Lignin
and Lignans
Advances in Chemistry

Lignin
and Lignans
Advances in Chemistry

Edited by
**Cyril Heitner
Donald R. Dimmel
John A. Schmidt**

CRC Press
Taylor & Francis Group
Boca Raton   London   New York

CRC Press is an imprint of the
Taylor & Francis Group, an **informa** business

First published in paperback 2024

First published 2010
by CRC Press
2385 NW Executive Center Drive, Suite 320, Boca Raton FL 33431

and by CRC Press
4 Park Square, Milton Park, Abingdon, Oxon, OX14 4RN

CRC Press is an imprint of Taylor & Francis Group, LLC

© 2010, 2024 Taylor & Francis Group, LLC

Library of Congress Cataloging-in-Publication Data

Lignin and lignans : advances in chemistry / editors, Cyril Heitner, Don Dimmel, John A. Schmidt.
    p. cm.
Includes bibliographical references and index.
ISBN 978-1-57444-486-5 (hardcover : alk. paper)
  1. Lignin. 2. Lignans. I. Heitner, Cyril, 1941- II. Dimmel, Don. III. Schmidt, John A. IV. Title.

TS933.L5L48 2010
572'.56682--dc22
    2010006628

ISBN: 978-1-57444-486-5 (hbk)
ISBN: 978-1-03-291781-8 (pbk)
ISBN: 978-0-429-13371-8 (ebk)

DOI: 10.1201/EBK1574444865

Visit the Taylor & Francis Web site at
http://www.taylorandfrancis.com

and the CRC Press Web site at
http://www.crcpress.com

Dedications

We dedicate this book to the memories of Gordon Leary and
Karl-Erik Eriksson, who both contributed chapters to this book

Gordon Leary coauthored Chapter 12, "The Chemistry of Lignin-Retaining Bleaching: Oxidative Bleaching Agents" with John Schmidt. He made many seminal contributions to our understanding of wood and lignin chemistry over a career that spanned more than 40 years. His approximately 90 publications described research on wood, lignin, bleaching, pulping, light-induced yellowing of paper and lignin, quinone methides, lignin-carbohydrate bonding, and the characterisation of lignin by NMR.

Shortly after receiving his PhD in 1965 from Canterbury University (New Zealand), Gordon pioneered the modern era of lignin photochemistry. In several elegant publications in *Nature,* he proposed quinones as the primary chromophores formed in photochemical yellowing of mechanical pulps and suggested a mechanism for their formation. These publications have become classic references cited in all publications on the photochemistry of wood fiber components. Much of the progress in the understanding of the reaction pathways of lignin yellowing and methods to stop this yellowing have their foundations in Gordon's pioneering research.

With his colleague R.W. Newman, Gordon published numerous papers on the use of NMR to characterize lignin in wood and in the various morphological regions of the wood fiber. This research has contributed to our knowledge of the structure of proto-lignin and the changes in lignin caused by the various extraction techniques.

Gordon's status as a leading wood chemist was recognized by various administrative appointments, the first as director of the Chemistry Division of the Department of Scientific and Industrial Research (DSIR), the equivalent of Canada's National Research Council or Australia's Commonwealth Scientific and Industrial Research Organization (CSIRO). He held this position from 1981 until the dismantling of DSIR in 1992. The Chemistry Division was the largest division in DSIR, with a very diverse range of activities.

After his career at DSIR, Gordon joined the Pulp and Paper Research Institute of Canada (Paprican) in September 1992 to carry out research into the bleaching of mechanical pulps. In recognition of his scientific achievements, he was elected a principal scientist (Paprican's highest scientific ranking) by his peers. He was later appointed executive director of the Canadian Mechanical Pulps Network of Centres of Excellence, a nationwide research association dedicated to enhancing the properties and value of mechanical pulps. In 1996 it included researchers from 15 universities, the National

Research Council of Canada, and Paprican; its annual budget of about $C8 million supported the work of approximately 65 university professors and 120 graduate students and postdoctoral fellows. The excellence of the scientists in this network was augmented by Gordon's exceptional leadership. During Gordon's tenure as executive director, the Canadian Mechanical Pulps National Network represented a renaissance in Canadian pulp and paper research.

Gordon left Paprican and the Network in 1996 to become the manager of the Pulp and Paper business unit at the Alberta Research Council (ARC) in Edmonton, which specialized in mechanical pulps, sensors, nonwood fibers, and papermaking. He built up the ARC laboratories, pilot plant, and staff from a skeleton of only eight staff and little more than a refiner and pulp-testing equipment. By the time he retired, this unit had 24 staff and pilot facilities for pressurized refining, chip production and impregnation, papermaking, pulp and paper testing, coating, print quality evaluation, and sensor development.

Under Gordon's leadership, ARC successfully developed a range of sensors for improved control of mill processes. The sensors were based mainly on spectroscopy or image sensing, and a number of them are operating in various mills. In 2001 he established a separate unit (Aquantix) to manufacture and market white water sensors for bleaching, dissolved solids and pitch control. The group had also an active technology development program in pulping and papermaking with fibers from agricultural waste.

Even in his "retirement," Gordon continued to contribute to wood and lignin chemistry as a member of the editorial boards of *Holzforschung, Appita,* and PAPTAC. In 2005, he served as program chair for the 13th International Symposium on Wood and Pulping Chemistry in Auckland, New Zealand, and was recognized for his lifetime achievements at the 2007 edition of this conference in Durban, South Africa. He continued to work as a visiting researcher in the Westermark group in Umea, Sweden, on what we would now recognize as the concept of the forest biorefinery.

Gordon possessed a rare combination of scientific, administrative, and leadership ability. But even this does not fully capture the man known to his friends, colleagues, and family and to those whom he mentored. To complete the picture, we need to add a sense of wonder about the natural world and scientific inquiry and a generosity of spirit towards his fellow humans. These remained undiminished by the tedium that often accompanies exacting experiments or never-ending committee meetings. Gordon enriched the lives of everyone that had the pleasure of working with him.

Karl-Erik Eriksson, who wrote Chapter 14, "Lignin and Lignan Biodegradation," received his BS in chemistry and PhD in biochemistry in 1958 and 1963, respectively, from the University of Uppsala, Sweden. He then completed his DSc in biochemistry in1967 at the University of Stockholm. He joined the Swedish Forest Products Research Laboratory (STFI) in Stockholm as a research assistant in 1958; in 1964 he was promoted as department head for Biochemical, Microbial, and Biotechnical Research, working closely with Börje Steenberg. He received a Fullbright Fellowship from 1968 to 1969 that allowed him to pursue postdoctoral studies at the California Institute of Technology with Norman Horowitz.

Karl-Erik pioneered the purification and characterization of fungal enzymes involved in lignocellulose degradation. He invented several processes using fungal enzymes to solve problems in pulp and paper manufacturing and recycling, published more than 280 research articles, and gave more than 250 lectures at universities, professional meetings, and companies. His success was due in part to his early career with scientists whose discoveries laid the foundations of modern approaches to protein purification–column chromatography using hydroxylapatite (Tiselius) and Sephadex (Flodin, Ingelman, Porath), as well as isoelectric focusing (Vesterberg). He was awarded along with T. Kent Kirk the 1985 Marcus Wallenberg Prize, for investigations into the fundamental biochemistry and enzymology of wood degradation by white-rot fungi, which arose from his work on fungal enzymes. Karl-Erik coauthored the book *Microbial and Enzymatic Degradation of Wood and Wood Components* with longtime friends and colleagues Robert Blanchette and Paul Ander in 1991.

In 1988, Karl-Erik joined the faculty of the University of Georgia as Eminent Scholar of Biotechnology and Professor of Biochemistry. He also served as an adjunct professor at the Institute of Paper Science and Technology (IPST) in Atlanta from 1990 to 1999. On his retirement from the University of Georgia in 1999, he was named Professor Emeritus.

He was a member of the Royal Swedish Academy of Engineering Sciences since 1978, sitting on its board as chairman of its Forestry and Forest Industry Sciences section from 1982 to 1985. He was elected to the World Academy of Art and Science in 1987, and was named a TAPPI Fellow in 2002.

Karl-Erik had numerous engagements consulting on behalf of United Nations Agencies and the governments of several countries, using his expertise in biotechnological applications of enzymes to industrial processing of biomaterials. He was also highly sought after as a consultant for companies around the world.

In addition to his scientific career, Karl-Erik founded a construction company and later cofounded the company Enzymatic Deinking Technologies (EDT) to exploit technology developed in his laboratory. After returning to Sweden, he became the board chairman of SweTree Genomics AB. He also served on the boards of directors of several additional companies.

Karl-Erik challenged the many students and researchers who passed through his laboratory to expand their horizons and continue learning throughout their lives, in the same way he would challenge himself. He was a kind, wonderful, and talented man with a big appetite for life.

We hope that this volume proves a fitting tribute to Gordon's and Karl-Erik's legacy.

Contents

Preface

Lignin, a constituent in almost all dry-land plant cell walls, is second only to cellulose in natural abundance. The purpose of this book is to provide an up-to-date compendium of the research on selected topics in lignin and lignan chemistry.

The structure and reactions of lignin have been studied for more than 100 years, and the extensive output of this research has been summarized in several comprehensive review texts. The first, *The Chemistry of Lignin,* written by F. E. Brauns in 1952, was followed by a supplemental volume in 1960 by F. E. Brauns and D. A. Brauns. Both Y. Hachihama and S. Jyodai in 1946 and I. A. Pearl in 1967 have written monographs with the same title. By the late 1960s, lignin chemistry had become so complex and covered such a large range of chemical and physical disciplines that authored chapters became the only way to provide authoritative coverage of all aspects of the field. In 1971, two prominent wood chemists, Kyösti V. Sarkanen and Charles H. Ludwig, edited the multiauthor reference textbook *Lignins.* Some of the contributors to this landmark text are still active in lignin research. This book has been used by both students and research scientists as the bible of lignin science.

Since the 1971 publication of *Lignins,* more than 14,000 papers have been published on the chemistry and physics of lignin. There has been immense progress in every area of lignin science. For example, advances in the understanding of the enzymology of lignin biodegradation led to the development of bioprocesses for the production of papermaking pulp. This has the potential for environmentally compatible industrial processes. A reliable determination of molecular weight distribution of lignin has come into its own since 1971. Also, there have been new processes developed in the area of pulping and bleaching. New areas of research have been developed in the field associated with environmentally friendly elemental chlorine–free and total chlorine–free bleaching processes.

When the 1971 edition of *Lignins* was published, spectroscopy of lignin was limited to degraded soluble lignins. The techniques of solid-state spectroscopy used today to characterize lignin in the plant fiber wall had not been developed. Today, UV-visible, infrared, and NMR spectroscopy are routinely used to characterise the changes in solid-state lignin *in situ* during and after various industrial processes. During the last 39 years, there have been considerable advances in the photochemistry of lignin. There is now a large body of research on the reaction pathways leading to the oxidative degradation and the formation of coloured chromophores.

This book is by no means a comprehensive treatise. The advances in the biosynthesis of lignin and lignans since 1971 have not been included in this volume. This should be the subject of a second book on the advances in lignin and lignan chemistry.

The editors thank the contributing authors for their dedicated effort in documenting the latest advances in their respective fields. Their cooperation and patience is greatly appreciated. In addition, we would like to thank those who spent countless hours reviewing the content and accuracy of each chapter. An effort was made by the editors to present a somewhat consistent writing style by exhaustively editing each chapter. We would like to thank the authors for their cooperation in this endeavor. We appreciate the kind support of FPInnovations, Paprican Division, and the Institute of Paper Science and Technology. Finally, the editors would like to thank their families for their cooperation in giving up time together to complete this book.

Editors

Donald Dimmel has been retired from professional life since 2002 and lives in Prescott, Arizona. He received a BS in chemistry from the University of Minnesota in 1962 and a PhD in organic chemistry from Purdue University in 1966. Following a postdoctoral position at Cornell University (New York), Dr. Dimmel was a faculty member at Marquette University (Milwaukee, Wisconsin) for several years, had a four-year stint in industry with Hercules Chemical Company (Wilmington, Delaware), and then went on to the Institute of Paper Chemistry, which in 1978 was located in Appleton, Wisconsin. He remained with the institute when it moved to Atlanta, Georgia, and changed its name to the Institute of Paper Science and Technology (IPST). He was a faculty member for 24 years, until his retirement in July 2002. At the end of his career, Dr. Dimmel was a professor, senior fellow, and the leader of the Process Chemistry group. He was a two-time winner of the IPST Teacher of the Year Award (1992 and 2001), was on the Editorial Advisory Board of the *Journal of Wood Chemistry and Technology,* and was a fellow in the International Academy of Wood Science. He has authored 100 refereed technical publications and patents. His research interests at IPST concerned reducing the energy and environmental impact associated with producing paper pulps from wood. Much of his research focused on developing a better understanding of the chemistry of lignin removal and carbohydrate degradation reactions that occur during pulping and bleaching.

John A. Schmidt is a principal scientist at FPInnovations, Paprican Division, in Pointe-Claire, Quebec, Canada. Dr. Schmidt earned a BSc in chemistry from the University of Western Ontario in 1979 and worked briefly for Dow Chemical of Canada before returning to Western for postgraduate studies. After earning a PhD in 1986, he joined Paprican and has remained there throughout his career. Dr. Schmidt is a member of the Chemical Institute of Canada, American Chemical Society, TAPPI, and the Pulp and Paper Technical Association of Canada. He has published 38 articles in peer-reviewed journals, holds five patents, and is a recipient of TAPPI's Best Research Paper Award. Dr. Schmidt's research interests are the photochemistry of lignocellulosic materials, pulp bleaching, aging and stabilization of paper, and wood-derived bioproducts.

Cyril Heitner retired from Paprican after a 36-year career. He received his BSc in chemistry from Sir George Williams University in 1963, his MSc in physical organic chemistry from Dalhousie University in 1966, and his PhD in organic photochemistry from McGill University in 1971. Dr. Heitner came to the institute as an Industrial Postdoctoral Fellow in 1970 and joined the staff in 1972.

The first of Dr. Heitner's scientific achievements were in the area of lignin modification to produce high-quality ultra-high-yield pulps. He discovered the effects of sulfonation on lignin softening, which has a profound effect on fiber length distribution and interfiber bonding of ultra-high-yield pulps. With R. Beatson, he was the first to determine the mechanism of lignin sulfonation in wood fiber. He and D. S. Argyropoulos also discovered that decreasing the pH of the sulfonation from 9 to 6 increased the amount of well-developed thin and flexible fibers and decreased the specific energy required. This discovery is now being used in most CTMP mills. Dr. Heitner has made significant scientific contributions in the area of chromophore chemistry of lignin-containing pulp and paper. He has developed a method for calculating the UV-visible absorption spectra of paper from reflectance values of thin ($10g/m^2$) sheets of paper. Using this technique, he studied both bleaching and light- and heat-induced reversion of lignin-containing pulps and paper. With J. A. Schmidt, he determined that multiplicity of the excited state leading to the cleavage of the phenacyl aryl ether bond using α-guaiacoxyacetoveratrone as a model was both singlet and triplet. It had been assumed by researchers that this group undergoes bond cleavage exclusively by the triplet excited state. This group also determined that an important reaction pathway leading light-induced yellowing involved cleavage of the β-O-4 aryl ether bond to a ketone and a phenol. This research has led to the development of a yellowing-inhibitor system that is close to commercial development.

Contributors

Umesh P. Agarwal
Fiber and Chemical Sciences Research
USDA Forest Products Laboratory
Madison, Wisconsin

Dimitris S. Argyropoulos
Department of Forest Biomaterials
North Carolina State University
Raleigh, North Carolina

Rajai H. Atalla
Cellulose Sciences International
Madison, Wisconsin

R. Bourbonnais
Biological Chemistry Group
FPInnovations, Paprican Division
Pointe-Claire, Quebec, Canada

Gösta Brunow
Department of Chemistry
University of Helsinki
Helsinki, Finland

Takeshi Deyama
Central Research Laboratories
Yomeishu Seizo Co., Ltd.
Nagano, Japan

Thomas Elder
Utilization of Southern Forest
    Resources
USDA Forest Service
Pineville, Louisiana

Karl-Erik L. Eriksson (Deceased)
Professor of Biochemistry &
    Molecular Biology Eminent
    Scholar in Biotechnology
University of Georgia
Athens, Georgia

Raymond C. Fort, Jr.
Department of Chemistry
University of Maine
Orono, Maine

Göran Gellerstedt
Department of Fibre and
    Polymer Technology
Royal Institute of Technology
Stockholm, Sweden

Hyoe Hatakeyama
Department of Applied Physics and
    Chemistry
Fukui University of Technology
Fukui, Japan

Tatsuko Hatakeyama
Lignocel Research
Fukui, Japan

Larry L. Landucci (Retired)
Chemistry and Pulping Group
US Forest Products Laboratory
USDA-Forest Service
Madison, Wisconsin

Catherine Lapierre
AgroParisTech
Thiverval-Grignon, France

and

Institut Jean-Pierre Bourgin
AgroParisTech-INRA
Versailles Cedex France

Gordon Leary (Deceased)
Pulp and Paper Business Unit
Alberta Research Council
Edmonton, Alberta, Canada

Knut Lundquist
Department of Chemical and
   Biological Engineering
Forest Products and Chemical
   Engineering
Chalmers University of Technology
Göteborg, Sweden

Sansei Nishibe
Faculty of Pharmaceutical Sciences
Health Sciences University
   of Hokkaido
Hokkaido, Japan

M.G. Paice
Biological Chemistry Group
FPInnovations, Paprican Division
Pointe-Claire, Quebec, Canada

John Ralph
DOE Great Lakes BioEnergy
   Research Center
University of Wisconsin
Madison, Wisconsin

and

Departments of Biochemistry and
   Biological Systems Engineering
University of Wisconsin
Madison, Wisconsin

I.D. Reid
Biological Chemistry Group
FPInnovations, Paprican Division
Pointe-Claire, Quebec, Canada

Sylvain Robert
Chemistry-Biology Department
University of Quebec at Trois-Rivieres
Trois-Rivieres, Quebec, Canada

1 Overview

Donald Dimmel

CONTENTS

INTRODUCTION

As the second most abundant natural polymer in our world, lignin has drawn the attention of many scientists for several centuries. Due to its complexity, nonuniformity, and conjunctive bonding to other substances, lignin has been difficult to isolate without modification and difficult to convert into useful consumer products, and its structure has been difficult to determine. The challenges presented in studying lignin have resulted in a vast amount of published literature. The goal of this volume is provide a resource that summarizes our present knowledge of lignin in certain key areas. The most inclusive description prior to this volume is best summarized in the book of K. V. Sarkanen and C. H. Ludwig, *Lignin* (Wiley-Interscience, 1971). This overview chapter takes much of its material from the aforementioned book and invites the reader to consult this book for greater depth. No references are presented in this first chapter; most discussion is supported by material in Sarkanen and Ludwig's book.

The biosynthesis of lignin is an important subject to understanding lignin structure, but it is not covered in this volume. It is a topic that deserves separate treatment, is steeped in controversy, and would add approximately 50% more pages to an already large volume. This volume focuses mainly on modern methods of lignin structure proof, on lignin reactivity, and on one aspect of lignan use. This brief introductory chapter is intended to familiarize the reader with a few basics, which are inherent to the discussions of the later chapters.

The discussion presented in this chapter will be expanded with the material given in Chapter 8. The layout of the volume is to present (i) a simple picture in this chapter; (ii) detailed chemical and spectral structural studies in Chapters 2–7; (iii) a coherent picture of lignin structure in Chapter 8; (iv) lignin/lignan reactions in Chapters 9–15, and then (v) pharmacological properties of lignans in Chapter 16. Since some readers will jump around from one chapter to another, we will (out of necessity) have some repetition of material. In general, a reader will better understand the chemistry pre-sented in later chapters by first reading the earlier chapters, especially Chapters 1 and 8.

OCCURRENCE

Nature is composed of minerals, air, water, and living matter. The latter contains polymers. The most abundant natural polymer is cellulose. It, together with lignin and hemicelluloses, are the principal components of plants. The principal function of lignin in plants is to assist in the movement of water; the lignin forms a barrier for evaporation and, thus, helps to channel water to critical areas of the plant.

Lignin is present in plants for which water conduction is important. Of greatest interest is its presence in trees. The lignin content depends on the type of tree: about 28% for softwoods and 20% for hardwoods. The cellulose content is approximately 45% in the wood of both types, while the hemicellulose content is roughly 17% in softwoods and 25% in hardwoods. Lignin structure can vary within the same plant, e.g., primary xylem, compression wood, early versus late wood, etc.

FORMATION AND STRUCTURE

Lignin is a polymer, built up by the combination of three basic monomer types, as shown in Figure 1.1. These building blocks, often referred to as phenylpropane or C_9 units, differ in the substitutions at the 3 and 5 positions. (Note: Typical phenols would have a numbering system that makes the phenol carbon #1; however, lignin nomenclature assigns the side-chain attachment to the aromatic ring as #1 and the phenol carbon as #4. Consequently, for the sake of consistency, we will use lignin nomenclature rules for the building blocks.)

Figure 1.2 outlines the main functional groups and numbering in lignin. The attachment of the aliphatic side chain to the aromatic ring is at C–1. The phenol oxygen is attached at C–4 and the numbering around the ring follows a rule that you use low numbers, which means that if there is only one methoxyl group it will be on C–3 (not C–5). The side-chain carbons are designated α, β, and γ, with C–α being the one attached to the aryl ring at its C–1 position. Not shown in Figure 1.2 are the possible occurrences of aliphatic and aryl ether linkages at C–α and C–γ, and ester

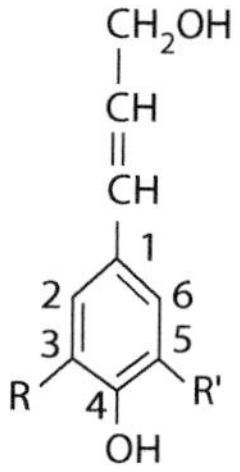

	Substituents	Name	Location
	$R = R' = H$	p-coumaryl alcohol	Compression wood, grasses
	$R = H, R' = OCH_3$	coniferyl alcohol	Hardwoods and softwoods
	$R = R' = OCH_3$	sinapyl alcohol	Hardwoods

FIGURE 1.1   Lignin monomeric building blocks.

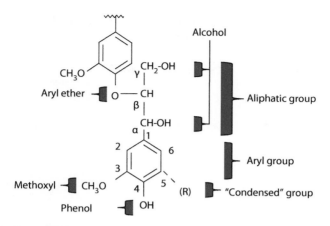

FIGURE 1.2 Lignin functional groups.

linkages at C–γ to non-lignin carboxylic acid groups. The existence of a carbon group at C–5 is often referred to a "condensed" structure. The term *condensed* is used rather loosely, being applied to both native and C–5 linkages formed during lignin reactions.

The principal monomer for softwood lignins is coniferyl alcohol, which has a methoxyl group on the C–3 position. Hardwood lignins have two main monomers: coniferyl alcohol and sinapyl alcohol, which has methoxyl groups on both the C–3 and C–5 positions. The third monomer, *p*-coumaryl alcohol, is more prominent in grasses and compression wood (branch conjunctures). The aromatic rings of the monomers are often referred to as follows: **guaiacyl** units have one aryl-OCH_3 group and are derived from coniferyl alcohol, **syringyl** units have two aryl-OCH_3 groups and are derived from sinapyl alcohol, and ***p*-hydroxyphenyl** units have no OCH_3 groups and are derived from *p*-coumaryl alcohol.

Native lignin arises via an oxidative coupling of the aforementioned alcohols with each other and (more important) with a growing polymer end unit. The oxidation produces a phenolic radical with unpaired electron density delocalized to positions O–4, C–1, C–3, C–5, and C–β; Figure 1.3 shows an example set of resonance forms for coniferyl alcohol. The lignin polymer can be initiated by coupling of two monomeric radicals, but more likely grows when monomer radicals couple with phenoxy radicals formed on the growing polymer. The phenoxy C–β position appears to be the most reactive, since the most abundant linkages in lignin involve this position (β–O–4, β–5, β–β).

An example of an oxidative coupling of coniferyl alcohol, which generates the abundant β–O–4 bond, is shown in Figure 1.4. The scheme is greatly simplified, since (a) only individual radical forms of the phenoxy radical are shown; (b) monomer-monomer coupling is shown, rather than the more prevalent monomer to polymer coupling process; and (c) the alcohols are likely conjugated with carbohydrates. Quinone methide intermediates from one coupling can participate in further coupling as the polymerization proceeds. Note, the term "quinone methide" refers to a nonaromatic structure that

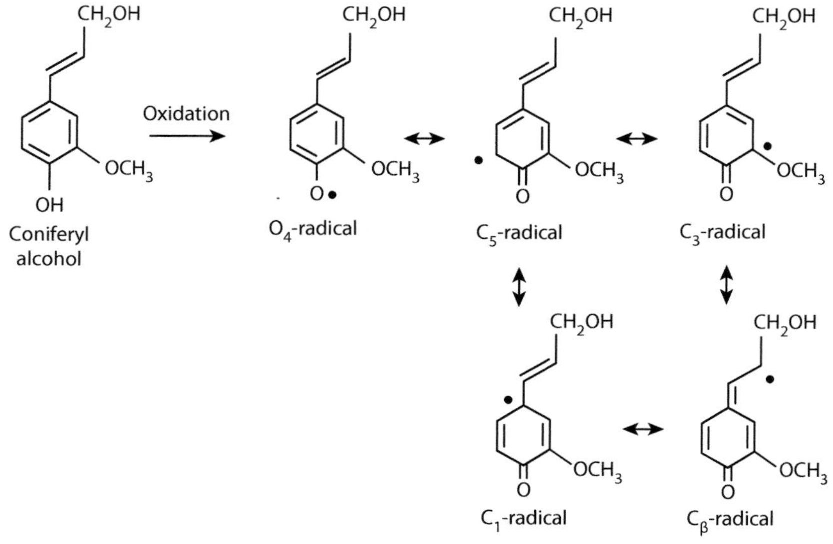

FIGURE 1.3 First step in lignin softwood polymerization.

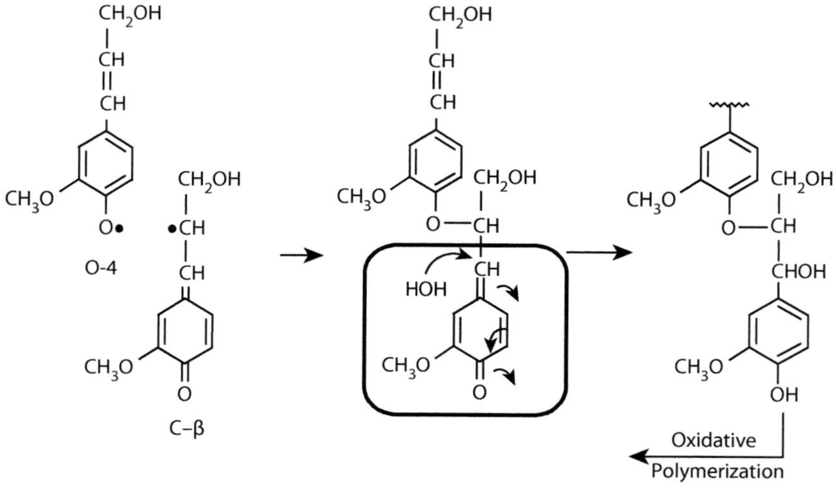

FIGURE 1.4 $C_\beta-O_4$ bond formation via radical coupling.

has two double bonds exiting the ring between $C_1 = C_\alpha$ and $C_4 = O_4$. These quinone methides are quite reactive and readily accept additions of nucleophiles to the $C_1 = C_\alpha$ double bond, resulting in regeneration of a much more stable aromatic ring, as shown by the chemistries presented in Figures 1.4 through 1.8.

In addition to the formation of β–O–4 and α–O–4 ether bonds, as shown in Figures 1.4 and 1.5, an ether linkage between C–5 and O–4 (a diphenyl ether) is

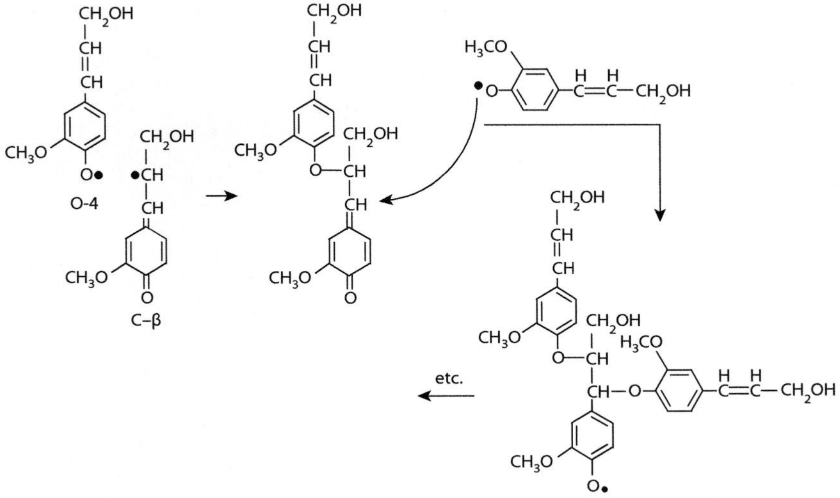

FIGURE 1.5 C_α–O_4 bond formation via radical coupling.

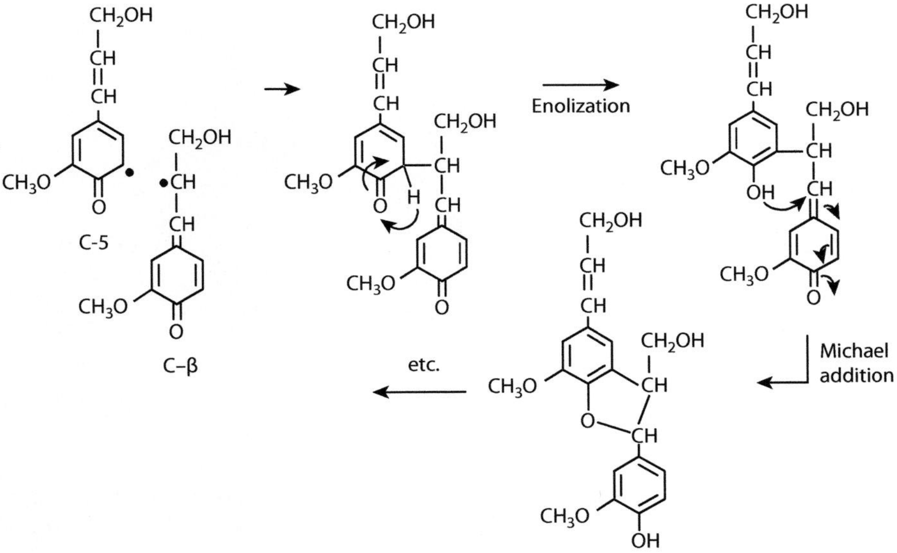

FIGURE 1.6 C_5–C_β bond formation via radical coupling.

also present to a small extent in lignin. Several C–C linkages also exist; Figure 1.6 outlines the chemistry for the production of a β–5 linkage. The latter is an example of a naturally occurring condensed structure. Another common C–C linkage in lignin exists in biphenyl units, which occur by the coupling of two phenoxy radicals at their C–5 positions. Coupling of a phenoxy radical at the C–1 position is also possible, an example of which produces a β–C–1 linkage (Figure 1.7). After the

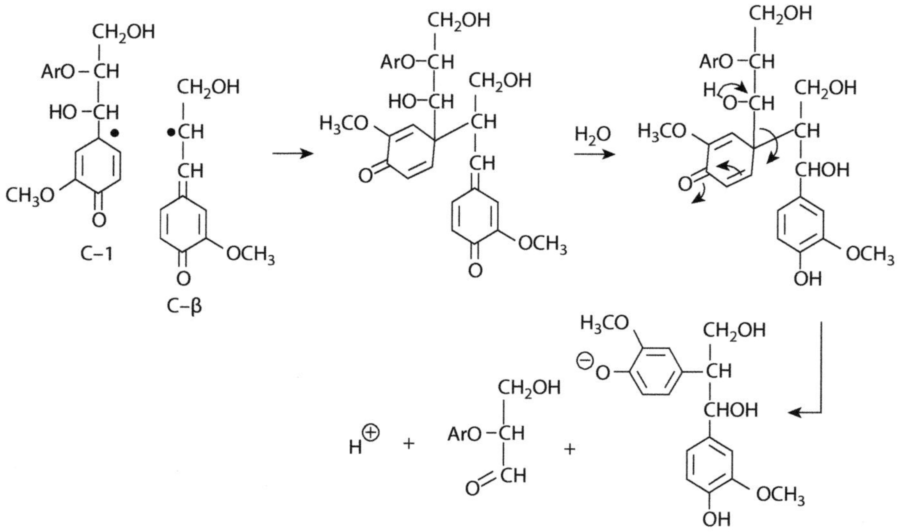

FIGURE 1.7 C_1–C bond formation via radical coupling.

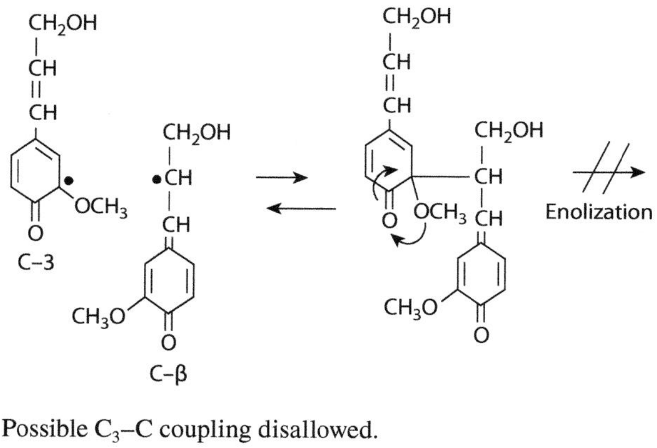

FIGURE 1.8 Possible C_3–C coupling disallowed.

initial coupling step, the side chain of one of the units is lost in order to regenerate aromaticity. The chemistry is facilitated by the relatively good stability of the aldehyde leaving group.

So far we have considered examples of coupling of all but one of the phenoxy radical density sites (O–4, C–1, C–3, C–5, and C–β [Figure 1.2]). Lignin linkages involving the C–3 site have not been observed. Coupling at this position likely occurs, but the process does not lead to a stable product (Figure 1.8). There is no good way for the aromatic ring to be regenerated with the methoxyl group present at C–3, since it is a poor leaving group. Consequently, the coupling likely reverses back to the individual radical species, which find other ways to couple, such as those shown in Figures 1.4 through 1.7. Since the building block sinapyl alcohol

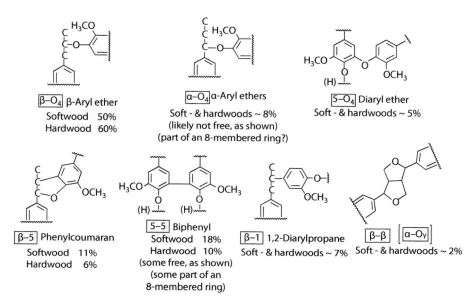

FIGURE 1.9 Lignin linkage types and amounts.

has methoxyl groups on both C–3 and C–5, coupling to both these positions will be inhibited and there will be little in the way of condensed structures formed. On the other hand, lignin derived from building block *p*-coumaryl alcohol, which has no methoxyl groups on C–3 and C–5, will have significantly more highly condensed structures. The proportion of condensed structures in a given lignin plays a major role in determining its reactivity, since C–C linkages are much less reactive than are C–O (ether) linkages.

Chapter 8 has much more detail on the structural units and linkage frequencies that exist in lignin. There are several variations of those presented here. However, of all the options that exist for generating interunit linkages, the β–O–4 linkage is the most predominant type in both softwoods and hardwoods (Figure 1.9). The second most abundant types involve linkages to the C–5 position (with linkages to C_β–, C_5–, or O_4-positions).

Figure 1.10 presents a partial representation of a softwood lignin. The main chain is shown by the combination of coniferyl alcohol units 1–10. Branching is shown by the units A and C attached to the main chain. The linkage types are color coded: dashed lines = more reactive ether bonds, dotted lines = low reactive C–C and O–4/ C–5 ether bonds, and dashed and dotted lines = the generally reactive α–O–4 and α–O–γ bonds. It should be pointed out that this picture is an oversimplification; the picture will be further refined as the reader delves into the various chapters. The message intended to be conveyed now is that lignin is a complex cross-linked polymer made up of different monomer units, linked in a variety of ways. Lignin exhibits a wide polydispersity, meaning that it has no characteristic molecular weight; values of 400 to more than a million weight average molecular weight have been reported.

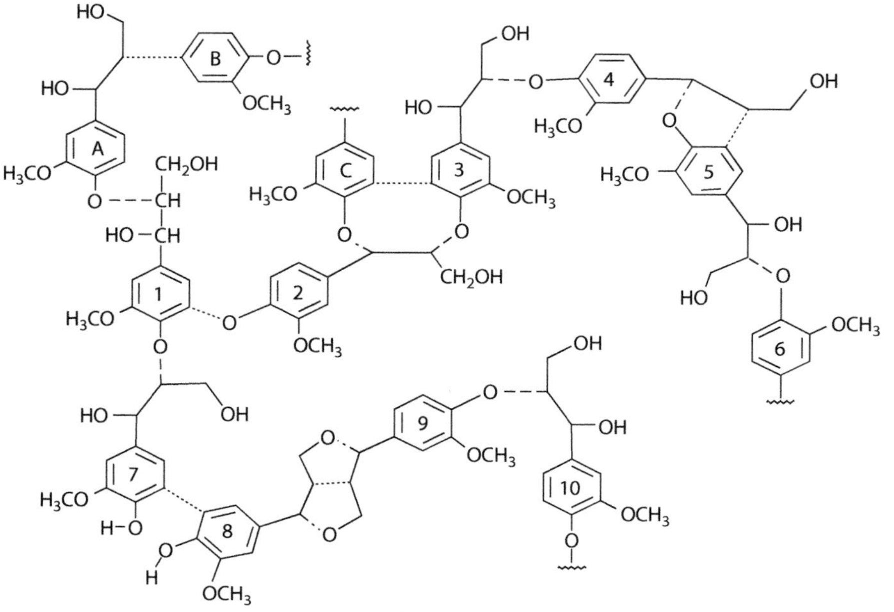

FIGURE 1.10   Example lignin structure.

ISOLATION AND STRUCTURE PROOFS

How have scientists defined such a complex substance? The answer is as complex as lignin itself and is still under active investigation. Early lignin studies involved isolating a lignin sample, degrading the polymer into small pieces, and deducing the polymer structure from the identity of the pieces. This was extremely tedious work. Newer methods, such as thioacidolysis, in combination with gas chromatography-mass spectroscopy, have been valuable tools in the determination of the lignin-derived monomers (Chapter 2). The advent of sophisticated nuclear magnetic resonance (NMR) techniques has greatly aided the understanding of lignin structure (Chapters 5 and 6). In addition, further structural insights are possible by the use of other spectral techniques (Chapters 3 and 4) and by thermal analysis (Chapter 7).

A real problem with all of these structural studies is to obtain a lignin sample that has not been significantly altered by its isolation from the other plant components. Research has made it clear that lignin is not a stand-alone polymer, but has linkages to polymeric carbohydrates. These unions are referred to as "lignin-carbohydrate complexes" (Chapter 8). Much will be said about the issues of isolation in the upcoming chapters.

To aid in the study of native lignin, researchers have prepared synthetic lignins, referred to as dehydrogenation polymers (DHP), by mixing lignin building blocks with oxidative enzymes. The DHPs can be obtained without interferences from other wood components, providing a baseline sample for comparison to structural analysis of native lignin.

REACTIVITY

The topic of lignin/lignan reactivity is the principal focus of Chapters 9 through 15. The correlation of molecular orbital calculations with lignin reactivity is taken up in Chapter 9. The geometry of a molecule markedly affects its energy and reactivity. Molecular orbital calculations give insight into lignin geometry, specifically conformational preferences, with respect both to ground states and to reaction intermediates. In addition, the calculations can pinpoint reactive sites by computing charge (and radical) densities of intermediates.

The greater reactivity of the various functional groups in lignin, as compared to those in carbohydrates, is the key to producing chemical pulps that are used in making high-quality paper products. Here the goal is to retain carbohydrate wood components and remove lignin components. The two primary steps in chemical pulp production are pulping (Chapter 10) and bleaching (Chapter 11). Chemical pulping largely involves alkaline processes that initiate reaction at the lignin's phenolic hydroxyl groups and give rise to cleavage of many of the aryl ether bonds; such chemistry is not possible with carbohydrates, since they lack both of these functionalities. Pulping can go only so far with this kind of chemistry; some lignin will still be resistant. This is where bleaching comes in. Here the chemistry involves breakdown of the lignin aromatic units, again a feature not present in carbohydrates. While this all seems simple enough, the reality is that there are many complexities, which will be the topics in Chapters 10 and 11.

Lignin does not have to be totally, or even partially, removed to give rise to paper products. Examples include pulps produced by the mechanical defibration of logs and steamed and/or chemically treated chips. Such pulps still contain significant quantities of lignin and suffer from lower bleachability and yellowing due to thermal- and light-induced oxidation. However, such pulps can be produced with double the yield and one-quarter the pollution than that obtained by chemical pulping. In addition, it is advantageous to use lignin-rich pulps because of higher bulk that permits lower basis weight and larger printing surface per ton of paper. Chapters 12 and 13 report on the advances in the chemistry of oxidative and reductive lignin-retaining bleaching. Brightness is one of the most important parameters that determine value. The oxidation and reduction of colored chromophores in lignin-containing pulp (stone and refiner groundwood), sometimes in sequence, increases the value of the paper produced. Chapter 14 reports on the photochemical processes of lignin and lignin model compounds, most of which cause yellowing of high-brightness lignin-containing papers. Lignin contains both moieties that absorb light to produce free-radicals, and react with oxygen, and moieties that react with photo-induced radicals. Also, there are lignin groups that sensitize the formation of reactive singlet oxygen ($^1O_2$), which in turn react with the various groups in lignin to cause color production and contribute to $\beta-O-4$ aryl etheir cleavage.

An evolving area is the use of biodegradation as a means to facilitate lignin removal (Chapters 15 and 16). Useful biodegradation chemistry again takes advantage of existing reactivity differences between lignin and carbohydrates. The employed enzymes often have high specificity for phenolic structures.

USES

By far, the principal use for lignin is as fuel in the production of pulp used for paper and corrugated board. High-quality paper products require that the lignin be separated from the cellulose in wood. The pulping process produces a pulp rich in cellulose and a liquor rich in degraded lignin. The liquor is partially evaporated and burnt in a furnace. Inorganic pulping chemicals are recovered, and the energy generated is used in the pulp production. Lignin has a high calorie content, which makes it an excellent fuel. In essence, the lignin in wood provides the energy needed to make the cellulose-rich pulp. This sentence can be changed to the following: Bleaching follows pulping when high-brightness products are desired. The lignin-derived fragments in the bleaching liquors have no value and disposal of these liquors is a problem.

For many decades, researchers have tried to find applications for uses of lignin derived from pulping liquors. This highly altered, complex lignin presents real challenges with respect to finding commercially valuable end uses. However, the future use of plants (including wood) as sources of chemicals, rather than just for making paper products, will generate large quantities of a new kind of lignin. Some plants are now being processed for ethanol fuel production (from their carbohydrate components); commercial uses of the lignin by-product will greatly enhance the processing costs.

Chapter 17, Pharmacological Properties of Lignans, is the only chapter in this volume that specifically addresses uses and characteristics of lignans. This chapter describes the activity of a wide variety of lignans derived from medicinal plants and used in traditional and folk medicines. The chapter also reports on the physiological changes in tumors in the digestive, reproductive, and endocrine systems caused by lignans and how these effects can be incorporated into various therapies.

2 Determining Lignin Structure by Chemical Degradations

Catherine Lapierre

CONTENTS

INTRODUCTION

One of the greatest challenges in the structural biochemistry of the lignified cell wall is to determine the nature and proportion of building units and interunit linkages in native lignin structures. Before the advent of powerful nuclear magnetic resonance (NMR) methods, chemical degradation reactions of lignins were the only viable ways to get structural information [1,2]. Among the first pioneering techniques, acidolysis [1,3], thioacetolysis [4], and hydrogenolysis [5] played an undisputed role in our current knowledge of lignin structure. However, as these methods have a low sample throughput capability and/or require a prolonged training to be mastered, they will not be presented in detail in this chapter.

The purposes of this chapter are twofold. The first is to provide an overview on the significance and comparative performances of the most commonly used chemical degradation methods, especially with regard to screening, informative, and quantitative capabilities. The second purpose is to provide lignin structural information based on the use of these methods, with a special focus on thioacidolysis. Since the detailed laboratory procedures are beyond the scope of this chapter, the reader is advised to consult the corresponding original papers mentioned in the reference section.

Before proceeding, we need to briefly remind the reader of the specific terminology used for lignin structure. Lignins are essentially composed of C_6C_3 units, namely p-hydroxyphenyl (H), guaiacyl (G), and syringyl (S) units, in various proportions according to botanical, physiological, and cytological criteria. These units are interconnected by a series of carbon–oxygen (ether) and carbon-carbon linkages in various bonding patterns [1]. The main ones have been established through a wealth of chemical and spectroscopic data obtained over the last 30 years and are outlined in Figure 2.1, with the conventional carbon numbering of the ring (C–1 to C–6) and the less conventional side chain carbon Greek marking (C–α, C–β, C–γ). In a few recent papers, carbon numbering is extended to the side chain (C–α, C–β, C–γ changed to C–7, C–8, and C–9).

Figure 2.1 is probably an oversimplified caricature of the complex lignin structure due to structural refinements of the outlined basic patterns and to the occurrence of other coupling modes between aromatic units, such as the β–6 one. There is a universal consensus for the predominance of the labile β–O–4 ether linkages (structure A, Figure 2.1) in native lignins, as well as the occurrence of the more resistant β–5, 5–5, β–β, 5–O–4, and β–1 interunit linkages (structures B through G, Figure 2.1). Some biphenyl 5–5 structures (structure C, Figure 2.1) are involved in dibenzodioxocin 8-membered ring structures (structure C', Figure 2.1), as discovered by Brunow and coworkers [6]. There is still some controversy about the relative importance of some interunit linkage types, which vary according to the employed characterization method. For instance, degradation products with diarylpropane β–1 skeleton (emanating from structure F or G, Figure 2.1) are recovered in substantial relative amount when native lignins are subjected to acidolysis [3] or thioacidolysis [7], whereas other studies [8] have revealed only small amounts of these structures. The origin of this controversy has been alleviated; in native lignins, the β–1 bonding pattern would occur as spirodienone precursors (structure F, Figure 2.1) converted to 1,2-diarylpropane structures when lignins are subjected to hydrolytic or acidic conditions [9,10]. The dienone hypothesis, already proposed in 1965 to account for the formation of β–1 bonding pattern [11], is now supported by NMR signals of spirodienone structures [9,10,12,13].

Lignin chemical degradation methods can be classified according to the mechanism underlying the depolymerization of the lignin network, namely oxidative, solvolytic, or hydrogenolytic reactions. In addition to these chemical lignin fragmentation procedures, analytical pyrolysis has been used to evaluate the lignin content and structure in lignocellulosic materials. The main advantages of analytical pyrolysis are (a) its high throughput, since this procedure does not involve any time-consuming wet chemistry, and (b) the low sample demand [14,15]. The pyrolytic depolymerization of lignins mainly proceeds by cleavage of labile ether bonds and thereby suffers the same limitations as most of the chemical degradations.

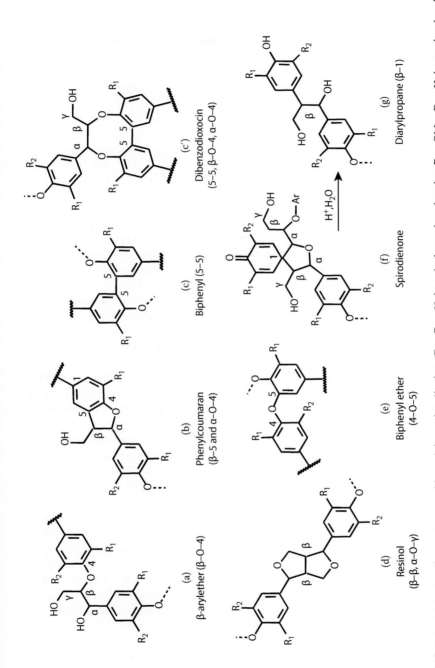

FIGURE 2.1 Main bonding patterns evidenced in native lignins ($R_1 = R_2 = H$ in *p*-hydroxyphenyl units; $R_1 = OMe$, $R_2 = H$ in guaiacyl units; $R_1 = R_2 = OMe$ in syringyl units). (Adapted from Ralph, J., Lundquist, K., Brunow, G., Lu, F., Kim, H., Shatz, P.F., Marita, J.M., Hatfield, R.D., Ralph, S.A., Christensen, J.H., and Boerjan, W., *Phytochem Rev.*, 3, 29–60, 2004.)

The extent to which the typical C_6C_3 phenylpropane skeleton of the building units is preserved in the recovered degradation products is a valuable guide to the selection of a chemical degradation procedure. For instance, most of the developed solvolytic or reductive procedures essentially preserve this C_6C_3 skeleton, which is not the case for oxidative procedures. Accordingly, the aromatic monomers and dimers released by the former techniques can be viewed as specific signatures of lignins in a plant material. In contrast, nonlignin phenolics may provide the same C_6C_1 degradation products as lignins when subjected to oxidative processes. This interference draw-back is carefully considered in the chapter.

THE OXIDATIVE DEGRADATION OF LIGNIN C_6C_3 UNITS INTO C_6C_1 MONOMERS AND DIMERS

ALKALINE NITROBENZENE OXIDATION: A 50-YEAR-OLD TECHNIQUE AND STILL A LEADERSHIP POSITION

Nitrobenzene oxidation (NBO) was introduced 50 years ago by Freudenberg to confirm the aromatic nature of lignins [reviewed in 16]. In this procedure, lignins are oxidized by nitrobenzene in alkaline medium (2M NaOH) and at high temperature (160–180°C for 2–3 hours). Benzaldehydes are recovered as the main products, together with lower amounts of the corresponding benzoic acids. Nitrobenzene was first considered to act as a two-electron-transfer oxidant and to produce quinone methide intermediates from free phenolic units. Later studies proposed that the oxidation was a free radical process involving a one-electron-transfer at the level of the benzylic alcohol group [17]. It is likely that both mechanisms coexist since NBO of model compounds with Me–O–4 [17] or with a benzylic methylene group [18] gives some benzaldehyde, albeit in low yield.

NBO is still probably the most commonly employed chemical degradation tech-nique for lignin analysis. This acceptance is related to the fact that NBO provides in satisfactory yield *p*-hydroxybenzaldehyde, vanillin, and syringaldehyde from lignin H, G, and S units, together with smaller amounts of the corresponding ben-zoic acids (Figure 2.2). In addition, NBO yields small amount of 5-carboxyvanillin, 5-formylvanillin, and dehydrodivanillin (Figure 2.2) from C–5 condensed guaiacyl units; however, these trace components are generally not included in the quantitative determination of lignin-derived compounds. The simple reaction mixture is analyzed either by high-performance liquid chromatography (HPLC) with UV detection or by gas chromatography (GC) after derivatization. The total yield and composition of aromatic aldehydes recovered from extracted wood has been considered as taxo-nomically specific [19]. For instance, normal conifer samples give rise to vanillin as a major product in the 17 to 28% range (by weight) based on the Klason lignin content of the sample [16,19]. In contrast and due to a lower proportion of carbon-carbon linkages between aromatic rings, higher yields of vanillin and syringaldehyde are obtained from hardwood samples (in the 30–50% range) [19]. As NBO depolymeriza-tion proceeds (by cleavage of the α–β benzylic bonds and of the R–O–4 ether bonds), the yield in monomeric products is indicative of the amount of lignin units without aryl, aryloxy or alkyl substituents at C–5 and or C–6 (referred to as the non condensed lignin units) [16].

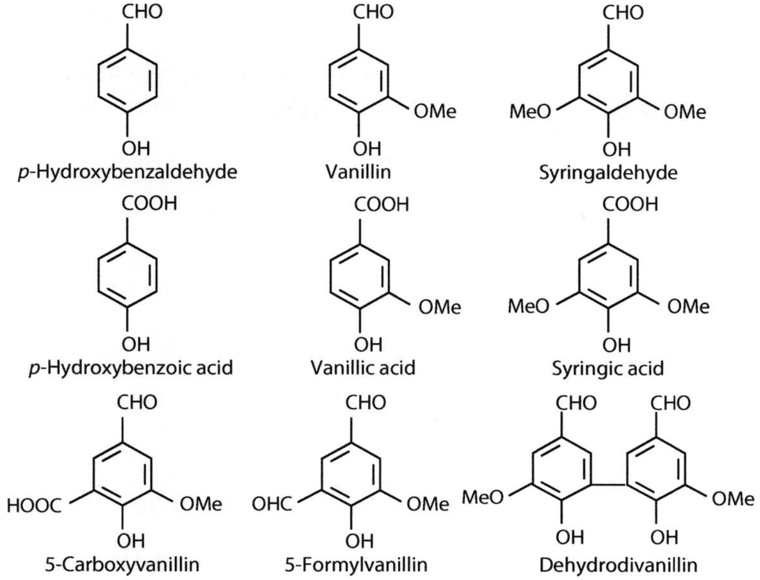

FIGURE 2.2 Main products recovered from the nitrobenzene oxidation of lignin structures. (Adapted from Chen, C.L., *Methods in Lignin Chemistry*, Berlin Heidelberg: Springer-Verlag, 301–321, 1992.)

The leadership position of NBO masks the difficulties of the method, such as its susceptibility to moderate experimental changes. As shown in Table 2.1, which reports on an interlaboratory comparative evaluation of lignin degradative analyses [20], the yield of monomeric products may differ to a very large extent, with a 20 to 30% standard deviation. This poor interlaboratory reproducibility may originate both from variations in reaction duration or temperature [21] and from analytical difficulties. After completion of the reaction, the classical procedure involves elimination of excess nitrobenzene and its reduction products from the alkaline reaction mixture; this is followed by the acidification of the hydrolysate, the extraction of the benzoic aldehydes and acids and their HPLC or GC analysis [16]. The possibility of incomplete extraction [22,23], as well interference from residual nitrobenzene derivatives [23], is often overlooked.

In contrast to the fluctuating NBO yield, the S/G ratio (e.g., the syringaldehyde to vanillin ratio, S/V in Table 2.1) displays a more satisfying interlaboratory reproducibility, with standard deviation in the 4–8% range. Due to this higher reproducibility, most of the discussion and interpretation of NBO data published in the literature focus on S/G ratios as an index of the so called lignin monomer composition. However, similar to other degradative methods, NBO probably gives an S/G ratio substantially higher than the actual proportion of S and G units in lignins. This is due to the fact that S lignin units are less involved in condensed interunit bonds than G units. This deviation does not decrease the value of the comparison of angiosperm lignin samples, provided that there is no interference from nonlignin components.

The recovery of a simple reaction mixture may be viewed as an advantage of NBO. Conversely, it can be considered as a drawback, in that little information is

TABLE 2.1
Comparative Nitrobenzene Oxidation of Isolated Lignin Samples Carried Out in Six Different Laboratories. The Total Yield in Vanillin V and Syringaldehyde S is Expressed in μmoles/g of Lignin

Milled Wood Lignin from Cotton Wood		Steam Explosion Lignin from Aspen		Organosolv Lignin from Mixed Hardwood		Kraft Lignin from Mixed Softwood
S + V	S/V	S + V	S/V	S + V	S/V	V
1672	2.27	985	2.57	595	2.44	560
2049	2.03	1041	2.41	425	2.76	649
1821	2.29	1328	2.47	803	2.29	872
2703	2.10	1776	2.61	1064	2.37	1082
1924	2.02	1397	2.42	1025	2.24	1007
2717	1.88	1730	2.35	789	2.30	900
Mean Value (Standard Deviation)						
2148 (453)	2.10 (0.16)	1376 (333)	2.47 (0.10)	784 (246)	2.40 (0.19)	845 (203)

Source: Gellerstedt, G., Rapporteur for degradative analyses of lignins: Modern methods of analysis of wood, annual plants and lignins. New Orleans: IEA Presymposium, 1991.

recovered on side chain functionality or bonding pattern of parent structures. Model compound studies reveal that benzaldehydes and benzoic acids can be generated from a variety of lignin structures, having β–O–4, β–1, β–β, and β–5 interunit bonds; however, the recovery yield vary with the parent structures [18,24]. The fact that diphenylmethane structures resist NBO, but are degraded by the phenyl nucleus exchange method [25], has been reassessed [18].

The main drawback of the NBO method, which has been repeatedly discussed in the literature, is its lack of specificity. As the lignin C_6C_3 units are degraded to C_6C_1 benzoic compounds, this method leads to ambiguous results when applied to samples containing nonlignin phenolics capable of generating the same benzaldehydes as lignin units. The case study of grass cell walls illustrates this interference limitation [26,27]. According to the data of Table 2.2, native wheat lignins appear as typical H–G–S lignins, with a substantial proportion of H units. However, a mild alkaline treatment induces a twofold decrease in the proportion of H monomers released from saponified straw (Table 2.2) or from the solubilized lignin fraction (data not shown). Such a phenomenon, first observed by Higuchi and coworkers [26,27], relates to the fact that most of the *p*-hydroxybenzaldehyde released from grass samples originates from *p*-coumaric esters. The extent to which the *p*-coumaric or ferulic units survive NBO and give rise to the free acids depends on the severity of the procedure [21]. The same interference between putative H lignin units and tyramine units was reported in the case of suberized cell walls [28]. Due to the same unsatisfying specificity, the occurrence of lignins in bryophyte samples is still the matter of debate [29,30].

TABLE 2.2

Yield of H, G, and S Monomers (Benzoic Aldehydes and Acids) Released by Nitrobenzene Oxidation (160°C, 3 Hours, 5 ml 2M NaOH, 0.5 ml Nitrobenzene, 25 mg Sample) of Wheat Straw Samples. Yields are Expressed in μmoles/g of Klason Lignin for the Extractive-Free Wheat Straw Before and After Mild Alkaline Hydrolysis (2M NaOH, 37°C, 2 Hours)

Sample	H + G + S	H/G/S
Extractive-free wheat straw	2518	22/41/37
Extractive-free wheat straw after alkaline hydrolysis	2752	12/44/44

Source: Lapierre, C., *Forage Cell Wall Structure and Digestibility,* Madison, WI: American Society of Agronomy Inc., 133–163, 1993.

Some misinterpretations of the NBO performances still occur in the literature. For instance, 5-hydroxyguaiacyl (5–OH–G) units have been tentatively searched for among the NBO products recovered from transgenic tobacco plants deficient in caffeic acid *O*-methyltransferase (COMT) activity [31]. Studies on COMT-deficient bm3 maize line [32] have shown that the labile 5–OH–G units cannot survive NBO. This situation was anticipated in the pioneering studies of Kuc and Nelson [33], in which they suspected "the occurrence of additional as yet undetected lignin units". By contrast and as explained in the next section, thioacidolysis provides an easy determination of 5–OH–G units in native or industrial lignins.

Cupric oxide oxidation [16], a companion method of NBO, provides a wider range of lignin-derived monomers and dimers [34,35]. In addition to the benzoic aldehydes and acids, acetophenones are recovered due to incomplete α–β cleavage. Moreover, a series of dimers with 5–5, 5–O–4, β–1, α–1, α–5, α–2 interunit bonds are obtained with carbonyl or carboxylic functions at Cα. Cupric oxide oxidation offers two advantages, relative to NBO. First, when applied to grass samples, the interference drawback is less severe since the skeleton of *p*-hydroxycinnamic units is preserved. Second and when CuO is used in anaerobic conditions (no trivial task) to avoid any superoxidation of the aldehydes, the acid/aldehyde ratio may be viewed as the signature content of α-carbonyl groups in lignin [34].

In spite of some limitations, it is very likely that NBO will maintain its position as a standard technique for lignins. In addition, there are still some developmental aspects for this technique [36,37].

PERMANGANATE OXIDATION: AN INFORMATIVE PROCEDURE WITH LOW THROUGHPUT CAPABILITIES

This analytical method, first developed by Freudenberg and coworkers in 1936 [reviewed in ref 38], was then comprehensively developed and improved by Miksche and coworkers [39]. This oxidative degradation conducted at alkaline pH involves an initial peralkylation of the phenolic hydroxyl groups (with diethylsulfate or dimethylsulfate), followed by two sequential oxidation steps with permanganate and

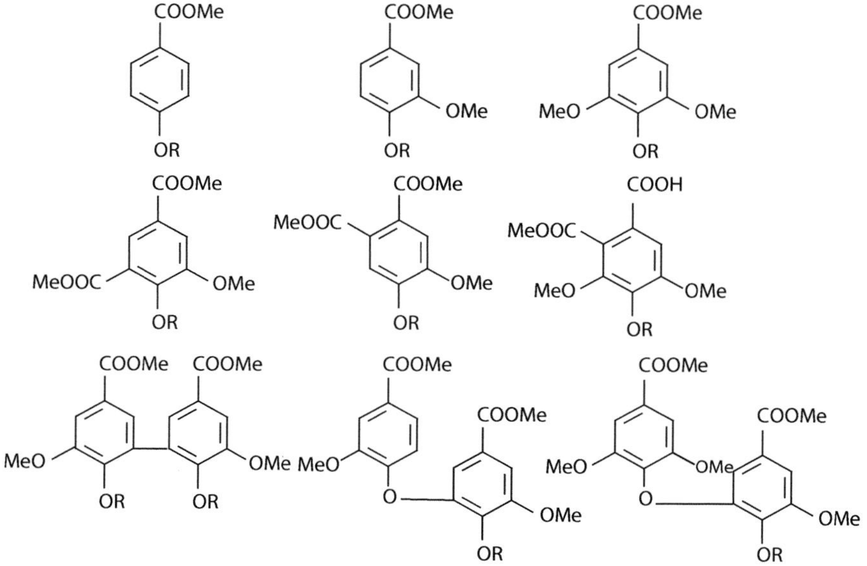

FIGURE 2.3 Main carboxylic acid methyl esters formed by permanganate oxidation of peralkylated lignin samples (R = OMe or OEt). (Adapted from Gellerstedt, G., *Methods in Lignin Chemistry*. Berlin Heidelberg: Springer-Verlag, 322–333, 1992.)

then hydrogen peroxide, to yield a variety of mono- and di-carboxylic acids [2,40]. These acids are finally methylated with diazomethane to obtain their methyl esters analyzed by gas chromatography (Figure 2.3).

The technique proceeds by oxidative cleavage of the side chains attached to the aromatic rings, thereby introducing a carboxylic substituent at each aromatic carbon provided with an aliphatic side chain. Monocarboxylic esters specifically originate from the so-called non condensed units, while dicarboxylic esters stem from C–5 or C–6 condensed units. The major drawback of this four-step method is that only structures in lignin carrying initially free, then methylated, phenolic groups can be analyzed; these account for 10 to 30 units per 100 C_9 units in native wood lignins [40]. The lignin aromatic units that are initially etherified at the 4-OH are, thereby, not considered in the analysis. Thus, the yield in aromatic carboxylic acids has been used to evaluate the frequency of free phenolic groups in lignins [40]. With the objective to increase this frequency and, thereby, this yield, researchers added another alkylation and a severe depolymerization of lignins (such as CuO oxidation) before the conventional method [2]. In other words, if six steps are employed, monomeric and dimeric aromatic esters are recovered in amounts that account for 60–70% of the lignin units [2]. Obviously, the permanganate oxidation method played an undisputable and important role in investigating lignin structure; however, the low acceptance of the method is probably related to its very low throughput capabilities (the complete sequence for one sample may require a whole week) and with its difficult, multi-step procedure that needs a prolonged experience to be mastered. Very few laboratories are still routinely using this technique.

THIOACIDOLYSIS: A MULTIFACETED METHOD WITH INFORMATIVE CAPABILITIES

Thioacidolysis is an acid-catalyzed reaction that results in β–O–4 cleavage. Similar in advantage to other solvolytic methods, it gives rise to C_6C_3 products that yield information about side chain functionality and interunit linkages of the parent polymer. The thioacidolysis method was adapted from acidolysis [3]. Acidolysis uses water and HCl, while thioacidolysis uses EtSH and BF_3 etherate; this change in conditions increases reaction yields. The rapid acceptance of the method is due to its simplicity and informative capabilities. Since its creation [41], thioacidolysis has evolved considerably [42–44]. The first procedure focused on the analysis of lignin-derived monomers, with the intent to evaluate the type and amount of units only involved in β–O–4 bonds. If a permethylation step is performed before thioacidolysis, the method gives an evaluation of free phenolic groups in β–O–4 linked units. Provided that a desulfurization step is carried out after thioacidolysis, lignin-derived dimers are obtained, which are representatives of lignin resistant bonding patterns. These various aspects are discussed in the following, with emphasis on the significance of the results.

LIGNIN-DERIVED MONOMERS: ORIGIN AND SIGNIFICANCE

Thioacidolysis proceeds by cleavage of β–O–4 bonds [42], combining the hard Lewis acid, boron trifluoride etherate, with a soft nucleophile, ethanethiol [45]. The first reaction step is thioethylation of any alcoholic or ether groups at C–α. After Cα–OH(R) replacement with Cα- SEt, subsequent thioethylations at C–β, then C–γ, proceed with participation of the neighboring thioethyl group. Lignin units only involved in arylglycerol-β-aryl ether substructures A (Figure 2.1) afford C_6C_3 phenylpropane monomers *1–3*, recovered as a ~50/50 *erythro/threo* mixture (Figure 2.4). Two other monomers, released from the same parent structures through minor reaction pathways, represent less than 10% of monomers *1–3*. Monomers *4* and *5,* with displaced side chains due to some rearrangements in acidic medium, are respectively eluted as downfield shoulders of the GC doublet peak corresponding to monomers *2* and *3* (Figure 2.5). Monomers *6* and *7* originate from the loss of terminal hydroxymethyl groups.

Products issued from the demethylation of *1–3* are generally recovered in trace amount (relative importance < 0.5% of the main monomers *1–3*), unless a high BF_3 concentration is used [42]. By contrast, catechol (C) or 5–OH–G monomers may be released from lignins that have been subjected to chemically- or physically-induced demethylation reactions [42]. In addition, 5–OH–G monomers are the signature of COMT deficiency in angiosperm lignins [32,46]. This is illustrated on the GC traces obtained from a COMT-deficient poplar line and the corresponding control (Figure 2.5) [46]. These traces reveal that COMT deficiency causes a severe shortage in S lignin units and the incorporation of 5–OH–G units, as shown by the corresponding monomers that are in trace amount in the control. Neither NBO nor acidolysis are capable of detecting these 5–OH–G units [32].

The efficiency of β–O–4 cleavage has been evaluated with arylglycerol GG, GS and SS β–O–4 dimers. Under standard conditions (4h, 100°C, dioxane/EtSH 9/1 mixture with 0.2M BF_3 etherate), these dimers provide the main monomers in 75–85%

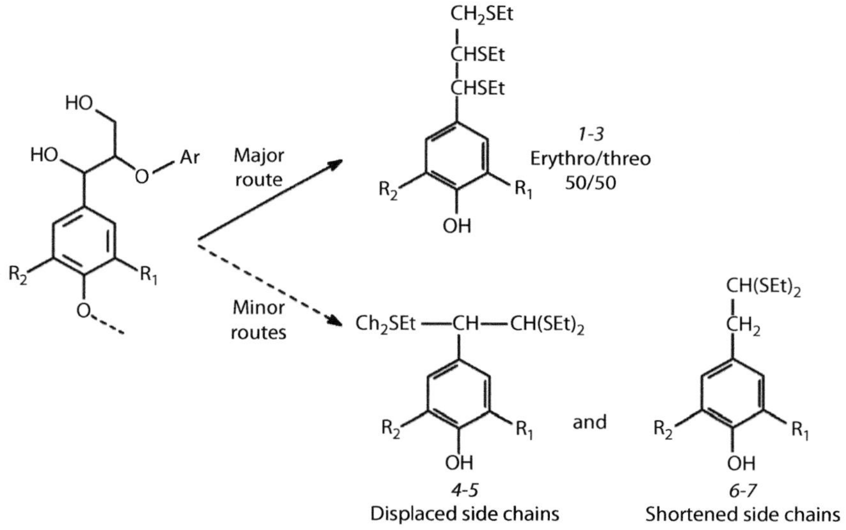

FIGURE 2.4 Thioethylated monomers recovered from the thioacidolysis of arylglycerol-β-O-4 structures. The *p*-hydroxyphenyl H *1* ($R_1 = R_2 = H$), guaiacyl G *2* ($R_1 = OMe, R_2 = H$) and syringyl S *3* ($R_1 = R_2 = OMe$) diastereomers are essentially recovered from the corresponding parent H, G, and S structures. Beside, minor G and S monomers with displaced side chains (*4* $R_2 = H$; *5* $R_2 = OMe$) or with shortened side chains (*6* $R_2 = H$; *7* $R_2 = OMe$) are recovered in low amount. (Adapted from Lapierre, C., Rolando, C., and Monties, B., *Holzforschung*, 40, 47–50, 1986.)

yield [43,47]. While the yield is high, it is not quantitative [48]. Some experimental changes have been suggested to improve this reaction yield. For instance, a swelling pretreatment of lignocellulosic samples before thioacidolysis has been shown to increase the recovery of lignin-derived monomers by about 25%, which was interpreted as an improved access of all of the lignin by the thioacidolysis reagent [49].

Dimers of the β-O–4 type with α–CO groups give rise to unsaturated monomers (Ar–CSEt = CHSEt) in low yield (about 20%) and to thioethylated dimers (Ar–CSEt = CHOAr). Accordingly, lignin samples with high α–CO content should be reduced by $NaBH_4$ before thioacidolysis. Vinyl ether structures (Ar–CH = CH–OAr) present in kraft lignins specifically give Ar–CH$_2$–CH(SEt)$_2$ in 40% yield [47]. These monomers are also released in minor amount from arylglycerol-β-aryl ether structures (compounds *6* and *7* in Figure 2.4). When recovered in high relative amount (e.g., more than 10–15% of the main monomers *1–3*, Figure 2.4), these C_6C_2 monomers are thus diagnostic for the occurrence of vinyl ethers in lignins.

Specific markers are released from special lignin side chains (Tables 2.3 and 2.4) such as cinnamyl alcohol, Ar–CH = CH–CH$_2$OH, and cinnamaldehyde, Ar–CH = CH–CHO, end-groups. Cinnamyl alcohol-derived monomers are obtained in low amount from cell wall lignins, in contrast to synthetic lignins (dehydrogenation polymers, or DHPs) [50] (Table 2.4). This difference reveals that natural lignins and synthetic lignins are formed by distinct coupling mechanisms [10,51]. Coniferaldehyde end-groups, responsible for the Wiesner lignin staining, give a specific thioacidolysis marker. The figure evaluated by thioacidolysis (about 0.5% of the main G monomer

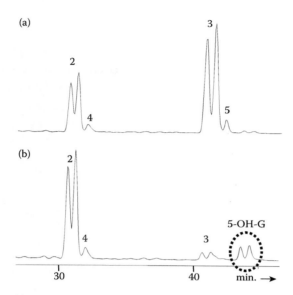

FIGURE 2.5 Partial GC chromatograms showing the separation of G and S thioacidolysis monomers (compounds *2–5* outlined in Figure 2.4 and analyzed as their trimethylsilylated [TMS] derivatives) recovered from (a) a one-year-old control poplar tree and (b) a one-year-old transgenic poplar tree severely depressed in COMT activity. (Based on Jouanin, L., Goujon, T., de Nadaï, V., Martin, M.T., Mila, I., Vallet, C., Pollet, B., Yoshinaga, A., Chabbert, B., Petit-Conil, M., and Lapierre, C., *Plant physiol.*, 123, 1363–1373, 2000.)

in hardwood lignins and 1–2% in softwood lignins, Table 2.4) is in agreement with functional group analysis [1]. This marker is also released from DHPs prepared from coniferyl alcohol, which suggests some reoxidation of the alcohol during or after polymerization. Sinapaldehyde or coniferaldehyde, incorporated into lignins by end-wise-type β–O–4 cross coupling (e.g., Ar–CH = C(OAr)–CHO in Table 2.3), specifically give rise to thioacidolysis indene derivatives. These indenes are recovered as trace components from normal lignins. By contrast, they are recovered in substantial amount from the lignins of cinnamyl alcohol dehydrogenase (CAD) deficient lines. As such, they are valuable marker compounds for CAD deficiency [52,53].

Let's now turn attention to the type of specific information that thioacidolysis provides about lignin structures in softwood, hardwood, and then grass samples (Tables 2.5 through 2.7). The yield in monomers *1–3* (Figure 2.4) can be converted into molar percentage of arylglycerol units only involved in β–O–4 bonds. The data of Table 2.5 reveal the following information about softwood lignins. Monomers *1* and *2* are recovered with yield ranging between 1000 and 1400 μmoles per gram Klason lignin (spruce wood and maritime pine wood samples in Table 2.5). This level suggests that about 30% of lignin units in the wood of common conifers are only involved in β–O–4 bonds. It conversely means that about 70% participate in carbon-carbon or biphenyl ether linkages, a percentage referred to as the "lignin condensation degree." H monomers *1* are recovered in low amount from mature softwood. In contrast, their relative frequency increases at early developmental

TABLE 2.3

Specific Monomers Released by Thioacidolysis of Various Lignin and Non Lignin Substructures. Monomers can be Recovered if the Aromatic Ring of the Parent Structure is not Involved in C–C or Biphenyl Ether Bond

Parent Structure and Side Chain	Recovered Monomers*
C_6C_3 arylglycerol-β-arylethers: –CHOH–CHOAr–CH$_2$OH (for H, G, S, C, or 5–OH G ring only involved in β–O–4)	R–CHSEt–CHSEt–CH$_2$SEt *erythro/threo* ca.50/50 high yield (75–85%)
C_6C_3 arylglycerol-β-arylethers with αCO:–CO–CHOAr–CH$_2$OH	R–CSEt = CHSEt Z and E, low yield (20%)
p-hydroxybenzaldehyde end-groups: –CHO	R–CH(SEt)$_2$
p-hydroxycinnamyl alcohol end-groups: –CH = CH–CH$_2$OH	R–CH = CH–CH$_2$SEt and R–CHSEt–CH$_2$–CH$_2$SEt
Coniferaldehyde end-groups (linked at C–4): G–CH = CH–CHO	G–CHSEt–CH$_2$–CH(SEt)$_2$
p-hydroxycinnamaldehydes linked at C$_β$: Ar–CH = C(OAr)–CHO	Thioethylated indene isomers
Vinyl ether structures in industrial lignins: –CH = COAr	R–CH$_2$–CH(SEt)$_2$ moderate yield (40%)
Benzoic esters and acids: –COOR' and –COOH	R–COOH (highly variable yield for esters)
p-hydroxycinnamic acids (ester and/or ether-linked)	R–CH = CH–COOH and R–CHSEt–CH$_2$–COOH

Source: Lapierre, C., *Forage Cell Wall Structure and Digestibility*, Madison, WI: American Society of Agronomy Inc., 133–163, 1993; Lapierre, C., Rolando, C., and Monties, B., *Holzforschung.*, 40, 47–50, 1986; Kim, H., Ralph, J., Lu, F., Pilate, G., Leplé, J.C., Pollet, B., and Lapierre, C., *J Biol Chem.*, 277, 47412–47419, 2002.

*R = aromatic ring of *p*-hydroxyphenyl, guaiacyl, or syringyl units and, in some industrial or transgenic samples, of catechol or 5-hydroxy-guaiacyl units.

stages (spruce seedlings, Table 2.5) [54]. Compression wood is also substantially enriched in H units and gives less monomers, which confirms that H units make the lignin condensation degree increase. S monomers *3* are generally recovered as trace components from the thioacidolysis of softwood lignins. In exceptional conifers, and in agreement with nitrobenzene data [19], S units occur in substantial amount, which makes their lignins richer in β–O–4 bonds (*Tetraclinis* and *Ephedra* samples in Table 2.5). The impact of physical treatments on softwood lignin structure can be evaluated by thioacidolysis. For instance, UV-photodegradation increases the condensation degree of spruce thermomechanical pulp (TMP) lignin, as well as the frequency of catechol units and vanillin end-groups (irradiated TMP in Table 2.5), signatures of demethylation reactions and of α–β cleavages [55]. A thermal treatment also increases the condensation degree of pine lignins (Table 2.5) and the relative frequency of the C_6C_2 monomer *4*, which is diagnostic for the formation of vinyl ether structures, a situation reminiscent of kraft lignins [43,47].

Relative to softwood lignins, native hardwood lignins provide thioacidolysis monomers in higher yield (between 2000 and 2800 µmoles per gram KL, Table 2.6).

TABLE 2.4

Yields of Thioacidolysis Major and Minor Monomers Released by Natural and Synthetic (DHPs) Lignins. Yields are Expressed in µmoles Per Gram Klason Lignin for Poplar and Spruce Woods and in µmoles Per Gram Sample for DHPs. The Molar Proportion H/G/S of the Major Monomers (Ar–CHR–CHR–CHR$_2$) is Given Together with Their Recovery Yield. The Relative Molar Frequencies of the Minor Monomers, as Compared to the Major Monomers, are Given Between Brackets. Tr: Trace Amounts

Monomer [R = SEt]	Poplar Wood Lignin	Spruce Wood Lignin	Bulk DHP HGS 39/33/28*	Endwise DHP HGS 33/36/31*	Bulk DHP GS 54/46*	Endwise DHP GS 70/30*	G DHP from Coniferin**
Ar–CHR–CHR–CH$_2$R	2310 (100)	1260 (100)	683 (100)	734 (100)	591 (100)	1035 (100)	766 (100)
H/G/S	Tr/41/59	2/98/Tr	45/48/7	45/25/30	-/93/7	-/50/50	
Ar–CH$_2$–CHR$_2$	155 (7)	85 (7)	21 (3)	25 (3)	19 (3)	39 (4)	45 (6)
Ar–CH = CH–CH$_2$R + Ar–CHR–CH$_2$–CH$_2$R	167 (7)	78 (6)	283 (41)	15 (2)	206 (35)	120 (12)	245 (32)
G–CHR–CH$_2$–CHR$_2$	8 (0.3)	18 (1.4)	24 (4)	13 (2)	28 (5)	10 (1)	19 (2)
Total	2712	1439	1011	787	844	1204	1045

Source: Lapierre, C., *Forage Cell Wall Structure and Digestibility*, Madison, WI: American Society of Agronomy Inc., 133–163, 1993; Jacquet, G., Pollet, B., Lapierre, C., Francesch, C., Rolando, C., and Faix, O., *Holzforschung*, 51, 349–354, 1997; Terashima, N., Atalla, R.H., Ralph, S.A., Landucci, L.L., Lapierre, C., and Monties, B., *Holzforschung*, 50, 9–14, 1996.

*Molar proportion of the employed H, G, and/or S cinnamyl alcohols.

**Prepared from coniferin with glucosidase, glucose oxidase, and peroxidase at pH 5.

TABLE 2.5
Total Yield and Relative Frequencies of the Main p-Hydroxyphenyl H, Guaiacyl G, Syringyl S, or Catechol C Monomers Released by Thioacidolysis of Softwood Lignins. The Data are Expressed Relative to the Lignin Content of the Extractive-Free Sample

Sample	Total Yield µmoles/gr Lignin	Type and Proportion of the Main Monomers (H, G, C, S = Monomers with Side Chain-CHSEt-CHSEt-CH₂SEt, Unless Specified)	Molar Percentage of Parent Structures*
Spruce wood *Picea abies* L.			
Normal wood area	1246	0.4% H, 99.6 % G	28
Compression wood area	1072	14% H, 86% G	22
Spruce thermomechanical pulp TMP			
Without irradiation	1080	Essentially G	24
With irradiation (24 h by a xenon lamp)	192	78% G, 11% G-CH(SEt)$_2$, 11% C	5
Spruce seedlings *Picea abies* (six-week old)			
Hypocotyls	1210	6% H, 94% G	27
Needles	1400	7% H, 93% G	31
Maritime pine wood *Pinus maritima*			
Control	1330	Essentially G	30
Heated 5 minutes at 260°C, under N$_2$	600	78% G, 22% G-CH$_2$-CH(SEt)$_2$	13
Wood from exceptional conifers			
Tetraclinis articulata (Vahl)	1790	7% H, 47% G, 46% S	45
Ephedra helvetica (C.M. Meyer)	1830	0.5% H, 35% G, 64.5% S	46

Source: Lapierre, C., *Forage Cell Wall Structure and Digestibility.* Madison, WI: American Society of Agronomy Inc., 133–163, 1993; Lapierre, C., Rolando, C., and Monties, B., *Holzforschung,* 40, 47–50, 1986; Lange, B.M., Lapierre, C., and Sandermann, H., *Plant Physiol.,* 108, 1277–1287, 1995; Pan, X., Lachenal, D., Lapierre, C., and Monties, B., *J Wood Chem Technol.,* 12, 135–147, 1992.

*This percentage corresponds to lignin units only involved in β–O–4 bonds. It is calculated with the assumption that thioacidolysis yield is 80% for the β–O–4 linked structures and that the average molecular weight of C$_6$C$_3$ units is 180.

This level means that the percentage of lignin units only involved in β–O–4 bonds exceeds 50%, a structural trait conventionally correlated to the frequency of S lignin units, since sinapyl alcohol has a pronounced tendency to be involved in β–O–4 bonds [56]. In other words, S-rich hardwood lignins have linear domains made of successive β–O–4 linked units [43,56]. This correlation between S units and β–O–4 bonds has been confirmed by the isolation of a β–O–4 and syringyl- rich lignin

TABLE 2.6
Total Yield and Relative Frequencies of the Main Monomers Released by Thioacidolysis of Native Hardwood Lignins. The Data are Expressed Relative to the Lignin Content of the Extractive-Free Sample

Sample	Total Yield μmoles/gr Lignin	Type and Proportion of the Main Monomers	Molar Percentage of Parent Structures*
Wood of Juvenile Poplar *Populus Tremula x Populus Alba* (six Month-old)			
Control line	2500	34% G, 66% S	62
COMT-deficient line	1220	84% G, 5% S, 11% 5-OH-G	30
CAD-deficient line	1370	54% G, 42% S, 3% SA**	34
Mature Wood			
Poplar *Populus euramericana* cv I214	2310 (± 132***)	37% G, 63% S	58
Beech *Fagus silvatica*	2400	29% G, 71% S	60
Eucalyptus *Eucalyptus urophylla*	2450	26% G, 74% S	61
Birch *Betula verrucosa*	2490	22% G, 78% S	62

Source: Lapierre, C., Rolando, C., and Monties, B., *Holzforschung*, 40, 47–50, 1986; Jouanin, L., Goujon, T., de Nadaï, V., Martin, M.T., Mila, I., Vallet, C., Pollet, B., Yoshinaga, A., Chabbert, B., Petit-Conil, M., and Lapierre, C., *Plant physiol.*, 123, 1363–1373, 2000; Kim, H., Ralph, J., Lu, F., Pilate, G., Leplé, J.C., Pollet, B., and Lapierre, C., *J Biol Chem.*, 277, 47412–47419, 2002.

*This percentage corresponds to lignin units only involved in β–O–4 bonds. It is calculated with the assumption that thioacidolysis yield is 80% for the β–O–4 linked structures and that the average molecular weight of C_6C_3 units is 200.

**SA = sinapaldehyde units linked to lignins at C_β.

***Standard deviation indicated for 55 independent analyses carried out by different analysts and over a 4-year period.

fraction from birch wood [57]. By contrast, a decrease of S units in hardwood lignins is associated with a higher condensation degree. For instance, the lignins of COMT-deficient transgenic poplars correlatively display a high condensation degree and a low frequency of S lignin units, together with the appearance of unusual amount of 5–OH–G units [46,58]. Such a higher condensation degree is also revealed by thioacidolysis and in the lignins of CAD-deficient plants [53,58]. In CAD-deficient poplars, the indene marker compounds originating from sinapaldehyde incorporated into lignins by β–O–4 coupling are recovered to an extent that closely mirrors the CAD deficiency level [53]. *p*-Hydroxyphenyl (H) thioacidolysis monomers, which are obtained as trace components from the thioacidolysis of mature wood lignins (e.g., with a relative frequency in the 0.1–0.3% range), are recovered in higher amount at early developmental stages and from foliar lignins (data not shown).

Similar to softwood lignins, grass lignins generally display moderate to low thioacidolysis yields, indicative of a high condensation degree, in spite of a substantial frequency in S units (Table 2.7) [7]. Grass lignins include a few percent H units (Table 2.7) and some S units acylated at C–γ by *p*-coumaric acid [59–61]. In the lignins of mature maize stems, up to 25% of S units might be *p*-coumaroylated [62]. This acylation does not seem to significantly lower the efficiency of the thioacidolysis depolymerization, since a mild alkaline hydrolysis performed without any lignin loss does not substantially increase the monomer yield [62]. By contrast, the acylation of sinapyl alcohol could alter its tendency to be involved in β–O–4 bonds. Such a hypothesis is supported by the fact that maize lignins have similar frequencies of thioacidolysis S monomers as hardwood lignins, but three to four times lower thioacidolysis yields (Tables 2.6 and 2.7).

TABLE 2.7
Total Yield and Relative Frequencies of the Main Monomers Released by Thioacidolysis of Native Grass Lignins (Mature Stems). The Data are Expressed Relative to the Lignin Content of the Extractive-Free Sample

Sample	Total Yield μmoles/gr Lignin	Type and Proportion of the Main Monomers	Molar Percentage of Parent Structures*
Wheat straw *Triticum aestivum* cv Capitole			
Control	1190	4.6% H, 43.2% G and 52.2% S	30
After a mild alkaline hydrolysis**	1380	3.7% H, 44.5% G and 51.8% S	34
Triticale *Secalotriticum* cv Montcalm	1610	3% H, 42% G and 55% S	40
Rye *Secale cereale* L. cv Dominant	1670	2% H, 44% G and 54% S	42
Corn *Zea mays* F292 line			
Control	900	2.1% H, 36.5% G, 61.4% S	22
bm3 COMT-deficient mutant	790	1.3% H, 58% G, 30.7% S, 10% 5-OH-G	20
bm1 CAD-deficient mutant	470	2.2% H, 34% G, 55% S, 6.6% CA***, 2.6% SA****	12

Source: Barrière, Y., Ralph, J., Grabber, J.H., Guillaumie, S., Argillier, O., Méchin, V., Chabbert, B., Mila, I., and Lapierre, C., *C.R. Biologies*, 847–860, 2004; Jacquet, G., Structure et réactivité des lignines de Graminées et des acides phénoliques associés. PhD Thesis Aix Marseille III University, 301, 1997.

*This percentage corresponds to lignin units only involved in β–O–4 bonds. It is calculated with the assumption that thioacidolysis yield is 80% for the β–O–4 linked structures and that the average molecular weight of C_6C_3 units is 200.

**NaOH 1 M, 2h, 40°C, followed with acidification, H_2O washing and drying of the residue. About half of the grass lignins are eliminated.

***CA = coniferaldehyde units (linked at C4–OH or at C–β).

****SA = sinapaldehyde units (linked at C–β).

Table 2.7's data show that, similar to other COMT-deficient angiosperms [46,63,64], the lignins of the COMT-deficient bm3 maize mutant have a reduced amount of S units and are enriched in condensed bonds and in 5–OH–G units [32,59]. The lignins of the CAD-deficient bm1 corn mutant (Table 2.7) display a higher condensation degree together with unusual amounts of coniferaldehyde and sinapaldehyde units [59].

The analysis of lignin-derived thioacidolysis monomers affords specific information on the type and amount of lignin units only involved in β–O–4 bonds. This robust and routine procedure provided the first evidence for the incorporation of unusual units in the lignins of transgenic or mutant plants. Further developments of the basic thioacidolysis procedure provided novel insights on lignin structure and biosynthesis as explained in the following.

Evaluation of Free Phenolic Units in Lignins by Thioacidolysis of Permethylated Samples

The content of free phenolic groups in lignins has a major biochemical and techno-logical significance. When thioacidolysis is performed on diazomethane-methylated samples, additional monomers **8–10** with methylated phenolic groups (Figure 2.6) are recovered [65,66]. The total yield of thioacidolysis monomers is similar before

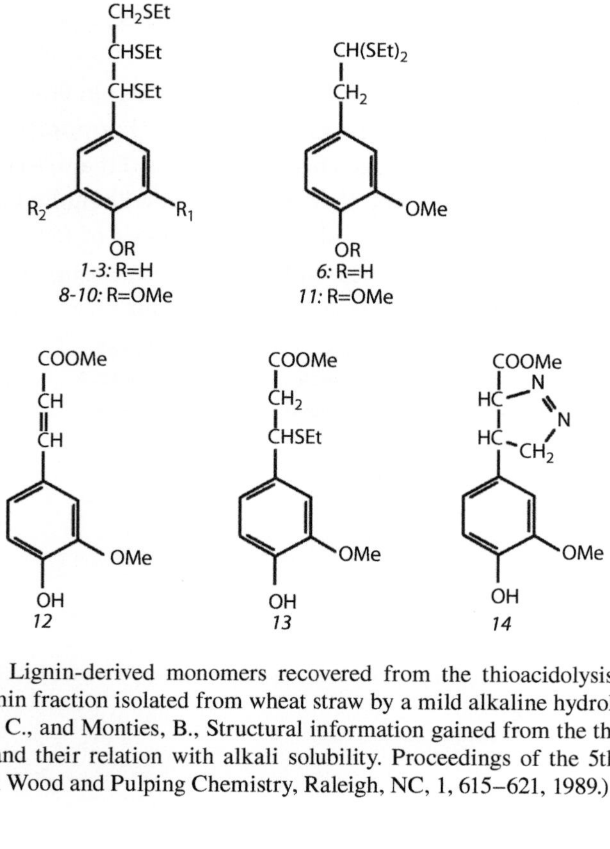

FIGURE 2.6 Lignin-derived monomers recovered from the thioacidolysis of a CH_2N_2-methylated lignin fraction isolated from wheat straw by a mild alkaline hydrolysis. (Adapted from Lapierre, C., and Monties, B., Structural information gained from the thioacidolysis of grass lignins and their relation with alkali solubility. Proceedings of the 5th International Symposium on Wood and Pulping Chemistry, Raleigh, NC, 1, 615–621, 1989.)

(monomers *1–3*, Figure 2.4) and after (sum of monomers *1–3* and monomers *8–10*, Figure 2.6) methylation of the sample; this is not the case for NBO treated lignins, due to Canizzaro reactions [67]. Beside the main monomers *1–3* and *8–10*, other monomers are recovered from methylated samples. When recovered in low relative amount, phenolic *4* and methylated *11* C_6C_2 G monomers, have the same origin as G C_6C_3 monomers *2* and *9*. Ferulate methyl esters *12–14*, which are released from alkali-soluble grass lignins methylated prior to thioacidolysis, originate from ferulic acid ether-linked to lignins. The pyrazoline derivative *14* is due to the 1-3 dipolar addition of CH_2N_2. The recovery of ferulate derivatives *12–14*, esterified at COOH and with a free phenolic group at C–4, establishes that (a) methyl esters partially survive thioacidolysis and (b) ferulic acid is ether-linked to grass lignin fractions isolated by a mild alkaline treatment [71].

The relative frequency of monomers *8–10* indicates the proportion of free phenolic groups within β–O–4 linked H, G, or S units. The data in Table 2.8, obtained for a series of soluble organosolv spruce lignins, points to the fact that this proportion closely follows the frequency of free phenolic units determined by NMR [68]. This correlation holds even for lignin fractions with low β–O–4 content that are recovered at the end of the cook. The frequency of free phenolic groups in insoluble cell wall lignins ranges from 15 to 30% for softwoods and 10 to 15% for hardwoods [69]. Thioacidolysis values (Table 2.9) are consistently found within these ranges. In other words, the frequency of free phenolic groups

---

TABLE 2.8

Evaluation of Free Phenolic Units in Five Successive Lignin Fractions Recovered by Organosolv Delignification of Spruce Thermomechanical Pulp TMP (29% Klason Lignin) in a Flow-Through Reactor (Ethanol–Water 1:1, v/v, 0.1M CH_3COOH, 175°C). The Percentage of Free Phenolic Groups in β–O–4 Linked G Units is Determined by Thioacidolysis of CH_2N_2 Methylated Samples. The Percentage of Free Phenolic Groups Per 100 C_6C_3 is Measured by ^{19}F NMR of Trifluoroacetylated Lignins

Sample	% Total Lignin	Thioacidolysis Yield in μmole Per Gram Lignin	% free Phenolic Groups in β–O–4 Linked G Units	% Free Phenolic Groups Per 100 C_9
Spruce TMP	100	1160	22	–
L1 (0–30 min)	9.5	840	31.4	33
L2 (30–60 min)	21.8	803	25	29
L3 (60–90 min)	19.4	808	20	24
L4 (90–120 min)	18.1	763	16	20
L5 (120–300 min)	31.2	624	12	16
Final residue	28	420	7	–

Source: Sjöström, E., *Wood Chemistry: Fundamental and Applications,* 2nd ed. San Diego: Academic Press, 71–89, 1992.

TABLE 2.9
Percentage of Free Phenolic Groups in β–O–4 Linked H, G, and S Lignin Units, as Determined by Thioacidolysis of CH_2N_2 Methylated Samples. The Reported Relative H/G/S Frequency is Based on Lignin-Derived Thioacidolysis Monomers (tr : Trace Amount). The Lignin-Derived Monomers *1–3* and *8–10* are Outlined in Figure 2.6

| | % Free Phenolic Groups in | | | Relative Frequency |
| | H : 100 × | G : 100 × | S : 100 × | H/G/S (in % |
Sample	8/[8 + 1]	9/[9 + 2]	10/[10 + 3]	Molar)
Spruce (*Picea abies*)				
Wood native lignin	90	25	–	2/98/tr
Cell culture lignin (extracellular)	93	54	–	8/92/-
Maritime pine (*Pinus maritima*)				
Normal wood native lignin	–	23	–	1/99/tr
Compression wood native lignin	91	22	–	18/82/tr
MWL lignin fraction from compression wood	90	33	–	15/85/tr
Poplar (*P. euramericana* cv I214)				
Wood native lignin	–	25	3.5	tr/33/67
MWL lignin fraction	–	31	6	tr/41/59
Native lignins of grass samples				
Wheat straw (*Triticum aestivum*, cv champlein)	92	43	5	5/49/46
Rye straw (*Secale cereale*, cv Montcalm)	91	38	5	2/44/54
Corn stem (*Zea mays*)	91	50	5	4/35/61
HGS synthetic lignins (DHPs)				
Bulk-type DHP	69	57	50	45/48/7
End wise-type DHP	80	20	6	45/25/30

Source: Lapierre, C., *Forage Cell Wall Structure and Digestibility*. Madison, WI: American Society of Agronomy Inc., 133–163, 1993; Jacquet, G., Pollet, B., Lapierre, C., Francesch, C., Rolando, C., and Faix, O., *Holzforschung*, 51, 349–354, 1997; Lange, B.M., Lapierre, C., and Sandermann, H., *Plant Physiol.*, 108, 1277–1287, 1995; Lapierre, C., Monties, B., and Rolando, C., *Holzforschung*, 42, 409–411, 1988; Lapierre, C., and Rolando, C., *Holzforschung*, 42, 1–4, 1988; Jacquet. G., Structure et réactivité des lignines de Graminées et des acides phénoliques associés. PhD Thesis Aix Marseille III University, 301, 1997.

in β–O–4 linked lignin units nicely approximates that of the whole polymer. This frequency is higher in milled wood lignin (MWL) samples relative to their parent native lignins (Table 2.9). This increase is probably related to the cleavage of β–O–4 bonds during the milling process.

A major outcome of the permethylation-thioacidolysis procedure was to provide the first evidence that H, G, and S units are not evenly distributed in native lignins [65,66]. Whenever present, more than 90% of H units are terminal units with free phenolic groups (Table 2.9). In contrast, S units essentially are internal units with etherified phenolic groups. The behavior of G units is intermediate, with a frequency of free phenolic groups ranging between 20 and 60%. These differences have been rationalized on the basis of the redox potentials of the species involved in the oxidatively driven lignin biosynthesis; the high oxidation potential of *p*-coumaryl alcohol account for the preferential distribution of H lignin units as terminal end-groups [70].

Thioacidolysis of permethylated samples also revealed that grass lignins have a twofold higher level of free phenolic G units, as compared with wood lignins (Table 2.9) [65,71]. This structural trait may be interpreted as a higher content in branching structures and/or as the occurrence of smaller lignin domains easily extracted from the walls. While wood lignins are poorly extractable in alkali at room temperature [72], this treatment solubilizes 50–75% of grass lignins from mature stems. The high frequency of free phenolic G units in grass lignins is essentially responsible for this alkaline solubility [73]. Grass lignins would be distributed in the cell wall as small domains rich in free phenolic groups and leachable by cold alkali. Relative to control samples, lignins of CAD-deficient plants also have a higher level of free phenolic groups (Figure 2.7). This higher level makes them unusually soluble in cold alkali [53] and more susceptible to kraft pulping [58].

Structural information gained from thioacidolysis of permethylated samples is provided by a routine technique. In contrast, the last development of thioacidolysis, the analysis of lignin-derived dimers after an additional desulfurization step, is not an easy technique to master.

DETERMINATION OF THIOACIDOLYSIS LIGNIN-DERIVED DIMERS: FURTHER INFORMATION FROM A NONROUTINE PROCEDURE

The analysis of lignin-derived dimers is possible after desulfurization of the thioacidolysis mixture over Raney nickel under mild conditions (in dioxane, 50°C, 4 h) in order to reduce their molecular weight and isomeric forms [43]. Gymnosperm lignins essentially give rise to GG dimers, together with trace amount of HG dimers. Angiosperm lignins additionally provide GS and SS dimers. The main dimers, their GC separation and their mass spectra are outlined in Figures 2.8 and 2.9 and in Table 2.10.

Dimers **15–19** represent the β–5 bonding pattern. Upon thioacidolysis, the benzylic ether linkage of phenylcoumaran-type models is cleaved with concomitant thioethylation. The C_γ-methylol group of β–5 structures is lost to an extent that depends on thioacidolysis duration. In contrast, the total amount of **16 + 17** or **18 + 19** is stable when this duration varies between 3 and 6 hours. Minor dimers of the β–5 series are observed as dimers with shortened side chains, similar to dimer **15** or with rearranged side chains similar to monomers **4–5** (Figure 2.4).

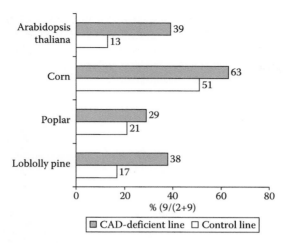

FIGURE 2.7 Frequency of #–O–4 linked G units with free phenolic groups, given as % (9/ (9 + 2) (Figure 2.6) in the mature stems of bm1 corn mutant (Based on Barrière, Y., Ralph, J., Grabber, J.H., Guillaumie, S., Argillier, O., Méchin, V., Chabbert, B., Mila, I., and Lapierre, L., *C.R. Biologies.*, 847–860, 2004); and A. thaliana mutant (Sibout, R., Eudes, A., Mouille, G., Pollet, B., Lapierre, C., Jouanin, L., and Seguin, A., *Plant Cell.*, 17, 2059–2076, 2005); or in the wood of loblolly pine null mutant (Lapierre, C., Pollet, B., MacKay, J.J., and Sederoff, R.R., *J Agric Food Chem.*, 48, 2326–2331, 2000); and transgenic poplar (Lapierre, C., Pilate, G., Pollet, B., Mila, I., Leplé, J.C., Jouanin, L., Kim, H., and Ralph, J., *Phytochemistry*, 65, 313–321, 2004); samples, relative to the control samples.

Thioacidolysis of β–1 lignin structures also causes thioethylation at C_α and partial loss of the C_γ-group, as revealed by model studies with appropriate β–1 dimers. Dimers **20–22** originate from β–1 structures of native lignins and from *p,p'*-stilbene structures of industrial kraft lignins. Dimers **20–26** are released in high amount from cell wall lignins in contrast to synthetic lignins [50]. The frequency of β–1 parent structures is higher in water-soluble lignin fragments associated with polysaccharides [76]. These results support the preferential formation of β–1 structures as end-groups at the interface between lignins and the polysaccharide matrix. The β–1 coupling mode may involve the detachment of a side chain as a glyceraldehyde-2 aryl ether [1,11], a mechanism confirmed by the identification of glycerol-2-aryl ethers from the thioacidolysis of borohydride-reduced samples [77]. The parent structures of dimers **20–26** would be (at least partially) dienone progenitors [70] undergoing acid-catalyzed degradation to form the corresponding 1,2-diarylpropanes upon thioacidolysis. This hypothesis is supported by acid hydrolytic experiments run on diazomethane methylated spruce wood [78] and by the identification of the spirodienone NMR signals [12]. As shown in Table 2.11, the borohydride reduction of a spruce MWL sample induces an important decrease in the recovery of the coniferaldehyde-derived monomer (as expected) and also in the recovery of the β–1 dimers, whereas the other dimers essentially are not affected. This β–1 decrease may be assigned to the elimination of the lignin fraction most closely associated to carbohydrates during $NaBH_4$ reduction [79] and/or to the reduction of the spirodienone structures.

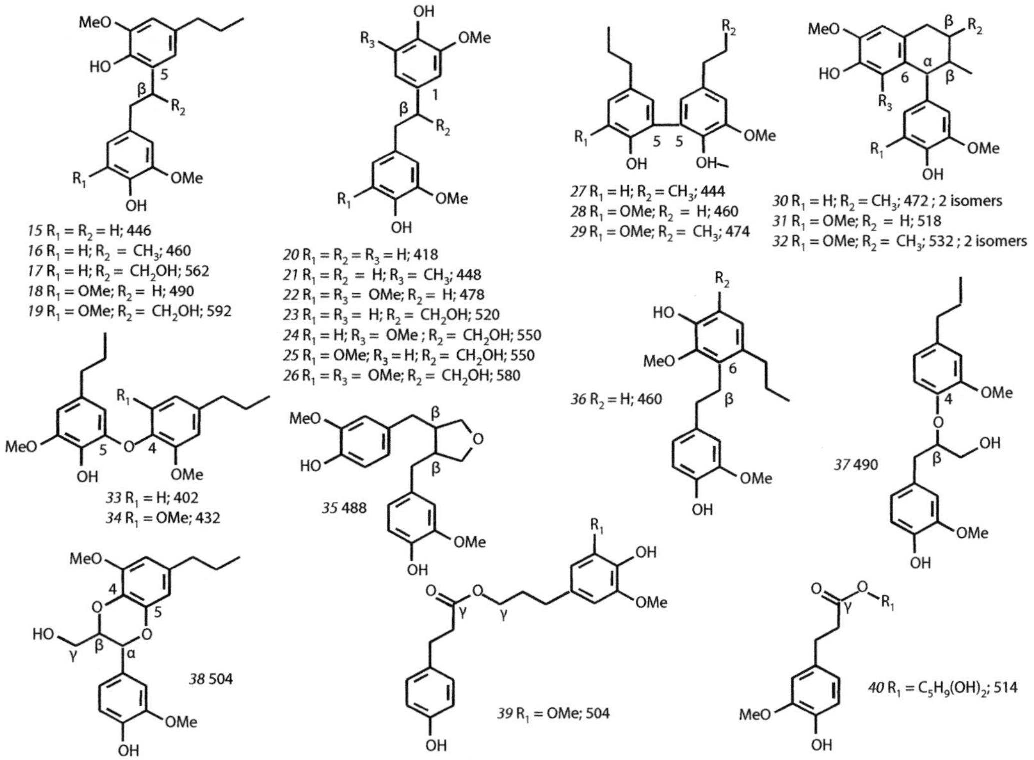

FIGURE 2.8 Dimers recovered from thioacidolysis then Raney nickel desulfurization of lignins. The nominal mass of the TMS derivative is indicated for each compound. (Adapted from Lapierre, C., *Forage Cell Wall Structure and Digestibility*, Madison, WI: American Society of Agronomy Inc., 133–163, 1993; Jacquet, G., Structure et réactivité des lignines de Graminées et des acides phénoliques associés. PhD Thesis Aix Marseille III University, 301, 1997.)

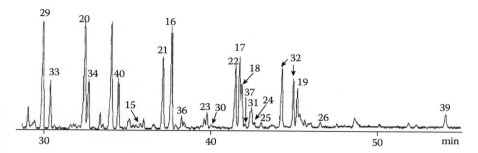

FIGURE 2.9 Partial GC chromatogram showing the separation of the main lignin-derived dimers (analyzed as their TMS derivatives) recovered from thioacidolysis, then desulfurization of a wheat straw sample. Peak numbers correspond to the assignments of Figure 2.8. (Jacquet, G., Structure et réactivité des lignines de Graminées et des acides phénoliques associés. PhD Thesis Aix Marseille III University, 301, 1997.)

Dimers *27–29* represent the biphenyl bonding pattern. Lignin biphenyl structures are involved in the dibenzodioxocin bonding pattern [6,70] to an extent that seems to vary with the origin of the sample [10,80]. Upon thioacidolysis, a dibenzodioxocin trimeric model gives rise to a biphenyl dimer in about 25% yield and to a seven-membered ring trimer as the main degradation product [81]. This observation may suggest that dibenzodioxocin structures are poorly characterized by thioacidolysis. However, steric effects have been shown to preclude the formation of the seven-membered ring structure in native lignins [80].

Upon thioacidolysis, resinol structures yield thioethylated tetralin-type dimers, which produce dimers *30–32* upon desulfurization. Lignin resinol structures are thus degraded through Cα–O–Cγ bond cleavage and Cα–C6 bond formation. The 3,4-divanillyltetrahydrofuran dimer *33*, specifically released by softwood samples, is already observed before desulfurization [78], suggesting that oxido-reductive mechanisms contribute to the assembly of lignins [3].

Some minor dimers can be assigned to β-6 structures on the basis of their mass spectra. For instance, dimer *36* is obtained in higher relative amount from kraft lignin samples [82]. Dimer *37*, diagnostic for incomplete β-O-4 cleavage [83], is recovered in trace amount in the standard thioacidolysis conditions (peak *37*, Figure 2.9). The benzodioxane dimer *38* is specifically released by COMT-deficient angiosperms. Upon thioacidolysis, the parent benzodioxane structures give 5–OH–G monomers as the main product (80%). They, nevertheless, partially survive (20%), thereby revealing the main bonding mode of the non conventional 5–OH–G units in lignins [46,84].

Dimers *39* and *40*, issued from grass samples, reveal the linkage type of *p*-coumaric and ferulic esters and show that some ester bonds survive thioacidolysis. Dimer *39* confirms that *p*-coumaric acid is ester-linked at Cγ of S lignin units [60,61]. Compound *40*, tentatively assigned from its mass spectral fragmentation pattern, may originate from ferulic units ester-linked to pentose units [85].

The relative frequencies of the main dimers released by thioacidolysis of lignins display important variations according to the sample origin (Tables 2.11 and 2.12). In

TABLE 2.10
Abbreviated Mass Spectra (Electronic Impact, 70 eV) of the TMS Derivatives of Lignin-Derived Dimers Recovered After Thioacidolysis Then Raney Nickel Desulfurization

TMS of*	m/z (Relative Intensity)
16	460 (25), 445 (5), 251 (20), 236 (7), 221 (5), 209 (100), 207 (20), 193 (7), 179 (5), 73 (40)
17	562 (25), 472 (15), 457 (5), 352 (10), 322 (5), 263 (15), 209 (40), 191 (50), 179 (15), 103 (5), 73 (100)
18	490 (30), 251 (10), 239 (100), 236 (5), 209 (25), 207 (5), 73 (30)
19	592 (30), 502 (5), 471 (5), 352 (5), 263 (10), 239 (50), 209 (10), 191 (60), 179 (5), 103 (5), 73 (100)
20	418 (10), 403 (3), 209 (100), 193 (5), 179 (25), 149 (5), 73 (25)
21	448 (25), 433 (5), 239 (100), 223 (10), 209 (50), 193 (5), 179 (15), 149 (5), 73 (30)
22	478 (25), 463 (5), 448 (5), 239 (100), 223 (10), 209 (25), 179 (5), 73 (20)
23	520 (25), 311 (100), 280 (10), 223 (20), 209 (40), 207 (30), 192 (10), 179 (20), 149 (40), 103 (5), 73 (80)
24	550 (25), 535 (5), 520 (10), 341 (100), 253 (20), 239 (30), 209, 50, 179 (10), 73 (80)
25	550 (25), 311 (75), 280 (5), 253 (5), 239 (50), 223 (25), 209 (25), 192 (10), 179 (15), 73 (100)
26	580 (25), 565 (5), 341 (100), 310 (5), 253 (25), 239 (25), 209 (25), 179 (5), 73 (60)
27	444 (50), 429 (40), 415 (40), 147 (30), 73 (100)
28	460 (50), 445 (40), 430 (30), 371 (20), 343 (10), 147 (5), 73 (100)
29	474 (100), 459 (50), 445 (40), 415 (15), 385 (10), 357 (8), 343 (10), 147 (5), 73 (100)

30	472 (100), 457 (8), 442 (5), 415 (8), 385 (40), 355 (5), 276 (5), 261 (8), 223 (5), 209 (8), 178 (10), 73 (75)
31	518 (80), 503 (15), 488 (20), 445 (5), 414 (5), 384 (5), 341 (10), 292 (70), 277 (25), 262 (25), 239 (8), 173 (20), 73 (100)
32	532 (70), 517 (5), 501 (15), 445 (10), 414 (5), 384 (8), 341 (5), 306 (60), 291 (8), 275 (15), 239 (8), 187 (20), 73 (100)
33	402 (100), 387 (40), 373 (40), 372 (50), 357 (25), 343 (25), 221 (15), 209 (10), 179 (5), 157 (5), 73 (70)
34	432 (100), 417 (40), 403 (25), 387 (8), 373 (8), 239 (5), 172 (8), 73 (50)
35	488 (70), 473 (5), 209 (100), 193 (20), 179 (75), 149 (20), 73 (70)
36	460 (20), 445 (5), 251 (100), 223 (10), 220 (15), 209 (20), 193 (5), 179 (15), 149 (5), 73 (25)
37	490 (5), 324 (5), 281 (5), 235 (8), 223 (5), 209 (100), 193 (5), 179 (10), 165 (3), 137 (5), 103 (7),73 (50)
38	504 (20), 489 (3), 414 (10), 324 (15); 293 (8), 234 (5), 209 (8), 294 (5), 193 (25), 103 (8), 73 (100).
39	504 (40), 239 (25), 237 (20), 209 (25), 193 (30), 179 (25), 166 (5), 163 (5), 73 (100)
40	514 (20), 397 (5), 340 (25), 268 (8), 251 (7), 231 (8), 222 (8), 209 (45), 192 (8), 179 (9), 147 (8), 117 (100), 73 (90)

Source: Lapierre, C., Pollet, B., and Rolando, C., *Res Chem Intermed.*, 21, 397–412, 1995; Lapierre, C., Pollet, B., Monties, B., and Rolando, C., *Holzforschung*, 45, 61–68, 1991; Ralph, J. and Grabber, J.H., *Holzforschung*, 50, 425–428, 1996; Jacquet, G., Structure et réactivité des lignines de Graminées et des acides phénoliques associés. PhD Thesis Aix Marseille III University, 301, 1997.

*The numbers refer to structures outlined in Figure 2.8 and to peaks quoted in Figure 2.9.

TABLE 2.11
Main Lignin-Derived Monomers and Dimers Recovered from Thioacidolysis
of (A) Spruce MWL and (B) Borohydride Reduced Spruce MWL. The
Reduced MWL Sample was Recovered in 80% Yield. Yields in Lignin-
Derived Monomers and Dimers are Expressed in μmole/gram Sample

Thioacidolysis Compound (R = SEt)	A	B
G–CHR–CHR–CH$_2$R	910	1050
G–CH$_2$–CHR$_2$	70	40
Coniferaldehyde-derived monomer	30	0
Coniferyl alcohol-derived monomer	60	110
Dihydroconiferyl alcohol	10	10
5–5 dimers	66	68
β–5 dimers	65	68
β–1 dimers	62	33
5–O–4 dimer	16	17
Tetrahydrofuran dimer	13	14

Source: Based on Lapierre, C., Pollet, B., Monties, B., *Phytochemistry*, 30, 659–662, 1991.

the case of softwood lignins, the mains lignin-derived dimers are representatives of biphenyl, β–5 and β–1 bonding patterns (Table 2.11) [82], whereas the β–β GG dimers originating from pinoresinol structures are recovered as trace components. The data reported in Table 2.12 for a few native or isolated angiosperm lignins are an oversimplification of dimer profiles. Detailed data, not reported herein for clarity, give the proportion of each possible GG, GS, and/or SS dimer-types for each bond type [7]. Not unexpectedly, the S/G ratio of lignin-derived monomers generally exceeds the one calculated from the dimers. Hardwood samples predominantly release β–1 dimers, mainly represented by SS and GS dimers. In addition, syringaresinol-derived SS dimers are recovered in substantial amount, whereas medioresinol- and pinoresinol-derived dimers are obtained in trace amount. Biphenyl dimers are recovered in low amounts from hardwoods, with the exception of COMT-deficient poplar lines, which display a softwood-type profile due to shortage in S units [46]. Biphenyl dimers are recovered in substantial amounts from the bm6 sorghum mutant, with altered CAD activity [85]. This trait is also observed in CAD-deficient pine [77]. In both plants, this structural peculiarity is linked to the unusually high incorporation of coniferaldehyde through a biphenyl coupling mode.

The determination of lignin-derived dimers illustrates that thioacidolysis scope is no more confined to the non condensed moiety of lignins. The GC-MS analysis of dimers recovered from the desulfurization of thioacidolysis-degraded lignins is a complex technique. However, a simpler and more comprehensive way to evaluate the condensed fraction of thioacidolysis-degraded lignins is by size exclusion chromatography (SEC) of thioacidolysis mixtures. Thioacidolysis combined with SEC analysis of the oligomeric, dimeric, and monomeric thioacidolysis products provides original information on the condensation degree of industrial lignins [86]. More recently, Gellerstedt and

TABLE 2.12
Relative Frequencies (% Molar) of the Main Dimers (Figure 2.8) Released by Thioacidolysis Then Desulfurization of Angiosperm Lignins. Their Recovery Yield Approximately Represents 10% of That of the Main Monomers (Tables 2.6 and 2.7). These Values are Given for Comparative Purposes. The Figures May Substantially Vary with the S Frequency and, Therefore, with Parameters Such as Age or Tissue. Tr: Trace Amount

Sample	5–5	β–5	β–1	4–O–5	Syringaresinol-Derived Dimers
Native Hardwood Lignins					
Poplar wood					
control line	10	26	29	8	26
COMT-deficient line* [46]	26	37	22	6	Tr
Birch wood	5	15	44	9	29
Native Grass Lignins					
Wheat straw	18	30	29	14	9
Sorghum stem					
control line	31	18	18	17	16
bm6 CAD-deficient mutant	60	10	12	12	4
Isolated Hardwood Lignins					
Poplar MWL	6	17	36	11	30
Poplar dioxan lignin	10	21	20	12	36
Aspen steam explosion lignin	3	17	23	6	51
Birch kraft lignin (39% delignification)	3	8	56	8	29

Source: Lapierre, C., Forage Cell Wall Structure and Digestibility. Madison, WI: American Society of Agronomy Inc., 133–163, 1993; Lapierre, C., Pollet, B., and Rolando, C., *Res Chem Intermed*, 21, 397–412, 1995; Jacquet, G., Structure et réactivité des lignines de Graminées et des acides phénoliques associés. PhD Thesis Aix Marseille III University, 301, 1997.

*This line gives 8% of the diether dimer *38* (Figure 2.8).

coworkers used this method to elegantly establish that lignin structure varies between different lignin carbohydrate complexes (LCCs) isolated from softwood [87].

DERIVATIZATION FOLLOWED BY REDUCTIVE CLEAVAGE (DFRC): A METHOD WITH UNIQUE FEATURES THAT PROVIDED NOVEL INFORMATION ON LIGNIN STRUCTURE

In 1997, Ralph and coworkers introduced a new analytical degradation method, the DFRC procedure (acronym for derivatization followed by reductive cleavage). Similar to thioacidolysis, this method selectively cleaves β-aryl ethers in lignins and releases analyzable monomers for quantification. It consists of three successive steps: (a) acetyl bromide treatment, which results in the bromination of the benzylic position and the acetylation of free hydroxyl groups, (b) reductive cleavage of the β–O–4 bonds with zinc dust, and (c) acetylation of newly formed free phenolic groups for GC determination of the lignin-derived monomers [88,89] (Figure 2.10a). The main recovered monomers, originating from lignin units only involved in arylglycerol-β–ethers structures are the acetylated derivatives of p-hydroxycinnamyl alcohols (Figure 2.10a).

When applied to β–O–4 dimers, this multistep method gives these monomers in quantitative yield [48]. However, when applied to lignins, and contrary to the first published data by Ralph and coworkers [89], it repeatedly seems that DFRC lignin-derived monomers are recovered in lower yield than their thioacidolysis analogues [90–92]. This lower depolymerization efficiency is supported by the comparative SEC analysis of thioacidolysis and DFRC mixtures recovered from a softwood MWL sample [92]. The fact that the DFRC method does not seem to quantitatively cleave the β–O–4 bonds of the lignin polymer has been assigned to the persistence of the α-brominated group when the bromine and the β–O–4 ether oxygen do not display a cis orientation or, by contrast, to the hydrolysis of the α-bromide prior to reductive cleavage [92].

When subjected to DFRC, lignins not only provide major monomers from arylglycerol-β–ethers structures, but also a series of minor monomers originating from other lignin structures [89,93]. However, the quantitative determination of these minor monomers seems to be difficult, since most parent structures give rise to several DFRC products [88]. For instance, vanillin end-groups give rise to three different monomers (Figure 2.10b). The detailed identification of DFRC dimers recovered from loblolly pine wood has been reported [94], without further published quantification and/or comparison of DFRC dimers released by a wider range of plant materials.

Ten years after its creation, the DFRC method does not seem to be so widely used, based on its initial promise. This is probably due to some difficulties in mastering its quantitative and routine use. This new method nevertheless has provided new and enhanced data about lignin chemistry and biosynthesis. For instance, one advantage of the method is that γ-ester groups on lignins remain largely intact. The recovery of diagnostic dimeric compounds (mainly the acetate derivative of p-coumaroylated sinapyl alcohol, Figure 2.10c) confirmed that p-coumarates are exclusively attached to the γ-position of grass lignin side chains [95].

More important, and with a modified DFRC procedure (all acetate-based reagents replaced with their propionate analogues), the recovery of a DFRC β–β syringyl dimer acetylated at C_γ has provided unique evidence that kenaf lignins are acetylated at C_γ and that this acetylation occurs at the monolignol stage [96]. The DFRC-based evidence of acetylated units in lignins has been obtained in a larger set of vascular plants [10,97,98].

Since DFRC does not scramble the β-carbon stereochemistry, the recovery of optically inactive β–5 (Figure 2.10d) and β–β DFRC dimers very elegantly confirmed the racemic nature of lignins [99]. The isolation of novel arylisochroman trimers from DFRC-degraded loblolly pine wood sample supports the occurrence of

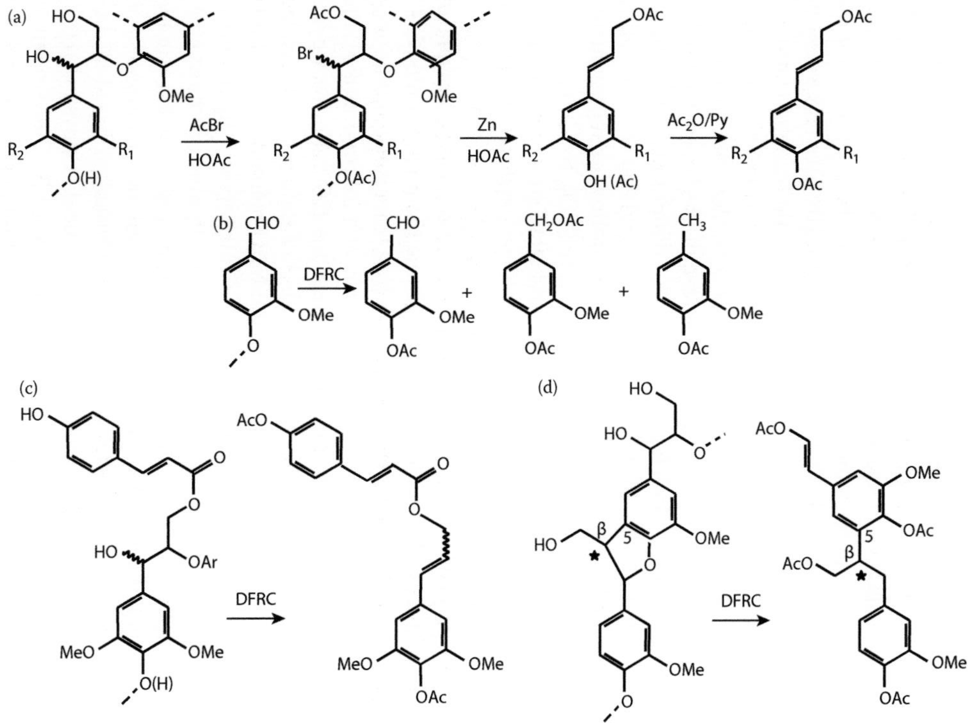

FIGURE 2.10 Basic DFRC reactions on lignin structures: (a) degradation of H ($R_1 = R_2 = H$), G ($R_1 = OMe$; $R_2 = H =$ and S ($R_1 = R_2 = OMe$) lignin units only involved in arylglycerol-β–ethers structures; (b) degradation of vanillin end-groups; (c) main dimer originating from p-coumaroylated syringyl lignin units; and (d) main dimer recovered from G units involved in phenylcoumaran structures (the optical center at β-carbon is denoted with a★). (Lu, F., and Ralph, J., *J Agr Food Chem.*, 45, 2590–2592, 1997; Lu, F., and Ralph, J., *J Agr Food Chem.*, 47, 1988–1992, 1999; Ralph, J., Lu, F., Hatfield, R.D., and Helm, R.F., *J Agr Food Chem.*, 47, 2991–2996, 1999.)

spirodienone structures in lignins [10,100]. These results, provided by a new degradative method, further illustrate the specific role of chemical degradations in lignin research.

OZONATION: AN OUTSTANDING TOOL TO EXPLORE THE STRUCTURE AND STEREOCHEMISTRY OF LIGNIN SIDE CHAINS

The ozonation technique is the only degradative method exclusively targeted at lignin side chains. The method preserves the side chains and destroys aromatic rings; the latter appear as carboxylic groups on the survived side chains (Figure 2.11) [1,8,101]. Importantly, the stereochemistry of the side chains is preserved in the recovered monocarboxylic or dicarboxylic aliphatic acids. For instance, the *erythro* and *threo* diastereomers of arylglycerol-β-aryl ether structures specifically give rise to erythronic and threonic acids (*E* and *T* in Figure 2.11a). The side chain oxidation products of β-1 diarylpropane structures, which occur in lignins as a mixture of *erythro* and *threo* configurations, are the *erythro* (*E'*) and *threo* (*T'*) isomers of 3-hydroxy-2-hydroxymethyl butanedioic acid (Figure 2.11b). By contrast, the β–5 phenylcoumaran structure, which essentially occurs with the trans (*erythro*) configuration, would only provide the *E'* isomer (Figure 2.11b) [8,102]. The ozonation yield and reproducibility have been recently improved by a mild reduction treatment following the ozonation step and by an optimization of the derivatization procedure [103]. By doing so, the researchers were able to conclude that arylglycerol-β-aryl ether structures comprise at least 35% of the C_6C_3 structure in birch wood meal, with an *erythro/threo* ratio of 2.8 [103].

With the improved procedure, both the *erythro/threo* ratio and the content of β–O–4 structures in lignin can be evaluated by determining the relative ratio (*E/T*) and absolute yield of erythronic and threonic acid recovered from isolated lignins or from wood meal samples (*E* and *T* in Figure 2.11a). In native hardwood lignins, the *erythro* form of β–O–4 structures predominates, whereas the *E/T* ratio from gymnosperms is approximately 1.0. In hardwood species, this *E/T* ratio is closely related to the S/G ratio of hardwood lignins (determined by the NBO method), a result that suggests that the type of aromatic ring may stereochemically control lignin structure [104]. Ozonation-based studies further confirmed that lignins are biosynthesized as racemic and optically inactive structures [105]. They also revealed that the *E/T* ratio of β–O–4 structures is higher in tension wood [106] and in compression wood [107] lignins, relative to the normal wood samples. Ozonation-based studies also provided information about the stereochemistry of chemical or physical lignin degradation procedures. For instance, they revealed that *erythro* β–O–4 structures are more rapidly degraded than *threo* structures during kraft pulping [108] or wood milling [109].

The β–1 diarylpropane structures and their companion glyceraldehyde-2-aryl ether structures have been reported to be very minor constituent of softwood lignins, based on the recovery of diagnostic ozonation compounds [8]. However, in more recent studies, ozonation analysis showed that the LCCs isolated from birch wood or from sugi wood have a relatively high number of both β–1 and glyceraldehyde-2-aryl ether structures [102,110,111]. This result is quite consistent with former

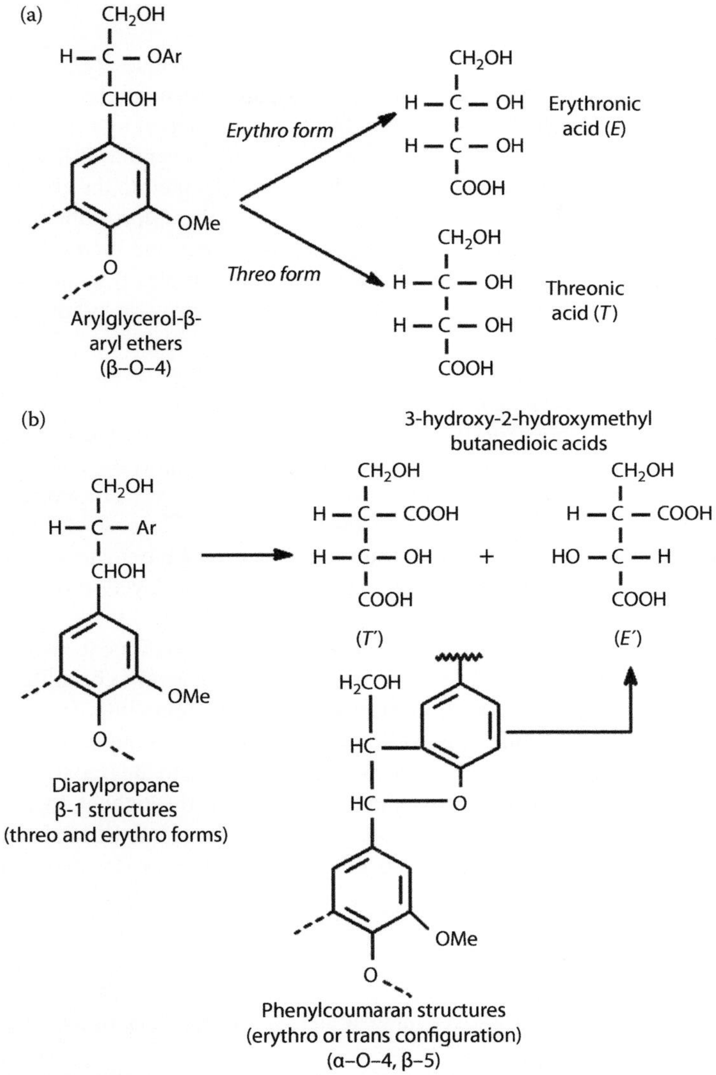

FIGURE 2.11 Basic ozonation reactions on lignin structures: (a) formation of erythronic and threonic acids from *erythro* and *threo* forms of lignin side chains involved in β–O–4 bonds and (b) degradation of diarylpropane and phenylcoumaran structures. (Adapted from Akiyama, T., Sugimoto, T., Matsumoto, Y., and Meshitsuka, G., *J Wood Sci.*, 48, 210–215, 2002; Aimi, H., Matsumoto, Y., and Meshitsuka, G., *J Wood Sci.*, 51, 252–255, 2005.)

thioacidolysis data revealing that the frequency of β–1 structures is higher in water-soluble lignin fragments associated with polysaccharides [76].

The ozonation method has provided novel insight into the stereo structure of lignin side chains. It has other important biological and industrial outcomes, such as the analysis of benzylic ether bonds established between lignins and polysaccharides [112].

CONCLUSION

The first evidence of the aromatic nature of lignin was provided at the beginning of the twentieth century by the recovery of veratric acid from methylated spruce wood subjected to permanganate oxidation, a method then developed by Freudenberg and, more comprehensively, by Miksche and coworkers in the seventies [40]. Over the last 50 years, considerable efforts and creativity have been devoted to the conception and development of innovative degradation methods or to the improvement of already established methods. However, as repeatedly reported, no one method can answer all questions related to lignins, which, because of their complex polymeric structure, call for the combined and critical use of various methods provided with complementary performances.

Some lignin degradation methods have received a low or even null acceptance, whatever their informative capabilities, mainly due to practical considerations, such as low reproducibility, poor routine capabilities, and/or high expertise to be mastered. This is why acidolysis, ethanolysis, thioacetolysis, hydrogenolysis, and permanganate oxidation, which play an outstanding role in our current knowledge on lignin structure, have been mainly confined to a few laboratories. The case of thioacidolysis is intermediate, since this method has routine aspects, which benefited an unusually rapid acceptance, and more sophisticated aspects when employed in the determination of lignin-derived dimers.

It is very likely that chemical degradation methods will continue to have an important role in lignin research. However, two main challenges need to be addressed. The first is to develop methods capable of providing novel insights into lignin structural features, whatever their routine capabilities. The second is to increase the throughput performances of established (or forthcoming) chemical degradation methods. Chemical degradation methods will thus assist the high-throughput spectral methods as ways to efficiently screen the genetic and environmental variability of lignin structures and to rationally design lignocellulosics suitable for downstream conversion processes.

REFERENCES

1. E Adler. Lignin chemistry—Past, present and future. *Wood Sci Technol* 11:169–218, 1977.
2. WG Glasser. Classification of lignins according to chemical and molecular structures. In WG Glasser, RA Northey, TP Schultz, eds. *Lignin : Historical, Biological and Material Perspectives.* ACS Symp Series, 742. Washington, DC: ACS, 1999, pp. 216–238.
3. K Lundquist. Low-molecular weight lignin hydrolysis products. *Appl Polymer Symp* 28:1393–1407, 1976.
4. H Nimz. Beech lignin—Proposal of a constitutional scheme. *Angew Chem Internat Edit* 13:313–321, 1974.
5. A Sakakibara. Hydrogenolysis. In SY Lin, CW Dence, eds. *Methods in Lignin Chemistry.* Berlin Heidelberg: Springer-Verlag, 1992, pp. 350–368.
6. P Karhunen, P Rummako, J Sipilä, G Brunow. The formation of dibenzodioxocin structures by oxidative coupling: A model reaction for lignin biosynthesis. *Tetrahedron Lett* 36:4501–4504, 1995.

7. C Lapierre. Application of new methods for the investigation of lignin structure. In HG Jung, DR Buxton, RD Hatfield, J Ralph, eds. *Forage Cell Wall Structure and Digestibility.* Madison, WI: American Society of Agronomy Inc., 1993, pp. 133–163.

8. N Habu, Y Matsumoto, A Ishizu, J Nakano. The role of the diarylpropane structure as a minor constituent in spruce lignin. *Holzforschung* 44:67–71, 1990.

9. E Ammälahti, G Brunow, M Bardet, D Robert, I Kilpeläinen. Identification of side-chain structures in a poplar lignin using three-dimensional HMQC-HOHAHA NMR spectroscopy. *J Agr Food Chem* 46:5113–5117, 1998.

10. J Ralph, K Lundquist, G Brunow, F Lu, H Kim, PF Shatz, JM Marita, RD Hatfield, SA Ralph, JH Christensen, W Boerjan. Lignins: Natural polymers from oxidative couplings of 4-hydroxyphenylpropanoids. *Phytochem Rev* 3:29–60, 2004.

11. K Lundquist, G Miksche. Nachweis Eines Neuen Verknupfungsprinzips von Guajacylpropaneheiten im Fichtenlignin. *Tetrahedron Lett* 25:2131–2136, 1965.

12. LM Zhang, G Gellerstedt. NMR observation of a new lignin structure, a spiro-dienone. *Chem Commun* 2744–2745, 2001.

13. EA Capanema, MY Balakshin, JF Kadla. Quantitative characterization of a hardwood milled wood lignin by nuclear magnetic resonance spectroscopy. *J Agr Food Chem* 53:9639–9649, 2005.

14. D Meier, O Faix. Pyrolysis-gas chromatography-mass-spectrometry. In SY Lin, CW Dence, eds. *Methods in Lignin Chemistry.* Berlin Heidelberg: Springer-Verlag, 1992, pp. 177–199.

15. J Ralph, RD Hatfield. Pyrolysis-GC-MS characterization of forage materials. *J Agr Food Chem* 39:1426–1437, 1991.

16. C-L Chen. Nitrobenzene and cupric oxide oxidation. In SY Lin, CW Dence, eds. *Methods in Lignin Chemistry.* Berlin Heidelberg: Springer-Verlag, 1992, pp. 301–321.

17. TP Schultz, MC Templeton. Proposed mechanism for the nitrobenzene oxidation of lignin. *Holzforschung* 40:93–97, 1986.

18. FD Chan, AFA Wallis, KL Nguyen. Contribution of lignin substructures to nitrobenzene oxidation products. *J Wood Chem Technol* 15:329–347, 1995.

19. KV Sarkanen, HL Hergert. Classification and distribution. In KV Sarkanen, CH Ludwig, eds. *Lignins: Occurrence, Formation, Structure and Reactions.* New York: Wiley-Interscience, 1971, pp. 43–94.

20. G Gellerstedt. Rapporteur for degradative analyses of lignins: Modern methods of analysis of wood, annual plants and lignins. New Orleans: IEA Presymposium, 1991.

21. E Billa, M-T Tollier, B Monties. Characterization of the monomeric composition of *in situ* wheat straw lignins by alkaline nitrobenzene oxidation : Effect of temperature and reaction time. *J Sci Food Agric* 72:250–256, 1996.

22. B Fernandez de Simon, J Perez-Ilzarbe, T Hernandez, C Gomez-Cordoves, I Estrella. HPLC study of the efficiency of extraction of phenolic compounds. *Chromatographia* 30:35–37, 1992.

23. G Galetti, R Piccaglia, G Chiavari, V Conciliani. HPLC/electrochemical detection of lignin phenolics by direct injection of nitrobenzene hydrolysates. *J Agr Food Chem* 37:985–987, 1989.

24. B Leopold. Aromatic keto and hydroxypolyethers as lignin models. *Acta Chem Scand* 4:1523–1537, 1950.

25. M Funaoka, I Abe, VL Chiang. Nucleus exchange reaction. In SY Lin, CW Dence, eds. *Methods in Lignin Chemistry.* Berlin Heidelberg: Springer-Verlag, 1992, pp. 385–391.

26. M Shimada, T Fukuzuka, T Higuchi. Ester Linkages of *p*-coumaric acid in bamboo and grass lignin. *Tappi* 54:72–78, 1971.

27. T Higuchi, Y Ito, I Kawamura. *p*-Hydroxyphenylpropane components of grass lignin and the role of tyrosine ammonia-lyase in its formation. *Phytochemistry* 6:875–881, 1967.

28. L B Davin, NG Lewis. Phenylpropanoid metabolism: Biosynthesis of monolignols, lignans and neolignans, lignins and suberins. In HA Stafford, RK Ibrahim, eds. *Phenolic Metabolism in Plants.* New York: Plenum Press, 1992, pp. 325–375.

29. KJ Logan, BA Thomas. Distribution of lignin derivatives in plants. *New Phytol* 99: 571–585, 1985.

30. R Ligrone, A Carafa, JG Duckett, KS Renzaglia, K Ruel. Immunocytochemical detection of lignin-related epitopes in cell walls in bryophytes and the charalean alga Nitella. *Pl Syst Evol* 270:257–272, 2008.

31. UN Dwivedi, WH Campbell, J Yu, RSS Datla, RC Bugos, VL Chiang, GK Poldila. Modification of lignin biosynthesis in transgenic *Nicotiana* through the expression of an antisense *O*-methyltransferase gene from *Populus. Plant Mol Biol* 26:61–71, 1994.

32. C Lapierre, M-T Tollier, B Monties. Occurrence of additional monomeric units in the lignins from the internodes of a brown-midrib mutant of maize bm3. *CR Acad Sci Paris* 307:723–728, 1988.

33. J Kuc, OE Nelson. The abnormal lignins produced by the brown-midrib mutants of maize. *Arch Biochem Biophys* 105:103–113, 1964.

34. MA Goni, B Nelson, RA Blanchette, JI Hedges. Fungal degradation of wood lignins : Geochemical perspectives from CuO-derived phenolic dimers and monomers. *Geochem Cosmochim Acta* 57:3985–4002, 1993.

35. JI Hedges, JE Ertel. Characterization of lignin by gas capillary chromatography of cupric oxide oxidation products. *Anal Chem* 54:174–178, 1982.

36. YZ lai, H Xu, R Yang. An overview of chemical degradation methods for determining lignin condensed units. In WG Glasser, RA Northey, TP Schultz, eds. *Lignin: Historical, Biological and Material Perspectives.* ACS Symp Series, 742. Washington, DC: ACS, 1999, pp. 239–249.

37. I Tomoda, Y Matsumoto, G Meshitsuka. Semi-quantitative method to evaluate the alpha -carbonyl content in lignin. *J Wood Sci* 51:172–175, 2005.

38. K Freudenberg, AC Neish. *Constitution and Biosynthesis of Lignin.* Berlin, Heidelberg, New York: Springer, 1968.

39. E Erikson, S Larsson, GE Miksche. Gaschromatographische Analyse von Ligninoxydationsprodukten: VIII. Zur Struktur des Lignins von Fichte. *Acta Chem Scand* 27:903–914, 1973.

40. G Gellerstedt. Chemical degradation methods: Permanganate oxidation. In SY Lin, CW Dence, eds. *Methods in Lignin Chemistry.* Berlin Heidelberg: Springer-Verlag, 1992, pp. 322–333.

41. C Lapierre, C Rolando, B Monties. Preparative thioacidolysis of spruce lignin: Isolation and identification of the main monomeric products. *Holzforschung* 40:47–50, 1986.

42. C Rolando, B Monties, C Lapierre. Thioacidolysis. In SY Lin, CW Dence, eds. *Methods in Lignin Chemistry.* Berlin Heidelberg: Springer-Verlag, 1992, pp. 131–147.

43. C Lapierre, B Pollet, C Rolando. New insights into the molecular architecture of hardwood lignins by chemical degradative methods. *Res Chem Intermed* 21:397–412, 1995.

44. A Boudet, C Lapierre, J Grima-Pettenatti. Biochemistry and molecular biology of lignification. *New Phytol* 129:203–236, 1995.

45. K Fuji, K Ichikawa, M Node, E Fujita. Hard and soft nucleophile system: New efficient method for removal of benzyl protecting groups. *J Org Chem* 44:1661–1664, 1979.

46. L Jouanin, T Goujon, V de Nadaï, M-T Martin, I Mila, C Vallet, B Pollet, A Yoshinaga, B Chabbert, M Petit-Conil, C Lapierre. Lignification in transgenic poplars with extremely reduced caffeic acid *O*-methyltransferase activity. *Plant physiol* 123:1363–1373, 2000.

47. MF Pasco, ID Suckling. Lignin removal during kraft pulping: An investigation by thioacidolysis. *Holzforschung* 48:504–508, 1994.

48. F Lu, J Ralph. DFRC method for lignin analysis. 1 New method for β-aryl ether cleavage: Lignin model studies. *J Agr Food Chem* 45:4655–4660, 1997.
49. H Onnerud, G. Gellerstedt G. Inhomogeneities in the chemical structure of spruce lignin. *Holzforschung* 57:165–170, 2003.
50. G Jacquet, B Pollet, C Lapierre, C Francesch, C Rolando, O Faix. Thioacidolysis of enzymatic dehydrogenation polymers from *p*-hydroxyphenyl, guaiacyl and syringyl precursors. *Holzforschung* 51: 349–354, 1997.
51. N Terashima, RH Atalla, SA Ralph, LL Landucci, C Lapierre, B Monties. New preparations of lignin polymer models under conditions that approximate cell wall lignification. *Holzforschung* 50:9–14, 1996.
52. H Kim, J Ralph, F Lu, G Pilate, J-C Leplé, B Pollet, C Lapierre. Identification of the structure and origin of thioacidolysis marker compounds for cinnamyl alcohol dehydrogenase deficiency in angiosperms. *J Biol Chem* 277:47412–47419, 2002.
53. C Lapierre, G Pilate, B Pollet, I Mila, J-C Leplé, L Jouanin, H Kim, J Ralph. Signatures of cinnamyl alcohol dehydrogenase deficiency in poplar lignins. *Phytochemistry* 65: 313–321, 2004.
54. BM Lange, C Lapierre, H Sandermann Jr. Elicitor-induced spruce stress lignin. *Plant Physiol* 108:1277–1287, 1995.
55. X Pan, D Lachenal, C Lapierre, B Monties. Structure and reactivity of spruce mechanical pulp lignins: Part I. Bleaching and photoyellowing of in situ lignins. *J Wood Chem Technol* 12:135–147, 1992.
56. KV Sarkanen. Precursors and their polymerization. In KV Sarkanen, CH Ludwig, eds. *Lignins: Occurrence, Formation, Structure and Reactions.* New York: Wiley-Interscience, 1971, pp. 95–164.
57. N Fukagawa, G Meshitsuka, A Ishizu. Isolation of a syringyl β-O-4' rich end-wise type lignin fraction from birch periodate lignin. *J Wood Chem Technol* 12:91–109, 1992.
58. G Pilate, E Guiney, K Holt, M Petit-Conil, C Lapierre, J-C Leplé, B Pollet, I Mila, EA Webster, HG Marstorp, DW Hopkins, L Jouanin, W Boerjan, W Schuch, D Cornu, C Halpin. Field and pulping performances of transgenic trees with altered lignification. *Nat Biotech* 20:607–612, 2002.
59. Y Barrière, J Ralph, JH Grabber, S Guillaumie, O Argillier, V Méchin, B Chabbert, I Mila, C Lapierre. Genetic and molecular basis of grass cell wall biosynthesis and degradability: II. lessons from brown-midrib maize mutants. *C.R. Biologies* 847–860, 2004.
60. JH Grabber, S Quideau, J Ralph. *p*-Coumaroylated syringyl units in maize lignin: Implication for β-ether cleavage by thioacidolysis. *Phytochemistry* 43:1189–1194, 1996.
61. FC Lu, J Ralph. Detection and determination of *p*-coumaroylated units in lignins. *J Agr. Food Chem* 47:1988–1992, 1999.
62. V Méchin, O Argillier, F Rocher, Y Hebert, I Mila, B Pollet, Y Barrière, C Lapierre. In search of a maize ideotype for cell wall enzymatic degradability using histological and biochemical lignin characterization. *J Agr Food Chem* 53:5872–5881, 2005.
63. G Pinçon, S Maury, L Hoffmann, P Geoffroy, C Lapierre, B Pollet, M Legrand. Repression of O-methyltransferase genes in transgenic tobacco affects lignin synthesis and plant growth. *Phytochemistry* 57:1167–1176, 2001.
64. T Goujon, R Sibout, B Pollet, B Maba, L Nussaume, N Bechtold, FC Lu, J Ralph, I Mila, Y Barriere, C Lapierre, L Jouanin. A new Arabidopsis thaliana mutant deficient in the expression of O-methyltransferase impacts lignins and sinapoyl esters. *Plant Mol Biol* 51:973–989, 2003.
65. C Lapierre, B Monties, C Rolando. Thioacidolyses of diazomethane-methylated pine compression wood and wheat straw in situ lignins. *Holzforschung* 42:409–411, 1988.

66. C Lapierre, C Rolando. Thioacidolysis of pre-methylated lignin samples from pine compression and poplar woods. *Holzforschung* 42:1–4, 1988.
67. H-M Chang, GG Allan. Oxidation. In KV Sarkanen, CH Ludwig, eds. *Lignins: Occurrence, Formation, Structure and Reactions.* New York: Wiley-Interscience, 1971, pp. 433–485.
68. X Pan, D Lachenal, C Lapierre, B Monties. Structure and reactivity of spruce mechanical pulp lignins: Part II. Organosolv fractionation of lignins in a flow-through reactor. *J Wood Chem Technol* 12:279–298, 1992.
69. E Sjöström. *Wood Chemistry: Fundamentals and Applications,* 2nd ed. San Diego: Academic Press, 1992, pp. 71–89.
70. G Brunow, I Kilpelaïnen, J Sipilä, K Syrjänen, P Karhunen, H Setälä, P Rummako. Oxidative coupling of phenols and the biosynthesis of lignin. In NG Lewis, S Sarkanen, eds. *Lignin and Lignan Biosynthesis.* ACS Symposium Series 697. Washington, DC, 1998, pp. 131–147.
71. C Lapierre, B Monties. Structural information gained from the thioacidolysis of grass lignins and their relation with alkali solubility. Proceedings of the 5th International Symposium on Wood and Pulping Chemistry, Raleigh, NC, 1989, Vol. 1, pp. 615–621.
72. E Beckman, O Liesche, F Lehman. Qualitative und quantitative Unterschiede des Lignine einiger Holz und Stroharten. *Biochem Z* 139:491–498, 1923.
73. C Lapierre, D Jouin, B Monties. On the molecular origin of the alkali solubility of gramineae lignins. *Phytochemistry* 28:1401–1403, 1989.
74. R Sibout, A Eudes, G Mouille, B Pollet, C Lapierre, L Jouanin, A Seguin. Cinnamyl alcohol dehydrogenase-C and -D are the primary genes involved in lignin biosynthesis in the floral stem of Arabidopsis. *Plant Cell* 17:2059–2076, 2005.
75. C Lapierre, B Pollet, JJ MacKay, RR Sederoff. Lignin structure in a mutant pine deficient in cinnamyl alcohol dehydrogenase. *J Agric Food Chem* 48:2326–2331, 2000.
76. C Lapierre, B Pollet, B Monties. Heterogeneous distribution of diarylpropane structures in spruce lignins. *Phytochemistry* 30:659–662, 1991.
77. C Lapierre, K Lundquist. Investigation of low molecular weight and high molecular weight lignin fractions. *Nordic Pulp Paper Res J* 14:158–162, 1999.
78. G Gellerstedt, L Zhang. Reactive structures in high yield pulping: Part 1. Structures of the 1,2-diarylpropane-1,3-diol type. *Nordic Pulp Paper Res J* 6:136–139, 1991.
79. K Lundquist, R Simonson, K Tingsvik. Studies of lignin-carbohydrate linkages in milled wood lignin preparations. *Svensk Papperstidning* 82:452–454, 1980.
80. DS Argyropoulos, L Jurasek, L Krtsitofova, Z Xia, Y Sun, E Palus. On the abundance and reactivity of dibenzodioxocins in softwood lignin. Proceedings of the 11th International Symposium on Wood and Pulping Chemistry, Nice, France, 2001, Vol. 1, pp. 175–179.
81. P Karhunen, P Rummakko, A Pajunene, G Brunow. Synthesis and crystal structure determination of model compounds for the dibenzodioxocin structure occurring in wood lignins. *J Chem Soc Perkin Trans* I:2305–2308, 1996.
82. C Lapierre, B Pollet, B Monties, C Rolando. Thioacidolysis of spruce lignin: GC-MS analysis of the main dimers recovered after Raney nickel desulphuration. *Holzforschung* 45:61–68, 1991.
83. J Ralph, JH Grabber. Dimeric beta-ether thioacidolysis products resulting from incomplete ether cleavage. *Holzforschung* 50:425–428, 1996.
84. J Ralph, C Lapierre, F Lu, JM Marita, G Pilate, J Van Doorsselaere, W Boerjan, L Jouanin. NMR evidence for benzodioxane structures resulting from incorporation of 5-hydroxyconiferyl alcohol into lignins of O-methyltransferase-deficient poplars. *J Agr Food Chem* 49:86–91, 2001.
85. G Jacquet. Structure et réactivité des lignines de Graminées et des acides phénoliques associés. PhD Thesis Aix Marseille III University, 301 pp., 1997.

86. ID Suckling, MF Pasco, B Hortling, J Sundquist. Assessment of lignin condensation by GPC analysis of lignin thioacidolysis products. *Holzforschung* 48:501–503, 1994.

87. M Lawoko, G Henriksson, G Gellerstedt. Structural differences between the lignin-carbohydrate complexes present in wood and in chemical pulps. *Biomacromolecules* 6:3467–3473, 2005.

88. F Lu, J Ralph. Derivatization followed by reductive cleavage (DFRC method), a new method for lignin analysis: Protocol for analysis of DFRC monomers. *J Agr Food Chem* 45:2590–2592, 1997.

89. F Lu, J Ralph. The DFRC method for lignin analysis: 2. Monomers from isolated lignins. *J Agr Food Chem* 46:547–552, 1998.

90. D Fournand, B Pollet, C Lapierre. Lignin evaluation by chemical degradative methods: Relative performances of the DFRC and thioacidolysis techniques. Proceedings of the 6th European Workshop on Lignocellulosic and Pulp, Bordeaux, France, 2000, pp. 313–316.

91. L Zhang, W Wafa-Al-Dajani, G Gellerstedt. Comparison between different methods for quantitative lignin analysis. Proceedings of the 7th European Workshop on Lignocellulosics and Pulp, Turku, Finland, 2002, pp. 521–524.

92. KM Holtman, HM Chang, H Jameel, JF Kadla. Elucidation of lignin structure through degradative methods: Comparison of modified DFRC and thioacidolysis. *J Agr Food Chem* 51:3535–3540, 2003.

93. F Lu, J Ralph. The DFRC method for lignin analysis: 7. Behavior of cinnamyl end groups. *J Agr Food Chem* 47:1981–1987, 1999.

94. J Peng, F Lu, J Ralph. The DFRC method for lignin analysis: 4. Lignin derived dimers isolated from DFRC-degraded loblolly pine wood. *J Agr Food Chem* 46:553–5560, 1998.

95. F Lu, J Ralph. Detection and determination of *p*-coumaroylated units in lignins. *J Agr Food Chem* 47:1988–1992, 1999.

96. F Lu, J Ralph. Preliminary evidence for sinapyl acetate as a lignin monomer in kenaf. *Chem Com* 90–91, 2002.

97. J Ralph, F Lu. The DFRC method for lignin analysis: 6. A simple modification for identifying natural acetates on lignins. *J Agr Food Chem* 46: 4616–4619, 1998.

98. JC DelRio, G Marques, J Rencoret, AT Martinez, A Gutierrez. Occurrence of naturally acetylated lignin units. *J Agr Food Chem* 55:5461–5468, 2007.

99. J Ralph, J Peng, F Lu, RD Hatfield, RF Helm. Are lignins optically active? *J Agr Food Chem* 47:2991–2996, 1999.

100. J Peng, F Lu, J Ralph. Isochroman lignin trimers from DFRC-degraded Pinus taeda. *Phytochemistry* 50:659–666, 1999.

101. KV Sarkanen, A Islam, CD Andersson. Ozonation. In SY Lin, CW Dence, eds. *Methods in Lignin Chemistry.* Berlin Heidelberg: Springer-Verlag, 1992, pp. 387–406.

102. H Aimi, Y Matsumoto, G Meshitsuka. Structure of small lignin fragment retained in water-soluble polysaccharide extracted from sugi MWL isolation residue. *J Wood Sci* 50:415–421, 2004.

103. T Akiyama, T Sugimoto, Y Matsumoto, G Meshitsuka. *Erythro/threo* ratio of beta -O-4 structures as an important structural characteristic of lignin: I. Improvement of ozonation method for the quantitative analysis of lignin side-chain structure. *J Wood Sci* 48:210–215, 2002.

104. T Akiyama, H Goto, DS Nawawi, W Syafii, Y Matsumoto, G Meshitsuka. *Erythro/threo* ratio of beta -O-4-structures as an important structural characteristic of lignin: Part 4. Variation in the *erythro/threo* ratio in softwood and hardwood lignins and its relation to syringyl/guaiacyl ratio. *Holzforschung* 59:276–281, 2005.

105. T Akiyama, K Magara, Y Matsumoto, G Meshitsuka, A Ishizu, K Lundquist. Proof of the presence of racemic forms of arylglycerol-beta-aryl ether structure in lignin: Studies on the stereo structure of lignin by ozonation. *J Wood Sci* 46:414–415, 2000.

106. T Akiyama, Y Matsumoto, T Okuyama, G Meshitsuka. Ratio of *erythro* and *threo* forms of beta-O-4 structures in tension wood lignin. *Phytochemistry* 64:1157–1162, 2003.

107. TF Yeh, JL Braun, B Goldfarb, HM Chang, JF Kadla. Morphological and chemical variations between juvenile wood, mature wood, and compression wood of loblolly pine (Pinus taeda L.). *Holzforschung* 60:1–8, 2006.

108. T Sugimoto, T Akiyama, Y Matsumoto, G Meshitsuka. The *erythro/threo* ratio of beta-O-4 structures as an important structural characteristic of lignin: Part 2. Changes in *erythro/threo* (E/T) ratio of beta-O-4 structures during delignification reactions. *Holzforschung* 56:416–421, 2002.

109. A Fujimoto, Y Matsumoto, H Chang, G Meshitsuka. Quantitative evaluation of milling effects on lignin structure during the isolation process of milled wood lignin. *J Wood Sci* 51:89–91, 2005.

110. H Aimi, Y Matsumoto, G Meshitsuka. Lignin fragments rich in detached side-chain structures found in water-soluble LCC. *J Wood Sci* 51:252–255, 2005.

111. H Aimi, Y Matsumoto, G Meshitsuka. Structure of small lignin fragments retained in water-soluble polysaccharides extracted from birch MWL isolation residue. *J Wood Sci* 51:303–308, 2005.

112. O Karlsson, T Ikeda, T Kishimoto, K Magara, Y Matsumoto, S Hosoya. Isolation of lignin-carbohydrate bonds in wood: Model experiments and preliminary application to pine wood. *J Wood Sci* 50:142–150, 2004.

3 Electronic Spectroscopy of Lignins

John A. Schmidt

CONTENTS

INTRODUCTION

Lignin ultraviolet (UV)-visible spectroscopy was last comprehensively reviewed by Goldschmid [1]. More recently, detailed descriptions of measurement techniques for solution absorbance [2] and diffuse reflectance [3] spectra of lignin-containing materials have appeared. The reader is referred to these works for experimental details. This chapter will concentrate on interpretation of spectra.

The electronic spectra of lignin and lignin model compounds can be adequately understood by examining the substituent effects on the three bands of benzene [4]. This chapter will consider such an analysis, along with a brief discussion of the basic concepts of light and the Beer–Lambert Law. More detailed theoretical treatments are available in any text on molecular spectroscopy, such as Jaffé and Orchin [5].

Earlier reviews considered only solution absorption spectra of lignocellulosic materials. However, in the thirty years since Goldschmid's review, lignin chemists have used several other techniques based on electronic transitions within molecules. This chapter will summarize the earlier data on the solution properties of lignins and lignin model compounds and, in somewhat more detail, the newer work on the application of diffuse reflectance spectroscopy, luminescence spectroscopy, and transient spectroscopy to problems in lignin chemistry.

The standard designations for the positions of substituents in lignin and lignin model compounds will be adopted for this chapter. The fundamental lignin structural unit is n-propyl benzene with substituents on both the aromatic ring and the alkyl side chain. The carbon atoms on the propyl side chain are designated α, β and γ, with the α-carbon being that which is attached to the aromatic ring. Ring carbons are numbered 1 through 6, with the 1-position being that where the propyl side chain is attached.

LIGHT, ENERGY, AND THE BEER–LAMBERT LAW

Electronic spectroscopy concerns transitions induced by UV and visible wavelengths of the electromagnetic spectrum. The commonly used wavelength unit in this range is nanometers (nm). Visible wavelengths extend from 400 nm to 750 nm and the UV wavelengths from 190 nm to 400 nm. The range between 300 nm and 400 nm, although not visible to the human eye, is present in terrestrial sunlight and is sometimes referred to as the near UV.

Light has both particle and wave characteristics. Energy is absorbed or emitted as discrete quanta, or photons. The Planck relation, Equation 3.1, connects the energy of a photon to its wavelength; E is the energy of the photon, υ the frequency, λ the wavelength, c the speed of light, and h Planck's constant (6.63×10^{-34} joule s). Equation 3.1 shows that photon energy is inversely proportional to wavelength.

$$E = h\nu = h\frac{c}{\lambda} \qquad (3.1)$$

In the absence of light scattering, the decrease in the intensity of light, I, passing through a solution is given by Equation 3.2, where c is the concentration of a homogeneously

$$dI = I\alpha c dl \qquad (3.2)$$

distributed absorber, α is the strength of the absorber and l is distance through the sample. After separating the variables and integrating, one obtains the familiar Beer–Lambert Law, Equation 3.3, where A is the absorbance, I_0 and I_t are the incident and transmitted light intensities, T is the transmittance, and ε is the extinction coefficient.

$$A = -\log\left(\frac{I_0}{I_t}\right) = -\log\left(\frac{1}{T}\right) = \frac{\alpha}{2.303}lc = \varepsilon lc \qquad (3.3)$$

ABSORPTION OF SUBSTITUTED BENZENES RELATED TO LIGNIN

The energy of UV-visible photons varies from about 170 kJ/mol to 598 kJ/mol. This corresponds to the energy required to cause transitions between electronic states; that is, states that differ in the spatial distribution of their electrons. The energy separation between the electronic states of saturated organic compounds is large, and occurs at shorter wavelengths than the conventional UV range. The presence of π orbitals is required for molecules to have easily observable absorptions in the UV-visible range. Thus, polysaccharides have very weak UV absorption. Lignin, with its aromatic structure, is responsible for almost all the UV-visible absorption of lignocellulosic materials.

Benzene

The three electronic absorption bands of benzene are listed in Table 3.1, identified by two common systems of notation. Platt developed the perimeter or free-electron model of the absorption bands of polycyclic aromatic hydrocarbons [5], of which benzene is the simplest. Others have reformulated the theory, but retained the system of nomenclature. According to this system, the three bands of benzene, in order of increasing energy, are designated ^1L_b, ^1L_a, and ^1B. Doub and Vandenbelt, in their exhaustive study of substituted benzenes [6], called the moderately strong band at 203 nm the primary band, the weak band at 256 nm the secondary band, and the highest energy band at 180 nm the second primary band. The latter band occurs at too low a wavelength to be observed under normal experimental conditions, but in substituted benzenes it is sufficiently red-shifted to fall in the normal UV range.

TABLE 3.1
Absorption Bands of Benzene

λ (nm)[a]	ε (L mol⁻¹ cm⁻¹)[a]	Designation in Platt Notation	Designation by Doub and Vandenbelt
180	46,000	^1B	second primary
203	7400	^1L_a	primary
256	220	^1L_b	secondary

a Values from Jaffé, H. H. and Orchin, M., *Theory and Applications of Ultraviolet Spectroscopy,* John Wiley and Sons, Inc., New York, 295–303, 1962. With permission.

The extinction coefficient (ε) at the band maximum is a rough indication of the probability of a transition and is governed by so-called selection rules. A transition between two states is forbidden if the initial and final states differ in spin multiplicity (singlet or triplet), if either of the states is antisymmetric (sign of the wave function changes upon inversion through a center of symmetry), or if the initial and final wave functions have zero overlap. The wave functions used in such calculations are quite simple, so the zero transition probability predicted for a forbidden transition is almost never observed. However, these calculations do usefully distinguish strong from weak transitions. Thus, the ^1L_a and ^1L_b transitions of benzene are formally symmetry-forbidden, but they have a finite intensity because they are vibronically coupled to the fully allowed ^1B transition at 180 nm.

Effect of Substituents

Substitution on the benzene ring shifts the absorption maxima of the forbidden ^1L_a and ^1L_b transitions by roughly equal amounts and further increases their intensity, largely because of resonance effects. The wavelength shifts are always to the red, regardless of whether the substituent is electron donating (*ortho, para* directing) or electron withdrawing (*meta* directing). Matsen [7] explains this in terms of the effect of the substituent electronegativity on Hückel π energy levels, while Duben [8] presents a simplified explanation in terms of Frontier Molecular Orbital Theory.

Doub and Vandenbelt [6] observed the following series of band displacements for the ^1L_a and ^1L_b transitions in monosubstituted benzenes:

Ortho-para directing:

$$CH_3 < Cl < Br < OH < OCH_3 < NH_2 < O^- < NHCOCH_3 < N(CH_3)_2$$

Meta directing:

$$NH_3^+ < SO_2NH_2 < CN = CO_2^- < CO_2H < COCH_3 < CHO < NO_2$$

The displacements correlated well with the Hammett substituent constants. Tomasik and Krygowski [9,10] extended this approach to include solvent effects on band displacement, and Uno and Kubota have attempted to put the application of Hammett constants to benzene electronic spectra on a more substantial theoretical foundation [11].

Two types of shift are observed in disubstituted compounds. If both substituents are of the same type (*o*, *p*-directing or *m*-directing) there is no additive effect due to the second substituent; the band displacements are those observed for the most displaced monosubstituted compound. Thus the ^1L_a band (203 nm in benzene) is shifted by 26.5 nm in benzoic acid, 65 nm in nitrobenzene, and 61 nm in *p*-nitrobenzoic acid [6].

If one of the substituents is electron donating and the other is electron withdrawing, further wavelength shifts are observed when the second substituent is added, as shown in Table 3.2. For example, in the hydroxybenzaldehydes the ^1L_a band is shifted by 53 nm in the *o*-isomer and 51 nm in the *m*-isomer, almost the sum of the shifts for phenol and benzaldehyde (7 nm and 46 nm, respectively). The wavelength shift in *p*-hydroxybenzaldehyde is greater than additive, 80 nm, and is large enough to obscure the ^1L_b band. In the *o*- and *m*-disubstituted compounds, the ^1L_b band is red shifted by an amount greater than the sum of the individual substituent shifts.

The larger-than-additive wavelength shifts have been interpreted as evidence for significant contribution of quinoid resonance structures to the excited state [12]. However, these generalizations should be regarded with care. When analyzed in terms of frequency, rather than wavelength, the shifts of the ^1L_a band upon adding a second substituent were proportional to the effect the same substituents had on unsubstituted benzene [5,14].

The spectra of trisubstituted benzenes with one electron-withdrawing and two electron-donating substituents contain the most red-shifted bands found in the constituent disubstituted compounds [4]. This is illustrated in Table 3.3 for

---

TABLE 3.2
Wavelength Shifts of the ^1L_a and ^1L_b Bands in Some Disubstituted Benzenes

	^1L_a		^1L_b		
	λ_{max} (nm)	$\varepsilon \times 10^{-3}$ (L mol^{-1} cm^{-1})	λ_{max} (nm)	$\varepsilon \times 10^{-3}$ (L mol^{-1} cm^{-1})	Reference
Benzene	203	7.40	256	0.22	[5]
Benzaldehyde	249	1.14	279*	–	[6]
Phenol	210	6.24	270	1.45	[6]
p-Hydroxy-benzaldehyde	283	16.0	–	–	[6]
o-Hydroxy-benzaldehyde	256	12.6	324	3.4	[12]
m-Hydroxy-benzaldehyde	254	10.1	314	2.58	[12]

* Doub and Vandenbelt (Doub, L. and Vandenbelt, J. M., *J. Am. Chem. Soc.*, 69, 2714–2723, 1947) did not observe this weak band, probably because of the broadening of the ^1L_a band in hydroxylic solvents. It is observed in hydrocarbon solvents such as cyclohexane (Morales, R.G.E. and Toporowicz, M., *Spectrochim. Acta.*, 37A, 11–15, 1981.)

TABLE 3.3

Absorption Maxima of Some Substituted Benzaldehydes in Cyclohexane

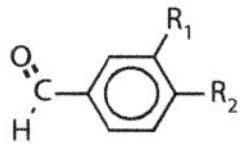

	^1L_a Band λ_{max} (nm)	^1L_b Band λ_{max} (nm)
$R_1 = R_2 = H$	241	279
$R_1 = H$, $R_2 = OMe$	267	281
$R_1 = OMe$, $R_2 = H$	248	306
$R_1 = R_2 = OMe$	272	301

Source: Extracted from Morales, R.G.E. and Toporowicz, M., *Spectrochim. Acta.*, 37A, 11–15, 1981.

3,4-dimethoxybenzaldehyde [13]. The ^1L_a band is the most red-shifted in *p*-methoxybenzaldehyde (267 nm), while the ^1L_b band is the most red-shifted in *m*-methoxybenzaldehyde (306 nm). Both these red shifted bands appear in 3,4-dimethoxybenzaldehyde.

LIGNIN MODEL COMPOUNDS

In a series of papers, Aulin-Erdtman et al. [15–17] compiled extensive data on UV-visible spectra of lignin model phenols and the corresponding phenolate ions. A selection of these is presented in Table 3.4.

The hypothesis of Doub and Vandenbelt that the spectra of substituted benzenes can be understood as a shifting of the three basic bands of benzene seems sufficient to understand qualitatively the spectra of these lignin model compounds. Absorptions below about 260 nm are ^1B transitions, while those above are due to ^1L_a and ^1L_b transitions. Depending on the exact substitution pattern, the ^1L_b band may be obscured under the more intense ^1L_a band.

For the series of *p*-hydroxybenzyl alcohols a single absorption maximum is observed above 260 nm, which is probably due largely to the ^1L_a transition. This band is hardly shifted from its position in phenol (see Table 3.4) because all the substituents are of the same character, electron donating. Similarly, in the series of coumaryl [15], coniferyl [17] and sinapyl [15] alcohols the longest wavelength band is only slightly shifted relative to that observed for the most shifted mono-substituted compound, cinnamyl alcohol, $\lambda_{max} = 292$ nm [195].

The benzaldehyde and acetophenone series follow the pattern observed by Doub and Vandenbelt for benzenes with one electron-withdrawing and two electron-donating substituents. Thus, in *p*-hydroxybenzaldehyde the ^1L_a band is shifted to 285 nm, and obscures the ^1L_b band, which is little affected by *p*-disubstitution. The additional *m*-methoxy groups in vanillin and syringaldehyde intensify and red-shift the ^1L_b band, which becomes visible as a discrete maximum at 310 nm.

TABLE 3.4
UV-Visible Data for Selected Lignin Model Phenols and Phenolate Anions*

Structure	R₁ = R₂ = H				R₁ = OMe, R₂ = H				R₁ = R₂ = OMe			
	λ_{max} (nm)	Solvent	$\varepsilon \times 10^{-3}$ (L mol^{-1}cm^{-1})	Reference	λ_{max} (nm)	Solvent	$\varepsilon \times 10^{-3}$ (L mol^{-1}cm^{-1})	Reference	λ_{max} (nm)	Solvent	$\varepsilon \times 10^{-3}$ (L mol^{-1}cm^{-1})	Reference
(benzyl alcohol; OH, R₁, R₂)	226	0.1N HCl in EtOH	8.71	[15]	230	Abs. EtOH	6.92	[16]	240[1]	95% EtOH	5.6	[17]
	277		1.66		280		2.95		271		1.12	
											9.33	
	247*	0.1N KOH in EtOH	13.5*	[15]	248*	pH 12 buffer	10.2*	[17]	258*	1N KOH in EtOH	1.8	[17]
	295*		2.69*		291*		4.27*		288*,2			
(aldehyde; O=, R₁, R₂, OH)	222	0.1N HCl in EtOH	12.3	[15]	231	0.01N HCl in EtOH	15.1	[17]	231	0.01N HCl in EtOH	16.2	[17]
	285		16.2		279		10.5		310		13.5	
					310		10.7					
	241*	0.1N KOH in EtOH	7.41*		296*	1N KOH in EtOH	2.82*		253*	0.1N KOH in EtOH	10.7*	
	336*		30.2*		349*		30.2*		368		27.5*	
(cinnamyl; OH, R₁, R₂)	263	0.1N HCl in EtOH	19.5	[15]	266	95% EtOH	15.1	[17]	222	95% EtOH	26.9	[15]
	291[3]		2.45		300[5]		6.3		276		14.1	
	290*	0.1N KOH in EtOH	22.4*		290*	1N KOH in EtOH	16.6*		300*,6	0.1N KOH in EtOH	13*	
	330*,4		7.9*		315*		15.1*		330*,6		16*	

(Continued)

TABLE 3.4
UV-Visible Data for Selected Lignin Model Phenols and Phenolate Anions* (Continued)

λ_{max} (nm)	Solvent	$\varepsilon \times 10^{-3}$ (L mol^{-1}cm^{-1})	Reference	λ_{max} (nm)	Solvent	$\varepsilon \times 10^{-3}$ (L mol^{-1}cm^{-1})	Reference	λ_{max} (nm)	Solvent	$\varepsilon \times 10^{-3}$ (L mol^{-1}cm^{-1})	Reference
	$R_1 = R_2 = H$				$R_1 = OMe, R_2 = H$				$R_1 = R_2 = OMe$		
(R_3 = Et)			[15]	(R_3 = Me)			[15, 17]	(R_3 = Et)			[15]
220	0.1N HCl in EtOH	10.2		228	0.01N HCl in EtOH	14.1		299	0.1N HCl in EtOH	10.5	
276		14.8		275		10.0					
				303		7.94					
237*	0.1N KOH in EtOH	6.31*		300*[7]	1N KOH in EtOH	3.2*		357*	0.1N KOH in EtOH	19.9*	
327*		24.5*		344*		23.4*					
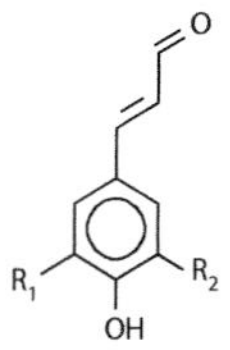 234	0.01N HCl	10.5	[15]	239	Water pH 5	9.77	[15]	242	0.01N HCl	16.2	[15]
323		27.5		300[8]		14.1		341		21.9	
				337		21.9					
249	1N NaOH	7.76		414*	0.2N NaOH in EtOH	35.5*		263	1N NaOH	12.9	
383*		36.3*						421*		30.9*	

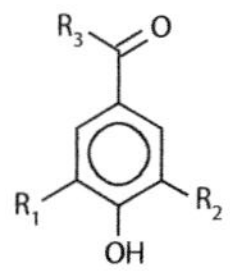

Notes: [1]Shoulder on a band with $\lambda_{max} < 210$ nm. [2]Shoulder on the 258 nm band. [3]Shoulder on the 263 nm band. [4]Shoulder on the 290 nm band. [5]Shoulder on the 266 nm band. [6]Aulin-Erdtman et al. give a single λ_{max} at 320 nm in their tabular data. However, to the author's eye, the spectrum seems better described as a coalescence of two bands at 300 and 330 nm. [7]Shoulder on 344 nm band. [8]Shoulder on the 337 nm band.

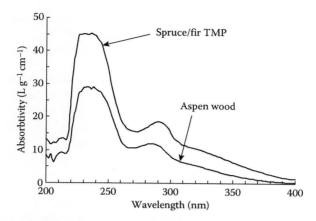

FIGURE 3.1   Typical UV-visible spectra of lignin solutions.

SOLUBLE LIGNINS

Figure 3.1 shows the UV-visible spectra of milled-wood lignin isolated from soft-wood thermomechanical pulp and from aspen wood. These are typical of soluble lignins. Discrete maxima occur at about 205 nm and 280 nm, and a shoulder appears at about 230 nm. Above 280 nm the absorbance declines continuously, without any clear features. A similar spectrum is observed for a number of other lignin preparations, such as milled-wood lignin in 2-methoxyethanol, Brauns lignin in ethanol, aqueous lignin sulfonic acid from spruce and sulfonated synthetic lignin [2].

The absorbance at either 205 or 280 nm is the basis of several techniques for the quantitative determination of lignin. The extinction coefficient depends on species and solvent, and varies from 10 to about 26 L g^{-1} cm^{-1}. Summaries of extinction coefficients are available in reviews by Dence [18], Lin [2], and Fengel and Wegener [19]. Lin [2] has provided guidelines for measurement of lignin absorption spectra, while Dence [18] provides instructions for determination of acid-soluble lignin and lignin solubilized by the acetyl bromide method.

Compared with the electronic spectra of individual model compounds, the bands in lignin spectra are broader, and have less distinct minima and maxima. This is expected for a material composed of a variety of structural units with similar, but not identical, absorption maxima and extinction coefficients. Several authors have attempted to resolve lignin spectra into their constituent bands using curve-fitting techniques. Early work by Hess [20] and Tiyama et al. [21] resolved the spectra of spruce milled-wood lignin and thiolignin into six bands. Norrström et al. [22–25] reported that the spectra of many technical lignins could be reconstructed as a combination of 13 overlapping bands; an additional band was required to reconstruct the spectra of milled-wood lignin [26].

A second derivative of the absorbance improves the resolution and discrimination of individual peaks in a spectrum consisting of several overlapping absorption bands. Lin [27] resolved the spectra of milled-wood lignin and lignosulfonates into

7–13 second derivative peaks between 200 and 300 nm. The technique was unable to resolve any maxima in the tailing absorption beyond 300 nm. The position and intensity of the second derivative peaks differ significantly from the peaks reported by the modeling procedures.

Garver [28] has proposed UV absorbance measurements of process liquors as a way to control operations where lignin is solubilized, such as during pulping and bleaching. The underlying principle is that parameters of interest, such as pulp brightness or residual bleaching chemical, are correlated with the ratio of the absorbances at two or more appropriately chosen UV wavelengths.

Difference Spectroscopy—The Δε Method

Difference spectroscopy has been useful in interpreting the UV-visible spectra of lignins. In this technique, the absorption spectrum measured before a given chemical treatment is subtracted from that recorded after the treatment (or vice versa). In the simplest case, a compound, C, is transformed to a product P. At any time during the reaction, these quantities are related by mass balance, Equation 3.4, where C_0 is the initial concentration of C. The difference in

$$P(t) + C(t) = C_0 \qquad (3.4)$$

absorbance ΔA is given by Equation 3.5, where l is the path length through the sample cell, and ε_P

$$\Delta A = l\left[\varepsilon_P P(t) + \varepsilon_C C(t) - \varepsilon_C C_0\right] \qquad (3.5)$$

and ε_C are the extinction coefficients for P and C, respectively. Equation 3.6 results when the mass balance in Equation 3.4 is substituted for C_0. Note that absorptions by groups that are unaffected by the chemical treatment will cancel in the subtraction.

$$\Delta A = l\left(\varepsilon_P - \varepsilon_C\right) P(t) = l \Delta\varepsilon P(t) \qquad (3.6)$$

Absorption difference spectra have been used to estimate the content of phenolic groups, carbonyl groups, ethylenic double bonds, noncondensed phenolic groups and phenylcoumarans in soluble lignins, and to study the photochemistry of lignin solutions. These experiments are summarized briefly in the following paragraphs.

As Table 3.4 shows, the absorption of a phenolate ion is shifted by at least 10 nm, and usually more, relative to the corresponding phenol. The difference between the UV spectrum of a lignin solution at neutral or lower pH and the spectrum measured at pH 11 or above (where the phenolic hydroxyl groups are ionized) is an ionization difference spectrum. This is the most commonly used lignin difference spectrum. Lin [2] and Zakis [29] have described procedures for the measurement of ionization difference spectra. Aulin-Erdtman et al. [30–32] compared the ionization difference spectra of a number of soluble lignins with a series of lignin model phenols to estimate the amounts of the different types of phenolic groups in each lignin preparation.

Milne et al. [33] reported the results of international round-robin tests that compared the phenolic hydroxyl content of five standard lignin samples determined by wet-chemical, FTIR, ^{1}H NMR, and $\Delta\varepsilon$ methods. The $\Delta\varepsilon$ method gave consistently lower values than the other techniques. Participants in the study recommended that the $\Delta\varepsilon$ method no longer be used for phenolic hydroxyl content.

Faix et al. [34] analyzed the phenolic hydroxyl contents of milled wood lignins from 42 plant species. Aminolysis [35,36] was the most accurate and reproducible technique. The ^{1}H NMR and FTIR techniques were similar to each other in their correlations with the aminolysis results (correlation coefficient $R = 0.77$ and 0.75, respectively). The correlation of the $\Delta\varepsilon$ technique (analysis wavelength 250 nm) was poorer, with $R = 0.57$.

Gärtner et al. [37] modified the $\Delta\varepsilon$ technique of Zakis [29] to determine the phenolic hydroxyl contents of lignins isolated from unbleached and oxygen-bleached softwood kraft pulps. The modified procedure requires measurements at 300 nm and at 350–370 nm and the use of an equation to account for interferences from conjugated structures such as stilbenes, enol ethers, and α-carbonyl groups. The authors do not recommend the technique for highly modified lignins or lignins of exotic origin that may contain structures not included in their model. Phenolic hydroxyl contents determined by this modified $\Delta\varepsilon$ method correlated well with those determined by aminolysis. However, as observed by Milne [33], the absolute values determined by the $\Delta\varepsilon$ method were consistently 15% lower. This consistent underestimation may occur because the second hydroxyl group of a structure containing two hydroxyl groups, such as biphenyl [37] or catechol [38], is difficult to ionize at the chosen pH due the formation of an internal hydrogen bond.

Adler and Marton [39] used a combination of ionization difference spectra and NaBH$_4$ reduction to estimate the amounts of the four conjugated carbonyl groups shown in Table 3.5 in spruce milled-wood lignin. Note that, although the difference

---

TABLE 3.5
Carbonyl Groups in Spruce Milled-Wood Lignin, Estimated by Borohydride Reduction

Structure	Observed λ_{max} (nm) in Alkaline NaBH$_4$ Difference Spectra	Proportion of Total C$_9$ Units
HO–C$_6$H$_3$(OMe)–CH=CH–CHO	400	<1
RO–C$_6$H$_3$(OMe)–CH=CH–CHO	340	3
HO–C$_6$H$_3$(OMe)–C(=O)–R	350	<1
RO–C$_6$H$_3$(OMe)–C(=O)–R	310	5–6

Source: Extracted from Adler, E. and Marton, J., *Acta. Chem. Scand.*, 13, 75–96, 1959.

TABLE 3.6

Estimation of Ethylenic Double Bonds in Lignin by Catalytic Hydrogenation

Structure	Observed λ_{max} (nm) in Hydrogenation Difference Spectra	Proportion of Total C_9 Units
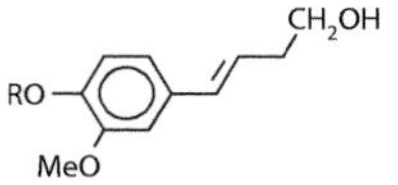 CH₂OH structure (RO—, MeO—)	262–269 295–298	3
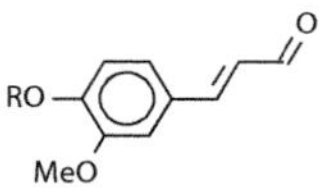 O aldehyde structure (RO—, MeO—)	250 343	3–4
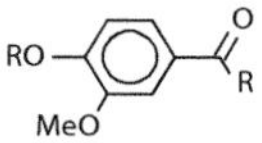 O, R ketone structure (RO—, MeO—)	270–280 305–308	5

Source: Extracted from Marton, J. and Adler, E., *Acta Chem. Scand.,* 15, 370–383, 1961.

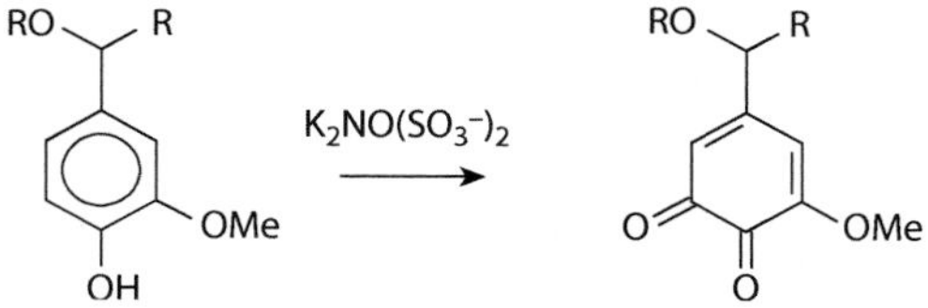

FIGURE 3.2   Reaction of uncondensed phenolic groups with Fremy's salt.

spectra for etherified γ-carbonyls and phenolic α-carbonyls both have maxima in the alkaline-borohydride difference spectrum at 340–350 nm, they could be easily distinguished because the phenolic carbonyls are reduced much more slowly.

Marton and Adler [40] estimated the content of α–β double bonds in spruce milled-wood lignin by comparing the difference spectra due to hydrogenation of the lignin with those for hydrogenation of the appropriate model compounds, as shown in Table 3.6. This experiment also produced estimates of the amounts of etherified conjugated α– and γ–carbonyl groups that agree well with those obtained by the borohydride reduction experiment.

Adler and Lundquist [41] used the reactions of Fremy's salt with phenols (Figure 3.2) to determine that 40–50% of the phenolic groups in spruce milled-wood lignin were the condensed type (those with a bond to C_5-aromatic carbon). The remaining phenolic groups were uncondensed. When the α-position has been etherified, uncondensed phenolic groups react with Fremy's salt to give an *o*-quinone with distinctive maxima at 365 nm and 465 nm ($\varepsilon = 2.5 \times 10^3$ and $0.7–0.8 \times 10^3$ L mol^{-1} cm^{-1}, respectively), while condensed phenolic units are unreactive.

These same authors also used difference spectroscopy to estimate the number of phenylcoumaran units in spruce milled-wood lignin at 9 per 100 C_9 units [42].

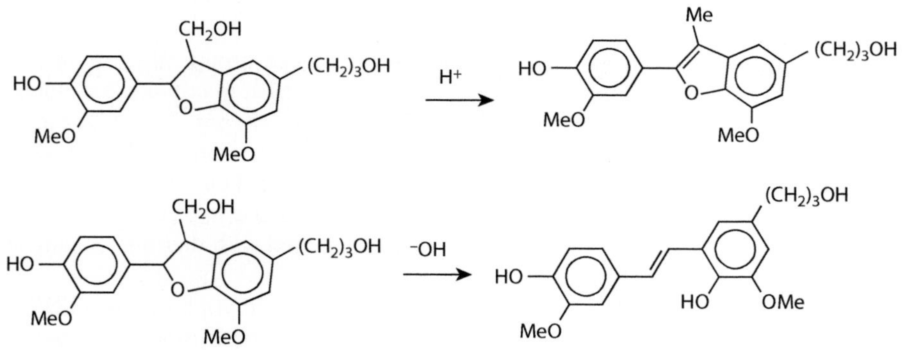

FIGURE 3.3 Chromophore formation from phenyl courmaran model compounds. Acidolysis gives a phenyl coumarone; alkaline hydrolysis gives a stilbene.

Acidolysis of phenylcoumarans, whether phenolic or etherified, yields the corresponding phenylcoumarone (Figure 3.3), which has $\lambda_{max} = 310$ nm in neutral solution and 330 nm in alkali ($\varepsilon = 26.4 \times 10^3$ and 31.8×10^3 L mol^{-1} cm^{-1}, respectively).

Under kraft cooking conditions, phenylcoumaran units undergo a retro-aldol condensation to give trans-stilbenes with $\lambda_{max} = 378$ nm in alkaline solution (Figure 3.3). Falkehag et al. [43] demonstrated that such stilbenes are present in kraft lignins, and in spruce milled-wood lignin subjected to kraft cooking conditions. Lignin samples that were reduced with LiAlH$_4$ to remove interferences from carbonyl compounds had ionization difference spectra that resembled those of the appropriate hydroxystilbene model compounds.

Lin and Kringstad [44,45] studied the UV solution photochemistry of spruce milled-wood lignin by difference spectroscopy. The difference spectra after irradiation showed a broad, featureless absorption increase at wavelengths above 400 nm, corresponding to the discoloration of the solutions. Absorption decreased at wavelengths below 400 nm, with two clear minima at 315 nm and 285 nm. These wavelengths corresponded to the absorptions of lignin α-carbonyl groups. The authors concluded that α-carbonyl groups were important sensitizers of the reactions and that they were destroyed during the course of the irradiation. Others have confirmed that loss of absorption occurs at these two wavelengths because of irradiation, both for milled-wood lignin in solution and for lignin immobilized in hydroxypropyl cellulose films [46]. However, rather different observations are made when lignin in high-yield pulp fibers are irradiated (see the Effect of Aromatic Carbonyl Groups section and Chapter 16).

COLOR OF WOOD, LIGNIN, AND HIGH-YIELD PULPS

Relatively weak absorptions at wavelengths above 400 nm are sufficient to cause the perception of color. The color of woody tissue varies widely from off-white (e.g., aspen, birch), yellow (e.g., spruce, pine, fir), amber-orange (e.g., western hemlock, Douglas fir, red cedar, teak) and rich browns (walnut). While some of this color is due to tannins and other extractives, in most species the color originates from chromophores in the lignin.

CONIFERALDEHYDE AND FERULIC ACID

Pew and Connors [47] proposed that coniferaldehyde end groups are responsible for some of the color of softwoods. The longest wavelength maximum for etherified coniferaldehyde model compounds occurs at around 340 nm. Although this is below the visible region, the band is very broad and there is nonzero absorption at wavelengths as high as 430 nm. Heitner and Min [48] concluded that removal of coniferaldehyde (by sulfonation) is partly responsible for the higher brightness of softwood chemithermomechanical pulp (CTMP) compared with thermomechanical pulp (TMP) prepared from the same wood.

Chandra et al. [49] recently reported that tris(hydroxymethyl)phosphine (THP, a novel reductive bleaching agent) removes the strong coniferaldehyde absorption at 340 nm more effectively than sodium hydrosulfite, a common bleaching agent for lignin-containing pulps. Mass spectal and NMR data suggest the zwitterionic phosphonium compound shown in Figure 3.4 as the product of the reaction.

Oxidation of the γ-aldehyde group in coniferaldehyde to a carboxylic acid yields ferulic acid. While they are not common in wood, ferulic acid and its 5,5' oxidative coupling product, diferulic acid, are important constituents of the cell walls of graminaceous plants such as wheat straw. They are thought to link polysaccharides to lignin through an ester linkage. Pan et al. [50] found that ferulic acid did not contribute to the color of wheat straw pulp. However, the absorption of diferulic acid in solution was sufficiently broadened to shift the tail of the absorption to above 400 nm. This effect was strengthened when the γ-carboxylic acid group was esterified, or when the compound was deposited on a solid cellulose matrix. Hence, the authors concluded that esterified diferulates are probably important contributors to the color of mechanical pulps from wheat straw and other annual plants.

BIPHENYLS

Soluble lignins and biphenyls have shallower minima at 250–270 nm than monomeric phenols. Pew and Connors [51,52] prepared a number of o-o'-dihydroxybiphenyl derivatives by dehydrogenative coupling of simple guaiacyl phenols. They concluded that much of the difference between the spectra of lignins and monomeric phenols

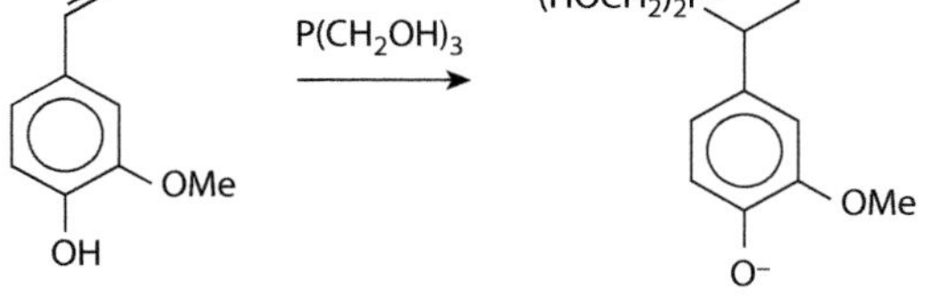

FIGURE 3.4   Reaction of coniferaldehyde with tris(hydroxymethyl)phosphine.

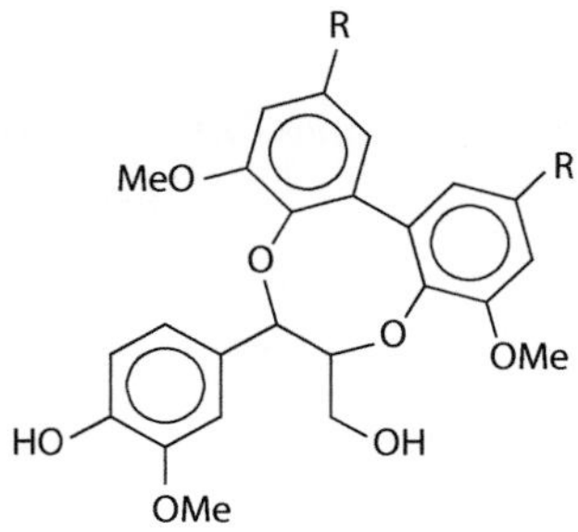

FIGURE 3.5 Biphenyl units bonded as dibenzodioxocin structures.

TABLE 3.7
Absorption Bands of Simple Benzoquinones

	λ_{max} (nm)	log ε	Assignment
para-Benzoquinones	240–300	3.8–4.5	π→π*
	285–485	2.4–3.5	π→π*
	420–460	1.2–2.3	n→π*
ortho-Benzoquinones	250–300	2.6–4.2	π→π*
	370–470	2.8–3.5	π→π*
	500–580	1.4–1.8	n→π*

Source: Reproduced from Berger, S., Hertl, P., and Rieker, A., *The Chemistry of Quinonoid Compounds,* *II*. London: John Wiley and Sons, 29–86, 1988. With permission.

could be accounted for by the presence of a substantial number of biphenyl structures in lignin. The recent work by Karhunen et al. [53–55], reporting substantial amounts of biphenyl groups bonded in dibenzodioxicin structures (Figure 3.5), would appear consistent with this view.

QUINONES

Quinones have long been considered contributors to the color of wood and high-yield pulps [56,57]. *o*-Quinones, in particular, are invoked in many reaction schemes purporting to explain pulp bleaching reactions, as well as reactions proposed to explain darkening under the influence of heat or light.

Berger et al. [58,59] have reviewed the UV-visible spectra of quinones. Generally, *p*- and *o*-quinones each show three characteristic bands, although the weaker bands are sometimes obscured under the stronger ones. The spectra of the two series are quite different, since *p*-quinones contain a cross-conjugated π system, while *o*-quinones are linearly conjugated. Table 3.7 shows the general characteristics of *p*- and *o*-quinones and the band assignments, while Table 3.8 gives specific data on some quinones of relevance to lignin. The differences between *o*- and *p*-quinones

TABLE 3.8

UV-Visible Absorption Data for Some Quinones Relevant to Lignin (sh = shoulder)

Compound	λ_{max} (nm)	Log ε	Solvent	Reference
(p-benzoquinone structure)	246	4.42	CHCl₃	[58]
	288	2.50		
	439	1.35		
(methoxy-p-benzoquinone structure, OMe)	254	4.26	CHCl₃	[58]
	357	3.21		
(lignin-related quinone structure; HO, MeO, HO, OMe)	286	3.64	95% EtOH	[47,51]
	328	3.55		
	400 (sh)	3.14		
(o-benzoquinone structure)	254	3.04	Diethyl ether	[58]
	368	3.28		
	587	1.35		
(methoxy-o-benzoquinone structure, MeO)	269	2.66	CH₂Cl₂	[58]
	465	3.26		
	545 (sh)	1.78		
	575 (sh)	1.56		
	370	3.31	water	[60]
	480 (sh)	2.8		
(ethyl-methoxy-o-benzoquinone structure, MeO)	470	3.19	CHCl₃	[61]
	570 (sh)	2.60		
(di-tert-butyl-o-benzoquinone structure)	256	3.5	CH₃OH	[58]
	402	3.28		
	550	1.77		

are readily apparent to the eye—*p*-quinones are yellow, while *o*-quinones tend to be dark red.

Imsgard et al. [62] determined that spruce milled-wood lignin contained 0.7% *o*-quinone structures by reducing the quinones to the corresponding catechols, and then measuring the visible spectrum of the complex formed between the catechols and ferric ion. This *o*-quinone level agrees well with other values in more recent literature. Zakis [63] determined a quinone content of 0.8% for spruce milled-wood lignin, based on the ability of quinones to oxidize hydrazine and release a stoichiometric amount of nitrogen. Based on ^{31}P NMR spectra of the trimethyl phosphite-quinone adducts, Argyropoulos and Heitner [64] found 0.7% quinone groups in black spruce TMP.

Pew and Connors [47,51] isolated a phenyl-substituted *p*-quinone (row 3, Table 3.8) from enzymatic oxidation of α-ethyl vanillyl alcohol. In addition to two discrete maxima at 328 and 286 nm, this compound has a strong shoulder at 400 nm. No further reports on this structure have appeared in the literature, however, and it is unlikely that this could be formed in solid lignin.

o-Quinones with an adjacent methoxy group have an unusual feature that is relevant to the color of wood and high-yield pulps. In chloroform (or other nonaqueous solvents), these compounds show a prominent maximum at 468 nm. In an aqueous solvent, the absorption at 468 nm is diminished (although still present), and a second, more intense maximum appears at about 370 nm. Adler et al. [60] attribute this behavior to an equilibrium between the *o*-quinone and a monohydrate adduct (Figure 3.6).

The phosphine bleaching agent THP (see the Coniferaldehyde and Ferulic Acid section) also reacts with quinones. When methoxy-*p*-benzoquinone is reacted with THP, a zwitterionic phosphonium ion (Figure 3.7) is formed [49]. The broad, 368 nm absorption of the quinone is replaced with a much sharper absorption band at 305 nm. This differs from the reaction with sodium hydrosulfite, which reduces quinones to the corresponding hydroquinones.

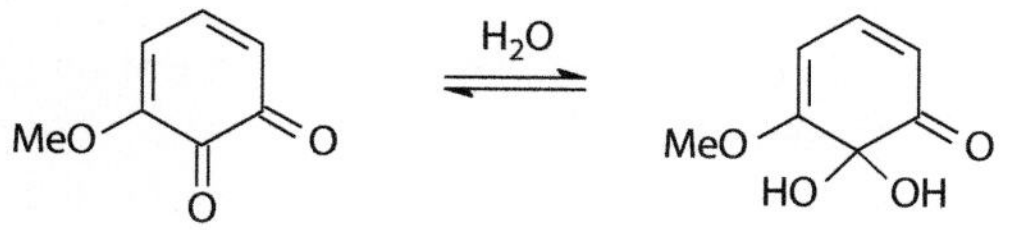

FIGURE 3.6 Equilibrium between 3-methoxy-1,2-benzoquinone and a monohydrate adduct.

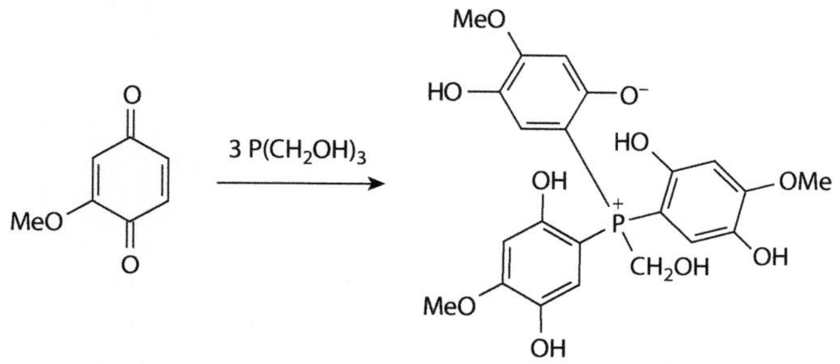

FIGURE 3.7 Reaction between methoxy-*p*-benzoquinone and tris(hydroxymethyl) phosphine.

METAL AND HYDROGEN-BONDED COMPLEXES

Some of the color of lignin is attributed to complexes between lignin units or between a lignin unit and a metal ion. The structures in Figure 3.8 have been proposed in the literature and are discussed in the following section.

Among lignin-metal complexes only those involving iron have significant color. Recent work [65–68] indicates that iron-lignin complexes also contribute to light-induced yellowing of lignin-containing fibers. Manganese can catalyze chromophore formation, especially in the presence of oxygen and alkali, and entrained MnO_2 can darken pulp [69,70]. However, manganese lignin complexes do not contribute to color.

Imsgard et al. [62] noted that spruce milled-wood lignin contained, in addition to 0.7 quinone groups, 1% catechol groups. Lignin-like catechols form complexes with ferric ions that have a broad absorption from 470–700 nm with λ_{max} around 560–800 nm and ε in the range of 1–2×10^3 L mol^{-1} cm^{-1}.

Aqueous ferric ion forms a complex with guaiacol with two discrete maxima and a similar extinction ($\lambda = 422$ nm, $\varepsilon = 2.4 \times 10^3$ Lmol^{-1}cm^{-1}; $\lambda = 461$ nm, $\varepsilon = 2.3 \times 10^3$ Lmol^{-1}cm^{-1}) as the catechol complexes [71]. However, 4-substituted phenols are better models than simple guaiacol for the guaiacyl function in lignin. The extinction coefficients for ferric complexes of such compounds range from 4 to 10^2 Lmol^{-1}cm^{-1}, which is 10–100 times smaller than those of catechol complexes [62,71].

There are sporadic reports of the contributions of complexes between different lignin structures to color. Steelink [72] proposed the existence of quinhydrones in pine kraft lignin. These donor-acceptor complexes between a quinone and a hydroquinone are held together by hydrogen bonds between the hydroxyl group of the hydroquinone and the quinone carbonyl group and can be particularly strong in the solid state. These complexes usually have an intense absorption in the visible region due to a charge-transfer transition [73,74].

Complexes between quinones and phenols have also been proposed as important lignin chromophores in pulping liquors. Furman and Lonsky [75,76] attributed two-thirds of the light absorption of kraft lignin to quinone-phenol charge-transfer complexes.

Incidental information suggests that similar donor-acceptor complexes contribute to the color of wood or high-yield pulp fibers. For example, solution spectra of

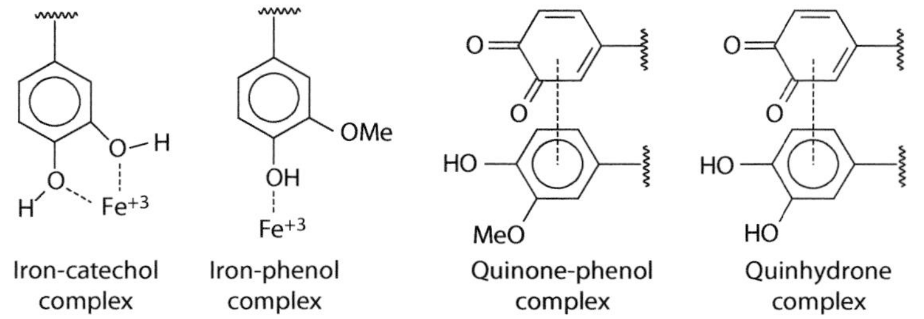

Iron-catechol   Iron-phenol   Quinone-phenol   Quinhydrone
complex   complex   complex   complex

FIGURE 3.8 Lignin metal and hydrogen-bonded complexes that contribute to the color of lignin.

etherified coniferaldehyde have $\lambda_{max} = 340$ nm. The maximum is red-shifted to about 350 nm when such a compound is incorporated into cellophane [47] or adsorbed onto fully bleached kraft pulp [77]. Broadening and red-shifting of electronic transitions is well-known when moving from fluid solution to the solid state. However, the absorption maximum is shifted a further 10 nm to 360 nm when methyl coniferaldehyde is adsorbed onto softwood TMP [70]. This may indicate the existence of other interactions, in addition to those that normally cause red-shifting and band broadening.

The spectra of simple *o*- and *p*-quinones adsorbed on high-yield pulps have been recorded [70,78,79]. Table 3.9 compares these spectra with the corresponding spectra in solution. For the *o*-quinones in particular, the solution spectra are significantly different from those recorded when the compound was adsorbed on high-yield pulp.

TABLE 3.9
Comparison of Absorption Spectra of Quinones in Solution and Adsorbed on Lignin-Containing Pulp

Compound	λ_{max} (nm) in Solution (log ε)	Reference	λ_{max} Adsorbed on Lignin-Free Pulp	λ_{max} Adsorbed on Lignin-Containing Pulp	Reference
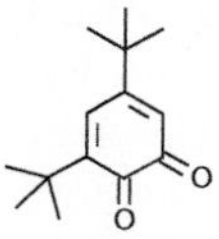	256 (3.50) 402 (328) 550 (1.77)	[58]	–	440	[70]
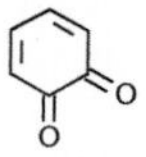	254 (3.04) 368 (3.28) 587 (1.35)	[58]	–	400 540	[78]
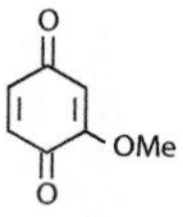	254 (4.26) 357 (3.21)	[58]	–	385	[70]
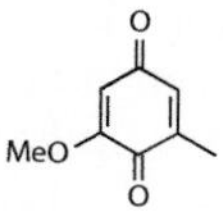	357	[79]	375	416	[79]
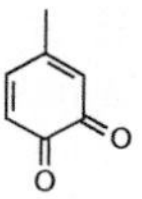	387	[79]	427	444	[79]
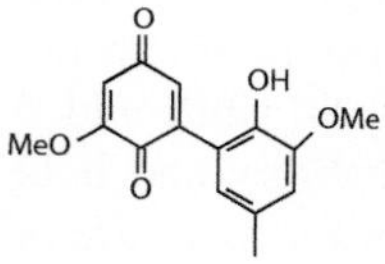	328	[79]	338	125	[79]

Barsberg et al. [80] have recently presented more systematic evidence for existence of lignin charge-transfer (CT) complexes in wood fibers. When deposited on filter paper, a series of lignin model donors (catechol, resorcinol and vanillyl, sinapyl, and coniferyl alcohols) formed complexes with p-benzoquinone. The characteristic absorptions at $\lambda > 500$ nm existed only when the donor and acceptor were deposited on filter paper. The forces holding the complexes together were evidently too weak to overcome solvent stabilization of the individual molecules. Further evidence for the formation of CT complexes comes from the result of adding p-benzoquinone to thermomechanical pulp from beech or Norway spruce. New absorptions with $\lambda_{max} = 400$ nm and 500 nm appeared, which tailed to well beyond 700 nm.

STILBENES

Stilbenes are claimed to exist in lignin-containing pulps. One mechanism for their formation is the retroaldol condensation of β–1 and β–5 lignin structures in the presence of alkali (see the Alkali Darkening section).

Mechanical action, which occurs during the refining of wood into mechanical pulp, has also been proposed to induce the formation of stilbenes. The basis of this proposal relates to the isolation of small yields of stilbenes after ball milling β–1 [81] or β–5 (phenylcoumaran) [82] lignin model compounds. Johansson et al. [83] searched for more direct evidence of this transformation by comparing milled-wood lignin isolated from first-stage refined thermomechanical pulp from Norway spruce with milled-wood lignin isolated from the wood before refining. The lignin isolated from the pulp was slightly darker than that isolated from the wood. However, ^{13}C NMR and UV-visible difference ionization spectra showed no evidence supporting transformation of β–1 or β–5 structures to stilbenes. These data suggested that the darkening was caused by formation of α-carbonyl structures during refining, as well as the formation of metal complexes.

DIFFUSE REFLECTANCE SPECTROSCOPY

Although studies of lignin model compounds in solution are useful guides, very different reactions can occur in the solid state. Diffuse reflectance spectroscopy is an ideal technique to monitor the chromophore changes that occur in lignin-containing fibers. Kubelka and Munk developed the basic theory of diffuse reflectance in the 1930s [84], and van den Akker [85] extended it to paper shortly after. However, the arduous calculations required to generate spectra severely limited its use. The development of microprocessor controlled spectrophotometers over the last decade has made diffuse-reflectance measurements much more accessible.

Since the 1970's, there has been heightened interest in using lignin-containing pulps in value-added paper grades, in addition to their traditional use in newsprint. The more stringent standards for whiteness and whiteness stability in value-added grades created a need for more detailed knowledge of the bleachability and light stability of the chromophores in mechanical pulp lignin. A significant literature of diffuse reflectance studies on lignin-containing pulps has thus appeared. In contrast, there are few diffuse reflectance studies of chemical pulp.

BASIC MEASUREMENTS

Figure 3.9 shows the reflectance spectra of unbleached and peroxide-bleached TMP from black spruce. Reflectance is the ratio of the intensity of reflected to incident light, and is thus mathematically analogous to the transmittance of the Beer–Lambert Law. Unlike transmittance, however, reflectance is not easily rendered to a quantity proportional to chromophore concentration. Diffuse reflectance is related to chromophore concentration by the Kubelka–Munk remission function, Equation 3.7, where R_∞ is the reflectance of a homogeneous sample that is sufficiently thick that no light can pass through, k is the absorption coefficient and s the scattering coefficient. The absorption coefficient is directly proportional to chromophore concentration, while the scattering coefficient is related to the surface area available to scatter light.

$$\frac{k}{s} = \frac{(1-R_\infty)^2}{2R_\infty} \tag{3.7}$$

Several approaches to interpret reflectance spectra have appeared in the literature:

1. Hirashima and Sumimoto [86,87] converted reflectance spectra to "apparent absorbance," $-\log(1/R_\infty)$, by analogy to the relationship between transmittance and absorbance in the Beer–Lambert Law. However, according to Equation 3.7, $-\log(1/R_\infty)$ is not proportional to k and, therefore, not proportional to chromophore concentration.
2. The remission function, k/s in Equation 3.7, is proportional to chromophore concentration. If the scattering coefficient remains constant, changes in the remission function will be proportional to changes in chromophore concentration. A common form of this treatment of reflectance data is the so-called post-color (PC) number, Equation 3.8, where the subscripts 0 and 1 indicate the remission function before and after a given treatment, respectively.

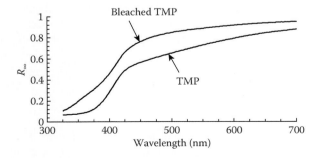

FIGURE 3.9 Diffuse reflectance spectrum of unbleached softwood TMP and TMP bleached with 4% alkaline hydrogen peroxide.

As proposed by Giertz [88], the reflectance used in Equation 3.8 is a standard brightness measurement, which is actually a normalized summation of reflectances (see Equation 3.9). Polcin and Rapson used the remission function to study the spectra of groundwood pulps [89–93].

$$PC = \left[\left(\frac{k}{s} \right)_1 - \left(\frac{k}{s} \right)_0 \right] \times 100 \qquad (3.8)$$

3. If the samples to be compared do not have similar scattering coefficients or do not maintain a homogeneous distribution of chromophores through the thickness of the sample, the PC number will not be strictly proportional to chromophore concentration. In such cases, Schmidt and Heitner [3,94,95] propose explicit calculation of absorption and scattering coefficients. This requires measurement of the reflectance of transmitting samples over white and black backgrounds of known reflectance, and additional computation. Further details are available in recent reviews [3,94,95].

Brightness is one of the most important commercial attributes of pulp. Research laboratories often do not possess the instrument specified by standards-setting bodies for this measurement and instead use a UV-visible spectrophotometer equipped with a diffuse reflectance attachment. The reflectance at 457 nm, the nominal brightness wavelength, is reported as a surrogate for brightness. Using this assumption, one would reasonably conclude that coniferaldehyde groups in pulp should not affect the brightness of mechanical pulps since coniferaldehyde does not absorb above 430 nm [77]. This reasoning, however, is based on a misunderstanding of this important industrial measurement. Brightness is not a single-wavelength measurement. Rather, it is calculated as the summation shown in Equation 3.9, where $R(\lambda)$ is the reflectance and $f(\lambda)$ is the normalized standard brightness function.

$$\text{ISO Brightness} \ = \sum_{\lambda} R(\lambda) f(\lambda) \qquad (3.9)$$

Tables of $f(\lambda)$ are given by various standard-setting organizations for the paper industry [96–98]. While $f(\lambda)$ is largest at 457 nm, the so-called central wavelength, it is nonzero from 400 to 510 nm. Thus, chromophores that absorb at wavelengths somewhat longer than 400 nm can affect brightness even if their absorption ceases well below 457 nm (e.g., coniferaldehyde).

Pulp-Bleaching Studies

Reductive Bleaching

Polcin and Rapson published a comprehensive series of papers describing diffuse reflectance studies of the chromophores in spruce and hemlock groundwood pulps [89–91], and their response to some common bleaching agents [92,93]. Based on either reflectance spectra or differences in remission function, they concluded that the major chromophores in both pulps were simple quinones and coniferaldehyde. They attributed the darker color of hemlock to larger amounts of condensed quinones and structures containing ring-conjugated carbonyl groups.

Sodium dithionite, $Na_2S_2O_4$, decreased the remission function at all wavelengths greater than 300 nm, but the largest decreases occurred in the visible region. Polcin and Rapson [93] attributed this to the reduction of simple quinones and coniferaldehyde. Kuys and Abbot [99] made similar observations in a spectroscopic study of dithionite bleaching of radiata pine TMP. However, they suggested that coniferaldehyde was reduced by sodium bisulfite formed as a dithionite oxidation product, not by dithionite itself.

Svensson et al. [100] compared dithionite bleaching of spruce TMP at pH 10 and pH 5, using principal components analysis to reduce reflectance spectra to contributing subspectra. The authors concluded that at pH 5 (the common pH for commercial dithionite bleaching) quinones were the major chromophores bleached. At pH 10, quinones were formed due to alkali darkening. In spite of the higher redox potential of dithionite at pH 10, bleaching was less efficient than at pH 5, and there was no evidence that any chromophores other than quinones were removed.

Polcin and Rapson also studied the effect of sodium borohydride and uranium (III) as reductive bleaching agents [93]. These two bleaching agents decreased the absorption coefficient in the visible region by about the same amount as dithionite. Unlike dithionite, borohydride and U^{3+} reduced the absorption coefficient much more in the ultraviolet region than in the visible region. The remission function difference spectra showed a strong maximum at about 360 nm, which the authors attributed to efficient reduction of aromatic ketone structures, in addition to quinones and coniferaldehyde.

Peroxide Bleaching

Polcin and Rapson [92] found that remission difference spectra for peroxide bleaching of groundwood pulp from spruce or hemlock were qualitatively similar to those recorded for reduction with borohydride or U^{3+}. The difference spectra showed broad, featureless changes in the visible and a discrete maximum at about 360 nm. Peroxide caused larger decreases in absorption coefficient than the reducing agents.

Despite having higher brightness than pulp bleached with reducing agents, pulp bleached with either alkaline (pH 10) hydrogen peroxide or peracetic acid (pH 7) still had a slight yellowish tint. This color could be removed by treating the oxidatively bleached pulp with a reducing agent in a second stage. The order of effectiveness of the reducing agents was dithionite < borohydride < U^{3+}. Polcin and Rapson [92] attributed this additional brightening to reduction of residual quinone or coniferaldehyde groups, and to reduction of new chromophores formed under the alkaline conditions of peroxide bleaching. In this regard, Gellerstedt [101] has noted that o-quinones treated with alkaline peroxide form small amounts of hydroxyquinones, which are resistant to further bleaching by peroxide.

Heitner and Min [48] compared the chromophore content of TMP and CTMP prepared from black spruce wood. The scattering coefficients of these two pulps were quite different; therefore, the remission function was unsuitable to compare the chromophore changes and the absorption coefficients were calculated explicitly.

The difference in absorption between TMP and CTMP showed a strong bleaching at $\lambda_{max} = 350$ nm, consistent with sulfonation of coniferaldehyde groups. Much smaller decreases in absorption occurred in the visible range, with no discernible maxima. The magnitude of the bleaching increased with increasing sulfite charge, to a maximum of 4% on oven-dried wood. Johansson and Gellerstedt observed similar spectral differences for TMP and CTMP prepared from Norway spruce [102]. Additionally, these authors noted that there is some chromophore formation during first-stage refining of TMP production.

The difference spectra for alkaline peroxide bleaching of black spruce TMP and CTMP also showed a maximum at $\lambda_{max} = 350$ nm, which was attributed to oxidative removal of coniferaldehyde [48]. The magnitude of the bleaching at 350 nm was largest for TMP, since it contained the largest amount of coniferaldehyde. For CTMP, the magnitude of the absorption decrease at $\lambda_{max} = 350$ nm after peroxide bleaching decreased with increasing sulfite charge (up to 4%). This is because the prior sulfite treatment removed increasing amounts of coniferaldehyde.

Holah and Heitner [103] gave further spectroscopic evidence that aromatic ketone groups contribute to the yellowish cast of softwood mechanical pulps, and further showed that they are removed by hydrogen peroxide. Figure 3.10 shows the absorption difference spectra obtained for treatment of softwood TMP with a 50% charge of alkaline peroxide. For reaction times less than five hours, the strong bleaching at $\lambda_{max} = 360$ nm and the decrease in absorption throughout the visible region is adequately explained by oxidation of coniferaldehyde and quinone chromophores. For reaction times greater than five hours, the shape of the difference spectrum changed.

Figure 3.11 shows the differences in absorption between pulp bleached for 5 hours and pulp bleached for up to 300 hours. Peroxide continued to bleach stone groundwood pulp and TMP at these long reaction times. However, λ_{max} was shifted

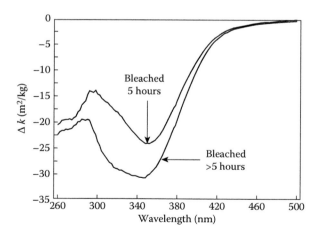

FIGURE 3.10 Absorption difference spectra for peroxide bleaching of softwood TMP, $\Delta k = k_t - k_{t=0}$. The shape of the difference spectrum for pulps bleached less than 5 hours is different from that for pulps bleached longer than 5 hours. (Reproduced from Holah, D.G. and Heitner, C., *J. Pulp Paper Sci.*, 18, J161–J165, 1992. With permission.)

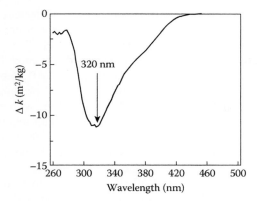

FIGURE 3.11 Difference in absorption between softwood TMP bleached with alkaline hydrogen peroxide for longer than 5 hours and that bleached for 5 hours, $\Delta k = k_{t > 5h} - k_{t=5h}$. (Reproduced from Holah, D.G. and Heitner, C., *J. Pulp Paper Sci.*, 18, J161–J165, 1992. With permission.)

to 320 nm, consistent with removal of aromatic ketone groups. This behavior was not observed for CTMP, perhaps because the aromatic ketones were transformed by the prior sulfite treatment to a chromophore resistant to peroxide.

Wright et al. [104,105] studied the kinetics of peroxide bleaching of *E. regnans* mechanical pulp by monitoring changes in the absorption coefficient at 457 nm. Their initial model proposed that two chromophores were bleached in independent reactions, either consecutively or in parallel. This model could not, however, adequately predict the effect of sudden changes in peroxide or alkali concentration. Instead, Wright et al. suggested a model where an initial chromophore is in equilibrium with a stable leucochromophore that is irreversibly transformed to a colorless product.

Alkali Darkening

Wood and lignin-containing pulps darken at pH greater than 10, a behavior important to both chemical and mechanical pulping. Kraft and soda pulping release cellulose fibers from wood by alkaline hydrolysis; the response of lignin to the alkali determines the chromophores that must be eliminated in subsequent bleaching stages. Peroxide bleaching of both chemical and mechanical pulps requires a pH greater than 10 to generate the major bleaching species, the hydroperoxide anion.

The absorption difference spectra for alkali-treated wood slices [106,107], TMP [70] or groundwood pulp [78] from softwoods all show significant absorption increases in the visible region, with λ_{max} at about 420 nm. The darkening can be diminished significantly, but not eliminated, by alkylation of phenolic groups [107] or reduction with borohydride [70,107] before exposure to alkali. Pretreatment with dithionite has no effect on alkali darkening [106,107]. Thus, phenolic and aromatic carbonyl groups would appear to be the most important contributors to alkali-darkening reactions.

High-yield (80%) sulfite pulp also darkens in alkali [78], but an additional chromophore is involved. Although the absorption increases significantly in the visible, the largest increase is in the ultraviolet, with λ_{max} at about 380 nm. This is due at least in part to formation of coniferaldehyde groups. Coniferaldehyde is sulfonated during production of high-yield sulfite, but the sulfonic acid groups are easily released in alkali, regenerating the original coniferaldehyde [108,109].

Leary and Giampaolo [70] found that unbleached softwood TMP behaved differently upon alkali treatment than peroxide-bleached TMP. Unbleached softwood TMP was bleached with peroxide (4% charge) to an ISO brightness of 78%; however, alkali-darkened pulp could only be bleached to 68% with the same peroxide charge. Borohydride reduction of the pulp before the alkali treatment inhibited the darkening and almost fully restored its bleachability. The absorption maximum for alkali darkening of unbleached pulp appeared at 420 nm, but was shifted to 360 nm if the pulp was pretreated with borohydride.

Peroxide-bleached TMP regained all the brightness lost on alkali darkening upon subsequent peroxide bleaching. Borohydride pretreatment inhibited the alkali darkening, but was not required to maintain full bleachability of the pulp, except at 120°C. The maximum in the absorption difference spectrum was at 360 nm.

Following the earlier suggestion of Gellerstedt [101], Leary and Giampaolo [70] attribute the irreversible darkening of unbleached TMP and the absorption maximum at 420 nm to the formation of hydroxylated o-quinones. The precursors of these hydroxylated quinones are lignin groups that can be removed by peroxide or borohydride, e.g., coniferaldehyde, quinones or aryl carbonyls.

The maximum that was observed at 360 nm for alkali darkening of peroxide or borohydride treated pulps was attributed to the formation of stilbenes [70]. Stilbenes, which are most likely formed by a retro-aldol condensation when β–1 or phenylcoumaran lignin structures are treated with alkali (Figure 3.3), have been isolated from peroxide-bleached mechanical pulps [110]. The action of alkali and oxygen removes such phenolic stilbenes [111].

Chemical Pulp Bleaching

Only three diffuse reflectance studies on the bleaching of chemical pulp have appeared over the period surveyed by this chapter. Ragnar [112] used difference spectra to monitor the chemical changes in birch kraft pulp during various stages of an elemental-chlorine free (ECF) bleaching sequence. Removal of hexenuronic acid, $\lambda_{max} = 232$ nm, could be easily distinguished from removal of lignin, $\lambda_{max} = 280$ nm. Ragnar thus proposes that diffuse-reflectance UV measurements could form part of a strategy to monitor and control removal of hexenuronic acid (which consumes bleaching chemicals without increasing brightness) during bleaching.

Wójciak et al. [113] used diffuse reflectance spectroscopy to determine that an acidic peroxide pretreatment of oxygen-delignified kraft pulp improved the removal of residual lignin in the subsequent alkaline extraction stage. da Silva Perez et al. [114] described a two-stage, chlorine-free bleaching sequence—active oxygen species were generated from O_2 using TiO_2 as a photosensitizer in the first stage, followed by photochemical decomposition of hydrogen peroxide in the second stage. Diffuse reflectance spectra indicated that the TiO_2 stage removed coniferaldehyde, but not quinones, from *E. regnans* pulp. The TiO_2 catalyst did not improve brightness by itself, but improved the response of the pulp to the subsequent peroxide stage.

PHOTOCHEMISTRY

A detailed treatment of lignin photochemistry is given in Chapter 17. This chapter will be restricted to the contributions of diffuse-reflectance and transient absorption studies.

Lignin-Containing Pulps

The rapid, light-induced yellowing of lignin is one of the most serious and recalcitrant problems preventing the wider use of high-yield pulps in value-added paper grades. Much research effort has been devoted to this problem over the last 40 years.

Action Spectra

An action spectrum plots the magnitude of the response to a fixed light exposure (in either J/cm^2 or photons/cm^2), as a function of wavelength. Andrady et al. [115] used yellowness index (ASTM D1925-70) to measure the response of newsprint made from unbleached loblolly pine TMP to irradiation between 260 nm and 600 nm. Wavelengths from 260 to 340 nm efficiently increased the yellowness index, although the wavelength dependence was much weaker for samples that had been yellowed by a previous exposure to broadband irradiation. Wavelengths between 500 nm and 600 nm caused a decrease in the yellowness index.

Forsskåhl and Tylli [116] constructed action spectra for irradiation of softwood chemimechanical pulp (CMP) and CTMP by measuring the changes in PC number at 457 nm and 557 nm. The spectra for peroxide-bleached chemimechanical pulp are shown in Figure 3.12. Both action spectra have a maximum response at 325 nm, although the response for PC number at 457 nm is significantly broader, perhaps due to a hidden maximum at 350 nm.

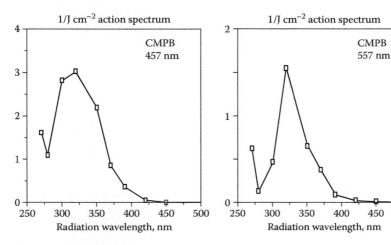

FIGURE 3.12   Action spectra for irradiation of peroxide-bleached softwood CMP. The *y* axis is the inverse of the incident energy required to produce a unit change in PC number at either 457 nm or 557 nm. (Reproduced from Forsskåhl, I. and Tylli, H., *Photochemistry of Lignocellulosic Materials*, Washington, DC: American Chemical Society, 45–59, 1993. With permission.)

These action spectra indicate that a lignin chromophore with λ_{max} at 325 nm initiates absorption increases at both 457 nm and 557 nm. This would seem to substantiate the proposal by many authors that aryl carbonyl groups are important sensitizers of photoyellowing (see Chapter 16, Photochemistry of Lignin and Lignans). These data also suggest that a second chromophore with λ_{max} at about 350 nm contributes to absorption increases at 457 nm, but not at 557 nm. The nature of this chromophore is less certain, although the position of the absorption is consistent with stilbenes, which, as noted earlier, may be formed under the conditions of peroxide bleaching [70,110].

Unlike newsprint, neither CMP nor CTMP were photobleached by any wavelength. However, pulp previously yellowed by broadband exposure was photobleached by almost all wavelengths of light, except for a narrow band around 350 nm.

Difference Spectra

Effect of Coniferaldehyde   Claesson et al. [117] and Leary [118] first reported the effect of ultraviolet light on the diffuse reflectance spectra of lignin-containing pulps. Wood or unbleached mechanical pulp from spruce or radiata pine showed decreases in reflectance at wavelengths above 400 nm, with a maximum at about 420 nm. At wavelengths below 400 nm, however, reflectance increased with a maximum at 360 nm. This photobleaching was not observed for high-yield bisulfite pulp. Instead, the reflectance decreased even more at wavelengths below 400 nm than at those above. Heitner [119] attributed this bleaching to photochemical oxidation of coniferaldehyde groups. The effect would thus not occur in high-yield bisulfite pulp, since sulfite pulping removes coniferaldehyde groups.

Although coniferaldehyde model compounds react rapidly and quantitatively with peroxide [120], the reaction in pulp fibers is much slower. Residual coniferaldehyde has been detected in peroxide-bleached mechanical pulps [87,110], and can have a profound effect on the absorption difference spectra for irradiation. Figure 3.13 shows difference spectra for ultraviolet irradiation of peroxide-bleached softwood TMP and CTMP [121]. The spectrum for CTMP has a broad absorption increase

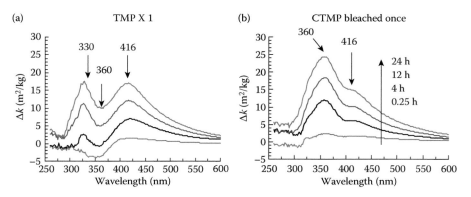

FIGURE 3.13   Absorption difference spectra for ultraviolet irradiation of peroxide-bleached softwood TMP and CTMP, calculated from diffuse reflectance measurements. (Reproduced from Schmidt, J.A. and Heitner, C., *J. Wood Chem. Technol.*, 15, 223–245, 1995. With permission.)

with λ_{max} at 360 nm and a clear shoulder at about 416 nm. In contrast, the spectrum of TMP bleached with 3% peroxide has two maxima at 330 nm and 416 nm, and a minimum at 360 nm. The minimum at 360 nm is attributed to photobleaching of residual coniferaldehyde, which retards the otherwise rapid increase in absorption that occurs at this wavelength. This interpretation is consistent with the observation that TMP that is given a more thorough treatment with peroxide shows the same difference spectrum as peroxide-bleached CTMP. For CTMP; the combination of sulfite treatment and peroxide bleaching removes all coniferaldehyde.

Davidson et al. [122] reported the difference spectra of spruce wood sections and groundwood pulp that had been irradiated and then stained with phloroglucinol. The irradiated and stained samples showed an absorption increase at $\lambda_{max} = 550$ nm, which the authors attributed to photochemical formation of coniferaldehyde, based on the well-known Wiesner reaction. However, in a subsequent publication [123], they found that phloroglucinol gives a similar color reaction with methoxy-p-quinone, which would be a more reasonable explanation of the earlier results.

Effect of Aromatic Carbonyl Groups   The action spectra of Forsskåhl and Tylli show that light of 320–330 nm induces yellowing most efficiently, consistent with the hypothesis that aromatic carbonyl groups are the most important sensitizers of the reactions. However, several authors [124,125] have noted that thorough borohydride reduction of mechanical pulps reduces light-induced yellowing only marginally.

Schmidt and Heitner [121,126] argued that aromatic carbonyl groups, even if not present initially, are formed by photochemical reactions. Figure 3.14 shows

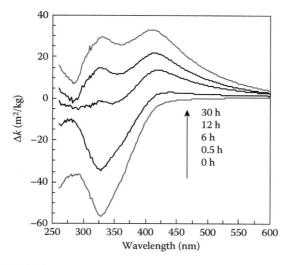

FIGURE 3.14   Absorption difference spectra for ultraviolet irradiation of peroxide-bleached, borohydride-reduced softwood TMP. (Reproduced from Schmidt, J. A. and Heitner, C., *J. Wood Chem. Technol.*, 13, 309–325, 1993. With permission.)

the absorption difference spectra for ultraviolet irradiation of borohydride-reduced peroxide-bleached TMP. Borohydride treatment without irradiation caused bleaching at 320 nm due to reduction of the carbonyl groups to the corresponding benzyl alcohols. This absorption recovered rapidly upon irradiation. Schmidt and Heitner proposed the cleavage of ketyl radicals of β–O–4 linked guaiacylglycerol structures, path b of Figure 3.15, to account for the recovery of the aromatic carbonyl absorption. This was based on an assumed rate constant of $>5 \times 10^7$ s^{-1} for cleavage of the ketyl radical to the corresponding enol [127]. However, it has been subsequently shown that the ketyl radical is much longer-lived than originally thought (rate constant $1.7 - 2.7 \times 10^3$ s^{-1}), so the bimolecular reaction of path a of Figure 3.15 seems a more likely route [128] (see Chapter 16, The Photochemistry of Lignin and Lignans, for more detail).

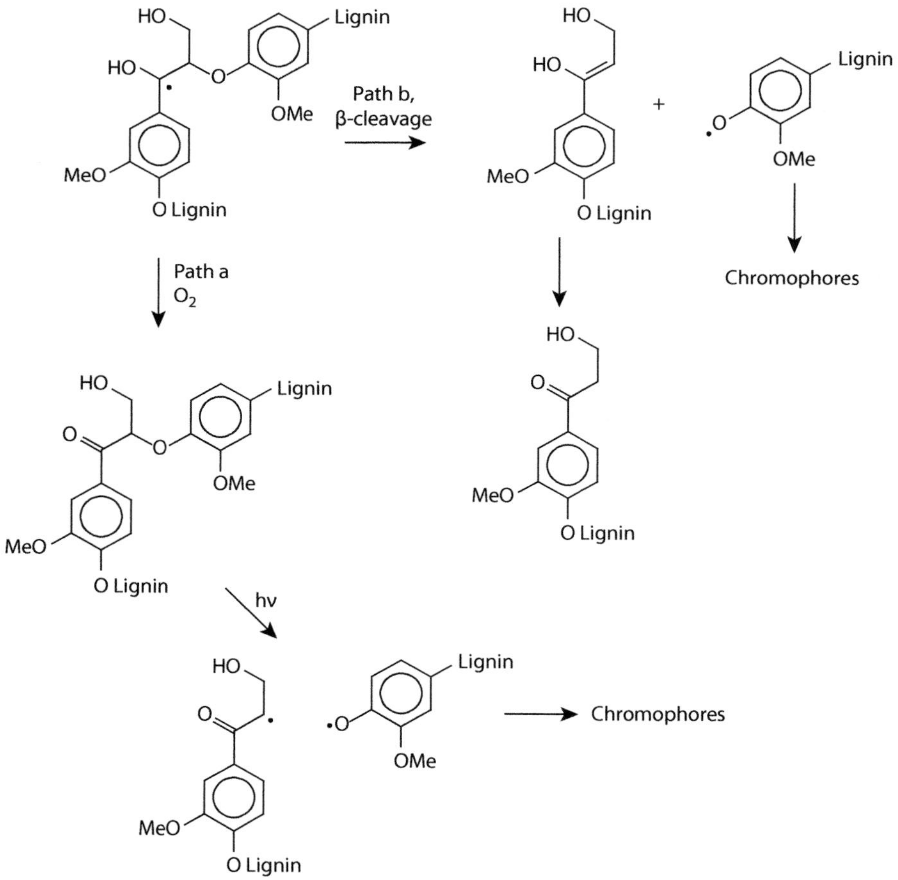

FIGURE 3.15 Two pathways proposed to account for yellowing of lignin-containing pulps and the formation of aromatic carbonyls from β–O–4 guaiacylglycerol ketyl radicals.

Summary of Light-Induced Chromophore Changes

Other difference spectra for irradiation of high-yield pulps have been reported [129–131], in addition to those already cited. In all cases, a minimum of three absorption peaks, with λ_{max} at 320–330 nm, 350–360 nm, and 420–430 nm, are required to adequately explain the spectra.

The absorption increase with λ_{max} = 320–330 nm in borohydride treated pulps is most probably due to formation of aryl carbonyl groups. This peak is obscured by the much stronger absorption increase at 350–360 nm in pulps where coniferaldehyde groups have been completely removed. However, in pulps where moderate coniferaldehyde photobleaching occurs (see the Effect of Coniferaldehyde section), it is clear that the absorption increase at 320–330 nm is independent of that at 350–360 nm.

The band with λ_{max} at 420–430 nm is most likely due to formation of *o*-quinone structures, at least during the early stages of photolysis. This is consistent with the absorption spectra of ligninlike *o*-quinones (see Table 3.8), and with the difference spectra observed after treatment of irradiated pulp with trimethyl phosphite, a reagent that selectively tags *o*-quinones for subsequent detection by ^{31}P NMR [132].

The chromophore changes occurring at 350–360 nm are the most complex. Strong photobleaching at this wavelength during irradiation of unbleached pulps is most likely due to the photooxidation of coniferaldehyde. In contrast, for highly bleached pulps that are free of coniferaldehyde residues the largest absorption increase occurs at this wavelength. Methoxy-*p*-quinones absorb in this region (Table 3.8). Their formation is consistent with the phloroglucinol staining experiments of Davidson et al. (see the Effect of Coniferaldehyde section), and with the observation that both borohydride and dithionite bleach this absorption [132]. Stilbenes have also been proposed as a contributor to the absorption at 350–360 nm, although there is no clear evidence supporting this.

Several authors have presented evidence of photochromism in lignin-containing pulps. Photochromism is a reversible change between two chemical species that is induced in a least one direction by a specific wavelength of light, as shown in Figure 3.16. Ek et al. [133] observed that the brightness was about one point lower immediately after a period of irradiation than after 24 hours storage in the dark. This effect was repeatable over several cycles of irradiation and storage. Forsskåhl and Maunier [134] noted that irradiation with monochromatic light at 370 nm decreased the brightness of bleached spruce TMP, but subsequent irradiation at 430 nm restored some of the brightness loss. This behavior was also repeatable over as many as 12 irradiation cycles. Both these groups suggest that transformation between quinones and their corresponding hydroquinones or semiquinones are the most probable chemical species involved. Choudhury et al. [135] also observed a photochromic

$$\text{Species A} \underset{\lambda_B \text{ or } \Delta}{\overset{\lambda_A}{\rightleftharpoons}} \text{Species B}$$

FIGURE 3.16   A general scheme for photochromism.

effect when they stimulated fluorescence of groundwood pulp alternately at 365 nm and then at 450–490 nm. They suggested that reversible transformation between phenyl coumarans and hydroxystilbenes might account for these effects.

Chemical Pulps

Chemical pulps bleached using Cl_2 and ClO_2 show very little light-induced yellowing because of their extremely low lignin contents. However, so-called ECF and totally chlorine-free (TCF) pulps contain somewhat more lignin, and are more sensitive to light. Research in this area is in its infancy, and only a few spectroscopic studies have appeared.

Castellan et al. [136] observed broad increases in absorption at wavelengths greater than 400 nm for irradiation of pine sulfite pulp bleached with a variety of oxygen bleaching agents. Only the raw spectra, not difference spectra, were reported; so, no clear absorption maxima were evident. Positive phloroglucinol staining indicated that either coniferaldehyde or methoxy-p-quinone is the main chromophore formed, although the latter would seem more likely given the discussion in the Effect of Coniferaldehyde section.

Forsskåhl et al. [137] used difference reflectance spectroscopy to study the role of 5-hydroxy-2-methylfuraldehyde (HMF) on light- and heat-induced yellowing of softwood TCF pulp. HMF is a carbohydrate-derived substance that forms under typical pulp bleaching conditions, and has been linked to color formation in cellulosic materials [138]. The reflectance difference spectrum of heat-aged TCF pulp showed an absorption maximum at 450 nm. The same pulp doped with HMF had a more intense, but otherwise similar difference spectrum, suggesting that HMF contributes to heat-induced yellowing of TCF pulps. Light-aged TCF pulp had a similar maximum as heat-aged samples, but the absorption band was much narrower.

TRANSIENT SPECTROSCOPY

Transient spectroscopy, also known as flash photolysis, makes possible the observation of short-lived chemical species. The award of the 1967 Nobel Prize in Chemistry to Norrish, Porter, and Eigen recognized the importance of this technique to the investigation of reaction mechanisms.

The basic principle is to observe the change in absorbance after an intense radiation pulse has created a significant population of short-lived reactive intermediates. Early experiments used xenon flash lamps as excitation sources and were able to detect intermediates with lifetimes $\geq 10^{-3}$ second. Modern experiments use pulsed lasers as excitation sources. The monochromatic output of a laser allows selective excitation; the narrow pulse width allows detection of species with lifetimes as low as 10^{-12} second. A complementary experiment uses a pulse of electrons from a linear accelerator to generate the reactive species. More experimental detail is available in many reviews [139].

This section deals primarily with the spectra of excited states and short-lived intermediates of lignin and lignin model compounds. More detailed discussion of reaction mechanisms is contained in Chapter 16, Photochemistry of Lignin and Lignans.

TRANSITIONS BETWEEN STATES—THE JABLONSKI DIAGRAM

An important application of transient spectroscopy is the characterization of the excited state involved in a photochemical reaction. The Jablonski, or energy-level, diagram in Figure 3.17 is useful to understand both transient absorption and emission spectra (see the Emission Spectroscopy section).

The ground electronic state of a closed shell molecule is always a singlet state, S_0 (spin quantum number $s = 0$), since all electron spins are paired in the molecular orbitals (MOs). Absorption of a photon promotes one electron from the highest-energy occupied molecular orbital (HOMO) to a higher-energy, singlet state MO, S_n. *Internal conversion*, which is a radiationless transition between states of the same multiplicity, rapidly converts S_n to the first excited singlet state, S_1. With rare exceptions [140], photochemistry and photophysics of organic molecules occur only from vibrationally relaxed S_1 (Kasha's Rule).

Since one electron is in a different MO in an excited state, the requirement of the Pauli principle for paired electron spins is relaxed. The excited state can therefore exist as either a singlet state S_1 ($s = 0$) or a triplet state T_1 ($s = 1$). The reactivity of singlet states and triplet states is generally very different. *Intersystem crossing*, which is a radiationless transition between states of different multiplicity, is a formally spin-forbidden transition; however, spin-orbit coupling removes this constraint in carbonyl compounds.

EXCITED STATES OF LIGNIN AROMATIC CARBONYL MODEL COMPOUNDS

The aromatic carbonyl is an important structure in lignin photochemistry. The electronic structure of aromatic carbonyl compounds has been exhaustively studied [141],

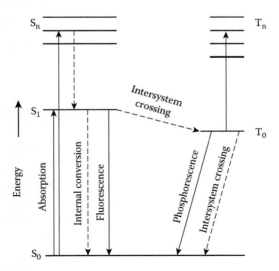

FIGURE 3.17 Jablonski (energy-level) diagram for a closed-shell molecule. Solid line: transitions that occur with absorption or emission of radiation; dashed line: radiationless transitions.

and is briefly summarized here. Aromatic carbonyl compounds have two different excited states that are close in energy, but have very different reactivity. So-called $\pi-\pi^*$ states result from promotion of an electron from a π bonding MO to a π^* antibonding MO, while $n-\pi^*$ states result from promotion of one of the lone-pair electrons on the carbonyl oxygen to the π^* MO. The energy order of the $\pi-\pi^*$ and $n-\pi^*$ states is sensitive to substituent effects, as is the observed photochemistry [141,142]. The lowest energy state of ligninlike compounds, e.g., *p*-methoxyacetophenone, is the $\pi-\pi^*$ state.

A favorite model compound for photochemical studies is $\alpha-(2\text{-methoxyphenoxy})$-3,4-dimethoxyacetophenone, Figure 3.18. Several groups have observed the transient absorption spectrum of this compound, which is shown in Figure 3.19 [143–145]. The spectrum is assigned to the $\pi-\pi^*$ triplet state, based on quenching by oxygen and by dienes, well-known triplet quenchers.

The lifetime of this triplet state is unusual for its longevity compared with other aromatic ketones and for its solvent dependence. The lifetime varies from about 200 ns in dioxane or acetonitrile, to 600 ns in 2-ethoxyethanol [144]. In contrast, the

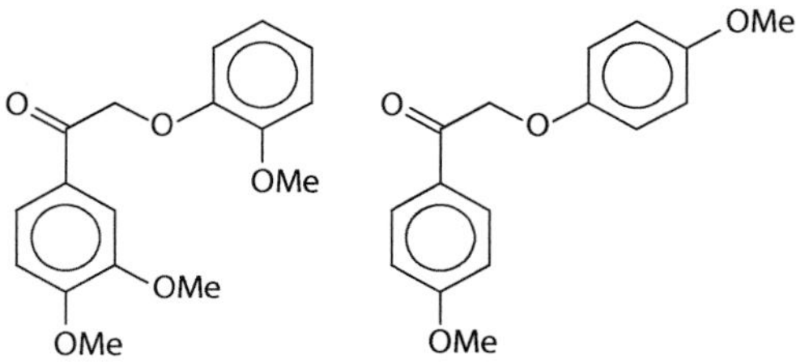

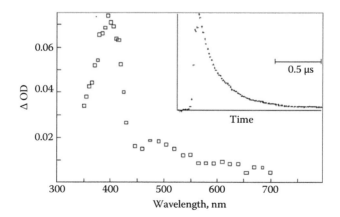

FIGURE 3.18   Structures of α-(2-methoxyphenoxy)-3,4-dimethoxyacetophenone (left) and α-(4-methoxyphenoxy)-4-methoxyacetophenone (right).

FIGURE 3.19   Transient absorption spectrum and decay (inset) for triplet state of α _ (2-methoxyphenoxy)-3,4-dimethoxyacetophenone. (Reproduced from Schmidt, J.A., Berinstain, A.B., de Rege, F., Heitner, C., Johnston, L.J. and Scaiano, J.C., *Can. J. Chem.*, 69, 104–107, 1991. With permission.)

triplet state of α-(4-methoxyphenoxy)-4-methoxyacetophenone (Figure 3.18), also a π–π* state, has a lifetime of only 0.5 ns. This difference in lifetimes arises from the significantly lower triplet energy of the 3,4-dimethoxyacetophenone moiety: 250 kJ mol^{-1} [145] compared to 301 kJ mol^{-1} for 4-methoxyacetophenone [146].

TRANSIENTS FROM CONIFERYL ALCOHOL

Flash photolysis of coniferyl alcohol in water and in acetic acid produces a transient species with λ_{max} at 350 nm [147]. It decays by first-order kinetics in both solvents, but with very different lifetimes: 500 s in water and 1.2 s in acetic acid. The transient is unreactive toward oxygen. Based on this reactivity pattern, Leary [147] assigned this transient as the corresponding quinone methide. Although this initial experiment used a flash-photolysis setup, the quinone methide is sufficiently long-lived that it can be detected with a modern UV-visible spectrophotometer using diode-array detection [148].

RADICAL SPECIES

The phenoxy radical is central to lignin chemistry. It is an important intermediate in lignin biosynthesis and in the degradation of lignin-containing pulps under the influence of heat or light. Das et al. [149] reported the spectra of a large number of substituted phenoxy radicals, and Shukla et al. [150] examined in detail the behavior of methoxy-substituted phenoxy radicals of relevance to lignin. Phenoxy radicals have a sharp absorption in the 400 nm region. Those with a *m*-methoxy substituent show, in addition, a broad visible absorption centered at about 650 nm. The intensity of this band is solvent dependant and is much weaker in hydroxylic solvents.

Upon irradiation, α-(aryloxy)acetophenones cleave at the C-phenoxy bond generating phenoxy and phenacyl radicals [143,151], as shown in Figure 3.20 for α-(4-methoxyphenoxy)-4-methoxyacetophenone. The phenacyl radical has two absorption bands; the most intense band occurs at about 380 nm and a broader, weaker band appears at about 540 nm [127]. The 540 nm band is attributed to the canonical form that has the unpaired electron on the carbonyl oxygen.

The role of ketyl radicals derived from the β–O–4 linkage in the light-induced yellowing of mechanical pulps has been established using transient methods

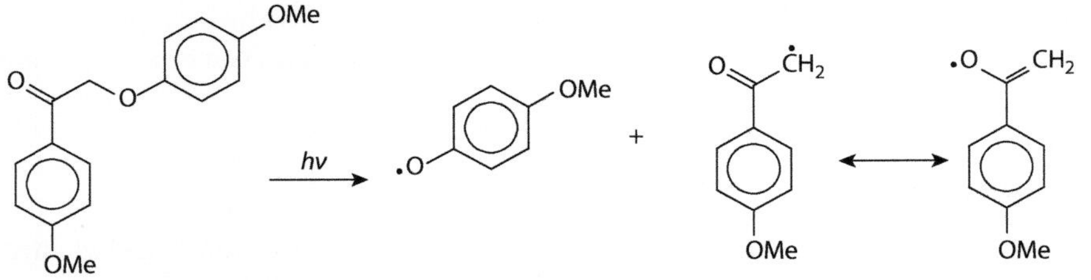

FIGURE 3.20 C-phenoxy cleavage of α-(aryloxy)acetophenones. The canonical form of the phenacyl radical with the unpaired electron on oxygen absorbs at about 540 nm.

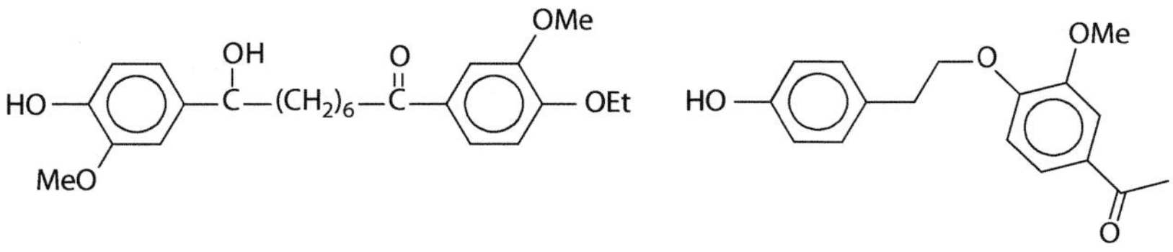

FIGURE 3.21 There is no ground-state interaction between the benzoyl and phenol chromophores of these molecules, even though transient studies show that the linking chains are sufficiently flexible to allow interaction of the triplet carbonyl and the phenol.

(Figure 3.15). Early studies reported rate constants for cleavage of the radical to an enol and a phenoxy radical (path a) $>5 \times 10^7$ s^{-1} [127]. This rate would be too fast to allow competition from bimolecular reaction of the ketyl radical with oxygen. However, a later study, which generated the β-O-4 ketyl radicals of a series of seven model dimers from two precursors, reported cleavage rate constants no higher than 2.7×10^3 s^{-1} and rapid quenching of the radical by oxygen [128]. The lower rate constants were also calculated by Huang et al. [152] and Anderson and Wayner [153] based on electrochemical measurements. Hence, the bulk of the evidence suggests that path b is the more likely pathway for photochemical decay of the β–O–4 linkage.

Neumann et al. [154] observed transient spectra of milled-wood lignins in deaerated solutions, using a xenon flash lamp for excitation. They interpreted the observed spectra as a combination of ketyl and phenoxy radicals, consistent with the conclusions of Fabbri et al. [128].

For bichromophoric molecules, such as those in Figures 3.16 and 3.19, UV-spectra are a sensitive probe of interactions between the two chromophoric groups, e.g., π–π complexes. Vanucci et al. [143] found that the solution absorption spectrum of α-(2-methoxyphenoxy)-3,4-dimethoxy-acetophenone (Figure 3.18) was essentially identical to the sum of the spectra of the monomeric reference compounds, indicating little or no ground-state interaction. This is not surprising, since the short methyloxy chain allows little opportunity for interaction between the two chromophores. However, there is also no evidence of ground-state interaction for the two compounds shown in Figure 3.21, which have longer ethyloxy and heptyl chains separating acetophenone and phenol chromophores [155,156]. Steric restrictions cannot explain the lack of interaction here, because transient studies show that in both these molecules the linking chain allows a folded conformation where the triplet aromatic ketone can abstract the phenolic hydrogen [155,156]. Thus, for linked bichromophoric molecules, the preferred conformations of ground and excited states can be substantially different.

TRANSIENT DIFFUSE REFLECTANCE STUDIES

In highly scattering materials, absorbance is not directly observable, and changes in diffuse reflectance after the laser flash are monitored instead. Wilkinson and Kelly have described details of the technique [157]. Hurrell et al. [151] used diffuse-reflectance laser-flash photolysis to study the behavior of several α-aryloxy-3,4-dimethoxyacetophenones adsorbed onto solid supports such as silica, zeolites, and microcrystalline cellulose. The spectra observed on the solid supports are similar to those observed in solution, with an absorption maximum appears at about 400 nm and a broad, tailing absorption extending to 600 nm. As in solution, the absorption is assigned to a $\pi-\pi^*$ triplet state, based on quenching by oxygen and dienes.

Scaiano et al. studied the photochemistry of crystalline α–(2-methoxyphenoxy)-3,4-dimethoxyacetophenone (Figure 3.18) using diffuse-reflectance laser-flash photolysis [158]. Surprisingly, no transient species formed within the 200 ns time resolution of their instrument, and the crystalline compound was stable to lamp irradiation. This contrasts to its behavior in solution or on surfaces, where triplet ketone spectra are easily observed and a variety of oligomeric and rearrangement products are formed [151]. This peculiar behavior arises from the crystal structure of the compound. In the unit cell, pairs of the 3,4-dimethoxybenzoyl group are stacked almost perfectly in a head-to-tail arrangement. This allows efficient formation of singlet excimers, which are complexes between ground-state molecules and excited-state molecules. The excimer forms rapidly enough to preempt both cleavage at the C-phenoxy bond and intersystem crossing to the triplet state, the usual behavior of ligninlike excited aromatic ketones. The excimer decays rapidly to the ground state. Hence, neither the triplet state nor any photoproducts are formed.

The crystal structure of ligninlike ketones is very sensitive to subtle changes in substitution. The unit cell of α-(4-methoxyphenoxy)-4-methoxyacetophenone (Figure 3.18) does not have any stacking of the benzoyl groups, and the ketone triplet state is easily observed in the solid state [127].

Schmidt et al. [159] and Wilkinson et al. [160] reported transient spectra from bleached softwood mechanical pulp. The transient spectrum has an absorption maximum at 450 nm. This is assigned to the carbonyl triplet state, even though it is redshifted by about 50 nm from the typical absorption of ketone triplets in solution. The basis of the assignment is the observed static quenching of the absorption by oxygen and by phenolic hydroxyl groups. The phenolic groups react with these triplets by a hydrogen abstraction reaction. Wójciak et al. [113] observed similar transient spectra and kinetics for oxygen-delignified softwood kraft pulp that had been treated with hydrogen peroxide followed by alkaline extraction. As did Schmidt et al. [159] and Wilkinson et al. [160], they assigned the transients to an aromatic carbonyl triplet state.

EMISSION SPECTROSCOPY

A molecule in an excited electronic state can return to the ground state by emitting a photon corresponding to the energy gap between the two states (Figure 3.17). A transition between states of the same multiplicity ($S_1 \rightarrow S_0$) results in fluorescence, while a transition between states of different spin multiplicity ($T_1 \rightarrow S_0$) gives phosphorescence. Fluorescence is a spin-allowed transition that typically occurs with a half-life in the picosecond to nanosecond range. Phosphorescence is spin-forbidden and occurs much more slowly, with half-lives ranging from microseconds to minutes. The two types of emission are therefore easily distinguished.

There are two major experiments in emission spectroscopy. An emission spectrum is obtained by exciting the sample at a fixed wavelength and scanning the emission wavelengths; an excitation spectrum is obtained by detecting the emission at a fixed wavelength while scanning the excitation wavelengths. If the observed emission is due only to one chromophore, the excitation spectrum will have the same peaks, with the same relative intensities and peak widths, as the absorbance spectrum. Olmstead and Gray have written a thorough review of the fluorescence spectroscopy of cellulose, lignin and mechanical pulps [161], to which the reader is referred for more detailed information.

FLUORESCENCE OF LIGNIN MODEL COMPOUNDS

Lundquist et al. [162] compiled a comprehensive collection of fluorescence data of lignin model compounds. Table 3.10 presents a representative selection of these data.

Guaiacyl compounds with saturated side chains have a fluorescence maximum of modest intensity at 316 nm. The excitation spectra correspond to the absorbance spectra expected for saturated guaiacyl compounds from the data in Table 3.4. The S_1 state for simple benzenes is a $\pi-\pi^*$ state, where intersystem crossing is inefficient and fluorescence is expected. The only exception to this behavior was the phenylcoumaran structure, which was only weakly fluorescent. The authors attributed this to a low-lying $n-\pi^*$ state, but no corroboration was given for this assignment.

Biphenyl compounds have fluorescence intensities similar to saturated guaiacyl compounds, but the emission maximum is shifted to 415 nm. Styrenic structures such as coniferyl alcohol are slightly more fluorescent and have their emission maximum at about 340 nm. Stilbene structures have the most intense fluorescence among lignin model compounds. Fluorescence is enhanced in molecules with extended conjugation and more rigid structures, since there are fewer vibrational modes available to dissipate the excited state energy. Thus, a phenyl coumarone, with its fused-ring structure, is the most highly fluorescent lignin model.

Unsubstituted acetophenone and benzophenone do not fluoresce. Their lowest energy excited state is $n-\pi^*$ and intersystem crossing to T_1 occurs with 100% efficiency. Methoxy substitution at the *para* position causes the $\pi-\pi^*$ state, where intersystem crossing is slowed, to become the lowest energy transition. Thus, guaiacyl compounds that have conjugated carbonyl groups show a weak fluorescence. However, there is uncertainty regarding the fluorescence of α-(2-methoxyphenoxy)-3,4-dimethoxyacetophenone (Figure 3.18). Okano et al. [163] found that this molecule does not fluoresce and attribute earlier observed fluorescence [158,164] to the photodegradation products acetoveratrone and guaiacol.

FLUORESCENCE OF LIGNIN PREPARATIONS

Lundquist et al. [162,165,166] reported that the peak shapes and positions in emission spectra of spruce milled-wood lignin in dioxane/water were independent of excitation wavelength. The emission maximum occurred at 358 nm. The excitation spectrum was consistent with the absorption spectrum of milled-wood lignin, indicating

TABLE 3.10
Fluorescence of Lignin Model Compounds

Compound	Excitation Maxima (nm)	Emission Maxima (nm)	Solvent	$\left(\dfrac{RI^*}{\varepsilon}\right) \times 10^3$
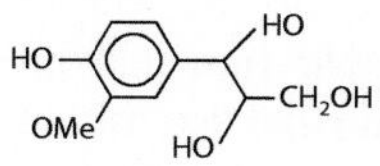 (structure: HO–, OMe, HO, CH₂OH, HO)	281, 233	315	Dioxane/Water	34
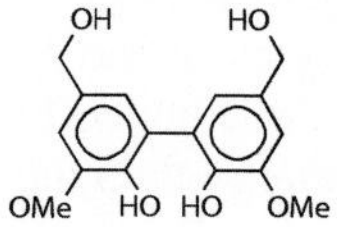 (structure: OH, HO, OMe, HO, HO, OMe)	289, 238	414	Dioxane/Water	10–50
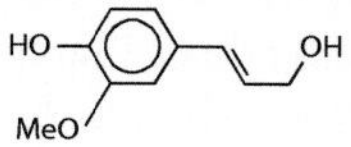 (structure: HO–, MeO, –OH)	264, 303	340	Dioxane/Water	50–70
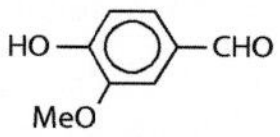 (structure: HO–, MeO, –CHO)	321	403	Dioxane/Water	<0.5
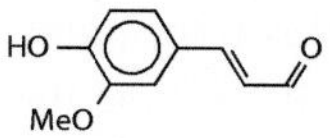 (structure: HO–, MeO, =O)	340	465	Water	<0.5
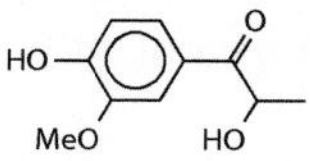 (structure: HO–, MeO, O, HO)	311	415	Water	<0.5
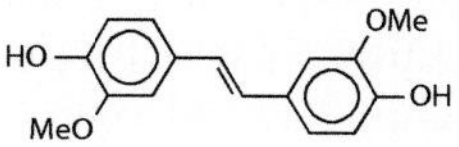 (structure: HO–, MeO, OMe, –OH)	338	390	Dioxane/Water	50–70
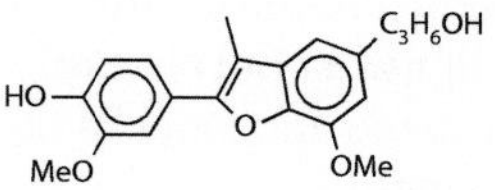 (structure: HO–, MeO, C₃H₆OH, O, OMe)	310	355	Dioxane/Water	350

Source: Data from Lundquist, K., Josefsson, B., and Nyquist, G., *Holzforschung,* 32, 27–32, 1978. With permission.

*RI = relative intensity. The fluorescence intensity of 1-(glyceryl)guaiacol was arbitrarily chosen as 100.

that all the lignin chromophores had an equal probability of causing fluorescence. Reduction of carbonyl groups with borohydride increased the intensity of the fluorescence, but the band shape and the position of absorption and emission maxima remained the same. They explained these results by proposing that lignin solutions behave as an energy sink; rather than fluorescing individually, excited lignin units transfer their singlet energy to some common structure from which emission occurs. Based on the position of the emission maximum, they proposed either coniferyl alcohol or phenylcoumarone as the emitting group.

The emission maxima of kraft lignin and of lignosulfonate appear at 400 and 390 nm, respectively [165]. The positions of these emission maxima are consistent with stilbene as the emitting group, which is known to exist in both these technical lignins. The excitation spectrum of lignosulfonate was identical to the UV spectrum, indicating that the energy sink hypothesis is applicable. The excitation spectrum of kraft lignin, however, resembled the absorption spectrum of a stilbene (λ_{max} = 334 nm), rather than the absorption spectrum of kraft lignin. In this case, emission arises mostly from direct excitation of the stilbene group; energy transfer within the polymer is limited.

Machado et al. reported the emission spectra of performic acid lignin from *Eucalytus grandis* [167]. The emission spectra were dependent on the excitation wavelength, indicating that the energy sink hypothesis is not applicable to this lignin. The authors used fluorescent probes to characterize the singlet energies of the *E. grandis* chromophores. In this technique, a fluorescent probe molecule is added to the lignin solution. Chromophores in the lignin with a lower singlet energy than the probe molecule will quench the probe's fluorescence by energy transfer. By using four probes of different singlet energies (biphenyl, naphthalene, phenanthrene, and pyrene), the authors determined that the singlet energies of the chromophores in *E. grandis* lignin were distributed as follows: 70% between 418 and 385 kJ/mol, 2% between 385 and 322 kJ/mol, and 28% < 322 kJ/mol.

Reduction of *E. grandis* lignin with sodium borohydride increased the fluorescence intensity and shifted the emission to shorter wavelengths. This was most likely a result of reducing aromatic carbonyl groups, which quench fluorescence, to the corresponding alcohols, which fluoresce. The singlet energies of the chromophores in reduced *E. grandis* lignin were distributed as follows: 52% 418–385 kJ/mol, 40% 385–346 kJ/mol, and only 8% <346 kJ/mol.

In contrast to the simple behavior of milled-wood lignin in solution, Castellan et al. [168] observed complex behavior for maritime pine milled-wood lignin incorporated into transparent hydroxypropylcellulose films. They were also able to measure the fluorescence quantum yields (ϕ_f = fluorescent photons emitted/ photons absorbed) for these films. Untreated milled-wood lignin is only weakly fluorescent, $\phi_f \approx 10^{-4}$, and the appearance of the fluorescence spectrum depends on the excitation wavelength. Oddly, excitation at shorter wavelength gives an emission maximum at longer wavelengths. Thus, excitation at 330 nm gives a broad emission centered at about 585 nm, excitation at 400 nm gives a somewhat sharper emission maximum at 570 nm, and excitation at 450 nm gives an emission maximum at 560 nm.

Bleaching the milled-wood lignin with peroxide shifted the emission excited by 330 nm irradiation to 400 nm, and increased ϕ_f by a factor of 10. Weak emissions occurred at 470 nm (400 nm excitation) and 520 nm (450 nm excitation). Reduction with either sodium borohydride or H_2/Pd eliminated the long-wavelength emissions—only emissions with λ_{max} = 400 nm remained. ϕ_f increased by an additional factor of 10 to about 10^{-1}. UV irradiation of these films caused the longer wavelength emissions to reappear. Castellan et al. [168] concluded that biphenyl groups were the major fluorophore in this lignin, and were responsible for the emission at 400 nm. The longer wavelength emissions (500–600 nm) were attributed to carbonyl-containing groups such as coniferaldehyde, quinones and methoxyacetophenones, which fluoresce only weakly. Quinones, and perhaps coniferaldehyde [162], may contribute to the 500–600 nm emission, based on their reported emission spectra. Methoxyacetophenones, however, fluoresce at about 415 nm [162,163], and are not likely to contribute to this longwave emission.

FLUORESCENCE OF LIGNOCELLULOSIC PULPS

Various celluloses, such as fully bleached kraft and sulfite pulps, Whatman filter paper and algal and bacterial celluloses, show a characteristic fluorescence. Excitation at wavelengths between 320 and 360 nm causes a broad emission with λ_{max} = 420–460 nm [161,169,170]. Definitive evidence for a universal absorbing group responsible for cellulose fluorescence is lacking. Despite the similarity of the emission spectra for many lignocellulosic materials, the identity of the fluorophore will likely depend on the nature and history of a specific pulp.

Castellan et al. [171] have recently suggested that dityrosine protein residues or p-hydroxycinnamic acid are possible candidates for the fluorphores in pure cellulose. Other recent studies [172,173] have shown that typical cellulose processing conditions (e.g., hot alkali) can induce the formation of small amounts of aromatic structures from reducing end groups or hemicelluloses. However, many of the structures identified in these studies are quinones, which are at best weakly fluorescent.

Liukko et al. [174] have examined the fluorescence of bleached softwood kraft pulps after each bleaching stage for six different bleaching sequences. For excitation at 350 nm, fluorescence intensity increased after each stage as more lignin was removed, except for ozone stages late in the bleaching sequence. The decreased fluorescence after a late ozone stage is attributed to the formation of carbonyl groups, which will quench fluorescence. Fluorescence excited at 430 nm behaved somewhat differently. In this case, fluorescence increased with decreasing lignin content down to a kappa number of about 7. At lower kappa numbers, further lignin removal caused fluorescence to decrease.

Wood and lignin-containing pulps fluoresce weakly [161,164,169,175–177]. As with cellulose, a broad emission occurs between 400 and 500 nm and the position of the emission maximum depends on the excitation wavelength. As with soluble lignins, chemical treatments that reduce the concentration of visible-light absorbing chromophores (peroxide bleaching, borohydride reduction) increase the fluorescence intensity and shift the emission maximum to the blue. Mechanical pulps with higher amounts of visible-light absorbing chromophores have weaker fluorescence and red-shifted emission spectra.

It is reasonable to assign the origin of this fluorescence to the lignin [176,177]. However, an alternative, although counterintuitive, hypothesis exists [169,175]. Olmstead and Gray [169] proposed that cellulose was the primary source of emission in mechanical pulps, with lignin acting as a quencher of this fluorescence. These authors attribute the emission of mechanical pulp only to cellulose. However, they do not attribute light-induced yellowing of high-yield pulps to cellulose, as suggested in a recent review [178]. Olmstead and Gray obtained a good approximation of the emission spectrum of softwood stone groundwood pulp by multiplying the cellulose emission spectrum by the reflectance of the groundwood pulp. In later experiments [161], they showed that reducing lignin content by an order of magnitude by acid chlorite delignification increased fluorescence. Adding milled-wood lignin to the acid-chlorite delignified sheet decreased the fluorescence.

Castellan et al. [176,177] vigorously disputed this interpretation. They determined that the fluorescence of wood and groundwood pulp from spruce excited at 350 nm decayed according to a biexponential rate law, with lifetimes of 0.5 and 1.8 ns. Cellulose fluorescence excited at 290 nm also decayed according to a biexponential function, with lifetimes of 0.9 and 6.7 ns. Based on these differing fluorescence decay times, they concluded that the fluorescence of lignin-containing pulp is due primarily to lignin. The longer fluorescence lifetime of cellulose rules out a mechanism where lignin quenches cellulose fluorescence and then fluoresces itself, but not one where lignin and cellulose compete for incident photons. However, there is some ambiguity in this measurement. The fluorescence lifetimes measured for wood and pulp varied by a factor of three as the excitation wavelength was changed from 350 to 455 nm. Unfortunately, the cellulose fluorescence lifetime was determined at only one excitation wavelength, 290 nm.

FLUORESCENCE FOR ANALYSIS AND MONITORING

As an analytical tool, fluorescence is from two to three orders of magnitude more sensitive than absorbance. This is because fluorescence is an absolute measurement, whereas absorbance is a relative measurement. Fluorescence is also more selective than absorbance, since not all compounds that absorb radiation will emit fluorescence. Despite this potential, no fluorometric lignin analyses have come into routine use.

Nonfluorescent species in a complex matrix can be observed by tagging them with a fluorophore. Thus, Zhu et al. [179] confirmed the formation of o-quinones during light-induced yellowing of high-yield pulps by treating the pulp with o-phenylenediamine. The diamine reacts with o-quinones to form phenazines, which fluoresce at about 460 nm under excitation at 350 and 375 nm. In other work, Kónya and Scaiano [180] tagged o-quinones with a trialkyl phosphite, where one of the alkyl chains was a pyrenyl-n-butyrate. The pyrene monomer fluoresces in the 370 to 400 nm region, with a well-resolved vibrational structure. Pyrene also forms an excimer, which has a broad, unresolved fluorescence in the 450–550 nm region. Using time-resolved techniques, the authors were able distinguish a pyrene excimer emission from the background emission of the pulp, even at low-levels of pyrene incorporation. The existence of the excimer emission implies that, during light-induced yellowing, at least some quinones are formed in islands or domains.

Bublitz et al. [181–183] examined the fluorescence of spent pulping liquors from kraft, sulfite and soda pulping of various softwoods. The fluorescence intensity at 430 nm increased linearly with the amount of lignin in the pulping liquor over a concentration range from 0 to 30 mg/L, after appropriate correction for Raman scattering of water. The authors determined that fluorescence was a useful indicator of the progress of a cook, but only if pulping is under good control.

Hanson and Wenzel [184,185] observed fluorescence from black liquor samples at 680 nm, using excitation wavelengths from 585 to 621 nm. In tetrahydrofuran solvent, the emission signal increased as the kraft cook progressed. However, for an aqueous alkaline liquor, the fluorescence intensity at 680 nm decreased as the cook progressed.

Liu et al. [186,187] used fluorescence microspectroscopy to determine the lignin content of single kraft pulp fibers. The stain Acridine Orange fluoresces strongly at about 630 nm in the presence of lignin. In the absence of lignin, the fluorescence occurs at 535 nm. The ratio of the fluorescence intensities at 630 nm and 535 nm is, therefore, a measure of delignification. The authors propose this as an on-line monitor of the uniformity of kraft pulping.

Bergström et al. [188] studied the time-resolved fluorescence of various newsprint samples. Newsprint containing no recycled fiber fluoresced weakly under both ultraviolet and visible (570–670 nm) excitation. They attributed both the ultraviolet and visible excited emissions to lignin. Newsprint containing recycled fiber had a much stronger ultraviolet-excited emission, which overwhelmed the native lignin fluoresence. This strong emission was attributed to optical brightening agents in the recycled component of the sheet. The authors suggest that fluorescence could be used to monitor the content of recycled content during production.

Two studies have attempted to use fluorescence to distinguish different wood species. Sum et al. [189] measured the fluorescence emission spectra excited at 308 nm in the heartwood and sapwood of jack pine (*Pinus banksiana*), white spruce (*Picea glauca*) and balsam fir (*Abies balsamea*). Very broad emission between 400 and 600 nm occurs for all these species. Pandey et al. [190] examined the fluorescence of several tropical hardwoods. The fluorescence spectra of the wood extracts were highly dependent on the excitation wavelength, while those of dry solid woods were not.

Barsberg and Nielson [191] used confocal laser scanning microscopy (CLSM) to obtain images of spruce TMP fibers before and after oxidative treatment with laccase enzyme and the mediator 2,2′-azobis-3-ethylbenzthiazoline-6-sulfonate (ABTS). Excitation was done using the 488 nm line of an argon-ion laser. Laccase-ABTS oxidation forms lignin carbonyl groups that will quench the lignin autofluorescence. This oxidation occurs initially at the most accessible fiber surfaces and is expected to progress slowly as the ABTS diffuses through the fiber wall. Thus, one might expect to observe diminished fluorescence intensity only at the fiber surfaces until the ABTS has had time to penetrate the secondary wall. What is observed instead is immediate, uniform quenching of fluorescence throughout the fiber wall. The authors interpret this as evidence that transfer of the excitation energy among individual lignin units is fast enough to compete effectively with fluorescence emission as a mode of excited state decay. Thus, a lignin excited state formed in the interior of the fiber wall can be rapidly transported to the oxidized fiber surface, where quenching occurs. This rapid excitation energy transfer is reminiscent of the behavior of lignin in solution observed by Lundquist et al. [162,165,166] (see the Fluorescence of Lignin Preparations section).

An underlying assumption of emission CLSM is that fluorescence occurs locally with respect to the initial chromophore excitation. Rapid excitation energy transfer; however, implies that an excited state can be transported far from the initial site of excitation before a fluorescence (or quenching) event occurs. This suggests caution in the interpretation of emission CLSM images of TMP and perhaps other lignocellulosic fibers, as local information in such images may be obscured [191].

PHOSPHORESCENCE

Phosphorescence is associated with formally spin-forbidden transitions such as $T_1 \rightarrow S_0$. It occurs rarely in fluid solution due to efficient quenching of triplet states by oxygen. Phosphorescence is usually observed from molecules that are frozen in glassy solvents or adsorbed on a solid substrate such as filter paper—so-called room-temperature phosphorescence.

Phosphorescence emission occurs in glassy ethanol (77 K) at 440 nm for acetoveratrone and α-(2-methoxyphenoxy)-3,4-dimethoxyacetophenone (Figure 3.18) [145,192], and at 460 nm for vanillin [193]. The excitation spectra observed for vanillin show an unusual feature. The fluorescence excitation spectrum is, as expected, identical to the absorption spectrum. The phosphorescence excitation spectrum, however, does not show the weak band at 350 nm and has enhanced intensity at 300 nm. This implies that intersystem crossing to T_1 occurs from S_2 rather than S_1 [193]. This seems to differ with *ab initio* calculations for benzaldehyde that predict maximal intersection between the potential surfaces for S_1 and T_2 [194].

Beyer et al. [164] observed that the emission intensity of mechanical pulp sheets decreased drastically when the sheets were moistened. They assigned the residual luminescence to phosphorescence from triplet states inaccessible to oxygen.

SUMMARY

Our understanding of the absorbance of lignins and lignin model compounds in solution is mature and has changed little since the earlier comprehensive review of Goldschmid [1]. The more refined theoretical calculations that have been made since the original work of Doub and Vandenbelt [4,6,12] have not significantly changed the electronic structure model that these earlier authors proposed to explain the solution spectra of di- and trisubstituted benzenes.

Improvements in instrument design and the wide availability of microcomputers have made possible the application of reflectance spectroscopy and the Kubelka–Munk equations to the study of lignin chromophores in wood fibers. Significant work has accumulated since 1970 on the response of lignin-containing pulps to bleaching agents, such as sodium hydrosulfite and hydrogen peroxide, and to degradation by light and heat. These studies have confirmed the importance of quinone structures to the color of wood and pulps and clarified their response to the common bleaching agents.

Challenging unresolved problems include clear identification of the lignin structures that limit the achievable brightness of lignin-containing pulps. For example, while sodium hydrosulfite and hydrogen peroxide both react with quinones, the latter gives up to 25 points gain in ISO brightness while the former achieves at the most 12. Even with hydrogen peroxide applications, the maximum achievable brightness is 5–10 ISO brightness points less than is possible by complete removal of lignin, e.g., fully bleached chemical pulp.

Transient spectroscopy and emission spectroscopy have clarified the spectra and chemistry of the excited electronic states of lignin. Lignin aromatic carbonyl groups show reactivity from their first excited singlet states, an unusual behavior for aromatic carbonyls. The fluorescence spectra of both soluble lignins and lignin-containing pulps show evidence of complex energy transfer within the polymer. Since its sensitivity, fluorescence spectroscopy shows promise as an analytical technique, although routine applications have yet to appear.

REFERENCES

1. O Goldschmid. Ultraviolet Spectra. In KV Sarkanen and CH Ludwig, eds. *Lignins: Occurence, Formation, Structure and Reactions.* New York: Wiley-Interscience, 1971, pp. 241–266.
2. SY Lin. Ultraviolet Spectrophotometry. In SY Lin and CW Dence, eds. *Methods in Lignin Chemistry.* Berlin: Springer-Verlag, 1992, pp. 217–232.
3. J Schmidt. UV-Visible Spectroscopy of Chromophores in Mechanical Pulps. In RA Meyers, ed. *Encyclopedia of Analytical Chemistry: Applications, Theory and Instrumentation,* 9. Chichester: John Wiley & Sons, 2000, pp. 8388–8407.
4. L Doub and JM Vandenbelt. The Continuity of the Ultraviolet Bands of Benzene with Those of Its Derivatives: Application to Certain Trisubstituted Derivatives. *J. Am. Chem. Soc.* 77:4535–4540, 1955.
5. HH Jaffé and M Orchin. *Theory and Applications of Ultraviolet Spectroscopy.* New York: John Wiley and Sons, Inc., 1962, pp. 295–303.
6. L Doub and JM Vandenbelt. The Ultraviolet Absorption Spectra of Simple Unsaturated Compounds: I. Mono- and *p*-Disubstituted Benzene Derivatives. *J. Am. Chem. Soc.* 69:2714–2723, 1947.
7. FA Matsen. Molecular Orbital Theory and the Spectra of Monosubstituted Benzenes. I. The Resonance Effect. *J. Am. Chem. Soc.* 72:5243, 1950.
8. AJ Duben. Frontier Orbital Analysis of the Bathochromic Shift of Monosubstituted Benzenes. *J. Chem. Ed.* 62:373–375, 1985.
9. P Tomasik, TM Krygowski, and T Chellathurai. The Hammett-Type Approach to the Substituent Effects in the UV Absorption Spectra of Aromatic Compounds. III. The Spectra of *ortho*-Disubstituted Benzene Derivatives. *Bull. Acad. Polon. Sér. Sci. Chim.* 22:1065–1074, 1974.
10. P Tomasik and TM Krygowski. The Hammett-Type Approach to the Substituent Effects in the UV Absorption Spectra of Aromatic Compounds. I. The Spectra of *para*-Disubstituted Benzene Derivatives. *Bull. Acad. Polon. Sér. Sci. Chim.* 22:443–456, 1974.
11. B Uno and T Kubota. New Description of the Substituent Effect on Electronic Spectra by Means of Substituent Constants. Part VII. Electronic Spectra of Substituted Benzenes. *J. Molec. Structure (Theochem)* 230:247–261, 1991.
12. L Doub and JM Vandenbelt. The Ultraviolet Absorption Spectra of Simple Unsaturated Compounds. II. *m*- and *o*-Disubstituted Benzene Derivatives. *J. Am. Chem. Soc.* 71:2414–2420, 1949.
13. RGE Morales and M Toporowicz. Assignation spectrale des bandes électroniques du 3-méthoxy, du 4-méthoxy et du 3,4-diméthoxybenzaldéhyde. *Spectrochim. Acta* 37A:11–15, 1981.
14. BG Gowenlock and KJ Morgan. Correlations in the Electronic Spectra of Substituted Aromatic Molecules. *Spectrochim. Acta* 17:310, 1961.
15. G Aulin-Erdtman and R Sanden. Spectographic Contributions to Lignin Chemistry IX. Absorption Properties of Some 4-Hydroxyphenyl, Guaiacyl, and 4-Hydroxy-3,5-dimeethoxyphenyl Type Compounds for Hardwood Lignins. *Acta Chem. Scand.* 22:1187–1209, 1968.
16. G Aulin-Erdtman. Spectographic Contributions to Lignin Chemistry III. Investigations on Model Compounds. *Svensk Papperstid.* 56:91–101, 1953.

17. G Aulin-Erdtman and L Hegbom. Spectrographic Contributions to Lignin Chemistry VII. The Ultra-Violet Absorption and Ionization $\Delta\varepsilon$-Curves of Some Phenols. *Svensk Papperstid.* 60:91–101, 1957.
18. CW Dence. The Determination of Lignin. In SY Lin and CW Dence, eds. *Methods in Lignin Chemistry.* Berlin: Springer-Verlag, 1992, pp. 33–61.
19. D Fengel and G Wegener. *Wood: Chemistry, Ultrastructure, Reactions.* Berlin: Walter de Gruyter, 1984, pp. 157–161.
20. CL Hess. An Investigation of the Homogeneity of Isolated Native Black Spruce Lignin. *Tappi* 35:312, 1952.
21. K Tiyama, J Nakano, and N Migita. The Resolution of Electronic Spectra of Lignin and Thiolignin. *J. Japan Wood Res. Soc.* 13:125, 1967.
22. H Norrström and A Teder. Absorption Bands in Electronic Spectra of Lignins. Part 1. Lignins from Alkaline Cooks on Spruce. *Svensk Papperstidn.* 74:85–93, 1971.
23. H Norrström and A Teder. Absorption Bands in Electronic Spectra of Lignins. Part 2. Band Intensities for Alkali Lignins from Spruce. *Svensk Papperstidn.* 74:337–344, 1971.
24. H Norrström. Absorption Bands in Electronic Spectra of Lignins. Part III. Lignins from Bisulphite Pulping of Spruce. *Svensk Papperstidn.* 75:611–618, 1972.
25. K Nilsson, H Norrström, and A Teder. Absorption Bands in Electronic Spectra of Lignins. Part IV. Lignins from Sulphate Pulping of Birch. *Svensk Papperstidn.* 75:733–738, 1972.
26. H Norrström and A Teder. Absorption Bands in Electronic Spectra of Lignins. Part V. Milled-Wood Lignins. *Svensk Papperstidn.* 76:458–462, 1973.
27. SY Lin. Derivative Ultraviolet Spectroscopy of Lignin and Lignin Model Compounds: A New Analytical Technique. *Svensk. Papperstidn.* 85:R162–R171, 1982.
28. TM Garver. Method and Apparatus for Monitoring and Controlling Characteristics of Process Effluents. International Patent WO 98/40721, 1998.
29. GF Zakis. *Functional Analysis of Lignins and Their Derivates.* Atlanta: TAPPI Press, 1994, pp. 43–49.
30. G Aulin-Erdtman and L Hegbom. Spectrographic Contributions to Lignin Chemistry VIII. $\Delta\varepsilon$-Studies on Braun's Native Lignins from Coniferous Woods. *Svensk Papperstidn.* 61:187–210, 1958.
31. G Aulin-Erdtman and L Hegbom. Spectrographic Contributions to Lignin Chemistry VI. Investigations on an Enzymatic Dehydrogenation Polymerisate of Coniferyl Alcohol (Freudenberg's DHP). *Svensk Papperstidn.* 59:363–371, 1956.
32. G Aulin-Erdtman. Spectrographic Contributions to Lignin Chemistry V. Phenolic Groups in spruce Lignin. *Svensk Papperstidn.* 57:745–760, 1954.
33. TA Milne, HL Chum, F Agblevor, and DK Johnson. Standardized Analytical Methods. *Biomass Bioenergy.* 2:341–366, 1992.
34. O Faix, C Grünwald, and O Beinhoff. Determination of Phenolic Hydroxyl Group Content of Milled Wood Lignins (MWLs) from Different Botanical Origins Using Selective Aminolysis, FTIR, ^{1}H NMR, and UV Spectroscopy. *Holzforschung* 46:425–432, 1992.
35. P Månsson. Quantitative Determination of Phenolic and Total Hydroxyl Groups in Lignin. *Holzforschung* 37:143–146, 1983.
36. G Gellerstedt and E-L Lindfors. Structural Changes in Lignin during Kraft Cooking. Part 4. Phenolic Hydroxyl Groups in Wood and Kraft Pulps. *Svensk. Papperstidn.* 87:R115–R118, 1984.
37. A Gärtner, G Gellerstedt, and T Tamminen. Determination of Phenolic Hydroxyl Groups in Residual Lignin Using a Modified UV-Method. *Nord. Pulp Pap. Res. J.* 14:163–170, 1999.
38. E Tiainen, T Drakenberg, T Tamminen, and A Hase. Determination of Phenolic Hydroxyl Groups in lignin by Combined Use of ^{1}H NMR and UV Spectroscopy. *Holzforschung* 53:529–533, 1999.

39. E Adler and J Marton. Zur Kenntnis der CarbonylGruppen im Lignin. I. *Acta. Chem. Scand.* 13:75–96, 1959.
40. J Marton and E Adler. Carbonyl Groups in Lignin III. Mild Catalytic Hydrogenation of Björkman Lignin. *Acta Chem. Scand.* 15:370–383, 1961.
41. E Adler and K Lundquist. Estimation of Uncondensed Phenolic Units in Spruce Lignin. *Acta Chem. Scand.* 15:223–224, 1961.
42. E Adler and K Lundquist. Spectrochemical Estimation of Phenylcourmaran Elements in Lignin. *Acta Chem. Scand.* 17:13–26, 1963.
43. SI Falkehag, J Marton, and E Adler. *Chromophores in Kraft Lignin. Advances in Chemistry: Lignin Structure and Reactions,* 59. Washington, DC: American Chemical Society, 1966, pp. 75–89.
44. SY Lin and KP Kringstad. Some Reactions in the Photoinduced Discoloration of Lignin. *Norsk Skogindustri* 9:252–256, 1971.
45. SY Lin and KP Kringstad. Photosensitive Groups in Lignin and Lignin Model Compounds. *Tappi* 53:658–663, 1970.
46. A Castellan, A Nourmamode, P Fornier de Violet, N Colombo, and C Jaeger. Photoyellowing of Milled Wod Lignin and Peroxide-Bleached Milled Wood Lignin in Solid 2-Hydroxypropylcellulose Films after Sodium Borohydride Reduction and Catalytic Hydrogenation in Solution: An UV/vis Absorption Spectroscopic Study. *J. Wood Chem. Technol.* 12:1–18, 1992.
47. JC Pew and WJ Connors. Color of Coniferous Lignin. *Tappi* 54:245–251, 1971.
48. C Heitner and T Min. The Effect of Sulphite Treatment on the Brightness and Bleachability of Chemitermomechanical Pulp. *Cellulose Chem. Technol.* 21:289–296, 1987.
49. R Chandra, TQ Hu, BR James, MB Ezhova, and DV Moiseev. A New Class of Bleaching and Brightness Stabilizing Agents. Part IV. Probing the Bleaching Chemistry of THP and BBHPE. *J. Pulp Paper Sci.* 33:15–22, 2007.
50. GX Pan, CI Thomson, and GJ Leary. UV-VIS: Spectroscopic Characteristics of Ferulic Acid and Related Compounds. *J. Wood Chem. Technol.* 22:137–146, 2002.
51. JC Pew and WJ Connors. New Structures from the Enzymic Dehydrogenation of Lignin Model *p*-Hydroxy-α-carbinols. *J. Org. Chem.* 34:580–584, 1969.
52. JC Pew. Evidence of a Biphenyl Group in Lignin. *J. Org. Chem.* 28:1048–1054, 1963.
53. P Karhunen, P Rummakko, J Sipilä, G Brunow, and I Kilpeläinen. The Formation of Dibenzodioxocin Structures by Oxidative Coupling: A Model Reaction for Lignin Biosynthesis. *Tetrahedron Lett.* 36:4501–4504, 1995.
54. P Karhunen, P Rummakko, J Sipilä, G Brunow, and I Kilpeläinen. Dibenzodioxocins: A Novel Type of Linkage in Softwood Lignins. *Tetrahedron Lett.* 36:169–170, 1995.
55. P Karhunen, P Rummakko, A Pajunen, and G Brunow. Synthesis and Crystal Structure Determination of Model Compounds for the Dibenzodioxocine Structure Occurring in Wood Lignins. *J. Chem. Soc., Perkin Trans.* 1:2303–2308, 1996.
56. J Gierer. Basic Principles of Bleaching. Part 2: Anionic Processes. *Holzforschung* 44:395–400, 1990.
57. J Abbot. Reactions of *ortho*-quinones and the Model Compound 4-*t*-butyl-1,2-benzoquinone in Alkaline Solution. *Res. Chem. Intermed.* 21:535–562, 1995.
58. S Berger and A Rieker. Identification and Determination of Quinones. In S Patai, ed. *The Chemistry of the Quinonoid Compounds, I.* London: John Wiley and Sons, 1974, pp. 163–229.
59. S Berger, P Hertl, and A Rieker. Physical and Chemical Analysis of Quinones. In S Patai and Z Rappoport, eds. *The Chemistry of Quinonoid Compounds, II.* London: John Wiley and Sons, 1988, pp. 29–86.
60. E Adler, R Magnusson, B Berggren, and H Thomelius. Periodate Oxidation of Phenols II. Oxidation of 2,6-Dimethoxyphenol and 3-Methoxycatechol: Formation of a 1,2-Naphthoquinone. *Acta Chem. Scand.* 14:515–528, 1960.

61. H-J Teuber and G Staiger. Reaktionen mit Nitrosodisulfonat. VIII. Mitteil. *ortho*-Benzochinone und Phenazine. *Chem. Ber.* 89:802–827, 1955.

62. F Imsgard, SI Falkehag, and KP Kringstad. On Possible Chromophoric Structures in Spruce Wood. *Tappi* 54:1680–1684, 1971.

63. GF Zakis, AA Melkis, and BY Nieberte. Methods for Determination of Quinone Carbonyl Groups in Lignins. *Khim. Drev.* 82–86, 1987.

64. DS Argyropoulos and C Heitner. ^{31}P NMR Spectroscopy in Wood Chemistry. Part IV. Solid State ^{31}P NMR of Trimethyl Phosphite Derivatives of Chromophores and Carboxylic Acids Present in Mechanical Pulps: A Method for the Quantitative Determination of *ortho*-Quinones. *Holzforschung* 48:112–116, 1994 Suppl.

65. Y Ni, A Ghosh, Z Li, C Heitner, and P McGarry. Photo-Stabilization of Bleached Mechanical Pulps with DTPA Treatment. *J. Pulp Paper Sci.* 24:259–263, 1998.

66. B-H Yoon, L-J Wang, and G-S Kim. Possible Formation of Lignin-Metal Complex during Photo-irradiation of TMP. 9th International Symposium on Wood and Pulping Chemistry, Montreal, 1997, 2, pp. 127-121–127-125.

67. Y Ni, Z Li, and ARP van Heiningen. Minimization of the Brightness Loss Due to Metal Ions in Process Water for Bleached Mechanical Pulps. *Pulp Paper Can.* 98:T396–T399, 1997.

68. J Janson and I Forsskåhl. Color Changes in Lignin-Rich Pulps on Irradiation by Light. *Nordic Pulp Paper Res. J.* 5:197–205, 1989.

69. GW Kutney and TD Evans. Peroxide Bleaching of Mechanical Pulps. Part 2. Alkali Darkening-Hydrogen Peroxide Decomposition. *Svensk Papperstidn.* 88:R84–R89, 1985.

70. G Leary and D Giampaolo. The Darkening Reactions of TMP and BTMP during Alkaline Peroxide Bleaching. *J. Pulp Paper Sci.* 25:141–147, 1999.

71. C Peart and Y Ni. UV-VIS Spectra of Lignin Model Compounds in the Presence of Metal Ions and Chelants. *J. Wood Chem. Technol.* 21:113–125, 2001.

72. C Steelink, T Reid and G Tollin. On the Nature of the Free-Radical Moiety in Lignin. *J. Am. Chem. Soc.* 85:4048–4049, 1963.

73. R Foster and MI Foreman. Quinone Complexes. In S Patai, ed. *The Chemistry of the Quinoid Compounds, I.* London: John Wiley and Sons Ltd., 1974, pp. 257–333.

74. MC Depew and JKS Wan. Quinhydrones and Semiquinones. In S Patai and Z Rappoport, eds. *The Chemistry of Quinonoid Compounds, II.* London: John Wiley and Sons Ltd., 1988, pp. 963–1018.

75. GS Furman and WFW Lonsky. Charge-Transfer Complexes in Kraft Lignin. Part 1. Occurrence. *J. Wood Chem. Technol.* 8:165–189, 1988.

76. GS Furman and WFW Lonsky. Charge-Transfer Complexes in Kraft Lignin. Part 2. Contribution to Color. *J. Wood Chem. Technol.* 8:191–208, 1988.

77. ID Suckling. The Effect of Coniferaldehyde Sulfonation on the Brightness and Absorption Spectra of Chemithermomechanical Pulps. 6th International Symposium on Wood and Pulping Chemistry, Melbourne, 1991, 1, pp. 587–593.

78. W Giust, F McLellan, and P Whiting. Alkali Darkening and Its Similarity to Thermal Reversion. *J. Pulp Paper Sci.* 17:J73–J79, 1991.

79. L Zhang and G Gellerstedt. Quinone Chromophores and Their Contribution to Photoyellowing in Lignin. 5th European Workshop on Lignocellulosics and Pulp, Aveiro, Portugal, 1998, pp. 285–289.

80. S Barsberg, T Elder, and C Felby. Lignin-Quinone Interactions: Implications for Optical Properties of Lignin. *Chem. Mater.* 15:649–655, 2003.

81. Z-H Wu, M Matsuoka, D-Y Lee, and M Sumimoto. Mechanochemistry of Lignin VI. Mechanochemical Reactions of β-1 Lignin Model Compounds. *Mokuzai Gakkaishi* 37:164–171, 1991.

82. D-Y Lee, M Matsuoka, and M Sumimoto. Mechanochemistry of Lignin IV. Mechanochemical Reactions of Phenylcoumaran Models. *Holzforschung* 44:415–418, 1990.

83. M Johansson, L Zhang, and G Gellerstedt. On Chromophores and Leucochromophores Formed during the Refining of Wood. *Nordic Pulp Paper Res. J.* 17:5–8, 2002.

84. P Kubelka. New Contributions to the Optics of Intensely Light-Scattering Materials. Part I. *J. Opt. Soc, Am.* 38:1948.

85. JA Van den Akker. Scattering and Absorption of Light in Paper and Other Diffusing Media. *Tappi* 32:498–457, 1949.

86. H Hirashima and M Sumimoto. Fundamental Properties of Mechanical Pulp Lignins. I. Visible Light Spectrophotometry and Nitrobenzene Oxidation. *J. Japan Wood Res. Soc.* 32:705–712, 1986.

87. H Hirashima and M Sumimoto. Fundamental Properties of Mechanical Pulp Lignins. II. Behaviors of Coniferyl Aldehyde Type Structures in Pulp Lignin. *J. Japan Wood Res. Soc.* 33:31–41, 1987.

88. HW Giertz. Om massans eftergulning. *Svensk Papperstidn.* 48:317–323, 1945.

89. J Polcin and WH Rapson. Spectrophotometric Study of Wood Chromophores In Situ. III. Determination of the Spectrum of Lignin by the Cotton-Dilution. *Tappi* 52:1970–1974, 1969.

90. J Polcin and WH Rapson. Spectrophotometric Study of Wood Chromophores In Situ. I. The Method of Differential $(K/S)_\lambda$ Curves Related to Bleaching and Color Reversion. *Tappi* 52:1960–1965, 1969.

91. J Polcin and WH Rapson. Spectrophotometric Study of Wood Chromophores In Situ. II. Determination of the Absorption Spectrum of Lignin from Reflectance and Reflectivity Measurements. *Tappi* 52:1965–1970, 1969.

92. J Polcin and WH Rapson. Effects of Bleaching Agents on the Absorption Spectra of Lignin in Groundwood Pulps. Part II. Oxidative Bleaching. *Pulp Paper Mag. Can.* 72:80–91, 1971.

93. J Polcin and WH Rapson. Effects of Bleaching Agents on the Absorption Spectra of Lignin in Groundwood Pulps. Part I. Reductive Bleaching. *Pulp Paper Mag. Can.* 72:69–80, 1971.

94. JA Schmidt and C Heitner. The Use of UV-Visible Diffuse Reflectance Spectroscopy for Chromophore Research on Wood Fibres: A Review. *Tappi* 76:117–123, 1993.

95. J Schmidt and C Heitner. Use of Diffuse Reflectance UV-Visible Spectroscopy to Characterize Chromophores in Wood Fibers. In D Argyropoulos, ed. *Progress in Lignocellulosics Characterization.* Atlanta: TAPPI Press, 1999, pp. 187–209.

96. International Standards Organization. ISO 2470: Measurement of Diffuse Blue Reflectance Factor (ISO Brightness).

97. Tappi. T 452: Brightness of Pulp, Paper and Paperboard.

98. Pulp and Paper Technical Association of Canada. E.1: Brightness of Pulp, Paper and Board.

99. K Kuys and J Abbot. Bleaching of Mechanical Pulps with Sodium Bisulfite. *Appita* 49:269–273, 1996.

100. E Svensson, H Lennholm, and T Iversen. Pulp Bleaching with Dithionite: Brightening and Darkening Reactions. *J. Pulp Paper Sci.* 24:254–259, 1998.

101. G Gellerstedt, HL Hardell, and EL Lindfors. The Reactions of Lignin with Alkaline Hydrogen Peroxide. Part IV. Products from the Oxidation of Quinone Model Compounds. *Acta Chem. Scand.* B 34:669–673, 1980.

102. M Johansson and G Gellerstedt. Chromophoric Content in Wood and Mechanical Pulps. *Nordic Pulp Pap. Res. J.* 15:282–286, 2000.

103. DG Holah and C Heitner. Alkaline Hydrogen Peroxide Bleaching of Mechanical and Ultra-High Yield Pulps. *J. Pulp Paper Sci.* 18:J161–J165, 1992.

104. PJ Wright and J Abbot. Kinetic Models for Peroxide Bleaching under Alkaline Conditions. Part 1. One and Two Chromophore Models. *J. Wood Chem. Technol.* 11:349–371, 1991.

105. PJ Wright, YA Ginting, and J Abbot. Kinetic Models for Peroxide Bleaching under Alkaline Conditions. Part 2. Equilibrium Models. *J. Wood Chem. Technol.* 12:111–134, 1992.
106. C Heitner, HI Bolker, and HG Jones. How Chromophores Are Generated by Alkaline Treatment of Wood. *Pulp Paper Can.* 76:T243–T247, 1975.
107. HI Bolker, BI Fleming, and C Heitner. Chromophores Generated by Treating Wood with Alkali. Part II. *Trans. Tech. Sec.* CPPA 4:30–32, 1978.
108. BO Lindgren and H Mikawa. The Presence of Cinnamyl Alcohol Groups in Lignin. *Acta Chem. Scand.* 11:826–835, 1957.
109. E Adler and L Ellmer. Coniferylaldehydgruppen im Holz und in Isolierten Ligninpräparaten. *Acta Chem. Scand.* 2:839–840, 1948.
110. G Gellerstedt and L Zhang. Formation and Reactions of Leucochromophoric Structures in High Yield Pulping. *J. Wood Chem. Technol.* 12:387–412, 1992.
111. DF Wong, G Leary, and G Arct. The Role of Stilbenes in Bleaching and Colour Stability of Mechanical Pulps. I. The Reaction of Lignin Model Stilbenes with Alkali and Oxygen. *Res. Chem. Intermed.* 21:329–342, 1995.
112. M Ragnar. A Novel Spectrophotometric Tool for Bleaching Studies and Determination of Hexenuronic Acid Removal. *Nordic Pulp Paper Res. J.* 16:68–71, 2001.
113. A Wójciak, M Sikorski, R Gonzalez Moreno, JL Bourdelande, and F Wilkinson. The Use of Diffuse-Reflectance Laser-Flash Photoysis to Study the Photochemistry of the Kraft Pulp Treated with Hydrogen Peroxide under Alkaline and Acidic Conditions. *Wood Sci. Technol.* 36:187–195, 2002.
114. D da Silva Perez, A Castellan, S Grelier, MGH Terrones, AEH Machado, R Ruggiero, and AL Vilarinho. Photochemical Bleaching of Chemical Pulps Catalyzed by Titanium Dioxide. *J. Photochem. Photobiol. A: Chem.* 115:73–80, 1998.
115. AL Andrady, Y Song, VR Parthasarathy, K Fueki, and A Torikai. Photoyellowing of Mechanical Pulp. Part 1. Examining the Wavelength Sensitivity of Light-Induced Yellowing Using Monochromatic Radiation. *Tappi* 74:162–168, 1991.
116. I Forsskåhl and H Tylli. Action Spectra in the UV and Visible Region of Light-Induced Changes of Various Refiner Pulps. In C Heitner and JC Scaiano, eds. *Photochemistry of Lignocellulosic Materials.* ACS Symposium Series, 531. Washington, DC: American Chemical Society, 1993, pp. 45–59.
117. S Claesson, E Olson, and A Wennerblom. The Yellowing and Bleaching by Light of Lignin-Rich Papers and the Re-Yellowing in Darkness. *Svensk Papperstidn.* 71:335–340, 1968.
118. GJ Leary. The Yellowing of Wood by Light: Part II. *Tappi* 51:257–260, 1968.
119. C Heitner. Light-Induced Yellowing of Wood-Containing Papers: An Evolution of the Mechanism. In C Heitner and JC Scaiano, eds. *Photochemistry of Lignocellulosic Materials.* ACS Symposium Series, 531. Washington, DC: American Chemical Society, 1993, pp. 2–25.
120. G Gellerstedt and R Agnemo. The Reactions of Lignin with Alkaline Hydrogen Peroxide. Part III. The Oxidation of Conjugated Carbonyl Structures. *Acta Chem. Scand.* B 34:275–280, 1980.
121. JA Schmidt and C Heitner. Light-Induced Yellowing of Mechanical and Ultra-High Yield Pulps. Part 3. Comparison of Softwood TMP, Softwood CTMP and Aspen CTMP. *J. Wood Chem. Technol.* 15:223–245, 1995.
122. RS Davidson, H Choudhury, S Origgi, A Castellan, V Trichet, and G Capretti. The Reaction of Phloroglucinol in the Presence of Acid with Lignin-Containing Materials. *J. Photochem. Photobiol. A: Chem.* 91:87–93, 1995.
123. S Origgi, V Trichet, A Castellan, and RS Davidson. The Action of Phloroglucinol in the Presence of Hydrochloric Acid on Photoaged, Totally-Chlorine-Free, Bleached Chemical Pulps. *J. Photochem. Photobiol. A: Chem.* 103:159–162, 1997.

124. M Ek, H Lennholm, and T Iversen. A Comment on the Effect of Carbonyl Groups on the Light-Induced Yellowing of Groundwood Pulp. *Nordic Pulp Paper Res. J.* 5:159–160, 1990.

125. JA Schmidt and C Heitner. Light-Induced Yellowing of Mechanical and Ultra-High Yield Pulps. Part 1. Effect of Methylation, NaBH$_4$ Reduction and Ascorbic Acid on Chromophore Formation. *J. Wood Chem. Technol.* 11:397–418, 1991.

126. JA Schmidt and C Heitner. Light-Induced Yellowing of Mechanical and Ultra-High Yield Pulps. Part 2. Radical-Induced Cleavage of Etherified Guaiacylglycerol-β-Arylether Groups Is the Main Degradative Pathway. *J. Wood Chem. Technol.* 13:309–325, 1993.

127. JC Scaiano, MK Whittlesey, AB Berinstain, PRL Malenfant, and RH Schuler. Pulse Radiolysis and Laser Flash Photolysis Studies of the Lignin Model α-(p-Methoxyphenoxy)-p-methoxyacetophenone and Related Compounds. *Chem. Mater.* 6:836–843, 1994.

128. C Fabbri, M Bietti, and O Lanzalunga. Generation and Reactivity of Ketyl Radicals with Lignin Related Structures. On the Importance of the Ketyl Pathway in the Photoyellowing of Lignin Containing Pulps and Papers. *J. Org. Chem.* 70:2720–2728, 2005.

129. I Forsskåhl and J Janson. Sequential Treatment of Mechanical and Chemimechanical Pulps with Light and Heat. Part 1. UV-VIS Reflectance Spectroscopy. *Nordic Pulp Paper J.* 6:118–126, 1991.

130. P Fornier de Violet, A Nourmamode, N Colombo, and A Castellan. Study of Brightness Reversion of Bleached Chemithermomechanical Pulp by Solid State Electronic Spectroscopy. *Cellulose Chem. Technol.* 23:535–544, 1989.

131. A Michell, PJ Nelson, and CWJ Chin. Diffuse Reflectance Spectroscopic Studies of the Bleaching and Yellowing of *Eucalyptus regnans* Cold Soda Pulp. *Appita* 42:443–448, 1989.

132. DS Argyropoulos, C Heitner, and JA Schmidt. Observation of Quinoid Groups during the Light-Induced Yellowing of Softwood Mechanical Pulp. *Res. Chem. Intermed.* 21:263–274, 1995.

133. M Ek, H Lennholm, G Lindblad, T Iversen, and DG Gray. Photochromic Behaviour of UV-Irradiated Mechanical Pulps. In C Heitner and JC Scaiano, eds. *Photochemistry of Lignocellulosic Materials,* 531. Washington, DC: American Chemical Society, 1993, pp. 147–155.

134. I Forsskåhl and C Maunier. Photocycling of Chromophoric Structures during Irradiation of High-Yield Pulps. In C Heitner and JC Scaiano, eds. *Photochemistry of Lignocellulosic Materials,* 531. Washington, DC: American Chemical Society, 1993, pp. 156–166.

135. H Choudhury, S Collins, and RS Davison. The Colour Reversion of Papers Made from High Yield Pulp: A Photochromic Process? *J. Photochem. Photobiol. A: Chem.* 69:109–119, 1992.

136. A Castellan, V Trichet, J-C Pommier, A Siohan, and S Armagnacq. Photo and Thermal Stability of Totally Chlorine Free Softwood Pulps Studied by UV/Vis Diffuse Reflectance and Fluorescence Spectroscopy. *J. Pulp Paper Sci.* 21:J291–296, 1995.

137. I Forsskåhl, H Tylli, and C Olkkonen. Participation of Carbohydrate-Derived Chromophores in the Yellowing of High-Yield and TCF Pulps. *J. Pulp Paper Sci.* 26:245–249, 2000.

138. O Theander. Non-Enzymic Conversion of Carbohydrates to Phenols and Enones. 4th International Symposium on Wood and Pulping Chemistry, Paris, 1987, 2, pp. 287–289.

139. LM Hadel. Laser Flash Photolysis. In JC Scaiano, ed. *Handbook of Organic Photochemistry,* 1. Boca Raton: CRC Press, 1989, pp. 279–292.

140. NJ Turro, V Ramamurthy, W Cherry, and W Farneth. The Effect of Wavelength on Organic Photoreactions in Solution: Reactions from Upper Excited States. *Chem. Rev.* 78:125–145, 1978.

141. NJ Turro. *Modern Molecular Photochemistry.* Menlo Park: Benjamin Cummings, 1978, pp. 375–382.
142. H Lutz, E Bréhéret, and L Lindqvist. Effects of Solvent and Substituents on the Absorption Spectra of Triplet Acetophenone and the Acetophenone Ketyl Radical Studied by Nanosecond Laser Photolysis. *J. Phys. Chem.* 77:1758–1762, 1973.
143. C Vanucci, P Fornier de Violet, H Bouas-Laurent, and A Castellan. Photodegradation of Lignin: A Photophysical and Photochemical Study of a Nonphenolic α-carbonyl β-O-4 Lignin Model Dimer, 3,4-dimethoxy-α-(2'-methoxyphenoxy)acetophenone. *J. Photochem. Photobiol., A:Chem.* 41:251–265, 1988.
144. JA Schmidt, AB Berinstain, F de Rege, C Heitner, LJ Johnston, and JC Scaiano. Photodegradation of the Lignin Model α-guaiacoxyacetoveratrone, Unusual Effects of Solvent, Oxygen, and Singlet State Participation. *Can. J. Chem.* 69:104–107, 1991.
145. AB Berinstain, MK Whittlesley, and JC Scaiano. Laser Techniques in the Study of the Photochemistry of Carbonyl Compounds Containing Ligninlike Moieties. In C Heitner and JC Scaiano, eds. *Photochemistry of Lignocellulosic Materials,* 531. Washington, DC: American Chemical Society, 1993, pp. 111–121.
146. I Carmichael and GL Hug. Spectroscopy and Intramolecular Photophysics of Triplet States. In JC Scaiano, ed. *Handbook of Organic Photochemistry,* 1. Boca Raton: CRC Press, 1989, pp. 369–403.
147. G Leary. The Chemistry of Reactive Lignin Intermediates. Part I. Transients in Coniferyl Alcohol Photolysis. *J. Chem. Soc. Perkin* II:640–642, 1972.
148. K Radotic, J Zakrzewska, D Sladic, and M Jeremic. Study of Photochemical Reactions of Coniferyl Alcohol. I. Mechanism and Intermediate Products of UV Radiation-Induced Polymerization of Coniferyl Alcohol. *Photochem. Photobiol.* 65:284–291, 1997.
149. PK Das, MV Encinas, S Steenken, and JC Scaiano. Reaction of *tert*-Butoxy Radicals with Phenols: Comparison with the Reactons of Carbonyl Triplets. *J. Am. Chem. Soc.* 103:4162–4166, 1981.
150. D Shukla, N Schepp, N Mathivanan, and LJ Johnston. Generaton and Spectroscopic and Kinetic Characterization of Methoxy-Substituted Phenoxyl Radicals in Solution and on Paper. *Can. J. Chem.* 75:1820–1829, 1997.
151. L Hurrell, LJ Johnston, N Mathivanan, and D Vong. Photochemistry of Lignin Model Compounds on Solid Supports. *Can. J. Chem.* 71:1340–1348, 1993.
152. Y Huang, D Pagé, DDM Wayner, and P Mulder. Radical-Induced Degradation of a Lignin Model Compound: Decomposition of 1-phenyl-2-phenoxyethanol. *Can. J. Chem.* 73:2079–2085, 1995.
153. ML Andersen and DDM Wayner. Electrochemistry of Electron Transfer Probes: α-Aryloxyacetoveratrones and Implications for the Mechanism of Photo-yellowing of Pulp. *Acta Chem. Scand.* 53:830–836, 1999.
154. MG Neumann, RAMC De Groote, and AEH Machado. Flash Photolysis of Lignin. Part 1. Deaerated Solutions of Dioxane-Lignin. *Polymer Photochemistry* 7:401–407, 1986.
155. WJ Leigh, EC Lathioor, and MJ St. Pierre. Photoinduced Hydrogen Abstraction from Phenols by Aromatic Ketones: A New Mechanism for Hydrogen Abstraction by Carbonyl n,π* and π,π* Triplets. *J. Am. Chem. Soc.* 118:12339–12348, 1996.
156. A Castellan, S Grelier, L Kessab, A Nourmamode, and Y Hannachi. Photophysics and Photochemistry of a Lignin Model Molecule Containing α–carbonyl guaiacyl and 4-hydroxy-3-methoxybenzyl Alcohol Moieties. *J. Chem. Soc. Perkin Trans.* 2:1131–1138, 1996.
157. F Wilkinson and G Kelly. Diffuse Reflectance Flash Photolysis. In JC Scaiano, ed. *Handbook of Organic Photochemistry,* 1. Boca Raton: CRC Press, 1989, pp. 293–314.
158. JC Scaiano, AB Berinstain, MK Whittlesey, PRL Malenfant, and C Bensimon. Lignin-like Molecules: Structure and Photophysics of Crystalline α-Guaiacoxyacetoveratrone. *Chem. Mater.* 5:700–704, 1993.

159. JA Schmidt, C Heitner, GP Kelly, and F Wilkinson. Diffuse-Reflectance Laser-Flash Photolysis of Mechanical Pulp. Part 1. Detection and Identification of Transient Species in the Photolysis of Thermomechanical Pulp. *J. Pulp Paper Sci.* 16:J111–J117, 1990.
160. F Wilkinson, A Goodwin, and DR Worrall. Diffuse Reflectance Laser Flash Photolysis of Thermomechanical Pulp. In C Heitner and JC Scaiano, eds. *Photochemistry of Lignocellulosic Materials,* 531. Washington, DC: American Chemical Society, 1993, pp. 86–98.
161. JA Olmstead and DG Gray. Fluorescence Spectroscopy of Cellulose, Lignin and Mechanical Pulps: A Review. *J. Pulp Paper Sci.* 23:J571–J581, 1997.
162. K Lundquist, B Josefsson, and G Nyquist. Analysis of Lignin Products by Fluorescence Spectroscopy. *Holzforschung* 32:27–32, 1978.
163. LT Okano, R Ovans, V Zunic, JN Moorthy, and C Bohne. Effect of Cyclodextrin Complexation on the Photochemistry of the Lignin Model α-Guaiacoxyacetoveratrone. *Can. J. Chem.* 77:1356–1365, 1999.
164. M Beyer, D Steger, and K Fischer. The Luminescence of Lignin-Containing Pulps: A Comparison with the Fluorescence of Model Compounds in Several Media. *J. Photochem. Photobiol. A: Chem.* 76:217–224, 1993.
165. K Lundquist, I Egyed, B Josefsson, and G Nyquist. Lignin Products in Pulping Liquors and Their Fluorescence Properties. *Cellulose Chem. Technol.* 15:669–679, 1981.
166. B Albinsson, L Shiming, K Lundquist, and R Stomberg. The Origin of Lignin Fluorescence. *J. Mol. Struct.* 508:19–27, 1999.
167. AEH Machado, DE Nicodem, R Ruggiero, D da Silva Perez, and A Castellan. The Use of Fluorescent Probes in the Characterization of Lignin: The Distribution, by Energy, of Flurophores in *Eucalyptus grandis* Lignin. *J. Photochem. Photobiol. A: Chemistry* 138:253–259, 2001.
168. A Castellan, A Nourmamode, C Noutary, C Belin, and P Fornier de Violet. Photoyellowing of Milled Wood Lignin and Peroxide-Bleached Milled Wood Lignin in Solid 2-Hydroxypropyl Cellulose Films after Sodium Borohydride Reduction and Catalytic Hydrogenation in Solution: A Fluorescence Spectroscopic Study. *J. Wood Chem. Technol.* 12:19–33, 1992.
169. JA Olmstead and DG Gray. Fluorescence Emission from Mechanical Pulp Sheets. *J. Photochem. Photobiol. A: Chem.* 73:59–65, 1993.
170. H Tylli, I Forsskähl, and C Olkkonen. Photochromic Behaviour of Ozonated and Photoirradiated Cellulose Studied by Fluorescence Spectroscopy. *Cellulose* 3:203, 1996.
171. A Castellan, R Ruggiero, E Frollini, LA Ramo, and C Chirat. Studies on Fluorescence of Cellulosics. *Holzforschung* 61:504–508, 2007.
172. T Rosenau, W Milacher, A Hofinger, and P Kosma. Isolation and Identification of Residual Chromophores in Cellulosic Materials. *Polymer* 45:6437–6443, 2004.
173. A Vikkula, J Valkama, and T Vuorinen. Formation of Aromatic and Other Unstarurated End Groups in Carrboxymethylcellulose during Hot Alkali Treatment. *Cellulose* 13:593–600, 2006.
174. S Liukko, V Tasapuro, and L Tiina. Fluorescence Spectroscopy for Chromophore Studies on Bleached Kraft Pulps. *Holzforschung* 61:509–515, 2007.
175. JA Olmstead, JH Zhu, and DG Gray. Fluorescence Spectroscopy of Mechanical Pulps. III. Effect of Chlorite Delignification. *Can. J. Chem.* 73:1955–1959, 1995.
176. A Castellan and RS Davidson. Steady-State and Dynamic Fluorescence Emission from Abies Wood. *J. Photochem. Photobiol. A: Chem.* 78:275–279, 1994.
177. A Castellan, H Choudhury, RS Davidson, and S Grelier. Comparative Study of Stone-Ground Wood Pulp and Native Wood. 2. Comparison of the Fluorescence of Stone-Ground Wood Pulp and Native Wood. *J. Photochem. Photobiol. A: Chem.* 81:117–122, 1994.

178. RS Davidson. The Photodegradation of Some Naturally Occurring Polymers. *J. Photochem. Photobiol. B. Biology* 33:3–25, 1996.

179. JH Zhu, JA Olmstead, and DG Gray. Fluorescent Detection of *o*–Quinones Formed in Lignin-Containing Pulps during Irradiation. *J. Wood Chem. Technol.* 15:43–64, 1995.

180. K Kónya and JC Scaiano. Development and Applications of Pyrene-Containing Fluorescent Probes for Monitoring the Photodegradation of Lignin-Rich Products. *Chem. Mater.* 6:2369–2375, 1994.

181. WJ Bublitz. Fluorescence of Pulping Liquors: A Tool for Digester Control? *Tappi* 64:73–76, 1981.

182. WJ Bublitz and TY Meng. The Fluorometric Behavior of Pulping Waste Liquors. *Tappi* 27:27–30, 1978.

183. WJ Bublitz and DC Wade. Applying Waste Liquor Fluorescence to Control Pulp Quality. *Svensk Paperstidn.* 82:535–538, 1979.

184. DM Hanson and DC Wenzel. Development of Size Exclusion Chromatography/Laser Induced Fluorescence Analysis of Isolated Lignins. *J. Wood Chem. Technol.* 9:189–200, 1989.

185. DM Hanson and DC Wenzel. Laser-Induced Fluorescence of Lignins with Excitation from 457 to 621 Nanometers. *J. Wood Chem. Technol.* 11:105–115, 1991.

186. Y Liu, R Gustafson, W McKean, and J Callis. Application of Single Fiber Fluorescence Staining to Assess Kappa Uniformity. TAPPI Pulping Conference, Montreal, Quebec, 1998, 3, pp. 1501–1507.

187. Y Liu, R Gustafson, and J Callis. Microspectroscopic Analysis and Kappa Determination of Single Pulp Fibres Stained with Acridine Orange. *J. Pulp Paper Sci.* 25:351–355, 1999.

188. H Bergström, J Carlsson, P Hellentin, and L Malmqvist. Spectrally and Temporally Resolved Studies of Fluorescence from Newsprint and from Substances Occurring in Newsprint. *Nordic Pulp Paper Res. J.* 11:48–55, 1996.

189. ST Sum, DL Singleton, G Parakevopoulos, RS Irwin, RJ Barbour, and R Sutcliffe. Laser-Excited Fluorescence Spectra of Eastern SPF Wood Species: An Optical Technique for Identification and Separation of Wood Species? *Wood Sci. Technol.* 25:405–413, 1991.

190. KK Pandey, NK Upreti, and VV Srinivasan. A Fluorescence Spectroscopic Study on Wood. *Wood Sci. Technol.* 32:309–315, 1998.

191. S Barsberg and KA Nielsen. Oxidative Quenching of Spruce Thermomechanical Pulp Fiber Autofluorescence Monitored in Real Time by Confocal Laser Scanning Microscopy: Implications for Lignin Autofluorescence. *Biomacromolecules* 4:64–69, 2003.

192. A Castellan, N Colombo, C Vanucci, P Fornier de Violet, and H Bouas-Laurent. A Photochemical Study of an O-methylated α-carbonyl β-1 Lignin Model Dimer: 1,2-di-(3',4'-dimethoxyphenyl)ethanone (deoxyveratroin). *J. Photochem. Photobiol., A:Chemistry* 51:451–467, 1990.

193. A Nishigaki, U Nagashima, A Uchida, I Oonishi, and S Ohshima. Hysteresis in the Temperature Dependence of Phosphorescence of 4-Hydroxy-3-methoxybenzaldehyde (Vanillin) in Ethanol. *J. Phys. Chem.* A 102:1106–1111, 1998.

194. V Molina and M Merchán. Theoretical Analysis of the Electronic Spectra of Benzaldehyde. *J. Phys. Chem.* A 105:3745–3751, 2001.

195. Beilstein. 3rd supplement, 6: 2401.

4 Vibrational Spectroscopy

Umesh P. Agarwal and Rajai H. Atalla

CONTENTS

INTRODUCTION

Vibrational spectroscopy is an important tool in modern chemistry. In the past two decades, thanks to significant improvements in instrumentation and the development of new interpretive tools, it has become increasingly important for studies of lignin. This chapter presents the three important instrumental methods—Raman spectroscopy, infrared (IR) spectroscopy, and near-infrared (NIR) spectroscopy— and summarizes their contributions to analytical, mechanistic and structural studies of lignin. The conceptual frameworks used to interpret vibrational spectra are first described in the following section.

VIBRATIONAL SPECTROSCOPY

Vibrational spectra are of two types [1], infrared and Raman, and arise from two different types of energy exchanges between the molecules under study and electromagnetic radiation. In infrared spectroscopy, a vibrational transition that involves a change in dipole moment results in absorption of an infrared photon. The energy of the absorbed photon is equal to the energy difference between the two vibrational states of the molecule.

In Raman spectroscopy, the electromagnetic field induces a dipole moment in the molecule, with the result that an exchange of energy occurs simultaneously with the vibrational transition. The energy of the exciting photons is higher than the energy difference between the two vibrational states, and the exchange with the field results in a scattered photon, shifted in frequency from the incident photon by an amount equal to the energy difference between the vibrational states.

With both types of vibrational spectroscopy, distinctive spectra and facility in interpretation are possible because only vibrational transitions corresponding to changes in the vibrational quantum number of ± 1 are allowed by the spectral selection rules. That is, $\Delta n = \pm 1$, where n is the vibrational quantum number. Due to this, the frequencies observed are usually the fundamental frequencies. In addition, because of analogies between the mathematical descriptions of classical and quantum mechanical vibrating molecular systems, it is possible to rationalize many spectral observations by analogy with classical vibrating systems that possess characteristic force constants and reduced masses. This rationalization has become the basis for systematizing much of the structural and chemical information derived from vibrational spectra.

A useful exception to the primary selection rule is that the overtones and combination bands associated with C–H and O–H stretching vibrations are, in fact, active in infrared absorption. Here $\Delta n = \pm 2$ or ± 3, which results in overtone and combination bands. The apparent violation of the selection rule arises because of anharmonicities in the potential energy function that govern the vibrations. These absorptions occur primarily in the near-infrared region because the selection rules are less rigorous in infrared absorption than in Raman scattering. These transitions are responsible for the spectral features in the NIR. Though the assignment of these transitions to particular vibrational modes is not as easy, the spectral features in this region are very sensitive to compositional variations and provide a valuable analytical tool when they are analyzed using chemometric methods.

In organic molecules, there are well-defined frequencies at which certain bond types of carbon, hydrogen, and oxygen are expected to absorb or scatter [2]. It is therefore possible to correlate frequencies and structures in very useful ways. For example, the most distinctive spectral features are the bands associated with the C–H and O–H stretching vibrations. Since the relatively low mass of the hydrogen atom, the frequencies associated with these stretching vibrations are quite removed from those of the other fundamental vibrational spectral features. Thus, they undergo vibrational transitions as local modes, where the majority of the energy is localized in the particular bond.

It is also anticipated that C=C and C=O double bonds will have distinctive frequencies because these bonds also undergo vibrational transitions as local modes. This is also true of some functional groups, such as carboxyl groups (COOH). In contrast, C–C and C–O single bonds, when adjacent to similar bonds, become involved in a high degree of coupling that can result in bands over a range of frequencies. The high degree of coupling arises because the bonds have similar energies and, therefore, force constants, and they also have similar reduced masses. Thus, skeletal vibrations in an organic molecule tend to be highly coupled, while functional group bands tend to be highly localized.

APPLICATION TO LIGNIN

It is important first to consider the types of functional groups that usually occur in lignin [3] and the influence of structural variations on these groups. Two classes of functional groups are notable. The first class are those that derive their character from a single type of bond and that result in characteristic absorption (infrared) [4] or scattering (Raman) [5] bands. Here it is expected that the frequencies would occur in a narrow range with slight variations according to the nature of the bonded atoms due to a second-order effect. Good examples of this category are the C–H and O–H stretching vibrations. Their general region (2800 to 3600 cm^{-1}) is determined by the mass of the hydrogen atom; their individual regions are determined by the force constants of the two types of bonds. The C=O bond also fits into this category; the small second-order effects depend on whether it is bonded to a hydrogen atom and carbon atom as in an aldehyde, two other carbon atoms as in carbonyl, or an oxygen atom and carbon atom as in carboxyl. For the C=C bond, the second-order effects would depend primarily on the degree of conjugation in adjacent structures.

The second important class of lignin functional groups consists of clusters of bonds that have collective properties and that derive their identity from the aromatic center. These groups are best discussed in terms of the characteristic vibrations of the aromatic ring and their variation with substitution on the ring. The most complete analyses of aromatic vibrational frequencies, not surprisingly, are those carried out for benzene [6]. Such analyses take advantage of the high symmetry of the benzene ring to carry out detailed normal mode analyses to identify the key internal coordinates contributing to particular vibrational frequencies. These analyses can then become the bases for analyses of the motions of aromatic centers substituted with different functional groups or attached to other substructures. In lignin, the most common functionalities are the variously modified propane substructures that can occur at C1, and the phenoxy, methoxy, and aryl oxygen linkages that can occur at C3, C4, and C5.

In a comprehensive analysis of the vibrational modes that can arise in lignin, Ehrhardt [7] identified the modes that are expected to have a high degree of frequency invariance. They include a cluster of five modes in the aromatic C–H stretching region between 3042 and 3061 cm^{-1}. Another cluster of coupled C–C stretching and C–H bending modes covers the region between 968 and 1594 cm^{-1}; this region includes nine modes. The other modes occur at lower frequencies that may not be readily observed in spectra of lignin. It is anticipated that bands characteristic of the aromatic center will be observed in lignin spectra, although their frequencies will be somewhat altered, and their relative intensities will be significantly different from those observed in the spectra of benzene. The types of modes observed in the spectra of lignin are expected to be similar to those observed in benzene, except that some stretching vibrations associated with bonds to functional groups will not have occurred in the benzene spectra. Moreover, bands that are disallowed in the spectra of benzene because of its high symmetry will not be forbidden in the spectra of aromatic centers in lignin.

Due to its high symmetry, benzene obeys the mutual exclusion rule, so that its infrared active bands are inactive in the Raman spectrum and vice versa. Since the aromatic centers in lignin do not have this symmetry, many more bands are active in both infrared and Raman spectra. However, the pattern remains that the highly polar vibrations are expected to be strongest in the infrared spectrum, whereas the least polar and most polarizable vibrations are expected to be most intense in the Raman spectrum.

The assignments of the spectra discussed in the following sections were undertaken with this background in mind. The analysis by Ehrhardt [7] provided the starting point, but this work was significantly complemented by investigations of the spectra of model compounds. Ehrhardt included studies of mono-substituted and di-substituted model compounds and one tri-substituted model compound. This work has been extended by Agarwal et al. [5,8–10] to a number of other model compounds that more closely approximate the C_9 units in lignin. The studies included consideration of the various effects that can enhance Raman spectra, as well as the infrared spectra in the fundamental region.

In the following sections, three methods for observing vibrational transitions are outlined together with overviews of their application in studies of lignin and lignocellulosics. The Raman spectral observations are presented in the next section. The instrumental methods applicable to lignin are summarized, the assignments of characteristic bands in the spectra of lignins and lignocellulosics are discussed, and the application in investigations of lignin is outlined. Following the Raman Spectroscopy section, the same classes of information are described for observations based on infrared absorption in the mid-infrared region. This is the region dominated by the fundamental vibrations. Finally, the application of NIR is described. This section does not focus on the assignments, which are not an issue, but rather on the ability to detect subtle variations that are not immediately obvious upon visual inspection. This is accomplished using chemometric spectral analyses.

RAMAN SPECTROSCOPY

Background

Raman spectroscopy has existed as an analytical technique for more than 70 years, but its applications to study lignin and lignin-containing materials did not begin until the early 1980s [11–15]. A number of factors were probably responsible for this outcome: user unfamiliarity with the technique, the belief that IR and Raman spectroscopy provided similar information (actually they are complementary), and the high cost of Raman instrumentation. Initial spectra [11,13] indicated that the Raman signal was almost completely obscured by laser-induced fluorescence (LIF). Even though a better signal-to-noise ratio could be obtained after acquiring multiple scans and subtracting the background, the spectra were of poor quality. The problem of LIF in lignocellulosics was first reported in 1984 [13]. It arose because lignin absorbed in the visible region, which was used ubiquitously in Raman spectroscopy for sample excitation. It was only after special methods were developed, for both macro [1,11] and micro [1,15] sampling, that the situation improved.

Although a number of visible-laser-based Raman techniques were used to study lignin and lignin-containing materials, a persistent need to suppress the accompanying fluorescence existed. This was true for most nonlignocellulosic materials as well. An important breakthrough in dealing with LIF occurred in 1986 when a new Raman instrument based on NIR excitation was developed [16]. This advance solved the problem of fluorescence for most samples and greatly revived interest in Raman spectroscopy. NIR Raman spectroscopy (also called NIR FT-Raman) has proven to be an effective technique in lignin research [5] and is rapidly becoming a widely used tool for analysis.

INSTRUMENTATION

Information on obtaining a Raman spectrum is abundant in the literature [17–19]. Raman instruments are either of a dispersive type [17,18] or are based on an interferometer [19]. A dispersive instrument consists of a source of monochromatic radiation (laser), an appropriate way of sampling, suitable gratings (for dispersion of the scattered radiation), and a detection device. In the past, gas lasers were the most common radiation sources; more recently, diode lasers are used. Commercial spectrometers consist of single, double, or triple monochromators, depending upon how efficiently the intensity of the excitation laser line needs to be suppressed to detect the weakly scattered Raman light. In a dispersive Raman spectrometer, a detector consists of either a photomultiplier tube or some multichannel device (e.g., charge-coupled device or photodiode array). A multichannel detector is particularly useful where high resolution is not necessary and rapid analysis is desired.

In 1986, a Raman instrument based on NIR excitation (1064 nm) and a Michaelson interferometer became available [16]. This development revolutionized Raman spectroscopy. In addition to the advantages of throughput and multiplex inherent to Fourier Transform (FT) techniques, this instrument overcame the obstacle of fluorescence. Fluorescence was eliminated by excitation at a NIR wavelength where electronic transitions in most samples are absent. Availability of such NIR FT-Raman instruments was particularly useful in the studies of lignin.

SPECIAL TECHNIQUES AND EFFECTS

Micro Raman

Raman microspectroscopy couples an optical microscope to the conventional Raman spectrometer [20]. The main advantage of this technique is that a sample can be investigated in a spatially resolved manner. This capability is especially important for heterogeneous materials (e.g., woody tissue) where composition and structure at the microscopic scale can be investigated. In the case of woods, chemical information from the morphologically distinct regions can be obtained. Depending upon the excitation wavelength and microscope-objective characteristics, lateral spatial resolution of the order of 1 µm has been achieved. Most Raman microscopes produced today are also confocal (do not detect out-of-focus scattering), which means that experiments requiring axial resolution (e.g., 2 µm at 633 nm) can be performed. Raman microspectroscopy has been used in investigations of the ultrastructure of woody tissues and other lignocellulosic materials.

Raman Imaging

Raman imaging maps the spatial distribution of a component in a sample, using a Raman frequency that is component-specific [21]. An image is produced using the Raman scattered photons at this frequency. Single-element detectors are used to create line (1D) images and multichannel detectors are used for 2D images. As useful advances in Raman imaging have occurred only recently, only few reports of applications to lignin-containing materials exist in the literature [22,23].

Resonance and Preresonance Raman

When a compound has an absorption band close to the sample excitation wavelength, either resonance Raman or preresonance Raman scattering occurs [17]. Compared to normal Raman scattering, the resonance Raman effect enhances scattering by as much as a million times. On occasion, overtones of a vibrational mode may be detected. Considering that laser wavelengths are available over most of the absorption frequency region (from NIR to ultraviolet (UV)), the effect can be observed by selecting an appropriate excitation wavelength. Although the preresonance Raman effect was previously reported to be present in the spectrum of native lignin [24], rigorous resonance Raman was observed in residual lignin only in 2001 [25], using UV excitation.

Conjugation Effect

The Raman intensities of certain vibrations depend upon conjugation [26]. Conjugation is defined as the presence of delocalized molecular orbitals caused by overlapping of adjacent atomic orbitals. In chemistry, the concept of conjugation has been used in determining bond lengths and the extent of π-charge transfer between groups of atoms. Raman studies of lignin model compounds indicated that the conjugation effect significantly enhanced the intensities of certain vibrations [9]; for instance, conjugated benzene ring modes, conjugated C=C bond stretching, and conjugated C=O bond stretching. The enhancement seemed to depend upon the extent of conjugation. The presence of conjugated structures in lignin has implications for quantitative work because conjugated structures contribute disproportionately (on a molar basis) to band intensity.

Surface Enhanced Raman

Another way to enhance the usually weak Raman scattering signal of a compound is by introducing the surface enhanced Raman (SER) effect [27] through the influence of small metal particles (usually silver, gold, or copper). In the SER technique, first reported in 1973, the enhancement effect depends upon the nature of the surface roughness and the chosen metal. The mechanisms that lead to the SER effect remain a subject of discussion [28] but are thought to arise from the interaction of adsorbate (chemical effect) and surface plasmons (electromagnetic effect). The SER effect has recently been induced in lignin [29]. The technique has the potential for *in situ* analysis.

SPECTRAL INTERPRETATION

No significant differences have been observed between the FT-Raman spectra of native and milled wood lignins (MWLs) from black spruce [30]. This is likely to be true for other lignocellulosics as well. Therefore, a MWL Raman spectrum can be considered to represent native lignin.

FT-Raman spectra of black spruce [30] and aspen [31] MWLs (representing guaiacyl and guaiacyl-syringyl lignin, respectively) are shown in Figure 4.1, and the

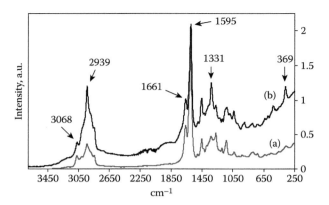

FIGURE 4.1 FT–Raman spectra of milled wood lignins (MWLs): (a) black spruce, (b) aspen. Annotated peaks are some of the peak positions in the spectrum of aspen MWL. Although the spectra appear to be similar, differences between peak positions and intensities were detected (Table 4.1). (Based on UP Agarwal and SA Ralph. FT-Raman Spectroscopy of Wood: Identifying Contributions of Lignin and Carbohydrate Polymers in the Spectrum of Black Spruce (Picea mariana). *Appl. Spectrosc.* 51, 1648–1655, 1997; UP Agarwal, JD McSweeny, and SA Ralph. An FT-Raman Study of Softwood, Hardwood, and Chemically Modified Black Spruce MWLs. 10th International Symposium on Wood and Pulping Chemistry, Yokohama, 2, 136–140, 1999.)

detected band frequencies are listed in Table 4.1. Relative intensities in the table are given with respect to other peaks in the spectrum.

Band assignments are based on the authors' previous work as well as the work of others in the field of vibrational spectroscopy. Useful information was obtained from the spectra of woods [30,32] mechanical pulps [33–36], MWLs [30,31,37], dehydrogenation polymer (DHP) lignins [38] and many lignin models [7,8,10]. Moreover, literature on vibrational assignment of benzene derivatives [2,6] was consulted. Except for the spectral assignments in the region below 925 cm^{-1}, where coupled modes involving deformation and torsional vibrations are present (vibrations of C–C–C, O–C–O, C–O–C, C–O–H, and aromatic ring deformation modes), the assignments have proven quite useful.

Displayed spectra were not processed in any way except that they are shifted on the y-axis with respect to one another for clarity. Contributions from O–H stretching and ester group modes were not detected because their Raman intensity is very weak. This is in keeping with the expectation that polar bonds are not easily detected in Raman spectroscopy because they have poor scattering cross-sections. Also, compared with IR spectra, fewer intense peaks were detected in Raman and they did not overlap as much.

3100–2800 cm^{-1}

In lignin, both aromatic and aliphatic C–H stretches are expected to contribute in this region. The medium intensity peak present at 3071/3068 cm^{-1} (the frequencies are indicated as spruce/aspen) is due to aromatic C–H stretches (Table 4.1). This is supported by the studies of benzene derivatives [2,6,7] and lignin models [10].

TABLE 4.1
Assignment of Bands in FT-Raman Spectra of Softwood and Hardwood Milled-Wood Lignins

Guaiacyl L.[a] (cm^{-1})	Guaiacyl-Syringyl L.[b] (cm^{-1})	Assignment[c]
3071 m[d]	3068 m	aromatic C–H stretch
3008 sh	3003 sh	C–H stretch in OCH$_3$, asymmetric
2940 m	2939 s	C–H stretch in O–CH$_3$, asymmetric
2890 sh	2893 sh	C–H stretch in R$_3$C–H
2845 m	2847 sh	C–H stretch in OCH$_3$, symmetric
1662 s	1661 s	ring conj. C=C stretch of coniferyl/sinapyl alcohol; C=O stretch of coniferaldehyde/sinapaldehyde
1621 sh	1620 sh	ring conjugated C=C stretch of coniferaldehyde/ sinapaldehyde
1597 vs	1595 vs	aryl ring stretching, symmetric
1508 vw	1501 vw	aryl ring stretching, asymmetric
1453 m	1455 s	O–CH$_3$ deformation; CH$_2$ scissoring; guaiacyl/syringyl ring vibration
1430 w	1426 w	O–CH$_3$ deformation; CH$_2$ scissoring; guaiacyl/syringyl ring vibration
1392 sh	1395 sh	phenolic O–H bend
1363 sh	1367 sh	C–H bend in R$_3$C–H
1334 m	1331 s	aliphatic O–H bend
1298 sh	---[e]	aryl-O of aryl-OH and aryl-O–CH$_3$; C=C stretch of coniferyl alcohol
1272 m	1272 m	aryl-O of aryl-OH and aryl-O–CH$_3$; guaiacyl/syringyl ring (with C=O group) mode
1226 vw	1224 w	aryl-O of aryl-OH and aryl-O–CH$_3$; guaiacyl/syringyl ring (with C=O group) mode
1192 w	1190 w	a phenol mode
---	1156 sh	unassigned
1136 m	1130 m	a mode of coniferaldehyde/sinapaldehyde
1089 w	1088 w	out of phase C–C–O stretch of phenol
1033 w	1037 m	C–O of aryl-O–CH$_3$ and aryl–OH
975 vw	984 sh	CCH and -HC=CH- deformation
928 vw	918 sh	CCH wag
895 vw	899 w	skeletal deformation of aromatic rings, substituent groups, and side chains
787 w	797 w	skeletal deformation of aromatic rings, substituent groups, and side chains
731 w	727 w	skeletal deformation of aromatic rings, substituent groups, and side chains
637 vw	638 w	skeletal deformation of aromatic rings, substituent groups, and side chains
---	597 m	skeletal deformation of aromatic rings, substituent groups, and side chains
588 vw	588 w	skeletal deformation of aromatic rings, substituent groups, and side chains
557 vw	---	skeletal deformation of aromatic rings, substituent groups, and side chains
534 vw	531 m	skeletal deformation of aromatic rings, substituent groups, and side chains
---	522 sh	skeletal deformation of aromatic rings, substituent groups, and side chains
---	503 vw	skeletal deformation of aromatic rings, substituent groups, and side chains
491 vw	490 vw	skeletal deformation of aromatic rings, substituent groups, and side chains

(Continued)

TABLE 4.1
Assignment of Bands in FT-Raman Spectra of Softwood and Hardwood Milled-Wood Lignins (Continued)

Guaiacyl L.[a] (cm⁻¹)	Guaiacyl-Syringyl L.[b] (cm⁻¹)	Assignment[c]
---	472 vw	skeletal deformation of aromatic rings, substituent groups, and side chains
457 vw	461 vw	skeletal deformation of aromatic rings, substituent groups, and side chains
---	447 vw	skeletal deformation of aromatic rings, substituent groups, and side chains
---	431 vw	skeletal deformation of aromatic rings, substituent groups, and side chains
---	417 vw	skeletal deformation of aromatic rings, substituent groups, and side chains
384 w	---	skeletal deformation of aromatic rings, substituent groups, and side chains
361 w	369 m	skeletal deformation of aromatic rings, substituent groups, and side chains

[a] Frequencies of black spruce MWL.
[b] Frequencies of aspen MWL; 43% of units are syringyl type.
[c] From UP Agarwal, SA Ralph, and RH Atalla, FT Raman Spectroscopic Study of Softwood Lignin, 9th International Symposium on Wood and Pulping Chemistry, Montreal, 1997.
[d] Note: vs is very strong; s is strong; m is medium; w is weak; vw is very weak; and sh is shoulder. Band intensities are relative to other peaks in spectrum.
[e] No corresponding band in guaiacyl-syringyl lignin

The contributions present at 2845/2847 cm⁻¹, 2890/2893 cm⁻¹, 2940/2939 cm⁻¹, and 3008/3003 cm⁻¹ are likely to arise from aliphatic C–H stretches. Considering the types of C–H groups present in the structure of lignin, it is proposed that the bands at 3008/3003 cm⁻¹ and 2940/2939 cm⁻¹ are due to the asymmetric C–H stretches in O–CH$_3$ groups. This is supported by published assignments [2,6,7] and our studies of lignin model compounds [10]. The C–H stretch in R$_3$C–H structures is expected to contribute at 2890/2893 cm⁻¹. Since several types of these structures are in lignin, small wavenumber shifts can also be expected in the frequency of this mode. Finally, based on the literature [2,6,7] and lignin model work [10], the peak at 2845/2847 cm⁻¹ can be assigned to the symmetric C–H stretch in the O–CH$_3$ group.

1800–1500 cm⁻¹

This is the most informative region of the Raman spectrum of lignin—contributions due to aromatic rings, ethylenic C=C, α– and γ– C=O, and some other chemical groups are detected here. Earlier work with woods [30,32], mechanical pulps [33–36], and MWLs [30,31,37] was useful in assigning the 1662/1661 cm⁻¹ band to the ethylenic C=C (in coniferyl alcohol/sinapyl alcohol units) and γ-C=O (in coniferaldehyde/sinapaldehyde) bond stretches in lignin. Similarly, the 1621 cm⁻¹ shoulder was associated with the ring-conjugated C=C bond stretch (in coniferaldehyde/sinapaldehyde). The aromatic ring stretch modes of lignin were detected at 1600/1595 and 1508/1501 cm⁻¹. Unlike IR, where the 1508/1501 cm⁻¹ band is strong, this mode is very weak in Raman.

1500–1000 cm⁻¹

In this region, several bands may represent mixed vibrations because many of the modes are coupled. In addition, there may be overlap as a result of CH_3, CH_2, and CH bending modes. The bands at 1453/1455 cm⁻¹ and 1430/1426 cm⁻¹ are assigned to $O–CH_3$ deformation and CH_2 scissoring modes. In addition, some scattering is expected from the guaiacyl/syringyl-ring vibration. Phenolic O–H bending is thought to be responsible for the shoulder at 1392/1395 cm⁻¹. C–H bending in $R_3C–H$ structures is likely to give intensity to the feature at 1363/1367 cm⁻¹. Aliphatic O–H bending seems to be the dominant contribution at 1334/1331 cm⁻¹. This feature is stronger in syringyl-rich aspen lignin (Figure 4.1). The triplet at 1298/--- cm⁻¹, 1272/1272 cm⁻¹, and 1226/1224 cm⁻¹ (Table 4.1) is likely to have contributions from aryl-O (in both aryl-O–H and aryl-O–CH_3) stretches. In addition, the 1272/1272 cm⁻¹ band intensity is in part due to the guaiacyl/syringyl ring (with C=O group) breathing. The peak at 1192/1190 cm⁻¹ is weak and is likely to be associated with the phenolic units in lignin. The medium intensity band at 1136/1130 cm⁻¹ is associated with the coniferaldehyde/sinapaldehyde unit, although further assignment within the unit is not yet clear. The weak peak at 1033/1037 cm⁻¹ seems to be due to the C–O of aryl-OCH_3 and aryl–OH.

1000–350 cm⁻¹

The lowest frequency region of the lignin Raman spectrum is difficult to assign because the number of contributions from skeletal vibrations is likely to increase. The bands are assigned to the skeletal modes of aromatic rings, substituent groups, and side chains. This includes out-of-plane vibrations of these structures. The aspen lignin band present at 369 cm⁻¹ is stronger than the 361 cm⁻¹ spruce lignin band (Figure 4.1 and Table 4.1). This difference could possibly be due to higher syringyl content (43%) of the hardwood lignin.

UV Resonance Raman Spectra

Lastly, the UV resonance Raman spectra of lignin model compounds were interpreted with the help of partial least squares (PLS) models [39].

APPLICATIONS

Lignin in Wood

Several differences have been detected in the FT-Raman spectra of the native lignin in hardwoods and softwoods [30–32,37]. In other research, spectra were used to classify wood type using a mathematical model [40,41]. The model was used for data processing and pattern recognition. In one study [40], a neural computing method was applied to extract key spectral features that were different between the hardwood and softwood species; in another, genetic algorithms were used to classify woods [41]. A chemometric technique called PLS regression has also been used to quantitate constituents, including lignin, of eucalyptus wood [42,43].

From the black spruce study [30], we concluded that the Raman spectra of native, enzyme, and milled wood lignins are very similar. Moreover, we found that most lignin Raman features were present in regions of the wood spectrum where carbohydrate components did not contribute. These findings are expected to be true for other species of wood as well.

Raman spectroscopy may be useful in determining the syringyl-to-guaiacyl (S/G) ratio in hardwoods. In a study that focused on two species of eucalyptus (samples varied in age and color) [43], a chemometric model was developed for this task and the prediction of the S/G ratio was quite good. Nevertheless, in another study in which a much larger number of hardwoods were sampled and no chemometric approach was used (author's unpublished results), the calculation of S/G ratios was not as accurate. Further work is needed to establish the wider applicability of the Raman/chemometric approach in calculating S/G ratios in wood lignins.

Lignin carbohydrate complexes (LCC) in mangrove, ohirugi [*Bruguiera gymnorrhiza* (L.) Lamk.] and buna [*Fagus crenata* Bl.] were studied using Raman spectroscopy; the LCCs were found to be similar [44].

In a study of fungus-induced changes in lignin responsible for darkening of wood chips [45], the authors concluded that quinones were largely responsible for the brightness loss of the fungus-treated chips.

Studies of native lignin in various morphological regions of wood have been carried out using a Raman microprobe [12,14,22–24,46]. One important result was evidence supportive of lignin orientation in the secondary walls of black spruce [12,14]. Although it was not clear what caused this orientation, this result highlighted the unique information that Raman spectroscopy can provide. Another investigation, which focused on studying corner middle lamellae in white birch and black spruce woods, indicated that the lignin concentration was not constant and could vary by as much as 100% [46]. Using confocal Raman, a small lignified border toward the lumen was observed in the gelatinous (G-) layer of poplar (*Populus nigra* x *P. deltoids*) tension wood [23]. These findings have important implications for understanding the ultrastructure of wood.

Lignin in Mechanical Pulp and Paper

Effects of bleaching [33–36], yellowing (both photo and thermal) [34,35,47], and specific chemical reactions [5,9] on lignin in lignocellulosics have been investigated using Raman spectroscopy. Bleaching studies of softwood thermomechanical pulp indicated that the lignin contained coniferyl alcohol, coniferaldehyde, and *p*-quinone units. The studies further demonstrated that *p*-quinones play a major role in determining pulp brightness [36]. For example, analysis of bleaching-related changes showed that whereas both coniferaldehyde and *p*-quinones were modified upon bleaching, the *p*-quinones were primarily responsible for pulp brightness.

Similarly, in photoyellowing, light-induced changes in lignin structure were ascertained. A time-resolved analysis of thermomechanical pulps showed that the light exposure led to decay of the 1654 cm^{-1} lignin band [5] and was responsible for a new Raman feature at 1675 cm^{-1} [35]. The former band contains contributions from both coniferyl alcohol and coniferaldehyde structures [35,36]. The new band, detected at 1675 cm^{-1} [35], was assigned to *p*-quinones [47]. The conclusion was that exposure of pulps to light caused the decay of coniferyl alcohol and coniferaldehyde units in lignin and caused the formation of *p*-quinones. It is noteworthy that prior to application of FT-Raman spectroscopy, *in situ* detection of *p*-quinones in photoexposed mechanical pulps was not possible.

Raman spectroscopy was also used to analyze lignin-containing pulps that were chemically modified (sulfonated, acetylated, hydrogenated, methylated, and acid hydrolyzed) [5,9]. In each case, the nature of the change and the extent to which pulp lignin was modified were discerned from the changes in the spectra.

Both lignin-containing and lignin-free coated and uncoated papers have been studied using Raman spectroscopy [48]. In addition to the spectrum of lignin, spectra of the coating mixture components such as latex and $CaCO_3$ were used to analyze these papers [48]. Raman features of coated papers were interpreted in terms of these components; the results showed strong contributions by latex and $CaCO_3$.

Lignin-containing printing and writing papers were aged by light and studied using Raman spectroscopy [49,50]. The changes in the spectrum of lignin for natural and accelerated aging were compared. Based in part on Raman information, experimental conditions for an accelerated test method that simulated natural aging were identified.

FT-Raman spectroscopy was used to differentiate between sulfate and sulfite wood papers [51]. The study indicated that in addition to differences due to hardwood and softwood, sulfite paper spectra showed a band at 510 cm^{-1}.

Residual Lignin in Chemical Pulp

Residual lignin in chemical pulp is difficult to analyze because of its low concentration. However, using FT Raman spectroscopy, the 1600 cm^{-1} band of lignin was easily detected [52–54]. This band has been used to determine pulp lignin content after research showed that its intensity was linearly correlated with pulp kappa numbers. More recently, UV resonance Raman has been applied to study residual lignin [25] with good results. The resonance technique has proven to be highly selective and sensitive for investigating residual lignin. In this case, the content of hexenuronic acid in pulps was determined *in situ*.

Lignin in Other Lignocellulosics

In addition to wood, pulp, and paper spectra, Raman spectra of numerous other lignin-containing materials have been obtained. These include bamboo [55], kenaf, jute, corn, wheat, and sugarcane bagasse [56] and flax [57]. In most cases, information could be obtained on the Raman features of lignin. In the spectrum of bamboo [55], strong lignin features were detected at 1604 and 1630 cm^{-1}. The bands were assigned to free and esterified phenolic units in lignin. Studies of flax and its parts indicated that major components of each could be detected using Raman spectroscopy [57].

Another lignocellulose, *Zinnia elegans*, has been analyzed for the presence of cellulose and lignin, and for the effect of the cellulose inhibitor on lignin biosynthesis [58]. Raman information suggested that the inhibitor 2,6-dichlorobenzonitrile has an effect on lignin formation.

Commercial Lignins

Analysis of commercial lignins by Raman spectroscopy remains a challenge. The problem is due the color of the samples, which produces a significant amount of fluorescence even when the samples are excited at 1064 nm, where most chromophores do not absorb. Since the fluorescence, even longer excitation wavelength is desirable. The other possibility is to modify the chromophores in lignin such that the absorption at 1064 nm is drastically reduced. In principle, it is possible to use resonance Raman and surface-enhanced Raman spectroscopy to study commercial lignins, but research remains to be done to determine if these approaches would be successful.

Chemical Modification Reactions of Lignin

Raman spectroscopy is being increasingly used to monitor chemical reactions [59]. In the chemical industry, online monitoring capability is being developed using fiber-optic-based Raman systems. Our laboratory has obtained useful information on numerous reactions, including bleaching, sulfonation, acetylation, hydrogenation, methylation, and acid hydrolysis.

For example, in photoyellowing, there was a need to determine whether aromatic-ring conjugated coniferyl alcohol C=C groups were completely hydrogenated in mechanical pulps. Using Raman spectroscopy, the hydrogenation reaction was followed by monitoring the intensity decline at 1654 cm^{-1}. Under modified reaction conditions, diimide completely hydrogenated this lignin double-bond in pulp [35].

Acetylation and deacetylation reactions of a mechanical pulp were also monitored using Raman spectroscopy [5]. The most prominent spectral change occurred at 2938 cm^{-1}. The acetylation reaction was considered complete when, upon further acetylation, the intensity of the 2938 cm^{-1} band (relative to 1095 cm^{-1} band) did not change. Analogously, the success of deacetylation was measured by the decline of Raman intensity at 2938 cm^{-1}.

Lignin Quantitation

The lignin spectrum has been used to carry out limited quantitative work. Both the amount of lignin in a sample and the concentration of a specific group within lignin (e.g., coniferyl alcohol) can be determined. The technique was first used to quantify lignin [60], using the 1595 cm^{-1} band, in southern pinewood that was progressively delignified. Due to the presence of a pre-resonance Raman effect, the amount of lignin in untreated wood could not be accurately determined. Nevertheless, a calibration curve for partly delignified samples was obtained. Further quantitative studies have been focused on quantifying lignin in chemical pulps, and the results have been encouraging.

Several factors are important in quantitative work. First, an internal reference band is needed for calculating relative intensity (either area or peak) of the band being used in quantitation. Although an external standard can be used, the internal band-ratio calculation is more reliable. However, if a chemometrics approach (e.g., principal components analysis [PCA], principal components regression [PCR], or PLS) is used, a standard is not required.

Second, considering that Raman intensity depends upon other factors besides concentration (discussed under Special Techniques and Effects), it is important to ensure that these intensity enhancement effects are either absent or their role is minimal. In certain cases, mild chemical treatments can modify a structure to minimize its enhanced contribution. Another problem in quantitative FT-Raman spectroscopy is "self absorption" [61], which is the absorption of the Raman photons by the sample. As the scattered photons are passing through the sample, they are absorbed and the spectrum obtained is a convolution of Raman and NIR absorption spectra. An investigation specifically designed to determine the importance of self-absorption in lignin and lignin-containing materials [62] found that the bands in the region of 2800–3100 cm^{-1} were reduced in intensity. The decline in intensity depended upon the moisture content of the sample, and the very strong NIR absorption of water beyond 2000 cm^{-1} (when excited by 1064 nm laser) was involved. Self-absorption can be avoided by either performing quantitation using a band that is not affected by self-absorption or by using D_2O instead of H_2O in the sample.

MID-INFRARED SPECTROSCOPY

BACKGROUND

Mid-infrared spectroscopy (also called FT-IR) has been used to analyze lignins for many years. The mid-infrared (mid-IR) spectral region is the frequency range from 4000 to 400 cm^{-1} (2.5 to 25 μm). Most lignin fundamental molecular vibrations fall in this range. The region below 400 cm^{-1} is the far infrared and that above 4000 cm^{-1} is the near infrared. As the needs of the instrumentation in the far-, mid-, and near-IR regions are different, the techniques in these regions have traditionally been treated differently. The mid- and near-IR differ in sampling techniques as well. Factors such as differences in the absorption in the two regions and the interaction with IR radiation as a function of wavelength are important considerations.

As early as 1948, Jones [63,64] conducted a comprehensive study of lignin using mid-IR spectroscopy. The IR studies by others that followed focused on, for example, synthetic lignin (DHP), lignin *in situ*, Braun's lignin, and enzyme lignin. Hergert [4] reviewed the early research on mid-IR spectroscopy of lignin and summarized the IR band assignments.

Early lignin (solid state) spectra were largely obtained in transmission mode using the potassium bromide (KBr) (or potassium chloride (KCl)) pellet sampling method. Since then, the development of a large number of sampling methods has permitted analytical measurements to be made in reflectance, emission, and photoacoustic absorption modes. New technological advances in instrumentation have benefited FT-IR spectroscopy as well. For example, IR microspectrometry allows the analysis of a small amount of a substance or of a small region of the material. Use of an imaging-capable IR-microscope in conjunction with two-dimensional (2D) detector arrays has allowed production of IR images of materials. Two-dimensional infrared (2 D-IR) spectroscopy, which records spectra at different levels of an applied external perturbation, has also been recently used to study lignin-containing materials.

Advances in data manipulation and availability of mathematical, statistical, and chemometric analytical software programs have greatly assisted extraction of useful information from IR spectra. This area, in particular, has greatly benefited from the wide spread use of personal computers.

In another review of lignin mid-IR spectroscopy [65], band assignments of guaiacyl, guaiacyl-syringyl, and coumaryl–guaiacyl-syringyl type milled wood lignins (MWLs) were summarized.

INSTRUMENTATION

Most modern, commercially available mid-IR instruments are built around an interferometer and are different in this respect from their predecessor, the diffraction IR spectrometer. A modern instrument consists of a source of IR radiation, an interferometer, a sample chamber, and a detector. Brief comments on various spectrometer components are provided in the following text. Details of topics related to instrumentation can be obtained from the literature [66].

As a source, most manufacturers use either a conducting ceramic or a wire heater coated with the ceramic. When heated, these devices emit IR radiation. Temperatures of 1000°C are fairly typical. The interferometer is the heart of the instrument as it analyzes the infrared radiation and enables generation of a spectrum. The Michelson interferometer is the one used most commonly. A computer is used to control the interferometer. Using the fast Fourier transform method, the interferogram is converted into wavelength absorbances. The detectors used in FT-IR instruments are photo resistors—they have very high resistance in the dark, but resistance falls upon exposure to light. Most frequently, deuterated triglycine sulfate (DTGS) is used as the pyroelectric detector of mid-IR radiation. For enhanced sensitivity, a cryogenically cooled mercury cadmium telluride (MCT) semi-conductor detector is used. A computer (data system) is used not only to control the interferometer and perform Fourier transform on collected data, but also to process the data to obtain the most useful information.

SPECIAL TECHNIQUES/INTERFACES

Although mid-IR spectroscopy can be applied to any kind of material in any physical state, sample preparation is an important consideration in FT-IR. For routine analysis of solid lignin or lignin-containing samples, the KBr transmission method is generally used. This sampling approach involves preparation of a compressed KBr pellet containing the sample in a manner that minimizes light scattering. Sometimes, solid samples should be analyzed directly to obtain the desired information. Occasionally, only information from the sample surface is desired. There are special methods to generate such information.

Diffuse Reflectance

Diffuse reflectance infrared Fourier transform (DRIFT) spectra are obtained when IR radiation is incident on a scattering sample at a specific angle and is reflected at all angles. The diffuse reflectance process involves transmission, scattering, and reflection [67]. The technique is used to analyze an intact lignin sample without modification [65]. To study a sample by DRIFT, the sample is either dispersed in KBr (e.g., MWL) or is analyzed directly (e.g., paper sheet) and placed at the focal point of the diffuse reflectance accessory. The scattered light from the sample is collected by a concave mirror and directed to the detector. Information in the spectrum is dominated by contributions from the surface. However, optical effects, occurring at the surface, can significantly affect the quality of the spectrum. For powdered samples, particle size is important. Further information on the DRIFT sampling method and its applications to study of lignin can be obtained elsewhere [65,68].

Attenuated Total Reflection, or Internal Reflectance

Like DRIFT, attenuated total reflection (ATR) [69] is used to study materials that are difficult to analyze by absorption methods, such as thin layers on nontransparent substrates, substances with very high absorption that are difficult to prepare as thin layers, and materials in which surface sampling needs to be carried out. The principle of measurement is based on the transmission of light through an optical element of high refractive index material (e.g., zinc selenide). The angle of incident light is such that it results in internal reflection. The geometry of the ATR element generates multiple internal reflections. When an IR-absorbing material is on the air-ATR element interface, the light and sample interact and the intensity of the internally reflected light is weakened. A spectrum is produced from the first one or two micrometers (into the sample) from the interface. Further information is available in the literature [69].

IR Microscopy

Infrared microspectrometry has been available for some time [70,71]. This accessory has made it possible to view the sample and select a specific region for chemical analysis. This is of particular interest for materials like wood where spatially resolved (10 μm or larger) information from chemically inhomogeneous regions is desired. In addition, an IR microscope provides high sensitivity and is very useful for studying small amounts of samples (e.g., single fibers, picograms of a substance, and specks in paper sheet). Compared to conventional DRIFT, micro-transmission mode was reported to be more sensitive in studies of kraft pulps [72]. No sample preparation is required and samples are studied directly.

Chemical Imaging

IR imaging is possible by coupling an IR microspectrometer (spectral information) with the focal plane array (FPA, spatial information) [73]. The introduction of FPA detectors allowed the mapping of a sample in practical time because thousands of detector elements are read during a spectral acquisition. Moreover, large sample areas can be mapped by using a programmable microscopic stage. Once obtained from a sample region, thousands of spectra can be evaluated simultaneously (by various ways). This means that a plot of single IR band intensity, representing a single component in a multicomponent sample, can be generated. Such a plot shows how that component is distributed in the sample. Such chemical information can be converted into an image and then compared to the visual image [74].

2-D IR

Introduced around 1990 [75,76], 2-D IR is a technique in which an external perturbation (e.g., temperature, pressure, or strain) is applied to the sample and time-dependent IR spectra are recorded. A 2-D correlation spectrum is then produced, plotting v_1 and v_2 (bands at two wavenumbers) in the two dimensions; the third dimension shows the correlation function of the spectral intensities observed at v_1 and v_2. The shape of the resulting surface shows whether or not the bands at the two wavenumbers are correlated. This allows assessment of the extent to which two parts of a molecule are linked (coupled) in their response to the applied perturbation. This technique has only recently been applied to the study of lignin [77].

Photoacoustic IR

Photoacoustic IR (PAS-IR) [78,79] involves direct measurement of the absorption of IR radiation. This spectroscopic technique is based on the conversion of absorbed infrared radiation into thermal energy, followed by the emission of sound produced by the thermal energy transfer from the sample into the surrounding gas phase. Carbon black is used as a reference material because of its excellent absorption characteristics. It is a good method for difficult-to-analyze samples. The technique has been applied to the study of lignin [80,81].

Transient IR

Transient infrared spectroscopy (TIRS) is a mid-infrared technique [82] that has been developed to obtain spectra of moving solids and viscous liquids. TIRS spectra are obtained from the generation of a thin, short-lived temperature differential that is introduced by means of either a hot or cold jet of gas. When a hot jet is used, an emission spectrum is obtained from the thin, heated surface layer. This technique is known as transient infrared emission spectroscopy (TIRES). When a cold jet is used, the blackbody-like thermal emission from the bulk of the sample is selectively absorbed as it passes through the thin, cooled surface layer. The result is a transmission spectrum convoluted with the observed thermal spectroscopy. This method is known as transient infrared transmission spectroscopy (TIRTS). TIRS is ideally suited for online analysis because it is a single-ended technique that requires no sample preparation. This technique has been applied to the lignin analysis of wood chips [83].

Fiber Optic Mid-IR Spectroscopy

Using optical fibers, mid-IR spectroscopy has been used for online analysis and remote sampling [84]. Fibers used in the mid-IR region are produced from oxides, chalcogenides, and halides of various elements. To be useful, the fibers must have IR radiation transmission capability over short distances. Like ATR crystals, fibers are based on the total reflection of radiation inside a material.

Spectral Interpretation

Since mid-IR spectroscopy has been used for a long time in lignin analysis, the IR features of lignin have been studied extensively. Most bands have been previously assigned [4,65], and it has been reported that the coupling of different vibrational modes exists in the spectra. Spectra of black spruce and aspen MWLs are typical of softwood and hardwood MWLs (Figure 4.2). These spectra were obtained in the

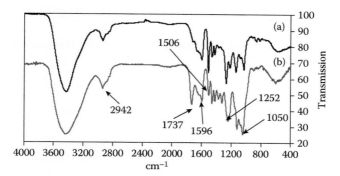

FIGURE 4.2 FT–IR spectra of MWLs obtained in the authors' laboratory: (a) black spruce; (b) aspen. Some IR bands in the aspen spectrum are annotated. The spectra differ with respect to band position, intensity, and shape.

authors' laboratory. Since there are wide variations in wood lignin structure and composition, wood spectra have significant differences (band positions, intensities, and shapes). The band assignments given in Table 4.2 were taken from Hergert [4] or Faix [65] unless otherwise indicated. Readers are referred to this original literature for details of band assignment.

In addition to the band assignments, mathematical approaches such as deconvolution [85,86], band fitting [87], and derivatization [88] have been used extensively in the literature as aids to interpreting spectra. Although these techniques are generally useful, some have limitations or potential for introducing artifacts [87]. The user of these approaches is cautioned and is advised to make informed choices.

APPLICATIONS

Lignin in Wood

A well-established use of mid-IR spectroscopy is to differentiate hardwoods and softwoods [90], based on the documented differences between hardwood and softwood lignin. It has also been used to quantitatively determine the amount of lignin in woods [83,91–93]. Examples of the use of FT-IR to study modification of wood lignin include weathering [94], photodegradation [95], fossilization [96,97], fungal treatment [98], and chemical reactions [99,100]. In addition, IR spectroscopy has been used *in situ* to determine functional groups in lignin. Studies of oak ray cell wall [101] and photodegradation were also carried out using infrared microscopy [102].

Lignin in Mechanical Pulps

Lignin-containing pulp and paper samples have been frequently analyzed by FT-IR (for a review see [103]). For instance, bleached [91,104,105], yellowed (both thermal and photo) [91,104–107]and biomechanical pulps [101] have been analyzed. Of the various methods of sampling, DRIFT has been used most often; photoacoustic, ATR, and micro methods have been used sparingly.

TABLE 4.2
Assignment of Bands in FT-IR Spectra of Softwood and Hardwood Milled-Wood Lignins

Guaiacyl L.[a] (cm⁻¹)	Guaiacyl Syringyl L.[b] (cm⁻¹)	Assignment[c]
3430 vs[d]	3440 vs	O–H stretch, H-bonded
2938 m	2942 m	C–H stretch methyl and methylene groups
2885 sh	2882 sh	C–H stretch in methyl and methylene groups
2849 sh	2848 sh	C–H stretch O–CH₃ group[e]
1717 sh	1737 vs	C=O stretch, unconjugated ketone, carboxyl, and ester groups
1667 sh	1670 sh	ring conj. C=O stretch of coniferaldehyde/sinapaldehyde
1645 sh	1643 sh	ring conjugated C=C stretch of coniferyl/sinapyl alcohol
1600s	1596 s	aryl ring stretching, symmetric
1513 vs	1506 vs	aryl ring stretching, asymmetric
1466 s	1464 s	C–H deformation, asymmetric
1458 sh	1425 m	O–CH₃ C–H deformation, asymmetric
1428 m	1379 m	Aromatic skeletal vibration combined with C–H in plane deformation
1375 w	1367 sh	O–CH₃ C–H deformation symmetric[e]
1331 sh	1330 m	Aryl ring breathing with C–O stretch
1270 vs	1252 vs	Aryl ring breathing with C=O stretch
1226 m	---[f]	C–C, C–O, and C=O stretches
1142 s	1159 sh	Aromatic C–H in plane deformation
---	1127 vs	Aromatic C–H in plane deformation
1085 w	1082 sh	C–O deformation, secondary alcohol and aliphatic ether
1035 s	1050 vs	Aromatic C–H in plane deformation
914 vw	905 w	C–H deformation out of plane, aromatic ring
878 sh	---	C–H deformation out of plane, aromatic ring
863 w	---	C–H deformation out of plane, aromatic ring
823 w	---	C–H deformation out of plane, aromatic ring
784 vw	---	CCH wag, mine from Raman
742 vw	---	skeletal deformation of aromatic rings, substituent groups, side chains[g]

[a] Frequencies of black spruce MWL.

[b] Frequencies of aspen MWL; 43% of units are syringyl type.

[c] From HL Hergert, in *Lignins: Occurrence, Formation, Structure and Reactions,* Wiley-Interscience, New York, 1971; O Faix, *Methods in Lignin Chemistry,* Springer-Verlag, Berlin, 1992, unless otherwise indicated.

[d] Note: *vs* is very strong; *s* strong; *m* medium; *w* weak; *vw* very weak; and *sh* shoulder. Band intensities are relative to other peaks in spectrum.

[e] From W Collier, VF Kalasinsky, and TP Schultz, *Holzforschung,* 51, 1997.

[f] No corresponding band was observed for guaiacyl-syringyl lignin.

[g] From D Lin-Vien, NB Colthup, WB Fateley, and JG Grasselli, *The Handbook of Infrared and Raman Characteristic Frequencies of Organic Molecules,* Academic Press, San Diego, 1991.

Both DRIFT and IR-PAS [81] have been used to investigate chemical changes in pulp lignin caused by the reductive or oxidative bleaching of mechanical pulps. The results showed a decline in aldehyde and ketone C=O band intensity (due to alkaline peroxide and sodium borohydride bleaching) and a reduction in contributions from conjugated carbonyl groups.

Light-induced modification of lignin in mechanical pulps has been thoroughly studied using FT-IR, both in pulps directly and in lignin isolated from yellowed pulps. Photoexposure resulted in the destruction of guaiacyl structures (in a softwood pulp) and led to production of carboxyl and/or ester groups [35,104]. IR-evidence supporting formation of p-quinone groups (upon photoyellowing) was also found [35]. Furthermore, the intensity of the 1727 cm^{-1} C=O band was linearly related to the post color number of the pulp [35]. FT-IR spectroscopy was also used to study chemical changes in the depth direction of paper exposed to light [106] on only one side. This study established that light-induced changes occurred to a depth of 50 to 100 μm, and that approximately 50% of the changes occurred at a depth of 20 μm. In addition to evaluating the effects of light- or heat-only treatments, FT-IR was applied to study mechanical pulps that were treated with both light and heat (light followed by heat and heat followed by light) [106].

Biomechanical straw pulps treated with enzyme mediator systems have been studied using mid-IR spectroscopy [108]. IR spectra indicated that a manganese-based enzyme system in combination with hydrogen peroxide generated the glucose oxidase couple, which removed aromatic ring structures.

A method based on an IR technique has been patented for determining lignin content (and other pulp properties) of pulp suspensions [109].

PAS-IR in conjunction with multivariate analysis has been applied to develop useful correlations to predict, among other things, methoxyl content and number of phenolic groups in a sample [110]. Another recently developed technique, 2D IR, was applied to study lignin-rich pulps [77]. The researchers concluded that lignin imparts strong viscoelastic behavior to pulp fibers, and the technique allows the study of lignin orientation.

Residual Lignin in Chemical Pulp

FT-IR has been used for both characterization [111] and quantitation [112,113] of residual lignin in chemical pulps. Due to the low lignin concentration, the infrared band at 1510 cm^{-1} is most suited for the latter purpose. Good correlations between kappa number and IR intensity were found for kraft pulps produced from a range of hardwoods and softwoods. The advantage of mid-IR analysis was that no sample preparation was required.

Isolated Lignins

When lignin is part of a multicomponent sample, it can often be analyzed *in situ*. However, the extent of information obtained is limited because of spectral overlap with other sample components and because the lignin is dispersed in the matrix. A spectrum of much higher quality can be obtained when isolated lignin is analyzed [114]. Depending on the source and the isolation procedure, the IR spectrum of lignin can vary significantly. For example, the spectrum of ball-milled enzyme lignin (isolated from cottonwood) is quite different from that of kraft indulin (mixed softwood) lignin [115,116]. The latter is isolated from the waste stream of the chemical pulping process and has thus undergone drastic structural changes.

Using IR band position, intensity, and shape, detailed studies of lignin have been conducted and the amounts of various functional groups (e.g., carbonyl, hydroxyl, and phenolic) have been determined. Isolated lignins have also been subjected to various chemical treatments (e.g., oxidation, acetylation, methylation, and oximation) and then analyzed by IR to understand how such treatments modify the IR spectrum of lignin.

In one investigation [117], 2-D correlation mid- and near-IR spectra of isolated lignins were used for qualitative spectral interpretation of the lignin from plant materials.

Lignin in Other Lignocellulosics

The determination of lignin content using IR techniques has also been conducted in plants other than wood. For example, the determination of kenaf lignin in plant materials using DRIFT has been described [118]. A linear relationship existed between the peak area at 1506 cm^{-1} and lignin content. The lignin content of four kenaf varieties (by weight of plant material) was 10.4–10.8% in the bark, 20.5–20.6% in the wood, and 14.9–15.3% in the pith.

Lignin in Solution

Infrared spectroscopy is seldom used to study lignin in solution, though the need for such a study occasionally arises; for example, for the analysis of spent pulping liquors or soluble lignin fractions. Faix [119] provides a review of sampling accessories and methods as well as information on the type of lignin samples analyzed.

Lignin Quantitation

FT-IR has played an important role in chemical analysis for decades. The current availability of numerous chemometric/multivariate techniques has also made IR spectroscopy a tool of quantitative measurement. Mid-IR spectroscopy answers the question of lignin content by performing either univariate or multivariate analysis. In the former case, the intensity of one lignin band is used for calculating the concentration. This is possible because observed IR band intensity is usually a linear function of lignin concentration. In such a calculation, the 1510 cm^{-1} band of lignin has almost always been used. In a multivariate analysis, multiple points in a spectrum are measured; this is done for each sample with varying concentration of lignin. Such data have been used to develop mathematical and statistical models capable of accurately predicting lignin concentration. The end goal of relating spectral changes to lignin concentration has thus been achieved.

NEAR-INFRARED SPECTROSCOPY

BACKGROUND

The region between the infrared and visible regions, from 4000 to 12500 cm^{-1}, is referred to as near-infrared. NIR vibrations of almost all materials are overtones and combination modes [120,121]. Most vibrations are modes of C–H, O–H, and N–H stretches. In a NIR spectrum, some absorption bands can be correlated with functional groups, although the bands are inherently weak compared to those of a mid-IR spectrum. However, an NIR spectrum also has information on composition, conformation, crystallinity, and inter- and intra-molecular interactions.

A review of NIR applications to lignin studies in 1992 [65] compared NIR spectra of milled softwoods and hardwoods and milled wood lignins (MWLs) (isolated from the same woods). There were numerous unresolved overlapping bands and there were only minor differences between the spectra. In addition, acetylation of MWL caused significant changes in the spectrum [65]. Another review of NIR spectroscopic studies of wood and wood products was published in 1995 [122]. Applications of NIR spectroscopy to lignin range from kappa number determination in chemical pulps [123] to lignin content in wood [124] and plant residues [125].

As for Raman and mid-IR spectroscopy, advances in data manipulation and the availability of chemometric analytical software have greatly assisted extraction of useful information from near-IR spectra. This is particularly important for NIR spectra because they are difficult to interpret in terms of chemical structure. An additional technical advance, fiber optics, has allowed the acquisition of NIR spectra by remote sensing.

INSTRUMENTATION

Early NIR work was performed using either a UV-Vis instrument with extension units for low wavenumbers or IR spectrometers with accessories for high wavenumbers. With these instruments, the quality of the collected spectrum was low. However, in modern times, good quality dispersive- and FT-NIR spectrometers exist that provide high quality spectra.

A NIR spectrometer can be built in a number of optical configurations. Analogous to the configuration of mid-IR, the basic configuration for NIR consists of a light source, a light dispersing element (grating), a sample, and a detector. In the region 1100–2500 nm, a lead sulfide (PbS) detector is usually used, whereas for the region below 1100 nm, PbS sandwiched with silicon photodiodes is used. Within a dispersive instrument, there are two modes of sample excitation, predispersive (light dispersed before falling on the sample) or post-dispersive. Moreover, spectrometers with Fourier transform (FT) capability are available in place of dispersive instruments. However, these spectrometers do not provide a spectrum with high signal-to-noise ratio. Filters (e.g., interference filter, turret-mounted filter) have also been used to obtain NIR radiation at various wavelengths. Another type of filter, the acousto-optic tuneable filter (AOTF), has been used as a diffraction grating. The AOTF relies on the acoustic diffraction of light in an anisotropic medium. Spectrometers are also developed with diode array detectors. Such instruments are very fast because they can detect multiple wavelengths simultaneously.

In the diffuse reflectance mode, two to four detectors can be used. The analog signal is averaged before being digitized. The computer records a signal representing wavelength and the reflectance or transmittance data, for both the sample and the reference. The final NIR spectrum is the difference between these two reflectance (transmittance) spectra.

Integrating spheres are also prevalent in NIR spectroscopy and have the advantage of improving the efficiency of signal collection. Workman and Burns [126] have reviewed NIR instrumentation.

SPECIAL TECHNIQUES AND INTERFACES

Samples are analyzed either in transmission or diffuse reflection mode. The latter is now a widely accepted method in quantitative analysis. Diffuse reflectance measurements penetrate about 1–4 mm of the front surface of samples; in nonhomogeneous samples, shallow penetration is the cause of variation in the spectrum. In transmission, on the other hand, the entire thickness of the sample is measured and error due to nonhomogeneity is minimized. These sampling methods are the same as those discussed in the section on mid-IR spectroscopy. For detailed information on diffuse reflection in NIR, the reader is referred to the literature [127].

Diffuse reflectance NIR spectra of black spruce and aspen MWLs are shown respectively in Figures 4.3 and 4.4. The two spectra are very similar.

APPLICATIONS

As direct spectral interpretation is limited in NIR spectroscopy, multivariate mathematical methods are used to obtain useful information. These techniques are used to develop mathematical models that correlate spectral features to properties of interest. For quantitative work, calibration models are needed that relate the concentration of a sample-analyte to spectral data. Information on developing calibration models and data analysis is provided elsewhere [128–132].

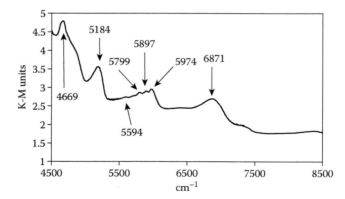

FIGURE 4.3 Near-infrared (NIR) spectrum, obtained in authors' laboratory, of black spruce milled wood lignin (MWL). Most peaks in the spectrum are annotated. The spectrum is similar to that of aspen (see Figure 4.4).

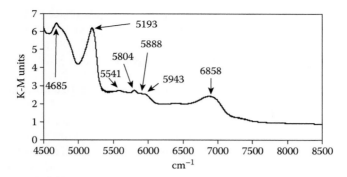

FIGURE 4.4 NIR spectrum of aspen MWL, obtained in authors' laboratory. Most peaks in the spectrum are annotated. The aspen and black spruce (Figure 4.3) spectra are very similar.

Lignin in Wood and Leaves

Schultz and Burns [133,134] applied NIR to determine lignin in pine and sweetgum (softwood and hardwood, respectively) and compared the two IR techniques for this application. They found that NIR was superior for determining lignin [134] with standard errors of calibration <1% over the lignin concentration range of 10–30%. FT-NIR has also been applied for wood identification in an examination of 90 samples of 12 different wood species [135]. The results showed that once the measuring unit had been calibrated, it was possible to identify new samples of the species previously tested. The authors concluded that differentiation between samples of a given wood species of different origins might be possible by NIR spectroscopy. A study for determining lignin content in wood by FT-NIR [136] concluded that this is an accurate, fast, and reliable method. The standard deviation was in the 0.11–0.14% range and root mean square error was <0.345%.

The quality of eucalypt woods for producing chemical pulps was evaluated using NIR spectra and chemometric methods [124]. NIR spectroscopy was used to predict pulp yield and cellulose content from spectra of powdered wood samples [137]. In another application, in addition to estimating lignin content, NIR spectra were used to quantify hardwood–softwood ratios in paperboard [138]. NIR spectra taken from solid European larch samples subjected to axial bending and compression tests revealed an excellent ability to model the variability of mechanical properties [139]. The study demonstrated that the model is based not only on the measurement of density, but also on surface geometry, composition, and, possibly, lignin content. The authors concluded that NIR spectroscopy shows considerable potential to become a tool for nondestructive evaluation of small clear wood specimens, e.g., increment cores.

NIR laboratory data of whole fresh leaves were evaluated with respect to leaf chemical composition [140]. NIR spectra were measured for 211 foliage samples, which included both broad- and needle-leafed species. Multiple linear regression analysis was used to determine if reflectance data from fresh leaf samples contained information on nitrogen, lignin, and cellulose concentrations. Calibration equations were developed for all leaf constituents, indicating that information on leaf biochemistry is present in the spectra of fresh as well as dried ground leaf samples.

Residual Lignin in Chemical Pulp

NIR spectroscopy has been used to investigate residual lignin and chemical pulps. In a study of delignification, birch chemical pulp was analyzed and a PLS model for Klason lignin was developed [141]. Compared to other methods, NIR was the most reliable (standard error <1%). The range of Klason lignin measurement was 0–25%. In a study of the kappa number of batch kraft pulp, different NIR sampling methods were compared [142]. The best results were obtained with homogenized, dried samples. The authors concluded that kappa could be measured (with a 95% confidence interval) consistently with ±2 kappa numbers. NIR was also used to predict kappa number, hexeneuronic acid, and Klason lignin in hardwood kraft pulp [143]. For the first two parameters, no difference was noted between dry and semidry samples; for Klason lignin, dry samples provided somewhat higher values. Finally, black liquors from alkaline pulping were analyzed and the concentration of the main liquor components (including lignin) were established [144].

Lignin in Other Lignocellulosics

NIR spectra of dry ground leaf samples were analyzed to obtain an unmixed spectrum of lignin and other forest foliage components [145]. In another application, NIR was used for remote sensing in the measurement of lignin, cellulose, and nitrogen concentrations in forest canopies [146]. The relative value of near- and mid-infrared diffused reflectance spectroscopy in determining the composition (including lignin) of forages and by-products has been evaluated [147]. Sixty-seven samples consisting of 15 alfalfa, 16 tall fescue, and 15 orchard grass hay samples, 10 corn stover samples, and 11 wheat straw samples at various stages of maturity were examined by these techniques. The results showed that diffuse mid-IR reflectance spectroscopy can perform as well as, and sometimes better than, diffuse NIR spectroscopy in determining the composition of forages and by-products. In addition, FT-NIR spectroscopy did not perform as well as either NIR spectroscopy using a scanning monochromator or FT-IR spectroscopy. Finally, diluting samples with potassium bromide was not beneficial for either of the Fourier-based determinations.

Other lignin applications of NIR included research on flax fiber [148], compost prepared from wheat straw and chicken litter [149], hard red winter and spring wheat [150], and degradation of plant cell walls by white-rot fungi [151].

CONCLUDING REMARKS

Raman, mid-IR, and near-IR techniques have played an important role in lignin analysis. These techniques provide fundamental knowledge at the molecular level, and will likely continue to be used in this way. Recent developments in Raman instrumentation have made this method even more versatile and user-friendly. Specific techniques such as FT-Raman, resonance Raman, SER, and Raman mapping are all expected to be widely used as compact, easy to use, integrated instruments become commercially available at reasonable cost. Historically, mid-IR spectroscopy has been used most extensively to analyze lignin due to the early development of the techniques, user familiarity and the existence of widely available, low-cost instrumentation. Between mid- and near-IR, the latter continues to see more vigorous developments in instrumentation and applications. Development of diffuse reflectance techniques and application of chemomertic methods have been two major advances in this area. Both are widely used in the quantitative analysis of lignocellulosic materials. The future of the lignin vibrational spectroscopy is bright and stimulating.

REFERENCES

1. RH Atalla, UP Agarwal, and JS Bond. Raman Spectroscopy. In C Dence and SY Lin, eds. *Methods in Lignin Chemistry*. Berlin: Springer-Verlag, 1992, pp. 162–176.
2. D Lin-Vien, NB Colthup, WB Fateley, and JG Grasselli. *The Handbook of Infrared and Raman Characteristic Frequencies of Organic Molecules*. San Diego: Academic Press, 1991.
3. WG Glasser. Classification of Lignin According to Chemical and Molecular Structure. In WG Glasser, RA Northey, and TP Schultz, eds. *Lignin: Historical, Biological and Materials Perspectives*. ACS Symposium Series, 742. Washington, DC: American Chemical Society, 2000, pp. 216–238.
4. HL Hergert. Infrared Spectra. In KV Sarkanen and CH Ludwig, eds. *Lignins: Occurrence, Formation, Structure and Reactions*. New York: Wiley-Interscience, 1971, pp. 267–297.
5. UP Agarwal. An Overview of Raman Spectroscopy as Applied to Lignocellulosic Materials. In DS Argyropoulos, ed. *Advances in Lignocellulosics Characterization*. Atlanta: TAPPI Press, 1999, pp. 201–225.
6. G Varsanyi. *Vibrational Spectra of Benzene Derivatives*. New York: Academic Press, 1969.
7. SM Ehrhardt. An Investigation of the Vibrational Spectra of Lignin Model Compounds. Ph.D. thesis, Georgia Institute of Technology, 1984.
8. UP Agarwal and RH Atalla. Raman Spectroscopic Evidence for Coniferyl Alcohol Structure in Bleached and Sulfonated Mechanical Pulps. In C Heitner and JC Scaiano, eds. *Photochemistry of Lignocellulosic Materials*. ACS Symposium Series, 531. Washington, DC: American Chemical Society, 1993, pp. 26–44.
9. UP Agarwal and RH Atalla. Using Raman Spectroscopy to Identify Chromophores in Lignin-Lignocellulosics. In WG Glasser, RA Northey, and TP Schultz, eds. *Lignin: Historical, Biological and Material Perspectives*. ACS Symposium Series, 742. Washington, DC: American Chemical Society, 2000, pp. 250–264.
10. UP Agarwal, RS Reiner, UP Agarwal, AK Pandey, SA Ralph, KC Hirth, and RH Atalla. Raman Spectra of Lignin Model Compounds. 13th International Symposium on Wood, Fibre and Pulping Chemistry, Auckland NZ, 2005, pp. 377–384.
11. U Agarwal and RH Atalla. Oxygen Sensitive Background in the Raman Spectra of Woody Tissue. 10th International Conference on Raman Spectroscopy, Eugene OR, 1986, pp. 14–46.
12. UP Agarwal and RH Atalla. In-Situ Raman Microprobe Studies of Plant Cell Walls: Macromolecular Organization and Compositional Variability in the Secondary Wall of Picea Mariana (Mill) B.S.P. *Planta* 169:325–332, 1986.

13. RH Atalla and UP Agarwal. Raman Microprobe Optimization and Sampling Technique for Studies of Plant Cell Walls. In AD Romig and DI Goldstein, eds. *Microbeam Analysis.* San Francisco: San Francisco Press, 1984, pp. 125–126.

14. RH Atalla and UP Agarwal. Raman Microprobe Evidence for Lignin Orientation in Cell Walls of Native Woody Tissue. *Science* 227:636–638, 1985.

15. RH Atalla and UP Agarwal. Recording Raman Spectra from Plant Cell Walls. *J. Raman Spectro.* 17:229–231, 1986.

16. T Hirschfeld and DB Chase. FT-Raman Spectroscopy: Development and Justification. *Appl. Spectrosc.* 40:133–137, 1986.

17. DA Long. *Raman Spectroscopy.* New York: McGraw Hill, 1977.

18. SK Freeman. *Applications of Laser Raman Spectroscopy.* New York: John Wiley, 1974, pp. 45.

19. PJ Hendra, C Jones, and G Warnes. *FT-Raman Spectroscopy.* Chicester, UK: Ellis-Horwood, 1991.

20. G Turrell, M Delhaye, and P Dhamelincourt. Characteristics of Raman Microscopy. In G Turrell and J Corset, eds. *Raman Microscopy: Developments and Applications.* San Diego: Academic Press, 1996, pp. 22–49.

21. J Barbillat. Raman Imaging. In G Turrell and J Corset, eds. *Raman Microscopy: Developments and Applications.* San Diego: Academic Press, 1996, pp. 175–200.

22. UP Agarwal. Raman Imaging to Investigate Ultrastructure and Composition of Plant Cell Walls: Distribution of Lignin and Cellulose in Black Spruce (Picea mariana). *Planta* 224:1141–1153, 2006.

23. N Gierlinger and M Schwanninger. Chemical Imaging of Poplar Wood Cell Walls by Confocal Raman Microscopy. *Plant Physiology* 140:1246–1254, 2006.

24. JS Bond. Raman Microspectroscopic Investigation of Patterns of Molecular Order in the Secondary Walls of Black Spruce and Loblolly Pine Tracheids. Ph.D. thesis, Georgia Institute of Technology, 1991.

25. M Halttunen, J Vyörykkä, B Hortling, T Tamminen, D Batchelder, A Zimmerman, and T Vuorinen. Study of Residual Lignin in Pulp by UV Resonance Raman Spectroscopy. *Holzforschung* 55:631–638, 2001.

26. ED Schmid and RD Topsom. Raman Intensity and Conjugation. 5. A Quantitative Relationship between Raman Intensity and the Length of Conjugation and an Analysis of the Raman Intensities of Some Substituted Benzenes and Biphenyls. *J. Am. Chem. Soc.* 103:1628–1625, 1981.

27. M Fleischmann, PJ Hendra, and AJ McQuillan. Raman Spectra from Electrode Surfaces. *Chem. Comm.* 80–81, 1973.

28. K Kneipp. Surface-Enhanced Raman Scattering and Related Effects. *Exp. Tech. Phys.* 38:3–28, 1990.

29. UP Agarwal, RS Reiner, and SA Ralph. Using Nano- and Micro-Particles of Silver in Lignin Analysis. TAPPI International Conference on Nanotechnology, Atlanta, 2006, pp. NanoCD-06.

30. UP Agarwal and SA Ralph. FT-Raman Spectroscopy of Wood: Identifying Contributions of Lignin and Carbohydrate Polymers in the Spectrum of Black Spruce (Picea mariana). *Appl. Spectrosc.* 51:1648–1655, 1997.

31. UP Agarwal, JD McSweeny, and SA Ralph. An FT-Raman Study of Softwood, Hardwood, and Chemically Modified Black Spruce MWLs. 10th International Symposium on Wood and Pulping Chemistry, Yokohama, 1999, 2, pp. 136–140.

32. PA Evans. Differentiating "Hard" from "Soft" Woods Using Fourier Transform Infrared and Fourier Transform Raman Spectroscopy. *Spectrochim. Acta* A 47:1441–1447, 1991.

33. UP Agarwal and RH Atalla. Raman Spectral Features Associated with Chromophores in High-Yield Pulps. *J. Wood Chem. Technol.* 14:227–241, 1994.

34. UP Agarwal, RH Atalla, and I Forsskähl. Sequential Treatment of Mechanical and Chemimechanical Pulps with Light and Heat. *Holzforschung* 49:300–312, 1995.
35. UP Agarwal and JD McSweeny. Photoyellowing of Thermomechanical Pulps: Looking beyond Alpha-Carbonyl and Ethylenic Groups as the Initiating Structures. *J. Wood Chem. Technol.* 17:1–26, 1997.
36. UP Agarwal and LL Landucci. FT-Raman Investigation of Bleaching of Spruce Thermomechanical Pulp. *J. Pulp Paper Sci.* 30:269–274, 2004.
37. UP Agarwal, SA Ralph and RH Atalla. FT Raman Spectroscopic Study of Softwood Lignin. 9th International Symposium on Wood and Pulping Chemistry, Montreal, 1997, pp. 8-1–8-4.
38. UP Agarwal and N Terashima. FT-Raman Study of Dehydrogenation Polymer (DHP) Lignins. 12th International Symposium on Wood, Fiber and Pulping Chemistry, Madison, WI, 2003, 3, pp. 123–126
39. AM Saariaho, DS Argyropoulos, AS Jääskeläinen, and T Vuorinen. Development of the Partial Least Squares Models for the Interpretation of the UV Resonance Raman Spectra of Lignin Model Compounds. *Vibrational Spectrosc.* 37:111–121, 2005.
40. IR Lewis, NW Daniel, NC Chaffin, and PR Griffiths. Raman Spectrometry and Neural Networks for the Classification of Wood Types. 1. *Spectrochim. Acta A* 50:1943–1958, 1994.
41. BK Lavine, CE Davidson, AJ Moores, and PR Griffiths. Raman Spectroscopy and Genetic Algorithms for the Classification of Wood Types. *Appl. Spectrosc.* 55:960–966, 2001.
42. T Ona, T Sonoda, M Shibata, T Kato, and Y Ootake. Non-destructive Determination of Wood Constituents by Fourier Transform Raman Spectroscopy. *J. Wood Chem. Technol.* 17:399–417, 1997.
43. T Ona, T Sonoda, K Ito, M Shibata, T Katayama, T Kato, and Y Ootake. Non-destructive Determination of Lignin Syringyl/Guaiacyl Monomeric Composition in Native Wood by Fourier Transform Raman Spectroscopy. *J. Wood Chem. Technol.* 18:43–51, 1998.
44. T Takei, T Iijima, M Higaki, and T Fukuzumi. The Properties and Chemical Components of Mangroves. V. Lignin Carbohydrate Complexes from Bruguiera Gymnorrhiza wood. *Mokuzai Gakkaishi* 40:868–873, 1994.
45. UP Agarwal and M Akhtar. Understanding Fungus-Induced Brightness Loss of Biomechanical Pulps. TAPPI Pulping/Process and Product Quality Conference, Boston, 2000, pp. 1229–1240.
46. VC Tirumalai, U Agarwal, and JR Obst. Heterogeneity of Lignin Concentration in Cell-corner Middle Lamella of White Birch. *Wood Sci. Tech.* 30:99–104, 1996.
47. UP Agarwal. Assignment of the Photoyellowing-Related 1675 Cm^{-1} Raman/IR Band to *P*-Quinones and Its Implications to the Mechanism of Color Reversion in Mechanical Pulps. *J. Wood Chem. Technol.* 18:381–402, 1998.
48. UP Agarwal and RH Atalla. Raman Spectroscopy. In TE Conners and S Banerjee, eds. *Surface Analysis of Papers.* Boca Raton: CRC Press, 1995, pp. 152–181.
49. JS Bond, RH Atalla, U Agarwal, and CG Hunt. The Aging of Lignin Rich Papers Upon Exposure to Light: Its Quantification and Prediction. 10th International Symposium on Wood and Pulping Chemistry, Yokohama, 1999, pp. 500–504.
50. JS Bond, X Yu, U Agarwal, RH Atalla, and CG Hunt. The Aging of Printing and Writing Papers upon Exposure to Light. Part I: Optical and Chemical Changes Due to Long-Term Light Exposure. 11th International Symposium on Wood and Pulping Chemistry, Nice, 2001, pp. 209–213.
51. AH Kuptsov. Fourier Transform Raman Spectroscopic Investigation of Paper. *Vibrational Spectrosc.* 7:185–190, 1994.

52. IA Weinstock, RH Atalla, U Agarwal, JL Minor, and CJ Petty. Fourier Transform Raman Spectroscopic Studies of a Novel Wood Pulp Bleaching System. *Spectrochim. Acta* A 49:819–829, 1993.

53. UP Agarwal, IA Weinstock, and RH Atalla. FT Raman Spectroscopy for Direct Measurement of Lignin Concentrations in Kraft Pulps. *Tappi J.* 2:22 –26, 2003.

54. A Ibrahim, PB Oldham, TE Connors, and TP Schultz. Rapid Characterization of Wood Pulp Lignin by Fourier Transform Raman Spectroscopy. *Microchemical J.* 56:393–402, 1997.

55. T Takei, T Kato, T Iijima, and M Higaki. Raman Spectroscopic Analysis of Wood and Bamboo Lignin. *Mokuzai Gakkaishi* 41:229–236, 1995.

56. UP Agarwal and RH Atalla. FT Raman Spectroscopy: What It is and What It Can do for Research on Lignocellulosic Materials. 8th International Symposium on Wood and Pulping Chemistry, Helsinki, 1995, pp. 67–72.

57. DS Himmelsbach and DE Akin. Near-Infrared Fourier-Transform Raman Spectroscopy of Flax (Linum usitatissimum L.) *Stems. J. Agric. Food Chem.* 46:991–998, 1998.

58. JG Taylor, TP Owen Jr, LT Koonce, and CH Haigler. Dispersed Lignin in Tracheary Elements Treated with Cellulose Synthesis Inhibitors Provides Evidence that Molecules of the Secondary Cell Wall Mediate Wall Patterning. *The Plant Journal* 2:959–970, 1992.

59. IR Lewis. Process Raman Spectroscopy. In IR Lewis and HGM Edwards, eds. *Handbook of Raman Spectroscopy*. New York: Marcel Dekker, 2001, pp. 919–973.

60. RH Atalla, JS Bond, and CP Woitkovich. Raman Spectroscopic Studies of Lignin in Native Woody Tissue. 4th International Symposium on Wood and Pulping Chemistry, Paris, 1987, pp. 437–439.

61. CJ Petty. Self-Absorption in Near-Infrared Fourier Transform Raman Spectrometry. *Vibrational Spectrosc.* 2:263–268, 1991.

62. UP Agarwal and N Kawai. Self-Absorption Phenomenon in Near-Infrared Fourier Transform Raman Spectroscopy of Cellulosic and Lignocellulosic Materials. *Appl. Spectrosc.* 59:385–389, 2005.

63. EL Jones. The Infrared Spectrum of Native Spruce Lignin. *Tappi J.* 32:167–170, 1949.

64. EL Jones. The Infrared Spectrum of Spruce Native Lignin. *J. Am. Chem. Soc.* 70:1984–1985, 1948.

65. O Faix. Characterization in the Solid State, Fourier Transform Infrared Spectroscopy. In SY Lin and C Dence eds. *Methods in Lignin Chemistry*. Berlin: Springer-Verlag, 1992, pp. 83–109.

66. PR Griffiths and JA deHasbeth. *Fourier Transform Infrared Spectroscopy*. New York: John Wiley, 1986.

67. G Kortüm. *Reflectance Spectroscopy*. Berlin: Springer-Verlag, 1969.

68. SR Culler. Sampling techniques For Qualitative/Quantitative Analysis of Solids. In PB Coleman, ed. *Practical Sampling Techniques: Techniques for Infrared Analysis*. Boca Raton: CRC Press, 1993, pp. 107–144.

69. FM Mirabella. *Internal Reflection Spectroscopy, Theory and Application*. Practical Spectroscopy Series, 15. New York: Marcel Dekker, 1993.

70. HJ Harthcock and RG Messerschmidt. *Infrared Microspectroscopy Theory and Applications*. New York: Marcel Dekker, 1988.

71. HJ Humecki. *Practical Guide to Infrared Microspectroscopy*. Practical Spectroscopy Series, 19. New York: Marcel Dekker, 1995.

72. N Duran and R Angelo. Infrared Microspectroscopy in the Pulp and Paper-Making Industry. *Appl. Spectrosc. Rev.* 33:219–236, 1998.

73. R Bhargava, T Ribar, and JL Koenig. Towards Faster FT-IR Imaging by Reducing Noise. *Appl. Spectrosc.* 53:1313–1322, 1999.

74. EN Lewis, PJ Treado, RC Reeder, GM Story, AE Dowrey, C Marcott, and IW Levin. Fourier Transform Spectroscopic Imaging Using an Infrared Focal-Plane Array Detector. *Anal. Chem.* 67:3377–3381, 1995.

75. I Noda. Two-Dimensional Infrared (2D IR) Spectroscopy: Theory and Applications. *Appl. Spectrosc.* 44:550–561, 1990.

76. C Marcott, I Noda, and AE Dowrey. Enhancing the Information Content of Vibrational Spectra through Sample Perturbation. *Anal. Chim. Acta* 250:131–143, 1991.

77. M Akerholm and L Salmen. The Oriented Structure of Lignin and Its Viscoelastic Properties Studied by Static and Dynamic FT-IR Spectroscopy. *Holzforschung* 57:459–465, 2003.

78. JF McClelland. Photoacoustic Spectroscopy. *Anal. Chem.* 55:89A–105A, 1983.

79. JF McClelland, RW Jones, S Lou, and LM Seaverson. A Practical Guide to FT-IR Photoacoustic Spectroscopy. In PB Coleman, ed. *Practical Sampling Techniques: Techniques for Infrared Analysis.* Boca Raton: CRC Press, 1993, pp. 107–144.

80. JF McClelland, RW Jones, and SJ Bajic. FT-IR Photoacoustic Spectroscopy. In JM Chalmers and PR Griffiths, eds. *Handbook of Vibrational Spectroscopy.* Chichester: John Wiley, 2002, pp. 1231–1251.

81. SGT St. Germain and DG Gray. Photoacoustic Fourier Transform Infrared Spectroscopic Study of Mechanical Pulp Brightening. *J. Wood Chem. Technol.* 7:33–50, 1987.

82. RW Jones and JF McClelland. Quantitative Analysis of Solids in Motion by Transient Infrared Emission Spectroscopy Using Hot-Gas Jet Excitation. *Anal. Chem.* 62:2074–2079, 1990.

83. RW Jones, RR Meglen, BR Hames, and JF McClelland. Chemical Analysis of Wood Chips in Motion Using Thermal-Emission Mid-Infrared Spectroscopy with Projection to Latent Structures Regression. *Anal. Chem.* 74:453–457, 2002.

84. MA Thomson, PJ Melling, and AM Slepski. Real Time Monitoring of Isocyanate Chemistry Using a Fiber-Optic FTIR Probe. *Polymer Preprints* 42:310–311, 2001.

85. JK Kauppinen, DJ Moffatt, HH Mantsch, and DG Cameron. Fourier Self-Deconvolution: A Method for Resolving Intrinsically Overlapped Bands. *Appl. Spectrosc.* 35:271–276, 1981.

86. W-J Yang, PR Griffiths, DM Byler, and H Susi. Protein Conformation by Infrared Spectroscopy: Resolution Enhancement by Fourier Self-Deconvolution. *Appl. Spectrosc.* 39:282–287, 1985.

87. WF Maddams. The Scope and Limitations of Curve Fitting. *Appl. Spectrosc.* 34:245–267, 1980.

88. JA Pierce, RS Jackson, W Van Every, PR Griffiths, and G Hongjin. Combined Deconvolution and Curve Fitting for Quantitative Analysis of Unresolved Spectral Bands. *Anal. Chem.* 62:477–484, 1990.

89. W Collier, VF Kalasinsky, and TP Schultz. Infrared Study of Lignin: Assignment of Methoxyl C–H Bending and Stretching Bands. *Holzforschung* 51:167–168, 1997.

90. NL Owen and DW Thomas. Infrared Studies of "Hard" and "Soft" Woods. *Appl. Spectrosc.* 43:451–455, 1989.

91. TP Schultz, MC Templeton, and GD McGinnis. Rapid Determination of Lignocellulose by Diffuse Reflectance Fourier Transform Infrared Spectrometry. *Anal. Chem.* 57:2867–2869, 1985.

92. J Rodrigues, O Faix, and H Pereira. Determination of Lignin Content of Eucalyptus Globulus Wood Using FTIR Spectroscopy. *Holzforschung* 46:46–50, 1998.

93. J Costa e Silva, BJ Nielsen, J Rodrigues, H Pereira, and H Wellendorf. Rapid Determination of the Lignin Content in Sitka Spruce (Picea sitchensis (Bong.) Carr.) Wood by Fourier Transform Infrared Spectrometry. *Holzforschung* 53:597–602, 1999.

94. EL Anderson, Z Pawlak, NL Owen, and WC Feist. Infrared Studies of Wood Weathering. Part I: Softwoods. *Appl. Spectrosc.* 45:641–647, 1991.

95. L Tolvaj and O Faix. Artificial Ageing of Wood Monitored by DRIFT Spectroscopy and CIE L[*]a[*]b[]Color Measurements. I: Effect of UV Light. *Holzforschung* 49:397–404, 1995.
96. JR Obst, NJ McMillan, DJ Blanchette, DJ Christensen, O Faix, JS Han, TA Kuster, L Landucci, RH Newman, RC Pettersen, VH Schwandt, and MG Wesolowski. Characterization of Canadian Arctic Fossil Woods. 5th International Symposium on Wood and Pulping Chemistry, Raleigh, 1989, Poster Sessions, pp. 289–308
97. JR Obst, NJ McMillan, DJ Blanchette, DJ Christensen, O Faix, JS Han, TA Kuster, L Landucci, RH Newman, RC Pettersen, VH Schwandt, and MG Wesolowski. Characterization of Canadian Arctic Fossil Woods. In RJ Christie and NJ McMillan, eds. *Tertiary Fossil Forests of the Geodetic Hills, Axel Heiberg Islands, Arctic Archipelago.* Bull. 403, Geological Survey of Canada, 1991, pp. 123–146.
98. S Backa, A Brolin, and T Nilsson. Characterisation of Fungal Degraded Birch Wood by FTIR and Py-GC. *Holzforschung* 55:225–232, 2001.
99. NL Owen and Z Pawlak. An Infrared Study of the Effect of Liquid Ammonia on Wood Surfaces. *J. Mol. Struct.* 198:435–449, 1993.
100. AJ Michell. FTIR Spectroscopic Studies of the Reactions of Wood and of Lignin Model Compounds with Inorganic Agents. *Wood Sci. Tech.* 27:69–80, 1993.
101. TP Abbott, FC Felker, and R Kleiman. FT-IR Microspectroscopy: Sample Preparation and Analysis of Biopolymers. *Appl. Spectrosc.* 47:180–189, 1993.
102. Y Kataoka and M Kiguchi. Depth Profiling of Photo-Induced Degradation in Wood by FT-IR Microspectroscopy. *J. Wood Sci.* 47:325–327, 2001.
103. JJ Workman. Infrared and Raman Spectroscopy in Paper and Pulp Analysis. *Appl. Spectrosc. Rev.* 36:139–168, 2001.
104. I Forsskähl and J Janson. Sequential Treatment of Mechanical and Chemimechanical Pulps with Light and Heat. Part 2. FTIR and UV-vis Absorption-Scattering Spectra. *Nord. Pulp Paper Res. J.* 7:48–54, 1992.
105. H Lennholm, M Rosenqvist, M Ek, and T Iversen. Photoyellowing of Groundwood Pulps. *Nord. Pulp Paper Res. J.* 9:10–15, 1994.
106. I Forsskähl, E Kentta, P Kyyronem, and O Sundstrom. Depth Profiling of a Photochemically Yellowed Paper. Part II. FT-IR Techniques. *Appl. Spectrosc.* 49:163–170, 1995.
107. F Kimura, T Kimura, and DG Gray. FT-IR Study of the Effect of Irradiation Wavelength on the Colour Reversion of Thermomechanical Pulps. *Holzforschung* 48:343–348, 1994.
108. T Vares, A Hatakka, J Dorado, G Almendros, P Bocchini, GC Galletti, and AT Martinez. Microhandsheet Evaluation and Chemical Analysis of Wheat-Straw Pulp Treated with Enzyme-Mediator Systems. 7th International Conference on Biotechnology in the Pulp and Paper Industry, Montreal, 1998, A, pp. 149–152.
109. H Furumoto, U Lampe, and C Roth. Analysis of Cellulose Characteristics. *Geman* 19, 613, 985, 1995.
110. G Gellerstedt and T Josefsson. Use of FT Spectroscopy and Chemometrics for Analysis of Wood Components. 3rd European Workshiop on Lignocellusics and Pulp, Stockholm, 1994, pp. 179–182.
111. B Hortling, T Tamminen, and E Kentta. Determination of Carboxyl and Non-conjugated Carbonyl Groups in Dissolved and Residual Lignins by IR Spectroscopy. *Holzforschung* 51:405–410, 1997.
112. S Berben, JP Rademacher, LO Sell, and DB Easty. Estimation of Lignin in Wood Pulp by Diffuse Reflectance Fourier-Transform Infrared Spectrometry. *Tappi J.* 70:129–133, 1987.
113. MA Friese and S Banerjee. Lignin Determination by FT-IR. *Appl. Spectrosc.* 46:246–248, 1992.

114. O Faix and O Beinhoff. Ftir Spectra of Milled Wood Lignins and Lignin Polymer Models (DHP's) with Enhanced Resolution Obtained by Deconvolution. *J. Wood Chem. Technol.* 8:505–522, 1988.

115. O Faix, DS Argyropoulos, D Robert, and V Neirinck. A Determination of Hydroxyl Groups in Lignins: Evaluation of ^{1}H-, ^{13}C-, ^{31}P-NMR, FTIR and Wet Chemical Methods. *Holzforschung* 48:387–394, 1998.

116. O Faix, B Andersons, and G Zakis. Determination of Carbonyl Groups of Six Round Robin Lignins by Modified Oximation and FTIR Spectroscopy. *Holzforschung* 52:268–274, 1998.

117. FE Barton II and DS Himmelsbach. Two-Dimensional Vibrational Spectroscopy II: Correlation of the Absorptions of Lignins in the Mid- and Near-Infrared. *Appl. Spectrosc.* 47:1920–1925, 1993.

118. C Pappas, PA Tarantilis, and M Polissiou. Determination of Kenaf (Hibiscus cannabinus L.) Lignin in Crude Plant Material Using Diffuse Reflectance Infrared Fourier Transform Spectroscopy. *Appl. Spectrosc.* 52:1399–1402, 1998.

119. O Faix. Characterization in Solution: Fourier Transform Infrared Spectroscopy. In SY Lin and C Dence, eds. *Methods in Lignin Chemistry.* Berlin: Springer-Verlag, 1992, pp. 233–241.

120. R Goddu and D Delker. Spectra-Structure Correlations for Near-Infrared Region. *Anal. Chem.* 32:140–141, 1960.

121. BG Osborne and T Fearn. *Near Infrared Analysis in Food Analysis.* New York: Wiley, 1986, pp. 200.

122. JM Pope. Near-Infrared Spectroscopy of Wood Products. In TE Conners and S Banerjee, eds. *Surface Analysis of Paper.* Boca Raton: CRC Press, 1995, pp. 142–151.

123. MD Birkett and MJT Gambino. Estimation of Pulp Kappa Number with Near-Infrared Spectroscopy. *Tappi J.* 72:193–197, 1989.

124. AJ Michell. Pulpwood Quality Estimation by Near-Infrared Spectroscopic Measurements on Eucalypt Woods. *Appita* 48:425–428, 1995.

125. JB Reeves III. Infrared Spectroscopic Studies on Forage and By-product Fibre Fractions and Lignin Determination Residues. *Vibrational Spectrosc.* 5:303–310, 1993.

126. JJ Workman and DA Burns. Commercial NIR Instrumentation. In DA Burns and EW Ciurczak, eds. *Handbook of Near Infrared Analysis.* New York: Marcel Dekker, 2001, pp. 53–71.

127. JM Olinger, PR Griffiths, and T Burger. Theory of Diffuse Reflection in the NIR Region. In DA Burns and EW Ciurczak, eds. *Handbook of Near Infrared Analysis.* New York: Marcel Dekker, 2001, pp. 19–52.

128. BG Osborne, T Fearn, and PH Hindle. *Practical NIR Spectroscopy with Applications in Food and Beverage Analysis.* Harlow, Essex, England.: Longman Scientific and Technical, 1993, pp. 99–144.

129. JJ Workman. NIR Spectroscopy Calibration Basics. In DA Burns and EW Ciurczak, eds. *Handbook of Near Infrared Analysis.* New York: Marcel Dekker, 2001, pp. 91–128.

130. H Mark. Multilinear Regression and Principal Component Analysis. In DA Burns and EW Ciurczak, eds. *Handbook of Near Infrared Analysis.* New York: Marcel Dekker, 2001, pp. 129–184.

131. H-R Bjorsvik and H Martens. Data Analysis: Calibration of NIR Instruments by PLS Regression. In DA Burns and EW Ciurczak, eds. *Handbook of Near Infrared Analysis.* New York: Marcel Dekker, 2001, pp. 185–208.

132. H Martens and T Naes. Multivariate Calibration by Data Compression. In P Williams and K Norris, eds. *Near-Infrared Technology in the Agriculture and Food Industries.* St. Paul: Amer. Assoc. Cereal Chem., 1987, pp. 57–105.

133. TP Schultz and DA Burns. Wood Chemistry: Rapid Secondary Analysis of Lignocellulose: Comparison of Near Infrared (NIR) and Fourier Transform Infrared (FTIR). *Tappi J.* 73:209–212, 1990.

134. DA Burns and TP Schultz. FT-IR versus NIR: A Study with Lignocellulose. In DA Burns and EW Ciurczak, eds. *Handbook of Near Infrared Analysis*. New York: Marcel Dekker, 2001, pp. 563–572.

135. M Brunner, R Eugster, E Trenka, and Bergamin-Strotz. FT-NIR Spectroscopy and Wood Identification. *Holzforschung* 50:130–134, 1996.

136. M Schwanninger and B Hinterstoisser. Determination of the Lignin Content in Wood by FT-NIR. 11th International Symposium on Wood and Pulping Chemistry, Nice, 2001, III, pp. 637–640.

137. JA Wright, MD Birkett, and MJT Gambino. Prediction of Pulp Yield and Cellulose Content from Wood Samples Using Near-Infrared Reflectance Spectroscopy. *Tappi J.* 73:164–166, 1990.

138. DB Easty, SA Berben, FA DeThomas, and PJ Brimmer. Near-Infrared Spectroscopy for the Analysis of Wood Pulping: Quantifying Hardwood-Softwood Mixtures and Estimating Lignin Content. *Tappi J.* 73:257–261, 1990.

139. W Gindl, A Teischinger, M Schwanninger, and B Hinterstoisser. The Relationship between Near Infrared Spectra of Radial Wood Surfaces and Wood Mechanical Properties. *Near Infrared Spectrosc.* 9:2001.

140. ME Martin and JD Aber. Analyses of Forest Foliage III: Determining Nitrogen, Lignin and Cellulose in Fresh Leaves Using Near Infrared Reflectance Data. *Near Infrared Spectrosc.* 2:25–32, 1994.

141. L Wallbacks, U Edlund, B Norden, and I Berglund. Multivariate Characterization of Pulp Using Solid-State C(13)-NMR, FTIR, and NIR [Near-IR Spectroscopy]. *Tappi J.* 74:201–206, 1991.

142. E Yuzak and C Lohrke. At-Line Kappa Number Measurement by Near-Infrared Spectroscopy. TAPPI Pulping Conference, Atlanta, 1993, pp. 663–672.

143. E Sjöholm, F Lunqvist, T Liljenberg, and S Backa. Prediction of Kappa Number, Hexenuronic Acid and Klason Lignon in Pulps by NIR Spectroscopy. 11th International Symposium on Wood and Pulping Chemistry, Nice, 2001, III.

144. U Lampe, H Meixner, J Mühlsteff, R Pastusiak, H Furomoto, H Hartenstein, M Amaral, A Brochado, and D Trancoso. Kappa Number Prediction Based on NIR Spectroscopy of Black Liquors from Alkaline Pulping for Process Optimisation. 11th International Symposium on Wood and Pulping Chemistry, Nice, 2001, I, pp. 367–370.

145. CA Hlavka and DL Peterson. Analysis of Forest Foliage Spectra Using a Multivariate Mixture Model. *Near Infrared Spectrosc.* 5:167–173, 1997.

146. JD Aber, KL Bolster, SD Newman, M Soulia, and ME Martin. Analyses of Forest Foliage II: Measurement of Carbon Fraction and Nitrogen Content by End-Member Analysis. *Near Infrared Spectrosc.* 2:15–23, 1994.

147. JB Reeves III. Near- versus Mid-Infrared Diffuse Reflectance Spectroscopy for the Quantitative Determination of the Composition of Forages and By-products. *Near Infrared Spectrosc.* 2:49–57, 1994.

148. GJ Faughey and HSS Sharma. A Preliminary Evaluation of Near Infrared Spectroscopy for Assessing Physical and Chemical Characteristics of Flax Fibre. *Near Infrared Spectrosc.* 8:61–69, 2000.

149. HSS Sharma, M Kilpatrick, and L Burns. Determination of Phase II Mushroom (Agaricus bisporus) Compost Quality Parameters by Near Infrared Spectroscopy. *Near Infrared Spectrosc.* 8:11–19, 2000.

150. FE Barton II, DS Himmelsbach, and DD Archibald. Two-Dimensional Vibration Spectroscopy. V. Correlation of Mid- and Near Infrared of Hard Red Winter and Spring Wheats. *Near Infrared Spectrosc.* 4:139–152, 1996.

151. FE Barton II, DS Himmelsbach, DE Akin, A Sethuraman, and K-E Eriksson. Two-Dimensional Vibrational Spectroscopy. III: Interpretation of the Degradation of Plant Cell Walls by White Rot Fungi. *Near Infrared Spectrosc.* 3:25–34, 1995.

5 NMR of Lignins

John Ralph and Larry L. Landucci

CONTENTS

INTRODUCTION

NMR in Lignin Research

NMR spectroscopy has enormously facilitated investigations into structural aspects of complex lignin polymers. Principally due to its high resolution and the high dispersion of chemical shifts, ^{13}C NMR has played a major role in the qualitative and quantitative understanding of lignin structure. Now modern two- and three-dimensional correlative methods are powerful tools for identifying structural units.

Nuclear Magnetic Resonance (NMR) spectroscopy has developed far beyond the simple acquisition of a proton or carbon spectrum. Today there are literally thousands of "NMR experiments" and variants that can be run to target the type of information sought. Fortunately, these experiments fall naturally into a handful of classes of useful experiments. Two-dimensional (2D) NMR, where more than a single NMR parameter is presented in a spectrum, is common and has been applied to the elucidation of lignin structure and bonding. Three-dimensional (3D) NMR has more recently been applied to problems in lignin chemistry. NMR hardware has become increasingly sophisticated and recent advances have improved both the sensitivity and the quality of data that can be acquired. Spectrometers operating at a 1000 MHz proton frequency are commercially available. The increase in magnetic field strength provided by such spectrometers offers enormously improved sensitivity and spectral dispersion over the 60 MHz instruments that were available at the time Sarkanen and Ludwig edited their famous book on Lignins [1].

This chapter will consider the basic aspects and findings of several forms of NMR spectroscopy, including separate discussions of proton, carbon, heteronuclear, and multidimensional NMR. Enhanced focus will be on ^{13}C NMR, because of its qualitative and quantitative importance, followed by NMR's contributions to our understanding of lignin structure by topic rather than by the type of NMR experiment. The more recent availability of lignin-biosynthetic-pathway mutant and transgenic plants has greatly aided the NMR analysis of perturbed and normal lignins, providing deeper insights into the lignification process. Solution-state NMR analysis of lignins, without the need for isolation from the other cell wall components, is also covered.

Sensitivity of NMR Spectroscopy

Compared to other spectral techniques, NMR may be orders of magnitude less sensitive. Ultraviolet and infrared spectroscopy involve transitions from an extensively populated ground state to an excited state that is not occupied under normal conditions because the energy levels are sufficiently far apart. Relatively large energies are involved. In NMR, the "ground state" and excited energy levels are so close in energy that thermal equilibrium assures that both states are essentially equally populated. In low-field spectrometers, for example, there may be only one in a million protons more in the lower energy state than in the high energy state. It is only the population *difference* that contributes to the NMR signal. This factor alone explains a sensitivity several orders of magnitude lower than other spectral techniques. The low energy involved, however, provides one striking advantage to NMR. Since the RF excitation frequency only acts on nuclear spins, and since the energy involved is too low to cause significant other physical or chemical changes in the molecule, NMR is the only spectroscopy that does not "visibly" perturb the system being measured. In other words, the structure and three-dimensional shape of a compound in solution being "viewed" by NMR is exactly the same as the structure when it is not being viewed. Other techniques are invasive; they perturb the system they are measuring.

Several nuclei of interest suffer from low natural abundance: protons are essentially 100% ^1H (~0.015% ^2H); phosphorus is 100% ^{31}P; the major carbon isotope, ^{12}C, is not NMR-active, while the ^{13}C isotope is only present in 1.11% relative abundance. Consequently, ^{13}C NMR at natural abundance loses another two orders of magnitude in sensitivity. Also, the energy levels are even closer in ^{13}C than in ^1H (roughly a factor of $^1/4$), resulting in an even smaller population difference between the levels and causing yet another decrease in sensitivity. Sample sizes typically in the milligram range are required to obtain good spectra in reasonable times. Despite this apparently abominable sensitivity, the information content of NMR spectra is significantly higher than for other spectral techniques. NMR alone can often fully identify compounds. Structure and bonding patterns are readily elucidated even in complex molecules.

Considerable gains in sensitivity have been realized through other approaches. A 2-dimensional (2D) $^{13}C–^1H$ correlation spectrum, for example, can now be acquired much more quickly than a 1-dimensional (1D) ^{13}C spectrum [2]. Although a relatively insensitive spectroscopic method, NMR still provides an amazingly detailed picture of lignin structure from a combination of experiments, if sufficient sample (typically 10–100 mg) is available.

Evolution of NMR Methods

Proton and FT-NMR

Proton NMR was the first type of NMR spectroscopy to become available to the researcher. The high natural abundance of the NMR-active ^1H-nucleus (~100%) and its relatively high sensitivity allowed early low-field continuous-wave (CW) spectrometers, with little in the way of signal-averaging capabilities, to successfully produce spectra from milligram quantities of small organic molecules. Such instruments, however, were far from ideal for studying a polymer such as lignin. One problem is to get a significant weight of material to dissolve in the small volume required for NMR (typically 0.5 ml). Even if this were accomplished, researchers were then faced with lignin's complexity, which meant that the concentration of any given structural type is rather low. Some kind of signal averaging was required to improve the signal-to-noise ratio of such spectra. Before the implementation of Fourier Transform NMR (FT-NMR, described below), this could only be accomplished by acquiring and adding frequency-domain data directly. A typical ^{1}H NMR scan on an early CW instrument might take up to 5 minutes. With the right accessories, a second scan could then be recorded and added to the first, and so on. The signal-to-noise improvement from such scan-averaging comes from the differing signal and noise properties; coherent signals from the sample will grow in proportion to the number of scans (NS), whereas the noncoherent noise grows only as $NS^{1/2}$. Therefore, the signal to noise of the spectrum grows as $NS/NS^{1/2} = NS^{1/2}$. Thus, the signal-to-noise ratio of a spectrum will be doubled if four scans are averaged; a 10-fold increase in signal-to-noise will require averaging 100 scans. With scans requiring as long as 5 minutes, the amount of signal averaging possible was limited.

The advent and implementation of Fourier Transform NMR (FT-NMR) allowed essentially unlimited signal averaging by digitally acquiring time-domain Free Induction Decay (FID) data at a far higher rate than could be accomplished by acquiring and adding frequency-domain data directly. In FT-NMR, the entire frequency response (amplitude vs time) is acquired following the application of a sharp, narrow duration (~10 μs) pulse. Once the sample has relaxed back toward its equilibrium state, the sample can be pulsed again, and the response acquired again, adding it to the former. Two aspects make this method superior. First, the experiment is ready to be repeated after typically just a few seconds (rather than the 5 minutes on the CW instrument). This means that acquiring even a hundred thousand scans becomes feasible. Secondly, the required frequency data can be extracted from the amplitude vs time response by a Fourier transformation. The Fourier transform itself was initially a computationally intensive task that could tie up a minicomputer for hours; however, advances in computing power have been such that 1D transform times are now subsecond. Another innovation was required to provide the stability required to allow accumulation of scans over an extended time. Independent RF circuitry was used to monitor the deuterium NMR signal supplied via the sample's solvent, allowing the instrument's electronics to track and compensate for any "drift" of the field within the sample.

Carbon NMR

The carbon skeleton of any organic molecule is a key feature; thus, ^{13}C NMR becomes an important diagnostic spectroscopic method. Fourier-transform-NMR was the enabling technology that allowed ^{13}C-NMR to be viable on real samples [3,4]. Due to the low sensitivity associated with the ^{13}C-nucleus, researchers needed rapid and extensive signal averaging. Once the sensitivity aspects could be overcome, ^{13}C NMR had some advantages over ^{1}H NMR. Primarily, the chemical shift range is significantly larger, ~200 ppm vs ~10 ppm for protons in typical organic molecules. Dispersion in ^{13}C NMR is considerable because the line widths are comparatively narrow; that is, it is relatively rare for two unrelated carbons in a small molecule or in a mixture to have identical chemical shifts. However, the complex nature of lignins leads to broad signals and to considerable overlap. A significant feature of ^{13}C-NMR is that chemical shifts are almost entirely dictated by through-bond, rather than through-space, interactions, except in cases of steric compression, where carbons are forced to occupy positions within the Van der Waals radii of other atoms [5,6]. The result is that a carbon chemical shift can reasonably be predicted by adding up contributions from its bonding environment [3,4,7]. For lignins, the chemical shifts in good model compounds are, therefore, the same as their counterparts in the polymer. In practice, side-chain carbon chemical shifts can be rather well modeled by dimeric lignin models such as guaiacylglycerol-β-guaiacyl ether, even though the "B-ring" (with no side-chain) is not well modeled [8,9]. Modeling side-chain carbon shifts in etherified units by using simple 4–O–Me derivatives is often satisfactory. The aromatic carbons can be much more difficult to model. Obviously, free-phenolic β-ether units can be modeled successfully using the A-ring of the same β-aryl ether model. However, to successfully model an etherified unit, it is necessary to use the internal unit of a trimeric model compound [10,11]. Data from β-aryl ether homopolymers are now also available [12]. Detailed analysis and assignment of model compound data provides good insight into the interpretation of the NMR signals of polymeric lignins, as will be explained below.

Another consideration was crucial to the development of the ^{13}C NMR method. Two NMR-active nuclei will normally interact by coupling interactions. The interactions of two ^{13}C nuclei can be neglected in most experiments, due to the low abundance of ^{13}C isotopes; most ^{13}C nuclei will find themselves within range of only ^{12}C nuclei that are not NMR active and that will, therefore, not couple. However, essentially all protons in a molecule are ^{1}H isotopes. Therefore, a ^{13}C nucleus will interact with any ^{1}H within range—usually within three bonds of the carbon. Such short- and long-range coupling interactions give rise to rather complex ^{13}C NMR spectra. This is the primary reason why, from the outset, protons are usually "decoupled" in ^{13}C NMR. Initially, ^{1}H-decoupling was via broadband irradiation of proton frequencies, but is now more efficiently accomplished via pulse sequences such as WALTZ [13,14] and MLEV [15]. The result is that all signals in ^{13}C NMR spectra of normal organic molecules (with normal isotopic distributions) are a series of single, sharp signals (peaks)—one peak for each (magnetically distinct) carbon. Another advantage of decoupling is that the carbon signal intensity is enhanced by irradiating attached protons; this is typically referred to as the nuclear Overhauser enhancement, or nOe [16]. Enhancements of up to a factor of 3 provide much needed sensitivity improvements for this insensitive nucleus. A disadvantage is that unequal enhancements produce peaks with areas that are not proportional to the number of carbons contributing to the peak. Quantitative (integrable) ^{13}C-NMR spectra can be acquired under special conditions, as will be covered in the Quantitative ^{13}C NMR of Lignins section.

Pulsed NMR Experiments

The power of NMR methods rose exponentially with the introduction of NMR pulse sequences. The first experiments simply pulsed the system and directly measured the response (a so-called single-pulse experiment). Later, pulse sequences strung together a series of pulses, each characterized by (a) the angle through which they flip the nuclear spin of the applied nucleus, and (b) the specific delays between the pulses and before the eventual acquisition of the signal. The intricacies of pulse-programming and the details of how such pulse experiments work are well beyond this review; a number of texts deal with these aspects [17–24]. The scope of such experiments and the ingenuity with which they were developed seemed unlimited; hundreds of experiments became available. One of the earliest experiments, the Attached Proton Test or APT experiment [25], allowed spectra to be acquired that distinguished the number of protons attached to each carbon, an extraordinarily useful feature. What followed were increasingly more sophisticated experiments, with increasingly bizarre acronyms such as INEPT, Insensitive Nuclei Enhanced by Polarization Transfer [26]. Unfortunately, not all of the new acronyms were relevant or accurate; some were proposed before researchers elucidated the mechanisms by which the experiments worked. The resulting jungle of fanciful, to silly acronyms remains. It is normal today to use the acronym and reference the experiment, but not to expand the acronym.

2D (and nD) NMR

Two-dimensional NMR resulted naturally from pulse programming concepts [20]. The basic idea is that a second dimension could be simulated by acquiring a series of spectra in which a delay period in the pulse sequence was incrementally increased. In what seemed at first like pure magic, it became possible, for example, to correlate a carbon's chemical shift in one dimension with the proton chemical shift of its attached proton in a second dimension. Again, the ingenuity for designing pulse programs to tease out the data of interest was boundless, and development of new experiments has continued unabated until the present.

A particular advantage of the ^{13}C–^{1}H correlative 2D NMR methods for lignin (described in more detail below) is the virtual dispersion that can be realized. Many of the protons in lignin from different units severely overlap, but their carbons are often distinct (or vice versa). The ability of experiments to separate overlapping peaks from one type of nucleus (e.g., ^{1}H) by correlating it with separated peaks from another nucleus (e.g., ^{13}C) produces spectra far better resolved than exhibited in 1D NMR. The initial impediments to 2D NMR were mainly associated with data size and computational difficulties, but advances in computing have obviated any such concerns.

Correlations anticipated in various homonuclear (^{1}H–^{1}H) and heteronuclear (^{13}C–^{1}H) 2D NMR experiments are conceptualized in Figure 5.1. A hypothetical model compound (the chemical shifts are not accurate and are for illustrative purposes only) with three aromatic protons and four side-chain protons on its three side-chain carbons is used to illustrate the information available from each experiment. A set of five experiments, in addition to the standard 1D proton and carbon spectra, are useful for characterizing any model compound or lignin. The correlation spectroscopy (COSY) experiment correlates directly coupled protons (Figure 5.1a).

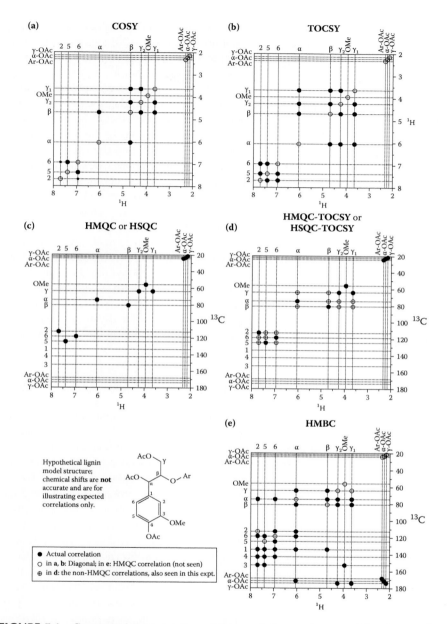

FIGURE 5.1 Conceptual representation of the correlation contours resulting from various homonuclear (^{1}H–^{1}H) a) COSY and b) TOCSY and heteronuclear (^{13}C–^{1}H) c) HMQC or HSQC, d) HMQC-TOCSY or HSQC-TOCSY, e) HMBC) 2D NMR experiments. A hypothetical lignin model compound (the chemical shifts are *not* accurate and are for illustrative purposes only) with 3 aromatic protons and 4 side-chain protons on its 3 side-chain carbons is used. (See text for details.)

Thus, Hα correlates with Hβ, which also correlates with the two Hγ-protons, etc. A weak H2–H6 correlation is shown; the coupling constant is ~2 Hz for this long-range correlation. In the TOCSY experiment (Figure 5.1b), a proton correlates with all other protons in its coupling network. Thus, all side-chain protons, for example, correlate with each other. The HMQC or HSQC experiment simply correlates each carbon with its directly attached proton (Figure 5.1c). Thus, Hα correlates with Cα, etc. The HMQC-TOCSY or HSQC-TOCSY experiment extends the HMQC by correlating a carbon not only with its attached proton but all other protons in the same coupling network (Figure 5.1d). Thus, Cα correlates with Hα, Hβ, and the two Hγ-protons. As a corollary, Hα correlates with the three side-chain carbons Cα, Cβ and Cγ. The heteronuclear multiple-bond correlation (HMBC) experiment correlates protons and carbons that are related by long-range coupling interactions over 2–3 bonds, providing extremely valuable connectivity information (Figure 5.1e). Thus, for example, Hα correlates with all carbons within 3 bonds; Cβ, Cγ, C1, C2, C6, and the α-acetate carbonyl. It is quickly apparent that carbon assignments can be made unambiguously from HMBC data. For example, C1 is the only aromatic carbon able to correlate with Hβ, C3 is the only carbon that can correlate with the methoxyl protons, and α- and γ-acetate carbonyls are distinguished by their correlations with α- vs γ-protons.

In the following section, which briefly describes the various experiments, spectra from various synthetic and isolated lignins will be used for illustration; synthetic lignins have the advantage of containing all of the major lignin unit types, without the complications of carbohydrates.

Homonuclear Correlation Spectroscopy, COSY, TOCSY, and INADEQUATE Classes

Proton-proton 2D experiments are an extraordinary improvement over 1D proton spectra (see Figure 5.7e later in the Broadness of Proton Spectra section). Broadness per se is not a significant deterrent in 2D spectra, which are taken at much lower resolution than 1D spectra. Homonuclear COSY experiments correlate protons that are coupled to each other with reasonable coupling constants, typically >2 Hz, as conceptualized in Figure 5.1a. Normally, these are vicinal protons (protons on adjacent carbons) and geminal protons (protons on the same carbon). The side-chain carbons of the common units in lignin all bear protons, so correlations are evident from H_α to H_β to the H_γ protons for each side-chain unit [2,27,28]. As with most other experiments, a gradient-edited COSY experiment allows acquisition of spectra without extensive phase cycling, so that even a single scan per increment is sufficient. For lignins, acquiring several scans per increment improves signal-to-noise, so remains worthwhile, particularly for identifying minor components. Multiple-quantum-filtered variants are useful to prevent strong (single-quantum) correlations from the methoxyl and acetate methyl groups from dominating the spectrum. An example of COSY experiment for a pine compression wood acetylated milled wood lignin (MWL) is shown in Figure 5.2a. Long-range coupling is typically under 2 Hz. The long-range variant of the COSY experiment can enhance correlations from protons associated by small coupling constants (down to ~0.1 Hz), particularly in the case of low-molecular mass lignin models and lignans [29].

Typical synthetic dehydrogenation polymers (DHPs), such as the one giving the data in Figure 5.2b, have an overabundance of units arising from dimerization of monomers—resinols **C** and cinnamyl alcohol end-groups **X1**, and also noncyclic α-aryl ethers **A2**. Resinols and cinnamyl alcohol end-groups show up at lower contour levels. One of two types of the minor dihydroconiferyl alcohol (DHCA) units **X5** is clearly evident in the pine lignin, Figure 5.2c; the other higher δ component seen in Figure 5.2b arises from 5-coupling of such units and can be seen in lignins with increased incorporation of DHCA.*

Although COSY type experiments are useful, the extra correlations provided by TOCSY experiments (Figure 5.1b) are particularly valuable [2,30,31]. Figures 5.2b and 5.2c show example TOCSY spectra for a synthetic lignin (DHP) and an isolated softwood lignin, both acetylated. Unlike typical DHP spectra, the spectra of isolated softwood lignins confirm the low abundance of products from monolignol coupling, having only minor contours corresponding to end-groups **X5** and resinols **C**. They also have a virtual absence of α,β-diethers **A2**. One type of DHCA product **X5** appears in the pine lignin, but two different products **X5** are in the DHP, which was synthesized from coniferyl alcohol containing ~2% DHCA. The second product comes from 5–5-coupling of at least one, and probably two, DHCA units. This component becomes significant in a pine mutant down-regulated in cinnamyl alcohol dehydrogenase (CAD), which is described later in the CAD-Deficient Pine: Incorporation of Dihydroconiferyl Alcohol and Guaiacylpropane-1,3-diol section and in Figure 5.12.

In TOCSY spectra, the mixing time determines the relative intensity of the various contours; mixing times of 60–120 ms are typically used. A short mixing time (10–30 ms) yields essentially COSY-type spectra, where protons correlate only with their directly coupled neighbors. The TOCSY (also called HOHOHA) experiments have some advantages over COSY and long-range COSY experiments for lignins [30]. During the mixing time, relaxation is less significant and TOCSY spectra are less sensitive to the initial state of the magnetization, which means that pulsing can be faster; full 2D TOCSY experiments can be acquired in as little as 2 minutes. Model compound data agreed well with corresponding data in lignins, obtained by TOCSY experiments [31]. TOCSY spectra have been reported often [2,27,30–40], and have been valuable in finding minor structures in lignins [34,41]. The older relayed coherence transfer experiments (where magnetization is transferred sequentially from one proton to the next coupled proton) provided similar data, but is seldom used today; these experiments were useful for assignments in the eight isomers of trimeric lignin β-ether models [11].

INADEQUATE experiments [42–44] correlate two neighboring ^{13}C-resonances and operate through direct ^{13}C–^{13}C coupling. They are consequently low-sensitivity experiments at natural ^{13}C-abundance (1.1%). Uniform ^{13}C-labeling at the 10–15% level provides an enormous sensitivity increase, while keeping carbon 1D spectra relatively free of homonuclear coupling peaks and minimizing long-range-coupling effects. Many versions of the experiment exist, including more sensitive implementations with INEPT-type transfer and with gradients [45,46]. However, some of these implementations only provide connectivity for protonated carbons. A version

---

* see later in Figure 5.12

providing symmetric COSY-like correlations [44] was used in the only previously published lignin spectra of this type [2,47–50]. A series of papers [47–49] illustrated the connectivity available for the major units in a uniformly ^{13}C-enriched poplar lignin [51]. INADEQUATE-type experiments provide diagnostic connectivity information on the carbon skeleton of lignins; however, connectivity

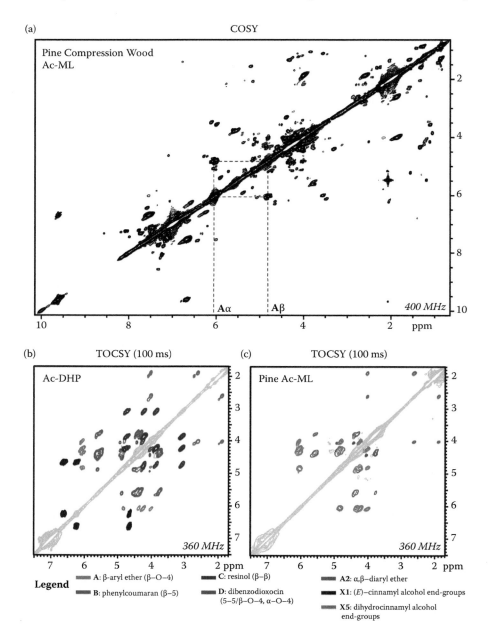

information can be more easily obtained by using long-range C–H correlations in the more sensitive HMBC experiment (Long-Range C–H Correlation, HMBC Class section).

J-Resolved Experiments

Chemical shifts in the acquired dimension and proton-proton coupling in the second dimension can be separated using "J-resolved experiments." Refocusing of relaxation effects from magnetic field inhomogeneity is an asset when seeking small (long-range) coupling constants in model compounds; however, J-resolved experiments are not widely used because of second-order artifacts that arise from tightly coupled protons. Higher magnetic fields can often simplify these spectra; tight coupling, where two distinct coupled protons have chemical shifts that differ by less than ~10J (where J is the coupling constant in Hz), becomes less frequent. Two protons which are, for example, only 20 Hz apart at 200 MHz are 75 Hz apart at 750 MHz. For lignins, J-resolved experiments have revealed useful information. The first-order α-protons in β-ether units of acetylated lignins allow *threo-* and *erythro*-isomers to be distinguished based on a combination of their chemical shifts and coupling constants [28]. Both etherified and free-phenolic *threo*-isomers can also be resolved [2]. In CDCl$_3$, there is further resolution of etherified vs nonetherified β-ether units [28].

Through-Space Interactions, NOESY and ROESY Class

NOESY or ROESY spectra correlate protons that are close to each other in space (rather than through-bond). The magnitude of the interaction falls off with the 6th power of the distance between the two protons, so such interactions are diagnostic for interproton distances of less than about 5 Å. Consequently, they are preeminent experiments for deducing 3D solution-state conformations for proteins, where the NOESY data provide sufficient numbers of distance constraints to allow constrained molecular modeling to produce essentially unambiguous 3D structures. However, lignin correlations are few and limited to intraunit responses, providing little conformational insight; this is largely due to the absence of regularly repeating macrostructures in lignins and perhaps to lignin's rather linear nature, with little chain folding. The major correlations seen are (a) between the methoxy protons and the aromatic 2-proton in guaiacyl units or the 2/6-protons in syringyl units and (b) between side-chain protons (Hα, Hβ and Hγ) and aromatic protons (H2 and H6) of the same unit [30,33]. Perhaps the limited correlations in NOESY/ROESY spectra provide the best evidence that MWLs and other solvent-extractable lignins have little in the way of structural regularity, a conclusion also supported by the complexity of

FIGURE 5.2   (See color insert following page 336.) (Opposite) COSY and TOCSY spectra. (a) COSY spectrum (gradient, double quantum filtered) from a pine compression wood acetylated milled wood lignin in acetone-d$_6$, at 400 MHz. Methoxyl and acetone (solvent) artifacts are severely suppressed by using the double quantum filtered variant shown here. (b) TOCSY spectrum, using a 100 ms mixing period, from an acetylated synthetic lignin (DHP, made from coniferyl alcohol containing a few percent of dihydroconiferyl alcohol, DHCA); and (c) an acetylated isolated lignin from a softwood (pine), in acetone-d$_6$, at 360 MHz. Contour assignment coloring scheme follows that to be established in Figure 5.3.

1D ^{13}C–NMR spectra of lignins. Useful interunit correlations are even not seen in a highly syringyl-rich lignin, where long arrays of β–O–4-coupled units exist, such as in birch [33] or the bast fibers of kenaf [37]. On low molecular mass lignin models and lignans, the NOESY experiment is, of course, valuable for identifying isomers and conformational insight. For example, isomers of the various 8–β-coupling products from ferulates and coniferyl alcohol can be nicely distinguished by NOESY and long-range COSY experiments [29].

Short-Range C–H Correlation, HMQC and HMQC-TOCSY Class

By far the most useful experiments are those providing correlations between protons and carbons in two dimensions. Extra apparent resolution is gained, which exceeds anything that can be achieved in 1D spectra with today's field strengths. Overlapping protons that are attached to carbons with different shifts are pulled apart by those carbon shift differences; overlapping carbons may be distinguished by their attachment to protons with different chemical shifts [2].

So great are the advantages of the so-called inverse-detected experiments that little use is now made of the older normal-mode experiments, except on older instruments. The former acquires data in the proton dimension, while the latter the carbon dimension. Satisfactory HMQC (or HSQC) spectra can be run in as little as 4 minutes with gradient selection [2], or about 15 minutes without. Longer experiments are still recommended for detailed work, since longer acquisitions obviously still give valuable signal-to-noise improvements and consequent interpretability enhancements.

HSQC or HMQC Spectra   Figure 5.3a shows a gradient-HSQC [52–54] spectrum of an acetylated synthetic lignin, with interpretation of major structural units; the structures, structure numbers, and color coding established here are used throughout this chapter. An HSQC contour implies that a proton at the proton frequency of the contour (x-axis) is directly attached to a carbon at the carbon frequency of the contour (y-axis) (Figure 5.1c). In all isolated lignins, such as those in Figures 5.3c and 5.3e, β-ether structures **A** dominate. The availability of an assigned spectrum, e.g., the clean synthetic lignin spectrum in Figure 5.3a, allows major structures in other spectra to be rapidly assigned. HMQC or HSQC spectra of lignins have been reviewed [2] and well reported [27,30,32–34,38–40,50,55–77,78–113].

The HMQC/HSQC experiments have been valuable in assigning major structures and have been indispensable in identifying new and minor units. The clear identification of dibenzodioxocins **D** as major new structures in lignins has been a significant finding [114–116]. In acetylated lignins, dibenzodioxocins are readily identified by unique and often well-resolved correlations in HMQC/HSQC spectra, Figures 5.3a and 5.3c (and later in Figure 5.10). Evidence provided by 2D NMR is far more diagnostic than 1D data, purely because of the simultaneous constraints that are revealed in the data. Thus, the observation that there is a proton at 4.9 ppm directly attached to a carbon at 84.4 ppm and a proton at 4.1 ppm attached to a carbon at 82.5 ppm is more revealing than just observing two new carbons at 84.4 and 82.5 ppm in a 1D spectrum. Also, dibenzodioxocins are not readily observed in 1D ^{13}C NMR spectra due to spectral overlap and their disperse (broad signal) nature.

Although 2D correlative spectra have the advantage over 1D of providing at least two pieces of simultaneous data, overlap and confusion can still occur in 2D. One pitfall in HMQC and HSQC spectra is found with structures **A2** (α,β-diaryl ethers) and **A3** (α-keto-β-aryl ethers). The α-proton/carbon contour in **A2** is located in almost exactly the same region (~5.6/81 ppm) as the β-proton/carbon contour in **A3** [117]. Keto-units **A3** occur mainly in syringyl-rich isolated lignins, and may be produced during ball-milling and lignin isolation [1,78]; however, synthetic lignins that have not been ball-milled also contain such oxidized units. The two components are best distinguished in HMBC spectra (see Long-Range C–H Correlation, HMBC Class section), where guaiacyl α-ethers signals for H–**A2**$_\alpha$ (5.6 ppm) correlate with carbons γ (63.5), β (81.5), 1 (137.5), 2 (112.3), 6 (120.0), and 4 (147.5) of the next unit [2]. In contrast, in units **A3**, H–**A3**$_\beta$ (5.6 ppm) correlates with the γ carbons (64.5) and, diagnostically, the α-carbonyl (194.8). One should confirm structural assignments from all obtainable data. None of the spectra in Figure 5.3 reveals any **A3** units, but these can be seen later in Figures 5.10f, 5.10g, 5.18e, and 5.18g.

The 2D HSQC pulse experiment contains more pulses and delays than its HMQC counterpart and requires more time during the pulse sequence before the actual signal acquisition. It can, therefore, be a less sensitive experiment for rapidly relaxing samples, but offers some advantages, including artifact reduction (although it is sensitive to the 180° pulse accuracy, which should be measured on each sample). Early versions were also reduced in sensitivity by a factor of two by selection of only one of the two coherence pathways [118,119]. More recent "sensitivity enhanced" or "sensitivity improved" HSQC schemes select both pathways and have become the standard experiments [52,53]. Phasing of HSQC spectra can be a problem. The older gradient-HMQC experiment may sometimes be preferable for rapidly relaxing samples. A useful discussion of the issues in HMQC vs HSQC pulse sequences has been published [120].

Multiplicity editing in 2D NMR is possible via INEPT- or DEPT-HMQC [121] experiments. Choosing the editing pulse or delay, as in 1D experiments, can yield spectra with CH and CH_3 contours positive and CH_2 contours negative [2,74]. Useful and potentially cleaner spectra can be obtained this way; however, these sequences have been little utilized in lignin work to date. They are experiencing renewed interest with the adoption of the latest shaped-pulse variants.

Improving the uniformity of pulse excitation has become more important, particularly with the expansion in use of cryogenically cooled probes and high-field instruments. Specialty shaped pulses, such as adiabatic pulses, provide significant improvements. Such pulses appear to be complex in terms of their rather extreme phase ramping or amplitude modulation, but are designed to produce broad and uniform excitation. Peaks near the edges of the acquisition windows in spectra acquired over wide sweep widths may be significantly more intense in adiabatic-pulse versions of HSQC spectra than in the normal (hard, rectangular pulse) versions, especially at high field (see Recent Advances in NMR: Inverse Detection, Gradients, and Cryoprobes section).

HMQC-TOCSY or HSQC-TOCSY Spectra Although less sensitive than their HMQC/HSQC counterparts (because of the extra correlations for each proton and

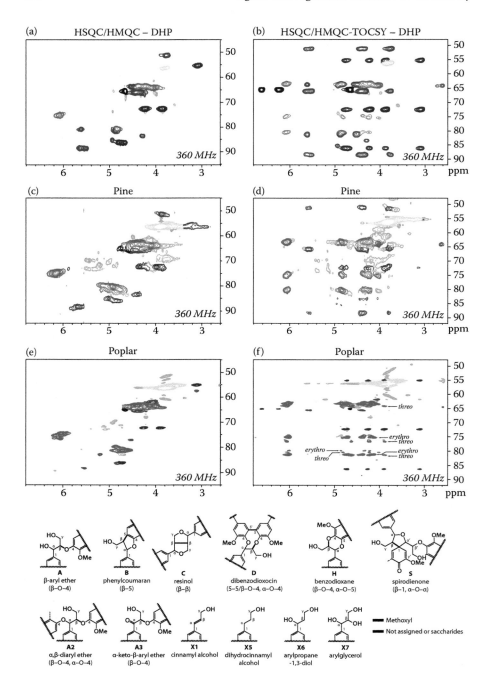

(a) HSQC/HMQC – DHP

(b) HSQC/HMQC-TOCSY – DHP

(c) Pine

(d) Pine

(e) Poplar

(f) Poplar

A β-aryl ether (β–O–4)

B phenylcoumaran (β–5)

C resinol (β–β)

D dibenzodioxocin (5–5/β–O–4, α–O–4)

H benzodioxane (β–O–4, α–O–5)

S spirodienone (β–1, α–O–α)

A2 α,β-diaryl ether (β–O–4, α–O–4)

A3 α-keto-β-aryl ether (β–O–4)

X1 cinnamyl alcohol

X5 dihydrocinnamyl alcohol

X6 arylpropane -1,3-diol

X7 arylglycerol

Methoxyl
Not assigned or saccharides

carbon), 2D HMQC-TOCSY spectra [122], or its relative, HSQC-TOCSY [118], enjoy deservedly more widespread use. They provide a useful redundancy in information and a clear identification of coupling networks (see Figure 5.1d) [2,33,35,37–39,41,57]. Assignments in HMQC/HSQC spectra, e.g., the synthetic lignin spectrum in Figure 5.3a, must be made with good knowledge of both carbon and proton chemical shifts for each structural type. In contrast, in HMQC-TOCSY spectra (Figure 5.3b), just knowledge of rough proton shifts allows for a rapid identification of units; this is because of the useful redundancy in the data. Knowing that acetylated β-ether units have α-protons at ~6 ppm, β-protons at ~5 ppm, and γ-protons between 4 and 4.6 ppm, allows us to readily assign the blue contours β-ether structures **A**, for example. It is further easy to see which carbons are in the same structure by simply looking for matching sets of correlations in the proton dimension. The DHP spectrum is particularly useful because the major inter-unit structures are present in a single sample. Due to the differences in proton coupling constants and the ability to transfer magnetization around the coupling network, the proton correlations for each carbon (or vice versa) will not all have the same intensity, and some may even be absent. It is, however, easy to recognize that all the blue peaks (β-ether structure **A**) belong together, as do all the green peaks (phenylcoumaran structure **B**), etc. The simple interpretability of the side-chain region of a 2D HMQC/HSQC-TOCSY spectrum and the valuable redundancy of correlation peaks make this experiment very useful.

As can be seen from the higher resolution Poplar spectra in Figures 5.3e and 5.3f, we can distinguish nicely between *erythro-* and *threo-*isomers of the dominant β-ether units in particular regions of the spectra. The *threo-*units always have one higher-field (lower δ) γ-proton; additionally, various carbon shifts are partially resolved. Although HSQC-TOCSY spectra are not quantitative, it remains apparent that hardwood lignins, with a substantial syringyl component, heavily favor the *erythro-*β-ether **A** isomers (see again later in Figure 5.18g vs 5.18f).

The HMQC-TOCSY experiment was the key to rapid identification of (a) dihydroconiferyl alcohol units **X5** and arylpropane diol units **X6** present in a mutant pine lignin (see later in Figure 5.12) [38], (b) tyramine units in genetically altered tobacco lignins [69,71], and (c) the tracing out of spirodienone side-chains [57], as noted in later sections.

HSQC or HSQC-TOCSY Difference Spectra   In cases where there is an obvious control sample and one with some perturbation, difference spectra can be

FIGURE 5.3 **(See color insert following page 336.)** (Opposite) Heteronuclear 2D correlation NMR spectra (HSQC/HMQC, HSQC/HMQC-TOCSY, side-chain regions) of (a–b) an acetylated synthetic lignin derived from coniferyl alcohol, and acetylated isolated lignins from (c–d) a softwood (loblolly pine) and (e–f) a hardwood (poplar). Prominent structures **A-D, S, X** are colored for ready identification. The left-hand spectra are HSQC/HMQCs; the right-hand spectra are HSQC/HMQC-TOCSYs with 80–100 ms TOCSY mixing times. The reason why the poplar spectra, Figures 5.3e and 5.3f, are more highly resolved is because they were more recently acquired at higher resolution in F_1 (the carbon dimension) and a single level of forward linear prediction was used in processing. This figure establishes the color coding for lignin units in all figures in this chapter (unless otherwise noted).

generated by electronically subtracting two spectra. Just as 1D difference spectra have tremendous value in delineating assignments [123–126], 2D difference spectra are now being similarly used. The first example is from synthetic lignins generated from specifically labeled coniferyl alcohol [111,127].

The 2D HSQC and HSQC-TOCSY difference spectra in Figure 5.4 illustrate how strikingly clean these spectra are and indicate how powerful this method may prove in the future. The spectra are unedited (apart from colorization of the contours) difference spectra from Terashima's DHPs [111], showing just the major units at this time. Delving into details from lower contour levels is the obvious next step to look into minor structures. Similar experiments on lignins or solubilized whole cell wall Gingko plant material, fed with labeled precursors, have been reported [111]. The difference spectra were readily generated using Bruker processing software. The quality is presumably dependent on machine stability and having samples of similar quality. Obviously all spectra do need to be run under the same conditions on a single instrument (or at least on instruments with the same field strength). Difference spectra have also now been used on isolated lignins and whole-cell-wall samples [97,99,104]. They can be particularly valuable for visualizing minor components, such as the *bis*-8–O–4-ethers that arise from the elevated incorporation of ferulic acid into lignins in plants deficient in cinnamoyl-CoA reductase (CCR) [97,104].

Long-Range C–H Correlation, HMBC Class

HMQC/HSQC-type spectra are valuable for their direct attachment information and because of the apparent extra dispersion they provide over 1D spectra. Long-range ^{13}C–^{1}H correlation experiments provide enormously valuable connectivity data (Figure 5.1e). Two- and 3-bond ^{13}C–^{1}H coupling constants are in the 2–15 Hz range, and HMBC [128] experiments are typically set with coupling evolution times of 60–120 ms, corresponding to coupling constants of 4–8 Hz. It is possible to miss correlations because they may be twice the set value; the response is a sine curve peaking at 1/2J. In some circumstances it may be appropriate to run spectra with different long-range coupling delays to allow for all correlations to become visible; however, as with the HSQC-TOCSY experiment, the useful redundancy of data often makes this unnecessary. If a given correlation from a particular structure is required, the experiment can be easily optimized for that correlation by appropriately choosing the coupling evolution delay as $0.5/J_{LRCH}$, using the long-range coupling constant of interest. For example, 110 ms corresponding to the measured $^{3}J_{C–O–C–H}$ of 4.5 Hz was chosen to observe ferulate arabinosyl moiety attachments [129]. What must be kept in mind with polymeric lignin samples, however, is that the longer this delay, the less overall intensity the spectrum will have due to relaxation losses. Obviously, if the (proton) FID from the lignin is reduced to the noise level in 100 ms, no signal will be produced in an HMBC experiment incorporating a 100 ms delay. Some isolated lignins, especially those prepared by steel ball-milling, can have very short relaxation times that can often be improved by EDTA washing the lignin to remove metals [56]. Acetylated samples can readily be EDTA-washed in an extraction step, and this is frequently worthwhile [71].

Applications of HMBC experiments to lignins have been documented [2,27,32,35,36,55–57,60,64,70,71,74,106,129–132]. The enormous power of this experiment can be seen in diverse applications that show unambiguously, for example, how *p*-coumarates are attached to γ-positions of lignins (Naturally Acylated Lignins section) [56], and how ferulates are responsible for lignin-polysaccharide cross-linking in grasses (Ferulates in Lignins; Cell Wall Cross-Linking in Grasses by Ferulates section) [55,129,130,133]. One of the most useful applications of HMBC experiments is in delineating the monolignol incorporation into lignins and the consequent distribution of syringyl, guaiacyl, and *p*-hydroxyphenyl (and other) structures (Figure 5.5) (see also the Lignins in Monolignol-Biosynthetic-Pathway Mutant or Transgenic Plants section and later in Figures 5.17, 5.20, and 5.21). The α-protons correlate with carbons β and γ of the side-chain and carbons 1, 2 and 6 of the aromatic rings; all of these carbons are within 3-bonds of the α-proton. What makes these correlation spectra diagnostic is that the equivalent syringyl S2/S6 carbons, resonating at ~105 ppm, are well separated from their guaiacyl counterparts (for which G2 and G6 are different, at ~112 and ~120 ppm), and from *p*-hydroxyphenyl H2/H6 carbons at ~129 ppm. Consequently, the partial HMBC spectra from, for example, the β-ether **A** α-protons gives an immediate impression of the distribution of such units in the lignin. Hardwoods (e.g., Figure 5.5d) obviously differ from softwoods (e.g., 5.5b). More compelling examples derive from the analysis of perturbations caused in compression wood (5.5c) and in the various mutants and transgenics (5.5a, 5.5g, 5.5h, and 5.5j) discussed in more detail in the Lignins in Monolignol-Biosynthetic-Pathway Mutant or Transgenic Plants section. The experiment can even be run on the whole cell wall component, without first isolating the lignin (Figure 5.5e) [64]. The information rivals that from the diagnostic ^{13}C–^{13}C-correlated INADEQUATE experiments; however, HMBC spectra are obtained with significantly higher sensitivity and allow correlations to be made over several bonds (and therefore through oxygen and nitrogen, for example), which the INADEQUATE experiments cannot.

Quantification in 2D NMR

Attempts have been made to utilize 2D NMR experiments for quantification, due to their improved signal dispersion over 1D experiments. The most logical experiments to utilize in a quantitative or semiquantitative fashion are those that depend only on one type of coupling interaction, and for which the sensitivity to that coupling constant is relatively low. The best candidate is, therefore, the (short-range) ^{13}C–^{1}H correlation experiments, HMQC or HSQC. A general method to limit the dependence of signal intensity on the 1-bond coupling constant, and on signal offset (from the center of the spectrum), has been described [72]. It has the advantage of being largely structure-independent, but is less sensitive than the modern "sensitivity-improved" HSQC experiment. Guidelines have been provided for acquiring lignin spectra for quantification using integrals from quantitative 1D ^{13}C NMR spectra as "internal standards" for deriving relative response factors between the various types of units [95]. The semiquantification available from the volume-integration of cross-peak contours in HMQC/HSQC spectra can be strikingly useful, especially for comparative purposes. Such integration is illustrated in the Lignins in Monolignol-Biosynthetic-Pathway Mutant or Transgenic Plants section.

For aromatic carbon-2 (or carbons-2/6 in symmetrical *p*-hydroxyphenyl or syringyl units) ^{13}C–^{1}H correlations, essentially no correction appears necessary to achieve quantification [60]. Quantification correction factors, like response factors in GC, can be determined for side-chain correlations most efficiently via model compounds that contain at least two types of structural units [60]. However, the authenticity and portability of such methods have not been determined. It should be emphasized that not all 2D spectra are equally quantitative. HSQC/HMQC may

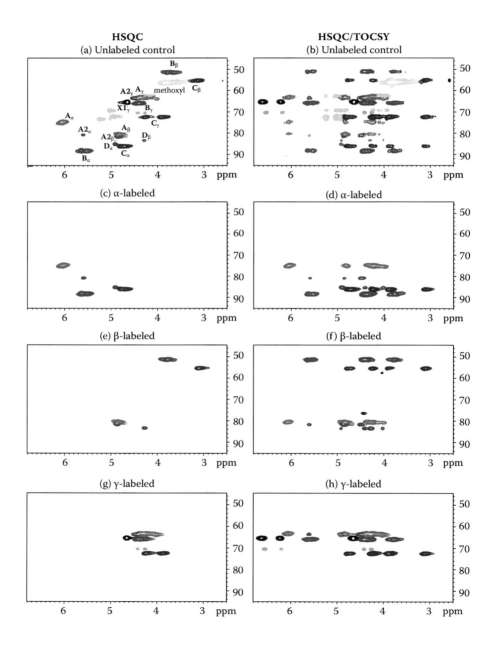

be quantitative, particularly if a small spectral window is acquired (to minimize off-resonance effects). However, HMBC experiments, which have correlation contour integrals that depend on a variety of coupling constants, are far from quantitative; some 2- or 3-bond $^{13}C-^{1}H$ correlations may not be evident at all, i.e., have essentially zero integral. This will happen, for example, if the actual long-range coupling constant J_{LR} is twice that corresponding to the chosen long-range coupling evolution period; for example, if the long-range coupling delay is chosen to be 100 ms (corresponding to an optimal J_{LR} of 5 Hz), a C–H pair having a coupling constant of 10 Hz will peak at 50 ms and be a null at 100 ms. Similarly COSY-type correlations depend on $^{1}H-^{1}H$ coupling constants. TOCSY integrals have a complex dependence on all the coupling constants in the proton network and on the mixing time.

3D Experiments for Lignins

NMR is not limited to one and two dimensions. Three-dimensional experiments are now commonplace, and 4D, 5D, and higher-D experiments have been applied to labeled proteins. Much of the value of these experiments comes from the further dispersion realized by correlating over the additional dimensions. This is particularly valuable in proteins where ^{13}C, ^{1}H, and ^{15}N dimensions are available. The expectation that 3D experiments necessarily are more time demanding overlooks the fact that signal-to-noise is gained on the *total number of scans in the entire experiment*. Lignins do not contain nitrogen; however, acquiring 3D spectra with one carbon and two proton dimensions is useful. The 3D HMQC-TOCSY experiment (with one ^{13}C and two ^{1}H axes) was first applied with some success using uniformly ^{13}C-enriched samples [2,50,77,78,134]. The increased complexity and data size, and the reduced resolution compared to 2D experiments are offset by the ability to "isolate" spin systems on their own planes. Although labeled materials facilitate 3D (as well as 1D and 2D) experiments, valuable spectra from unlabeled materials are readily obtained. For example, the spectra shown later in Figure 5.19f were acquired from unenriched samples in 24 h at 750 MHz. At 360 MHz, 48 h produces very satisfactory 3D spectra. Figure 5.6 is particularly impressive because it is on ~70 mg of a whole cell wall sample that is only ~30% lignin, but was acquired using a 500 MHz cryoprobe system in 36 h [61].

The two experiments used with lignin to date are variations of 3D HSQC-TOCSY and the reverse TOCSY-HSQC experiments (or their HMQC analogs). The

FIGURE 5.4 (See color insert following page 336.) (Opposite) HSQC and HSQC-TOCSY spectra and difference-spectra of acetylated DHPs (400 MHz, acetone-d$_6$) from specifically ^{13}C-enriched coniferyl alcohols (Adapted from Terashima, N., Akiyama, T., Ralph, S.A., Evtuguin, D., Pascoal Neto, C., Parkås, J., Paulsson, M., Westermark, U., and Ralph, J., *Holzforschung*, **63**, 379–384, 2009). (a–b) unenriched control; (c–h) 2D Difference spectra obtained by subtraction of the spectrum of the unenriched control from the spectra of DHPs specifically ^{13}C-enriched at side-chain carbon, (c–d) Cα, (e–f) Cβ, (g–h) Cγ. Note that there are two components in these DHPs that are different from usual DHPs and may have been a result of the preparation mode (Adapted from Terashima, N., Atalla, R.H., Ralph, S.A., Landucci, L.L., Lapierre, C., and Monties, B., *Holzforschung*, 50(1), 9–14, 1996); one is a glucose or glucoside component (light green), the other appears to be perhaps a γ-ether (light pink). Contour assignment coloring scheme otherwise follows that established in Figure 5.3.

former provides planes of 2D HSQC-TOCSY spectra at given proton frequencies, while the latter gives particularly clean HSQC spectra at planes corresponding to any of the protons of the coupling network; other planes may also be useful—see caption to Figure 5.6. Both experimental types have been illustrated and reviewed [2,50,61,68,77,78,134]. Researchers in this field love the ready interpretability of

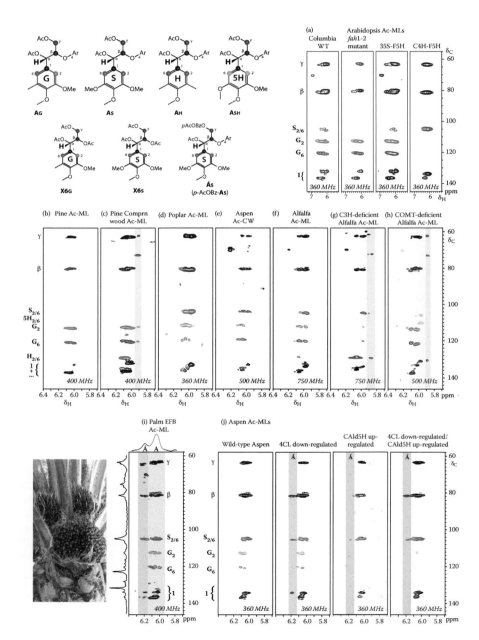

the 3D HSQC-TOCSY experiment (for the same reason that 2D HSQC-TOCSY is valuable—the appearance of the whole spin system is pleasing). The 3D TOCSY-HSQC experiment is, however, more sensitive, since only HSQC correlations are seen after magnetization is first transferred to the entire proton system (before the HSQC-part of the experiment). It is possible to observe HSQC spectral planes that are purely from a single substructure in lignin, even at natural ^{13}C abundance [2,68,135] (see Figures 5.19e and 5.19f, the COMT-Deficiency: Benzodioxanes from Incorporation of 5-Hydroxyconiferyl Alcohol section), and even when the sample is not an isolated lignin but the entire cell wall component (following ball milling and cell wall dissolution) (Figure 5.6) [61].

FIGURE 5.5 (See color insert following page 336.) (Opposite) Gradient-selected 2D HMBC subspectra showing α-proton correlations in β-aryl ether units **A** (and glycerols **X7**), illustrating their value in delineating H:G:S differences (and more). (a) Spectra from wild-type arabidopsis, the *fah*1-2 mutant deficient in ferulate 5-hydroxylase (F5H, see caption to Figure 5.16), and two transgenics upregulated in F5H, the C4H-F5H one spectacularly so (Marita, J., Ralph, J., Hatfield, R.D., and Chapple, C., *Proc. Natl. Acad. Sci.*, 96(22), 12328–12332, 1999), all in acetone-d_6. The wild-type and 35S–F5H lignins are guaiacyl-rich syringyl/guaiacyl copolymers. In the fah1–2 mutant, syringyl units **As** are essentially absent. When F5H is expressed under the control of the C4H-promoter, the lignin that is deposited is comprised almost exclusively of syringyl units **As**. (b–h) Selection of similar HMBC regions of acetylated samples (in acetone-d_6, unless otherwise indicated) for H:G:S elucidation: (b) pine ML, showing only **AG** and no **As** (nor **AH** at this contour level); (c) Pine compression wood ML (from Ian Suckling and Bernadette Nanayakkara, SCION) showing increase **AH** level; (d) Poplar ML, a typical S/G lignin; (e) An Aspen dissolved whole cell wall sample, in CDCl$_3$, showing that valuable HMBC spectra can be acquired without the need for lignin isolation; (f) a control alfalfa ML in CDCl$_3$, a typical G-rich S/G lignin, but with minor **AH** levels; (g) a heavily C3H-down-regulated alfalfa ML, in CDCl$_3$, with ~65% **AH** units, and showing that β-ethers are derived from addition to the growing polymer of coniferyl, sinapyl, and especially *p*-coumaryl alcohol monomers; (h) a COMT-deficient alfalfa, showing an almost complete absence of S–β-ethers **As**, and the appearance of novel 5-hydroxyguaiacyl β-ethers **A5H**, derived from lignification using 5-hydroxyconiferyl alcohol. Note that glycerols **X7** are most prominently seen at the lower proton chemical shift (~5.9 ppm) in c, g, and h. (i) Lignin isolated from the empty fruit bunches of oil palm (Sun, R. C., Mott, L., and Bolton, J., *Journal of Agri. Food Chem.*, 46(2), 718–723, 1998) in which normal β-ethers **A** are S/G but those naturally acylated with γ-*p*-hydroxybenzoate groups, in which the proton chemical shifts are higher (~6.2 ppm) are S-only **Ás**. (j) Aspen acetylated lignins from a study in which the interactions between down-regulation of 4CL and up-regulation of F5H to successfully produce reduced-lignin transgenics with higher S:G ratios were explored (Li, L., Zhou, Y., Cheng, X., Sun, J., Marita, J.M., Ralph, J., and Chiang, V.L., *Proc. Natl. Acad. Sci.*, 100(8), 4939–4944, 2003). The 4CL-down-regulated transgenic had a lower lignin content with similar structure and S:G as the wild-type, but a higher level of *p*-hydroxybenzoates in units **Ás**. (F5H- or) CAld5H-upregulated aspen, like the C4H-F5H alfalfa in (a), was extremely S-rich. The doubly manipulated transgenic in which 4CL was down-regulated and F5H was up-regulated showed that the effects were essentially additive—the lignin level was lower, the **As** content high (indicating a high flux toward syringyl lignin), and the *p*-hydroxybenzoate on lignin level was elevated as indicated by **Ás**-β-ethers. Contour assignment coloring scheme is for S:5H:G:H delineation and does not follow that established in Figure 5.3; the colors are the same as those used for similar purposes in Figures 5.16a through c and 5.17.

The 2D slices from the 3D TOCSY-HSQC spectrum (Figure 5.6), show how elegantly individual unit types in lignin can be "isolated." Although these spectra are from a polysaccharide-dominated sample (as seen in Figure 5.6a), even isomers of lignin units may be resolved onto their own planes. Individual planes representing HSQC spectra at a given proton shift (Figures 5.6b through 5.6d), HSQC-TOCSY

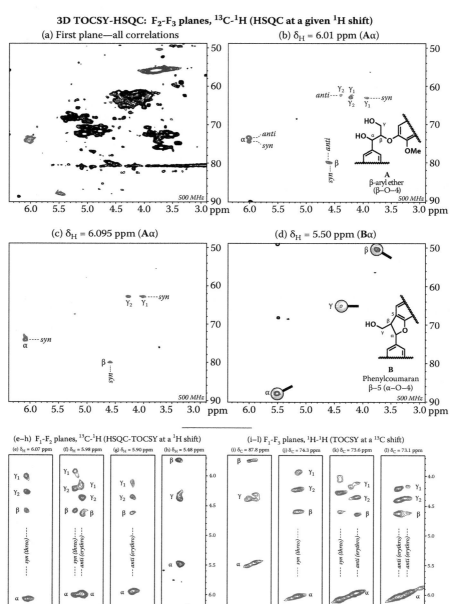

3D TOCSY-HSQC: F_2-F_3 planes, ^{13}C-^{1}H (HSQC at a given ^{1}H shift)

(a) First plane—all correlations

(b) δ_H = 6.01 ppm (Aα)

(c) δ_H = 6.095 ppm (Aα)

(d) δ_H = 5.50 ppm (Bα)

(e–h) F_1-F_2 planes, ^{13}C-^{1}H (HSQC-TOCSY at a ^{1}H shift)

(i–l) F_1-F_3 planes, ^{1}H-^{1}H (TOCSY at a ^{13}C shift)

(e) δ_H = 6.07 ppm   (f) δ_H = 5.98 ppm   (g) δ_H = 5.90 ppm   (h) δ_H = 5.48 ppm

(i) δ_C = 87.8 ppm   (j) δ_C = 74.3 ppm   (k) δ_C = 73.6 ppm   (l) δ_C = 73.1 ppm

sub-spectra at a given proton shift (5.6e through h), and TOCSY subspectra at a given carbon shift (5.6i through l) can all be viewed. (It is actually also possible to project onto planes other than these three orthogonal ones defined by the acquisition axes, and useful information may be obtained from those.)

Frankly, the biggest "problem" with 3D NMR is not its acquisition or processing but its simplicity. To the uninitiated, seeing only a few spots on a plane seems unimpressive. The simplicity, however, belies its incredible power of isolating structural units, from a polymer as complex as lignin, in a sample as complex as the entire cell wall, on their own planes. How much easier would Freudenberg's task have been with such a powerful tool?

Recent Advances in NMR: Inverse Detection, Gradients, and Cryoprobes

Recent advances in NMR technology have resulted in features so essential that it is now difficult to contemplate serious structural studies on complex molecules without them. "Inverse detection" came from a recognition that, no matter what nucleus you were interested in, physically acquiring data from the most sensitive nucleus in the system, usually proton, brought large sensitivity gains. Further developments allowed the more sensitive nucleus to be initially excited, with another significant gain in sensitivity. These gains are powers of the gyromagnetic ratios of the nuclei involved. For example, a polarization transfer experiment is γ_H/γ_X (~4 for carbon) times more sensitive than direct observation (without nOe). The enhancements are even larger for the lower-γ nuclei such as ^{15}N and ^{29}Si. A 2D ^{13}C–^1H correlation experiment carried out by acquiring proton data (rather than ^{13}C-data, as was conventional) experiences an incredible $4^{5/2}$ or 31.6-fold gain in sensitivity over a direct observation experiment and $4^{3/2}$ or 7.9-fold gain over the polarization transfer experiment (in which the protons are excited initially, but carbon data is still acquired).

The time required to obtain a given signal to noise (S/N) is proportional to the square of the *S/N*. A 2D ^{13}C–^1H correlation experiment that required 24 h. in the "normal mode" experiment would, therefore, take only 23 minutes using the inverse mode to acquire a spectrum with essentially the same sensitivity. Obviously, experiments that require a weekend to run in inverse mode cannot even be contemplated using the older scheme. Before these inverse experiments, it would have been impossible to study the minor (~1% of the lignin) ferulate cell wall cross-linking elements in grass lignins. This operating method alone allows 2D ^{13}C–^1H experiments to be acquired far more quickly than a 1D ^{13}C NMR spectrum. For many samples (and for proteins in particular) the carbon data are therefore obtained without ever acquiring an actual 1D ^{13}C NMR spectrum. Significant difficulties with such experiments have

FIGURE 5.6 (See color insert following page 336.) (Opposite) Slices from a 3D NMR spectrum of acetylated pine ball-milled cell walls (Lu, F. and Ralph, J., *Plant J.*, 35(4), 535–544, 2003), illustrating the spectacular lignin substructural editing possible in this complex sample (Ralph, J. and Lu, F., *Org. Biomol. Chem.*, 2(19), 2714–2715, 2004). The 3D spectrum was acquired with the acquisition restricted to the lignin side-chain region. (a–d) F_2-F_3 (HSQC) slices at a given ^1H shift; (e–h) F_1-F_2 (HSQC-TOCSY) slices at a given ^1H shift; (i–l) F_1-F_2 (TOCSY) slices at a given ^{13}C shift. Contours in figure d are artificially enlarged to enhance their visibility. Partial substructures of β-ether units **A** and phenylcoumaran units **B** are given with conventional side-chain labeling.

been overcome by improvements in hardware and clever pulse programming and phase cycling. An example will serve to illustrate the issues involved.

To acquire a "normal mode" $^{13}C-^{1}H$ correlation spectrum, researchers traditionally acquired, with proton decoupling, a set of 1D ^{13}C NMR spectra. Proton decoupling was already well established with the advent of 1D ^{13}C NMR. If it is desired to acquire proton spectra instead, we need a method to detect only those protons that are attached to ^{13}C-carbons, not ^{12}C-carbons. Therefore, some way of getting rid of the 99% unwanted signals had to be devised. This was initially via phase cycling—ensuring that the pulse sequence produced additive in-phase signals for the desired $^{13}C-^{1}H$ pairs, but opposite and canceling signals (in alternate scans, for example) for $^{12}C-^{1}H$ pairs. Also, instead of decoupling protons from ^{13}C, the instrument was now faced with the problem of decoupling ^{13}Cs from protons. Unfortunately, the large ^{13}C-spectral width required huge power levels in conventional decoupling schemes, sufficient to easily boil the sample. The advent of more efficient decoupling schemes such as GARP-decoupling [136], and the hardware to implement them were the keys to making such experiments routinely available. Again, the enormous improvement in sensitivity for these "inverse-detected" experiments over their traditional "normal-mode" counterparts, and the rapid evolution of NMR instrumentation to confront the rigors of ever more complex and more variable pulse sequences, has allowed these "inverse" experiments to almost completely replace the original ones. The term *inverse* is being phased out, since it has now become the "normal" mode.

The latest innovation to completely alter the way spectra are acquired is in the application of large pulsed field gradients during an NMR pulse sequence [137]. For years, NMR spectroscopists sought the strongest and most uniform stable magnetic fields possible in which to place their samples. Then theoretical specialists began advocating strategic use of "homospoil" pulses (to temporarily wreck the homogeneity of the magnetic field) and field gradients (where the field is ramped rapidly). Again, these methods have become accepted in "normal" experiments due to the beautiful way they can clean up spectra (by "coherence selection") and reduce artifacts without requiring phase cycling. Long-range $^{13}C-^{1}H$ correlation experiments have particularly benefited from the application of pulsed field gradients [118,138]. For lignins, the use of gradients and inverse detection allows acquisition of strikingly good 2D spectra that allow substantive interpretation in as little as 4 minutes [2].

The move to all-digital instrument components and digital oversampling (as used in CD players, for example) added further benefits. Commercial systems now offer cryogenic probes, which improve signal-to-noise (S/N) by reducing the operating temperature of the coil and the preamplifier. The increase in the S/N ratio by a factor of ~4, as compared to conventional probes, leads to a possible reduction in experiment time of 16-fold or a reduction in required sample concentration by a factor of 4. An NMR system with a cryoprobe may have sensitivity exceeding a traditional system with a much higher-field magnet.

Finally, as we alluded to above (HSQC or HMQC Spectra section), advances in pulse-shaping hardware and software have led to pulse-shaped variants of many pulse programs becoming the standard [139,140]. For example, the more energy-efficient adiabatic-pulse variants of HSQC-type experiments are recommended as standard experiments on newer high-field instruments, particularly those equipped with cryogenically cooled probes. They offer the advantages of a wide inversion bandwidth meaning minimized ^{13}C-pulse offset effects, lower power, and a wider decoupling range using CHIRP or WURST decoupling [140–142].

PROTON AND CARBON NMR

Arranging lignin studies under NMR-specific headings (such as ^{1}H, ^{13}C, 2D etc) leads to considerable redundancy (with the exception of multinuclear NMR, which is treated in its own section). Consequently, after a brief introduction to proton and carbon NMR of lignins, we provide the arguably more important contributions of NMR to various aspects of lignin investigation in the following sections. Where possible, the techniques used are delineated.

INTRODUCTION TO PROTON NMR OF LIGNINS

Broadness of Proton Spectra

NMR practitioners know that increasing molecular weight broadens resonances due to reduced relaxation times. But lignins appear broader and more featureless than, for example, comparably sized proteins (Figure 5.7e vs 5.7f). It is easily recognized that much of this broadness comes from the huge range of environments in the lignin polymer. Since isolated lignins are polydisperse and have no regularly repeating macrostructures, each unit finds itself in a unique magnetic environment. But complexity in lignin extends far beyond the complexities caused by irregular bonding of one unit to the next. Even a β–O–4-ether homopolymer is a stereochemical nightmare [11]. A simple trimeric β-ether model compound (Figures 5.7c and 5.7d), has four optical centers and therefore $2^4 = 16$ optical isomers or $2^4/2 = 8$ physically different (and NMR-different) isomers. A pure random β-ether hexamer, such as might be isolated, especially from syringyl-rich lignins, has 10 optical centers and therefore $2^{10} = 1024$ optical isomers and half that number, 512, of physically distinct isomers. (In general a molecule with n optical centers has 2^n optical isomers and $2^n/2 = 2^{n-1}$ real isomers.) In reality, although β-aryl ether units in guaiacyl lignins are essentially random (~50:50 *threo:erythro*), such units in syringyl lignins are mainly (~75:25) *erythro*-isomers [143,144]. Also, units such as phenylcoumarans and resinols are found as only single isomers in nature. Therefore, the stereochemical complexity in oligomers is reduced, but still becomes overwhelming with increasing molecular size. What is often overlooked by those unfamiliar with basic chemical stereochemistry is that, for example, a phenylcoumaran-phenylcoumaran trimeric unit still has two physically distinct isomers, even though there is only a single isomer of each dimeric moiety.

The stereochemical complexity causes significant broadening due to the overlapping of NMR signals, as illustrated in Figure 5.7e. Even in a trimer, with 8 real (NMR-distinct) isomers [11], the α-, β-, and γ-protons become complex (Figures 5.7c and 5.7d). The lines in these spectra where sharpened for presentation and assignment purposes by careful Gaussian resolution-enhancing apodization of the data. Even so, broadening is present due to the increased molecular weight. Obviously, spectra of the trimer will be markedly broader than those of the dimer. It is not, therefore, surprising that proton spectra of lignins (Figure 5.7e), are so

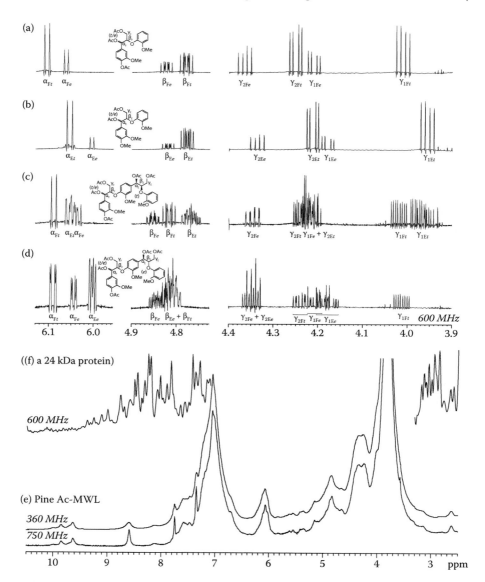

FIGURE 5.7 1D spectra of side-chain regions of β-ether dimers and trimers, and lignin showing how lignins are broadened by its stereochemical complexity. Figures a to d are 600 MHz resolution-enhanced spectra of acetylated β-ether models: (a) a free-phenolic dimer; (b) an etherified dimer; (c) 4 of the 8 isomers of a trimer (etherified dimeric unit is *threo*); (d) the other 4 of the 8 isomers of a trimer (etherified dimeric unit is *erythro*). (e) Acetylated pine isolated lignin, at 360 and 750 MHz. (f) A section of a 24 kDa protein at 600 MHz showing how much more resolved and detailed the proton NMR spectrum of even a much higher molecular mass protein is due to its being a single molecular structure (Cornilescu, G., Cornilescu, C.C., Zhao, Q., Frederick, R.O., Peterson, F.C., Thao, S., and Markley, J.L., *J. Biomol. NMR,* 29(3), 387–390, 2004).

apparently featureless, especially when compared with spectra of even much larger polymers with a single structure, e.g., the protein in Figure 5.7f [410]. Careful interpretation of lignin ^{1}H NMR spectra can reveal important details and, fortunately, the broadness is not an overbearing deterrent to producing valuable 2D correlations, as will be demonstrated.

1D Proton NMR Studies of Isolated and Synthetic Lignins

A comprehensive review of proton NMR of lignin model compounds, synthetic, isolated, and degraded lignins and their derivatives prior to 1968 was made in Ludwig's chapter 8 [145] in Lignins: Occurrence, Formation, Structure and Reactions [1]. Much of the work reported stemmed from the seminal studies by Ludwig, Nist, and McCarthy published in 1964 [146,147], and the 1968 review by Lenz [148].

More recently, Lundquist has provided systematic and comprehensive studies on lignins by proton NMR at magnetic field strengths ranging up to 500 MHz. Formyl groups in isolated lignins were found at a level of ~7% [126]. They were primarily due to coniferaldehyde (~4%) and vanillin end-groups. The aldehyde signals completely disappeared following borohydride reduction, as was elegantly shown in the first application of difference spectroscopy reported in lignin literature. Birch [149] and spruce [150] lignins, purified to low carbohydrate level, were examined in considerable detail with the help of spectra from model compounds [151]. Carbohydrates, present either as contaminants or linked to lignins, were primarily xylans in birch; evidence for benzyl ester linkages was presented [152,153].

Comparisons were made between synthetic coniferyl alcohol dehydrogenation polymer (DHP) lignins and MWL from spruce [154]. MWLs contained more aldehydes, possibly attributable to oxidative lignin aging reactions. DHPs contained more β–5- and β–β-units as well as cinnamyl alcohol end-groups and less β–O–4-units; the DHPs had a consequently lower hydroxyl content. Differences in interunit distributions between DHPs and isolated lignins have come to be cited as a justification for a theory of carefully controlled lignification [155]. However, the result is simply explained by the relatively large amount of dimerization that occurs in DHPs compared with the endwise polymerization that occurs during lignification when the coniferyl alcohol supply or the radical generation capacity is more limited, as reviewed [107]. These ideas, and the notion of "bulk" vs "endwise" polymerization were lucidly discussed at the end of chapter 4 [156] in the original *Lignins* book [1], and in Adler's 1977 review [157]. Enhancing cross-linking of a monolignol with a dilignol (model), using only simple methods to limit diffusion, elegantly demonstrated that such simple explanations were all that were required to explain high β-ether levels [158].

Compositional differences between various fractions of isolated lignins, in studies also using ^{13}C-NMR, suggested that differences in monomeric composition relate to *in situ* heterogeneity of poplar lignins [159]. Proton NMR was applied to an examination of the ratios of *threo-* and *erythro-*isomers of β-aryl ether units in softwoods and hardwoods, using a set of guaiacyl–guaiacyl, guaiacyl–syringyl, and syringyl–syringyl model compounds [160]. The study extended earlier proton [150,151] and ^{13}C NMR work [161–163] that showed that the two isomers were present in roughly equal amounts in softwood guaiacyl lignins, but that *erythro*-isomers predominated in hardwood syringyl-guaiacyl lignins.

The solvent DMSO has particular value in NMR. This is due to its being a poor facilitator of proton exchange that, therefore, limits the normal broadening seen in the resonances of exchangeable protons, such as in hydroxyl groups. The various phenolic hydroxyl groups in lignin are well dispersed in this solvent (8.1–9.3 ppm) and can be quantified [123,124]. Aldehydes from 9.3–10.3 ppm (as in other solvents) are also quite well resolved. In acetylated lignins, the various phenolic and aliphatic acetates have acetate methyl protons that are reasonably well resolved. Lignin is sufficiently complex that absolute quantification is difficult, but reasonable quantification of various units and functional groups is possible.

Proton 1D NMR continues to be extremely useful for lignin structural studies, as reviewed more recently [164]. Its role is being augmented by 2D methods.

Quantification in 1D Proton NMR

Proton NMR spectra are inherently quantitative, providing the sample is allowed to relax sufficiently between scans (as is normally the case, since proton T_1-relaxation times in polymeric lignins are typically well under 1 s). The sensitivity is considerably higher than in ^{13}C NMR. Some groups, such as methoxyls and acetate methyl groups in acetylated lignins, receive enhancements due to the multiplicity of identical protons. Proton NMR and quantitative aspects were extensively treated in the previous treatise on lignins [145] and will not be reviewed here. The emphasis on ^{13}C NMR quantification in the Quantitative ^{13}C NMR of Lignins section of this chapter is a reflection more of the need to understand the issues of such quantification, and should not be interpreted as implying any superiority over ^{1}H NMR quantification. In both cases, the inability to fully resolve resonances of interest provides the most significant quantification problems.

Introduction to ^{13}C NMR of Lignins

The introduction of ^{13}C NMR opened up a powerful new spectroscopic method for the analysis of complex polymers, such as lignins. Its value lay mainly in qualitative aspects; unlike ^{1}H NMR, ^{13}C NMR requires special time-consuming methods for quantification (see Quantitative ^{13}C NMR of Lignins section). A series of seminal papers established the necessary model compound database and led to signal assignments to synthetic and isolated lignins from a variety of plants [8,9,161,162,165–170]. Most of the assignments made in subsequent papers on lignin ^{13}C NMR spectra are based directly on the data in this series, which were, in turn, based on numerous monomeric and dimeric model compounds. The most important paper in this era, published by Nimz et al., is a listing of about 40 ^{13}C chemical shift assignments of MWLs isolated from maple, beech, oak, cherry, compression wood, bamboo, wheat straw, a GSH-DHP, and an H-DHP (these terms are defined below) [166]. It is worth noting, however, that assignments in many subsequent papers are often not fully authenticated and there is still considerable overlap. Suspected structures must have all of their peaks present. For some structures it is difficult to unambiguously identify diagnostic peaks that are separated sufficiently from those of other structures. The previous limitations of the model compounds are perhaps the main reason why typically only 35–50 chemical shift assignments were made in subsequent reports prior to the recent characterization of G-, GS-, and S-DHPs, in which over 100 assignments were made [171]. Work continues on improving the accuracy and reliability of assignments, based primarily on the data of model compounds that more accurately represent structural entities in lignin.

In a 1D ^{13}C NMR spectrum of underivatized lignin, almost all signals from side-chain carbons are found in a rather narrow chemical shift range (52–90 ppm) and there are few signals that can be attributed to single lignin structures. Carbon-13 NMR signals are broadened by the structural and stereochemical complexity noted in the Broadness of Proton Spectra section. Without complete signal resolution, ^{13}C NMR suffers from many of the same problems as ^{1}H NMR.

Researchers use a variety of conditions for acquiring lignin (and model compound) NMR data. Underivatized lignins are usually run in either DMSO-d_6 or acetone-d_6/ D$_2$O (typically ~9:1) solvent. In both cases, chemical shifts are somewhat sensitive to solute concentrations, temperature, and, in the latter case, the exact make-up of the solvent. Tables of NMR data in these solvents appear in several references [171–173]. Acetylated lignins provide more ideal samples. They are quite soluble in common organic solvents and solute interactions are less severe. Nevertheless, there are significant differences between samples run in CDCl$_3$ vs acetone-d_6, and attempts to relate data between the two solvents have only been partially successful. From extensive solvent effect studies of lignin model compounds, researchers have concluded that acetone-d_6 is a better solvent choice if the intent is to correlate model chemical shifts with lignin chemical shifts [174].

In an attempt to address the issues of solvent on chemical shift, researchers have established a database of lignin model compounds in which they attempted to run all compounds in three solvents, acetone-d_6, CDCl$_3$, and DMSO-d_6 [175]. However, the massive undertaking of unequivocally assigning spectral peaks has really only been met for samples in acetone-d_6, the solvent upon which our groups have standardized for acetylated lignins. Others, with their own databases of chemical shifts, have standardized on CDCl$_3$ or DMSO. Each solvent system offers its own small advantages (and personal preferences); a convergence to one solvent is not likely in the near future. The electronic model compound database is updated regularly and is freely available [175].

Since ^{13}C NMR of lignins has been reviewed a number of times [176,177], we will not provide particularly much detail here. A comprehensive table of chemical shift assignments for fully acetylated lignins was recently compiled by comparing ^{13}C NMR data from numerous G, GS, and S trimeric model compounds with data from DHPs and lignins isolated from softwoods and hardwoods [10,178,179]. Table 5.1 indicates the authors' best attempt to reliably assign peaks in representative acetylated softwood and hardwood isolated lignins. The assignments are based mainly on dimeric and trimeric model compounds and synthetic lignins, aided by 2D studies.

TABLE 5.1
^{13}C-NMR Chemical Shift Assignments for Acetylated Lignins in Acetone-d$_6$

Chemical Shift		Assignment[#]
From	**To**	
20.2	20.3	AcMe on phenolic OH of **S**-rings
20.5	20.6	AcMe on Bβ in **S–b–S**-glycerol
20.6	20.7	AcMe on primary (γ) OH
20.8	20.9	AcMe on phenolic OH of **G**-rings & on benzylic OH (α)
	29.83	Center signal of acetone-d$_6$ multiplet (solvent and internal standard)
50.7	51.1	β in **EG–b1–GF**
51.1	51.4	β in **G–c–G**
55.3	55.5	β in **G–r–G, G–r–S, S–r–S**
56.3	56.5	OMe
62.8	63.3	γ in **G–b–G**(*e*), **S–b–G**(*e*),* **G–c–G**-glycerol
63.2	63.6	γ in **G–b–S**(*e*), **S–b–S**(*e*), **G–a–G**
63.4	63.8	γ in **G–b–G**(*t*), **S–b–G**(*t*)*
64.2	64.5	γ in **G–b–S**(*t*), **S–b–S**(*t*), γ in dibenzodioxocin moiety
65.0	65.2	γ in **EG–b1–GF**
65.2	65.5	γ in **C**-alc & **S**-alc end-groups
65.9	66.1	γ in **G–c–G**
72.4	72.7	γ in **G–r–G, G–r–S, S–r–S**
72.9	73.4	Bβ in **G–c–G**-glycerol, **S–b–S**-glycerol
74.4	74.7	α in **G–b–G**(*e*)
74.9	75.4	α in **G–b–S**(*e*)
75.2	75.7	α in **G–b–G**(*t*), **S–b–S**(*e*)
75.7	75.8	α in **EG–b1–GF**
76.3	76.7	α in **G–b–S**(*t*)
76.5	77.0	α in **S–b–S**(*t*)
79.9	80.4	β in **G–b–G**(*e*), **S–b–G**(*e*)*
80.3	80.7	β in **G–b–G**(*t*); **S–b–G**(*t*)*
80.6	81.0	α in **G–a–G**
81.0	81.4	β in **G–b–S**(*e*), **S–b–S**(*e*)
81.4	81.8	β in **G–b–S**(*t*), **S–b–S**(*t*), **G–a–G**
83.3	83.6	β in dibenzodioxocin moiety
85.3	85.5	α in dibenzodioxocin moiety
86.1	86.3	α in **G–r–G**
86.3	86.	α in **S–r–S**
88.3	88.6	α in **G–c–G**
88.6	88.9	α in **S–c–G**
103.0	103.3	2,6F in **S–r–S**
103.4	103.6	2,6F in **G–r–S**
103.6	103.9	2,6E in **S–r–S, S–c–G**
104.5	105.0	2,6E,F in **S–b–S**(*e*)

TABLE 5.1
^{13}C-NMR Chemical Shift Assignments for Acetylated Lignins in Acetone-d_6 (Continued)

Chemical Shift		Assignment[#]
From	To	
105.2	105.5	2,6E in S–b–S(t)
106.6	107.3	2,6 in S-rings with $\alpha C = O$
111.0	111.3	2F in G–c–G, G–r–G, G–r–S
111.3	111.7	2E in G–c–G, G–r–G, C-alc, vanillin
111.7	112.0	2F in G–55–G
112.2	112.4	2F in G–b–S; B2 in G–c–C-alc
112.4	112.5	2 in EG–b1–GF
112.5	113.1	2E,F in G–b–G, G–c–G; 2E in G–b–S, cinnamaldehydes; 2F in G–a–G
114.2	114.4	2F in EG–b1–GF
115.7	115.9	5E in vanillin
116.2	116.4	B6 in G–c–C-alc
118.2	118.4	5E in cimmanaldehydes
118.5	118.6	6F in G–r–G, G–r–S
118.6	118.7	6F in G–c–G
118.5	119.1	5E in G–b–G, G–b–S, G–r–G, EG–b1–GF
119.0	119.2	5E in C-alc; 6E in G–c–G
119.3	119.5	5E in G–c–G, 6E in G–r–G
119.9	120.1	6F in G–b–S
120.1	120.7	6E,F in G–b–G; 6E in G–b–S, EG–b1–GF, C-alc; 6F in G–a–G, in dibenzodioxocin moiety
121.5	121.7	6F in G–55–G
122.1	122.3	βE in C-alc; 6F in EG–b1–GF
122.4	122.6	5F in G–r–S, in dibenzodioxocin moiety
122.7	122.9	3,5E in p-OH-benzoate
123.0	123.2	5F in EG–b1–GF
123.2	123.4	5F in G–b–G, G–b–S, G–a–G; βE in C-alc
123.6	123.8	5F in G–c–G, G–r–G; 6E in cinnamaldehydes
124.0	124.2	βE in S-alc
125.9	126.1	6E in vanillin
127.8	128.0	1F in p-OH-benzoate
128.2	128.4	βE in C-ald
128.7	128.9	4F in S–r–S
128.9	129.3	B5 in G–c–G
129.3	129.5	4F in S–b–S
129.9	130.1	1E in cimmamaldehydes
131.0	131.2	B1 in G–c–G-glycerol
131.5	131.7	5F in dibenzodioxocin moiety; B1 in G–c–C-alc; 2,6F in p-OH-benzoate

(Continued)

TABLE 5.1

^{13}C-NMR Chemical Shift Assignments for Acetylated Lignins in Acetone-d$_6$ (Continued)

Chemical Shift		Assignment[#]
From	**To**	
131.7	132.0	5F in G–55–G
132.0	132.2	1E in vanillin
132.2	132.6	1E in C-alc
132.6	133.0	1E in G–b–G
133.0	133.3	B5 in dibenzodioxocin moiety
133.3	133.4	1E in S-alc
133.8	134.4	1E in S–b–S
134.0	134.2	αE in C-alc
134.3	134.5	αE in S-alc
134.5	134.7	Bα in G–c–C-alc
134.7	134.9	1 in EG–b1–GF
135.5	135.7	4 in ES–r–S
135.8	136.0	1 in FG–55–G
136.0	136.6	1 in FG–b–G; 1F & 4E in S–b–S; 4 in ES–c–G;
136.6	136.8	1 in EG–r–G
137.0	137.3	1 in FG–b–S; 1 in EG–c–G
137.5	137.7	1 in FG–a–G
137.7	137.9	B1 in EG–b1–GF
138.0	138.2	1F in dibenzodioxocin moiety
138.3	138.5	4 in FG–55–G
138.5	138.9	1 in ES–r–S
140.0	140.2	4 in FG–r–G, EG–b1–GF
140.5	140.7	4F in G–b–S, G–a–G
140.7	141.0	4F in G–b–G, in dibenzodioxocin moiety; 1,4F in G–c–G
141.2	141.4	1 in FS–r–S
141.7	142.0	1 in FG–r–G
145.0	145.4	B3 in G–c–G
147.0	147.6	4 in EG–r–G
147.8	148.0	4 in EG–b1–GF
148.0	148.5	4E in G–b–G, C-alc
149.2	149.4	B4 in G–c–C-alc, S–c–C-alc
150.9	151.0	4E in C-ald
151.4	151.7	3E in G–b–G, G–b–S, vanillin, EG–b1–GF
151.8	152.2	3E in C-alc, cinnamaldehydes, G–r–G, G–c–G; 3F in G–b–G, G–b–S, G–a–G, EG–b1–GF, in dibenzodioxocin moiety; 4E in G–b–S
152.3	152.7	3F in G–r–G, G–c–G, G–55–G
152.8	153.3	3,5F in S–b–S, S–r–S; αE in C-ald
153.4	153.6	4E in vanillin
153.7	154.1	3,5E in S–b–S, S–r–S
155.6	155.8	4F in p-OH-benzoate

TABLE 5.1
^{13}C-NMR Chemical Shift Assignments for Acetylated Lignins in Acetone-d$_6$ (Continued)

Chemical Shift		Assignment#
From	To	
165.5	166.5	Benzoate C=O in p-OH-benzoate
168.2	168.3	Phenolic AcC=O on S-rings with α C=O
168.4	168.7	Phenolic AcC=O on S-rings
168.9	169.0	Phenolic AcC=O on G-rings
169.2	169.4	Phenolic AcC=O on p-OH-benzoate entities
169.9	170.0	Benzylic (α) AcC=O in G- & S-rings
170.2	170.4	Bβ AcC=O in G–c–G-glycerol, S–b–S-glycerol
170.6	170.8	Primary (γ) AcC=O in G–b–G, G–b–S, G–a–G, G–c–G-glycerol
170.9	171.0	Primary (γ) AcC=O in G–c–G
180.5	180.9	p-Quinone C=O in G-rings
191.1	191.5	C=O in vanillin
193.8	194.2	C=O in cinnamaldehydes

Legend: **G** = guaiacyl, **S** = syringyl, **C** = coniferyl, **S** = sinapyl, **V** = vanillyl, E = etherified phenolic, F = free phenolic, a = α–O–4, b = β–O–4, r = resinol (β–β), c = coumaran (β–5), b1 = β–1, 55 = 5–5 (biphenyl), (*e*) *erythro*, (*t*) = *threo*, alc = alcohol, ald = aldehyde. #All assignments refer to the A-ring of the dimer fragment unless otherwise noted. *Shifts are from model compound **S–b–G** only and were not found in lignins or DHPs.

LIGNIN MODEL STUDIES

Low molecular weight model compounds, particularly dimers and trimers, are at the heart of NMR studies on lignins. Nearly all studies that attempt to assign new structures to lignin spectra incorporate model syntheses to provide key NMR data and for structural authentication; these are, therefore, dealt with under the lignin NMR sections. To accumulate much of the required data in a single repository, researchers have established an NMR database of lignin model (and related) compounds, which continues to be regularly updated [175]. A great deal of literature is associated with the painstaking synthesis, structural authentication, and spectral interpretation of a range of lignin model compounds.

POLYMERIC LIGNIN MODELS (DHPs)

Synthetic polymers can be made from monolignols using single-electron oxidants under a variety of conditions. Perhaps the most biomimetic approach is using a peroxidase and hydrogen peroxide. The polymerization in bulk solution does not mimic all aspects of lignification, and the polymers produced can vary considerably from those in plants [154,166,169,177,180–186]. Even when considerable efforts are made to limit the rate at which radicals encounter each other [156,158,187–189], dimerization reactions may be more prevalent than in lignification, where the predominant reaction is cross-coupling of the monolignol (radical) with the growing polymer (radical) [158]. Nevertheless, the similar reaction chemistry provides all of the major products observed in lignins; thus, DHPs occupy a place of enormous value in structural and spectroscopic studies.

Utility of DHPs

Dehydrogenation polymers (DHPs) of *p*-hydroxycinnamyl alcohols have been used as oligomeric/polymeric models for the lignin structure for decades. There are many reasons for using DHPs rather than natural isolated lignins, such as a MWL. A DHP is free of extraneous wood components, such as extractives, tannins, and proteins. It can be difficult to remove all of the carbohydrate material from isolated lignins, especially because a fraction is thought to be chemically bound to the lignin. All of the major subunit types are represented in DHPs, although the linkage type distribution is quite different from native or isolated lignins. Another significant advantage of DHPs is that they are relatively simple to label isotopically in specific positions by using appropriately labeled *p*-hydroxycinnamyl alcohols [111,180,190–192] (see also the HSQC or HSQC-TOCSY Difference Spectra section and Figure 5.4). The use of isotopically labeled DHPs is important in the study of enzymatic delignification (pulping) and bleaching, and in microbial degradation studies. Use of a DHP prepared from [α–^{13}C]-coniferyl alcohol facilitated the interpretation of the ^{13}C NMR spectra of the product resulting from the oxidation of the labeled DHP by a lignin peroxidase [193]. Labeled DHPs were essential in studies of lignin/polysaccharide complexes, which are important in forage digestibility research [36,133].

It is also possible to incorporate into DHPs minor structures (labeled or unlabeled) that are known to be present in natural lignins but are not formed from *p*-hydroxycinnamyl alcohols during dehydropolymerization. For example, *p*-hydroxybenzoic acid with a ^{14}C-labeled carboxyl carbon was incorporated into a DHP for the purpose of determining the nature of this unit on lignins and its fate during alkaline hydrolysis [194]. Incorporation of strategically ^{13}C-labeled ferulates and diferulates into synthetic lignins and the application of various 2D experiments, primarily HMBC, provided the range of ferulate incorporation pathways [55,129,133]. Synthetic DHPs incorporating coniferyl *p*-coumarate and coniferyl *p*-hydroxybenzoate were used to elucidate mechanisms of lignin acylation [195]. Synthetic lignins from coniferaldehyde were used to model lignins in plants deficient in CAD [65,196]. Synthetic lignins incorporating small amounts of ^{13}C-labeled ferulic acid were used to provide confirmatory evidence for the incorporation of ferulic acid into lignin in transgenic plants deficient in cinnamoyl-CoA reductase (CCR) [97,104].

Experiments that utilize labeled DHPs supplement the more difficult and time-consuming experiments involving the labeling of natural lignins by feeding labeled precursors to the plant [197–202]. Nevertheless, DHPs are, like lignins, not homogeneous compounds and their structures are not completely determinable. Some of the problems in lignin structural elucidation are also evidenced with DHPs.

An extension of the DHP methods is to prepare lignins within preformed cell walls [203,204]. The walls may have their own peroxidases and other features, such as ferulates that may act as lignin nucleation sites in grass walls [130]. The lignins produced in this manner appear to be much closer in composition/structure to natural lignins, presumably due to the slow diffusion of monomer (radicals) into the lignifying zone [203]. Again, minor or novel monomers can be incorporated into lignins [112]. In this way, a case was made for incorporating coniferyl (or sinapyl) ferulate into lignins to improve delignification and enzymatic degradability of walls [113].

Enzyme-Initiated Dehydropolymerization

DHPs prepared from coniferyl alcohol are the most common and give guaiacyl DHPs (G-DHP) that are intended as models of lignins isolated from softwoods. Polymers prepared from a mixture of coniferyl and sinapyl alcohols (GS-DHP) are used as models of hardwood lignins, and those prepared from a mixture of all three alcohols (GSH-DHPs) are generally considered as models for grass lignins and some wood lignins with significant contents of p-hydroxyphenyl moieties [166]. DHPs derived from pure sinapyl [205] and pure p-coumaryl alcohol [206] have been prepared even though it was suspected that neither one relates to any natural polymer. In fact, lignins in certain monolignol-biosynthetic-pathway mutants or transgenics can come remarkably close (Lignins in Monolignol-Biosynthetic-Pathway Mutant or Transgenic Plants section). A DHP's main utility is as a mechanistic and characterization tool to obtain data on minor structural entities.

The key step in both natural lignification processes and conventional laboratory dehydrogenations is the enzyme-initiated generation of phenoxy radicals of the p-hydroxycinnamyl alcohols. This simple oxidation leads to a complex variety of nonenzymatic processes that includes radical coupling reactions, nucleophilic additions to quinone methides, rearrangements, side-chain oxidations, and side-chain eliminations to generate the complex lignin macromolecule [207]. Peroxidase and/or a phenol oxidase (such as laccase) have been postulated to be involved in lignification [208]. A horseradish peroxidase/hydrogen peroxide system is commonly used in laboratory dehydrogenations [205,209].

Guaiacyl DHPs were also prepared from coniferin by the action of β-glucosidase and peroxidase, with hydrogen peroxide generated *in situ* through the action of oxygen and glucose oxidase on the glucose liberated from the coniferin [188]. Examination of these DHPs by ^{13}C NMR spectroscopy indicated a closer approximation to the structure of a softwood lignin than DHPs prepared in the conventional manner from coniferyl alcohol.

A number of studies addressed the impact of reaction conditions, including the presence of various saccharides on the types and stereochemistries of various lignin units [143,210–212]. The potential for production of novel β–γ-linkages from sinapyl alcohol in the presence of pectins was illustrated [211]. *Pinus taeda* suspensions fed with labeled phenylalanine demonstrated that lignins resembling native lignins could be produced *in situ* [213]. In addition, DHPs synthesized within primary cell walls of suspension-cultured maize, using native peroxidases within the walls, but exogenously supplied H_2O_2 and monolignols, produced lignins that were remarkably similar to maize lignins, as determined by ^{13}C NMR as well as degradative methods [203]. An extracellular lignin produced by suspension cultures of *Picea abies* had intermediate character [186].

A DHP synthesized using NMR-invisible methoxyl groups, prepared from coniferyl alcohol synthesized via $^{12}CD_3I$ (^{13}C-depleted, trideuterated) allowed NMR spectra to be acquired without the complication of the overwhelming methoxyl group that can produce artifacts particularly in 2D spectra run without the use of gradient-edited pulse sequences [2,74]. That DHP is used for 2D illustrations here. The possibility of incorporation of *cis*-coniferyl alcohol into lignins was examined via a DHP [214]. Unfortunately, NMR was not used to characterize the nature of residual cinnamyl alcohol end-groups or the stereochemistries of the resulting units. A more recent study showed that the *cis*-double bonds remained in the product and that most of the units were produced as the same isomers; however, a number of currently unidentified units were also produced [39]. No evidence for *cis*-cinnamyl alcohol side-chains has been found in isolated lignins to date, suggesting that *cis*-hydroxycinnamyl alcohols are not significantly present in the cell wall during lignification.

Dehydropolymerization Initiated by Metal Oxidants

Dehydrodimerizations and polymerizations have alternatively been carried out using metallic single-electron oxidants in a variety of solvents. This nonenzymatic approach to synthesis of DHPs can result in polylignols that, in some aspects (such as β-ether frequency), have a closer resemblance to corresponding MWLs than conventionally prepared DHPs [179,215]. Using metal one-electron oxidants (iron, manganese, and copper salts) is more flexible, in that the reaction conditions are not constrained within the temperature, solvent, and pH limits required by the enzyme. Other examples of metal salt-initiated polymerizations of *p*-hydroxycinnamyl alcohols reported in the literature are the preparation of DHPs using ferric chloride in various organic-aqueous solvents [216–218]. However, no ^{13}C NMR data were obtained. β-Ether coupling of sinapyl alcohol was shown to be favored by increasing the dioxane content or acidity of the reaction medium [218].

Comparisons of DHPs with Natural Lignins

In general, synthetic lignins have excessive monolignol-monolignol coupling resulting in too frequent β–β-units (**B**, resinols) and unsaturated cinnamyl alcohol side-chains **X1** (from β–5- or β–O–4-dehydrodimerization). Attempts to improve the resemblance have met with variable success. Brunow found that low pH favors β-ether formation [219]. The closer resemblance of a typical Mn(III)-derived coniferyl alcohol guaiacyl G-DHP to a guaiacyl MWL isolated from spruce wood is seen in the ^{13}C NMR spectrum of the side-chain region (Figure 5.8). In contrast, in the spectrum of the conventional enzymatic G-DHP, signals due to β–5- **B** and β–β-units **C** are much more intense than in the MWL or the Mn(III)-derived DHP. A feature in common with both of the DHPs is the presence of coniferyl alcohol end units **X1**. This unsaturation is less prominent in a MWL. A comparison of a similarly prepared guaiacyl/syringyl GS-DHP and a conventional DHP with a MWL isolated from elm wood is shown in Figure 5.9. As with the G-DHP, the overabundance of the β–5- **B** and β–β-units **C** and unsaturated end-groups **X1** is apparent in the conventional DHP. The Mn(III)-derived GS-DHP has an even greater resemblance to the GS-MWL because, unlike the corresponding G-DHPs, unsaturated end units **X1** are almost absent.

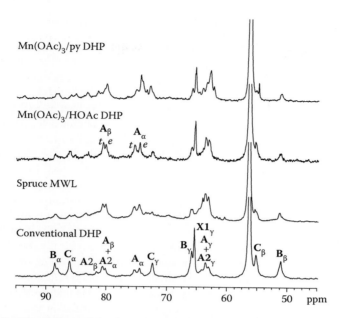

FIGURE 5.8 Comparison of the side-chain regions of 1D ^{13}C NMR spectra of various acetylated G-DHPs with a softwood lignin. Peak assignment follows that established in Figure 5.3 (Landucci, L.L., *J. Wood Chem. Technol.*, 20(3), 243–264, 2000).

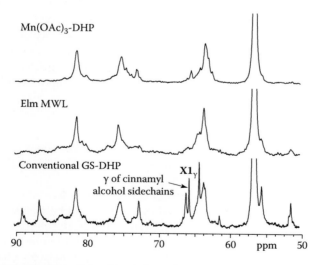

FIGURE 5.9 Comparison of the side-chain regions of 1D ^{13}C NMR spectra of two acetylated GS-DHPs with a hardwood lignin (Landucci, L.L., *J. Wood Chem. Technol.*, 20(3), 243–264, 2000).

NATURAL LIGNINS

A requirement of solution-state NMR is that samples must be soluble in a suitable organic solvent. Lignins isolated following solvent extraction (usually in 96:4 dioxane:water) of ball-milled plant cell wall materials provide perhaps the best, and most frequently used, materials of this type [220–222]. Unfortunately, only a fraction of the cell wall lignin is represented this way. MWL yields from woods are typically around 15–20% [221], although higher yields are easily obtained for some samples, particularly if crude polysaccharidases are used following ball milling and prior to lignin extraction. Milling time affects both the yield and the β-ether frequency in the product [223]. Poplar lignin was recently obtained in ~70% yield in NMR-suitable form this way [224]. In Lapierre's method, where no lyopholizing is done during the isolation, lignins from immature poplar are recovered essentially quantitatively [225,226]. These lignins are soluble in DMSO, but resolubilization is difficult following lyopholization. Some grasses have more soluble lignins; an isolated mature maize lignin was 67% of the total lignin [56]. Isolated lignins are typically dissolved in acetone-d_6:D_2O (~9:1) or DMSO-d_6 for NMR studies of underivatized lignins. More commonly, the isolated lignins are acetylated, which leads to good solubility in acetone-d_6 or $CDCl_3$, and less-viscous solutions. For NMR experiments that involve long delays in the pulse sequences, such as in long-range ^{13}C–^1H correlation experiments (e.g., HMBC), treating lignins with metal chelators can be beneficial [2,56].

Selectively labeling lignins by feeding specifically labeled precursors is an approach toward unambiguously assigning NMR resonances. This has been particularly successful with ginkgo, in which cut stems continue to grow for a month or more when the cut section is immersed in water. Exogenously administering coniferin does not appear to perturb lignification and provides a "natural selective enrichment" [50].

Quality lignins can also be isolated via mild acidolysis, including in conjunction with cellulolytic treatments [227–232].

A relatively recent development, aided by the high sensitivity of modern NMR instruments equipped with cryogenic probes, is to obtain lignin data from spectra run on the entire cell wall component; i.e., without lignin fractionation/isolation [61,64,98,100]. Ball-milled cell walls can be dissolved or simply swelled in various solvents to yield strikingly high quality spectra [102,103].

Elucidation of Lignin Substructures

NMR provides unparalleled methods for elucidating the structures of organic molecules. It is difficult to envision determining reaction mechanisms via model compound studies, for example, without the use of NMR.

Although lignin is complex, its derivation from only a few monomers suggests that the number of structures to be found in lignin should be relatively small, and should have all easily succumbed to identification. It was, therefore, a surprise that, as recently as 1995, a major new structure common to all lignins was discovered [114–116]. Since then, other pathways during lignification have also been revealed by less prevalent structures. More mysteries abound, as many peaks in 1D ^{13}C (or contours in 2D NMR) spectra of lignins remain unidentified. Not all of these can necessarily to attributed to in vivo lignins; structural changes certainly occur during lignin isolation. Nevertheless, reliably assigning all of the peaks or contours in lignin spectra remains a formidable challenge. Delineating the changes to lignin during various treatments, such as chemical pulping, requires as complete a characterization of the original lignin as possible.

Early proton and ^{13}C NMR spectra indicated that the already well-known major lignin structural units could be identified diagnostically from their side-chain proton and carbon shifts [9,146–149]. These included β–O–4- **A**, β–5- **B** (or phenylcoumaran), β–β- **C** (or resinol) units, as well as cinnamyl alcohol **X1** and cinnamaldehyde and benzaldehyde end-units. The major units were also revealed in suspension-cultured *Pinus taeda* walls by solid-state NMR following feeding of side-chain-labeled precursors [213]. Two types of units, α-aryl ethers **A2** (actually α,β-diaryl ethers) and β–1-units (see next section) were not readily observed (see below). Since aromatic shifts are more difficult to assign, and are severely overlapped in both proton and carbon NMR, researchers have more trouble detecting biphenyl (5–5-) and biaryl ether (4–O–5-) units. A great number of papers and reviews have dealt with establishing the nature and content of these standard units. Salient features are noted under the headings below.

β–O–4-ethers (β-aryl Ethers, **A**)

By virtue of its importance in lignins and in pulping mechanisms, the β-ether unit **A** has been the most studied. Various methods for assigning *threo-* and *erythro-* isomers in models and in lignins have appeared based on ^{1}H, ^{13}C, ^{19}F, and ^{31}P NMR [8,151,233–235]. Enormous numbers of isomers must be dealt with as molecules become more complex [107,236]. Although there are only two isomers of the popular β-ether dimeric model guaiacylglycerol-β-guaiacyl ether, there are eight isomers of the analogous trimer (Figure 5.7). Many of the proton chemical shifts are resolvable at high magnetic field, e.g., 600 MHz [11]. Fortunately, the isomers cluster into groups for *threo-* and *erythro-*isomers of each unit type, etherified or free-phenolic, and syringyl or guaiacyl. The various combinations of S and G units have been studied using a series of dimers [237], and more recently using all of the possible S/G trimers, for which detailed NMR data have been presented [10].

Isomer Distributions   Lignin β-guaiacyl ether units **A** are essentially 50:50 *erythro:threo* [143]. Syringyl lignins have a much higher proportion of *erythro-*isomers [238], as beautifully evident in NMR spectra (see for example Figures 5.3f, 5.18f, and 5.18g) [2,70,144,239]. This is because, under a wide range of aqueous conditions, water adds to β-guaiacyl quinone methides essentially randomly, whereas water addition to β-syringyl quinone methides favors *erythro-*isomers by about 75:25 [143,144]. Natural lignins are analogous in this regard to synthetic lignins and to β–O–4-dimers produced from the monolignols via peroxidase in aqueous systems. The product distribution is logically a result of simple chemical control [236], an observation that is inconsistent with recent proposals advocating absolute structural control of the polymer produced during lignification [155].

Syringyl and guaiacyl *threo-* and *erythro-*β-ether protons and carbons overlap somewhat. The *threo-*isomers always have a higher-field γ-proton, which makes them particularly easy to identify in 2D spectra, especially TOCSY and HMQC-TOCSY.

Figure 5.3f shows an example 2D HMQC-TOCSY spectrum of a syringyl-rich poplar acetylated MWL, in which the predominance of the *erythro*-isomers is clearly revealed, particularly in the correlation with C-α (~75 ppm). The Cα–Cβ correlation clusters for *erythro*- and *threo*-isomers and for syringyl and guaiacyl units are beautifully disperse in a 2D INADEQUATE spectrum [50]. A plot of model shifts for all 16 of the β-ether combinations (*erythro/threo*, syringyl/guaiacyl, and etherified/free-phenolic), correlated well with major features of the contour complex, on the same spectrum.

Chemical cross-coupling propensities of monolignols with guaiacyl and syringyl units in lignin are becoming available [10,158,210]. Sinapyl alcohol does not appear to readily cross-couple with guaiacyl units in vitro. Thus, in Table 5.1, shifts from model compound **S–b–G** (b = β–O–4) are listed, but such structures have not been authenticated in lignins or DHPs. The presence or absence of hetero-dimeric entities (**G–b–S** or **S–b–G**) in natural lignins could not be confirmed by ^{13}C NMR spectroscopy because the side-chain signals from both *erythro*- (*e*-) and *threo*- (*t*-) GS entities overlap the corresponding signals from *e*- and *t*-**S–b–S** entities; signals from *e*-**S–b–G** entities overlap those from *e*-**G–b–G** entities; and signals from *t*-**S–b–G** entities overlap those from both *e*-**S–b–S** (α-carbon) and *t*-**G–b–G** (β- and γ-carbons). The difficulty of preparing the **S–b–G** entity in trimers by oxidative coupling of sinapyl alcohol with **G–b–G** or **G–b–S** dimers suggests that **S–b–G**-units might not be found in the polymer. However, there is some evidence for such structures; all four combinations, including **G–b–S**, of β-ether dimers can be released by incomplete thioacidolysis [240]. In contrast, the ease of preparing the **G–b–S** entity in trimers by coupling of coniferyl alcohol with **S–b–G** or **S–b–S** dimers suggests that the **S–b–G**-entity is likely in lignin. Unfortunately the INADEQUATE Cα–Cβ contours noted above cannot unambiguously resolve this issue, although it appears to reveal **S–b–G** ethers [50]. Revealing such details of the lignin structure requires extensive use of carefully synthesized model compounds, for which authentic NMR data are available in several solvents.

β–5-Units (Phenylcoumarans, Arylcoumarans, **B**)

Phenylcoumarans **B** are typically recognized as the second most abundant interunit linkage type in lignins; they will, obviously, be at low to negligible levels in syringyl-rich lignins. Although several arylcoumaran natural products were reported as having a *cis*-ring, it is likely that most such products in nature are in fact *trans* [241,242]. The most extensive model compound data is from Miyakoshi and Chen [243,244], although these authors used mainly γ-methyl models that do not model lignins well; substituents effects were also examined earlier [245]. More appropriate model data is found in several references [8,34,175,241]. Model data derives primarily from synthetic schemes by Nakatsubo [246] or Brunow [247]. A clarification of the stereochemistry of various model compounds was addressed [248]. Trimeric compounds containing phenylcoumaran units were described [249–251].

β–β-Units (Resinols, **C** and Other Products)

Resinols **C** are distinguished from the more prominent units in lignin by their formation mechanism. They are generally thought to result from monomer–monomer coupling, although the dearth of releasable (i.e., etherified) pinoresinol units in softwood lignins has led to suggestions that they may be formed during chain extension rather than just dehydrodimerization [252,253]; however, no compelling evidence has been presented. Coupling of monolignols is relatively rare in lignification, where monomer–polymer coupling is the principal manner in which polymerization occurs. Since monolignols strongly prefer to dehydrodimerize than to cross-couple, the paucity of these structures (and of the cinnamyl alcohol end-groups) led researchers to conclude long ago that monomer supply (or radical supply) is carefully limited during lignification to encourage "endwise polymerization" [1,157]. Their levels are about 2–3% in softwoods and up to about 8% in syringyl-rich angiosperms. This difference is possibly due to the higher stability and longer lifetimes of the sinapyl alcohol radicals, and the dimerization favoring β–β-coupling by about 95% for sinapyl alcohol vs about 40% for coniferyl alcohol in aqueous systems with peroxidase [217,254]. The high β–β-coupling propensity of sinapyl alcohol in dimerization reactions also explains why lower amounts of unsaturated side-chains occur in syringyl-rich lignins.

Resinol structures **C** are readily seen in 1D and 2D NMR spectra because of the relative insensitivity of the side-chain proton and carbon chemical shifts to the limited ring substitution that can occur. Those derived from *p*-coumaryl alcohol, however, have a significantly higher α-proton shift [60]. Resinols must be judiciously quantified since, in 2D HMQC or HSQC spectra, the contours for the resinol unit are twice as intense as for any other (unsymmetrical) unit; two coniferyl alcohol monolignols, for example, yield one pinoresinol unit with two of each type of carbon/proton whereas β–O–4-coupling results in only a single β–O–4-unit (with only one set of β–O–4-carbons and protons contributing). Resinol structures **C** are readily seen in the spectra of Figures 5.2, 5.3, 5.4, 5.8, 5.9, 5.10, 5.17, 5.18, 5.19, and 5.24. Only one isomer has been detected in lignins that have not been subjected to acidolytic conditions [255]; *epi*-pinoresinol, for example, has not been verified in lignins.

Crossed resinols are apparently possible, as revealed by degradative analyses such as thioacidolysis [256,257], and as discovered in low-molecular weight oligomers during metabolic profiling [258]. However, NMR has not been used to address this issue in the polymer. The β–β-cross product of ferulate with both sinapyl and coniferyl alcohols has been validated in vivo. HMBC spectra of a ryegrass lignin convincingly revealed both syringyl and guaiacyl cross-products [130].

A recent finding was unanticipated. In spruce isolated lignin it is possible to detect a putative β–β-coupled α–O–α-tetrahydrofuran that appears must result from addition of water to the intermediary quinone methide [259]. A similar structure is found in Kenaf, when trapping of the quinone methide by the γ-OH is prevented due to its natural acylation [260]—see Naturally Acylated Lignins section and later in Figure 5.11. The whole range of such structures bearing γ-acetate, γ-*p*-coumarate, and γ-*p*-hydroxybenzoate groups has been rather unambiguously identified by NMR [101]. Analogs have also been found in ferulate coupling in grasses and cereal grains [261–263]. However, β–β-structures, other than the resinols **C**, were not expected monolignol coupling products, since it has always been assumed that internal trapping of the quinone methide by the γ-OH would be rapid and complete. This data suggests that lignification may be consistent with low pH conditions where the protonation-dependent addition of water is more rapid; the paucity of acyclic α-aryl ethers in lignins has been noted to be consistent with low-pH conditions [264].

5–5-Units (Biphenyls)

Biphenyls were difficult to authenticate and quantify by NMR, since only severely overlapped aromatic resonances were thought to be involved; the side-chains were assumed to be insignificantly different from their counterparts in non-5–5-coupled structures [177]. In fact, the side-chains of the biphenyl units themselves remain difficult to assign, but units coupled to them may be highly diagnostic, since 5–5-units are frequently involved in cyclic structures, dibenzodioxocins **D** (see next section). Prior assignments of 5–5-units as fully vs partially etherified need to be reexamined. Some model data are available [175,265,266] and substituent effects have been examined [267]. Trimers containing biphenyl units have recently had NMR data reported [265]. Unfortunately the β–O–4-dimers that were 5–5-linked in tetrameric models [268], and the trimers linked in hexameric models [269] did not have full NMR assignments.

Dibenzodioxocins **D**   The elucidation of dibenzodioxocins **D** in lignins has been an exciting recent development [114,115]. It provides a new pathway for 5–5-linked units to be incorporated into cyclic ether structures. Despite being ethers, such structures cannot necessarily be fully released by solvolytic methods. NMR was crucial to the identification of the novel units and to the unambiguous demonstration of their presence, at significant levels, in isolated lignins [2,76] and whole-cell-wall preparations [64]. A survey of plants reveals that dibenzodioxocins are present in all lignin classes (from softwoods and hardwoods as well as grasses and legumes).

Dibenzodioxocins **D** show up easily in HMQC or HSQC spectra, where the Cα–Hα and Cβ–Hβ correlations are unique and well-resolved (Figure 5.10). Due to diverse proton-proton coupling constants, the full side-chain correlation matrix does not always show up well in HSQC-TOCSY experiments with typical (80–100 ms) TOCSY mixing times. Figure 5.10a shows an HSQC-TOCSY spectrum of a guaiacyl DHP acquired using a 125 ms TOCSY mixing time, which emphasizes the full **D** coupling array. The peaks or correlations can be difficult to find in unacetylated samples, yet readily show up in acetylated samples. The reason appears to be that the chemical shifts are more variable in unacetylated samples, smearing out the required correlation peaks. They sharpen a little in DMSO. The contours in 2D spectra are also easier to find than significant peaks in 1D ^{13}C NMR spectra. Obviously, since 5–5-units arise only from guaiacyl units (and typically minor *p*-hydroxyphenyl, but not syringyl units), dibenzodioxocin levels can be very low in syringyl-rich lignins from angiosperms (hardwoods, grasses, or legumes). At lower contour levels, such as those shown in the spectra from ryegrass (Figure 5.10f) or aspen (Figure 5.10g), they can almost always be detected. An exception is in the strikingly guaiacyl-depleted lignins in an Arabidopsis mutant upregulated in ferulate 5-hydroxylase (see later in Figure 5.18e), **F5H (or Cald-5H)** section.

An array of dibenzodioxocins with chemical shifts differing from those seen previously in normal guaiacyl-syringyl lignins are found in the *p*-hydroxyphenyl-rich lignins from C3H-down-regulated alfalfa plants [60], as seen later in Figure 5.16e. Synthesis of a range of dibenzodioxocins incorporating *p*-hydroxyphenyl units is required to fully assign these data.

4–O–5-Units (Biaryl Ethers)

Biaryl ethers are also difficult to identify and quantify in lignins. Due to the severe overlap of aromatic resonances, they are not readily identified in syringyl–guaiacyl lignins. Few appropriate model compounds are available for this unit; simple models have been described [270–272], and more continue to appear in the lignin model compound database [175]. One interesting conundrum is that the anticipated aromatic correlations for 4–O–5-linked units cannot be detected in softwood lignins [273], despite there being purportedly some 4% of such units [157]. The ^{13}C–^1H correlations, which should be readily evident in softwood spectra, have not, to date, been found, either in isolated lignins or in whole-cell-wall samples.

α-Aryl Ethers (or α,β-Diaryl Ethers) A2

Such structures **A2**, easily visible in spectra of synthetic lignins (e.g., Figures 5.2b, 5.3a, 5.3b, 5.4, 5.8, 5.10a, 5.10b, and 5.10e), are difficult to detect in most isolated lignins, despite prior estimates that such units were in the 6–8% range [157]. A significant proportion of such estimates can presumably now be attributed to dibenzodioxocins **D**. In studies on pine lignins, noncyclic α-ethers **A2** were concluded to be less than 0.3% of their isolated lignins [40]. It is likely that some of the assignments reported in the literature, from the carbon at ~81 ppm, are actually attributable to α-keto-β-ethers **A3** (see α-Keto-β-Ethers A3 section).

One notable exception to the paucity of noncyclic α-aryl ethers in isolated lignins may occur in tobacco [71]. A tobacco lignin isolate was more similar to a synthetic lignin than any other isolated lignin reported to date. It had significant amounts of β–β-units (resinols **C**) and a substantial α-ether **A2** component characteristic of a bulk polymer. It appears that lignification is not controlled by slow diffusion-limited monomer supply as seen in other plants that have distinctly endwise lignins. Lignins from suspension-cultured spruce also appear more similar to synthetic DHPs, presumably because the lignins are formed outside the cell wall [274].

Diarylpropanes (β–1-Units)

Although β–1-units feature prominently in products from various acidolytic treatments, they have been extraordinarily difficult to find by NMR of lignins isolated by dioxane:water extraction, but have been reported [28,30,34,59,134,275,276]. In part this was because of overlapping signals in 1D and 2D experiments. Ede et al. initially estimated that the level was <2% in pine [28]. It was later found that β–1-units were more efficiently extracted into acetone–water fractions, or by direct derivatization and extraction into acetic anhydride [276]. TOCSY spectra clearly authenticated β–1-units for the first time. The structure is rather convincingly revealed in spectra of an acetylated poplar MWL [50,78,134]. It is likely that this acidic extraction converted other β–1-precursors in lignins to the conventionally described β–1-products, as discussed further below. The most extensive model data is from Miyakoshi and Chen [243,244]. Other data derives primarily from reports on the synthesis of such model compounds [250,277,278].

The occurrence of dienone structures was suggested as an explanation for the discrepancy between degradation and NMR spectral studies [275,279–281]. Mild acidolysis of methylated wood produced β–1-dimers with an unmethylated B-ring; the monomethylated β–1-dimers were always methylated in the A ring and the B-ring was free,

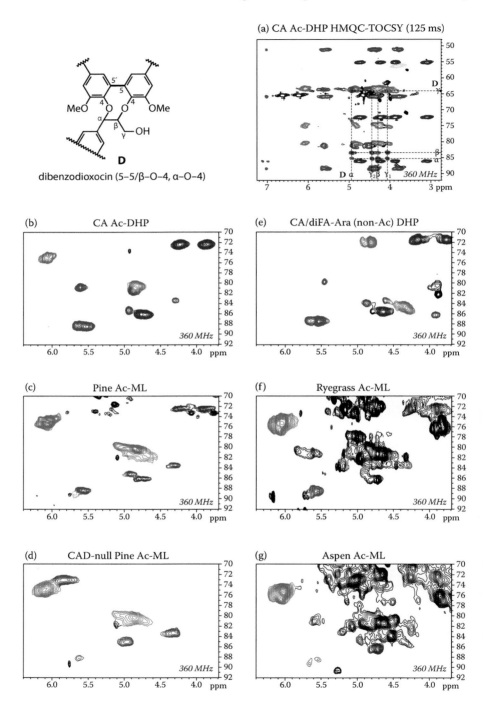

(a) CA Ac-DHP HMQC-TOCSY (125 ms)

dibenzodioxocin (5–5/β–O–4, α–O–4)

D

(b) CA Ac-DHP

(c) Pine Ac-ML

(d) CAD-null Pine Ac-ML

(e) CA/diFA-Ara (non-Ac) DHP

(f) Ryegrass Ac-ML

(g) Aspen Ac-ML

suggesting that it is not actually in the form of a phenolic-OH in lignin [282]. In addition to the dienones postulated from water addition to the quinone methide [279,280], spirodienones **S** were proposed as logical intermediates [283]; sinapate esters can form isolable analogs upon oxidative coupling. NMR evidence for spirodienones was found in poplar lignins [78]. Carbonyl peaks at ~182 ppm units have been detected [59]; although small (and most readily seen in quantitative spectra, presumably because of their long T_1-relaxation times), these carbons had long-range $^{13}C–^1H$ correlations reasonable for spiro-hexadienone structures. Further compelling NMR evidence for spirodienones in lignins from spruce, aspen, poplar kenaf, alfalfa, and even various fruits, has now been reported [57,59,60,103,104,106,107]. High-syringyl lignins appear to contain elevated amounts of such units (see the Poplar lignin in Figure 5.3e, for example). Remarkable amounts of syringyl spirodienones **S** appear in Kenaf bast fiber lignins, where the amount is comparable to that of the traditional resinol unit **C** (Figure 5.11f) [57].

The isolation of arylisochromans, following DFRC degradation, and the observation by NMR of trace amounts in isolated lignins, also suggests that dienone β–1-coupling products are formed during lignification [41]. Aryl isochromans can be identified in some softwood isolated lignins by their diagnostic HMQC or HSQC correlation at δ_C 41.3, δ_H 3.60 (acetylated units, in acetone). The complete side-chain of the crucial unit is seen in TOCSY spectra [41,284]. Whether the structures are present as such in native lignins is not yet clear, but, even if not, the internal trapping of a β–1-quinone methide intermediate is presumably operating *in vivo*. Compounds assigned as having β–6-linkages resulting from degradative procedures [256,285–287] should be carefully reexamined to determine if they are in fact their β–1/α–6 arylisochroman isomers (that are difficult to authenticate from mass spectrometry alone) and, therefore, also result from β–1-coupled units.

Revealingly, no β–1-coupling products are detectable in the highly *p*-hydroxyphenyl-rich lignins in C3H-deficient alfalfa, yet spirodienones are readily detected in the

FIGURE 5.10 (**See color insert following page 336.**) (Opposite) Dibenzodioxocins **D** are elegantly revealed in HMQC or HSQC spectra, where the Cα–Hα and Cβ–Hβ correlations, red, are unique and readily identified. (b–g) 360 MHz HSQC spectra representing various classes of lignins: (b) from the acetylated "methoxy-less" DHP (Ralph, J., Zhang, Y., and Ede, R.M., *J. Chem. Soc., Perkin Trans.*, 1 (16), 2609–2613, 1998); (c) a *Pinus taeda* acetylated milled wood lignin; (d) from the CAD-deficient *Pinus taeda* mutant (Ralph, J., MacKay, J.J., Hatfield, R.D., O'Malley, D.M., Whetten, R.W., and Sederoff, R.R., *Science.*, 277, 235–239, 1997)—note the enhanced dibenzodioxocin levels due to the high levels of 5–5-linked DHCA units from the incorporation of dihydroconiferyl alcohol DHCA into this lignin; (e) a nonacetylated (the reason for the different peaks of structure **D**!) synthetic DHP derived from coniferyl alcohol and 5% [9-^{13}C]5–5-diFA-Ara, a 5–5-coupled dimer of a ferulate-arabinofuranoside ester (Quideau, S. and Ralph, J., *J. Chem. Soc., Perkin Trans.*, 1(16), 2351–2358, 1997); (f) a ryegrass acetylated lignin known to contain diferulates and therefore likely to contain dibenzodioxocins derived from 5–5-coupled diferulate (Ralph, J., Grabber, J.H., and Hatfield, R.D., *Carbohydr. Res.*, 275(1), 167–178, 1995); (g) an acetylated aspen lignin. Due to diverse proton-proton coupling constants, the full side-chain correlation matrix does not always show up well in HSQC-TOCSY experiments with typical (80–100 ms) TOCSY mixing times, e.g., Figures 5.3b, 5.3d. Figure 5.10a shows an HSQC–TOCSY acquired using a 125 ms TOCSY mixing time, which beautifully emphasizes the full **D** coupling array. Contour assignment coloring scheme follows that established in Figure 5.3.

wild-type control [60]. It is logically concluded that *p*-coumaryl alcohol does not favor β–1-coupling, and/or that coniferyl and sinapyl alcohol do not favor cross-coupling with *p*-hydroxyphenyl-β-ethers, although this remains to be validated. Wild-type control alfalfa was useful for demonstrating that spirodienones **S** cleave to the traditional β–1-products (and the glyderaldehyde 2-aryl ether) upon acidic treatment; mild acidolysis of the residue from the milled lignin preparation produced a further NMR-analyzable fraction that had no spirodienones, but the conventional β–1-unit was evident [60].

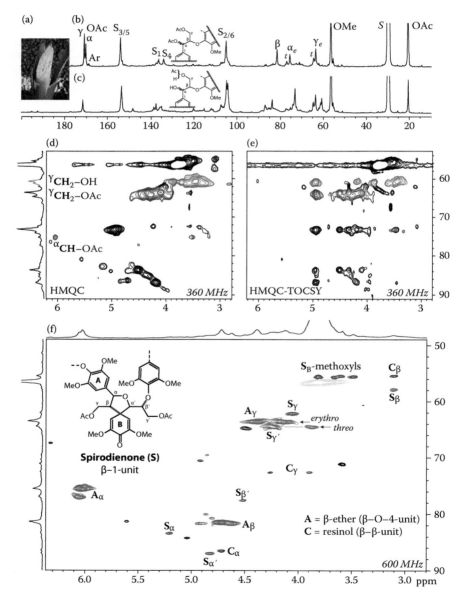

It appears that the product of β–1-coupling in lignins is the spirodienone **S** and that other β–1-products are likely the result of isolation. Thus NMR spectroscopy has again proven to be a revealing and diagnostic structural tool, with ramifications for the biosynthesis of lignin.

Cinnamyl Alcohol End-Groups *X1*

Cinnamyl alcohol end-groups **X1** result almost exclusively from monolignol dimerization reactions and are consequently relatively minor in lignins (which are formed primarily from endwise coupling of monolignols with the growing lignin polymer). Due to the relative invariance of particularly the γ-proton and carbon signals, they are, however, readily identified in proton [149,150] and carbon NMR spectra [288]. For the same reason, they stand out strongly in 2D spectra, particularly HMQC/ HSQC and their TOCSY variants (see Figures 5.2b, 5.3, 5.4, 5.8, 5.9, 5.10a, 5.16d, 5.16e, 5.18, and 5.19a through 5.19c).

The possibility of *(Z)*-cinnamyl alcohols being involved in lignification has been addressed [39,214]. Synthetic lignins prepared using *(Z)*-coniferyl alcohol retained much of the *(Z)*-geometry in the cinnamyl alcohol side-chains [39]. No evidence of such *(Z)*-cinnamyl alcohol end-groups in lignin has been observed, however.

Aldehydes

Formyl groups were easily identified in proton [126,149,150] and carbon NMR spectra [9]. Benzaldehyde and cinnamaldehyde units can be distinguished; they have chemical shift differences and the coupling with other side-chain protons allows their proton networks to be traced in COSY spectra (for example, see the correlations of protons at δ 9.6 in Figure 5.2a). A further discussion of aldehyde structures in lignins appears later in a section (CAD-Deficiency) that deals with the elevated incorporation of aldehydes into lignins in CAD-deficient plants.

Dihydrocinnamyl Alcohols *X5* and Derivatives

Softwood isolated lignins contain small amounts of dihydrocinnamyl alcohol units **X5** [27,289,290], most readily seen in TOCSY and HMQC-TOCSY spectra [2,38,66], e.g., Figures 5.2, 5.3, and 5.12. The low level coupling is predominantly via crosscoupling with normal lignin units (Figure 5.2c). In cross-coupling with coniferyl

FIGURE 5.11   (See color insert following page 336.) (Opposite) Spectra from *Tainung* Kenaf isolated lignin. (a) The lignin was isolated from the bast fibers, not the core; (b) 1D ^{13}C-NMR spectrum of acetylated kenaf lignin shows it to be particularly syringyl-rich; (c) 1D ^{13}C-NMR spectrum of unacetylated kenaf shows that the lignin bears its own natural acetates ((a–c) Ralph, J., *J. Nat. Prod.*, 59(4), 341–342, 1996); (d) a gradient HMQC spectrum of unacetylated kenaf lignin shows clear acylation at the γ-position; a low amount of α-acylation comes from natural acetyl group migration; (e) a gradient HMQC-TOCSY (100 ms mixing time) spectrum again shows the clearly acetylated γ-position ((d–e) Ralph, J., Marita, J.M., Ralph, S.A., Hatfield, R.D., Lu, F., Ede, R.M., Peng, J., Quideau, S., Helm, R.F., Grabber, J.H., Kim, H., Jimenez-Monteon, G., Zhang, Y., Jung, H.-J.G., Landucci, L.L., MacKay, J.J., Sederoff, R.R., Chapple, C., and Boudet, A.M., *Advances in Lignocellulosics Characterization*, TAPPI Press, Atlanta, GA, 1999); (f) 600 MHz HSQC spectrum showing how elegantly β–1-spirodienone structures **S** are revealed (Zhang, L., Gellerstedt, G., Ralph, J., and Lu, F., *J. Wood Chem. Technol.*, 26(1), 65–79, 2006). Colors in spectra (b–e) do not relate to those in other figures.

alcohol, β–O–4– and β–5–products will obviously be the most preponderant. A phenylcoumaran dimer with a propanol side-chain was released following dioxane-water hydrolysis of preextracted *Quercus mongolica* wood [291]. It was optically inactive, suggesting that it arose from lignin and not from lignans; the aglycone of an analogous glucoside isolated from Scotch pine (*Pinus sylvestrus*), for example, was optically active [292].

The derivation of the unsaturated side-chain is not clear. An NADPH-dependent enzyme has been reported to produce the β–5–compound from the coniferyl alcohol dimer β–5–dehydrodiconiferyl alcohol [293]. However, the lignin product is racemic. DHCA units occur at high levels in pine mutant trees that are deficient in CAD (see section CAD-Deficiency, Figure 5.2b) [294]. It is likely that DHCA is directly incorporated at low levels during softwood lignification, rather than arising from some post-lignification transformation. Coniferyl alcohol is apparently reduced to DHCA by the NADPH-dependent enzyme. The CAD-deficient mutant tree appears to have reduced capability of producing coniferyl alcohol, suggesting that DHCA may be formed by an alternative reduction pathway from coniferaldehyde [38].

When DHCA levels become elevated, such as in a CAD-deficient pine mutant (CAD-Deficiency section), or in synthetic lignins made from coniferyl alcohol containing a few percent of DHCA (Figure 5.2b), significant homo-dimerization can also occur, producing mainly 5–5-structures and elevated dibenzodioxocin levels (Figure 5.12d) [38]. These 5-coupled DHCA-derived units **X5** have higher side-chain proton chemical shifts and can be distinguished in TOCSY spectra (Figure 5.2b, for example). The presence of DHCA monomers during peroxidase-H_2O_2-promoted lignification also leads to other products in lignins [135]. Following two sequential formal H• abstractions (or, more likely, radical disproportionation), DHCA monomers will generate quinone methides that will produce guaiacylpropane-1,3-diol (GPD) upon re-aromatization by water addition (Figure 5.12d). Derived units **X6** also appear in lignins; their tell-tale propan-1,3-diol side-chains have been identified in pine MWLs, along with oxidized analogs that may have resulted during lignin isolation [135]. Figure 5.12 shows the identification of three structures derived from DHCA in the lignin of a CAD-deficient pine mutant. It is important to note, however, that these structures also show up at lower levels in lignins from normal pine (see, for example, the **X6** Cα–Hα correlation in Figure 5.12c, and the DHCA **X5** proton coupling network identified in the TOCSY spectrum of pine lignin in Figure 5.2c). The appearance of these structures in lignins provides compelling evidence that DHCA is a minor monomer in lignification. Figure 5.12 is discussed in more detail in CAD-Deficiency section.

Arylglycerols *X7*

Glycerol side-chains can result from oxidizing conditions and have been detected in enzymatic dehydrogenation products [60,295], but their presence in natural lignins remains in question [296]. A dimeric phenylcoumaran structure, containing an arylglycerol side-chain, is produced upon oxidation of coniferyl alcohol with Ag_2O (compound #262 in the "Database of Lignin and Cell Wall Model Compounds" [175]). The guaiacyl and syringyl monomers are also in the database (compounds #240, #272). The ^{13}C NMR assignments appropriate for lignins are given in Table 5.1. Considerable amounts have been found in DHPs made from *p*-coumaryl alcohol [60]; glycerol units **X7** show

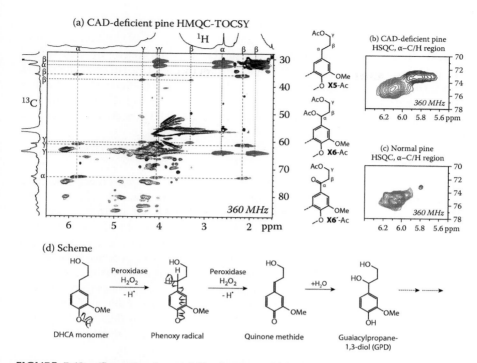

FIGURE 5.12 **(See color insert following page 336.)** Dihydroconiferyl alcohol and its derived products in pine lignins (Ralph, J., Kim, H., Peng, J., and Lu, F., *Org. Lett.*, 1(2), 323–326, 1999). (a) Partial 2D HMQC-TOCSY NMR spectra showing dihydroconiferyl alcohol (DHCA) units **X5**, derived guaiacylpropane-1,3-diol (GPD) units **X6**, and the benzylic ketone **X6′** derived from GPD units in an acetylated lignin isolated from a CAD-deficient pine mutant (Ralph, J., MacKay, J.J., Hatfield, R.D., O'Malley, D.M., Whetten, R.W., and Sederoff, R.R., *Science*, 277, 235–239, 1997). Data from the acetylated model 4-benzyloxy-3-methoxyphenyl-propane-1,3-diol are at the center of the yellow circles. Note also that dibenzodioxocin correlations are strong in this sample; see Figure 5.10d and caption. (b-c) The β-ether A and GPD **X6** Cα–Hα correlations in HSQC spectra from acetylated isolated lignins from the CAD-deficient pine and normal pine; GPD structures **X6** (along with DHCA units **X5**) are present at low levels in normal softwoods (Ralph, J., Kim, H., Peng, J., and Lu, F., *Org. Lett.*, 1(2), 323–326, 1999). (d) Scheme showing the generation of GPD monomer from DHCA monomer. Contour assignment coloring scheme, except for structure **X6′**, follows that established in Figure 5.3.

up readily in lignins derived from high levels of *p*-coumaryl alcohol [60], including in pine compression wood lignins—see Figures 5.5c, 5.5g, 5.16d, 5.16e, and 5.17. They are also elevated in lignins derived from the novel monomer 5-hydroxyconiferyl alcohol [63]—see Figures 5.5h and 5.20. Arylglycerols are now readily identified in "normal lignins"—see for example the HSQC spectra from alfalfa (Figure 5.16d) [60].

α-Keto-β-Ethers *A3*

Mainly found in high-syringyl isolated lignins, α-keto-β-ethers (structures **A3**) are thought to arise during isolation steps, presumably during ball milling; however, they are also found in synthetic lignins that have not been subjected to milling. Syringyl units are more prone to such oxidation. α-Keto-β-ethers **A3** have been found, for example, in aspen (Figure 5.10g) and arabidopsis (Figure 5.18e and 5.18g) lignins [70,117]. They are most readily identified in 1D ^{13}C NMR spectra by their shifted S2/6 peaks at ~107 ppm, and in 2D HSQC spectra by their Cβ–Hβ correlation peak at ~81/5.6. It should be noted that the Cα–Hα peak of α,β-diaryl ethers **A2** also occurs in this region. α-Keto-β-ethers **A3** can, however, be readily distinguished by correlating this contour with its α-carbonyl carbon (at 194.8 in acetylated samples in acetone-d$_6$) in HMBC spectra [70].

Quinone Methides

Although quinone methides are not structural units found in lignins themselves, they are important intermediates in lignification and in pulping reactions. They were initially considered to be too unstable to be subject to NMR but are in fact stable for considerable periods. The first relevant model spectra, both proton and carbon, were published in 1983 [297]. Guaiacyl quinone methides were shown to exist as a mixture of orientational isomers, in an approximately 70:30 ratio; the isomers derive from the relative orientation of the side-chain with respect to the methoxyl side of the ring. Further studies utilized 2D NMR methods to complete or improve assignments [298] and even provided evidence for differences in steric compression and reactivity between the two guaiacyl isomers [299]. An unexpectedly stable quinone methide resulting from β–β-coupling of sinapyl *p*-hydroxybenzoate was authenticated by NMR [300]. Further aspects of lignin quinone methides appear in a recent review [301].

Hydroxyl Groups

Hydroxyl groups in lignins are characteristic. Various types of phenols are recognized in proton spectra run in DMSO [123]. DMSO is a solvent that does not favor proton exchange; phenolic hydroxyl groups are therefore relatively sharp. In ^{13}C NMR, the acetate carbonyl resonances for phenolic, primary (γ) and secondary benzylic (α) acetates are resolved and can be used for quantifying such units [185,302]. Derivatizing with ^{13}C-labeled acetate (via [1-^{13}C]-acetyl chloride) allows for rapid quantification, via ^{13}C NMR, of hydroxyl environments in lignins [303].

Guaiacyl MWLs

Guaiacyl lignins appear to be remarkably similar to one another, regardless of their origin. For example, Figure 5.13 compares the ^{13}C signals from aromatic carbons from MWLs obtained from three different genera, *Ginkgo*, *Pinus*, and *Picea*. The similarity of this "fingerprint" region illustrates the uniformity amongst guaiacyl MWLs. The only differences generally observed between guaiacyl MWLs isolated from different wood species are in the side-chain region of the NMR spectra. These differences are mainly due to varying amounts of carbohydrates that are not separated from the lignins during isolation. The apparent relative homogeneity of guaiacyl lignin structure even across the cell wall, was indicated by a ^{13}C NMR comparison of lignin isolated from the middle lamella and secondary wall of Norway spruce (*Picea abies*) [304].

A study in which significant differences were found amongst guaiacyl lignins was the ^{13}C NMR comparison of MWLs isolated from fossil woods from three different

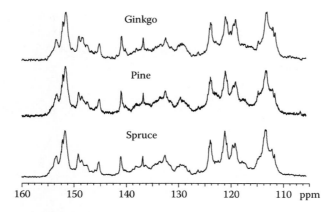

FIGURE 5.13 Comparison of the aromatic regions of 1D ^{13}C NMR spectra of three acetylated softwood lignins (Landucci, L.L., *J. Wood Chem Technol.*, 20(3), 243–264, 2000).

locations in the Canadian Artic [305]. The approximately 50-million-year-old lignins were found to be more condensed, demethylated, and side-chain-degraded than recent lignin from pine. The differences amongst the three fossil MWLs were likely due to varying degrees of degradation during the long burial time, and not because of initial differences.

The first published application of a 2D proton-carbon correlation of lignin [306] confirmed that loblolly pine contained no detectable (by NMR) syringyl content. The weak signals (at about 105 ppm) previously assigned to syringyl units by others were found to be due to residual carbohydrates. Low syringyl contents have been reported for loblolly pine (0.015%), radiata pine (0.001%), and ginkgo (0.060%) as measured by pyrolysis-gas chromatography/mass spectrometry or of the syringaldehyde formed upon nitrobenzene oxidation [307]. Generally, substructures of abundances less than about 0.5% in lignins are not observable by ^{13}C NMR [308]. Contamination artifacts perhaps cannot be ruled out in the above studies; our investigations, using high-sensitivity GC-MS spectra from DFRC degradation products of pine, failed to reveal even traces of syringyl monomers from loblolly pine.

Guaiacyl/Syringyl MWLs

In contrast to the similarity of G-lignins, GS-lignins are frequently dissimilar. For example, the ^{13}C NMR spectra (aromatic region) of GS lignins isolated from willow, beech, birch, elm, and sweetgum are shown in Figure 5.14. One common difference between hardwood MWLs is the relative amount of guaiacyl and syringyl units. This varies from very guaiacyl-rich lignins, such as those from elms and poplars, to syringyl-rich lignins such as those from birch and sweetgum. In addition, hardwood lignins are rather heterogeneous, so lignin isolated from various parts of the wood cell can have different proportions of G and S units [159,306]. Due to the unique chemical shift of the syringyl signals, especially those from the 2,6-carbons on the aromatic rings, it is generally simple to determine the relative abundances of G and

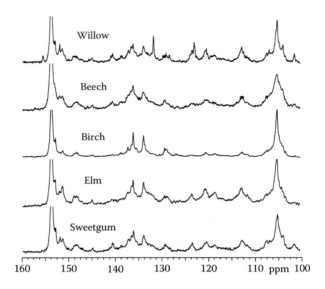

FIGURE 5.14   Comparison of the aromatic regions of 1D ^{13}C NMR spectra of five acetylated hardwood lignins (Landucci, L.L., *J. Wood Chem. Technol.*, 20(3), 243–264, 2000).

S units by quantitative ^{13}C NMR spectroscopy (Quantitative ^{13}C NMR of Lignins section) or from 2D HSQC spectra (Quantification in 2D NMR section). Other differences between GS lignins are frequently the presence of "unique" entities, such as incorporated hydroxycinnamates (Ferulates in Lignins; Cell Wall Cross-Linking in Grasses by Ferulates section), and γ-acylated units (*p*-Hydroxybenzoates in Various Hardwoods section) [309].

A study similar to one mentioned in the Guaiacyl MWLs section on ancient wood has also been conducted. However, the wood in this case was much younger (~6600 years) and could be identified as the hardwood species *Bischofia polycarpa*. [310]. Similar to the ancient guaiacyl lignin, the ancient guaiacyl/syringyl MWL had a higher degree of condensation, a lower methoxyl content and side-chain degradation, but not nearly to the same extent as the much older guaiacyl MWLs.

Naturally Acylated Lignins

Esters of various types are found in certain lignins. Hardwood lignins contain acetate groups [311,312], and all grasses contain varying levels of *p*-coumarates [36,56,313–320], and some plants contain *p*-hydroxybenzoates on lignin [194,309,321]. NMR is not only able to confirm such lignin acylation, but provides unambiguous regiochemical elucidation; the regiochemistry (or site of attachment) provides insights into the biochemistry of such acylation.

Acetates

Kenaf bast fiber lignins (Figure 5.11a), contain striking amounts of acetate [37,312]; over 50% of the units appear to be acetylated. Figures 5.11b through 5.11e show the

1D and 2D spectra identifying such acetylation and establishing that the acetates are primarily at the γ-side-chain position. Acetates will migrate from the γ- to the α-position [322], so strict regiochemistry is not expected. Experiments based on the DFRC (Derivatization Followed by Reductive Cleavage) degradative method confirm the regiochemistry and further reveal that acetates are almost entirely on syringyl units [312].

Such products implicate the biosynthesis of sinapyl γ-acetate as a precursor of Kenaf lignin (see Novel Mono-tetrahydrofurans Provide Evidence that Acylated Monolignols are True Lignin Precursors section) [101,260]. The functional role of such high levels of acetates in kenaf lignins awaits explanation. Although present at much lower levels, except in palms where they may be substantial [323], acetates on the γ-position have been confirmed in aspen by a modified DFRC analysis [312]. Detailed NMR studies elucidating the nature of these minor acetate levels have not been carried out.

p-Coumarates on Grass Lignins

All grasses contain lignins that are partially acylated by p-coumarate. The p-coumarate units remain free-phenolic (as most easily revealed by the anticipated shifts of resonances when the lignins are acetylated), and do not participate to a significant extent in the radical coupling reactions occurring during lignification. The functional role has not been fully elucidated, although some hypotheses are emerging [324–326]. The elucidation of attachment regiochemistry by NMR is elegant and unambiguous [2,56]. The application of HMQC experiments on maize lignins established that lignin γ-units were acylated and that α-acylation was undetectable. Subsequent HMBC experiments confirmed that it was p-coumarate that was acylating the lignin γ-positions. Experiments with other grasses, derived from either C_3 and C_4 biosynthetic pathways, revealed the same information—that acylation is entirely at the γ-position [2,75,76]. The additional finding of p-coumarates on many types of lignin units (both isomers of β-ethers, phenylcoumarans, and even cinnamyl alcohol end-groups) suggests that such acylation results from incorporation into lignin of pre-acylated monolignols [56]. More conclusive evidence is the recent discovery of nonresinol β–β-tetrahydrofuran units (see Novel Mono-tetrahydrofurans Provide Evidence that Acylated Monolignols are True Lignin Precursors section).

p-Hydroxybenzoates in Various Hardwoods

Aspen, poplar, and the willows all have significant amounts of p-hydroxybenzoates on their lignins [194,309,321]. p-Hydroxybenzoates have been found in significant quantities in lignins from oil palm empty fruit bunches [327,409]. Like p-coumarates, the p-hydroxybenzoates remain as free-phenolic pendant groups and appear as sharp resonances due to their increased mobility compared with the polymer backbone (see CCR-Deficiency section). The ^{13}C NMR chemical shifts of the p-hydroxybenzoate units are, therefore, predictably different in underivatized vs acetylated lignins [224,309]. Again, NMR experiments indicate that it is the lignin γ-positions that are acylated [323]. Furthermore, the HMBC experiments again show that it is syringyl units that are acylated by p-hydroxybenzoate; acylated guaiacyl units cannot be detected (Figure 5.5i); the γ-p-hydroxylated β-ethers Á conveniently have a higher α-proton chemical shift, allowing the S:G distribution on such units to be determined.

Novel Mono-tetrahydrofurans Provide Evidence that
Acylated Monolignols are True Lignin Precursors

There is one lignification pathway that is significantly altered by preacylation of the monolignols [101,260]. That is the pathway in which the γ-OH on the monolignol becomes involved in postcoupling reactions, i.e., the pathway normally leading to β–β-coupled (resinol) units. With the γ-position acetylated, β–β-coupling or cross-coupling can still presumably occur, but the rearomatization reactions following the radical coupling step can no longer be driven by the internal attack of the γ-OH on the quinone methide intermediates—the γ-acetylation prevents such a reaction. The acyl group remains attached in nonresinol β–β-coupling products; such products could not have arisen from postcoupling acetylation reactions. Therefore, finding these nonresinol syringyl β–β-structures in lignins establishes beyond reasonable doubt that lignin acylation arises through incorporation of preacylated sinapyl alcohol, as a lignin precursor, via radical coupling mechanisms [101,260].

The HSQC NMR spectra of the Kenaf lignin in Figure 5.15 (left column) show that all diagnostic side-chain ($^{13}C_\alpha$–H_α, $^{13}C_\beta$–H_β, $^{13}C_\gamma$–H_γ) correlations corresponding to the novel β–β-structures are observed and identified by comparison with the synthesized model compounds (right column).

Further evidence in the case of *p*-hydroxybenzoates comes from the discovery of a novel β–β-coupling product bearing a *p*-hydroxybenzoate substituent in actively lignifying poplar xylem [258]. It was identified by NMR as a cross-coupling product between sinapyl alcohol and sinapyl *p*-hydroxybenzoate, demonstrating that sinapyl *p*-hydroxybenzoate is the "monomer" in the coupling reaction *in planta* [258,300].

Ferulates in Lignins; Cell Wall Cross-Linking in Grasses by Ferulates

All grasses also contain ferulate that is intimately incorporated into the cell wall, as has been reviewed [133]. Polysaccharide OH groups (primarily the C5-OH of arabinosyl units in arabinoxylans) are acylated by (activated) ferulic acid to form ester bonds; dehydrodimerization of two ferulate esters can result in polysaccharide-polysaccharide cross-linking [133,272]. Both ferulates and ferulate dehydrodimers, which are bound to polysaccharides, can "attach" to lignins via radical coupling reactions with mono- or oligolignols, resulting polysaccharide-lignin cross-linking [55,129,133,203]. The radical nature of the coupling is indicated by the type of ferulate-lignin bonding that is present. Bonding to ferulate can occur, in the typical combinatorial sense, to the Cβ side-chain carbon, or the aromatic 5- or 4–O-positions and to analogous carbons on lignin, as has been elegantly revealed by NMR studies [129,130,133].

A synthetic lignin from coniferyl alcohol was made in the presence of ~5% of a labeled ferulate model; the single carbonyl resonance in the model produced some five different environments in the DHP [129]. Coupling modes were readily identified using inverse-detected long-range ^{13}C–^1H correlation (HMBC) spectra of the synthetic lignin, along with model compounds for the various possibilities [129]. In particular, there were products in which the 8-position of ferulate (analogous to the β-position of monolignols) was involved in radical coupling reactions to produce 8–5-, 8–O–4-, and 8–β-structures [129].

With the database of cross-coupling products from the synthetic lignin studies, researchers attempted to look for these components in real lignins. Unfortunately, the

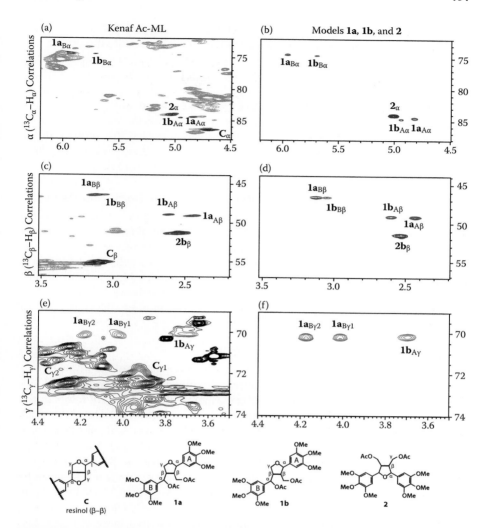

FIGURE 5.15 **(See color insert following page 336.)** HSQC NMR spectra showing the (a) α- ($^{13}C_\alpha–^1H_\alpha$), (b) β- and (c) γ-correlations of novel β–β-coupled structures produced from γ-acylated sinapyl alcohol in Kenaf Ac-ML. (d–f) HSQC NMRs of methylated and acetylated model compounds (right column; individual spectra of compounds **1a**, **1b**, and **2** are overlaid) are compared with MLs (left column). There are two isomers of **1**; stereochemical assignments have not yet been made. Only the color of contours **C** follows that established in Figure 5.3.

levels are so low (a few percent of the lignin), and the number of products so diverse, that spectra cannot easily show these products. However, NMR has sufficient dispersion and resolution to find the products if the signal to noise can be raised. Therefore, plants were grown in ^{13}C-enriched (~15%) CO_2 to provide uniformly labeled material [130]. Enrichment of ^{13}C in a sample to 15% provides a 15-fold sensitivity gain in

^{13}C NMR (and any 2D method that involves carbon), or a savings of $15^2 = 225$ in time to the same signal-to-noise. The incorporated ferulates were readily discerned.

Ferulate-monolignol 8–β-cross-products were readily identified in ryegrass lignin, providing compelling evidence that ferulates cross-couple with normal monolignols, both sinapyl and coniferyl alcohols, during lignification in grasses. The evidence that both monolignols reacted with ferulate was unexpected, since ferulates were suggested to be involved in early wall development, whereas syringyl units resulted later [319,328,329]. The NMR spectral data further suggested that ferulate in the wall reacts with monolignols and not preformed oligomers, suggesting that ferulates therefore act as nucleation sites for lignification in grasses [130,133].

Lignins in Monolignol-Biosynthetic-Pathway Mutant or Transgenic Plants

Recent advances in genetic engineering have allowed researchers to perturb the monolignol biosynthetic pathway [330–333], allowing the abundance of normally minor components to be substantially enhanced and, therefore, structurally analyzed [38,60,62,63,65–69,71,135,226,239,334–339]. This approach provides valuable insights into the control of lignification and into the remarkable biochemical flexibility of the lignification system [294,330]. Studies to date indicate that up- or down-regulating any of the enzymes of the major pathways can affect the course and/or extent of lignification.

Down-Regulation of Early Pathway Enzymes

General Lignin Down-Regulation   General down-regulation of the whole lignification pathway can be achieved by targeting enzymes early in the monolignol biosynthetic scheme. Chiang's group obtained aspen with significantly lower lignin contents by down regulation of an early CoA Ligase, Pt-4CL1 [117,239]. Such down-regulation had little effect (a) on the syringyl/guaiacyl ratio, as determined by examination of the HMBC spectra, shown in Figure 5.14j, and (b) on alteration of lignin structure. However, the *p*-hydroxybenzoate level on β-ether units was elevated. Meyermans et al. obtained a lesser reduction in lignin levels in aspen by downregulating the early *O*-methyl transferase, CCoA-OMT (CCOMT) [224], in poplar. A similar lignin reduction was found in antisense-CCOMT tobacco [340] and alfalfa [63,341]. Despite having significantly lower than normal lignin levels, the plants in these cases produce essentially normal lignins, as delineated by NMR [63,68,117]; the levels of *p*-hydroxybenzoates on the aspen/poplar lignins increased, however, suggesting that the pathways to the monolignol esters might not be fully coupled to general monolignol synthesis [342]. In tobacco transgenics down-regulated in both CCR and CAD, a general down-regulation of lignification was also observed. Again, the lignins were comparatively normal, showing only small indications of the previously substantial changes noted when each gene, singly, was down-regulated [69,343]. Down-regulation of PAL, the first committed enzyme in the phenylpropanoid pathway, also results in lower lignin levels [344].

Down-Regulation of C3H; Inhibition of Both G and S Lignin Formation   Richly detailed NMR spectra revealed both striking and subtle structural differences between the syringyl/guaiacyl lignins in normal wild-type alfalfa vs the *p*-hydroxyphenyl-rich lignins in the heavily C3H-down-regulated plants [60,345]. The alfalfa transgenics became the first plants available in sufficient quantities for lignin isolation and NMR; a stunted Arabidopsis mutant with no G- or S-lignin detectable via nitrobenzene oxidation [346] was not available in sufficient quantities for early studies.

The anticipated effect of C3H-deficiency, an enhancement of the relative level of p-hydroxyphenyl (H) units in the lignin, was compellingly demonstrated in the aromatic profiles revealed by HMBC spectra (Figures 5.5f and 5.5g), and HSQC NMR spectra (Figures 5.16b and 5.16c). Wild-type plants have syringyl/guaiacyl lignins with only low levels of H-units (Figures 5.5f and 5.16b). A severely down-regulated line (Figures 5.5g and 5.16c) was G- and S-depleted and strikingly H-rich. Reasonable quantification, gained by volume-integrating the contours in the HSQC spectra (Figures 5.16b and 5.16c), indicated that the H-content went from ~1% in the wild-type to ~65% in the transgenic (Table 5.2) [60].

High-field HSQC spectra of the side-chain regions were more revealing regarding the manner in which the monomeric units are assembled. The most immediately noticeable feature from Figure 5.16d and 5.16e is that the minor, but important, spirodienones **S**, resulting from β–1-coupling reactions, are absent in the C3H-deficient lignin (Figure 5.16e). It seems logical from the absence of β–1-coupling products that p-coumaryl alcohol simply does not efficiently β–1-couple with H-units in the lignin, nor probably with G- or S-units. As has been discovered with other units, this is likely due to a simple chemical incompatibility. The second striking difference is the multitude of dibenzodioxocins **D** that appear in the C3H-deficient lignins due to the coupling of the full range of hydroxycinnamyl alcohols with traditional guaiacyl-guaiacyl 5–5-biphenyl units, as well as the biphenyls formed from p-hydroxyphenyl units. Less readily seen before quantification (Table 5.2), is that β-ether units **A** were lower in the transgenic (~56% vs ~75% of the units quantified); this was compensated for by the two other major units, phenylcoumarans **B** and resinols **C**, each of which nearly doubled in relative proportion. The higher resinol concentration suggested that more monomer-monomer coupling reactions occurred during the lignification in the H-rich plants. Although still quite low, the relative dibenzodioxocin **D** level was about double that in wild-type plants. Glycerol structures **X7** were also at significantly higher (~9-fold) levels.

Significant details regarding lignification via enhanced levels of p-coumaryl alcohol came from analysis of the HMBC spectra (Figure 5.17). The data from these spectra allow a determination of which monomers are involved in forming each type of interunit linkage. From Figure 5.17, for example, essentially all phenylcoumarans **B** were revealed to be formed by coupling reactions involving p-coumaryl alcohol; this is the kind of information that is required to understand the cross-coupling propensity of p-coumaryl alcohol. The spectra, unfortunately, do not reveal whether the coupling was with a guaiacyl or another p-hydroxyphenyl unit (it obviously could not have been a syringyl unit, since the 5-position is not available for coupling).

Similarly, the change in resinols **C** from being almost entirely derived from sinapyl alcohol in the wild-type to being derived from all three monomers, and p-coumaryl alcohol in particular, in the transgenic plants, is striking. It is logical that sinapyl alcohol monomers find themselves only rarely able to dimerize (with other sinapyl alcohol monomers) in this sinapyl-alcohol-depleted plant, so presumably β–β-cross-couple with coniferyl, and possibly p-coumaryl alcohol; studies are

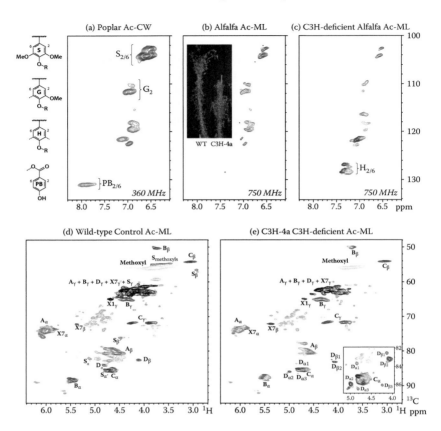

FIGURE 5.16 **(See color insert following page 336.)** HSQC spectra of acetylated lignins, in CDCl₃, with different H:G:S ratios. (a–c) Aromatic regions of acetylated isolated lignins showing quantifiable H:G:S levels in (a) Poplar (which also shows how γ-*p*-hydroxybenzoates can be evidenced in this spectral region); (b) Alfalfa wild-type control, a guaiacyl-rich S/G lignin with only traces of H-units (~2%); (c) a C3H-deficient alfalfa, extraordinarily enriched in H-units (~65%) from the typically minor monolignol *p*-coumaryl alcohol (Ralph, J., Akiyama, T., Kim, H., Lu, F., Schatz, P.F., Marita, J.M., Ralph, S.A., Reddy, M.S.S., Chen, F., and Dixon, R.A., *J. Biol. Chem.*, 281(13), 8843–8853, 2006). (d–e) Side-chain regions from the alfalfa lignins: (d) Wild-type control, (e) C3H-deficient. C3H-deficiency, and the incorporation of higher levels of *p*-coumaryl alcohol into the lignin, produces significant changes in the distribution of interunit linkage types. The absence of spirodienone **S** units in the transgenic reveals that *p*-coumaryl alcohol does not apparently favor β–1-cross-coupling reactions. Several types of new dibenzodioxocins **D** are more readily seen at the lower contour levels in the more highly resolved partial spectrum in the inset. Note that the contour levels used to display the two spectra were chosen to highlight the structural similarities and differences; with no internally invariant peaks, interpretation of apparent visual quantitative differences needs to be cautious. Photographic inset shows WT and C3H-deficient transgenic plants at the WT flowering stage (Reddy, M.S.S., Chen, F., Shadle, G.L., Jackson, L., Aljoe, H., and Dixon, R.A., *Proc. Natl. Acad. Sci.*, 102(46), 16573–16578, 2005). Semi-quantitative data derived from corrected volume integrals are given in Table 5.2. Contour assignment coloring scheme in d-e follows that established in Figure 5.3; contour assignment coloring for S:G:P:PB delineation is the same as used for similar purposes in Figures 5.5 and 5.17.

TABLE 5.2

S:G:H Measurement and Linkage Data from Corrected Volume Integrals for Contours in the Side-Chain Region of $^{13}C-^{1}H$ Correlation Spectra of Acetylated Isolated Stem Lignins from Control (Wild-Type) and C3H-Deficient Alfalfa

Lignin↓ Unit→	A	B	C	D	S	X1	X7		S	G	H
Alfalfa											
WT*	75	9	9	1.1	0.6	4.8	0.5		41	58	1
C3H-antisense*	56	18	16	2.6	–	2.9	4.6		18	17	65

Source: From Ralph, J., Akiyama, T., Kim, H., Lu, F., Schatz, P.F., Marita, J.M., Ralph, S.A., Reddy, M.S.S., Chen, F., and Dixon, R.A., *J. Biol. Chem.*, 281(13), 8843–8853, 2006.

Note: **A** = β-ether (β–O–4), **B** = phenylcoumaran (β–5), **C** = resinol (β–β), **D** = dibenzodioxocin (β–O–4/α–O–4), **S** = spirodienone (β–1); **X5** = cinnamyl alcohol endgroup; **X7** = arylglycerol end-group. These percentages are corrected ratios, via responses from model dimers, from the volume integrals of contours (Figures 5.16d and 5.16e), arise from units **A-D**, **S**, **X1** and **X7**, and are therefore better attempts at true ratios. **S:G:H** values are from uncorrected volume integrals of the S2/6, G2, and H2/6 correlations from HSQC spectra (Figures 5.16b and 5.16c). More details, and the quantification of other lignin fractions, have been reported.

required to determine which of these cross-coupling reactions are chemically feasible, i.e., if sinapyl alcohol and *p*-coumaryl alcohol, for example, are compatible for radical cross-coupling. Examination of thioacidolysis or DFRC dimers may eventually shed light on the occurrence of mixed resinols; mixed dimers can be elucidated by GC-MS [257].

Although it is less clear since dibenzodioxocins **D** did not show useful correlations in the wild-type lignins, it appears that the coupling in the C3H-deficient plant also involved coupling of *p*-coumaryl alcohol with the dibenzodioxocin-precursor 5–5-end-unit. The crucial β-aryl ether units derive from all three monolignols, illustrating that *p*-coumaryl alcohol is able to function in the most important endwise-coupling reactions that allow polymer growth.

Down-Regulation of Later Pathway Enzymes; Lignin Compositional and Structural Change

Down-regulating enzymes later in the pathway frequently did not significantly alter the quantity of "lignin" produced, but caused more extensive structural alterations.

Syringyl Lignin Pathways   Two key enzymes (beyond the coniferyl alcohol pathway) are required for the production of sinapyl alcohol, which is the essential monolignol for production of syringyl units in lignins. Ferulate 5-hydroxylase, FH5, now perhaps more appropriately called CAld-5H to reflect coniferaldehyde as the preferred substrate [347,348], effects the 5-hydroxylation. Caffeate *O*-methyltransferase,

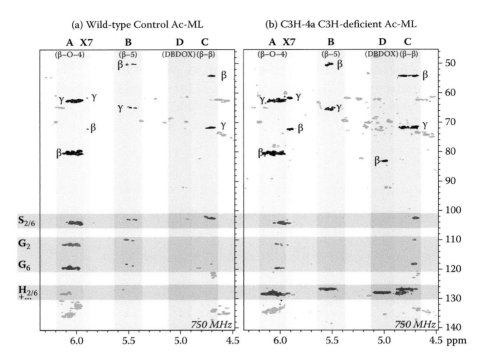

FIGURE 5.17 **(See color insert following page 336.)** Partial HMBC spectra of lignins, in $CDCl_3$, isolated from alfalfa, (a) the wild-type control and (b) a C3H-deficient line (Ralph, J., Akiyama, T., Kim, H., Lu, F., Schatz, P.F., Marita, J.M., Ralph, S.A., Reddy, M.S.S., Chen, F., and Dixon, R.A., *J. Biol. Chem.*, 281(13), 8843–8853, 2006). These spectra allow a determination of the monolignol involved in forming each type of structural unit (see text). The correlations highlighted are from α-protons to the carbons within three bonds, most diagnostically those to the 2- and 6-postions on the aromatic rings of *p*-hydroxyphenyl (H), guaiacyl (G) and syringyl (S) units. Contour assignment coloring scheme does not follow that established in Figure 5.3 (except for the subunit designations along the top); contour assignment coloring for S:G:H delineation is the same as used for similar purposes in Figures 5.5 and 5.16a through c.

COMT, for which the preferred substrate now appears to be 5-hydroxyconiferalde-hyde [342], catalyzes the methylation of the new 5-OH.

F5H (or Cald-5H)   The most striking examples of compositional flexibility in the two normally predominant monolignols came from the F5H (ferulate 5-hydroxylase) *Arabidopsis* mutants and transgenics produced by Chapple's group. An F5H-deficient mutant, derived from chemical mutagenesis [349] had (essentially) no syringyl component in its lignin, as shown by NMR [70] and other methods [349]. When a suitably promoted *F5H* gene was introduced into the F5H-deficient mutant, the researchers observed that up-regulation of F5H proved to be strikingly effective at diverting the monolignol pool beyond coniferyl alcohol. With an approximately 3% guaiacyl component, these plants' lignins have a far lower guaiacyl contents than any plants reported to date [70].

The incredible compositional change, from extremely guaiacyl-rich to essentially guaiacyl-depleted, is dramatically illustrated in the NMR spectra given in Figures 5.18 and 5.5a [2,70]. Figure 5.5a shows the HMBC correlations between the α-protons of the major β-aryl ether units in lignins from wild-type control, the down-regulated mutant, and a C4H-promoted up-regulated transgenic. The control contains guaiacyl and syringyl (β-ether) units. The mutant has no detectable syringyl component. The lignin in the most highly up-regulated C4H-F5H transgenic is extraordinarily guaiacyl-deficient; only weak guaiacyl peaks can be discerned at lower levels. In fact, the syringyl:guaiacyl ratio in this transgenic is higher than has been recorded in any plant to date; but, as shown below and in Figure 5.18e, the methylation of 5-hydroxyguaiacyl precursors cannot apparently keep pace and 5-hydroxyconiferyl alcohol becomes a significant component of the lignin from this transgenic [66].

Similarly remarkable compositional changes were effected in Aspen by manipulation of the analogous gene and the enzyme, now called CAld5H to reflect the preferred substrate specificity for coniferaldehyde by the enzyme [347,348]. The HMBC spectra in Figures 5.5j show that CAld5H up-regulation is again strikingly effective at ramping up the monolignol pool to sinapyl alcohol, producing high-syringyl lignins. The F5H-upregulated poplars had up to ~94.5% S, and had a consequently dramatic benefit on pulping efficiency [350,351].

COMT-Deficiency: Benzodioxanes from Incorporation of 5-Hydroxyconiferyl Alcohol COMT is one of two enzymes required to 5-methoxylate guaiacyl monomeric units to produce sinapyl alcohol and eventually produce syringyl units in angiosperm lignins. If COMT is down-regulated, 5-hydroxyconiferyl aldehyde might build up and be expected to be reduced to 5-hydroxyconiferyl alcohol if the next enzyme, CAD, is sufficiently nonspecific. In fact, it appears that 5-hydroxyconiferyl alcohol is indeed formed, shipped out to the wall, and incorporated into lignin analogously to other lignin monomers (although it produces some novel structures in the final lignin).

Again, NMR provides compelling evidence that benzodioxane structures are produced in lignins that incorporate 5-hydroxyconiferyl alcohol. Figure 5.19b shows the side-chain region of an HMQC spectrum from an (acetylated) isolated lignin from a COMT-deficient gene-silenced poplar (*Populus tremula* x *Populus alba*) described recently [336]. The degree of COMT suppression was higher than in other trials using anti-sense suppression [226,334]. As with the dibenzodioxocins **D** [114,115], the benzodioxanes **H** are readily apparent in short-range $^{13}C-^1H$ correlation spectra (HMQC or HSQC) of acetylated isolated lignins. Well separated contours at δ_C/δ_H of 76.8/4.98 (α), and 75.9/4.39 (β) are diagnostic for the benzodioxanes **H**; the γ-correlations overlap with those in other lignin units. Anticipated shifts occur following acetylation; in the unacetylated lignins (not shown), the correlations are centered at δ_C/δ_H of 76.5/4.87 (α), and 78.9/4.06 (β). As seen in Figures 5.19b and 5.19d, the side-chain correlations are consistent with those in a model compound for the *trans*-benzodioxane, synthesized by biomimetic cross-coupling reactions between coniferyl alcohol and a 5-hydroxyguaiacyl unit [67]. The α-proton shift deviated the most, presumably since this is an acetylated (and therefore originally free-phenolic) model, rather than a phenol-etherified structure, which would correspond to most of the units in the lignin polymer. Lignins from COMT antisense poplars also contain benzodioxane units [67].

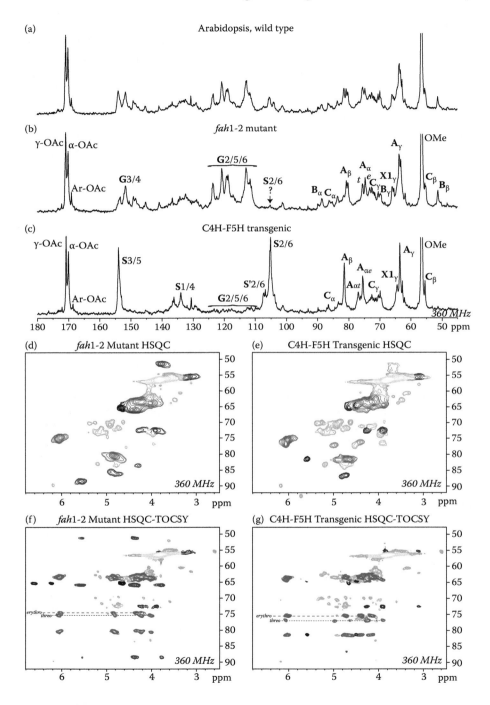

Figure 5.19e shows the first 3D spectrum of a natural ^{13}C-abundance lignin. It is a 3D gradient-selected TOCSY-HSQC taken on a 750 MHz instrument. In less than 24 h, the 3D experiment provided ample sensitivity to authenticate all of the major units in the natural ^{13}C-abundance sense-suppressed COMT transgenic poplar lignin. The $F_2–F_3$ slice of the new benzodioxane structure **H** is spectacularly clean with its α-, β-, and γ-correlations fully resolved (Figure 5.19f). In the 3D experiment, the γ-correlations of the benzodioxane in the $F_2–F_3$ plane are nicely resolved and isolated from the plane at its α-proton frequency (5.01 ppm), as well as at the β-proton frequency (~4.47 ppm; not shown) along the F_1 axis (Figure 5.18f). Obtaining a "pure" slice, leaves little ambiguity between correlations of different structures. For example, the γ_1- and γ_2-correlations of the benzodioxane units that overlap with the γ_1- and γ_2-correlations of the β-aryl ether **A** units and the dibenzodioxocin **D** units in the 2D spectra are unique to their respective 3D slices. This was the first reported "isolation" of the side-chain $^{13}C–^{1}H$ coupling network of a new lignin component using 3D NMR experiments at natural abundance. Again, the data in this slice agree well with those from a benzodioxane model compound [67].

Another detail revealed regarding benzodioxane units **H** in the 3D and 2D spectra is the degree of etherification. In acetylated lignins, units that were free-phenolic in lignin become phenol-acetylated, whereas those that were originally etherified remain so. Phenol acetylation causes Hα in **H** units to move to a lower field (higher ppm). Thus, unetherified units have Hα at 5.04 ppm, whereas etherified units are at 4.96 ppm. The 2D slice for the 3D experiment shown in Figure 5.19f has only a trace of correlations for the acetylated component. Other slices reveal slightly more. It appears that the benzodioxane units **H** are substantially etherified and, therefore, have been fully integrated into the polymer by further monolignol coupling reactions during lignification.

A reasonable quantification of units **H** (and other units) can be accomplished by measuring volume integrals in the 2D spectra [95], particularly if the similar Cα–Hα correlations are used. As can be seen in Table 5.3, the 5-hydroxyconiferyl alcohol-derived benzodioxane **H** units can become the second most abundant

FIGURE 5.18   (See color insert following page 336.) (Opposite) Spectra from wild-type Arabidopsis, a mutant deficient in ferulate 5-hydroxylase (F5H), and a transgenic upregulated in F5H (Marita, J., Ralph, J., Hatfield, R.D., and Chapple, C., *Proc. Natl. Acad. Sci.*, 96(22), 12328–12332, 1999). In the *fah1–2* mutant, syringyl units are almost completely absent (Marita, J., Ralph, J., Hatfield, R.D., and Chapple, C., *Proc. Natl. Acad. Sci.*, 96(22), 12328–12332, 1999; Meyer, K., Shirley, A.M., Cusumano, J.C., Bell-Lelong, D.A., and Chapple, C., *Proc. Natl. Acad. Sci.*, 95(12), 6619–6623, 1998). When F5H is expressed under the control of the C4H-promoter, the lignin that is deposited is comprised almost exclusively of syringyl units. G = guaiacyl, S = syringyl. (a–c) 1D ^{13}C NMR spectra of *Arabidopsis* lignins from (a) the wild type, (b) the *fah1–2* mutant, and (c) the C4H–F5H transgenic. Major peaks are shown with their assignments; G is used to represent a general guaiacyl unit, S a general syringyl unit; *e = erythro*-isomer, *t = threo*-isomer; oligosaccharides are present in these lignins. (d–g) Gradient-selected 2D HSQC spectra of *Arabidopsis* lignins from d, the *fah1–2* mutant, and e, the C4H–F5H upregulated transgenic, as well as higher F_1-resolution (carbon dimension) gradient-selected 2D HSQC-TOCSY spectra from (f) the *fah1–2* mutant and (g) the C4H–F5H transgenic. Contour assignment coloring scheme follows that established in Figure 5.3.

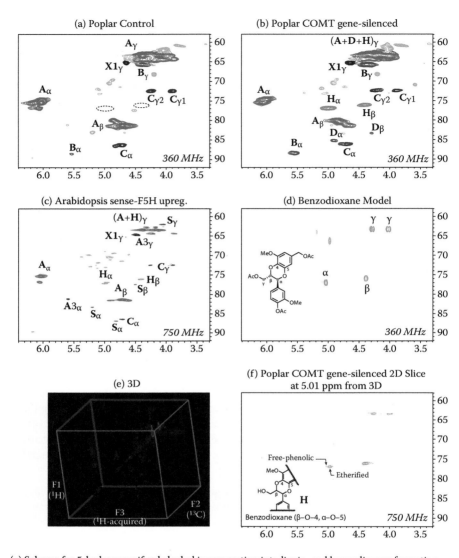

(a) Poplar Control

(b) Poplar COMT gene-silenced

(c) Arabidopsis sense-F5H upreg.

(d) Benzodioxane Model

(e) 3D

(f) Poplar COMT gene-silenced 2D Slice at 5.01 ppm from 3D

(g) Scheme for 5-hydroxyconiferyl alcohol incorporation into lignin, and benzodioxane formation

interunit type in transgenics and mutants with low COMT levels. Since the lignins analyzed by NMR represent 65% of the total lignin in the poplar transgenic, it is logical that the benzodioxane structures would remain a significant component, even if the lignins were drastically partitioned by the isolation process. The total β-ether frequency (normal β-ether **A** plus dibenzodioxocins **D** plus benzodioxanes **H**) in the transgenic is around 77%, lower than in the control because of the higher guaiacyl content. However, there are a few units not covered by these percentages, since they have no resonances in the aliphatic side-chain region of the NMR spectra (cinnamaldehyde end-groups and β–1-structures). The COMT-deficient brown-midrib-3 mutant of maize contains very significant levels of benzodioxanes **H** [352]. Although the data are limited at present, it appears that 5-hydroxyconiferyl alcohol may quantitatively make up for the sinapyl alcohol deficiency.

The incorporation of 5-hydroxyconiferyl alcohol into lignins has been documented several times [353–356]. Benzodioxanes have been proposed to be in lignins previously [336,353,355], and analogous structures occur in lignans [357,358]; however, until NMR spectra were obtained, the occurrence of benzodioxanes in lignins had not been observed directly. The likely mechanism for formation of the novel benzodioxane units is shown in Figure 5.19g, as described in the caption. Since etherified benzodioxanes don't cleave under alkaline pulping conditions, pulping efficiency would, therefore, likely be reduced. Recent pulping trials with COMT-deficient poplars confirm a lower pulping efficiency [336].

FIGURE 5.19   (See color insert following page 336.) (Opposite) Partial spectra from gradient HMQC NMR experiments highlighting new peaks for benzodioxane units **H**. Acetylated lignins were from (a) a control poplar, (b) a COMT-down-regulated poplar transgenic (Ralph, J., Lapierre, C., Marita, J., Kim, H., Lu, F., Hatfield, R.D., Ralph, S.A., Chapple, C., Franke, R., Hemm, M.R., Van Doorsselaere, J., Sederoff, R.R., O'Malley, D.M., Scott, J.T., MacKay, J.J., Yahiaoui, N., Boudet, A.-M., Pean, M., Pilate, G., Jouanin, L., and Boerjan, W., *Phytochem.,* 57(6), 993–1003, 2001; Marita, J.M., Ralph, J., Lapierre, C., Jouanin, L., and Boerjan, W., *J. Chem. Soc., Perkin Trans.,* 1(22), 2939–2945, 2001), (c) an F5H-upregulated Arabidopsis (cf. Figure 5.18d, but acquired at higher field and with higher resolution in the F_1 (carbon) dimension) (Ralph, J., Lapierre, C., Marita, J., Kim, H., Lu, F., Hatfield, R.D., Ralph, S.A., Chapple, C., Franke, R., Hemm, M.R., Van Doorsselaere, J., Sederoff, R.R., O'Malley, D.M., Scott, J.T., MacKay, J.J., Yahiaoui, N., Boudet, A.-M., Pean, M., Pilate, G., Jouanin, L., and Boerjan, W., *Phytochem.,* 57(6), 993–1003, 2001), and (d) similar correlations from a benzodioxane model compound (Ralph, J., Lapierre, C., Lu, F., Marita, J.M., Pilate, G., Van Doorsselaere, J., Boerjan, W., and Jouanin, L., *J. Agr. Food Chem.,* 49(1), 86–91, 2001). (e–f) The first natural ^{13}C-abundance lignin 3D NMR spectrum (Marita, J.M., Ralph, J., Lapierre, C., Jouanin, L., and Boerjan, W., *J. Chem. Soc., Perkin Trans.,* 1(22), 2939–2945, 2001). (e) A 3D gradient-selected TOCSY-HSQC spectrum (70 ms TOCSY) of a natural ^{13}C-abundance lignin from the sense-suppressed COMT transgenic; (f) 2D F_2-F_3 slices for the structural unit **H**. (g) Scheme for production of benzodioxanes **7** in lignins via incorporation of 5-hydroxyconiferyl alcohol **1** into a guaiacyl lignin (Ralph, J., Lapierre, C., Marita, J., Kim, H., Lu, F., Hatfield, R.D., Ralph, S.A., Chapple, C., Franke, R., Hemm, M.R., Van Doorsselaere, J., Sederoff, R.R., O'Malley, D.M., Scott, J.T., MacKay, J.J., Yahiaoui, N., Boudet, A.-M., Pean, M., Pilate, G., Jouanin, L., and Boerjan, W., *Phytochem.,* 57(6), 993–1003, 2001). Cross-coupling of 5-hydroxyconiferyl alcohol **1**, via its radical **1·**, with a guaiacyl lignin unit **3G·**, via its radical **3G**, produces a quinone methide intermediate **4** which rearomatizes by water addition to give the β-ether structure **5** (possessing a 5-hydroxyguaiacyl end-unit). This unit is capable of further incorporation into the lignin polymer via radical coupling reactions of radical **5·**. Reaction with the monolignol coniferyl alcohol **2G**, via its radical **2G·**, produces a quinone methide intermediate **6**. This time, however, quinone methide **6** can be internally trapped by the 5-OH phenol, forming a new 5-O-α-bond, and creating the benzodioxane ring system in **7**. Compound numbers relate only to this figure. Contour assignment coloring scheme follows that established in Figure 5.3.

TABLE 5.3

Volume Integrals for Acetylated Lignin Units

Lignin↓ Unit→	A	B	C	D	H	X5	Σ(β-ethers)
Poplar							
WT	88	3	7	0	0	2	88
COMT-silenced	53	13	5	6	18	5	78
COMT-antisense	65	12	5	3	10	4	78
Maize							
WT	86	7	0	0	0	6	86
bm3-mutant	60	7	0	0	25	8	85
Arabidopsis							
WT	67	13	6	6	0	7	73
F5H-sense	81	<1	6	0	10	4	91
Alfalfa							
WT	81	8	6	1	0	3	83
COMT-antisense	44	8	3	4	38	4	85
CCOMT-antisense	88	5	6	0	< 1	2	88

Source: Marita, J.M., Vermerris, W., Ralph, J., and Hatfield, R.D., *J. Agr. Food Chem.*, 51(5), 1313–1321, 2003; Marita, J.M., Ralph, J., Hatfield, R.D., Guo, D., Chen, F., and Dixon, R.A., *Phytochem.*, 62(1), 53–65, 2003; Marita, J.M., Ralph, J., Lapierre, C., Jouanin, L., and Boerjan, W., *J. Chem. Soc.*, Perkin Trans., 1 (22), 2939–2945, 2001; Marita, J., Ralph, J., Hatfield, R.D., and Chapple, C., *Proc. Natl. Acad. Sci.*, 96(22), 12328–12332, 1999; Marita, J.M., Lignin in Lignin-Biosynthetic-Pathway Mutants and Transgenics: Structural Studies. Ph.D., U. Wisconsin-Madison (University Microfilms International #3012586), 2001.

Note: Uncorrected Volume Integral Ratios of Contours in the Side-Chain Region of ^{13}C–^{1}H Correlation Spectra of Acetylated Isolated Lignins.
A = β-ether (β–O–4), **B** = phenylcoumaran (β–5), **C** = resinol (β–β), **D** = dibenzodioxocin (β–O–4/α–O–4/5–5), **H** = benzodioxane (β–O–4/α–O–5); the last column is the sum of all β–O–4 components, Σ(β-ethers) = **A** + **D** + **H**. *Note:* percentages are *not* traditional interunit linkage percentages since no account is taken of units that do not have specific contours in the side-chain region of their HSQC spectra (e.g., 5–5, 4–O–5, β–1 units). These percentages are uncorrected ratios of the volume integrals of contours arising from units **A–D**, **H**, and **X** and are therefore for comparative purposes only, not true quantification.

The levels of benzodioxanes in a lignin isolated from COMT-down-regulated alfalfa were higher than in other samples analyzed; however, partitioning of this unit into the soluble fraction was evident, implying that is was not necessarily higher in the whole lignin [63]. Spectral (HMBC) evidence elegantly reveals the presence of sequential benzodioxane units (Figure 5.20), indicating that 5-hydroxyconiferyl alcohol, along with the traditional monolignols, could β–O–4-couple to the 5-hydroxyguaiacyl groups in these lignins.

There is another sample in which benzodioxanes are evident, an interesting variant in the COMT-deficiency class. Arabidopsis transgenics with sense-upregulated F5H have previously been shown to have only a minor guaiacyl component [70,349], as noted above. Contours not immediately identified in the 2D NMR spectra of their lignins [70] later were shown to result from benzodioxane structures (Figure 5.18e) [66]. The observations here imply that, whereas syringyl production was enormously up-regulated in these transgenics, the methylation could apparently not keep pace with the accelerated production of 5-hydroxy units (e.g., 5-hydroxyconiferaldehyde and 5-hydroxyconiferyl alcohol). The result is a significant incorporation of 5-hydroxyconiferyl alcohol into the monolignol pool for the most heavily F5H-upregulated transgenics, ~10% as measured from contour volumes in the HMQC spectrum in Figure 5.18e and Table 5.3.

Evidence from degradative and NMR methods indicates that coniferyl and sinapyl alcohols, as well as further 5-hydroxyconiferyl alcohol, couple with the new terminus created by incorporation of 5-hydroxyconiferyl alcohol monomers into the growing polymer [63,66,336]. It has, therefore, become clear that 5-hydroxyconiferyl alcohol can be used as a lignin monomer by plants to offset the deficiency in sinapyl alcohol monomers [107,236,359]. The recent statement that "There is, however, no known precedent for the free interchange of monomeric units in any biopolymer assembly, then or now, and no biochemical evidence..." [360] is clearly incorrect [107]. In fact, plants and other organisms have long used loosely ordered chemistry to enhance their viability and resistance [361]. Indeed, even in more highly ordered polysaccharide biosynthesis, monomer substitution can occur, as seen in the substitution of α-L-fucose by α-L-galactose in xyloglucans in a fucose-deficient Arabidopsis mutant [362].

CAD-Deficiency

CAD is the last enzyme on the pathway to the monolignols coniferyl and sinapyl alcohols **2G** and **2s**, from which lignins are normally derived (Figure 5.21). When CAD is down-regulated, the hydroxycinnamaldehyde precursors to the monolignols, coniferaldehyde **1G** and sinapaldehyde **1s**, build up and may be incorporated into the polymer by radical coupling [38,71,196,363–368]. Indeed, coniferaldehyde was readily incorporated into synthetic lignins under biomimetic conditions [196].

Down-regulation of CAD in tobacco and poplar caused a marked increase in aldehyde components [71,363,369]; incorporation of aldehyde precursors to the monolignols into synthetic lignins established that coupling and cross-coupling reactions were possible [65,196]. Aldehydes were also increased in some brown-midrib maize and sorghum mutants [370], although differences can be hard to detect by NMR [365]. A pine CAD mutant identified in Sederoff's group had markedly increased aldehyde levels [38]. This latter lignin was also characterized by a rather massive incorporation of dihydroconiferyl alcohol monomer.

CAD-Deficient Angiosperms   The availability of a ^{13}C-enriched lignin from a CAD-deficient tobacco transgenic [71] and a series of model compounds [335] allows NMR methods to be used to ascertain hydroxycinnamaldehyde in vivo

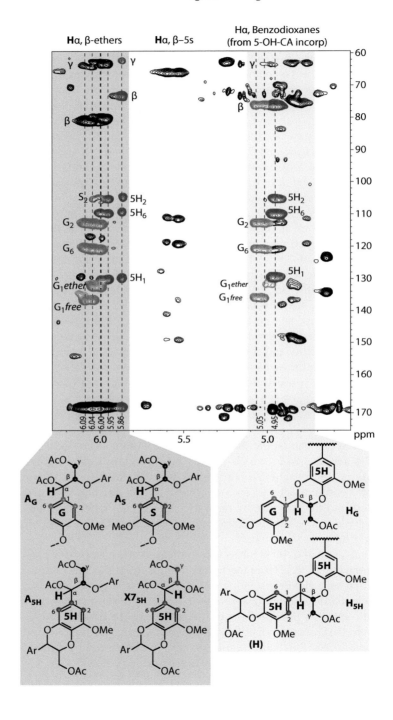

cross-coupling propensities. Data revealing such details of plant lignification are rarely obtained.

Radical cross-coupling of monolignols **2** (Figure 5.21), with the growing lignin oligomer/polymer **3** is the major reaction occurring during lignification. Thus, the hydroxycinnamyl alcohol **2** (primarily at its β-position) couples with phenolic units **3** (at the 4–O- or 5-position for guaiacyl units **3G**, and almost exclusively at the 4–O-position for syringyl units **3S**) to form a chain-extended oligomer/polymer, also with the general formula **3**. Figure 5.21 (right) shows a selected region of a long-range $^{13}C-^{1}H$ correlation (gradient-HMBC) spectrum from a CAD-deficient tobacco lignin, with peaks in the ^{13}C-projection and the resultant contours in the HMBC spectrum colored to match structures on the left for easy identification. In fact, the aldehyde structures in the lignins likely have been further incorporated into the polymer by primarily 4–O-coupling with the next monolignol, so they are not strictly the phenolic compounds **4** shown. Available coupling sites are indicated in Figure 5.21 by the dashed arrows.

Correlation of the aldehyde carbonyl carbons in cross-products **4** to the H7-protons 3-bonds away identifies the type of lignin units involved (Figure 5.21). Protons H7 resonate at ~7.3 ppm for hydroxycinnamaldehydes (either **1G** or **1s**) coupled 8–O–4 to guaiacyl units **3G** (compounds **4GG** and **4sG**, δ_C = 188.1 ppm); whereas, they resonate considerably upfield, at ~6.7 ppm when coupled 8–O–4 to syringyl units **3s** (compounds **4Gs** and **4ss**, δ_C = 186.8 ppm). Correlations from these H7 protons into the ring identify the hydroxycinnamaldehyde involved in the coupling; 3-bond correlations identify equivalent S2/6 carbons derived from sinapaldehyde **1s** units at higher field than the nonequivalent G2 and G6 carbons from units derived from coniferaldehyde **1G**. Therefore, for 8–O–4 cross-coupled units **4** in lignins, the G/S nature of both the hydroxycinnamaldehyde component (coupled 8-) and the lignin unit (coupled 4–O-) are diagnostically revealed. Other aromatic protons in the complex lignin polymer resonate in the H7 regions; so, correlations that are not of interest here will result. An absence of correlations is, therefore, more diagnostic, revealing the absence of a component.

Of the four possible aldehyde 8–O–4 incorporation products **4** (Figure 5.21), only three can be detected in the antisense-CAD tobacco lignin by NMR methods. Product **4GG** is conspicuously absent. The data imply that sinapaldehyde **1s** cross-couples 8–O–4 with both guaiacyl and syringyl units (to form cross-coupled structures **4sG** and **4ss**), but that coniferaldehyde **1G** cross-couples 8–O–4 *only* with syringyl units **3s** and not with guaiacyl units **3G**. Thus cross-coupling product **4Gs** is

FIGURE 5.20   (See color insert following page 336.) (Opposite) Partial 2D HMBC of lignin isolated from the COMT-deficient alfalfa transgenic (Marita, J. M., Ralph, J., Hatfield, R.D., Guo, D., Chen, F., and Dixon, R.A., *Phytochem.*, 62(1), 53–65, 2003), run on a 500 MHz cryoprobe instrument. The left columns reveal that β-ethers **A** are largely guaiacyl **AG** as expected from the low syringyl content, and as seen by the low level of syringyl β-ether **As** component. 5-Hydroxyguaiacyl β-ethers **A5H** are clearly evident, as are new 5-hydroxyguaiacyl glycerol structures **X75H**. The right columns reveal which monolignols have coupled to the new 5-hydroxyguaiacyl units. In addition to the expected coniferyl alcohol addition, giving **HG** units, it is clear that "chains" of benzodioxane units may occur, as seen by the **H5H** units, which demonstrate the presence of two sequential benzodioxanes.

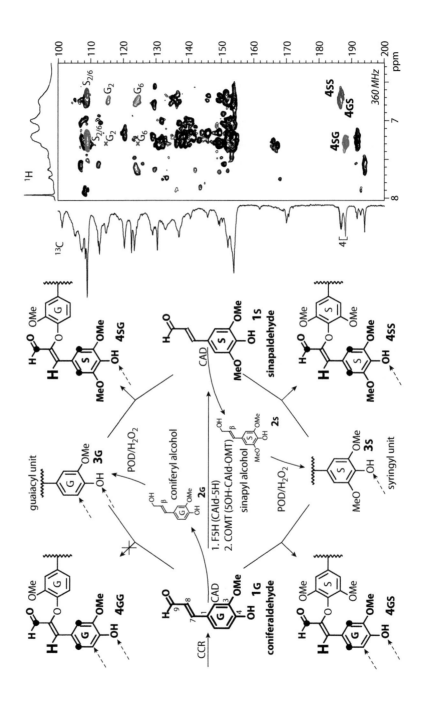

FIGURE 5.21 **(See color insert following page 336.)** (Opposite) Cross-coupling of hydroxycinnamaldehydes **1G** and **1s** with guaiacyl and syringyl lignin units **3G** and **3s** can potentially produce four 8–O–4-cross-coupled structures **4**. Only three of the four cross-products (**4Gs**, **4sG**, and **4ss** but not **4GG**) are significantly represented, as determined by analysis of the gradient-HMBC partial spectrum to the right. The HMBC section shows, in a lignin isolated from CAD-deficient tobacco stems, correlations from the aldehyde carbonyl carbons (185–195 ppm) to the H7 protons within 3-bonds in structures **4** that have been further incorporated into lignin at positions indicated by the dashed arrows. These H7 protons diagnostically correlate with the 2- and 6-carbons of the cinnamaldehyde ring; the G2 and G6 carbons for guaiacyl units, and the higher-field equivalent S2/6 (syringyl) carbons are identified and colored correspondingly. Colored dots on the structures indicate the carbons correlated by $^3J_{C-H}$ to the H7-protons (which are shown in a larger font size) shown in the HMBC spectrum. The aldehydes at ~188.1 ppm, with their corresponding H7s at ~7.3 ppm result from hydroxycinnamaldehydes **1** 8–O–4-coupled to guaiacyl units **3G** in lignin; correlations from H7 indicate, however, that sinapaldehyde **1s**, but not coniferaldehyde **1G**, couples with guaiacyl units **3G**—the green X's show where correlations are present in a model for **4GG**. The aldehydes at ~186.7 ppm, with their corresponding H7s at ~6.7 ppm result from hydroxycinnamaldehydes **1** 8–O–4-coupled to syringyl units **3s** in lignin; correlations from H7 indicate that sinapaldehyde **1s** and coniferaldehyde **1G** each couple with syringyl units **3s**. S = syringyl, G = guaiacyl, CCR = cinnamoyl-CoA reductase, CAD = cinnamyl alcohol dehydrogenase, F5H = ferulate 5-hydroxylase, which has recently also been termed CAld5H = coniferaldehyde 5-hydroxylase, to reflect the preferred in vivo substrate (Osakabe, K., Tsao, C.C., Li, L., Popko, J.L., Umezawa, T., Carraway, D.T., Smeltzer, R.H., Joshi, C.P., and Chiang, V.L., *Proc. Natl. Acad. Sci.*, 96(16), 8955–8960, 1999; Humphreys, J.M., Hemm, M.R., and Chapple, C., *Proc. Natl. Acad. Sci.*, 96(18), 10045–10050, 1999), COMT = caffeoyl O-methyl transferase (for which the preferred substrate appears to be 5-hydroxyconiferaldehyde (Li, L., Popko, J.L., Umezawa, T., and Chiang, V.L., *J. Biol. Chem.*, 275(9), 6537–6545, 2000), POD = peroxidase. Convention for compound numbering: e.g., **4Gs** = a guaiacyl aldehyde (coniferaldehyde **1G**) coupled 8–O–4 to a syringyl lignin unit (**3s**). Compound numbers and coloring relate only to this figure.

readily detected, whereas **4GG** cannot be detected even when spectra are viewed at close to the baseplane noise level. This cross-coupling propensity is presumably the reason why 8–O–4-coupled structures are not found to be prominent in CAD-deficient softwood lignins [38].

These observations require no departure from the existing theory of radical coupling of phenols into polymeric lignins, since they appear to reflect simple chemical coupling propensities [236]. Attempts to prepare compounds modeling **4GG** by biomimetically cross-coupling coniferaldehyde **1G** (at the 8-position) with coniferyl alcohol **2G** or with a simple guaiacyl model 1-(4-hydroxy-3-methoxy)-ethanol (at the 4–O-position) also failed to produce 8–O–4-coupled cross-products. Apparently the same factors that promote cross-coupling in vitro apply in vivo, suggesting that these radical coupling reactions and those involved in "normal" lignification are likely to be under simple chemical control in the plant [236,294]. In the cross-products, the chemical shifts match those in model compounds synthesized by traditional means [335]. Also available from the HMBC spectrum in Figure 5.21 is some information regarding the incorporation of the hydroxycinnamaldehydes **1** and their derived hydroxybenzaldehydes at ring 4–O-positions, as described elsewhere [335]. The novel coniferaldehyde and sinapaldehyde 8–O–4-cross-coupling structures in lignins have been validated by thioacidolysis marker compounds diagnostically derived from them [371,372].

CAD-Deficient Pine: Incorporation of Dihydroconiferyl Alcohol and Guaiacylpropane-1,3-diol   Preliminary NMR observations [38] on the novel lignins from a CAD-deficient pine (*Pinus taeda*) mutant were being strengthened as more diverse evidence accumulated [373,374]. The plant, despite a reduction in CAD levels to less than 1% of normal levels [375], appeared to be producing normal levels of lignin (Klason lignin values were almost identical). The wood was highly colored and stained strongly with phloroglucinol, suggesting a higher aldehyde component. In addition, NMR studies on isolated lignins indicated that aldehydes and DHCA were substantial components [2,38].

New aldehyde components were originally misidentified [38], but corrected in subsequent publications [2,71,135]. A book chapter on NMR details how the incorrect assignment was made [2]. The aldehyde components all stem from incorporation into the lignin of more significant quantities of coniferaldehyde (and its product vanillin) than in normal pine. The aldehyde levels seen in normal lignins could possibly result from postlignification oxidation, but the levels seen in this and other CAD-deficient plants suggest that incorporation of hydroxycinnamaldehydes (and hydroxybenzaldehydes) into lignins is a normal process that becomes more significant when the pathway to the final monolignols is down-regulated.

The CAD-deficient pine lignin had another feature that may or may not be attributable directly to the CAD-deficiency. Since this was a natural mutant, other unknown enzymes may have also been impacted by the mutation(s); however, this is unlikely because the genetic determinant of the modified wood colocalized with the coding sequence for the enzyme CAD [375]. Substantial levels of an unexpected unit derived from dihydroconiferyl alcohol (DHCA, Figure 5.12) were identified in the mutant's lignin [2,38,294]. Although found at lower levels in normal plant lignins [2,27,77,290], DHCA's derivation is not completely clear. Claims that the unit could only arise as a modified metabolic product *following* dimerization of traditional lignin monomers [376,377] were incorrect; there is ample evidence that DHCA is formed as a monomer in the CAD-deficient pine and that the monomer is incorporated into the lignin via radical coupling processes [2,38,294,374]. Evidence from NMR indicates that about half of the DHCA units in the soluble lignin fractions were involved in 5–5-coupled structures [2,38]; coniferyl alcohol has never been shown to 5–5-dimerize. Dissolution of the (dioxane-water insoluble) residual lignin in acetyl or propionyl bromide led to acylated and α-brominated lignin and caused some other structural changes; but, other than acylation, dissolution did not affect the saturated arylpropanol side-chain. Experiments employing HMQC and COSY NMR then unambiguously determined that DHCA was a component of that residual lignin, although quantification was not possible [2].

More recently, GPDs were also identified in lignins isolated from the CAD-deficient pine mutant (Figure 5.12). This helped to unravel the nature of weak correlations often seen in pine lignin ^{13}C–^1H correlation NMR spectra (Figure 5.12b); smaller amounts of GPDs are now recognized to be present in lignins from normal pines and presumably other softwoods. The structures in the complex lignin polymers are readily proven by NMR of isolated lignins.

Figure 5.12a shows a subplot of the side-chain region from a 2D HMQC-TOCSY NMR experiment on an acetylated lignin isolated from the mutant pine, highlighting the new (acetylated) guaiacylpropane-1,3-diols **X6**, along with the DHCA **X5** units. Data from a model compound, 1,3-diacetoxy-1-(4-*O*-benzyl-3-methoxyphenyl)-propane [135], are at the center of the yellow circles and obviously match well.

Invariably, traces of the novel components found in lignins from transgenic or mutant plants, can also be found in lignins from control plants. Figure 5.12b shows the α-C/H region of an HSQC spectrum of the mutant's lignin, where the β-aryl ether units (blue) and the strong new GPD unit **X6** (red) appear most cleanly. Figure 5.12c shows the same region in lignin from a normal pine control. Although the GPD peak is weak, it is diagnostic and, with other correlations evident (not shown), well authenticated.

The GPD monomeric units are derived from DHCA monomer **5** via the action of peroxidase and hydrogen peroxide (Figure 5.12d). The mechanism, via a vinylogous quinone methide, involves two H-radical abstractions. Abstraction from a benzylic CH_2 to produce quinone methides from phenoxy radicals has been noted previously [378]. When DHCA is subjected to peroxidase-H_2O_2, we found monomeric GPD **2**, as well as the range of homo- and crossed-dimers involving DHCA and GPD.

Also evident in the HMQC-TOCSY spectrum (Figure 5.12a), are (acetylated) ketones **X6′** (magenta contours). Products of benzylic alcohol oxidation are seen in various isolated lignins, notably from syringyl (3,5-dimethoxy-4-hydroxy-phenyl) units; they may arise during lignin isolation (particularly in the ball-milling step). Ketones **X6′** provide additional confirmatory evidence for the GPD structures **X6** described above.

In summary, a previously unidentified unit present in small quantities in normal lignins was identified by NMR as a major component of the hydroxyphenylpropanoid polymeric component of a pine mutant deficient in CAD. The guaiacylpropane-1,3-diol units arise from conversion of dihydroconiferyl alcohol monomers by radical reactions and also incorporate into lignins as monomers. Like DHCA, however, GPD does not possess an unsaturated side-chain and is, thus, limited to coupling on the ring 4–O-, 5-, and possibly 1-positions, and necessarily forming terminal units in lignin (except via 5–O–4-coupling). High MW polymers cannot be created without the incorporation of authentic hydroxycinnamyl alcohols (or other cinnamyl derivatives) as well. The CAD-deficient pine lignin is more soluble in alkali and has reduced molecular weight [373].

CCR-Deficiency

Tobacco plants down-regulated in CCR, the prior enzyme to CAD on the pathway, caused a reduction in lignin levels, but at the expense of structural integrity and plant vigor [379]. Such "lignins" incorporated significant amounts of tyramine ferulate, a known wound-response product in tobacco [71]. Hydroxycinnamoyl amide structures were established by NMR experiments, which provided preliminary evidence that the ferulate moiety incorporated into the polymer (similarly to its incorporation into lignins in grasses).

Criticisms [380] that the tyramine units are not part of the polymer, because they are sharp peaks (when the remaining peaks are broader), are invalid [236]. The lignins were MW-fractionated, but that aside, the comment belies a concept worth stating regarding NMR of various lignin polymers. Higher MW compounds or polymers do indeed yield generally broader peaks than smaller molecules. But mobile components on those polymers act far more like smaller molecules and appear sharper. This is the case for not only the tyramine units in these samples [71], but also for *p*-coumarates on grass lignins, and *p*-hydroxybenzoates on aspen, poplar, willow, and palm lignins (Naturally Acylated Lignins section). Such units are all free-phenolic (i.e., uncoupled) terminal appendages on the polymer that are free to rotate more rapidly than the bulk polymer. Rapid rotation in solution is associated with long relaxation times and sharp peaks. Also, in all these cases, the carbons and protons are relatively invariant, being rather remote from the nearest structural differences. The comment [380] that the tyramine peaks have to be from low-MW components of the sample can be dismissed when it is realized that the ferulate moieties to which they are attached would also then be sharp; in fact, it is hard to find them since they are radically coupled in the typical variety of ways in a complex set of structures embedded fully into the polymer [71]. Broad peaks come from the high MW, but more importantly from the many environments in which units find themselves (Broadness of Proton Spectra section).

A more general response, and particularly in plants that do not normally make tyramine ferulate (such as poplar and arabidopsis) appears to be the direct incorporation of ferulic acid itself into lignins. In sharp contrast with claims in another report [381], we readily detected thioacidolysis marker compounds for 4–O-etherified ferulic acid, as well as for a novel *bis*-(β–O–4-ether) that derives from ferulic acid radical coupling reactions, in natural grasses and in CCR-deficient plant materials; decarboxylation truncates the sidechain providing a pathway for a second β–O–4-coupling reaction [97,104]. Both types of structures are evidenced in NMR spectra [97,104].

Conclusions Regarding NMR Studies of
Lignin-Biosynthetic-Pathway Mutants and Transgenics

Perturbing lignification is an ideal way to increase or reduce components entering lignification. The value in structurally assessing the resultant lignins has been a tremendously enhanced understanding of lignins' structure and the lignification process itself. With the variety of useful experiments available, NMR is uniquely suited to delineating often novel new structures in complex polymers. Also, NMR studies have shown that manipulating specific lignin-biosynthetic-pathway genes can produce profound alterations in plant lignins. Although the lignins in mutant and transgenic plants may appear to be strikingly different from normal lignins, they often represent merely broad compositional shifts; most of the novel units that have been found to date appear to be (often previously unidentified) minor units in lignins from "normal" plants. A salient observation is that the process of lignification appears to be flexible enough to readily incorporate phenolic phenylpropanoids other than the traditional monolignols. The incorporation of hydroxycinnamaldehyde monomers, as well as 5-hydroxyconiferyl alcohol, also implies that the plant is apparently sending these products of incomplete monolignol biosynthesis out to the cell wall for incorporation. Such data require no alterations in the current model for lignification, since a solely chemical polymerization process remains the best theory [107,236,338].

The resultant modified lignins in some transgenics evidently have properties sufficient to accommodate the water transport and mechanical strengthening roles of lignin and to allow the plant to be viable, if not vigorous. Whether such plants will be able to confront the rigors of a natural environment replete with a variety of pathogens remains to be determined. However, the plants' flexible approach toward lignification, i.e., polymerizing monolignol precursors and derivatives along with the traditional monolignols, is a testament to their flexible strategy; in a single generation, these plants have circumvented genetic obstacles to remain viable. The recognition that novel units can incorporate into lignin provides significantly expanded opportunities for engineering the composition and consequent properties of lignin for improved utilization of valuable plant resources in processes ranging from natural ruminant digestion to industrial chemical pulping.

Pulping Mechanisms, and Lignins from Pulp and Pulping Liquors

Anthraquinone Pulping

In the early '80s, ^{13}C NMR was just coming on stream to aid in the elucidation of the mechanism of action of a newly discovered novel organic catalyst for alkaline pulping, anthraquinone. Both ^{13}C and ^{1}H NMR of isolated adducts between lignin model quinone methides and anthrahydroquinone suggested that carbon–carbon bonding had taken place between the α-position of the quinone methide and C-10 of the anthracenyl system [382–384]. Interesting shielding effects in proton NMR spectra, and hindered rotation occurring on the NMR timescale suggested a 3D conformation of the adducts that was compellingly verified by a subsequent X-ray crystal structure [385]. Preparation of adducts between anthranol and isolated lignins confirmed that similar structures, and even the same isomers, could be produced in the polymer [386].

Soda Pulping

The application of ^{13}C NMR techniques to the characterization of soda lignin confirmed that formaldehyde was released from the γ-hydroxymethyl groups of C9 units during pulping via reverse aldol reactions with subsequent formation of vinyl ether structures in the lignin matrix [387]. Furthermore, in the same study, it was confirmed by means of specific ^{13}C labeling that the released formaldehyde condensed with lignin fragments to form diarylmethylene structures, which was counterproductive to delignification.

Kraft Pulping

Comprehensive ^{13}C studies of underivatized kraft lignins in DMSO [388] and acetylated kraft lignins in acetone-d_6 [389], along with over 100 corresponding model compounds, provided a solid foundation for signal assignments. The NMR spectra of lignins isolated from kraft pulping liquors are much more complex than the corresponding spectra of MWLs because of all of the new structures formed, which are not originally present in the lignin. Main characteristics of lignins isolated during kraft pulping are a degraded side-chain, more unsaturation due to stilbenes and vinyl ethers, many new signals in the 25–50 ppm region due to fully reduced side-chains, and a much higher phenolic content.

Oxygen/Peracid Treatments

Researchers used ^{13}C NMR spectroscopy to examine products from MWLs from rice straw, Japanese beech (*Fagus crenata* Bl.) and Yezo spruce (*Picea jezoensis* Carr.) following oxygen pulping treatment [390] and from spruce MWL and a G-DHP following peracetic acid treatment [391].

Other Lignin Treatments

Spruce MWL was depolymerized with dry hydrogen iodide and then analyzed by ^{13}C NMR to reveal that ether bonds were cleaved and substitution of α-hydroxyl and α-alkoxy groups by iodine occurred [392]. Light-induced degradation of lignin was studied using spruce MWL impregnated on handsheets prepared from cotton linters. Results of NMR examination showed that α-ether, β-ether, and α–β-bond cleavage were the main reactions [393]. Lignins isolated from steam-exploded aspen [394], *Pinus radiata* wood [395], *Eucalyptus* [396], and beech [397] were studied by ^{13}C NMR spectroscopy. Extensive β-aryl ether cleavage, high phenolic content, and an extensively modified propanoid side-chain were common features in this type of lignin.

Lignin Side-Chain Proton/Carbon Assignment Table for 2D and 3D Spectra

In 2D and 3D NMR spectra, resonances are spread over a considerable volume; the contours are not sharp. However, the reliability of the assignments improve by utilizing at least two chemical shift parameters (e.g., the chemical shifts of a carbon and its attached proton) to define a contour as belonging to a given structural unit. Table 5.4 provides, for many of the units to be found in lignins, side-chain ^{13}C and ^{1}H chemical shifts that can be used to assign 2D and 3D spectra. All shifts are from acetylated lignins in the favored solvent, acetone-d_6, unless noted.

QUANTITATIVE ^{13}C NMR OF LIGNINS

THE NUCLEAR OVERHAUSER EFFECT AND T_1 RELAXATION

Both conventional and somewhat unconventional techniques have been used for liquid-state ^{13}C NMR quantification of lignins. The two main problems that must be

TABLE 5.4
2D Correlation Chemical Shifts of Acetylated Lignins in Acetone-d_6

Structure	Type	H_α	H_β	$H_{\gamma 2}$		$H_{\gamma 1}$	$C\alpha$	$C\beta$	$C\gamma$
A	Ge	6.03	4.84	4.33		4.22	74.5	80.2	63.2
	Gt	6.07	4.82	4.23		4.01	75.3	80.4	63.3
	Se	6.03	4.73	4.39		4.15	75.4	81.4	63.4
	St	6.08	4.63	4.24		3.90	76.8	81.6	64.3
B	G	5.66	3.77		4.36		88.4	51.1	65.8
C	G/S	4.8	3.1	4.26		3.90	86.5	55.2	72.4
D	G/S-GG	4.94	4.30	4.45		4.07	85.1	83.3	64.0
	G/S-GGc	4.81	4.11	-		-	84.0	82.5	-
H	S/G	4.98	4.39	4.28		4.02	76.8	75.9	63.4
S	SA	5.20	3.10		4.06		83.2	57.6	61.9
	SB	4.82	4.50	4.30		4.18	86.9	77.4	64.7
A2	G	5.61	4.89	4.56		4.45	80.9	81.5	63.5
A3	S	-	5.59		4.48		194.8	81.3	64.7
X1	G	6.63	6.23		4.65		134.2	122	65.3
X5a	G	2.62	1.90		4.04		32.1	30.9	64.0
X5b	G	2.73	1.89		4.11		32.2	30.9	64.1
X6	G	5.83	2.18		4.10		73.1	35.9	61.2
X6'	G	-	3.33		4.41		198.3	37.6	60.4
X7	G	5.91	~5.4		~4.1		~74	72.9	61.2
	Sc	5.88	5.37	-			73.5	72.2	61.5
	Pc	5.84	5.38	-			73.7	72.2	62.2
	5H	5.86						73.7	62.2
Benzaldehyde	G/S	9.86	–		–		191.6	–	–
Cinamaldehyde	G/S	7.60	6.70		9.68		153.1	129.8	194.0
Ald-b-G	S		7.31	–	–	9.57	136.1	149.8	188.3
Ald-b-S	G	6.71		–	9.32		126.7	152.1	187.0
Ald-b-S	S	6.69		–	9.32		126.8	152.1	187.0

Note: Data is presented to show how 2D correlation chemical shifts of acetylated lignins in acetone-d_6 (unless denoted with a superscript c) from various 2D spectra (author-generated from various in-house sources) can be used as a guide for assigning COSY, TOCSY, HMQC/HSQC, HMQC/HSQC-TOCSY, HMBC, etc. spectra. the author-generated data include components that may be easily seen only in various lignin-biosynthetic-pathway mutants and transgenics.
A = β–*O*–4, **B** = β–5, **C** = β–β, **D** = dibenzodioxocin, **A2** = α,β-di-ether, **A3** = α-keto-β-ether, **H** = benzodioxanes (α,β-ethers from 5-hydroxyguaiacyl units), **S** = spirodienones (β–1, note that shifts are given for the A- and B-ring moieties; see Figure 5.11), **X1** = cinnamyl alcohol, **X5** = dihydrocinnamyl alcohol (**X5a** = 5–β and 4–*O*–β-linked units, **X5b** = 5–5-linked units), **X6** = arylpropanediol, **X6'** = benzylic-oxidized arylpropanediol, **X7** = arylglycerol. See structures in Figures 5.3 and 5.12. **Ald-*b*-G/S** = the cinnamaldehyde 8–*O*–4-**G/S** products, c.f. structures **4** in Figure 5.21. cData from sample in CDCl₃. Similar data tables for acetylated lignins in CDCl₃ have been provided (Kilpeläinen, I., Sipilä, J., Brunow, G., Lundquist, K., and Ede, R.M., *J. Agr. Food Chem.*, 42(12), 2790–2794, 1994; Ämmälahti, E., Brunow, G., Bardet, M., Robert, D., and Kilpeläinen, I., *J. Agr. Food Chem.*, 46(12), 5113–5117, 1998). Chemical shifts in italics are from model data (where data from actual lignins overlaps or is not clear).

dealt with when making the transition from qualitative to quantitative ^{13}C NMR are the nuclear Overhauser enhancement, or nOe and the difference in relaxation rates of carbon nuclei [16].

The nOe is observed as a change in intensity of the signal of one nucleus when another nearby nucleus is perturbed in some way. One of the most important manifestations of the nOe is the carbon signal enhancement that results upon irradiating the proton nuclei, such as is done during decoupling. Although the nOe can result in a threefold increase in S/N, it is a complication in quantitative NMR, because the enhancement is generally not equal for all types of carbon signals. Therefore, reliable quantification over the full spectral range of carbon signals required that the nOe be minimized or eliminated. This is accomplished by the classical inverse-gated experiment in which the proton decoupling is gated off during the interpulse delay. In conventional proton decoupled qualitative ^{13}C NMR experiments the nOe builds up during this delay. The major disadvantage of eliminating this irradiation and consequent nOe is the lower resultant sensitivity.

The second problem in quantification is the relatively slow relaxation rate of carbon nuclei. This problem is simply addressed by providing a relatively long delay between scans so the carbon nuclei are completely relaxed. The major relaxation process is the spin-lattice relaxation defined by the time constant T_1 and complete relaxation requires a delay of at least $5xT_1$ seconds [398]. Fortunately, because lignin is a large molecule and tumbles relatively slowly in solution, the carbon nuclei relax very rapidly. This is a disadvantage when attempting to resolve close peaks, or in some experiments (such as 2D long-range coupling, HMBC), when longer signal persistence is desired; however, rapid relaxation is an advantage in quantification. The T_1 relaxation times can be determined by use of the inversion recovery pulse sequence or by the progressive saturation method although the latter is more demanding and subject to systematic errors [165,398].

The T_1 ranges of selected carbon types in various lignins are listed in Table 5.5 [308]. The data in Table 5.5 provide sufficient information to estimate the delay times required for valid quantification for almost any lignin type. It should be readily

TABLE 5.5
Spin-Lattice Relaxation Times (T_1)

T_1 Ranges of Selected Carbon Types (seconds)			
Lignin sample[a]	Methoxyl	Ternary	Quaternary
Cottonwood MWL (CU)	0.79	0.11–0.18	1.1–2.8
Poplar SE lignin (PU)	0.67	0.18–0.21	1.6–2.1
Acet. Poplar SE lignin (PA)	0.76	0.21–0.34	1.3–2.5
Indulin (IU)	0.82	0.16–0.23	1.3–2.8
Organosolv lignin (RU)	0.87	0.23–0.26	1.3–3.5

Source: Landucci, L.L., *Holzforschung*, 45(6), 425–432, 1991.

[a] Standard samples supplied for Round-Robin Lignin Analysis at IEA Symposium, New Orleans, 1991.

TABLE 5.6
Effect of Temperature on Spin-Lattice Relaxation Times (T_1)
of Selected Signals from Aspen MWL in DMSO-d_6

Signal	T_1 (seconds)		
(ppm)	30°C	50°C	70°C
169.5	1.52	2.19	2.23
169.2	1.35	2.30	2.88
131.3	—	0.21	0.23
106.7	0.09	0.13	0.15
104.3	0.14	0.15	0.15
63.2	0.06	0.08	0.09
55.9	0.48	0.76	0.74
20.9	0.55	0.76	0.79

Source: Landucci, L.L., *Holzforschung*, 45(6), 425–432, 1991.

apparent that slow relaxation of the quaternary carbons is what limits the scan repetition rate of the inverse-gated experiment. The minimum delay to be used for
these lignins is based on the quaternary carbon relaxation at ambient temperature
when a 90° tip angle is used. For example, the 5T_1 delay for poplar steam-exploded
lignin should be $5 \times 2.1 = {\sim}11$ s, whereas with the organosolv lignin it should be
$5 \times 3.5 = {\sim}18$ s. With the increasing popularity of the use of DMSO as a lignin solvent, it is a common practice to heat NMR samples to 50° or more during data acquisition. This is acceptable for qualitative NMR, because sharper signals and greater
resolution are obtained, especially if the sample solution is relatively concentrated.
However, sharper signals translate to slower relaxation, so it is disadvantageous to
heat samples during quantification. On the contrary, if more signal is desired over a
given period of time, cooling the sample would be advantageous, because the more
viscous the sample solution is, the faster the relaxation will be.

The effect of temperature on relaxation rate of selected signals from aspen MWL
in DMSO has been reported and is shown in Table 5.6 [399]. As indicated from the
T_1 values, heating the sample from 30° to 70° doubles the required relaxation delay,
if valid quantification is desired over the entire spectral range. This viscosity effect
is much more noticeable in DMSO than in acetone or chloroform.

Finally, even when all these parameters are taken into account, spectral baselines
are never flat. The situation has improved considerably as instruments have gone
fully digital and taken advantage of efficient digital filtering, but setting the baseline
remains a significant source of error.

PRECISION AND ACCURACY

The concepts of precision and accuracy are sometimes misunderstood as they relate
to lignin analysis. The most important of these concepts is precision, as this is a prerequisite to accuracy. Simply defined, precision is the ability to reproduce a result. If

this cannot be done, then accuracy is simply unobtainable. In the lignin literature, replications of identical experiments are seldom reported due to the excessive instrument time required for such studies. However, it has been shown that with large amounts (>500 mg) of lignin and multiple replications on older (predigital) instruments, precisions of ± 1% are possible [400]. More realistic values with smaller amounts of lignin (100–500 mg) are generally in the range of ± 10 to ± 25%. Only after the desired precision is obtained can the accuracy be improved. Accuracy generally depends on a set of assumptions, because, unfortunately, true values in lignin analysis are unknown. A comprehensive examination of the effects of various operations during NMR quantification on precision has been reported [400]. No systematic study using digital instrumentation, nor the recently available cryogenically cooled probes (in which both 1H and 13C coils are now cooled to reduce electronic noise), has been reported.

Approaches to Quantification

Signal Cluster Analysis

Shown in Figure 5.22 is a quantitative spectrum of a willow (Salix) MWL with various spectral regions highlighted [309]. These regions generally correspond to integral limits that are predefined and remain constant throughout a particular study. For high-precision work, this step is essential. Information obtained by measuring these signal clusters results in values of useful ratios, as listed in Table 5.7. The accuracy of these ratios generally depends upon certain assumptions. For example, the OMe/Aryl ratio assumes no unsaturation in the side-chain of C9 units. This ratio has been determined for a large number of MWL types by use of the inverse-gated experiment [306]. If unsaturation is present and the signal(s) position is known, a correction factor can be determined, as was reported in the quantitative analysis of a guaiacyl DHP [171]. In the case of the aliphatic carbon/C6 aryl ratio of lignin carbons, an additional assumption of a carbohydrate-free sample is made. With acetylated samples, the ratio of primary to secondary to phenolic hydroxyl groups can easily be determined with relatively high precision by quantifying the signals from the acetate carboxyl groups [185,401]. A

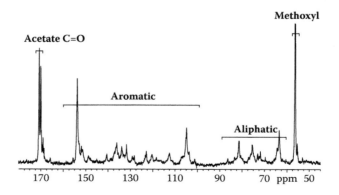

FIGURE 5.22 Quantitative ¹³C NMR spectrum of a willow lignin with highlighted spectral regions (Landucci, L.L., *Holzforschung*, 39(6), 355–359, 1985).

TABLE 5.7
Useful Ratios for Lignin Characterization[a]

Ratio	Assumptions
Methoxyl/Aryl(C6)	No unsaturation in side-chain
Aliphatic/Aromatic(C6)	No unsaturation in side-chain No carbohydrate impurity
Pri/Sec/Phenolic OH (acetylated samples)	No carbohydrate impurity
Ternary/Quaternary (aromatic region)	None, but a spectral editing experiment such as DEPT is needed to determine CH/C boundary.

Values for Selected Round-Robin Samples

Lignin sample[b]	Methoxy/Aryl(C6)	Alip/Aryl(C6)	Tern/Quat
Cottonwood MWL	1.4	2.7	2.4/3.6
Poplar SE lignin	1.2	1.2	2.1/3.9
Acet. Poplar SE lignin	1.2	1.3	2.1/3.9
Indulin	0.7	0.8	2.8/3.2
Organosolv lignin	1.0	0.7	2.4/3.6

Source: Landucci, L.L., *IEA Presymposium on Analysis of Wood, Annual Plant Lignins,* New Orleans, LA., 1991.

[a]   Area of aryl region (95–160 ppm) is set to 6.0 units.

[b]   Standard samples supplied for Round-Robin Lignin Analysis at IEA Symposium, New Orleans, 1991.

modification of this type of quantification using ^{13}C-labeled acetyl chloride for derivatization was reported [303]. Determination of the hydroxyl distribution in lignin also requires carbohydrate-free samples, since secondary hydroxyls from carbohydrates contribute to the central signal of the primary, secondary, and phenolic trio. However, it is possible to quantify carbohydrates, such as xylans, to some extent from the anomeric carbon (C1) that appears at about 100 ppm. Then a correction can be made [309].

There are two common approaches to quantification, each of which has advantages and disadvantages. The first approach is to set the aromatic + aliphatic regions (minus the methoxyl group) to nine carbons. This approach is unaffected by unsaturation in the side-chain. Disadvantages are that it is not valid for highly degraded lignins such as kraft lignin and not valid if carbohydrates are present. The second approach is based on the assumption that there are six aromatic carbons per C9 unit and that the signals all appear in the region between 100 and 160 ppm. Advantages of this approach are that it can be applied to a lignin material that is suspected of having considerable side-chain degradation—that is, one in which a C9 unit is unrealistic. Also, this approach is not seriously affected by the presence of carbohydrates. A disadvantage is that it is not valid if there is extensive aliphatic unsaturation in the side-chain such as that from vinyl ethers and stilbenes. Therefore, this approach would not be a good one for a kraft lignin.

Application of both of these approaches to a pine MWL is illustrated in Table 5.8 with some selected values [402]. Clearly, there is no significant difference in the two

TABLE 5.8
¹³C NMR Derived Values for Pine Wood Ac-MWL

	Approach 1	Approach 2
Functional Unit	(content/monomeric unit)	
Methoxyl	0.93	0.93
Acetoxy Me	1.86	1.86
Acetoxy C=O		
primary	0.85	0.86
secondary	0.61	0.61
phenolic	0.35	0.35
Aromatic	5.97	6.00
Aliphatic		
60–100 ppm	2.78	2.79
35–55 ppm	0.24	0.24
Total	3.02	3.03

Source: Netzel, D.A., *Anal. Chem.*, 59, 1775–1779, 1987.

TABLE 5.9
¹³C NMR Derived Values for Fossil Wood Ac-MWLs

MWL =>	GDH-80		Resolute		Grattan	
Functional Unit	**Approach**		**Approach**		**Approach**	
	1	**2**	**1**	**2**	**1**	**2**
Methoxyl	0.88	0.73	0.91	0.80	0.73	0.71
Acetoxy Me	1.52	1.27	0.96	0.85	1.37	1.20
Acetoxy C=O						
primary	0.52	0.43	0.46	0.41	0.51	0.45
secondary	0.35	0.29	0.12	0.10	0.26	0.23
phenolic	0.58	0.48	0.38	0.34	0.55	0.48
Aromatic	7.17	6.00	6.77	6.00	6.85	6.00
Aliphatic						
60–100 ppm	1.50	1.25	1.84	1.62	1.63	1.43
35–55 ppm	0.33	0.28	0.41	0.37	0.49	0.43
Total	1.83	1.53	2.25	1.86	2.12	1.86

Source: Netzel, D.A., *Anal. Chem.*, 59, 1775–1779, 1987.

columns because, for a relatively intact lignin with a low carbohydrate content, both approaches are valid. A contrasting situation is illustrated in Table 5.9 where the same two approaches are applied to three different MWLs that were isolated from wood that was found in tertiary fossil forests in the Canadian Arctic [402]. These lignins are thought to be between 25 and 65 million years old. Visually comparing these MWLs to the recent pine MWL on the top indicates extensive side-chain degradation and more aromatic character than the pine lignin. Also, it is interesting that the Resolute lignin was almost completely lacking in secondary acetate carbonyls and, with the other two lignins, this functionality was less predominant than in the recent pine MWL. The values generated from the two approaches are listed in Table 5.9. Clearly, they are very different. It is not really clear which approach is better with these unusual samples. If the degradation mechanisms involved appreciable condensation and aromatization then approach 1, which assumes a C9 entity might not be a bad assumption, because it would simply mean a shift of the aliphatic carbons into the aromatic pool.

Use of Internal Standards

Theoretically, the best approach to quantification would be the use of an internal standard because no assumptions concerning lignin structure need be made. However, this is usually not practical for lignin analysis because standards, which are generally simple and low molecular weight substances, have long relaxation rates. Therefore, the $5T_1$ requirement for the pulse delay would be unrealistically long. However, for specific applications in which only protonated carbons in the internal standard are utilized, the reliable "spiking" method can be useful. As an example, a vinyl ether model compound was added and used to quantify such structures in a lignin isolated from a soda cook [404]. Using this type of quantitative experiment, with conditions based on the protonated carbons of the model compound, the researchers were able to generate a fairly accurate value of the content of this functionality in the lignin. In this case, the long relaxation rate of quaternary carbons in the low MW compound was not a problem because the quaternary carbon signals were not needed for the quantification. Comparison of one protonated carbon in the model with the corresponding protonated carbon in the polymer required a delay of 5–10 s, sufficient for complete relaxation of the β-carbon in the model.

Quantification by Curve Fitting

Comparison of an experimental inverse-gated spectrum with a simulated spectrum (of the signals of interest) generated by a curve fitting program can provide more precise quantification. Advantages of this method are that there is no noise in the simulated spectrum and individual signals, rather than signal clusters, can be quantified. In fact, this is the only practical method to quantify individual signals in crowded regions of a spectrum, since signal overlap is generally not a problem. An example of this technique, schematically illustrated in Figure 5.23, is the quantification of the p–acetoxybenzoate group in a willow MWL. Based on the OMe/aryl ratio, the abundance of this entity was determined to be 0.04–0.06/C9 units in a series of six similar willow MWLs [309]. With this method it is desirable to simulate at least two or three signals of the entity of interest, so as to provide an internal check of the method. In this case, the 1/1/2 ratio of the C = O/C4/C2,6 provided such a check. This application involved the relatively discrete and sharp resonances from a substituent on lignin side-chains. Unfortunately, the backbone lignin units themselves (e.g., β–5, β–O–4) comprise differently populated substructures. The signals from a particular lignin structure can, therefore, not be expected to be symmetric. Curve fitting methods must therefore be applied with caution.

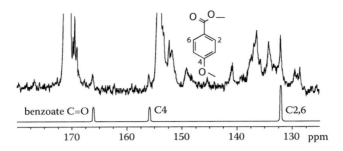

FIGURE 5.23 Quantification of *p*-acetoxybenzoate groups in acetylated willow ML (Landucci, L. L., Deka, G. C., and Roy, D. N., *Holzforschung*, 46(6), 505–511, 1992).

SEMIQUANTITATIVE METHODS

Comparison of the Same Signal(s) from Two or More Spectra

One technique is simply to compare signal intensities or areas of a particular signal or signals among different spectra. With this technique the same sample concentration, and acquisition and processing parameters must be maintained from sample to sample. The primary advantage is that the time-consuming inverse-gated technique is not necessary as identical carbon nuclei are being compared. Also, the signal enhancement feature of the nOe can be utilized. This technique is most frequently used when following the fate of a particular entity during some reaction or treatment of a given lignin.

Comparison of Signals from Protonated Carbons of Similar Type

Another technique that does not require an inverse-gated spectrum is illustrated in Figure 5.24 with the measurement of relative signal areas of the protonated carbons in the side-chain region of a guaiacyl DHP [215]. All of the signals were assigned by comparison with authentic dimeric and trimeric lignin models. This spectrum indicates that this DHP is relatively simple compared to a guaiacyl lignin and contains only the four major unit types shown here. The accuracy of this method depends upon the similarity of the relaxation behavior of these protonated carbons. This technique has also been applied to the determination of syringyl/guaiacyl ratio in poplar lignin [405].

Spectral Editing Experiments

The DEPT experiment detects only protonated carbons and depends upon polarization transfer or energy transfer from protons to adjacent carbons [17]. The efficiency of this transfer process depends largely upon a delay in the pulse sequence that is related to the proton-carbon coupling constant, but is relatively insensitive to coupling constant variation [406,407]. Therefore, the process can only be perfect, or quantitative, for one coupling constant. Since coupling constants for the various proton/carbon pairs can vary widely in lignin, as in most other materials, an average coupling constant must be used. For best results, it has been demonstrated by investigators in coal research that relatively long relaxation delays must be used [403]. Therefore, much of the advantage of the DEPT sequence disappears. However, in

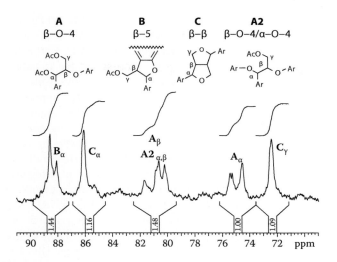

FIGURE 5.24 Integration of signals in the side-chain region of a guaiacyl DHP. Peak assignment follows that established in Figure 5.3. (Landucci, L.L., *Holzforschung*, 39(6), 355–359, 1985).

spite of all of these limitations, it has been shown to be a viable method for approximating the ternary/quaternary ratio in kraft lignins, which are notoriously difficult materials because of their appreciable unsaturated side-chain content in the aromatic region and extensive loss of carbon from the side-chain [408].

CONCLUSION

In summary, true quantification in lignins is difficult due to signal overlap and other factors. The best quantification method remains the relatively tedious inverse-gated technique, along with an internal standard substance. There are other options, but all are imperfect to some degree. With substances like lignins there are few guidelines and there are no absolutes; so, any value obtained from a nonquantitative technique must be compared to a corresponding value obtained from an inverse-gated experiment to determine its quantitative validity. Fortunately, most of the semiquantitative methods that have been described are adequate when the researcher wants to follow changes in structure over time during a particular treatment provided that the desired precision is maintained.

ACKNOWLEDGMENTS

The authors are grateful to many people for the data presented in this chapter. Jane Marita, Fachuang Lu, Hoon Kim, Takuya Akiyama, Paul Schatz, Stéphane Quideau, Richard Helm, Rick Ede, Noritsugu Terashima, and Sally Ralph isolated many of the lignins, produced compounds for the database, provided data, etc. Collaborators supplied many of the lignins from mutant and transgenic plants; these include Catherine Lapierre, Clint Chapple, Rochus Franke, Matt R. Hemm, Jan Van Doorsselaere, Wout Boerjan, Ronald R. Sederoff, David M. O'Malley, Jay T. Scott, John J. MacKay,

Nabila Yahiaoui, Matthieu Chabannes, Alain-M. Boudet, Michel Pean, Gilles Pilate, Lise Jouanin, Srinu Reddy, Fang Chen, Richard Dixon, Vincent Chiang, Runcang Sun, Ian Suckling, Wilfred Vermerris, and others. Ronald D. Hatfield, Sally Ralph, Gösta Brunow, Knut Lundquist, and others are thanked for valuable discussions and collaborations. High-field and cyroprobe NMR studies were carried out at the National Magnetic Resonance Facility at Madison with support from the NIH Biomedical Technology Program (RR02301) and additional equipment funding from the University of Wisconsin, NSF Academic Infrastructure Program (BIR-9214394), NIH Shared Instrumentation Program (RR02781, RR08438), NSF Biological Instrumentation Program (DMB-8415048), and U.S. Department of Agriculture. In addition to funding sources for the many collaborators, the following directly funded the work in our labs: USDA-NRI (Plant Growth and Development #96-35304-3864; Improved Utilization of Wood and Wood Fiber #94-03465, #97-02208, #99-02351, #2001-35103-10869, #2003-35103-13834; Food Characterization #2003-35503-13820), the DOE Great Lakes Bioenergy Research Center (DOE Office of Science BER DE-FC02-07ER64494), and the U.S. Dept. of Energy, Division of Energy Biosciences (#DE-AI02-00ER15067).

REFERENCES

1. Sarkanen, K. V., and Ludwig, C. H. (1971) *Lignins, Occurrence, Formation, Structure and Reactions*, Wiley-Interscience, New York.
2. Ralph, J., Marita, J. M., Ralph, S. A., Hatfield, R. D., Lu, F., Ede, R. M., Peng, J., Quideau, S., Helm, R. F., Grabber, J. H., Kim, H., Jimenez-Monteon, G., Zhang, Y., Jung, H.-J. G., Landucci, L. L., MacKay, J. J., Sederoff, R. R., Chapple, C., and Boudet, A. M. (1999) Solution-state NMR of lignins. In: Argyropoulos, D. S. (ed). *Advances in Lignocellulosics Characterization*, TAPPI Press, Atlanta, GA.
3. Stothers, J. B. (1972) *Carbon-13 NMR Spectroscopy*, Academic Press, New York.
4. Levy, G. C., and Nelson, G. L. (1972) *Carbon-13 Nuclear Magnetic Resonance for Organic Chemists*, Wiley Interscience, New York.
5. Winstein, S., Carter, P., Anet, F. A. L., and Bourn, A. J. R. (1965) The effects of steric compression on chemical shifts in half-cage and related molecules. *J. Am. Chem. Soc.* **87**(22), 5247–5249.
6. Cheney, B. V. (1968) Magnetic deshielding of protons due to intramolecular steric interactions with proximate hydrogens. *J. Am. Chem. Soc.* **90**(20), 5386–5390.
7. Breitmaier, E., and Voelter, W. (1987) *Carbon-13 NMR Spectroscopy. High-Resolution Methods and Applications in Organic Chemistry and Biochemistry*, 3 Ed., VCH Publishers, New York.
8. Lüdemann, H.-D., and Nimz, H. (1974) ^{13}C-Kernresonanzspektren von Ligninen, 1. Chemische Verschiebungen bei Monomeren und Dimeren Modelsubstanzen. *Makromol. Chem.* **175**, 2393–2407.
9. Lüdemann, H.-D., and Nimz, H. (1974) ^{13}C-Kernresonanzspektren von Ligninen, 2. Buchen- und Fichten-Björkman-Lignin. *Makromol. Chem.* **175**, 2409–2422.
10. Landucci, L. L., and Ralph, S. A. (2001) Reaction of *p*-hydroxycinnamyl alcohols with transition metal salts. IV. Tailored syntheses of β-O-4 trimers. *J. Wood Chem. Technol.* **21**(1), 31–52.
11. Ralph, J. (1993) ^{1}H NMR of acetylated β-ether/β-ether lignin model trimers. *Magn. Reson. Chem.* **31**(4), 357–363.

12. Kishimoto, T., Uraki, Y., and Ubukata, M. (2008) Synthesis of β-O-4-type artificial lignin polymers and their analysis by NMR spectroscopy. *Org. Biomol. Chem.* **6**(16), 2982–2987.

13. Shaka, A. J., Keeler, J., Frenkiel, T., and Freeman, R. (1983) An improved sequence for broadband decoupling: WALTZ-16. *J. Magn. Reson.* **52**(2), 335–338.

14. Shaka, A. J., Keeler, J., and Freeman, R. (1983) Evaluation of a new broadband decoupling sequence: WALTZ-16. *J. Magn. Reson.* **53**(2), 313–340.

15. Levitt, M. H., Freeman, R., and Frenkiel, T. (1982) Supercycles for broadband heteronuclear decoupling. *J. Magn. Reson.* **50**(1), 157–160.

16. Noggle, J. H., and Schirmer, R. E. (1971) *The Nuclear Overhauser Effect*, Academic Press, London.

17. Sanders, J. K., and Hunter, B. K. (1993) *Modern NMR Spectroscopy. A Guide for Chemists.*, 2 Ed., Oxford University Press, Oxford.

18. Ernst, R. R. (1987) *Principles of Nuclear Magnetic Resonance in One and Two Dimensions*, Clarendon Press, Oxford.

19. Chandrakumar, N., and Subramanian, S. (1986) *Modern Techniques in High-Resolution FT-NMR*, Springer-Verlag, New York.

20. Bax, A. (1982) *Two-Dimensional Nuclear Magnetic Resonance in Liquids*, D. Reidel Publishing Co., Dodrecht, Holland.

21. Croasmun, W. R., and Carlson, R. M. K. (eds). (1987) *Two-Dimensional NMR Spectroscopy. Applications for Chemists and Biochemists*, VCH, Weinheim, Germany.

22. Derome, A. E. (1987) *Modern NMR Techniques for Chemistry Research*, Permagon Press, Oxford.

23. Cavanagh, J., Fairbrother, W. J., Palmer, A. G., and Skelton, N. J. (1996) *Protein NMR Spectroscopy, Principles and Practice*, Academic Press, San Diego.

24. Günther, H. (1995) *NMR Spectroscopy: Basic Principles, Concepts, and Applications in Chemistry*, 2 Ed., John Wiley and Sons, Chichester.

25. Patt, S. L., and Shoolery, J. N. (1982) Attached proton test for carbon-13 NMR. *J. Magn. Reson.* **46**, 535–539.

26. Morris, G. A., and Freeman, R. (1979) Enhancement of nuclear magnetic resonance signals by polarization transfer. *J. Am. Chem. Soc.* **101**(3), 760–762.

27. Fukagawa, N., Meshitsuka, G., and Ishizu, A. (1991) A two dimensional NMR study of birch milled wood lignin. *J. Wood Chem. Technol.* **11**(3), 373–396.

28. Ede, R. M., Brunow, G., Simola, L. K., and Lemmetyinen, J. (1990) Two-dimensional proton-proton chemical shift correlation and J-resolved NMR studies on isolated and synthetic lignins. *Holzforschung* **44**(2), 95–101.

29. Quideau, S., and Ralph, J. (1993) Synthesis of 4,8-bis(4-hydroxy-3-methoxyphenyl)-3,7,dioxabicyclo[3.3.0]octan-2-ones and determination of their relative configuration via long-range proton couplings. *J. Chem. Soc., Perkin Trans. 1* (6), 653–659.

30. Ede, R. M., and Brunow, G. (1992) Application of two-dimensional homo- and heteronuclear correlation NMR spectroscopy to wood lignin structure determination. *J. Org. Chem.* **57**(5), 1477–1480.

31. Ede, R. M., and Ralph, J. (1996) Assignment of 2D TOCSY spectra of lignins: the role of lignin model compounds. *Magn. Reson. Chem.* **34**(4), 261–268.

32. Fukagawa, N., Meshitsuka, G., and Ishizu, A. (1992) 2D NMR study of residual lignin in beech kraft pulp combined with selective cleavage with pivaloyl iodide. *J. Wood Chem. Technol.* **12**(4), 425–445.

33. Fukagawa, N., Meshitsuka, G., and Ishizu, A. (1992) Isolation of a syringyl-β-O-4 rich end- wise type lignin fraction from birch periodate lignin. *J. Wood Chem. Technol.* **12**(1), 91–109.

34. Kilpeläinen, I., Sipilä, J., Brunow, G., Lundquist, K., and Ede, R. M. (1994) Application of two-dimensional NMR spectroscopy to wood lignin determination; Identification of some minor structural units of hard- and softwood lignins. *J. Agr. Food Chem.* **42**(12), 2790–2794.

35. Quideau, S., and Ralph, J. (1994) A biomimetic route to lignin model compounds *via* silver (I) oxide oxidation. 2. NMR characterization of non-cyclic benzyl aryl ether trimers and tetramers. *Holzforschung* **48**(2), 124–132.

36. Ralph, J., and Helm, R. F. (1993) Lignin/hydroxycinnamic acid/polysaccharide complexes: Synthetic models for regiochemical characterization. In: Jung, H. G., Buxton, D. R., Hatfield, R. D., and Ralph, J. (eds). *Forage Cell Wall Structure and Digestibility*, American Society of Agronomy, Crop Science Society of America, Soil Science Society of America, Madison, WI.

37. Ralph, J. (1996) An unusual lignin from Kenaf. *J. Nat. Prod.* **59**(4), 341–342.

38. Ralph, J., MacKay, J. J., Hatfield, R. D., O'Malley, D. M., Whetten, R. W., and Sederoff, R. R. (1997) Abnormal lignin in a loblolly pine mutant. *Science* **277**, 235–239.

39. Ralph, J., and Zhang, Y. (1998) A new synthesis of *(Z)*-coniferyl alcohol, and characterization of its derived synthetic lignin. *Tetrahedron* **54**, 1349–1354.

40. Ede, R. M., and Kilpeläinen, I. (1995) Homo- and hetero-nuclear 2D NMR techniques: unambiguous structural probes for non-cyclic benzyl aryl ethers in soluble lignin samples. *Res Chem Intermediat* **21**(3–5), 313–328.

41. Ralph, J., Peng, J., and Lu, F. (1998) Isochroman structures in lignin: A new β-1 pathway. *Tetrahedron Lett.* **39**(28), 4963–4964.

42. Bax, A., Freeman, R., and Kempsell, S. P. (1980) Natural abundance carbon-13-carbon-13 coupling observed via double-quantum coherence. *J. Am. Chem. Soc.* **102**(14), 4849–4851.

43. Bax, A., Freeman, R., and Frenkiel, T. A. (1981) An NMR technique for tracing out the carbon skeleton of an organic molecule. *J. Am. Chem. Soc.* **103**(8), 2102–2104.

44. Marcei, T. H., and Freeman, R. (1982) Echoes and antiechoes in coherence transfer NMR: determining the signs of double-quantum frequencies. *J. Magn. Reson.* **48**(1), 158–163.

45. Braun, S., Kalinowski, H.-O., and Berger, S. (1996) *100 and More Basic NMR Experiments: A Practical Course*, VCH, New York.

46. Braun, S., Kalinowski, H.-O., and Berger, S. (1998) *150 and More Basic NMR Experiments: A Practical Course*, VCH, New York.

47. Bardet, M., Gagnaire, D., Nardin, R., Robert, D., and Vincendon, M. (1986) Use of carbon-13 enriched wood for structural NMR investigation of wood and wood components, cellulose and lignin, in solid and in solution. *Holzforschung* **40**(Suppl.), 17–24.

48. Guittet, E., Lallemand, J. Y., Lapierre, C., and Monties, B. (1985) Applicability of the carbon-13 NMR "INADEQUATE" experiment to lignin, a natural polymer. *Tetrahedron Lett.* **26**(22), 2671–2674.

49. Lapierre, C., Monties, B., Guittet, E., and Lallemand, J. Y. (1987) Two-dimensional carbon-13 NMR of poplar lignins: study of carbon connectivities and reexamination of signal assignments by means of the INADEQUATE technique. *Holzforschung* **41**(1), 51–58.

50. Robert, D., Ämmälahti, E., Bardet, M., Brunow, G., Kilpeläinin, I., Lundquist, K., Neirinck, V., and Terashima, N. (1998) Improvement in NMR structural studies of lignin through two- and three-dimensional NMR detection and isotopic enrichment. In: Lewis, N. G., and Sarkanen, S. (eds). *Lignin and Lignan Biosynthesis*, American Chemical Society, Washington, DC.

51. Lapierre, C., Gaudillere, J. P., Monties, B., Guittet, E., Rolando, C., and Lallemand, J. Y. (1983) Enrichissement photosynthetique en carbone 13 lignines de peuplier: caractérisation préliminaire par acidolyse et RMN 13C. *Holzforschung* **37**(5), 217–224.

52. Kay, L. E., Keifer, P., and Saarinen, T. (1992) Pure absorption gradient enhanced heteronuclear single quantum correlation spectroscopy with improved sensitivity. *J. Am. Chem. Soc.* **114**(26), 10663–10665.

53. Palmer, A. G., Cavanagh, J., Wright, P. E., and Rance, M. (1991) Sensitivity improvement in proton-detected two-dimensional heteronuclear correlation NMR spectroscopy. *J. Magn. Reson. Ser. A* **93**, 151–170.

54. Davis, A. L., Keeler, J., Laue, E. D., and Moskau, D. (1992) Experiments for recording pure-absorption heteronuclear correlation spectra using pulsed field gradients. *J. Magn. Reson.* **98**(1), 207–216.

55. Quideau, S., and Ralph, J. (1997) Lignin-ferulate cross-links in grasses. Part 4. Incorporation of 5–5-coupled diferulate into lignin. *J. Chem. Soc., Perkin Trans. 1* (16), 2351–2358.

56. Ralph, J., Hatfield, R. D., Quideau, S., Helm, R. F., Grabber, J. H., and Jung, H.-J. G. (1994) Pathway of *p*-coumaric acid incorporation into maize lignin as revealed by NMR. *J. Am. Chem. Soc.* **116**(21), 9448–9456.

57. Zhang, L., Gellerstedt, G., Ralph, J., and Lu, F. (2006) NMR studies on the occurrence of spirodienone structures in lignins. *J. Wood Chem. Technol.* **26**(1), 65–79.

58. Zhang, L., Henriksson, G., and Gellerstedt, G. (2003) The formation of β-β structures in lignin biosynthesis - are there two different pathways? *Org. Biomol. Chem.* **1**(20), 3621–3624.

59. Zhang, L., and Gellerstedt, G. (2001) NMR observation of a new lignin structure, a spiro-dienone. *Chem. Commun.* (24), 2744–2745.

60. Ralph, J., Akiyama, T., Kim, H., Lu, F., Schatz, P. F., Marita, J. M., Ralph, S. A., Reddy, M. S. S., Chen, F., and Dixon, R. A. (2006) Effects of coumarate-3-hydroxylase down-regulation on lignin structure. *J. Biol. Chem.* **281**(13), 8843–8853.

61. Ralph, J., and Lu, F. (2004) Cryoprobe 3D NMR of acetylated ball-milled pine cell walls. *Org. Biomol. Chem.* **2**(19), 2714–2715.

62. Marita, J. M., Vermerris, W., Ralph, J., and Hatfield, R. D. (2003) Variations in the cell wall composition of maize *brown midrib* mutants. *J. Agr. Food Chem.* **51**(5), 1313–1321.

63. Marita, J. M., Ralph, J., Hatfield, R. D., Guo, D., Chen, F., and Dixon, R. A. (2003) Structural and compositional modifications in lignin of transgenic alfalfa down-regulated in caffeic acid 3-*O*-methyltransferase and caffeoyl coenzyme A 3-*O*-methyltransferase. *Phytochem.* **62**(1), 53–65.

64. Lu, F., and Ralph, J. (2003) Non-degradative dissolution and acetylation of ball-milled plant cell walls; high-resolution solution-state NMR. *Plant J.* **35**(4), 535–544.

65. Kim, H., Ralph, J., Lu, F., Ralph, S. A., Boudet, A.-M., MacKay, J. J., Sederoff, R. R., Ito, T., Kawai, S., Ohashi, H., and Higuchi, T. (2003) NMR Analysis of Lignins in CAD-deficient Plants. Part 1. Incorporation of hydroxycinnamaldehydes and hydroxybenzaldehydes into lignins. *Org. Biomol. Chem.* **1**, 268–281.

66. Ralph, J., Lapierre, C., Marita, J., Kim, H., Lu, F., Hatfield, R. D., Ralph, S. A., Chapple, C., Franke, R., Hemm, M. R., Van Doorsselaere, J., Sederoff, R. R., O'Malley, D. M., Scott, J. T., MacKay, J. J., Yahiaoui, N., Boudet, A.-M., Pean, M., Pilate, G., Jouanin, L., and Boerjan, W. (2001) Elucidation of new structures in lignins of CAD- and COMT-deficient plants by NMR. *Phytochem.* **57**(6), 993–1003.

67. Ralph, J., Lapierre, C., Lu, F., Marita, J. M., Pilate, G., Van Doorsselaere, J., Boerjan, W., and Jouanin, L. (2001) NMR evidence for benzodioxane structures resulting from incorporation of 5-hydroxyconiferyl alcohol into lignins of *O*-methyl-transferase-deficient poplars. *J. Agr. Food Chem.* **49**(1), 86–91.

68. Marita, J. M., Ralph, J., Lapierre, C., Jouanin, L., and Boerjan, W. (2001) NMR characterization of lignins from transgenic poplars with suppressed caffeic acid *O*-methyltransferase activity. *J. Chem. Soc., Perkin Trans. 1*(22), 2939–2945.

69. Chabannes, M., Barakate, A., Lapierre, C., Marita, J., Ralph, J., Pean, M., Danoun, S., Halpin, C., Grima-Pettenatia, J., and Boudet, A.-M. (2001) Strong decrease in lignin content without significant alteration of plant development is induced by simultaneous down-regulation of cinnamoyl CoA reductase (CCR) and cinnamyl alcohol dehydrogenase (CAD) in tobacco plants. *Plant J.* **28**(3), 257–270.

70. Marita, J., Ralph, J., Hatfield, R. D., and Chapple, C. (1999) NMR characterization of lignins in *Arabidopsis* altered in the activity of ferulate-5-hydroxylase. *Proc. Natl. Acad. Sci.* **96**(22), 12328–12332.

71. Ralph, J., Hatfield, R. D., Piquemal, J., Yahiaoui, N., Pean, M., Lapierre, C., and Boudet, A.-M. (1998) NMR characterization of altered lignins extracted from tobacco plants down-regulated for lignification enzymes cinnamyl-alcohol dehydrogenase and cinnamoyl-CoA reductase. *Proc. Natl. Acad. Sci.* **95**(22), 12803–12808.

72. Heikkinen, S., Toikka, M. M., Karhunen, P. T., and Kilpeläinen, I. A. (2003) Quantitative 2D HSQC (Q-HSQC) via suppression of J-dependence of polarization transfer in NMR spectroscopy: Application to wood lignin. *J. Am. Chem. Soc.* **125**(14), 4362–4367.

73. Al-Dajani, W. W., and Gellerstedt, G. (2002) On the isolation and structure of softwood residual lignins. *Nordic Pulp Paper Res. J.* **17**(2), 193–198.

74. Ralph, J., Zhang, Y., and Ede, R. M. (1998) Preparation of synthetic lignins with superior NMR characteristics via isotopically labeled monolignols. *J. Chem. Soc., Perkin Trans. 1* (16), 2609–2613.

75. Crestini, C., and Argyropoulos, D. S. (1997) Structural analysis of wheat straw lignin by quantitative ^{31}P and 2D NMR spectroscopy: The occurrence of ester bonds and a-O-4 substructures. *J. Agr. Food Chem.* **45**(4), 1212–1219.

76. Galkin, S., Ämmälahti, E., Kilpeläinen, I., Brunow, G., and Hatakka, A. (1997) Characterization of milled wood lignin from Reed Canary grass (*Phalaris aruncinacea*). *Holzforschung* **51**(2), 130–134.

77. Brunow, G., Ämmålahti, E., Niemi, T., Sipilä, J., Simola, L. K., and Kilpeläinen, I. (1998) Labelling of a lignin from suspension cultures of *Picea abies*. *Phytochem.* **47**(8), 1495–1500.

78. Ämmälahti, E., Brunow, G., Bardet, M., Robert, D., and Kilpeläinen, I. (1998) Identification of side-chain structures in a poplar lignin using three-dimensional HMQC-HOHAHA NMR spectroscopy. *J. Agr. Food Chem.* **46**(12), 5113–5117.

79. Vivas, N., Saint-Cricq de Gaulejac, N., Bourgeois, G., Vitry, C., Pianet, I., Barbe, B., and Glories, Y. (1998) Structural study of lignins extracted from oak heartwood. *J. Chim. Phys. Phys.-Chim. Biol.* **95**(2), 430–436.

80. Balakshin, M. Y., Capanema, E. A., Chen, C. L., and Gracz, H. S. (2003) Elucidation of the structures of residual and dissolved pine kraft lignins using an HMQC NMR technique. *J. Agr. Food Chem.* **51**(21), 6116–6127.

81. Chen, C. L., Capanema, E. A., and Gracz, H. S. (2003) Comparative studies on the delignification of pine kraft-anthraquinone pulp with hydrogen peroxide by binucleus Mn(IV) complex catalysis. *J. Agr. Food Chem.* **51**(21), 6223–6232.

82. Chen, C. L., Capanema, E. A., and Gracz, H. S. (2003) Reaction mechanisms in delignification of pine Kraft-AQ pulp with hydrogen peroxide using Mn(IV)-Me4DTNE as catalyst. *J. Agr. Food Chem.* **51**(7), 1932–1941.

83. Capanema, E. A., Balakshin, M. Y., Chen, C. L., Gratzl, J. S., and Gracz, H. (2001) Structural analysis of residual and technical lignins by H-1-C-13 correlation 2D NMR-spectroscopy. *Holzforschung* **55**(3), 302–308.

84. Balakshin, M. Y., Capanema, E. A., Goldfarb, B., Frampton, J., and Kadla, J. F. (2005) NMR studies on Fraser fir Abies fraseri (Pursh) Poir. lignins. *Holzforschung* **59**(5), 488–496.

85. Kishimoto, T., Ueki, A., Takamori, H., Uraki, Y., and Ubukata, M. (2004) Delignification mechanism during high-boiling solvent pulping. Part 6: Changes in lignin structure analyzed by H-1-C-13 correlation 2-D NMR spectroscopy. *Holzforschung* **58**(4), 355–362.

86. Xie, Y. M. (2004) Difference of condensed lignin structures in eucalyptus species. *Nordic Pulp & Paper Res. J.* **19**(1), 18–21.

87. Holtman, K. M., Chang, H. M., and Kadla, J. F. (2004) Solution-state nuclear magnetic resonance study of the similarities between milled wood lignin and cellulolytic enzyme lignin. *J. Agr. Food Chem.* **52**(4), 720–726.

88. Holtman, K. M., Chang, H. M., Jameel, H., and Kadla, J. F. (2003) Elucidation of lignin structure through degradative methods: Comparison of modified DFRC and thioacidolysis. *J. Agr. Food Chem.* **51**(12), 3535–3540.

89. Lin, L., Zhou, X. T., Zhao, D. S., and Qiu, Y. G. (2002) Structural changes of lignins caused by oxidative ammonolysis. *Acta Chimica Sinica* **60**(1), 176–179.

90. Lin, L. Z., Yao, Y. G., and Shiraishi, N. (2001) Liquefaction mechanism of beta-O-4 lignin model compound in the presence of phenol under acid catalysis. Part 1. Identification of the reaction products. *Holzforschung* **55**(6), 617–624.

91. Balakshin, M., Capanema, E., Chen, C. L., Gratzl, J., Kirkman, A., and Gracz, H. (2001) Biobleaching of pulp with dioxygen in the laccase-mediator system: Reaction mechanisms for degradation of residual lignin. *J. Mol. Catalysis B-Enzymatic* **13**(1–3), 1–16.

92. Liitia, T. M., Maunu, S. L., Hortling, B., Toikka, M., and Kilpeläinen, I. (2003) Analysis of technical lignins by two- and three-dimensional NMR spectroscopy. *J. Agr. Food Chem.* **51**(8), 2136–2143.

93. Al-Dajani, W. W., and Gellerstedt, G. (2002) On the isolation and structure of softwood residual lignins. *Nordic Pulp & Paper Res. J.* **17**(2), 193–198.

94. Capanema, E. A., Balakshin, M. Y., and Kadla, J. F. (2005) Quantitative characterization of a hardwood milled wood lignin by nuclear magnetic resonance spectroscopy. *J. Agr. Food Chem.* **53**(25), 9639–9649.

95. Zhang, L. M., and Gellerstedt, G. (2007) Quantitative 2D HSQC NMR determination of polymer structures by selecting suitable internal standard references. *Magn. Reson. Chem.* **45**(1), 37–45.

96. Ibarra, D., Chavez, M. I., Rencoret, J., del Rio, J. C., Gutierrez, A., Romero, J., Camarero, S., Martinez, M. J., Jimenez-Barbero, J., and Martinez, A. T. (2007) Lignin modification during *Eucalyptus globulus* kraft pulping followed by totally chlorine-free bleaching: A two-dimensional nuclear magnetic resonance, Fourier transform infrared, and pyrolysis-gas chromatography/mass spectrometry study. *J. Agr. Food Chem.* **55**(9), 3477–3490.

97. Ralph, J., Kim, H., Lu, F., Grabber, J. H., Leplé, J.-C., Berrio-Sierra, J., Mir Derikvand, M., Jouanin, L., Boerjan, W., and Lapierre, C. (2008) Identification of the structure and origin of a thioacidolysis marker compound for ferulic acid incorporation into angiosperm lignins (and an indicator for cinnamoyl-CoA reductase deficiency). *Plant J.* **53**(2), 368–379.

98. Yelle, D. J., Ralph, J., and Frihart, C. R. (2008) Characterization of non-derivatized plant cell walls using high-resolution solution-state NMR spectroscopy. *Magn. Reson. Chem.* **46**(6), 508–517.

99. Wagner, A., Ralph, J., Akiyama, T., Flint, H., Phillips, L., Torr, K. M., Nanayakkara, B., and Te Kiri, L. (2007) Modifying lignin in conifers: The role of HCT during tracheary element formation in Pinus radiata *Proc. Natl. Acad. Sci.* **104**(28), 11856–11861.

100. Yelle, D. J., Ralph, J., Lu, F., and Hammel, K. E. (2008) Evidence for cleavage of lignin by a brown rot basidiomycete. *Environ. Microbiol.* **10**(7), 1844–1849.

101. Lu, F., and Ralph, J. (2008) Novel tetrahydrofuran structures derived from β–β-coupling reactions involving sinapyl acetate in kenaf lignins. *Org. Biomol. Chem.* (6), 3681–3694.

102. Kim, H., and Ralph, J. (2009) Solution-state 2D NMR of ball-milled plant cell wall gels in DMSO-d6/pyridine-d5. *Org. Biomol. Chem.* **8**(3), 576–591.

103. Kim, H., Ralph, J., and Akiyama, T. (2008) Solution-state 2D NMR of ball-milled plant cell wall gels in DMSO-d_6. *BioEnergy Res.* **1**(1), 56–66.

104. Leplé, J.-C., Dauwe, R., Morreel, K., Storme, V., Lapierre, C., Pollet, B., Naumann, A., Gilles, Kang, K.-Y., Kim, H., Ruel, K., Lefèbvre, A., Josseleau, J.-P., Grima-Pettenati, J., De Rycke, R., Andersson-Gunnerås, S., Erban, A., Fehrle, I., Petit-Conil, M., Kopka, J., Polle, A., Messens, E., Sundberg, B., Mansfield, S. D., Ralph, J., Pilate, G., and Boerjan, W. (2007) Downregulation of cinnamoyl coenzyme A reductase in poplar: Multiple-level phenotyping reveals effects on cell wall polymer metabolism and structure. *Plant Cell* **19**, 3669–3691.

105. Holmgren, A., Brunow, G., Henriksson, G., Zhang, L., and Ralph, J. (2006) Non-enzymatic reduction of quinone methides during oxidative coupling of monolignols: Implications for the origin of benzyl structures in lignins. *Org. Biomol. Chem.* **4**(18), 3456–3461.

106. Bunzel, M., and Ralph, J. (2006) NMR characterization of lignins isolated from fruit and vegetable insoluble dietary fiber. *J. Agr. Food Chem.* **54**(21), 8352–8361.

107. Ralph, J., Lundquist, K., Brunow, G., Lu, F., Kim, H., Schatz, P. F., Marita, J. M., Hatfield, R. D., Ralph, S. A., Christensen, J. H., and Boerjan, W. (2004) Lignins: Natural polymers from oxidative coupling of 4-hydroxyphenylpropanoids. *Phytochem. Revs.* **3**(1), 29–60.

108. Guerra, A., Régis, M., Ferraz, A., Lu, F., and Ralph, J. (2004) Structural characterization of lignin during *Pinus taeda* wood treatment with *Ceripotiopsis subvermispora*. *Appl. Environ. Microbiol.* **70**(7), 4073–4078.

109. Akim, L. G., Colodette, J. L., and Argyropoulos, D. S. (2001) Factors limiting oxygen delignification of kraft pulp. *Can. J. Chem.* **79**(2), 201–210.

110. Balakshin, M. Y., Capanema, E. A., and Chang, H. M. (2007) MWL fraction with a high concentration of lignin-carbohydrate linkages: Isolation and 2D NMR spectroscopic analysis. *Holzforschung* **61**(1), 1–7.

111. Terashima, N., Akiyama, T., Ralph, S. A., Evtuguin, D., Pascoal Neto, C., Parkås, J., Paulsson, M., Westermark, U., and Ralph, J. (2009) 2D-NMR (HSQC) difference spectra between specifically ^{13}C-enriched and unenriched protolignin of *Ginkgo biloba* obtained in the solution-state of whole cell material. *Holzforschung* **63**, 379–384.

112. Grabber, J. H., Mertens, D. R., Kim, H., Funk, C., Lu, F., and Ralph, J. (2008) Cell wall fermentation kinetics are impacted more by lignin content and ferulate cross-linking than by lignin composition. *J. Sci. Food Agr.* **89**(1), 122–129.

113. Grabber, J. H., Hatfield, R. D., Lu, F., and Ralph, J. (2008) Coniferyl ferulate incorporation into lignin enhances the alkaline delignification and enzymatic degradation of maize cell walls. *Biomacromolecules* **9**(9), 2510–2516.

114. Karhunen, P., Rummakko, P., Sipilä, J., Brunow, G., and Kilpeläinen, I. (1995) Dibenzodioxocins; a novel type of linkage in softwood lignins. *Tetrahedron Lett.* **36**(1), 169–170.

115. Karhunen, P., Rummakko, P., Sipilä, J., Brunow, G., and Kilpeläinen, I. (1995) The formation of dibenzodioxocin structures by oxidative coupling: A model reaction for lignin biosynthesis. *Tetrahedron Lett.* **36**(25), 4501–4504.

116. Karhunen, P., Rummakko, P., Pajunen, A., and Brunow, G. (1996) Synthesis and crystal structure determination of model compounds for the dibenzodioxocine structure occurring in wood lignins. *J. Chem. Soc., Perkin Trans. 1*, 2303–2308.

117. Hu, W.-J., Lung, J., Harding, S. A., Popko, J. L., Ralph, J., Stokke, D. D., Tsai, C.-J., and Chiang, V. L. (1999) Repression of lignin biosynthesis in transgenic trees promotes cellulose accumulation and growth. *Nature Biotechnol.* **17**(8), 808–812.

118. Willker, W., Leibfritz, D., Kerssebaum, R., and Bermel, W. (1992) Gradient selection in inverse heteronuclear correlation spectroscopy. *Magn. Reson. Chem.* **31**, 287–292.

119. Bodenhausen, G., and Ruben, D. J. (1980) *Chem. Phys. Lett.* **69**, 185–188.

120. Mandal, P. K., and Majumdar, A. (2004) A comprehensive discussion of HSQC and HMQC pulse sequences. *Concepts Mag. Reson. Part A* **20A**(1), 1–23.

121. Kessler, H., Schmieder, P., and Kurz, M. (1989) Implementation of the DEPT sequence in inverse shift correlation: The DEPT-HMQC. *J. Magn. Reson.* **85**(2), 400–405.
122. Lerner, L., and Bax, A. (1986) Sensitivity-enhanced two-dimensional heteronuclear relayed coherence transfer NMR spectroscopy. *J. Magn. Reson.* **69**(2), 375–380.
123. Li, S., and Lundquist, K. (2001) Analysis of hydroxyl groups in lignins by ¹H NMR spectrometry. *Nordic Pulp Paper Res. J.* **16**(1), 63–67.
124. Li, S., and Lundquist, K. (1994) A new method for the analysis of phenolic groups in lignins by ¹H NMR spectrometry. *Nordic Pulp and Paper Res. J.* **9**, 191–195.
125. Lewis, N. G., Eberhardt, T. L., and Luthe, C. E. (1988) Developing new methodology to determine changes to the lignin macromolecule during delignification. *TAPPI J.* **71**(1), 141–142.
126. Lundquist, K., and Olsson, T. (1977) NMR studies of lignins. 1. Signals due to protons in formyl groups. *Acta Chem. Scand. B* **31**(9), 788–792.
127. Terashima, N., Atalla, R. H., Ralph, S. A., Landucci, L. L., Lapierre, C., and Monties, B. (1996) New preparations of lignin polymer models under conditions that approximate cell wall lignification. II. Structural characterization of the models by thioacidolysis. *Holzforschung* **50**(1), 9–14.
128. Bax, A., and Summers, M. F. (1986) Proton and carbon-13 assignments from sensitivity-enhanced detection of heteronuclear multiple-bond connectivity by 2D multiple quantum NMR. *J. Am. Chem. Soc.* **108**(8), 2093–2094.
129. Ralph, J., Helm, R. F., Quideau, S., and Hatfield, R. D. (1992) Lignin-feruloyl ester cross-links in grasses. Part 1. Incorporation of feruloyl esters into coniferyl alcohol dehydrogenation polymers. *J. Chem. Soc., Perkin Trans. 1* (21), 2961–2969.
130. Ralph, J., Grabber, J. H., and Hatfield, R. D. (1995) Lignin-ferulate crosslinks in grasses: Active incorporation of ferulate polysaccharide esters into ryegrass lignins. *Carbohydr. Res.* **275**(1), 167–178.
131. Argyropoulos, D. S., Jurasek, L., Kristofova, L., Xia, Z. C., Sun, Y. J., and Palus, E. (2002) Abundance and reactivity of dibenzodioxocins in softwood lignin. *J. Agr. Food Chem.* **50**(4), 658–666.
132. Ralph, J., Zhang, Y., and Ede, R. M. (1996) Preparation of synthetic lignins with superior NMR characteristics. In *211th National Meeting of the Am. Chem. Soc.*, American Chemical Society, Washington, DC, New Orleans, LA.
133. Ralph, J., Hatfield, R. D., Grabber, J. H., Jung, H. G., Quideau, S., and Helm, R. F. (1998) Cell wall cross-linking in grasses by ferulates and diferulates. In: Lewis, N. G., and Sarkanen, S. (eds). *Lignin and Lignan Biosynthesis*, American Chemical Society, Washington, DC.
134. Kilpeläinen, I., Ämmälahti, E., Brunow, G., and Robert, D. (1994) Application of three-dimensional HMQC-HOHAHA NMR spectroscopy to wood lignin, a natural polymer. *Tetrahedron Lett.* **35**(49), 9267–9270.
135. Ralph, J., Kim, H., Peng, J., and Lu, F. (1999) Arylpropane-1,3-diols in lignins from normal and CAD-deficient pines. *Org. Lett.* **1**(2), 323–326.
136. Shaka, A. J., Barker, P. B., and Freeman, R. (1985) Computer-optimized decoupling scheme for wideband applications and low-level operation. *J. Magn. Reson.* **64**(3), 547–552.
137. Hurd, R. E. (1990) Gradient-enhanced spectroscopy. *J. Magn. Reson.* **87**(2), 422–428.
138. Ruiz-Cabello, J., Vuister, G. W., Moonen, C. T. W., Van Gelderen, P., Cohen, J. S., and Van Zijl, P. C. M. (1992) Gradient-enhanced heteronuclear correlation spectroscopy: Theory and experimental aspects. *J. Magn. Reson.* **100**(2), 282–302.
139. Bradley, S. A., and Krishnamurthy, K. (2005) A modified CRISIS-HSQC for band-selective IMPRESS. *Magn. Reson. Chem.* **43**(2), 117–123.
140. Boyer, R. D., Johnson, R., and Krishnamurthy, K. (2003) Compensation of refocusing inefficiency with synchronized inversion sweep (CRISIS) in multiplicity-edited HSQC. *J. Magn. Reson.* **165**(2), 253–259.

141. Zhang, S. M., Wu, J., and Gorenstein, D. G. (1996) "Double-WURST" decoupling for N-15- and C-13-double-labeled proteins in a high magnetic field. *J. Magn. Reson. Series A* **123**(2), 181–187.

142. Kupce, E., and Freeman, R. (1997) Compensation for spin-spin coupling effects during adiabatic pulses. *J. Magn. Reson.* **127**(1), 36–48.

143. Brunow, G., Karlsson, O., Lundquist, K., and Sipilä, J. (1993) On the distribution of the diastereomers of the structural elements in lignins: The steric course of reactions mimicking lignin biosynthesis. *Wood Sci Technol* **27**(4), 281–286.

144. Bardet, M., Robert, D., Lundquist, K., and von Unge, S. (1998) Distribution of *erythro* and *threo* forms of different types of β-O-4 structures in aspen lignin by carbon-13 NMR using the 2D INADEQUATE experiment. *Magn. Reson. Chem.* **36**(8), 597–600.

145. Ludwig, C. H. (1971) Magnetic resonance spectra. In: Sarkanen, K. V., and Ludwig, C. H. (eds). *Lignins, Occurrence, Formation, Structure and Reactions*, Wiley-Interscience, New York.

146. Ludwig, C. H., Nist, B. J., and McCarthy, J. L. (1964) Lignin. XIII. The high resolution nuclear magnetic resonance spectroscopy of protons in acetylated lignins. *J. Am. Chem. Soc.* **86**, 1196–1202.

147. Ludwig, C. H., Nist, B. J., and McCarthy, J. L. (1964) Lignin. XII. The high resolution nuclear magnetic resonance spectroscopy of protons in compounds related to lignin. *J. Am. Chem. Soc.* **86**, 1186–1196.

148. Lenz, B. D. (1968) Application of nuclear magnetic resonance spectroscopy to characterization of lignin. *Tappi* **51**(11), 511–519.

149. Lundquist, K. (1979) NMR studies of lignins. 2. Interpretation of the ¹H NMR spectrum of acetylated birch lignin. *Acta Chem. Scand. B* **33**(1), 27–30.

150. Lundquist, K. (1980) NMR studies of lignins. 4. Investigation of spruce lignin by ¹H NMR spectroscopy. *Acta Chem. Scand.* **B34**(1), 21–26.

151. Lundquist, K. (1979) NMR studies of lignins. 3. Proton NMR spectroscopic data for lignin model compounds. *Acta Chem. Scand. B* **B33**(6), 418–420.

152. Lundquist, K., Simonson, R., and Tingsvik, K. (1979) On the occurrence of carbohydrates in milled wood lignin preparations. *Sven. Papperstidn.* **82**(9), 272–275.

153. Lundquist, K., Simonson, R., and Tingsvik, K. (1980) Studies on lignin carbohydrate linkages in milled wood lignin preparations. *Sven. Papperstidn.* **83**(16), 452-454

154. Brunow, G., and Lundquist, K. (1980) Comparison of a synthetic dehydrogenation polymer of coniferyl alcohol with milled wood lignin from spruce, using ¹H NMR nuclear magnetic resonance spectroscopy. *Pap Puu. Helsinki, Suomen Paperi ja Puutavaralehti Oy* **62**(11), 669–670.

155. Davin, L. B., and Lewis, N. G. (2000) Dirigent proteins and dirigent sites explain the mystery of specificity of radical precursor coupling in lignan and lignin biosynthesis. *Plant Physiol.* **123**(2), 453–461.

156. Sarkanen, K. V. (1971) Precursors and their polymerization. In: Sarkanen, K. V., and Ludwig, C. H. (eds). *Lignins, Occurrence, Formation, Structure and Reactions*, Wiley-Interscience, New York.

157. Adler, E. (1977) Lignin chemistry: Past, present and future. *Wood Sci Technol* **11**(3), 169–218.

158. Syrjänen, K., and Brunow, G. (2000) Regioselectivity in lignin biosynthesis: The influence of dimerization and cross-coupling. *J. Chem. Soc., Perkin Trans.* **1**(2), 183–187.

159. Lapierre, C., Lallemand, J. Y., and Monties, B. (1982) Evidence of poplar lignin heterogeneity by combination of carbon-13 and proton NMR spectroscopy. *Holzforschung* **36**(6), 275–282.

160. Hauteville, M., Lundquist, K., and Von Unge, S. (1986) NMR studies of lignins. 7. Proton NMR spectroscopic investigation of the distribution of *erythro* and *threo* forms of β-O-4 structures in lignins. *Acta Chem. Scand. B* **B40**(1), 31–35.

161. Nimz, H. H., Tschirner, U., Stähle, M., Lehmann, R., and Schlosser, M. (1984) Carbon-13 NMR spectra of lignins. 10. Comparison of structural units in spruce and beech lignin. *J. Wood Chem. Technol.* **4**(3), 265–284.

162. Nimz, H. H., and Lüdemann, H.-D. (1976) Kohlenstoff-13-NMR-Spektren von Ligninen, 6. Lignin- und DHP-Acetate. *Holzforschung* **30**(2), 33–40.

163. Ralph, J. (1982) Reactions of lignin model quinone methides and NMR studies of lignins. Ph.D. thesis, U. Wisconsin–Madison, University Microfilms #DA 82-26987.

164. Lundquist, K. (1992) Characterization in solution: Spectroscopic methods. 5.3. Proton (1H) NMR spectroscopy. In: Lin, S. Y., and Dence, C. W. (eds). *Methods in Lignin Chemistry*, Springer-Verlag, Berlin.

165. Nimz, H., Nemr, M., Schmidt, P., Margot, C., Schaub, B., and Schlosser, M. (1982) Carbon-13 NMR spectra of lignins. 9. Spin-lattice relaxation times (T1) and determination of interunit linkages in three hardwood lignins (*Alnus glutinosa, Corylus avellanus* and *Acer pseudoplatanus*). *J. Wood Chem. Technol.* **2**(4), 371–382.

166. Nimz, H. H., Robert, D., Faix, O., and Nemr, M. (1981) Carbon-13 NMR spectra of lignins, 8. Structural differences between lignins of hardwoods, softwoods, grasses and compression wood. *Holzforschung* **35**(1), 16–26.

167. Nimz, H., Ludemann, H. D., et al. (1975) Carbon-13 NMR spectra of lignins, 4. Lignins of the European mistletoe (Viscum album L.). **73**, 226–233.

168. Nimz, H. H., and Tutschek, R. (1977) Carbon-13 NMR spectra of lignins, 7. The question of the lignin content of mosses (Sphagnum magellanicum Brid.). *Holzforschung* **31**(4), 101–106.

169. Nimz, H., and Lüdemann, H.-D. (1974) ¹³C-Kernresonanzspektren von Ligninen, 5. Oligomere Ligninmodellsubstanzen. *Makromol. Chem.* **175**, 2577–2583.

170. Nimz, H., Mogharab, I., and Lüdemann, H.-D. (1974) ¹³C-Kernresonanzspektren von Ligninen, 3. Vergleich von Fichtenlignin mit Künstlichem Lignin nach Freudenberg. *Makromol. Chem.* **175**, 2563–2575.

171. Landucci, L. L., Ralph, S. A., and Hammel, K. E. (1998) ¹³C-NMR characterization of guaiacyl, guaiacyl/syringyl, and syringyl dehydrogenation polymers. *Holzforschung* **52**(2), 160–170.

172. Lapierre, C., Monties, B., Guittet, E., and Lallemand, J. Y. (1984) Photosynthetically carbon-13 labeled poplar lignins: Carbon-13 NMR experiments. *Holzforschung* **38**(6), 333–342.

173. Bardet, M., Foray, M. F., and Robert, D. (1985) Use of the DEPT pulse sequence to facilitate the carbon-13 NMR structural analysis of lignins. *Makromol. Chem.* **186**(7), 1495–1504.

174. Landucci, L. L., and Ralph, S. A. (1997) Assessment of lignin model quality in lignin chemical shift assignments: Substituent and solvent effects. *J. Wood Chem. Technol.* **17**(4), 361–382.

175. Ralph, S. A., Landucci, L. L., and Ralph, J. (2005) NMR database of lignin and cell wall model compounds. Available at http://ars.usda.gov/Services/docs.htm?docid = 10449 (previously http://www.dfrc.ars.usda.gov/software.html), updated at least annually since 1993.

176. Robert, D. (1992) Characterization in solution: Spectroscopic methods. 5.4. Carbon-13 nuclear magnetic resonance spectrometry. In: Lin, S. Y., and Dence, C. W. (eds). *Methods in Lignin Chemistry*, Springer-Verlag, Berlin.

177. Chen, C.-L. (1998) Characterization of milled wood lignins and dehydrogenative polymerisates from monolignols by carbon-13 NMR spectroscopy. In: Lewis, N. G., and Sarkanen, S. (eds). *Lignin and Lignan Biosynthesis*, American Chemical Society, Washington, DC.

178. Landucci, L. L., Ralph, S. A., and Hammel, K. E. (1997) ¹³C NMR characterization of guaiacyl, guaiacyl/syringyl, and syringyl dehydrogenation polymers (DHPs). *Holzforschung* **52**(2), 160–170.

179. Landucci, L. L. (2000) Reaction of *p*-hydroxycinnamyl alcohols with transition metal salts. 3. Preparation and NMR characterization of improved DHPs. *J. Wood Chem. Technol.* **20**(3), 243–264.
180. Lewis, N. G., Newman, J., Just, G., and Ripmeister, J. (1987) Determination of bonding patterns of carbon-13 specifically enriched dehydrogenatively polymerized lignin in solution and solid state. *Macromolecules* **20**(8), 1752–1756.
181. Brunow, G., and Wallin, H. (1981) Studies concerning the preparation of synthetic lignin. In: *Ekman-Days 1981, International Symposium on Wood and Pulping Chemistry*, SPCI (Svenska Pappers- och Cellulosaingeniörsföreningen, The Swedish Association of Pulp and Paper Engineers), Stockholm, Sweden.
182. Gagnaire, D., and Robert, D. (1978) Carbon-13 NMR study of a polymer model of lignin, DHP, carbon-13 selectively labeled at the benzylic positions. In. *Proc. Eur. Conf. NMR Macromol.*, Lerici, Rome, Italy.
183. Faix, O., and Beinhoff, O. (1988) FTIR spectra of milled wood lignins and lignin polymer models (DHPs) with enhanced resolution obtained by deconvolution. *J. Wood Chem. Technol.* **8**(4), 505–522.
184. Faix, O. (1986) Investigation of lignin polymer models (DHPs) by FTIR spectroscopy. *Holzforschung* **40**(5), 273–280.
185. Robert, D. R., and Brunow, G. (1984) Quantitative estimation of hydroxyl groups in milled wood lignin from spruce and in a dehydrogenation polymer from coniferyl alcohol using carbon-13 NMR spectroscopy. *Holzforschung* **38**(2), 85–90.
186. Brunow, G., Kilpeläinen, I., Lapierre, C., Lundquist, K., Simola, L. K., and Lemmetyinen, J. (1993) The chemical structure of extracellular lignin released by cultures of *Picea abies*. *Phytochem.* **32**(4), 845–850.
187. Freudenberg, K. (1956) Beiträge zur Erforschung des Lignins. *Angew. Chem.* **68**(16), 508–512.
188. Terashima, N., Atalla, R. H., Ralph, S. A., Landucci, L. L., Lapierre, C., and Monties, B. (1995) New preparations of lignin polymer models under conditions that approximate cell well lignification. I: Synthesis of novel lignin polymer models and their structural characterization by ^{13}C NMR. *Holzforschung* **49**(6), 521–527.
189. Tanahashi, M., Aoki, T., and Higuchi, T. (1982) Dehydrogenative polymerization of monolignols by peroxidase and hydrogen peroxide in a dialysis tube. II. Estimation of molecular weights by thermal softening method. *Holzforschung* **36**(3), 117–122.
190. Newman, J., Rej, R. N., Just, G., and Lewis, N. G. (1986) Synthesis of (1,2-^{13}C), (1-^{13}C), and (3-^{13}C) coniferyl alcohol. *Holzforschung* **40**(6), 369–373.
191. Terashima, N., and Seguchi, Y. (1988) Heterogeneity in formation of lignin. IX. Factors influencing the formation of condensed structures in lignins. *Cellulose Chem. Technol.* **22**, 147–154.
192. Gagnaire, D., and Robert, D. (1977) A polymer model of lignin (D.H.P.) carbon-13 selectively labelled at the benzylic positions: Synthesis and NMR study. *Makromol. Chem.* **178**(5), 1477–1495.
193. Hammel, K. E., Jensen, K. A., Mozuch, m. D., Landucci, L. L., Tien, M., and Pease, E. A. (1993) Ligninolysis by a purified lignin peroxidase. *J. Biol. Chem.* **268**(17), 12274–12281.
194. Okabe, J., and Kratzl, K. (1965) On the origin of bonds between *p*-hydroxybenzoic acid and lignin: Experiments with *p*-hydroxybenzoic acid-[[14]COOH] and dehydrogenation-polymerizates. *Tappi* **48**(6), 347–354.
195. Nakamura, Y., and Higuchi, T. (1978) Ester linkage of *p*-coumaric acid in bamboo lignin. III. Dehydrogenative polymerization of coniferyl *p*-hydroxybenzoate and coniferyl *p*-coumarate. *Cellul. Chem. Technol.* **12**(2), 209–221.
196. Higuchi, T., Ito, T., Umezawa, T., Hibino, T., and Shibata, D. (1994) Red-brown color of lignified tissues of transgenic plants with antisense CAD gene: Wine-red lignin from coniferyl aldehyde. *J. Biotechnol.* **37**(2), 151–158.

197. Araki, H., and Terashima, N. (1981) Radiotracer Experiments on Lignin Reactions VII. The origins of low molecular weight compounds formed during kraft cooking. *Mokuzai Gakkaishi* **27**(5), 414–418.

198. Lewis, N. G., Yamamoto, E., Wooten, J. B., Just, G., Ohashi, H., and Towers, G. H. N. (1987) Monitoring biosynthesis of wheat cell-wall phenylpropanoids in situ. *Science* **237**(4820), 1344–1346.

199. Lewis, N. G., Razal, R. A., Yamamoto, E., Bokelman, G. H., and Wooten, J. B. (1989) Carbon-13 specific labeling of lignin in intact plants. In. *Plant Cell Wall Polymers*, ACS, Washington, DC.

200. Tomimura, Y., Sasao, Y., Yokoi, T., and Terashima, N. (1980) Heterogeneity in formation of lignin VI. Selective labeling of guaiacyl-syringyl lignin. *Mazuzai Gakkaishi* **26**(8), 558–563.

201. Xie, Y., and Terashima, N. (1991) Selective carbon 13-enrichment of side chain carbons of ginkgo [Ginkgo biloba] lignin traced by carbon 13 nuclear magnetic resonance. *J. Japan Wood Res. Soc.* **37**(10), 935–941.

202. Yimin, X., and Terashima, N. (1993) Selective carbon-13 enrichment of side chain carbons of rice stalk lignin traced by carbon-13 nuclear magnetic resonance. *Mokuzai Gakkaishi* **39**(1), 91–97.

203. Grabber, J. H., Ralph, J., Hatfield, R. D., Quideau, S., Kuster, T., and Pell, A. N. (1996) Dehydrogenation polymer-cell wall complexes as a model for lignified grass walls. *J. Agr. Food Chem.* **44**(6), 1453–1459.

204. Grabber, J. H., Ralph, J., and Hatfield, R. D. (1998) Modeling lignification in grasses with monolignol dehydropolymerisate-cell wall complexes. In: Lewis, N. G., and Sarkanen, S. (eds). *Lignin and Lignan Biosynthesis*, American Chemical Society, Washington, DC.

205. Tanahashi, M., and Higuchi, T. (1981) Dehydrogenative Polymerization of Monolignols by Peroxidase and H_2O_2 in a dialysis Tube I. Preparation of highly polymerized DHPs. *Wood Res.* **67**, 29–42.

206. Nakatsubo, F. (1981) Enzymic dehydrogenation of *p*-coumaryl alcohol and syntheses of oligolignols. *Wood Research* **67**, 59–118.

207. Harkin, J. M. (1967) Lignin: A natural polymeric product of phenol oxidation. In: Taylor, W. I., and Battersby, A. R. (eds). *Oxidative Coupling of Phenols*, Marcel Dekker, New York.

208. Dean, J. F. D., and Eriksson, K.-E. (1992) Biotechnological modification of lignin structure and composition in forest trees. *Holzforschung* **46**(2), 135–147.

209. Sarkanen, K. V. (1971) Dehydrogenative polymerization to lignin. In: Sarkanen, K. V., and Ludwig, C. H. (eds). *Lignins, Occurrence, Formation, Structure and Reactions*, Wiley-Interscience, New York.

210. Syrjänen, K., and Brunow, G. (1998) Oxidative cross coupling of *p*-hydroxycinnamic alcohols with dimeric arylglycerol b-aryl ether lignin model compounds: The effect of oxidation potentials. *J. Chem. Soc., Perkin Trans. 1* (20), 3425–3429.

211. Yoshida, S. (1994) Effect of hemicelluloses on dehydrogenative polymerization of sinapyl alcohol II. Formation of β-γ linkage in the presence of pectin. *Mokuzai Gakkaishi* **40**(9), 966–973.

212. Tollier, M. T., Lapierre, C., Monties, B., Francesch, C., and Rolando, C. (1991) Structural variations in synthetic lignins (DHPs) according to the conditions of their preparation. In: *Sixth International Symposium of Wood and Pulping Chemistry*, Australian Pulp and Paper Industries Technical Association, Melbourne, Australia.

213. Eberhardt, T. L., Bernards, M. A., He, L., Davin, L. B., Wooten, J. B., and Lewis, N. G. (1993) Lignification in cell suspension cultures of *Pinus taeda*: In situ characterization of a gymnosperm lignin. *J. Biol. Chem.* **268**(28), 21088–21096.

214. Morelli, E., Rej, R. N., Lewis, N. G., Just, G., and Towers, G. H. N. (1986) cis-Monolignols in Fagus grandifolia and their possible involvement in lignification. *Phytochem.* **25**(7), 1701–1705.

215. Landucci, L. L. (1995) Reactions of *p*-hydroxycinnamyl alcohols with transition metal salts. 1. Oligolignols and polylignols (DHPs) from coniferyl alcohol. *J. Wood Chem. Technol.* **15**(3), 349–368.

216. Kawai, S., Ohashi, H., Hirai, T., Okuyama, H., and Higuchi, T. (1993) Degradation of syringyl lignin model polymer by laccase of *Coriolus versicolor*. *Mokuzai Gakkaishi* **39**(1), 98–102.

217. Tanahashi, M., Takeuchi, H., and Higuchi, T. (1976) Dehydrogenative polymerization of 3,5-disubstituted *p*-coumaryl alcohols. *Wood Res.* **61**, 44–53.

218. Tanahashi, M., and Higuchi, T. (1990) Effect of the hydrophobic regions of hemicelluloses on dehydrogenative polymerization of sinapyl alcohol. *Mokuzai Gakkaishi* **36**(5), 424–428.

219. Brunow, G. (1998) Oxidative coupling of phenols and the biosynthesis of lignin. In: Lewis, N. G., and Sarkanen, S. (eds). *Lignin and Lignan Biosynthesis*, American Chemical Society, Washington, DC.

220. Björkman, A. (1956) Studies on finely divided wood. Part I. Extraction of lignin with neutral solvents. *Sven. Papperstidn.* **59**(13), 477–485.

221. Obst, J. R., and Kirk, T. K. (1988) Isolation of lignin. *Methods Enzymol., Biomass, Pt. B* **161**, 3–12.

222. Lundquist, K. (1992) Isolation and Purification. 3.1. Wood. In: Lin, S. Y., and Dence, C. W. (eds). *Methods in Lignin Chemistry*, Springer-Verlag, Berlin.

223. Sugimoto, T., Akiyama, T., Matsumoto, Y., and Meshitsuka, G. (2002) The erythro/threo ratio of β-O-4 structures as an important structural characteristic of lignin. Part 2. Changes in erythro/threo (E/T) ratio of β-O-4 structures during delignification reactions. *Holzforschung* **56**(4), 416–421.

224. Meyermans, H., Morreel, K., Lapierre, C., Pollet, B., De Bruyn, A., Busson, R., Herdewijn, P., Devreese, B., Van Beeumen, J., Marita, J. M., Ralph, J., Chen, C., Burggraeve, B., Van Montagu, M., Messens, E., and Boerjan, W. (2000) Modifications in lignin and accumulation of phenolic glucosides in poplar xylem upon down-regulation of caffeoyl-coenzyme A *O*-methyltransferase, an enzyme involved in lignin biosynthesis. *J. Biol. Chem.* **275**(47), 36899–36909.

225. Lapierre, C., Monties, B., and Rolando, C. (1986) Thioacidolysis of poplar lignins: Identification of monomeric syringyl products and characterization of guaiacyl-syringyl lignin fractions. *Holzforschung* **40**(2), 113–118.

226. Lapierre, C., Pollet, B., Petit-Conil, M., Toval, G., Romero, J., Pilate, G., Leple, J. C., Boerjan, W., Ferret, V., De Nadai, V., and Jouanin, L. (1999) Structural alterations of lignins in transgenic poplars with depressed cinnamyl alcohol dehydrogenase or caffeic acid *O*-methyltransferase activity have an opposite impact on the efficiency of industrial kraft pulping. *Plant Physiol.* **119**(1), 153–163.

227. Pepper, J. M., Baylis, P. E. T., and Adler, E. (1959) The isolation and proprieties of lignins obtained by the acidolysis of spruce and aspen woods in dioxane-water medium. *Can. J. Chem.* **37**, 1241–1248.

228. Gellerstedt, G., Pranda, J., and Lindfors, E. L. (1994) Structural and molecular-properties of residual birch kraft lignins. *J. Wood Chem. Technol.* **14**(4), 467–482.

229. Wu, S., and Argyropoulos, D. S. (2003) An improved method for isolating lignin in high yield and purity. *J. Pulp Paper Sci.* **29**(7), 235–240.

230. Argyropoulos, D. S., Sun, Y., and Palus, E. (2002) Isolation of residual kraft lignin in high yield and purity. *J. Pulp and Paper Sci.* **28**(2), 50–54.

231. Jaaskelainen, A. S., Sun, Y., Argyropoulos, D. S., Tamminen, T., and Hortling, B. (2003) The effect of isolation method on the chemical structure of residual lignin. *Wood Sci Technol* **37**(2), 91–102.

232. Zoia, L., Orlandi, M., and Argyropoulos, D. S. (2008) Microwave-assisted lignin isolation using the enzymatic mild acidolysis (EMAL) protocol. *J. Agr. Food Chem.* **56**(21), 10115–10122.

233. Ralph, J., and Helm, R. F. (1991) Rapid proton NMR method for determination of *threo:erythro* ratios in lignin model compounds and examination of reduction stereochemistry. *J. Agr. Food Chem.* **39**(4), 705–709.

234. Ralph, J., and Wilkins, A. L. (1985) Rapid NMR method for determination of *threo:erythro* ratios in lignin model compounds. *Holzforschung* **39**(6), 341–344.

235. Arkhipov, Y., Argyropoulos, D. S., Bolker, H. I., and Heitner, C. (1991) Phosphorus-31 NMR spectroscopy in wood chemistry. Part I. Model compounds. *J. Wood Chem. Technol.* **11**(2), 137–157.

236. Ralph, J., Brunow, G., Harris, P. J., Dixon, R. A., Schatz, P. F., and Boerjan, W. (2008) Lignification: Are lignins biosynthesized via simple combinatorial chemistry or via proteinaceous control and template replication? In: Daayf, F., El Hadrami, A., Adam, L., and Ballance, G. M. (eds). *Recent Advances in Polyphenol Research*, Wiley-Blackwell Publishing, Oxford, UK.

237. Sipilä, J., and Syrjänen, K. (1995) Synthesis and ^{13}C NMR spectroscopic characterization of six dimeric arylglycerol-β-aryl ether model compounds representative of syringyl and *p*-hydroxyphenyl structures in lignins. On the Aldol reaction in b-ether preparation. *Holzforschung* **49**(4), 325–331.

238. Akiyama, T., Goto, H., Nawawi, D. S., Syafii, W., Matsumoto, Y., and Meshitsuka, G. (2005) *Erythro/threo* ratio of β-O-4-structures as an important structural characteristic of lignin. Part 4: Variation in the *erythro/threo* ratio in softwood and hardwood lignins and its relation to syringyl/guaiacyl ratio. *Holzforschung* **59**(3), 276–281.

239. Li, L., Zhou, Y., Cheng, X., Sun, J., Marita, J. M., Ralph, J., and Chiang, V. L. (2003) Combinatorial modification of multiple lignin traits in trees through multigene cotransformation. *Proc. Natl. Acad. Sci.* **100**(8), 4939–4944.

240. Ralph, J., and Grabber, J. H. (1996) Dimeric β-ether thioacidolysis products resulting from incomplete ether cleavage. *Holzforschung* **50**(5), 425–428.

241. Li, S., Iliefski, T., Lundquist, K., and Wallis, A. F. A. (1997) Reassignment of relative stereochemistry at C-7 and C-8 in arylcoumaran neolignans. *Phytochem.* **46**(5), 929–934.

242. Wallis, A. F. A. (1998) Structural diversity in lignans and neolignans. In: Lewis, N. G., and Sarkanen, S. (eds). *Lignin and Lignan Biosynthesis*, American Chemical Society, Washington, DC.

243. Miyakoshi, T., and Chen, C. L. (1991) ^{13}C NMR spectroscopic studies of phenylcoumaran and 1,2 diarylpropane type lignin model compounds. Pt. 1. Synthesis of model compounds. *Holzforschung* **45**(Suppl), 41–47.

244. Miyakoshi, T., and Chen, C. L. (1992) ^{13}C NMR spectroscopic studies of phenylcoumaran and 1,2-diarylpropane type lignin model compounds. Pt.2. Substituent effects on ^{13}C chemical shifts of aromatic carbons. *Holzforschung* **46**(1), 39–46.

245. Hassi, H. Y., Aoyama, M., Tai, D., Chen, C. L., and Gratzl, J. S. (1987) Substituent effects on carbon-13 chemical shifts of aromatic carbons in β-O-4 and β-5 type lignin model compounds. *J. Wood Chem. Technol.* **7**(4), 555–581.

246. Nakatsubo, F., and Higuchi, T. (1979) Syntheses of phenylcoumarans. *Mokuzai Gakkaishi* **25**(11), 735–742.

247. Brunow, G., and Lundquist, K. (1984) A new synthesis of model compounds for the b-5 structural unit in lignins. *Acta Chem. Scand.* **B38**(4), 335–336.

248. Ede, R. M., Ralph, J., and Wilkins, A. L. (1987) The stereochemistry of β-5 lignin model compounds. *Holzforschung* **41**(4), 239–245.

249. Nakatsubo, F., and Higuchi, T. (1980) Synthesis of trimeric lignin model compound composed of phenylcoumaran and b-O-4 structures. *Mokuzai Gakkaishi* **26**(1), 31–36.

250. Nakatsubo, F., and Higuchi, T. (1980) Synthesis of trimeric lignin model compound composed of phenylcoumaran and β-1 structures. *Mokuzai Gakkaishi* **26**(2), 107–111.

251. Ralph, J., Ede, R. M., and Wilkins, A. L. (1986) Synthesis of trimeric lignin model compounds composed of β-aryl ether and phenylcoumaran structures. *Holzforschung* **40**(1), 23–30.

252. Lundquist, K. (1992) ^{1}H NMR spectral studies of lignins. Results regarding the occurrence of b-5 structures, β-β-structures, non-cyclic benzyl aryl ethers, carbonyl groups and phenolic groups. *Nordic Pulp Paper Res. J.* **7**(1), 4–8, 16.

253. Önnerud, H., and Gellerstedt, G. (2003) Inhomogeneities in the chemical structure of hardwood lignins. *Holzforschung* **57**(3), 255–265.

254. Katayama, Y., and Fukuzumi, T. (1978) Enzymic synthesis of three lignin-related dimers by an improved peroxidase-hydrogen peroxide system. *Mokuzai Gakkaishi* **24**(9), 664–667.

255. Lindberg, B. (1950) *Epi*-pinoresinol. *Acta Chem. Scand.* **4**, 391–392.

256. Lapierre, C., Pollet, B., Monties, B., and Rolando, C. (1991) Thioacidolysis of spruce lignin: Gas chromatography-mass spectroscopy analysis of the main dimers recovered after Raney nickel desulfurization. *Holzforschung* **45**(1), 61–68.

257. Lapierre, C. (1993) Application of new methods for the investigation of lignin structure. In: Jung, H. G., Buxton, D. R., Hatfield, R. D., and Ralph, J. (eds). *Forage Cell Wall Structure and Digestibility*, American Society of Agronomy, Crop Science Society of America, Soil Science Society of America, Madison, WI.

258. Morreel, K., Ralph, J., Kim, H., Lu, F., Goeminne, G., Ralph, S. A., Messens, E., and Boerjan, W. (2004) Profiling of oligolignols reveals monolignol coupling conditions in lignifying poplar xylem. *Plant Physiol.* **136**(3), 3537–3549.

259. Zhang, L., and Gellerstedt, G. (2004) Observation of a novel b–b-structure in native lignin by high resolution 2D NMR techniques. In: *Eighth European Workshop on Lignocellulosics and Pulp*, Latvian State Institute of Wood Chemistry, Riga, Latvia, Riga, Latvia.

260. Lu, F., and Ralph, J. (2002) Preliminary evidence for sinapyl acetate as a lignin monomer in kenaf. *Chem. Commun.* (1), 90–91.

261. Grabber, J. H., Ralph, J., and Hatfield, R. D. (2002) Model studies of ferulate-coniferyl alcohol cross-product formation in primary maize walls: Implications for lignification in grasses. *J. Agr. Food Chem.* **50**(21), 6008–6016.

262. Ralph, J., Bunzel, M., Marita, J. M., Hatfield, R. D., Lu, F., Kim, H., Schatz, P. F., Grabber, J. H., and Steinhart, H. (2004) Peroxidase-dependent cross-linking reactions of *p*-hydroxycinnamates in plant cell walls. *Phytochem. Revs.* **3**(1), 79–96.

263. Ralph, J., Bunzel, M., Marita, J. M., Hatfield, R. D., Lu, F., Kim, H., Grabber, J. H., Ralph, S. A., Jimenez-Monteon, G., and Steinhart, H. (2000) Diferulates analysis: New diferulates and disinapates in insoluble cereal fibre. *Polyphénols Actualités* (19), 13–17.

264. Brunow, G., Sipilä, J., and Mäkelä, T. (1989) On the mechanism of formation of noncyclic benzyl ethers during lignin biosynthesis. Part 1: The reactivity of β-O-4 quinone methides with phenols and alcohols. *Holzforschung* **43**(1), 55–59.

265. Alves, V. L., Drumond, M. G., Stefani, G. M., Chen, C. L., and Pilo-Veloso, D. (2000) Synthesis of new trimeric lignin model compounds containing 5-5' and β-O-4' substructures, and their characterization by 1D and 2D NMR techniques. *J. Brazilian Chem. Soc.* **11**(5), 467–473.

266. Drumond, M. G., Veloso, D. P., Cota, S. D. S., Lemos de Morais, S. A., Do Nascimento, E. A., and Chen, C. L. (1992) Biphenyl-type lignin model compounds: Synthesis and carbon-13 NMR substituent chemical shift additivity rule. *Holzforschung* **46**(2), 127–134.

267. Drumond, M., Aoyama, M., Chen, C. L., and Robert, D. (1989) Substituent effects on carbon-13 chemical shifts of aromatic carbons in biphenyl type lignin model compounds. *J. Wood Chem. Technol.* **9**(4), 421–442.

268. Hyatt, J. A. (1987) Synthesis of some tetrameric lignin model compounds containing β-O-4 and 5,5'-interunit linkages. *Holzforschung* **41**(6), 363–370.

269. Kilpeläinen, I., Tervilae-Wilo, A., Peraekylae, H., Matikainen, J., and Brunow, G. (1994) Synthesis of hexameric lignin model compounds. *Holzforschung* **48**(5), 381–386.

270. Kuroda, K., and Inoue, Y. (1989) Synthesis of diphenyl ethers. II. Convenient synthesis of diphenyl ethers comprised of 1-O-4, 4-O-6 and 4-O-5 structures in lignin. *Mokuzai Gakkaishi* **35**(7), 6408.

271. Kuroda, K., and Inoue, Y. (1986) Synthesis of diphenyl ethers composed of 4-*O*-5 structures. *Mokuzai Gakkaishi* **32**(4), 285–288.

272. Ralph, J., Quideau, S., Grabber, J. H., and Hatfield, R. D. (1994) Identification and synthesis of new ferulic acid dehydrodimers present in grass cell walls. *J. Chem. Soc., Perkin Trans. 1* (23), 3485–3498.

273. Ralph, S. A. (2005) Conundrums regarding 5-O-4-linkages in softwood lignins. In. *Thirteenth International Symposium on Wood, Fiber, and Pulping Chemistry*, Auckland, New Zealand.

274. Brunow, G., Ede, R. M., Simola, L. K., and Lemmetyinen, J. (1990) Lignins released from *Picea abies* suspension cultures: True native spruce lignins? *Phytochem.* **29**(8), 2535–2538.

275. Lundquist, K. (1987) On the occurrence of β-1 structures in lignins. *J. Wood Chem. Technol.* **7**(2), 179–185.

276. Ede, R. M., Ralph, J., Torr, K. M., and Dawson, B. S. W. (1996) A 2D NMR investigation of the heterogeneity of distribution of diarylpropane structures in extracted *Pinus radiata* lignins. *Holzforschung* **50**(2), 161–164.

277. Ahvonen, T., Brunow, G., Kristersson, P., and Lundquist, K. (1983) Stereoselective syntheses of lignin model compounds of the β-O-4 and β-1 types. *Acta Chem. Scand. B* **B37**(9), 845–849.

278. Namba, H., Nakatsubo, F., and Higuchi, T. (1980) Synthesis of trimeric lignin model compound composed of b-O-4 and b-1 structures. *Mokuzai Gakkaishi* **26**(6), 426–431.

279. Lundquist, K., and Miksche, G. E. (1965) Nachweis eines neuen Verknüpfungsprinzips von Guajacylpropaneinheiten im Fichtenlignin. *Tetrahedron Lett.* (25), 2131–2136.

280. Lundquist, K., Miksche, G. E., Ericsson, L., and Berndtson, L. (1967) Über das Vorkommen von Glyceraldehyd-2-arylätherstrukturen im Lignin (On the occurrence of glyceraldehyde 2-aryl ethers in lignin). *Tetrahedron Lett.* (46), 4587–4591.

281. Brunow, G., and Lundquist, K. (1991) On the acid-catalyzed alkylation of lignins. *Holzforschung* **45**(1), 37–40.

282. Gellerstedt, G., and Zhang, L. (1991) Reactive lignin structures in high yield pulping. Part 1. Structures of the 1,2-diarylpropane-1,3-diol type. *Nordic Pulp Paper Res. J.* **6**(3), 136–139.

283. Setälä, H., Pajunen, A., Rummakko, P., Sipilä, J., and Brunow, G. (1999) A novel type of spiro compound formed by oxidative cross-coupling of methyl sinapate with a syringyl lignin model compound. A model system for the β-1 pathway in lignin biosynthesis. *J. Chem. Soc., Perkin Trans. 1* (4), 461–464.

284. Peng, J., Lu, F., and Ralph, J. (1999) The DFRC method for lignin analysis. Part 5. Isochroman lignin trimers from DFRC-degraded *Pinus taeda*. *Phytochem.* **50**(4), 659–666.

285. Sudo, K., and Sakakibara, A. (1974) Hydrogenolysis of protolignin. XI. *Mokuzai Gakkaishi* **20**(8), 396–401.

286. Yasuda, S., and Sakakibara, A. (1976) Hydrogenolysis of protolignin in compression wood. II. *Mokuzai Gakkaishi* **22**(11), 606–612.

287. Yasuda, S., and Sakakibara, A. (1977) Hydrogenolysis of protolignin in compression wood. III. Isolation of four dimeric compounds with carbon to carbon linkage. *Mokuzai Gakkaishi* **23**(2), 114–119.

288. Nimz, H. H., and Ludemann, H. D. (1976) Carbon 13 nuclear magnetic resonance spectra of lignins. 6. Lignin and DHP acetates. *Holzforschung* **30**(2), 33–40.

289. Sakakibara, A. (1980) A structural model of softwood lignin. *Wood Sci Technol* **14**(2), 89–100.

290. Lundquist, K., and Stern, K. (1989) Analysis of lignins by ^{1}H NMR spectroscopy. *Nordic Pulp Paper Res. J.* **4**, 210–213.

291. Aoyama, M., and Sakakibara, A. (1978) Isolation of two new dilignols from hydrolysis products of hardwood lignin. *Mokuzai Gakkaishi* **24**(6), 422–423.

292. Popoff, T., and Theander, O. (1975) Two glycosides of a new dilignol from *Pinus silvestris*. *Phytochem.* **14**, 2065–2066,

293. Gang, D. R., Kasahara, H., Xia, Z. Q., Vander Mijnsbrugge, K., Bauw, G., Boerjan, W., Van Montagu, M., Davin, L. B., and Lewis, N. G. (1999) Evolution of plant defense mechanisms - Relationships of phenylcoumaran benzylic ether reductases to pinoresinol-lariciresinol and isoflavone reductases. *J. Biol. Chem.* **274**(11), 7516–7527.

294. Sederoff, R. R., MacKay, J. J., Ralph, J., and Hatfield, R. D. (1999) Unexpected variation in lignin. *Curr. Opin. Plant Biol.* **2**(2), 145–152.

295. Higuchi, T., Nakatsubo, F., and Ikeda, Y. (1974) Enzymic formation of arylglycerols from p-hydroxycinnamyl alcohols. *Holzforschung* **28**(6), 189–192.

296. Sakakibara, A. (1991) Chemistry of lignin. In: Hon, D. N.-S., and Shiraishi, N. (eds). *Wood and Cellulosic Chemistry*, Marcel Dekker, New York.

297. Ralph, J., and Adams, B. R. (1983) Determination of the conformation and isomeric composition of lignin model quinone methides by NMR. *J. Wood Chem. Technol.* **3**(2), 183–194.

298. Ralph, J., and Ede, R. M. (1988) NMR of lignin model quinone methides. Corrected carbon-13 NMR assignments via carbon-proton correlation experiments. *Holzforschung* **42**(5), 337–338.

299. Ede, R. M., Main, L., and Ralph, J. (1990) Evidence for increased steric compression in *anti* compared to *syn* lignin model quinone methides. *J. Wood Chem. Technol.* **10**(1), 101–110.

300. Lu, F., Ralph, J., Morreel, K., Messens, E., and Boerjan, W. (2004) Preparation and relevance of a cross-coupling product between sinapyl alcohol and sinapyl *p*-hydroxybenzoate. *Org. Biomol. Chem.* **2**, 2888–2890.

301. Ralph, J., Schatz, P. F., Lu, F., Kim, H., Akiyama, T., and Nelsen, S. F. (2008) Quinone methides in lignification. In: Rokita, S. (ed). *Quinone Methides*, Wiley-Blackwell, Hoboken, NJ.

302. Faix, O., Argyropoulos, D. S., Robert, D., and Neirinck, V. (1994) Determination of hydroxyl groups in lignins evaluation of ^{1}H-, ^{13}C-, ^{31}P-NMR, FTIR and wet chemical methods. *Holzforschung* **48**(5), 387–394.

303. Orejuela, L. M., and Helm, R. F. (1996) Rapid quantitative ^{13}C-NMR analysis of hydroxyl environments in lignins. *Holzforschung* **50**(6), 569–572.

304. Sorvari, J., Sjöstrom, E., Klemola, A., and Lanine, J. E. (1986) Chemical characterization of wood constituents, especially lignin, in fractions separated from middle lamella and secondary wall of Norway spruce (*Picea abies*). *Wood Sci. Technol.* **20**, 35–51.

305. Obst, J. R., McMillan, N. J., Blanchette, R. A., Christensen, D. J., Faix, O., Han, J. S., Kuster, T. A., Landucci, L. L., Newman, R. H., Pettersen, R. C., Schwandt, V. H., and Wesolowsky, M. F. (1991) Characterization of Canadian arctic fossil woods. In: Christie, R. L., and McMillan, N. J. (eds). *Tertiary Fossil Forests of the Geodetic hills Axel Heiberg Island, Arctic Archipelago*, Geological Survey of Canada, Ottawa, Canada.

306. Obst, J. R., and Landucci, L. L. (1986) Quantitative carbon-13 NMR of lignins: Methoxyl:aryl ratio. *Holzforschung* **40**(Suppl.), 87–92.
307. Obst, J. R., and Landucci, L. L. (1986) The syringyl content of softwood lignin. *J. Wood Chem. Technol.* **6**(3), 311–327.
308. Landucci, L. L. (1991) One-dimensional liquid-state ^{13}C NMR characterization of lignins. In: *IEA Presymposium on Analysis of Wood, Annual Plant Lignins*, New Orleans, LA.
309. Landucci, L. L., Deka, G. C., and Roy, D. N. (1992) A ^{13}C NMR study of milled wood lignins from hybrid *Salix* Clones. *Holzforschung* **46**(6), 505–511.
310. Pan, D.-R., Tai, D.-S., and Chen, L.-L. (1990) Comparitive studies on chemical composition of wood components in recent and ancient woods of Bischofia polycarpa. *Holzforschung* **44**(1), 7–16.
311. Sarkanen, K. V., Chang, H.-M., and Allan, G. G. (1967) Species variation in lignins. III. Hardwood lignins. *Tappi* **50**(12), 587–590.
312. Ralph, J., and Lu, F. (1998) The DFRC method for lignin analysis. Part 6. A modified method to determine acetate regiochemistry on native and isolated lignins. *J. Agr. Food Chem.* **46**(11), 4616–4619.
313. Smith, D. C. C. (1955) Ester groups in lignin. *Nature* **176**, 267–268.
314. Scalbert, A., Monties, B., Lallemand, J. Y., Guittet, E., and Rolando, C. (1985) Ether linkage between phenolic acids and lignin fractions from wheat straw. *Phytochem.* **24**(6), 1359–1362.
315. Shimada, M., Fukuzuka, T., and Higuchi, T. (1971) Ester linkages of *p*-coumaric acid in bamboo and grass lignins. *Tappi* **54**(1), 72–78.
316. Monties, B., and Lapierre, C. (1981) Donnés récentes sur l'hétérogénéite de la lignine. *Physiologie Végétale* **19**(3), 327–348.
317. Atsushi, K., Azuma, J., and Koshijima, T. (1984) Lignin-carbohydrate complexes and phenolic acids in bagasse. *Holzforschung* **38**(3), 141–149.
318. Azuma, J., Nomura, T., and Koshijima, T. (1985) Lignin–carbohydrate complexes containing phenolic acids isolated from the culms of bamboo. *Agric. Biol. Chem.* **49**, 2661–2669.
319. Terashima, N., Fukushima, K., He, L.-F., and Takabe, K. (1993) Comprehensive model of the lignified plant cell wall. In: Jung, H. G., Buxton, D. R., Hatfield, R. D., and Ralph, J. (eds). *Forage Cell Wall Structure and Digestibility*, American Society of Agronomy, Crop Science Society of America, Soil Science Society of America, Madison, WI.
320. Lam, T. B. T., Iiyama, K., and Stone, B. A. (1992) Changes in phenolic acids from internode walls of wheat and Phalaris during maturation. *Phytochem.* **31**(8), 2655–2658.
321. Nakamura, Y., and Higuchi, T. (1978) Ester linkage of *p*-coumaric acid in bamboo lignin. II. Syntheses of coniferyl *p*-hydroxybenzoate and coniferyl *p*-coumarate as possible precursors of aromatic acid esters in lignin. *Cellul. Chem. Technol.* **12**(2), 199–208.
322. Helm, R. F., and Ralph, J. (1993) Stereospecificity for zinc borohydride reduction of a-aryloxy-b-hydroxy ketones. *J. Wood Chem. Technol.* **13**(4), 593–601.
323. Ralph, J., Sun, R., Kuroda, K.-I., Jimenez-Monteon, G., Lu, F., Schatz, P. F., Ralph, S. A., Hill, S., Jouanin, L., Lapierre, C., Boerjan, W., and Chiang, V. L. (2010) Naturally *p*-hydroxybenzoylated lignins. *J. Biol. Chem.*, in preparation.
324. Takahama, U., and Oniki, T. (1994) Effects of ascorbate on the oxidation of derivatives of hydroxycinnamic acid and the mechanism of oxidation of sinapic acid by cell wall-bound peroxidases. *Plant Cell Physiol.* **35**(4), 593–600.
325. Takahama, U., Oniki, T., and Shimokawa, H. (1996) A possible mechanism for the oxidation of sinapyl alcohol by peroxidase-dependent reactions in the apoplast: Enhancement of the oxidation by hydroxycinnamic acids and components of the apoplast. *Plant Cell Physiol.* **37**(4), 499–504.

326. Hatfield, R. D., Ralph, J., and Grabber, J. H. (2008) A potential role of sinapyl *p*-coumarate as a radical transfer mechanism in grass lignin formation. *Planta* **228**, 919–928.

327. Sun, R. C., Fang, J. M., Tomkinson, J., and Bolton, J. (1999) Physicochemical and structural characterization of alkali soluble lignins from oil palm trunk and empty fruit-bunch fibers. *J. Agr. Food Chem.* **47**(7), 2930–2936.

328. Terashima, N., Fukushima, K., Tsuchiya, S., and Takabe, K. (1986) Heterogeneity in formation of lignin. VII. An autoradiographic study on the formation of guaiacyl and syringyl lignin in poplar. *J. Wood Chem. Technol.* **6**(4), 495–504.

329. Terashima, N., Fukushima, K., and Takabe, K. (1986) Heterogeneity in formation of lignin. VIII. An autoradiographic study on the formation of guaiacyl and syringyl lignin in *Magnolia kobus* DC. *Holzforschung* **40**(Suppl.), 101–105.

330. Whetten, R. W., MacKay, J. J., and Sederoff, R. R. (1998) Recent advances in understanding lignin biosynthesis. *Annu. Rev. Plant Physiology Plant Mol. Biol.* **49**, 585–609.

331. Boudet, A.-M. (1998) A new view of lignification. *Trends Plant Sci.* **3**(2), 67–71.

332. Baucher, M., Monties, B., Van Montagu, M., and Boerjan, W. (1998) Biosynthesis and genetic engineering of lignin. *Crit. Rev. in Plant Sci.* **17**(2), 125–197.

333. Dixon, R. A., and Ni, W. (1996) Genetic manipulation of the phenylpropanoid pathway in transgenic tobacco: New fundamental insights and prospects for crop improvement. *Biotech. & Biotechnol. Equipment* **10**(4), 45–51.

334. Van Doorsselaere, J., Baucher, M., Chognot, E., Chabbert, B., Tollier, M.-T., Petit-Conil, M., Leplé, J.-C., Pilate, G., Cornu, D., Monties, B., Van Montagu, M., Inzé, D., Boerjan, W., and Jouanin, L. (1995) A novel lignin in poplar trees with a reduced caffeic acid/5-hydroxyferulic acid *O*-methyltransferase activity. *Plant J.* **8**(6), 855–864.

335. Kim, H., Ralph, J., Yahiaoui, N., Pean, M., and Boudet, A.-M. (2000) Cross-coupling of hydroxycinnamyl aldehydes into lignins. *Org. Lett.* **2**(15), 2197–2200.

336. Jouanin, L., Goujon, T., de Nadaï, V., Martin, M.-T., Mila, I., Vallet, C., Pollet, B., Yoshinaga, A., Chabbert, B., Petit-Conil, M., and Lapierre, C. (2000) Lignification in transgenic poplars with extremely reduced caffeic acid *O*-methyltransferase activity. *Plant Physiol.* **123**(4), 1363–1373.

337. Goujon, T., Sibout, R., Pollet, B., Mabra, B., Nussaume, L., Bechtold, N., Lu, F., Ralph, J., Mila, I., Barrière, Y., Lapierre, C., and Jouanin, L. (2003) A new *Arabidopsis thaliana* mutant deficient in the expression of O-methyltransferase impacts lignins and sinapoyl esters. *Plant Mol. Biol.* **51**(6), 973–989.

338. Boerjan, W., Ralph, J., and Baucher, M. (2003) Lignin biosynthesis. *Annu. Rev. Plant Biol.* **54**, 519–549.

339. Morreel, K., Ralph, J., Lu, F., Goeminne, G., Busson, R., Herdewijn, P., Goeman, J. L., Van der Eycken, J., Boerjan, W., and Messens, E. (2004) Phenolic profiling of caffeic acid *O*-methyltransferase-deficient poplar reveals novel benzodioxane oligolignols. *Plant Physiol.* **136**(4), 4023–4036.

340. Ni, W., Paiva, N. L., and Dixon, R. A. (1994) Reduced lignin in transgenic plants containing a caffeic acid O-methyltransferase antisense gene. *Transgenic Res* **3**(2), 120–126.

341. Guo, D., Chen, F., Inoue, K., Blount, J. W., and Dixon, R. A. (2001) Downregulation of caffeic acid 3-*O*-methyltransferase and caffeoyl CoA 3-*O*-methyltransferase in transgenic alfalfa: impacts on lignin structure and implications for the biosynthesis of G and S lignin. *Plant Cell* **13**, 73–88.

342. Li, L., Popko, J. L., Umezawa, T., and Chiang, V. L. (2000) 5-Hydroxyconiferyl aldehyde modulates enzymatic methylation for syringyl monolignol formation, a new view of monolignol biosynthesis in angiosperms. *J. Biol. Chem.* **275**(9), 6537–6545.

343. Chabannes, M., Ruel, K., Yoshinaga, A., Chabbert, B., Jauneau, A., Josseleau, J.-P., and Boudet, A.-M. (2001) *In situ* analysis of specifically engineered tobacco lignins reveals a differential impact of individual transformations on the spatial patterns of lignin deposition at the cellular and sub-cellular levels. *Plant J.* **28**(3), 271–282.

344. Sewalt, V. J. H., Ni, W., Blount, J. W., Jung, H. G., Masoud, S. A., Howles, P. A., Lamb, C., and Dixon, R. A. (1997) Reduced lignin content and altered lignin composition in transgenic tobacco down-regulated in expression of L-phenylalanine ammonia-lyase or cinnamate 4-hydroxylase. *Plant Physiol.* **115**, 41–50.

345. Reddy, M. S. S., Chen, F., Shadle, G. L., Jackson, L., Aljoe, H., and Dixon, R. A. (2005) Targeted down-regulation of cytochrome P450 enzymes for forage quality improvement in alfalfa (*Medicago sativa* L.). *Proc. Natl. Acad. Sci.* **102**(46), 16573–16578.

346. Franke, R., Hemm, M. R., Denault, J. W., Ruegger, M. O., Humphreys, J. M., and Chapple, C. (2002) Changes in secondary metabolism and deposition of an unusual lignin in the *ref8* mutant of Arabidopsis. *Plant J.* **30**(1), 47–59.

347. Osakabe, K., Tsao, C. C., Li, L., Popko, J. L., Umezawa, T., Carraway, D. T., Smeltzer, R. H., Joshi, C. P., and Chiang, V. L. (1999) Coniferyl aldehyde 5-hydroxylation and methylation direct syringyl lignin biosynthesis in angiosperms. *Proc. Natl. Acad. Sci.* **96**(16), 8955–8960.

348. Humphreys, J. M., Hemm, M. R., and Chapple, C. (1999) Ferulate 5-hydroxylase from Arabidopsis is a multifunctional cytochrome P450-dependent monooxygenase catalyzing parallel hydroxylations in phenylpropanoid metabolism. *Proc. Natl. Acad. Sci.* **96**(18), 10045–10050.

349. Meyer, K., Shirley, A. M., Cusumano, J. C., Bell-Lelong, D. A., and Chapple, C. (1998) Lignin monomer composition is determined by the expression of a cytochrome P450-dependent monooxygenase in *Arabidopsis. Proc. Natl. Acad. Sci.* **95**(12), 6619–6623.

350. Huntley, S. K., Ellis, D., Gilbert, M., Chapple, C., and Mansfield, S. D. (2003) Significant increases in pulping efficiency in C4H-F5H-transformed poplars: Improved chemical savings and reduced environmental toxins. *J. Agr. Food Chem.* **51**(21), 6178–6183.

351. Stewart, J. J., Akiyama, T., Chapple, C. C. S., Ralph, J., and Mansfield, S. D. (2009). The effects on lignin structure of overexpression of ferulate 5-hydroxylase in hybrid poplar. *Plant Physiol.* **150**(2), 621–635.

352. Marita, J. M. (2001) Lignin in Lignin-Biosynthetic-Pathway Mutants and Transgenics: Structural Studies. Ph.D., U. Wisconsin–Madison (University Microfilms International #3012586).

353. Hwang, B. H., and Sakakibara, A. (1981) Hydrogenolysis of protolignin. XVIII. Isolation of a new dimeric compound with a heterocycle involving a,b-diether. *Holzforschung* **35**(6), 297–300.

354. Lapierre, C., Tollier, M. T., and Monties, B. (1988) A new type of constitutive unit in lignins from the corn *bm3* mutant. *Comptes rendus de l'Académie des sciences. Série III* **307**(13), 723–728.

355. Jacquet, G. (1997) Structure et réactivité des lignines de graminées et des acides phénoliques associés: développement des méthodologies d'investigation. Ph.D., Institut National Agronomique.

356. Suzuki, S., Lam, T. B. T., and Iiyama, K. (1997) 5-Hydroxyguaiacyl nuclei as aromatic constituents of native lignin. *Phytochem* **46**(4), 695–700.

357. Ishikawa, T., Seki, M., Nishigaya, K., Miura, Y., Seki, H., Chen, I.-S., and Ishii, H. (1995) Studies on the chemical constituents of *Xanthoxylum nitidum* (Roxb.) D. C. (*Fagara nitida* Roxb.). III. The chemical constituents of the wood. *Chem. Pharm. Bull.* **43**(11), 2014–2018.

358. Su, B.-N., Li, Y., and Jia, Z.-J. (1997) Neolignan, phenylpropanoid and iridoid glycosides from *Pedicularis verticillata. Phytochem.* **45**(6), 1271–1273.

359. Ralph, J. (2006) What makes a good monolignol substitute? In: Hayashi, T. (ed). *The Science and Lore of the Plant Cell Wall Biosynthesis, Structure and Function*, Universal Publishers (BrownWalker Press), Boca Raton, FL.

360. Lewis, N. G. (1999) A 20th century roller coaster ride: A short account of lignification. *Curr. Opin. Plant Biol.* **2**(2), 153–162.

361. Denton, F. R. (1998) Beetle Juice. *Science* **281**, 1285.
362. Zablackis, E., York, W. S., Pauly, M., Hantus, S., Reiter, W. D., Chapple, C. C. S., Albersheim, P., and Darvill, A. (1996) Substitution of L-fucose by L-galactose in cell walls of Arabidopsis *mur1*. *Science* **272**(5269), 1808–1810.
363. Halpin, C., Knight, M. E., Foxon, G. A., Campbell, M. M., Boudet, A.-M., Boon, J. J., Chabbert, B., Tollier, M.-T., and Schuch, W. (1994) Manipulation of lignin quality by downregulation of cinnamyl alcohol dehydrogenase. *Plant J.* **6**(3), 339–350.
364. Stewart, D., Yahiaoui, N., McDougall, G. J., Myton, K., Marque, C., Boudet, A. M., and Haigh, J. (1997) Fourier-transform infrared and Raman spectroscopic evidence for the incorporation of cinnamaldehydes into the lignin of transgenic tobacco (Nicotiana tabacum L.) plants with reduced expression of cinnamyl alcohol dehydrogenase. *Planta* **201**(3), 311–318.
365. Provan, G. J., Scobbie, L., and Chesson, A. (1997) Characterisation of lignin from CAD and OMT deficient *bm* mutants of maize. *J. Sci. Food Ag.* **73**(2), 133–142.
366. Yahiaoui, N., Marque, C., Myton, K. E., Negrel, J., and Boudet, A.-M. (1998) Impact of different levels of cinnamyl alcohol dehydrogenase down-regulation on lignins of transgenic tobacco plants. *Planta* **204**(1), 8–15.
367. Vailhe, M. A. B., Besle, J. M., Maillot, M. P., Cornu, A., Halpin, C., and Knight, M. (1998) Effect of down-regulation of cinnamyl alcohol dehydrogenase on cell wall composition and on degradability of tobacco stems. *J. Sci. Food Agr.* **76**(4), 505–514.
368. Halpin, C., Holt, K., Chojecki, J., Oliver, D., Chabbert, B., Monties, B., Edwards, K., Barakate, A., and Foxon, G. A. (1998) Brown-midrib maize (*bm1*): A mutation affecting the cinnamyl alcohol dehydrogenase gene. *Plant J.* **14**(5), 545–553.
369. Baucher, M., Chabbert, B., Pilate, G., VanDoorsselaere, J., Tollier, M. T., Petit-Conil, M., Cornu, D., Monties, B., Van Montagu, M., Inze, D., Jouanin, L., and Boerjan, W. (1996) Red xylem and higher lignin extractability by down-regulating a cinnamyl alcohol dehydrogenase in poplar. *Plant Physiol.* **112**(4), 1479–1490.
370. Pillonel, C., Mulder, M. M., Boon, J. J., Forster, B., and Binder, A. (1991) Involvement of cinnamyl-alcohol dehydrogenase in the control of lignin formation in *Sorghum bicolor L. Moench. Planta* **185**(4), 538–544.
371. Kim, H., Ralph, J., Lu, F., Pilate, G., Leplé, J. C., Pollet, B., and Lapierre, C. (2002) Identification of the structure and origin of thioacidolysis marker compounds for cinnamyl alcohol dehydrogenase deficiency in angiosperms. *J. Biol. Chem.* **277**(49), 47412–47419.
372. Lapierre, C., Pilate, G., Pollet, B., Mila, I., Leplé, J. C., Jouanin, L., Kim, H., and Ralph, J. (2004) Signatures of cinnamyl alcohol dehydrogenase deficiency in poplar lignins. *Phytochem.* **65**(3), 313–321.
373. Dimmel, D. R., MacKay, J. J., Althen, E., Parks, C., and Boon, J. J. (2001) Pulping and bleaching of CAD-deficient wood. *J. Wood Chem. Technol.* **21**, 1–18.
374. Lapierre, C., Pollet, B., MacKay, J. J., and Sederoff, R. R. (2000) Lignin structure in a mutant pine deficient in cinnamyl alcohol dehydrogenase. *J. Agr. Food Chem.* **48**(6), 2326–2331.
375. MacKay, J. J., O'Malley, D. M., Presnell, T., Booker, F. L., Campbell, M. M., Whetten, R. W., and Sederoff, R. R. (1997) Inheritance, gene expression, and lignin characterization in a mutant pine deficient in cinnamyl alcohol dehydrogenase. *Proc. Natl. Acad. Sci.* **94**(15), 8255–8260.
376. Gang, D. R., Fujita, M., Davin, L. D., and Lewis, N. G. (1998) The 'abnormal lignins': mapping heartwood formation through the lignan biosynthetic pathway. In: Lewis, N. G., and Sarkanen, S. (eds). *Lignin and Lignan Biosynthesis*, Amer. Chem. Soc., Washington, DC.
377. Lewis, N. G., Davin, L. B., and Sarkanen, S. (1998) Lignin and lignan biosynthesis: distinctions and reconciliations. In: Lewis, N. G., and Sarkanen, S. (eds). *Lignin and Lignan Biosynthesis*, Amer. Chem. Soc., Washington, DC.

378. Zanarotti, A. (1982) Preparation and reactivity of 2,6-dimethoxy-4-allylidene-2,5-cyclohexadien-1-one (vinyl quinone methide). A novel synthesis of sinapyl alcohol. *Tetrahedron Lett.* **23**(37), 3815–3818.

379. Piquemal, J., Lapierre, C., Myton, K., O'Connell, A., Schuch, W., Grima-Pettenati, J., and Boudet, A.-M. (1998) Down-regulation of cinnamoyl-CoA reductase induces significant changes of lignin profiles in transgenic tobacco plants. *Plant J.* **13**(1), 71–83.

380. Anterola, A. M., and Lewis, N. G. (2002) Trends in lignin modification: A comprehensive analysis of the effects of genetic manipulations/mutations on lignification and vascular integrity. *Phytochem.* **61**(3), 221–294.

381. Laskar, D. D., Jourdes, M., Patten, A. M., Helms, G. L., Davin, L. B., and Lewis, N. G. (2006) The Arabidopsis cinnamoyl CoA reductase *irx4* mutant has a delayed but coherent (normal) program of lignification. *Plant J.* **48**(5), 674–686.

382. Landucci, L. L., and Ralph, J. (1982) Adducts of anthrahydroquinone and anthranol with lignin model quinone methides. 1. Synthesis and characterization. *J. Org. Chem.* **47**(18), 3486–3495.

383. Ralph, J., and Landucci, L. L. (1983) Adducts of anthrahydroquinone and anthranol with lignin model quinone methides. 2. Dehydration derivatives. Proof of *threo* configuration. *J. Org. Chem.* **48**(3), 372–376.

384. Ralph, J., and Landucci, L. L. (1983) Adducts of anthrahydroquinone and anthranol with lignin model quinone methides. 3. Independent synthesis of *threo* and *erythro* isomers. *J. Org. Chem.* **48**(22), 3884–3889.

385. Ralph, J., Landucci, L. L., Nicholson, B. K., and Wilkins, A. L. (1984) Adducts of anthrahydroquinone and anthranol with lignin model quinone methides. 4. Proton NMR hindered rotation studies. Correlation between solution conformations and x-ray crystal structure. *J. Org. Chem.* **49**(18), 3337–3340.

386. Ralph, J., and Landucci, L. L. (1986) Adducts of anthrahydroquinone and anthranol with lignin model quinone methides. 9,10-^{13}C labeled anthranol-lignin adducts; examination of adduct formation and stereochemistry in the polymer. *J. Wood Chem. Technol.* **6**(1), 73–88.

387. Landucci, L. L. (1989) Search for lignin condensation reactions with modern NMR techniques. In. *Adhesives from renewable resources*, Amer. Chem. Soc., Washington, DC.

388. Kringstad, K. P., and Moerck, R. (1983) Carbon-13 NMR spectra of kraft lignins. *Holzforschung* **37**(5), 237–244.

389. Mörck, R., and Kringstad, K. P. (1985) ^{13}C-NMR-spectra of kraft lignins. 2. Kraft lignin acetates. *Holzforschung* **39**(2), 109–119.

390. Chen, K. L., and Hayashi, J. (1993) Change of rice straw lignin with NaOH-oxygen pulping, 5: ^{13}C NMR analysis of stirred oxidized rice straw lignin and wood lignin. *J. Japan Wood Res. Soc.* **39**(9), 1069–1076.

391. Nimz, H. H., and Schwind, H. (1981) Oxidation of lignin and lignin model compounds with peracetic acid. *Ekman-Days 1981, Int. Symp. Wood Pulping Chem., Volume 2, 105–11. SPCI,* Stockholm.

392. Akim, L. G., Shevchenko, S. M., and Zarubin, M. Y. (1993) ^{13}C NMR studies on lignins depolymerized with dry hydrogen iodide. *Wood Sci Technol* **27**, 241–248.

393. Sjoholm, R., Holmbom, B., and Akerback, N. (1992) Studies of the photodegradation of spruce lignin by NMR spectroscopy. *J. Wood Chem. Technol.* **12**(1), 35–52.

394. Marchessault, R. H., Coulombe, S., Morikawa, H., and Robert, D. (1982) Characterization of aspen exploded wood lignin. *Can. J. Chem.* **60**(18), 2372–2382.

395. Hemmingson, J. A. (1983) The structure of lignin from Pinus radiata exploded wood. *J. Wood Chem. Technol.* **3**(3), 289–312.

396. Hemmingson, J. A. (1985) Structural aspects of lignins from Eucalyptus regnans wood steam exploded by the Iotech and Siropulper processes. *J. Wood Chem. Technol.* **5**(4), 513–534.

397. Sudo, K., Shimizu, K., and Sakurai, K. (1985) Characterization of steamed wood lignin from beechwood. *Holzforschung* **39**(5), 281–288.

398. Wehrli, F. W., and Wirthlin, T. (1980) *Interpretation of Carbon-13 NMR Spectra*. In., Heyden, London.

399. Landucci, L. L. (1991) Application of modern liquid-state NMR to lignin characterization. 2. ^{13}C signal resolution and useful techniques. *Holzforschung* **45**(6), 425–432.

400. Landucci, L. L. (1985) Quantitative carbon-13 NMR characterization of lignin 1. A methodology for high precision. *Holzforschung* **39**(6), 355–359.

401. Robert, D. R., Bardet, M., Gellerstedt, G., and Lindfors, E. L. (1984) Structural changes in lignin during kraft cooking: part 3. On the structure of dissolved lignins. *J. Wood Chem. Technol.* **4**(3), 239–263.

402. Obst, J. R., McMillan, N. J., Blanchetter, R. A., Christensen, D. J., Faix, O., Han, J. S., Kuster, T. A., Landucci, L. L., Newman, R. H., Pettersen, R. C., Schwandt, V. H., and Wesolowski, M. F. (1991) Characterization of Canadian Arctic (Arctic Archipelago, Northwest Territories Canada) fossil woods. *Geological Survery of Canada Bulletin* (403), 123–146.

403. Netzel, D. A. (1987) Quantification of carbon types using DEPT/QUAT NMR pulse sequences: Application to fossil-fuel-derived oils. *Anal. Chem.* **59**, 1775–1779.

404. Landucci, L. L. (1995) Search for lignin condensation reactions with modern NMR techniques. In: Hemmingway, R. W., Connor, A. H., and Branham, S. J. (eds). *Adhesives from Renewable Resources*, Amer. Chem. Soc., Washington, DC.

405. Lapierre, C., Monties, B., Guittet, E., and Lallemand, J. Y. (1985) The quantitative measurements in hardwood lignin carbon-13 NMR spectra. *Holzforschung* **39**(6), 367–368.

406. Bendall, M. R., Doddrell, D. M., and Pegg, D. T. (1981) Editing of C-13 NMR-Spectra: A pulse sequence for the generation of subspectra. *J. Am. Chem. Soc.* **103**(15), 4603–4605.

407. Doddrell, D. M., Pegg, D. T., and Bendall, M. R. (1982) Distortionless enhancement of NMR signals by polarization transfer. *J. Magn. Reson.* **48**(2), 323–327.

408. Gellerstedt, G., and Robert, D. (1987) Structural changes in lignin during kraft cooking. Part 7. Quantitative carbon-13 NMR analysis of kraft lignins. *Acta Chem. Scand. B* **B41**(7), 541–546.

409. Sun, R. C., Mott, L., and Bolton, J. (1998) Isolation and fractional characterization of ball-milled and enzyme lignins from oil palm trunk. *J. Agri. Food Chem.* **46**(2), 718–723.

410. Cornilescu, G., Cornilescu, C. C., Zhao, Q., Frederick, R. O., Peterson, F. C., Thao, S., and Markley, J. L. (2004) Letter to the editor: Solution structure of a homodimeric hypothetical protein, At5g22580, a structural genomics target from Arabidopsis thaliana. *J. Biomol. NMR* **29**(3), 387–390.

6 Heteronuclear NMR Spectroscopy of Lignins

Dimitris S. Argyropoulos

CONTENTS

INTRODUCTION

Nuclear magnetic resonance (NMR) spectroscopy is a powerful tool for examining lignin structure. Earlier chapters covered NMR techniques that involve viewing proton (H) and carbon-13 (C-13) nuclei, which are naturally present in lignin. Additional structural information can be gained by applying derivatization procedures that covalently link other NMR active nuclei to lignin and observing the resulting NMR spectrum of that nuclei.

Important considerations in selecting NMR-active nuclei for labeling functional groups in lignins are the sensitivity of the nuclei in an NMR experiment, the availability of suitable derivatizing reagents, and the ease of obtaining quantitative derivatization under mild conditions. Several heteronuclear NMR cases are discussed in this chapter, with primary emphasis given to the most informative one, phosphorus-31 NMR.

SILICON-29 NMR

Trimethylsilylation of hydroxyl groups has long been used to facilitate gas chromatographic separation. Its use to characterize the labile protons in lignin was not explored until the late eighties.

The application of silicon-29 (^{29}Si) NMR spectra to trimethylsilylated lignin allows aromatic, aliphatic and carboxylic acid hydroxyl groups to be distinguished

[1,2]. Silylation can be used after methylation with diazomethane (to eliminate signal overlap with phenolic groups) to detect different alcoholic groups in kraft lignin and humic acids [3]. In theory, the observation of the ^{29}Si nuclei (natural abundance of 4.7%) is twice as sensitive as a comparable ^{13}C experiment; the negative gyromagnetic ratio and rather long spin-lattice relaxation times of this technique, however, cause signal reduction when proton decoupled spectra are acquired. Consequently, the acquisition of ^{29}Si NMR spectra requires high sample concentrations and long delay times. Alternatively, the ^{29}Si NMR signals can be considerably enhanced using the INEPT (Insensitive Nuclei Enhanced by Polarization Transfer) sequence, but this poses major limitations in quantitatively interpreting the spectra [4].

NITROGEN-15 NMR

Nitrogen-15 (^{15}N) NMR spectroscopy has found only limited utility in lignin chemistry due to difficulties with recording ^{15}N NMR spectra.

The low natural abundance (0.365%) of the ^{15}N isotope dictates a low sensitivity for the method. The negative magnetic moment of the ^{15}N nucleus results in a negative Nuclear Overhauser Effect (nOe), which can lead to complete disappearance of the ^{15}N NMR signal under incorrectly chosen acquisition conditions. The significant variation in the relaxation time of ^{15}N atoms with different chemical environments may also result in the elimination of certain signals in ^{15}N NMR spectra. These factors limit the applicability of this technique to the analysis of nitrogen-incorporated lignins when functional group detection of the nitrogen-containing moieties is required. ^{15}N NMR spectroscopy in solution has been applied to the study N-containing products after oxidative ammonolysis of lignin model compounds, DHP, and Organocell lignin [5]. ^{15}N-enriched NH$_3$ was used in the ammonolysis reaction to ensure sensitive and reliable detection of ^{15}N NMR signals. The formation of formamide, acetamide, substituted benzamides, and urea was detected by ^{15}N NMR spectroscopy.

A possible application of ^{15}N NMR spectroscopy to lignin analysis is the study of lignin biodegradation products. The incorporation of nitrogen-containing functional groups can be detected and identified by this technique. Kniker et al. analyzed sulphonated lignin and organosolv lignin incubated with ^{15}N-enriched ammonium sulfate for 600 days to demonstrate the influence of ammonia fertilizer on pathways of microbial decomposition of lignin [6]. The introduction of amide-peptide structures into lignin was detected. Kögel-Knabner et al. [7] provided more examples of using ^{15}N NMR spectroscopy for the structural analysis of lignin-related humic substances.

MERCURY-199 AND TIN-119 NMR

The selective mono-mercuration of aromatic rings with mercury acetate followed by ^{199}Hg NMR spectroscopic analysis has been used to elucidate the aromatic substitution patterns in lignins [8]. Direct and indirect (HMQC 2D ^{199}Hg-^{1}H spectra) ^{199}Hg NMR spectra of the formed derivatives were recorded; $J(^{199}$Hg, ^{1}H) coupling constants provided information about the position of the mercury substitution on the aromatic ring. Mercury acetate was found to introduce Hg at the C-5 position of guaiacyl lignin model compounds and C-2 position of syringyl compounds.

Tin-119 (^{119}Sn) NMR spectroscopy has been used to characterize bis(tri-butyltin) oxide derivatives of various lignin-related phenols and lignin-containing material isolated from spent bleach liquors [9]. Signals of the model derivatives were well resolved, but the broad signals displayed by actual spent liquors limited the utility of the spectra.

FLUORINE-19 NMR

Florine-19 (^{19}F) NMR spectroscopy has also been explored for determining various hydroxyl groups in lignin; the OH groups are first converted into the corresponding fluorobenzoic acid alkyl and aryl esters and/or flurobenzyl ethers [10] under phase transfer catalysis conditions. While relatively small amounts of lignin sample (100 mg) and short acquisition protocols (one hour) are required for this technique, the method suffers from a somewhat cumbersome workup procedure that could introduce complexities in quantification and an inability to distinguish between primary and secondary hydroxyl groups. Despite these limitations, this approach has been applied successfully for determining the total phenolic and carboxylic acid groups in oxidized [11] and residual [12] lignins using 2-fluorobenzoic acid as an internal standard.

Another promising application of ^{19}F NMR spectroscopy, reported by Sevillano et al. [13], is the use of trifluoromethyl and trifluoromethoxyphenylhydrazine for the quantitative detection of carbonyl groups in lignin. The reliability of the technique, however, depends on the efficiency and the quality of the purification steps required after derivatization. A new method, capable of quantitatively detecting different classes of carbonyl groups for a series of model compounds [14] and lignins [15], employs a selective fluoride-induced trifluoromethylation of carbonyl groups (using trifluorotrimethylsilane), followed by ^{19}F NMR spectral analysis of the derivatives. These studies have shown that the trifluoromethyl ^{19}F-NMR chemical shifts vary significantly and consistently for derivatives of various aldehydes, ketones and quinones that may be present in complex lignocellulosic materials.

PHOSPHORUS-31 NMR

The use of phosphorus-containing derivatizing reagents for lignin analysis has grown in importance. Phosphorus-31 is a nucleus that is 100% naturally abundant. The sensitivity of a ^{31}P NMR experiment is about 15 times less than that of a proton NMR experiment. The range of ^{31}P chemical shifts is more than 1000 ppm for a variety of phosphorus compounds and the average line width is about 0.7 Hz [16]. Various types of organophosphorus compounds give signals within narrow ranges, characteristic of the oxidation state of the phosphorus nuclei. Furthermore, relationships have been identified between phosphorus chemical shifts and structure that in some instances even reveal stereochemical information [17,18]. All of these factors make phosphorus an ideal reporter group for NMR studies of labile

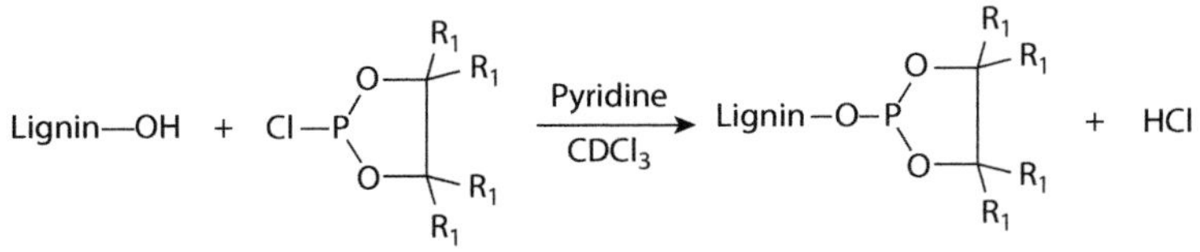

SCHEME 6.1 The phosphitylation reaction of labile protons in model compounds and lignins with reagents I (R_1 = H, 2-chloro-1,3,2-dioxaphospholane) and II (R_1 = CH_3, 2-chloro-4,4,5, 5-tetra-methyl-1,3,2-dioxaphospholane) (Based on Lucas, H. J., Mitchell, F. W., and Scully, C. N., *J. Am. Chem. Soc.*, 72, 5491, 1950.)

groups in lignin. The reactions of various phospholane chlorides with labile centers present in coal samples were investigated by Verkade's group in the mid 1980's [19–23].

Early work focused on the phospholanes produced after the derivatization of OH groups with 1,3,2-dioxaphospolanyl chloride (I), Scheme 6.1 [24,25,27,28]:

A detailed review of the advantages and limitations of this form of spectroscopy has been published [24]. In part I of series of papers *31P NMR in Wood Chemistry,* we evaluated this technique's potential by investigating a large variety of model compounds with structures likely to occur in lignins [25]. This research showed that this technique could distinguish not only most forms of phenolic hydroxyls (ArOH) but also primary and secondary aliphatic hydroxyls (R^1-OH and R^2-OH) and *erythro-* and *threo*-forms of β–O–4-structures [25]. Supporting evidence for the latter assignment was sought by using the lignin alkylation studies of Adler et al. [26]. Methylation of black spruce milled wood lignin resulted in the complete elimination of the two broad signals at 135.0 and 134.2 ppm attributed to the *erythro-* and *threo*-forms, respectively, of the phosphitylated *alpha*-hydroxyl groups in β–O–4 structures [27]. Early observations revealed a slight concentration-dependence of the ^{31}P NMR chemical shifts among a variety of phosphorus derivatives. The effect was explored to better understand the behavior of the signals from derivatized lignins [28]. Since chloroform and pyridine are used as solvents for the derivatization reaction, their role was examined separately. The resulting knowledge permitted spectra with greater resolution to be obtained and facilitated the assignment of ^{31}P signals in lignins (Table 6.1).

The derivatization of lignins with reagent I allowed visualization of the overall distribution of hydroxyl groups. However, signal overlap between the syringyl phenolic structures and those belonging to condensed phenolic groups limited its capacity for distinction and accurate determination of these moieties. Another phosphitylation reagent, 2-chloro-4,4,5,5-tetramethyl-1,3,2-dioxaphospholane (II, Scheme 6.1), was developed; it was found to be particularly good at resolving this region at the expense of fine resolution between the primary and secondary hydroxyls [29]. The different patterns of ^{31}P NMR spectra of ball milled cottonwood lignin phosphitylated with reagents I and II are illustrated in Figure 6.1.

The ^{31}P NMR signals, after use of reagent II, were well resolved for the free phenolic hydroxyls belonging to guaiacyl, syringyl, *p*-hydroxyphenyl units and most C5 and C6 related condensed phenolic forms. In addition, signals due to carboxylic

TABLE 6.1
^{31}P NMR Chemical Shifts Ranges of Various Functionalities in Lignins after Derivatization with Reagent I

Chemical Shift, ppm	Lignin Functionality
136.5–135.8	OH group in xylan
136.8–135.2	*Erythro* α-OH in β-O-4 syringyl units
135.2–135.4	*Erythro* α-OH in β-O-4 guaiacyl units
134.5–133.7	*Threo* α-OH in β-O-4 syringyl and guaiacyl units
133.7–133.2	γ-OH in α-carbonyl containing units, cinnamyl alcohols
133.2–132.7	γ-OH in β-O-4 units
132.7–132.1	Primary OH (probably phenylcoumaran type)
132.1–131.6	Phenolic OH in syringyl units
131.6–131.0	Phenolic OH in biphenyl units, cinnamyl aldehydes
130.4–129.7	Phenolic OH in guaiacyl units
129.7–129.3	Phenolic OH in guaiacyl units and catechol structures
127.1–126.5	COOH groups in aliphatic acids and cinnamic acids

Source: Based on Argyropoulos, D. S., Bolker, H. I., Heitner, C., and Archipov, Y., *J. Wood Chem. Technol.*, 13(2), 187–212, 1993.

acids were well separated from all other signals, allowing direct access to this important information related to the fundamental changes occurring within lignins under oxidative conditions [30]. The signal assignments for reagent II with lignin are given in Table 6.2. The structure/chemical shift relations of phosphitylated phenols of more than sixty lignin model compounds were explored using Hammett principles. This provided a set of empirical parameters that permits the accurate prediction of ^{31}P NMR chemical shifts of lignin-related phenolic compounds derivatized with II [31].

The ^{31}P spin-lattice relaxation behavior of phosphitylated lignins has also been investigated [28]. Further work, using two-dimensional ^{31}P NMR spectroscopic techniques, clarified the assignments of one-dimensional ^{31}P NMR spectra of lignins [32]. This research confirmed the absence of through-bond and through-space ^{31}P-^{31}P and ^{31}P-^{1}H couplings from within a ^{31}P NMR spectrum of phosphitylated lignins. Detailed measurements of the phosphorus spin-lattice and spin-spin relaxation times at various static magnetic fields and temperatures found that the predominant spin relaxation mechanism of phosphorus in phosphitylated lignins was due to chemical shift anisotropy [32]. This background information was then used to design an experimental protocol for obtaining quantitative ^{31}P NMR spectra of phosphitylated lignins [33]. The quantitative reliability of this methodology was verified by an international round-robin effort designed to validate analytical techniques; ^{31}P NMR data were compared to amminolysis [36] and permanganate oxidation [36,37] of lignin standards [34,35].

Recently, cellulose was dissolved in a variety of imidazolium chloride-based ionic liquids (IL's) bearing a series of substituents, which imparted varying degrees

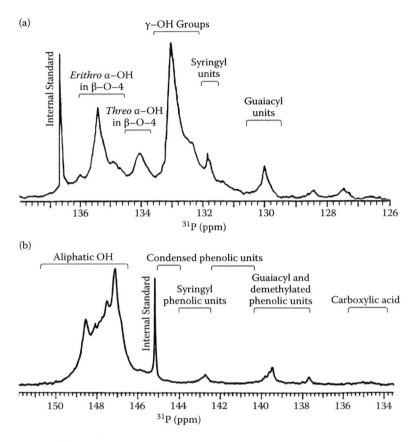

FIGURE 6.1 ^{31}P NMR spectra and assignments for ball milled cottonwood lignin phosphitylated with reagent I (a) and reagent II (b) ((a) Based on Argyropoulos, D. S., Bolker, H. I., Heitner, C., and Archipov, Y., *J. Wood Chem. Technol.*, 13(2), 187–212, 1993; Argyropoulos, D. S., *J. Wood Chem. Technol.*, 14(1), 65–82, 1994. (b) Based on Granata, A. and Argyropoulos, D. S., *J. Agric. Food Chem.*, 43(6), 1538–1544, 1995.)

of hydrophobicity [75]. While in solution, the cellulose was esterified with the hydrophobic phosphitylation reagent 2-chloro-4,4,5,5-tetramethyl-1,3,2-dioxaphospholane (II), inducing phase separation. Ionic liquids bearing hydrophobic substituents were found to create environments that reduced the phase separation. A similar, but less pronounced phase separation, was also observed during the phosphitylation of isolated enzymatic mild acidolysis lignin (EMAL) (74a–c), which is the major non-polysaccharide constituent of wood. A careful choice of IL and cosolvent addition resulted in homogenization and the advancement of quantitative ^{31}P NMR analysis procedures on lignocellulosics. Ultimately, this methodology provided a means of probing the structure of lignin on the fiber, without its prior isolation; the whole wood cell wall is made soluble in ionic liquids [76], allowing for the acquisition of detailed quantitative ^{31}P NMR spectra [75].

TABLE 6.2
^{31}P NMR Chemical Shifts Ranges of Various Functionalities in Lignins after Derivatization with Reagent II

Chemical Shift, ppm	Lignin Functionality
150.8–146.3	Aliphatic OH group
144.3–142.8	Condensed phenolic units: diphenylmethane type
143.7–142.2	Syringyl phenolic units
142.8–141.7	Condensed phenolic units: 4-O-5' type
141.7–140.2	Condensed phenolic units: 5-5' type
140.2–138.4	Guaiacyl and demethylated phenolic units
138.6–136.9	*p*-Hydroxyphenolic units
135.6–133.7	Carboxylic acids

Source: Argyropoulos, D. S., *Research on Chemical Intermediates,* 21(3–5), 373–395, 1995; Argyropoulos, D. S., Bolker, H. I., Heitner, C., and Archipov, Y., *J. Wood Chem. Technol.,* 13(2), 187–212, 1993.

APPLICATIONS OF QUANTITATIVE ^{31}P NMR

Early efforts to apply quantitative ^{31}P NMR were centered around quantitatively determining ArOH groups and the two diastereomeric forms of the α–OH in β–O–4 linkages in lignins [38]. The technique was applied to fractionated guaiacyl (G) and guaiacyl/syringyl (GS) DHPs prepared by continuous ("Zutropf," ZT) and discontinuous ("Zulauf," ZL) dehydrogenation schemes. Comparison of the data to that of milled wood lignin samples from softwood and hardwood species revealed that GS-DHPs resemble GS milled wood lignins to a greater extent than G milled wood lignins. The total phenolic OH contents of ZT-DHPs were always lower than those of ZL-DHPs, which is in agreement with the theory that predicts more β–O–4-linkages in the ZT-DHPs. Secondary aliphatic-OH groups in G-DHPs were extremely low. This underlies again the principal differences of G-DHPs compared to GS-DHPs and MWLs. The e*rythro:threo* ratios of the G-type samples were found to vary between 1 and 1.5, indicating only minor dependence on molar mass or mode of preparation. The *erythro:threo* ratio in GS-type samples was found to vary from 1.6 to 4.3, showing the highest value for low molar mass ZL-DHPs and lowest values for the cherry tree MWL. A significant amount of the quantitative functional group data collected for both types of synthetic lignins was later found to correlate well with the results of a "Lignin" computer simulation model developed by Jurasek [39].

Early spectroscopic evidence indicated that quantitative solution-state ^{31}P NMR could be used to follow and quantify the formation of stable carbon-carbon and carbon–oxygen bonds during kraft pulping. The use of reagent II, with four bulky methyl groups on the glycol bridge of the dioxaphospholane (Scheme 6.1), dramatically increased the resolution of the ^{31}P spectra in the region responsible for the uncondensed and condensed phenolic structures (Figure 6.2a).

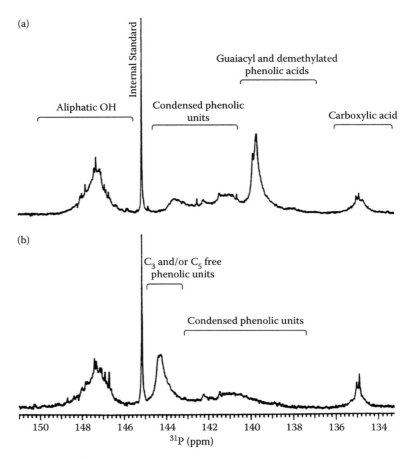

FIGURE 6.2 Quantitative ^{31}P NMR spectra and region assignments of (a) a kraft solubilized black spruce (*Picea mariana*) lignin sample and (b) its Mannich product (Jiang, Z. and Argyropoulos, D. S., *Can J. Chem.*, 76(5), 612–622, 1998.)

The phenolic hydroxyl groups belonging to condensed structures could be quantified by integrating the region between 140.3–144.4 ppm in the spectrum of Figure 6.2a. However, no information about condensed units lacking free phenolics can be obtained by this method. A significant amount of condensed structures was found to accumulate within solubilized kraft lignins at about 16% delignification. These species were found to further accumulate at subsequent levels of delignification [29].

To better understand the nature of the condensed structures in lignins, researchers examined the application of the Mannich reaction to probe these structures [40]. The Mannich reaction was applied to a large variety of lignin-model compounds to selectively and quantitatively block the available aromatic C3 and C5 positions [40]. Quantitative ^{31}P NMR provided a determination of the units that bear no substituents at the aromatic C3 and/or C5 positions.

Smit et al. combined thioacidolysis and phosphitylation with reagent II, followed by ^{31}P NMR spectroscopy, to measure the total amount of condensed and uncondensed units in wood and lignin [41]. Application of the method to *Picea radiata* wood and milled wood lignin showed that approximately 77% of the C9 units in the milled wood lignin and 71% of the C9 units in wood could be quantified. The amount of condensed structures/C9 units in MWL was determined to be: as high as 8% β–5, 5% 4–O–5, and 16% 5–5 structures.

Tohmura and Argyropoulos have developed an analytical method for lignins that involves the combination of DFRC with quantitative ^{31}P NMR [45a]. More specifically the methodology involves: (a) derivatization followed by reductive cleavage (DFRC) [42–44], (b) depolymerization, and (c) quantitative ^{31}P NMR spectroscopy [45]. This technique was shown to detect and quantify the various ether linkages present in softwood residual kraft and milled wood lignins. In addition, the technique supplied new quantitative information about β-aryl ethers linked to condensed and noncondensed aromatic moieties, including dibenzodioxocins. Within residual kraft lignin, β-aryl ether bonds connected to condensed phenolic moieties predominated over those connected to noncondensed phenolic moieties. In addition, the amount of DFRC monomers determined by GC was very small in the residual kraft lignin, but large in the milled wood lignin. This indicates that almost all noncondensed β-aryl ether linkages were cleaved during kraft pulping. The method offers new avenues for the detailed investigation of the bonding patterns of native and technical lignins. The same approach was used later by the same research group to define the abundance of dibenzodioxocin moieties in lignin [45b].

Recently, the Argyropoulos group has applied the combination of DFRC with quantitative ^{31}P NMR to determination the arylglycerol-β-aryl ether linkages in enzymatic mild acidolysis lignins (EMAL) [74a–c] and further compare the DFRC/^{31}P NMR protocol with thioacidolysis [45c]. More specifically, enzymatic mild acidolysis lignins [74a–c], isolated from different species of softwood and *Eucalyptus globules,* were submitted to comparative analysis that included thioacidolysis, DFRC, and DFRC followed by quantitative ^{31}P NMR (DFRC/^{31}P NMR) [45a]. Gas chromatography (GC) was used to determine the monomer yields from both thioacidolysis and DFRC; ^{31}P NMR was used to quantified the various phenolic hydroxyl groups released by DFRC [45a]. The monomer yields from thioacidolysis and DFRC were substantially different; thioacidolysis resulted in higher yields. In contrast, an excellent agreement was obtained in the total number of β-aryl ether structures determined by thioacidolysis and DFRC/^{31}P NMR. These results indicate that the lower monomer yields derived from DFRC are due to a limitation of using GC alone for detecting DFRC monomers, rather than an inefficiency in the DFRC protocol's chemistry.

To further validate this data, we applied both thioacidolysis and DFRC/^{31}P NMR to better understand the lignin isolation process from wood. The results show that mild rotary ball milling minimizes, but does not prevent the degradation of β–O–4 structures, during the early stages of wood pulverization. The extent of such degradation was found to be higher for *E. globulus* than for a variety of softwoods examined. However, the combination of enzymatic hydrolysis and mild acid hydrolysis [74a–c] with low-intensity ball milling protocols resulted in high lignin isolation yields, with less degradation, than traditional lignin isolation protocols. Furthermore, the structures of the EMALs isolated at yields ranging from 20 to 62% were very similar, indicating structural homogeneity in the lignin biopolymer within the secondary wall [45c].

Another acidolytic method of residual lignin isolation, proposed in the mid-1990s [46], has been widely accepted for the isolation of residual lignins from kraft and other pulps. However, it is not known to what extent the isolated lignins are affected by the process of acidolysis. To address this issue, Jiang and Argyropoulos [47] isolated residual lignin from various softwood kraft pulps using a dioxane acidolysis process in batch and flow-through reactors. The lignin was characterized using a variety of wet-chemical and ^{31}P NMR spectroscopic techniques. The dioxane acidolysis lignins, isolated from kraft pulps using the flow-through reactor, were found to have similar functional group distributions and elemental compositions as those of the acidolysis lignins obtained from the same pulps using the batchwise process. The data offers evidence that the structure of residual lignin in kraft pulps is not altered significantly during a dioxane acidolysis isolation process.

The formation of diphenylmethane (DPM) moieties in lignin during conventional kraft, soda and modified extended modified continuous cooking (EMCC) pulping conditions has also been probed using ^{31}P NMR [48a]. This effort confirmed the assignment of a ^{31}P NMR signal due exclusively to the presence of phosphitylated DPM phenolic hydroxyl groups. Softwood milled-wood lignin was subjected to kraft pulping conditions in the presence and absence of varying amounts of formaldehyde; the ^{31}P NMR spectra of the recovered lignins revealed selective signal growth in the region between 142.8 and 144.3 ppm. These signals were assigned to DPM ArOH groups, in accordance with previous model compound work and calculations based on the Hammett principles [31,40]. The DPM moieties in lignin accumulate in significantly higher proportions in an isothermal (120°C) soda pulping experiment than in a comparable kraft pulping experiment. DPM structures were also found to prevail among the condensed phenolic units of a conventionally cooked pulp compared to one produced using a modified cooking protocol (EMCC), providing additional evidence that modern modified pulping technologies alter the structure of residual kraft lignin beneficially [48a]. This effort was extended further with detailed computational analyses [48b]. Similar conclusions were also reached by Froass et al., who used quantitative ^{31}P NMR, together with ^{1}H and ^{13}C NMR spectroscopies, to investigate the structure of residual lignins from conventional and modified cooking protocols [49].

The stereoselective cleavage of the β–O–4-structures under conventional kraft pulping conditions has also been studied using quantitative ^{31}P NMR [50]. Quantitative spectra of residual and dissolved softwood kraft lignins, isolated at various degrees of delignification, confirmed earlier observations that the *erythro* isomers cleave faster than their *threo* counterparts [50]. In subsequent work, Ahvazi and Argyropoulos [51] measured the remaining β–O–4 ether structures present in softwood milled-wood lignin at various points along a kraft cook. This effort was focused at deriving the various absolute fundamental thermodynamic parameters that govern the stereoselective degradation of β-ethers under homogeneous kraft pulping conditions. The absolute rates of scission of the two diastereomers followed two kinetic regimes: an initial fast phase, followed by a slower phase. In agreement with previous accounts, the rate constants for these scission reactions were found to follow a pseudo-first-order rate law during both phases. Rate constant data invariably indicated that the *erythro*-isomers of the β–O–4 units of softwood milled-wood lignins cleave faster than their *threo* counterparts during both phases.

A number of research efforts have been focused at using mixtures of organic solvents with water as the pulping medium. The Alcell® process represents one such process, with the inherent limitation that it cannot be used for the pulping of softwoods. In recent work by Liu et al.[52], quantitative ^{31}P NMR was used for the comparative analysis of dissolved and residual lignins after conventional kraft and Alcell cooking of softwood and hardwood species. At all degrees of delignification, the ArOH group content for both wood species was: (a) higher for the solubilized kraft lignins than for the solubilized Alcell lignins and (b) lower in the residual kraft lignins than in the residual Alcell lignin. The authors rationalized these findings on the basis of the greater solvating abilities of alkaline aqueous media (toward inducing solubilization of the phenolic moieties) as opposed to those of ethanol. The condensed phenolic units in residual lignins formed in greater abundance in softwood rather than hardwood pulps for both processes. Condensed phenolic structures were formed most rapidly in the residual lignins of the softwood Alcell® pulps and in particular during the early and later phases of delignification. The condensation reactions, induced under the acidic conditions and elevated temperatures, may be responsible for the deceleration of the delignification observed for softwoods during the Alcell® process.

The fundamental changes of softwood and hardwood kraft lignins during high-pressure oxygen and low-pressure oxygen-peroxide reinforced delignification (Eop) stages have also been elucidated using quantitative ^{31}P NMR [53]. Conventional kraft pulps from a hardwood and a softwood, which were oxygen delignified using similar conditions, were subjected to residual and solubilized lignin isolation procedures. The formation of carboxylic acids and the degradation of condensed phenolic structures were found to be the main pathways that account for lignin dissolution during the oxygen or the Eop delignification stages. Significant degradation of the uncondensed and condensed phenolic units occurred during the oxygen and the Eop stages. Presumably, the formation of carboxylic acids compensated for the elimination of phenolic hydroxyl groups by keeping the lignin fragments hydrophilic enough for dissolution in alkaline media. In general, the chemical and physicochemical considerations that affect the efficiency of oxygen and Eop stages were found to be similar. However, due to the milder nature of the Eop stage, the resultant residual lignins were found to be less affected than their high-pressure oxygen counterparts [53].

An isolated softwood residual kraft lignin was systematically oxidized at different times and temperatures in an attempt to understand the complex interactions of this lignin with oxygen at elevated temperatures, pressures, and pH [54]. Quantitative ^{31}P NMR spectra of oxidized lignins, obtained after phosphitylation with reagent II, showed the formation and/or elimination of the various functional groups as a function of time and temperature. For all temperatures studied, two regimes described the rate of carboxylic acid group formation: an initial rapid phase that dominates the process for the first 20 minutes (at all temperatures), followed by a slower phase. This is not an unusual finding, since similar observations were also reported by Renard et al. [55,56], who studied the kinetics of oxidation of cuoxam lignin. The rate of carboxylic acid group formation was found to dramatically increase as the reaction temperature increased. At reaction temperatures, typical of conventional commercial oxygen delignification installations (80–100°C), only minor oxidation occurred within the residual kraft lignin. In general, the efficiency of oxidation of residual kraft lignin was found to significantly increase above 100°C, even though these experiments were conducted under idealized two-phase homogeneous conditions. The technological ramifications of these data imply that a three-phase oxygen delignification system, operating at temperatures below 100°C, causes only minor oxidative changes in the structure of the lignin on the fiber.

Quantitative ^{31}P NMR spectra of the oxidized lignin samples described above [54] also showed characteristic signals of catechols in the region 138.6–139.1 ppm [29,31,50]. Quantification of the catechol signals, as a function of time and temperature, showed that the catechol concentration was relatively constant over the whole temperature range examined, implying that catechols were oxidation intermediates. Their participation in an oxygen delignification was previously suggested by Gellerstedt and Lindfors [57] and Renard et al. [55].

Sun and Argyropoulos [58], using quantitative ^{31}P NMR, mapped the reactivity and efficiency of several bleaching agents: chlorine dioxide, ozone, dimethyldioxirane and alkaline hydrogen peroxide. Residual lignin isolated from a conventional softwood kraft pulp was reacted with varying charges of these reagents, followed by quantitative ^{31}P NMR analyses. Chlorine dioxide and ozone were found to be the most efficient reagents in causing the formation of carboxylic acids; alkaline hydrogen peroxide was less efficient. Guaiacyl phenolic units were found to be the major sites of attack of all the oxidative treatments. The results confirmed the reactivity of ozone toward both free and etherified phenolic structures; the reactivity of chlorine dioxide and hydrogen peroxide was mainly directed toward free phenolic structures. The elimination of condensed phenolic structures was also examined for all oxidative treatments. At a given reagent charge, the relative efficiencies of elimination of condensed phenolic moieties were ozone > chlorine dioxide > alkaline hydrogen peroxide. Ozone and chlorine dioxide exhibited relatively high reactivity toward condensed phenolic units. For these cases, the rate of elimination/unit charge of reagent was similar to that observed for the elimination of guaiacyl phenolic units.

Senior et al. used quantitative ^{31}P NMR spectroscopy to compare the effectiveness of chlorine dioxide (D) and hydrogen peroxide (P) in the DEDP and DEPD bleaching sequences (E = extraction stage) as applied to softwood and hardwood kraft pulps [59]. The amount of condensed and uncondensed phenolic units in the residual lignins after different bleaching stages was determined and an attempt was made to correlate the units with the Kappa number and the brightness of the pulp. The analysis demonstrated that the second D (D_2) stage reduced the content of the free phenolic units in lignin from softwood pulp by 60%, while the P stage had no such effect. For the hardwood pulp, the D_2 stage reduced the content of free phenolic units in lignin by 90–95%. Based on these findings, Senior et al. concluded that the chlorine dioxide stage should precede hydrogen peroxide. This complies with the findings of van Lierop et al. [60] that a peroxide bleaching stage can decrease the efficiency of a subsequent chlorine dioxide stage by degrading phenolic hydroxyl units, which are considered to be sites of attack by chlorine dioxide.

Quantitative ^{31}P NMR spectroscopy has also been used to study the mechanism of light-induced yellowing. Photochemically induced degradation, condensation, and rearrangement reactions were observed during irradiation [61]. Milled wood lignin produced from alkaline hydrogen peroxide bleached softwood TMP fibers was adsorbed on pure cellulose and irradiated for various times under oxygen and nitrogen. The absolute amounts of β–O–4 ethers, phenolic hydroxyl groups, carboxylic acids and various condensed phenolic units were nondestructively quantified, using ^{31}P NMR spectroscopy. The ability of quantitative ^{31}P NMR to determine the concentration of both diastereomeric forms of the β–O–4 linkages in lignins provided a unique opportunity to further investigate the kinetics of this scission and of any possible salient stereochemical effects. Irradiation of milled wood lignin caused severe cleavage of the β–O–4 structures. These structures were eliminated faster in the presence of excess oxygen compared to nitrogen. This research provided additional evidence on the extreme reactivity of phenoxy radicals formed by homolytic bond scission within the β–O–4 structures during irradiation. These radicals may further condense or disproportionate to yield chromophoric centers.

The kinetic profiles of formation and/or elimination of various condensed phenolic structures formed *via* radical coupling reactions in irradiated milled wood lignin were also followed using quantitative ^{31}P NMR [61]. A net increase of the C5-related condensed phenolic units occurs during irradiation of milled wood lignin. The build up of the condensed phenolic units was more rapid in the presence of excess oxygen than in the presence of nitrogen. This work also revealed a well-resolved signal at 144.2 ppm that appeared in the ^{31}P NMR spectra of irradiated milled wood lignins. The identity of this signal was found to be due to the formation of Cα–C5 and/or Cβ–C5 phenolic moieties in lignin during irradiation. The relative amounts of these units were found to gradually increase and did not diminish in the presence of oxygen or nitrogen.

^{31}P NMR SPECTROSCOPY FOR THE DETECTION OF QUINONES

Carbonyl groups of simple aliphatic aldehydes, *ortho* and *para* quinones, α,β-unsaturated carbonyls and cyclic aromatic anhydrides are known to condense with trimethyl phosphite, producing phosphite esters [62]. Lebo and Lonsky used trimethyl phosphite to determine the light-induced formation of *ortho*-quinonoid functional groups in refiner mechanical pulp [63].

Solid state ^{31}P NMR spectra of mechanical and ultra-high-yield pulps treated with trimethyl phosphite showed useful information on the type and amount of carbonyl groups. A thorough investigation of this reaction by Argyropoulos's group [64,65] revealed a number of salient features, which has led to the development of a technique for the quantitative determination of *ortho*-quinones present in solid lignocellulosic materials [65] and a semiquantitative protocol for their detection in soluble lignin [66].

A strong signal at 10 ppm in the ^{31}P solid state cross polarization/magic angle spinning (CP/MAS) NMR spectrum of stone-ground wood pulp (SGW) treated with trimethyl phosphite, has been attributed to a cyclic phosphite ester formed *via* hydrolysis of an oxyphosphorane adduct of trimethyl phosphite and *ortho*-quinones present in lignin, as shown in Scheme 6.2 [64–66].

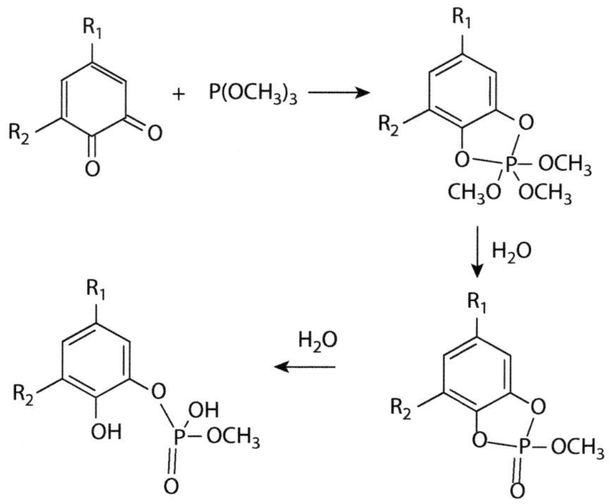

SCHEME 6.2 Reactions of *ortho*-quinones with trimethyl phosphite and the possible transformation products (Argyropoulos, D. S. and Heitner, C., *Holzforschung,* 48 (Suppl.), 112–116, 1994; Argyropoulos, D. S. and Zhang, L., *J. Agric. Food Chem.,* 46(11), 4628–4634, 1998.)

The oxyphosphorane adducts of *ortho*-quinones with trimethyl phosphite, when synthesized in the absence of water, are known to give rise to a signal at −45.8 ppm [64]. Such an adduct, however, was found to be unstable in a typical pulp sample because traces of water and hydroxyl groups seem to hydrolyze oxyphosphoranes to cyclic phosphite esters, resulting in a signal at 10 ppm [65,66]. The assignment the ^{31}P chemical shift at 10 ppm to cyclic phosphite esters formed from adducts of trimethyl phosphite and *ortho*-quinones present in lignin is supported by the significant increase in signal intensity when SGW was oxidized with potassium nitrosodisulphonate (Fremy's salt).

The broad signal extending over the 20–40 ppm region in the ^{31}P solid state NMR spectra of pulps treated with trimethyl phosphite has particular interest. It is known that trimethyl phosphite reacts with 3-benzylidene-2,4-pentanedione to rapidly form a stable adduct at room temperature and give rise to a ^{31}P NMR signal 27.9 ppm [64]. The 20–40 ppm broad signal in the pulps was assigned to conjugated structures (e.g., α,β-unsaturated aldehydes and ketones) present in lignin within high-yield pulps [66]. Further independent evidence supporting this assignment was obtained from the solid state diffuse reflectance UV/visible difference spectra of thermomechanical pulp (TMP) and TMP treated with trimethyl phosphite [66]. Carboxylic acids are known to be present in high-yield pulps and the amount increases after alkaline hydrogen peroxide treatment [69]. Some signals in the ^{31}P NMR spectra of bleached pulps may be due to trimethyl phosphite reacting with these carboxylic acid groups. Carboxylic acids and trialkyl phosphites adduct formation has been reported by a number of other workers [68–70].

Carboxylic acids in mechanical pulp also can undergo oxyphosphorylation with trimethyl phosphite; the signals from the adducts may interfere with those attributed to *ortho*-quinones. The adduct formation between carboxylic acids and trimethyl phosphite in oxyphosphorylated mechanical pulp was investigated with solid state ^{31}P NMR spectroscopy [65]; it was shown that complete ionization of the carboxylic acid groups in pulp prior to oxyphosphorylation prevents adduct formation and, thus, NMR signal overlap with *ortho*-quinone derivatives.

The oxyphosphorylation technique was used to probe the light-induced brightness reversion of black spruce SGW samples irradiated with a weak light source as a function of time [71]. The intensity of the central ^{31}P NMR signal at around 10 ppm due to *ortho* quinones was found to increase substantially after one and two days of irradiation. However, the rapid formation of *ortho*-quinones during early irradiation was followed by a reduction of these species at later phases. The data suggests that during early photochemistry, *ortho*-quinones are produced in high-yield pulps and subsequently react, creating other more complex chromophores that do not possess the quinone character.

To extend the technique for the detection of quinones in soluble lignins, Argyropoulos and Zhang [66] carried out detailed measurements and observations with model *ortho*- and *para*-quinones. These compounds, in dry organic solvents, were shown to form adducts with trimethylphosphite in quantitative yield and have ^{31}P NMR signals around −46 ppm and −2 ppm for *ortho*- and *para*-quinones, respectively. The adducts of *ortho*-quinones, in the presence of moisture and lignin, were shown to hydrolyze to the open ring product, dimethylphenylphosphate (−2 ppm), at an overall yield of about 70%. Similarly, the yield of the hydrolysis of *para*-quinone adducts with trimethylphosphite was about 70%. Consequently, a number of important issues (internal standard, and spin-lattice relaxation considerations for model compounds and lignins) were investigated, which allowed for the development of an experimental semiquantitative protocol recommended for spectral acquisition [66].

Lignin samples were then subjected to the developed protocol to determine the reliability of the technique in semiquantitatively detecting the changes in quinone content in accord with known chemistry. The accumulated data are tabulated in Table 6.3. The quinone content in MWL was 0.4 per 100 C9 units, which is lower than results obtained in other previous studies [67,72]. Furthermore, oxygen delignification was found to increase the quinone content within kraft lignin, possibly via demethylation reactions as recently demonstrated by Asgari and Argyropoulos [54].

TABLE 6.3
Total Quinone Content of Different Lignin Samples Detected in This Work

Lignin Sample	Quinone Content (mmol/g)
Black spruce MWL	0.020
Softwood solubilized kraft lignin	0.029
Indulin	0.023
Dimethyldioxirane-treated hardwood residual kraft lignin	0.24
Oxygen-treated kraft lignin	0.055

Source: Ramirez, F., *Pure and Appl. Chem.*, 9, 337–369, 1964.

Finally, a treatment of a hardwood (aspen) solubilized kraft lignin with dimethyl-dioxirane resulted in a dramatic increase in its quinone content as expected from its actual reactivity with lignin model compounds [60,73].

^{31}P NMR SPECTROSCOPY FOR THE DETECTION OF FREE RADICALS

In recent years our group has also emabrked at developing quantitative ^{31}P NMR spin trapping techniques that can be used as effective tools for the detection and quantification of many free radical species. Free radicals react with a nitroxide phosphorus compound, 5-diisopropoxy-phosphoryl-5-methyl-1-pyrroline-N-oxide (DIPPMPO) to form stable radical adducts, which are suitably detected and accurately quantified using ^{31}P NMR [77]. Initially, the ^{31}P NMR signals for the radical adducts of oxygen-centered ($^{•}$OH, $O_2^{•-}$) and carbon-centered ($^{•}CH_3$, $^{•}CH_2OH$, $CH_2^{•}CH_2OH$) radicals were assigned [77]]. Subsequently, the quantitative reliability of the developed technique was demonstrated under a variety of experimental conditions. The ^{31}P NMR chemical shifts for the hydroxyl and superoxide reaction adducts with DIPPMPO were found to be 25.3 and 16.9, 17.1 ppm (in phosphate buffer), respectively. The ^{31}P NMR chemical shifts for $^{•}CH_3$, $^{•}CH_2OH$, $^{•}CH(OH)CH_3$ and $^{•}C(O)CH_3$ spin adducts were 23.1, 22.6, 27.3 and 30.2 ppm, respectively [77].

The same system was also applied for the detection of phenoxy radicals, as an alternative to traditional EPR techniques. More specifically, the phenoxy radicals were produced via the oxidation of different phenols by $K_3Fe(CN)_6$. The ^{31}P NMR signals for the radical adducts of phenoxy radicals (PhO$^{•}$) were assigned and found to be located at 25.2 ppm [78]. Subsequently, this spin trapping system was applied to the oxidation of various phenols in the presence of peroxidases and 1-hydroxybenzotriazole (HBT) as a mediator: the 2,4,6-trichlorophenol and 2,4,6-tri-tert-butylphenol were oxidized and only phenoxy radical adducts were detected, whereas during the oxidation of 2,4-dimethylphenol and isoeugenol, other adducts were detected and related to radical delocalization [78].

This powerful system was also applied for the trapping of ketyl radicals, which are very difficult intermediates to be detected and quantified with traditional techniques (i.e. EPR). Ketyl radicals were initially produced using photochemical reactions of acetophenone, whose excited triplet state is able to abstract hydrogen from an H donor [79]. As such, the ^{31}P NMR signals for the radical adducts of the DIPPMPO spin trap with the ketyl radicals were assigned. Furthermore in efforts to confirm the structure of these adducts, their mass spectra and fragmentation patterns were carefully examined under GC-MS conditions [80]. Subsequently, the DIPPMPO spin trapping system was applied to the oxidation of 1-(3,4-dimethoxyphenyl)ethanol in the presence of horseradish peroxidase, hydrogen peroxide and 1-hydroxybenzotriazole as the electron carrier (mediator) [79]. Our work confirmed that the mechanism consists of a hydrogen abstraction reaction from the α position, involving the ketyl radical: during the oxidation, the hydroxyl, hydroperoxyl and ketyl radical intermediates were all detected.

These efforts demonstrate the efficacy of our methodology that provides for the first time a facile means for the detection of otherwise elusive radical species, with important implications in biology, chemistry and biochemistry.

CONCLUDING REMARKS

The selective and quantitative tagging of lignin with various heteronuclear spectroscopic reporter groups, followed by quantitative NMR data acquisition, has the potential to provide some truly unique insights toward understanding the complex and variable reactions of lignin under conditions of natural or commercial transformations. Many of the selective lignin chemical transformations reviewed here permit specific enquiries to be made on well-defined lignin functional groups. The described heteronulcear NMR approach has distinct advantages over its carbon and proton counterparts, since it allows for rapid quantitative acquisitions for well defined lignin moieties. This information, when coupled with careful ^{13}C NMR acquisitions, may offer a comprehensive detailed understanding of the complex lignin macromolecule. The remaining challenges of probing structural information of lignin present on the fiber, without the need of its prior isolation, may also be eventually addressed by modern advances of quantitative ^{31}P NMR of wood and pulps dissolved in ionic liquids.

REFERENCES

1. Brezny, R., and Schraml, J., "Silicon-29 NMR spectral studies of kraft lignin and related model compounds," *Holzforschung* **41**(5), 293–298 (1987).
2. Nieminen, M.O.J., Pukkinen, E., and Rahkamaa, E., "Determination of hydroxyl groups in kraft pine lignin by silicon-29 NMR spectroscopy," *Holzforschung* **43**(5), 303–307 (1989).
3. Herzog, H., Burba, P., and Buddrus, J., "Quantification of hydroxylic groups in a river humic substance by Si-29-NMR," *Fresenius J. Anal. Chem.*, **V354**, N3, 375–377 (1996).
4. Chen, C.-L., and Robert, D., in *Methods in Enzymology*, 161, Part B, pp. 137–174, W.A. Woods and S.T. Kellogg (eds.), Academic Press, New York (1988).
5. Potthast, A., Schiene, R., and Fischer, K., "Structural investigations of n-modified lignins by N-15-NMR spectroscopy and possible pathways for formation of nitrogen-containing compounds related to lignin," *Holzforschung* **50**(6), 554–562 (1996).
6. Knicker, H., Ludemann, H. D., and Haider, K., "Incorporation studies of NH4 + during incubation of organic residues by N-15-CPMAS NMR-spectroscopy," *Eur. J. Soil Sci.* **4**(3), 431–441 (1997).
7. Kögel-Knabner, I., "C-13 and N-15 NMR spectroscopy as a tool in soil organic matter studies," *Geoderma* **80**(3–4), 243–270 (1997).
8. Neirinck, V., Robert, D., and Nardin, R., "Use of indirect Hg-199 NMR detection for aromatic compounds of biological interest," *Mag. Reson. in Chem.* **31**(9), 815–822 (1993).
9. Kolehmainen, E., Paasivirto, J., Kauppinen, R., Otollinen, T., and Kasa, S., "^{119}Sn-NMR study on bis(tri-butiltin)oxide derivatives of phenolic compounds," *Intern. J. Environ. Anal. Chem.* **43**, 19–24 (1990).
10. Barrelle, M., "A new method for the quantitative fluorine-19 NMR spectroscopic analysis of hydroxyl groups in lignins," *Holzforschung* **47**(3), 261–267 (1993).
11. Barrelle, M., Fernandes, J. C., Froment, P., and Lachenal, D., "An approach to the determination of functional groups in oxidized lignins by fluorine-19 NMR," *J. Wood Chem. Technol.* **12**(4), 413–424 (1992).
12. Lachenal, D., Fernandes, J. C., and Froment, P., "Behavior of residual lignin in kraft pulp during bleaching," *J. Pulp Paper Sci.* **21**(5), J173–J177 (1995).

13. Sevillano, R. M., Mortha, G., Froment, P., Lachenal, D., and Barrelle, M., "^{19}F NMR spectroscopy for the quantitative analysis of carbonyl groups in lignin," 4th European Workshop Lignocellulosicsa and Pulp, Stresa, Italy, 292 (1996).

14. Ahvazi, B., and Argyropoulos, D. S., "^{19}F Nuclear magnetic resonance spectroscopy for the elucidation of carbonyl groups in lignins: Part I: Model compounds," *J. Agric. Food Chem.* **44**(8), 2167–2175 (1996).

15. Ahvazi, B., Crestini, C., and Argyropoulos, D. S., "^{19}F Nuclear magnetic resonance spectroscopy (NMR) for the quantitative detection and classification of carbonyl groups in lignins," *J. Agric. Food Chem.* **47**(1), 190–201 (1999).

16. Verkade, J. G., and Quin, L. D., "Phosphorus-31 NMR spectroscopy in stereochemical analysis of organic compounds and metal complexes," in *Methods of Stereochemical Analysis,* M. Grayson and E. J. Griffith (eds.), **7**, VCH Publ. (1987).

17. Anderson, R. C., and Shapiro, M. Y., "2-Chloro-4(R), 5(R)-dimethyl-2-oxo-1,3,2,-dioxaphospholane, a new chiral derivatizing agent," *J. Org. Chem.* **49**, 1304 (1984).

18. Johnson, C. R., Elliot, R. C., and Penning, T. D., "Determination of enantiomeric purities of alcohols and amines by a ^{31}P NMR technique," *J. Am. Chem. Soc.* **106**, 5019 (1984).

19. Schiff, D. E., Verkade, J. G., Metzler, R. M., Squires, T. G., and Vernier, C. G., "Determination of alcohols, phenols, and carboxylic acids using phosphorus-31 NMR spectroscopy," *Applied Spectroscopy* **40**(3), 348 (1986).

20. Wroblewski, A. E., Markuszewski, R., and Verkade J. G., "A novel application of ^{31}P NMR spectroscopy to the analysis of organic groups containing -OH, -NH, and -SH functionalities in coal extracts and condensates," *Am. Chem. Soc. Div. of Fuel Chem.* **32**(3), 202 (1987).

21. Wroblewski, A. E., Lensink, C., Markuszewski, R., and Verkade, J. G., "^{31}P NMR spectroscopic analysis of coal pyrolysis condensates and extracts for heteroatom functionalities possessing labile hydrogen," *Ener. Fuels* **2**, 765 (1988).

22. Lensink, C., and Verkade, J. G., "Identification of labile hydrogen functionalities in coal derived liquids by ^{31}P NMR spectroscopy," *Am. Chem. Soc. Div. of Fuel Chem.* **33**(4), 906 (1988).

23. Lucas, H. J., Mitchell, F. W., and Scully, C. N. "Cyclic phosphites of some aliphatic glycols," *J. Am. Chem. Soc.* **72**, 5491 (1950).

24. Argyropoulos, D. S., "^{31}P NMR in wood chemistry: A review of recent progress," *Research on Chemical Intermediates* **21**(3–5), 373–395 (1995).

25. Archipov, Y., Argyropoulos, D. S., Bolker, H. I., and Heitner, C., "^{31}P NMR spectroscopy in wood chemistry. Part I. Lignin model compounds," *J. Wood Chem. Technol.* **11**(2), 137–157 (1991).

26. Adler, E., Brunow, G., and Lundquist, K. "Investigation of the acid-catalysed alkylation of lignin by means of NMR spectroscopic methods," *Holzforschung* **41**, 199 (1987).

27. Argyropoulos, D. S., Bolker, H. I., Heitner, C., and Archipov, Y. "^{31}P NMR spectroscopy in wood chemistry. Part V. Qualitative analysis of lignin functional groups." *J. Wood Chem. Technol.* **13**(2),187–212 (1993).

28. Argyropoulos, D. S., Archipov, Y., Bolker, H. I., and Heitner, C., "^{31}P NMR spectroscopy in wood chemistry. Part IV. Lignin models: Spin lattice relaxation times and solvent effects in ^{31}P NMR," *Holzforschung* **47**, 50–56 (1993).

29. Granata, A., and Argyropoulos, D. S, "2-Chloro-4,4,5,5-Tetramethyl-1,3,2-Dioxaphospholane a reagent for the accurate determination of the uncondensed and condensed phenolic moieties in lignins," *J. Agric. Food Chem.* **43**(6), 1538–1544 (1995).

30. Sun, Y., and Argyropoulos, D. S., "Fundamentals of high pressure oxygen and low pressure oxygen-peroxide (Eop) delignification of softwood and hardwood kraft pulps: A comparison," *J. Pulp Paper Sci.* **21**(6), J185–190 (1995).

31. Jiang, Z. H., Argyropoulos, D. S., and Granata, A., "Correlation analysis of ^{31}P NMR chemical shifts with substituent effects of PHENOLS," *Magnetic Resonance in Chemistry* **33**, 375–382 (1995).
32. Mazúr, M., and Argyropoulos, D. S., "^{31}P NMR spectroscopy in wood chemistry. Part VII. Studies toward elucidating the phosphorus relaxation mechanism of phosphitylated lignins," *Cellulose Chem. Technol.* **29**(5), 589–601 (1995).
33. Argyropoulos, D. S., "Quantitative phosphorus-31 NMR analysis of lignins: A new tool for the lignin chemist," *Journal of Wood Chemistry and Technology* **14**(1), 45–63 (1994).
34. Faix, O., Argyropoulos, D. S., Robert, D., and Neirinck, V., "Determination of hydroxyl groups in lignins: Evaluation of ^{1}H-, ^{13}C-, ^{31}P-NMR, FTIR and wet chemical methods," *Holzforschung* **48**(5), 387–394 (1994).
35. Argyropoulos, D. S., "Quantitative phosphorus-31 NMR analysis of six soluble lignins," *J. Wood Chem. Technol.* **14**(1), 65–82 (1994).
36. Heuts, L., PhD Thesis, Royal Institute of Technology (1998).
37. Argyropoulos, D. S., Hortling, B., Poppius-Levlin, K., Sun, Y., and Mazur, M., "MILOX pulping, lignin characterization by ^{31}P NMR spectroscopy and oxidative degradation," *Nordic Pulp and Paper Research J.* **10**(1), 68–73 (1995).
38. Saake, B., Argyropoulos, D. S, and Faix, O., "Structural investigations of synthetic lignins (DHP's) as a function of molar mass and mode of preparation," *Phytochemistry* **43**(2), 499–507 (1996).
39. Jurasek, L., "Molecular modelling of fibre walls," *J. Pulp Paper Sci.* **24**(7), 209–212 (1998).
40. Jiang, Z., and Argyropoulos, D. S., Coupling ^{31}P NMR with the Mannich reaction toward the quantitative analysis of lignins*" Can J. Chem.* **76**(5), 612–622 (1998).
41. Smit, R., Suckling, I. D., and Ede, R. M., "A new method for the quantification of condensed and uncondensed softwood lignin structures." Proceedings of 9th International Symposium on Wood and Pulping Chemistry. Montreal, Canada, June 1997, Vol. 1, L4-1–L4-6.
42. Lu, F., and Ralph, J., "Derivatization followed by reductive cleavage (DFRC method), new method for lignin analysis: protocol for analysis of DFRC monomers," *J. Agric. Food Chem.* **45**, 2590–2592 (1997).
43. Lu, F., and Ralph, J., "DFRC method for lignin analysis. 1. new method for β-aryl ether cleavage: lignin model studies," *J. Agric. Food Chem.* **45**, 4655–4660 (1997).
44. Lu, F., and Ralph, J., "The DFRC method for lignin analysis. 2. Monomers from isolated lignins," *J. Agric. Food Chem.* **46**, 547–552 (1998).
45. (a) Tohmura, S., and Argyropoulos, D. S., "Determination of arylglycerol-β-aryl ethers and other linkages in native and technical lignins," *J. Agric. Food Chem.* **49**(2), 536–542 (2001); (b) Argyropoulos, D. S., Jurasek, L., Krištofová, L., Xia, Z., Sun, Y., and Paluš, P., "Abundance and reactivity of dibenzodioxocins in softwood lignin," *J. Agric. and Food Chem.* **50**, 658–666 (2002); (c) Guerra, A., Norambuena, M., Freer, J., and Argyropoulos, D. S., " Determination of arylglycerol-β-aryl ether linkages in enzymatic mild acidolysis lignins (EMAL): Comparison of DFRC/^{31}P NMR with thioacidolysis," *Journal of Natural Products* **71**(5), 836–841 (2008).
46. Gellerstedt, G., Pranda, J., and Lindfors, E.-L., "Structural and molecular properties of residual birch kraft lignins," *J. Wood Chem. Technol.* **14**(4), 467–482 (1994).
47. Jiang, Z., and Argyropoulos, D. S., "Isolation and characterization of residual lignin in kraft pulp," *J. Pulp Paper Sci.* **25**(1), 25–29 (1999).
48. (a) Ahvazi, B., Pageau, G. and Argyropoulos, D. S., "On the formation of diphenyl-methane structures in lignin under kraft, EMCC, and soda pulping conditions," *Can. J. Chem.* **76**, 506–512 (1998); (b) Jurasek, L., Kristofova, L., Sun, Y., and Argyropoulos, D. S., "Alkaline oxidative degradation of diphenylmethane structures: Activation energy and computational analysis of the reaction mechanism," *Can. J.Chem.* **79**, 1394–1401 (2001).

49. Froass, P. M., Ragauskas, A. J., and Jiang, J.-E., "Chemical structure of residual lignin from kraft pulp," *J. Wood Chem. Technol.* **16**(4), 347–365 (1996).

50. Jiang, Z., and Argyropoulos, D. S., "The stereoselective degradation of arylglycerol-β-aryl ethers during kraft pulping," *J. Pulp Paper Sci.* **20**(7), 183–188 (1994).

51. Ahvazi, B., and Argyropoulos, D. S., "Thermodynamic parameters governing the stereoselective degradation of arylglycerol-β-aryl ether bonds in milled wood lignin under kraft pulping conditions," *Nordic Pulp and Paper Res. J.* **12**(4), 282–288 (1997).

52. Liu, Y., Carriero, S., Pye K., and Argyropoulos, D. S., "A comparison of the structural changes occurring in lignin during alcell and kraft pulping of hardwoods and softwoods," in *Lignin: Historical, Biological and Materials Perspectives,* W. G. Glasser, R. A. Northey, T.P. Schultz (eds.), ACS Symposium Series No. 742, Chapter 22, 447–464 (1999).

53. Sun, Y., and Argyropoulos, D.S., "Fundamentals of high-pressure oxygen and low-pressure oxygen-peroxide (Eop) delignification of softwood and hardwood kraft pulps: A comparison," *J. Pulp Paper Sci.* **21**(6), J185–J190 (1995).

54. Asgari, F., and Argyropoulos, D. S., "Fundamentals of oxygen delignification. Part II. Kinetics of functional group formation/elimination in residual kraft lignin," *Can. J. Chem.* **76**, 1606–1615 (1998).

55. Renard, J. J., Mackie, D. M., and Bolker, H. I., *Paperi Ja Puu* **11**(1) (1975).

56. Renard, J. J., Mackie, D. M., Bolker, H. I., and Clayton, D. W., "Delignification of wood using pressurized oxygen. II. Mechanism of transfer of the reagents into the wood structure," *Cellulose Chem. Technol.* **9**(4), 341–352 (1975).

57. Gellerstedt, G., and Lindfors, E.-L., "Hydrophilic groups in lignin after oxygen bleaching," *Tappi J.* **70**(6), 119–122 (1987).

58. Sun, Y., and Argyropoulos, D. S., "A comparison of the reactivity and efficiency of ozone, chlorine dioxide, dimethyldioxirane and hydrogen peroxide with residual kraft lignin," *Holzforschung* **50**(2), 175–182 (1996).

59. Senior, D.J., Hamilton, J., Ragauskas, A.J., Sealey, J., and Froass, P., "Interaction of hydrogen peroxide and chlorine dioxide stages in ECF bleaching," *Tappi J.* **81**(6), 170–177 (1988).

60. Van Lierop, B., Jiang, Z. H., Chen, J., Argyropoulos, D. S., and Berry, R. M., "On the efficiency of hydrogen peroxide use in ECF bleaching," *J. Pulp Paper Sci.* **26**(7), 255–259 (2000).

61. Argyropoulos, D. S., and Sun, Y., "Photochemically induced solid-state degradation, condensation, and rearrangement reactions in lignin model compounds and milled wood lignin," *Photochem. and Photobiol.* **64**(3), 510–517 (1996).

62. Ramirez, F., "Condensations of carbonyl compounds with phosphite esters," *Pure and Appl. Chem.* **9**, 337–369 (1964).

63. Lebo, S. E., and Lonsky, W.F.W., "The occurrence and light induced formation of *ortho*-quinonoid lignin structures in white spruce refiner mechanical pulp," *J. Pulp Paper Sci.* **16**(5), J139–J143 (1990).

64. Argyropoulos, D. S., Heitner, C., and Morin, F. G., "^{31}P NMR spectroscopy in wood chemistry. Part III. Solid state ^{31}P NMR of trimethyl phosphite derivatives of chromophores in mechanical pulps," *Holzforschung* **46**(3), 211–218 (1992).

65. Argyropoulos, D. S., and Heitner, C., "^{31}P NMR spectroscopy in wood chemistry. Part VI. Solid state ^{31}P NMR of trimethyl phosphite derivatives of chromophores and carboxylic acids present in mechanical pulps: A method for the determination of *ortho*-quinones," *Holzforschung* **48** (Suppl.), 112–116 (1994).

66. Argyropoulos, D. S., and Zhang, L., "On the semiquantitative determination of quinonoid structures in soluble lignins by ^{31}P NMR," *J. Agric. Food Chem.* **46**(11), 4628–4634 (1998).

67. Gierer, J., and Imsgard, F., "The reactions of lignins with oxygen and hydrogen peroxide in alkaline media," *Svensk. Papperstidn.* **80**(16), 510–518 (1977).
68. Pudovik, A. N., Gazizov, T. K., and Kharlamov, V. A., "The reaction of trialkyl phosphites with mineral and carboxylic acids," *Dolkady Academii Nauk SSSR* **227**(2), 376–379 (1976).
69. Krutikov, V. I., Aleinikov, S. F., Kalaverdova, E. A., and Lavrent'ev, A. N., "Some characteristics of the formation of adducts of carboxylic acids with phosphorus acid esters," *Zh Obshch. Khim.* **58**(11), 2490–2493 (1988).
70. Huyser, E. S., and Dietter, J. A., "Synthesis and acetolysis of mixed trialkyl phosphites," *J. Org. Chem.* **33**, 4205–4210 (1968).
71. Argyropoulos, D. S., Heitner, C., and Schmidt, J. A., "Observation of quinonoid groups during the light-induced yellowing of softwood mechanical pulp," *Res. Chem. Intermed.* **21**(3–5), 263–274 (1995).
72. Imsgard, F., Falkehag, S. I., and Kringstad, K. P., "On possible chromophoric structures in spruce wood," *Tappi* **54**(10), 1680 (1971).
73. Argyropoulos, D. S., Sun, Y, Berry, R. M., and Bouchard, J., "Reactions of dimethyldioxirane with lignin model compounds," *J. Pulp Paper Sci.* **22**(3), J84–J90 (1996).
74. (a) Guerra, A., Filpponen, I., Lucia, L., and Argyropoulos, D. S., "A comparative evaluation of three lignin isolation protocols, with different wood species" *J. Agric. Food Chem.* **54**(26), 9696–9705 (2006); (b) Guerra, A., Filpponen, I., Lucia, L., Saquing, C., Baumberger, S., and Argyropoulos, D. S., "Toward a better understanding of the lignin isolation process from wood," *J. Agric. Food Chem.* **54**(16), 5939–5947 (2006); (c) Wu, S., and Argyropoulos, D. S., "An improved method for isolating lignin in high yield and purity," *J. Pulp Paper Sci.* **29**(7), 235–240 (2003).
75. King, A.W.T., Zoia, L., Filpponen, I., Olszewska, A., Xie, H., Kilpeläinen, I., Argyropoulos, D.S. *In-situ* Determination of Lignin Phenolics and Wood Solubility in Imidazolium Chlorides using ^{31}P NMR"; *J. Agric. Food Chem.* 2009, 57, 8236–8243.; King, A., Järvi, P., Kilpeläinen, I., Heikkinen, S., Argyropoulos., D. S., "Hydrophobic Interactions Determining Functionalized Lignocellulose Solubility in Dialkylimidazolium Chlorides, as Probed by ^{31}P NMR"; DOI: 10.1021/bm8010159, Biomacromolecules, *10*, 458–463 (2009).
76. Kilpelainen, I., Xie, H., King, A., Granström, M., Heikkinen, S., and Argyropoulos, D. S., "Dissolution of wood in ionic liquids," *J. Agric. Food Chem.* **55**(22), 9142–9148 (2007).
77. Argyropoulos, D. S., Li, H., Gaspar, A. R., Smith, K., Lucia, L., Rojas, O. J., "Quantitative ^{31}P NMR Detection of Oxygen-centered and Carbon-centered Radical Species." *Bioorganic & Medicinal Chemistry*, Volume 14, Issue 12, 15 June, pp. 4017–4028, (2006).
78. Zoia, L. Argyropoulos., D. S., "Phenoxy Radical Detection Using Quantitative ^{31}P NMR Spin Trapping", *J. Phys. Org. Chem.*, 22 1070–1077, (2009); www.interscience.wiley.com) DOI 10.1002/poc.1561, 2009.
79. Zoia, L. Argyropoulos., D. S., "Ketyl Radical Detection Using Quantitative ^{31}P NMR Spin Trapping"; *J. Phys. Org. Chem.* 2009, DOI:10.1002/poc. 1561.
80. Zoia, L., Argyropoulos D. S., "Characterization of free radical spin adducts of the DIPPMPO using Mass Spectrometry and ^{31}P NMR", European Journal of Mass Spectrometry, To appear January 2010, DOI:10.1155/ejms. 1062.

7 Functional Groups and Bonding Patterns in Lignin (Including the Lignin-Carbohydrate Complexes)

Gösta Brunow and Knut Lundquist

CONTENTS

INTRODUCTION

Phenylpropane units of types **1** (guaiacylpropane), **2** (syringylpropane), and **3** (*p*-hydroxyphenylpropane) are the main building blocks in lignins; these are shown in Figure 7.1, along with the designation of the atoms in the phenylpropane units. The proportions of **1–3** differ with the botanical origin of the lignin. The biosynthesis of lignins is considered to proceed via an oxidative polymerization of three primary precursors: the *p*-hydroxycinnamyl alcohols **4–6**. (Chapter 1 takes up the chemistry of several common oxidative polymerization pathways.) It is noteworthy that practically all the types of structural elements detected in lignins are formed *in vitro* in enzymatic oxidation experiments with the *p*-hydroxycinnamyl alcohols **4–6** and phenolic lignin model compounds [1–6].

The two main routes (reaction routes A and B) for the incorporation of **4–6** in the growing lignin polymer are shown in Figure 7.2. For simplicity only coniferyl alcohol (**4**) and guaiacylpropane units (**1**) are considered in Figure 7.2. Route A depicts the dimerization of monolignols, presumably the first step in lignin biosynthesis; route B represents the further growth of the polymer by coupling of monolignol radicals to the phenolic units in the lignin [3,4]. Studies of lignin degradation products provide evidence for both reaction routes but reaction route B is in all probability the

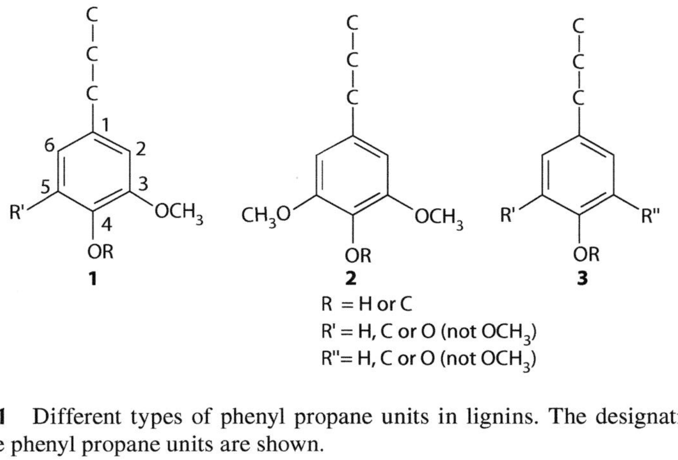

FIGURE 7.1 Different types of phenyl propane units in lignins. The designations of the atoms in the phenyl propane units are shown.

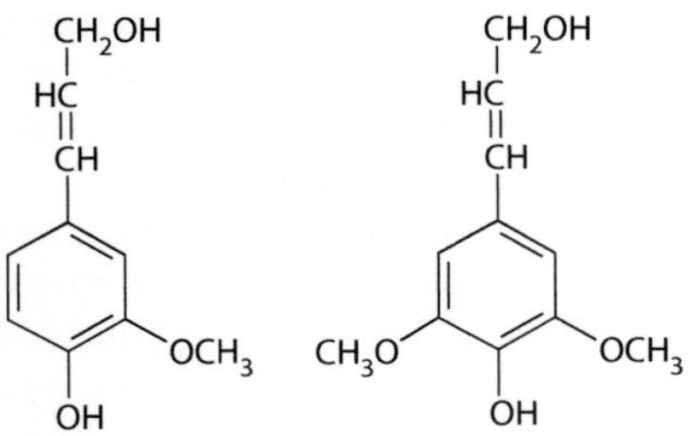

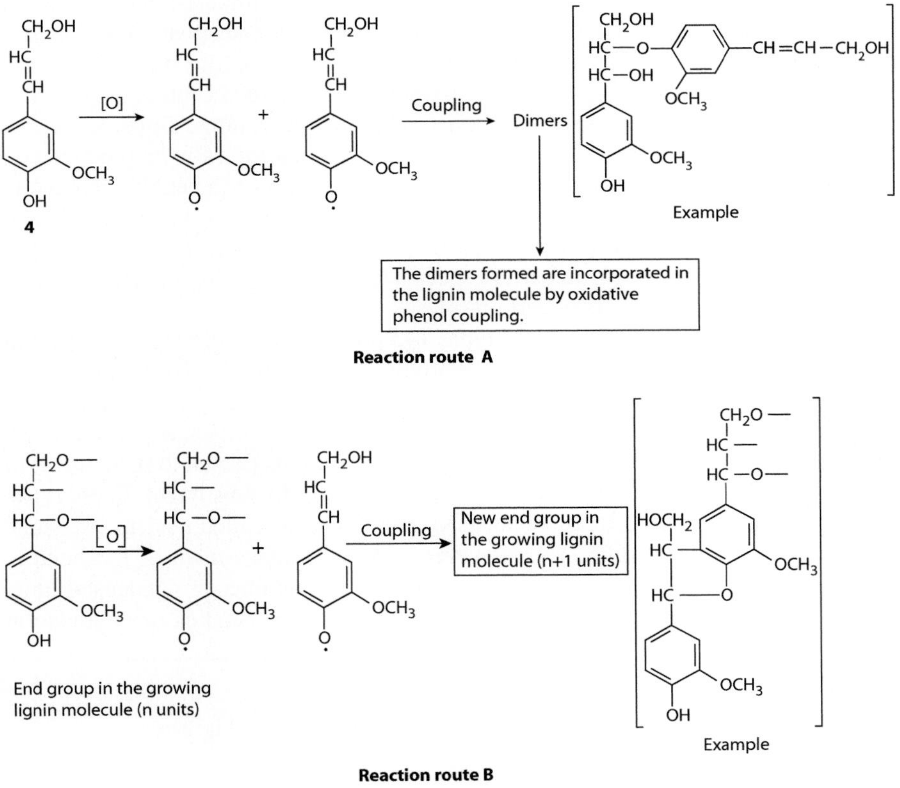

FIGURE 7.2   The two main routes (reaction routes A and B) for the incorporation of **4–6** in the growing lignin polymer.

most important one; the term "end-wise polymerization" for this reaction route was introduced by Sarkanen [3]. The prevalence of reaction route B (Figure 7.2) strongly influences the proportions of different types of bonding patterns and functional groups in lignins. The isolation of numerous dimeric lignin degradation products derived from lignin structures consisting of different types of units shows that cross-coupling of units **1–3** also occurs during the biosynthesis.

It follows from the types of reactions shown in Figure 7.2 that the structural elements in lignins are not linked to each other in any particular order, although there is the possibility that certain sequences of units are more probable than others. In addition, lignins will not exhibit optical activity, a fact that has recently been demonstrated [7,8]. It is likely that the particular reaction conditions prevailing in the plants influence the outcome of the polymerization (e.g., the relative importance of routes A and B, Figure 7.2). These conditions would be, among other things, the pH, the oxidation potential of the polymerizing enzyme(s), and the relative concentrations of 1–3, which is dependent on plant species and morphological regions.

Furthermore the presence of carbohydrates in the plant cells is not taken into account. There are strong indications of the occurrence of linkages between lignin and carbohydrates (see The Lignin-Carbohydrate Complex section). Also, certain types of lignins are esterified with phenolic acids, e.g., grass lignins with p-coumaric acid and aspen lignin with p-hydroxybenzoic acid (see Esterified Units section). It has been found that there are some types of units in lignins, for example dihydroconiferyl alcohol units (4, with a saturated side chain), which cannot logically be produced on oxidation of p-hydroxycinnamyl alcohols. The participation of phenolic precursors other than 4–6 in the oxidative polymerization leading to lignins has been stressed in recent studies [9–13]. However, the participation of such precursors seems to be quantitatively important only in certain species (e.g., incorporation of ferulic acid in grass lignins [11]), or mutants of certain species.

The proportions of structural units of types 1–3 provide a basis for the classification of lignins. The compositions of some important classes of lignins based on this criterion are listed in Table 7.1.

Tracer studies [14] indicate that roughly 50% of the 5-positions, and a small percentage of the 2- and 6-positions, in the guaiacyl units (units related to structure 1) in softwood lignin, as well as hardwood lignin, carry a C or O-substituent. These types of units are often referred to as "condensed units." The so-called nucleus exchange method gives similar results [15]. However, a recent reassessment of the nucleus exchange method [16] suggests that the values for the number of condensed units estimated by this method can be expected to be too low. Studies of degradation

---

TABLE 7.1
Approximate Composition of Some Important Classes of Lignins

	1	2	3
Softwood lignin	95%	1%	4%
Hardwood lignin*	≈ 50%	≈ 50%	≈ 2%
Grass lignin**	70%	25%	5%
Compression wood lignin	70%	≈ 0%	30%

Source: Based on Sarkanen, K.V. and Ludwig, C.H., *Lignins: Occurrence, Formation, Structure and Reactions.* New York: Wiley-Interscience, 1971; Adler, E., *Wood Sci Technol*, 11, 169–218, 1977.

*Most hardwood lignins are composed of about equal amounts of 1 and 2, but quite a few exceptions are known. Some hardwood lignins are esterified with p-hydroxybenzoic acid.

**p-Coumaric acid attached by ester linkages not included.

products obtained on permanganate oxidation of lignins yield results that point to a smaller proportion of condensed guaiacyl units, namely condensed guaiacyl units/ total number of guaiacyl units ≈ 0.4 [17,18]. A combination of degradation (thioacidolysis) and subsequent examinations by gel permeation chromatography constitutes an alternative way to determine of the extent of condensation [19].

The lignin polymer is expected to be branched and cross-linking may occur to some extent. If the lignin were a linear polymer the number of interconnections per unit in a molecule consisting of n units is $(n-1)/n$ and cannot exceed 1. This is true even if branching occurs. Cross-linking leads to the formation of rings of units. A consequence of this is that the number of interconnections per unit increases (values larger than 1 are possible). Available lignin data [3,4] suggest that the number of rings is small and it can therefore be assumed that the number of interconnections per unit is close to 1.

Lignins have been studied both *in situ* and as isolated samples. An isolated lignin sample is not chemically but operationally defined, i.e., it is defined by the starting material and the operations applied to it in order to obtain the sample. This implies that lignin preparations isolated from botanically and morphologically equivalent wood/plant samples using different methods may have different properties, such as the number of phenolic groups and its molecular-weight distribution. It also means that materials other than lignins may be present in the samples. As a concrete example of this type of ambiguity, proteins and tannins both give rise to so-called Klason lignin [3]. Furthermore, attention has to be paid to the possibility that chemical modifications have occurred during the isolation procedures.

Some of the drawbacks faced in connection with the use of isolated lignin samples can be avoided by studying the lignin *in situ*. In this case other constituents in the samples may interfere. Furthermore, study of solid wood/plant materials excludes the use of several of the most powerful analytical techniques, since they are only applicable to examinations of solutions. In some cases the biological microstructure would be another complication when lignins are studied *in situ*. In many such studies of lignins, scientists have based their results on particular color reactions; materials giving UV absorption at specific wavelengths and materials giving rise to certain NMR signals. This is adequate in some contexts but, in other cases, may lead to erroneous results. In most cases elucidation of the structures of lignins requires studies of lignins *in situ* as well as of isolated lignin samples.

In this chapter we have tried to amalgamate the results from studies of lignin *in situ* with those obtained in studies of isolated lignins. However, most of the data given emerge from examinations of milled wood lignin (MWL) [20,21]. Comparative studies of wood and MWL [17,22–24] suggest that MWL in most respects is representative for the lignin in wood. The nature of the functional groups and bonding patterns in lignins are currently well established. However, many results regarding the quantitative contribution of different types of structural elements are contradictory and controversial. This is not surprising, since it is very difficult to obtain reliable quantitative lignin data.

An example of the types of difficulties encountered in quantitative lignin analysis can be found in the calculation of the abundance of condensed units in lignins based on ^{1}H NMR spectroscopy [25,26]. The calculation is based on an integration of the signal due to aromatic protons (corrected for the contribution of certain vinyl protons), together with a determination of the number of phenylpropane units and consideration of the distribution of units **1–3** in the sample. A rather optimistic estimate of the experimental error in a determination of the number of aromatic protons/unit by this method would be ± 5%. If, for instance, the number of aromatic protons/unit is found to be 2.5 in a softwood lignin sample (composition, see Table 7.1), a 5% error makes the determination of the extent of condensation uncertain (50% ± 12.5%). In this, and in many other cases, quantification of functional groups and bonding patterns in lignins leads to uncertain results, in spite of basically sound approaches. The problems related to lignin analysis are outlined in [27].

BONDING PATTERNS AND FUNCTIONAL GROUPS

GENERAL

Biosynthetic considerations are in many cases one of the arguments for the existence of particular structural units in lignins. This has been briefly commented on in the introduction to this chapter and in Chapter 1. Biosynthetic arguments are only pointed out in some cases in the following. Each one of the types of units discussed below can be expected to be present in all the different classes of lignins listed in Table 7.1. For simplicity the different types of lignin units (framed in the formulas) are in general depicted as noncondensed guaiacylpropane units (**1**) but when applicable the formulas also represent the other types of phenylpropane units. Data given for hardwood lignins refer to such lignins that are composed of similar amounts of units of types **1** and **2**. The attachment of lignin units to carbohydrates is treated in The Lignin-Carbohydrate Complex section.

SURVEY OF DIFFERENT TYPES OF LIGNIN UNITS

Arylglycerol Units Attached to an Adjacent Unit by a β–O–4 Linkage

Evidence for the occurrence of a β–O–4 linkage (**7**) as an important lignin structural feature was obtained during the fifties by the synthesis of appropriate model compounds and comparisons of their reactions with lignin reactions. Both models and lignins gave, for instance, so called Hibbert ketones on acid degradation [28]. The existence of lignin structures of type **7** was later confirmed in numerous studies. It has been shown that all the three types of phenylpropane units (**1–3**) participate in structures of the arylglycerol β-aryl ether type. This is, for instance, demonstrated by the isolation of compounds **8** from acidolysis [29], compounds **9** from thioacidolysis [23], and compounds **10** from DRFC degradations [30].

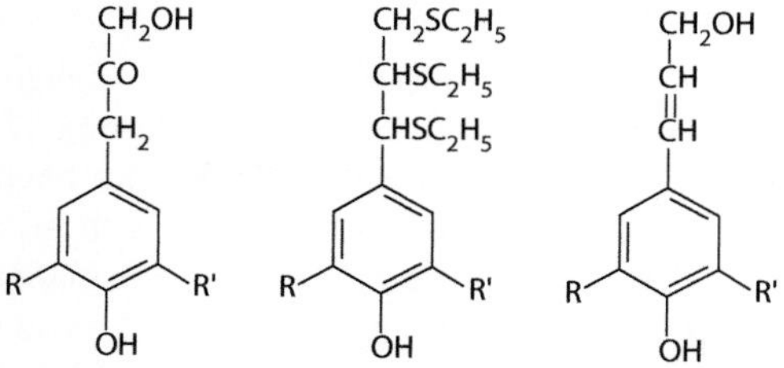

CH$_2$OH   CH$_2$SC$_2$H$_5$ CH$_2$OH
CO   CHSC$_2$H$_5$ CH
CH$_2$ CHSC$_2$H$_5$ CH

R — (ring) — R'   R — (ring) — R'   R — (ring) — R'

OH   OH   OH

8   **9**   **10**

R = R' = H
or
R = H, R' = OCH$_3$
or
R = R' = OCH$_3$

The distribution of the diastereomeric forms (**7a** and **7b**) of **7** in lignins has been studied by NMR spectral methods. The results concur in the respect that about equal amounts of the diastereomeric forms are found in softwood lignins and that there is a predominance of the *erythro* form (**7a**) in hardwood lignins [26,31–34]. Stereochemical studies, based on the formation of threonic acid (**11**) and erythronic acid (**12**) after ozonation [35–37], are largely in accordance with the NMR spectroscopic examinations. The distribution of diastereomeric forms of arylglycerol units attached to guaiacylpropane (**1**) units and syringylpropane (**2**) units in a hardwood lignin has been studied using the 2D INADEQUATE experiment [38]. The results show that the predominance of *erythro* forms is due to the presence of large amounts of *erythro* forms attached to syringylpropane units (**13a**). NMR studies suggest that there are 30–40% units of type **7** in softwood lignins [39–41] and 40–50% such units in hardwood lignins [40,42]. Most of the units of type **7** in hardwoods are linked to syringylpropane units (**13**) [38].

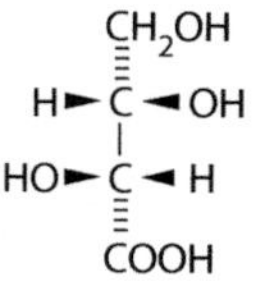

CH$_2$OH   CH$_2$OH
H — C — OH   H — C — OH
HO — C — H   H — C — OH
COOH   COOH

11 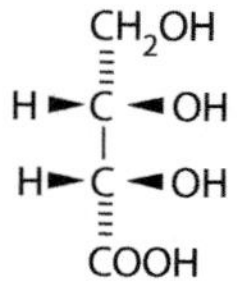   **12**

OCH$_3$ OCH$_3$
CH$_2$OH   CH$_2$OH
H — C — O — (ring) — C—C—C   H — C — O — (ring) — C—C—C
H — C — OH   OCH$_3$ HO — C — H   OCH$_3$

OCH$_3$ OCH$_3$
O   O

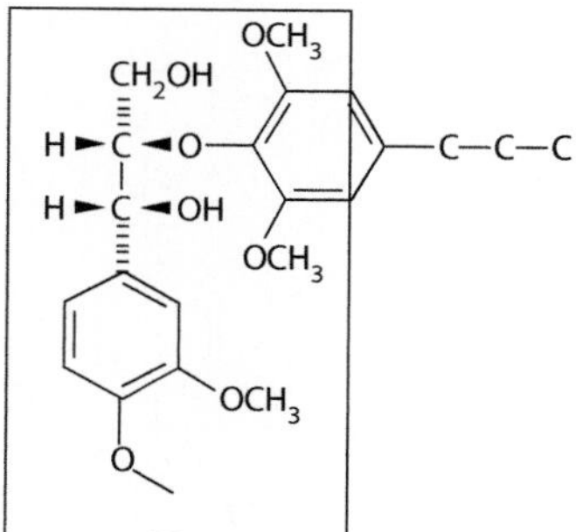

13a 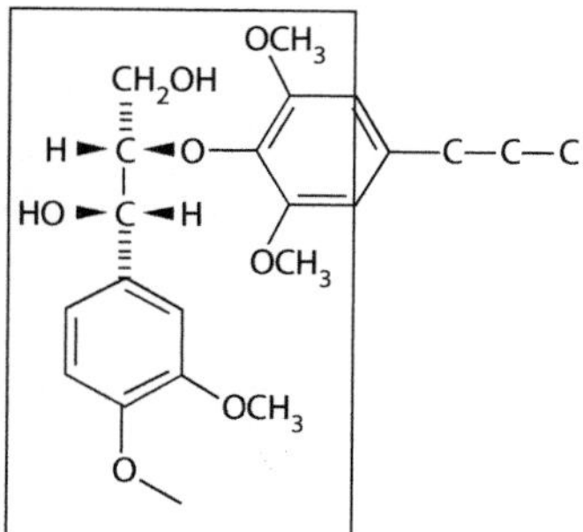   **13b**

Noncyclic Benzyl Aryl Ethers

Oxidation experiments with coniferyl alcohol lead to the conclusion that noncyclic benzyl aryl ethers (**14**, Figure 7.3) should be present in lignins [2]. The liberation of phenolic groups on mild acidolytic treatments of MWL has been taken as evidence for the occurrence of units of type **14** (most of them with an etherified phenolic group) in lignins [43,44]. There are, however, alternative explanations for the liberation of phenolic groups [45,46]. Studies using ^{1}H NMR show that units of type **14** are present in very small amounts, if at all, in MWL [47,48]. Evidence for their existence in MWL has been obtained in recent 3D NMR spectroscopic studies [49]. Studies

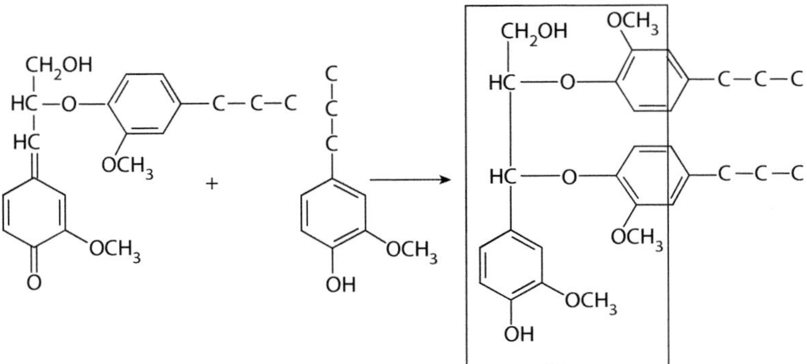

FIGURE 7.3   Formation of non-cyclic benzyl aryl ethers by addition of a phenolic unit to a quinone methide intermediate formed during lignin biosynthesis.

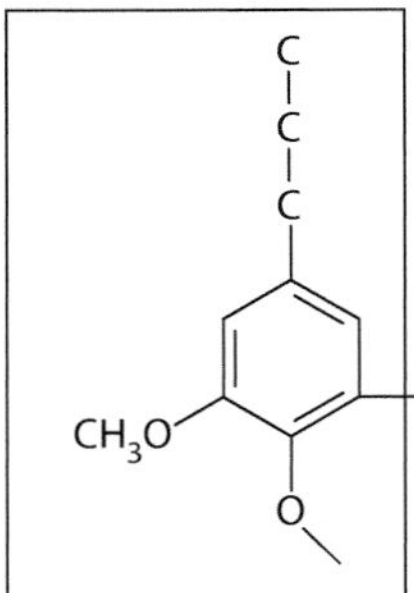

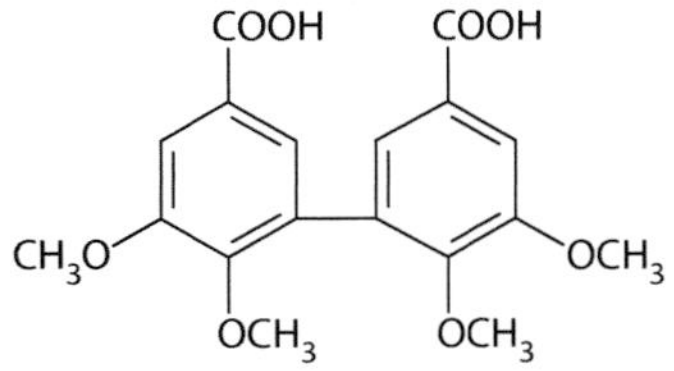

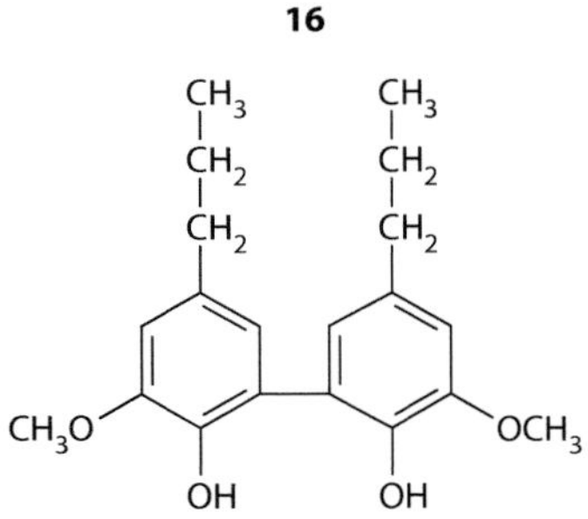

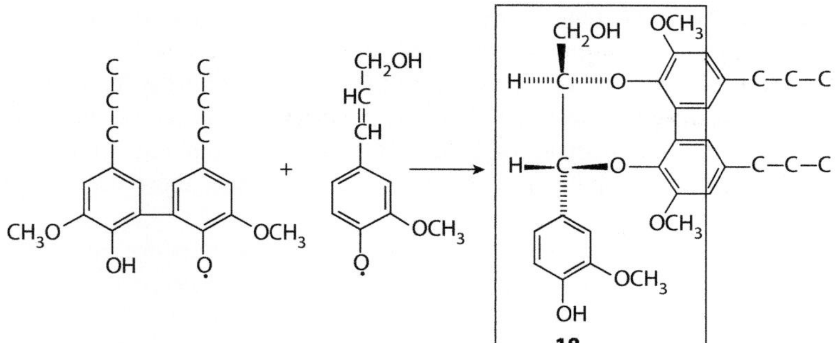

FIGURE 7.4 Formation of dibenzodioxocin structures during lignin biosynthesis. The phenol group in unit **18** may be etherified.

of lignin *in situ* also point to the presence of only a small number of type **14** units [50]. The pH-dependence of addition reactions to quinone methides may provide an explanation for the low number of type **14** units in lignins [51].

Biphenyl, Dibenzodioxocin, and Diaryl Ether Structures

The occurrence of biphenyl units (**15**) in lignins has been demonstrated in studies of lignin degradation products. Methylation followed by permanganate oxidation gives dehydrodiveratric acid (**16**) [17]. Degradation by reductive methods gives dehydrodicoerulignol (**17**) [22,52,53]. This provides unambiguous proof of the occurrence of phenylpropane units in lignins attached to each other by a biphenyl linkage. Pew [54] attempted to estimate the number of biphenyl units in softwood lignin by a UV spectrometric method. He concluded, "coniferous lignin may well contain 25% or more of biphenyl-linked units." Permanganate oxidation studies suggested that 19% of the lignin units are of the biphenyl type (**15**) in softwood lignin [17]. This estimate is based on results obtained on permanganate oxidation of lignin pretreated by cupric oxide oxidation.

Biphenyl coupling occurs to some extent on cupric oxide oxidation of lignin model compounds [55,56]. This indicates that the value derived from permanganate oxidation studies is slightly too high. On the basis ^{13}C NMR examinations, Drumond et al. [57] arrived at a higher value for the frequency of biphenyl units (24–26%). The accuracy of this estimate can be questioned, since no separate signals from biphenyl units can be discerned in the spectra.

The estimation of biphenyl units (**15**) is complicated by the recent finding [58] that there are significant amounts of dibenzodioxocin structures in lignins; the formation of such structures during lignin biosynthesis is shown in Figure 7.4. Conclusive evidence of the occurrence of units of type **18** has been obtained by NMR spectral studies [46]. Based on studies of the liberation of phenolic groups in biphenyl structures on treatments leading to cleavage of β-ethers Argyropoulos et al. [59] concluded that the number of units of type **18** was 4% in a softwood

MWL sample. The determination relies on the correctness of a number of assumptions. This implies that about 8% biphenyl units may be present in dibenzodioxocin structures. Model compound studies [46,60] indicate that the biphenyl units in dibenzodioxocin structures are included in the estimates of biphenyl units by NMR spectroscopy and permanganate oxidation. The number of biphenyl units is smaller in hardwood lignins. Permanganate oxidation suggests that there are 9% units of type 15 in birch lignin [18].

The occurrence of diaryl ether structures in lignin is indicated by the formation of acid 19 on permanganate oxidation [2] and was later confirmed in studies of lignin degradation by reductive methods [22,52,53]. Permanganate oxidation studies suggest 3.5% units of type 20 in softwood lignin [17] and significantly larger amounts of such units (6.5%) in hardwood lignin [18].

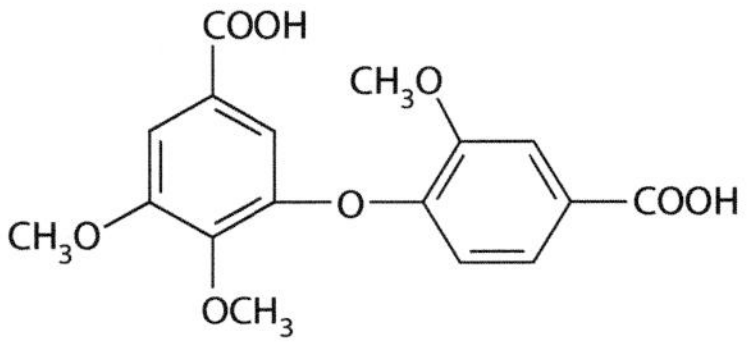

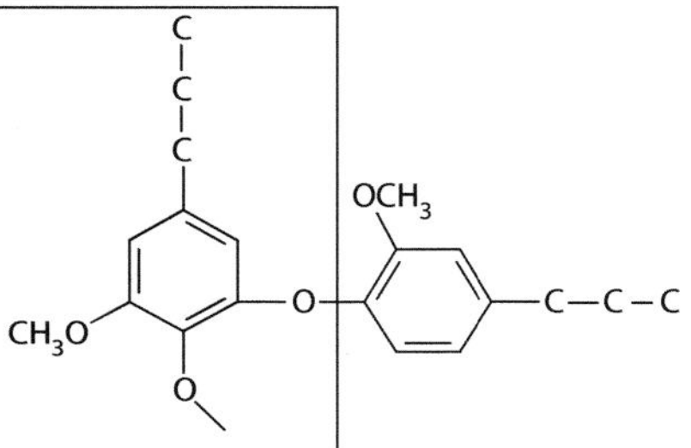

Phenylpropane Units Attached to an Adjacent Unit by a β–5 Linkage

Units linked to an adjacent unit by a β–5 linkage (21) are present in phenylcoumaran structures. Other types of lignin structures containing a β–5 linked unit have been speculated to occur in lignin [61,62], but conclusive evidence for this has not been obtained.

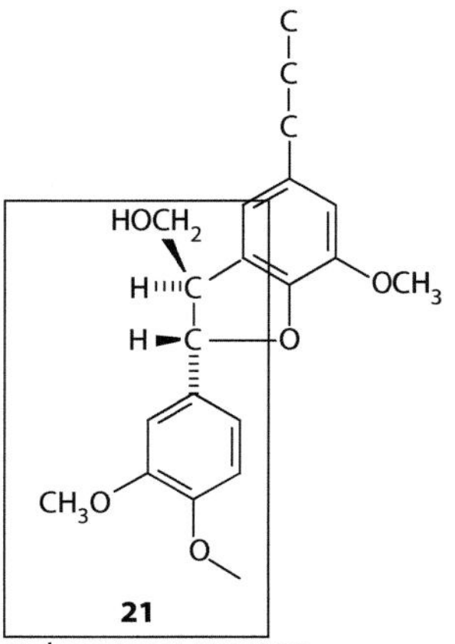

Phenylcoumaran structure

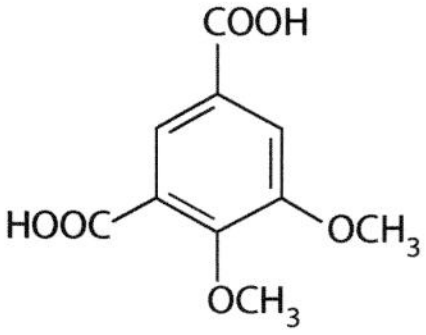

Freudenberg and coworkers [2] have shown that permanganate oxidation of methylated lignin gives isohemipinic acid (**22**). This acid can be expected to originate from β–5 linked units; labeling studies have shown that this is true to some extent [2,63]. Acidolysis studies provided evidence for the occurrence of a significant number of phenylcoumaran structures in softwood lignin [61]. Ozonolysis studies [64] have shown that phenylcoumaran structures in lignins have primarily the *trans*-configuration, as shown in unit **21**. Attempts to detect the *cis*-form in spruce lignin by ^{1}H NMR spectroscopy based on model compound data [65] have failed [66].

The occurrence of phenylcoumaran structures involving units of type **21** in lignins has been confirmed in a large number of studies of degradation products [e.g., 22, 67] and in several ^{1}H NMR and ^{13}C NMR spectroscopic studies. In addition, UV spectroscopic acidolysis studies [61] indicated that there are about 10% units of type **21** in spruce lignin. This figure should be adjusted somewhat downwards because of interference of diguaiacylstilbene formed from β–1 structures on acid treatment [29, 67]. Recent acidolysis studies point to a frequency of 6–9% units in spruce lignin [68]. Permanganate studies suggest 12% β–5 linked units [17], which agrees fairly well with results obtained from ozonolysis studies [64]. The number of units of type **21** in hardwood lignins is comparatively small [18,47,69].

Phenylpropane Units Attached to an Adjacent Unit by a β–β Linkage

Units linked to an adjacent unit by a β–β linkage [23] are present in pinoresinol structures and in analogous structures **24** and **25**. The isolation of dilactone **26** from lignins degraded by nitric acid provides proof of the occurrence of structures of the pinoresinol type [70]. Solvolytic [29,71,72] and reductive [22,52,53] degradation studies provide evidence of the occurrence of different types of pinoresinol-related structures [**24** and **25**] in lignins.

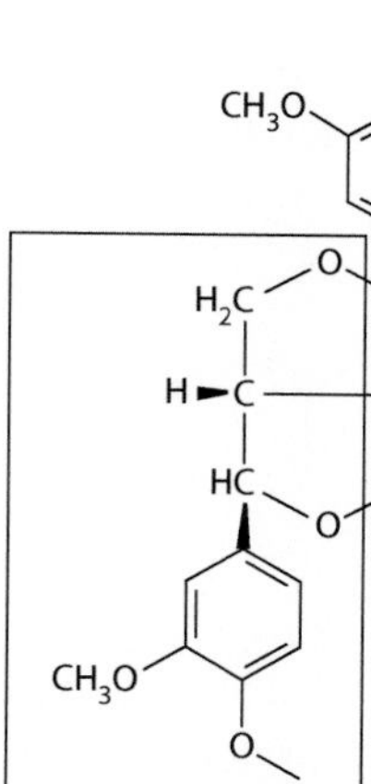

23
Pinoresinol structure

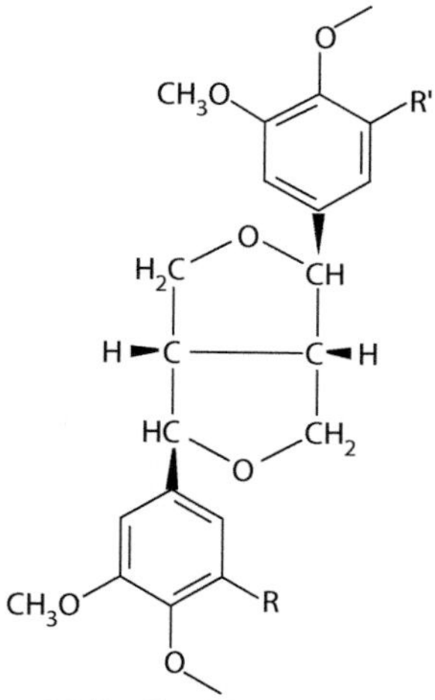

24 R = R' = OCH₃
25 R = OCH₃, R' = H

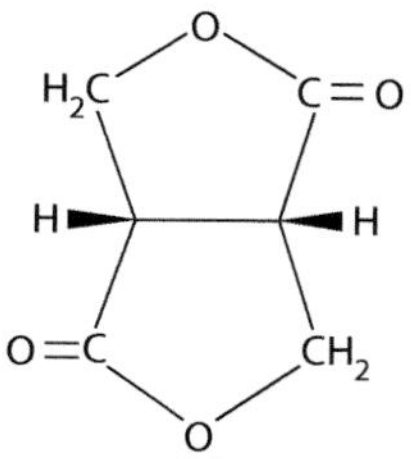

26

Both pinoresinol and syringaresinol structures have been detected in lignins by a variety of NMR spectroscopic techniques [see, e.g., 40,47,73,74]. Ogiyama and Kondo have determined the abundance of units of type **23** in a softwood lignin to 5–10% [70]. From the reported dilactone yield [70], one can estimate that the number of units of type **23** is about 10% in hardwood lignin. NMR spectral studies suggest somewhat lower values (≈ 7%) [40,69].

The occurrence of β–β structures of types **27** and **28** has been considered [2]. Epipinoresinol structures (**27**) have not been detected in lignins [75], while structures of type **28** have been found to be present in a grass lignin [76]. The formation of 3,4-divanillyltetrahydrofuran on acidolysis [75] or thioacidolysis of softwood lignin [24] suggests the occurrence of β–β linked units of types **29** or **30**. These structures cannot be formed by *oxidation* of coniferyl alcohol (see the discussion of end-groups of dihydroconiferyl alcohol type below).

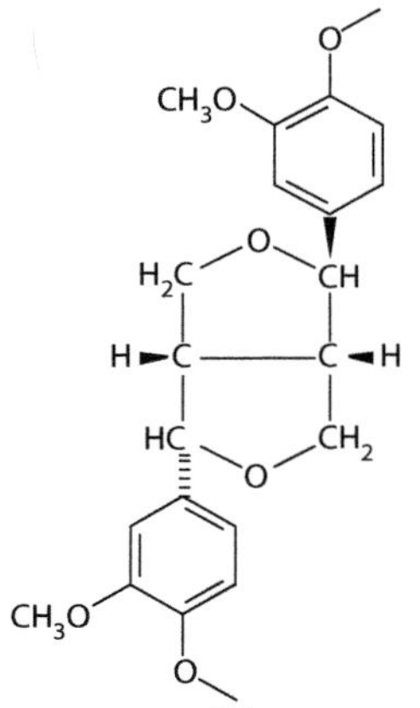

27

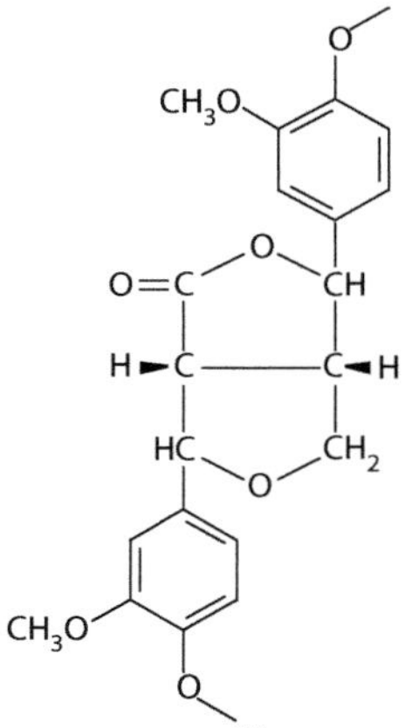

28

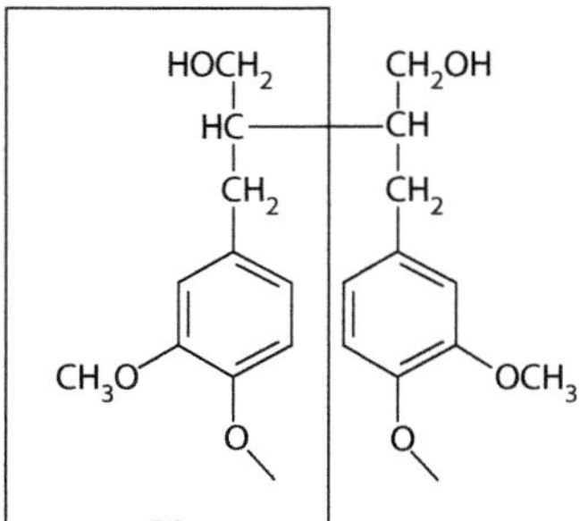

29

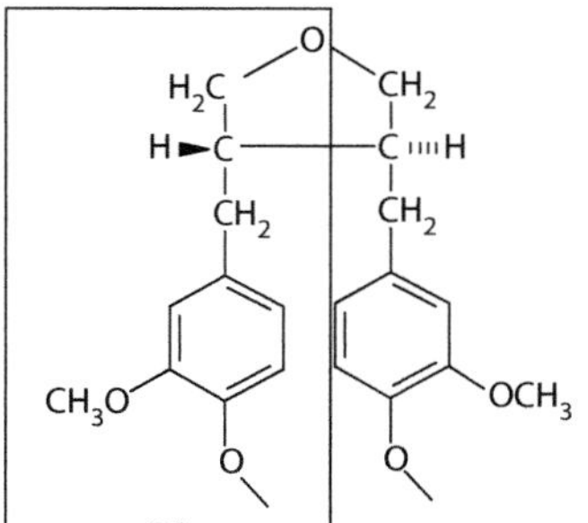

30

1-Aryl-2-aryloxy-1-propanone Units

The occurrence of arylketone units that may be of the 1-aryl-2-aryloxy-1-propanone type (**31**) in softwood lignin has been suggested on the basis of UV spectrometric studies [77]. Spectroscopic studies (^{1}H NMR) suggest that only small amounts of such units are present in lignins [78,79]. However, degradation studies provide support for the occurrence of units of type **31** in lignins [24,29]. Proof of their existence in both softwood [74,80] and hardwood (1-aryl-2-syringyloxy-1-propanone units, 74) lignins has recently been obtained in NMR spectroscopic studies. They are probably more frequent in hardwood lignins than in softwood lignins.

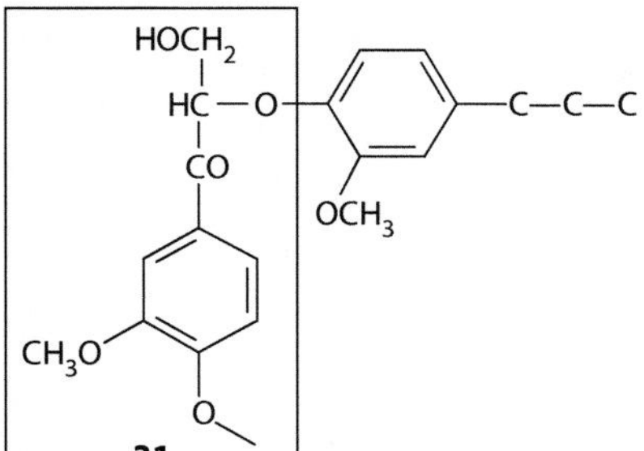

31

Phenylpropane Units Attached to an Adjacent Unit by a β–1 Linkage

The occurrence of β–1 structures of the 1,2-diaryl-1, 3-propanediol type (Figure 7.5) in lignins was derived from studies of degradation products [81,82]. Almost all types of solvolytic and reductive degradations of lignins yield comparatively large amounts of dimeric products that can be envisioned to originate from structures of this type [22,67]. Only recently has it been possible to obtain spectroscopic evidence for the occurrence of β–1 structures in lignins [49,74,83]. Degradation studies suggest, however, much larger amounts of such structures than do the spectroscopic studies. Two possible explanations of the discrepancy have been suggested: uneven distribution of the β–1 structures in lignins [84,85] or formation of β–1 structures during lignin degradation from cyclohexadienone precursors of type **32** [45,83] [or, as judged from model experiments, of type **33** [86].

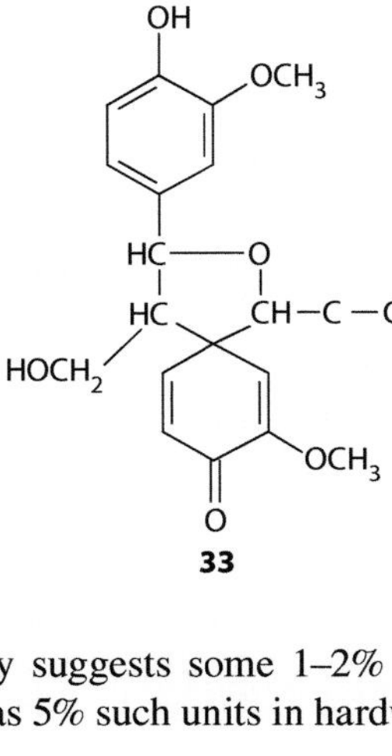

33

Proton NMR spectroscopy suggests some 1–2% units of type **34** in softwood lignins and perhaps as much as 5% such units in hardwood lignins [83]. The number

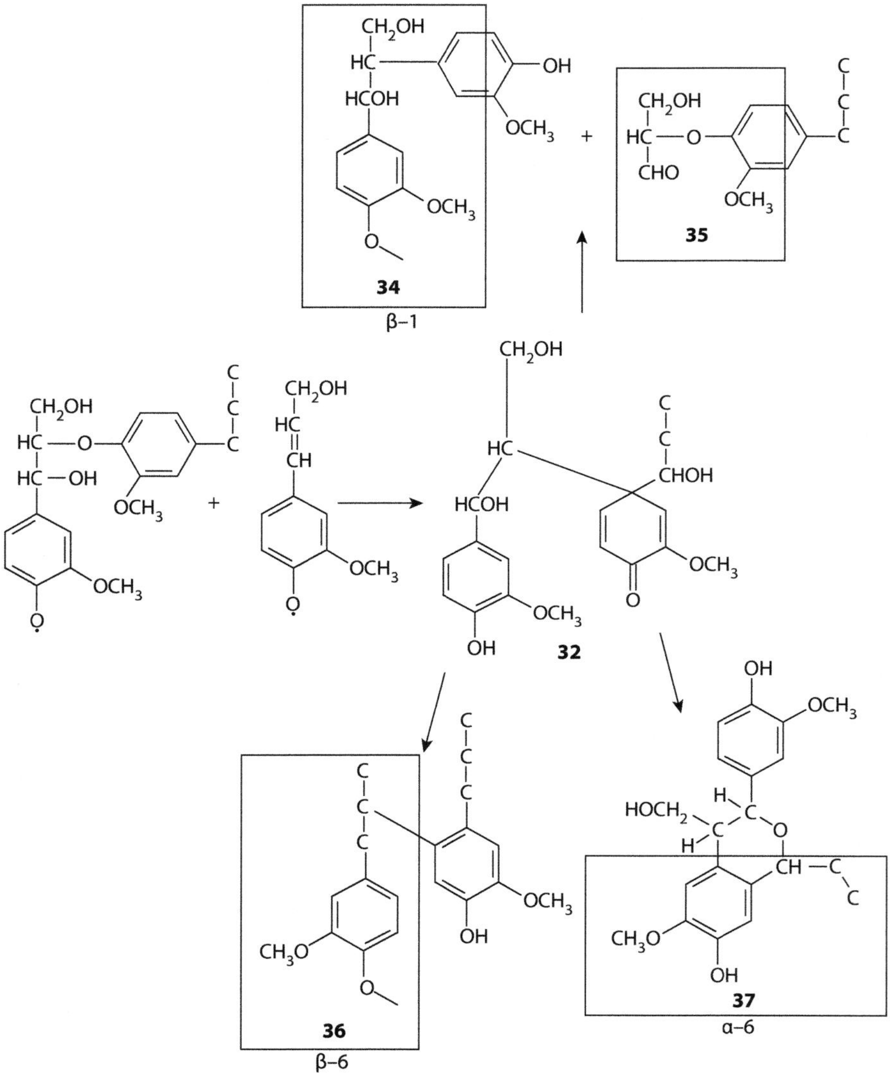

FIGURE 7.5 Biosynthesis of β-1, β-6 and α-6 structures and end groups of glyceraldehyde 2-aryl ether type (**35**).

of cyclohexadienone units of types **32** and **33** has been estimated to be about 1% in softwood lignin in a recent study [80].

Phenylpropane Units Linked to an Adjacent Unit by a β–6 or α–6 Linkage

The detection of lignin degradation products that are proposed to originate from lignin structures of β–6 type [22,53] or α–6 structures [87] has been described; conceivable biosynthetic routes leading to such structures are shown in Figure 7.5.

Calculating from the yield of metahemipinic acid (**38**) on degradation of lignins by permanganate oxidation, Miksche and coworkers estimated that there are about 3% units of types **36** or **37** in softwood lignin [17] and about 2% such units in hardwood lignins [18] [Note: the 2- and 6-positions are equivalent in syringylpropane units (**2**)].

The occurrence of α–6 structures of the diphenylmethane type has been suggested by Nimz [52]; however, permanganate oxidation studies [17] indicated that the number of such structures is very small.

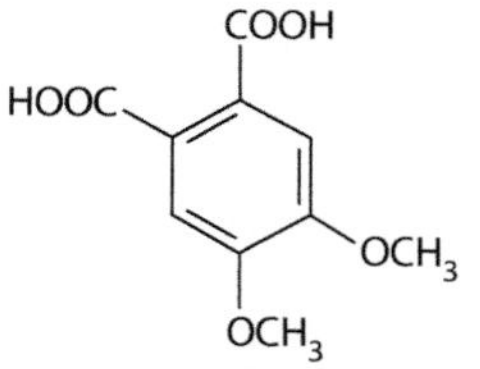

38

Cyclohexadienone Units and Quinoid Units

Different types of cyclohexadienone units and quinoid units are shown in Figure 7.6. Phenol oxidation often results in the formation cyclohexadienones and quinones [88]. It could, therefore, be expected that the phenol oxidation involved in the biosynthesis of lignin results in the formation of quinoid lignin units (**39, 40**) and lignin units of cyclohexadienone type (**41–46**) (see also the discussion of β–1 structures in lignins above). It is notable that compounds of dioxepin type (a structural element of type **41** is present in such compounds) are formed from units of type **7** on enzymatic oxidation under conditions similar to those prevailing during lignin biosynthesis [89]. Most of the quinoid and cyclohexadienone units (**39–43** and possibly **44**), which conceivably are present in lignins, carry a methoxyl group that is hydrolyzed on acid treatment. Studies of the liberation of methanol, suggest that there are a few percent units of quinone or cyclohexadienone type in lignins [90,91]. Measurements of the decolorization on treatment with SO_2-water indicate that there are 1–2% quinoid units in softwood lignin [91]. Derivatization with a phosphorus reagent and subsequent examination by ^{31}P NMR spectroscopy revealed the presence of 0.8% o-quinoid units in softwood lignin [92]. The possibility that quinoid groups are formed during the preparation of MWL should be considered [93].

End-Groups

Different types of end-groups are shown in Figure 7.7. End-groups of the cinnamyl alcohol (**47**) and cinnamaldehyde (**48**) types have been detected in both isolated lignins and lignins in wood. Quantitative analysis of MWL from spruce wood by ^{1}H NMR spectroscopy suggests the presence of about 1% units of type **47** [40] and 3–4% units of type **48** [47,77]. Lindgren and Mikawa [94] examined spruce wood using a color reaction and arrived at the conclusion that there was about 2% of each

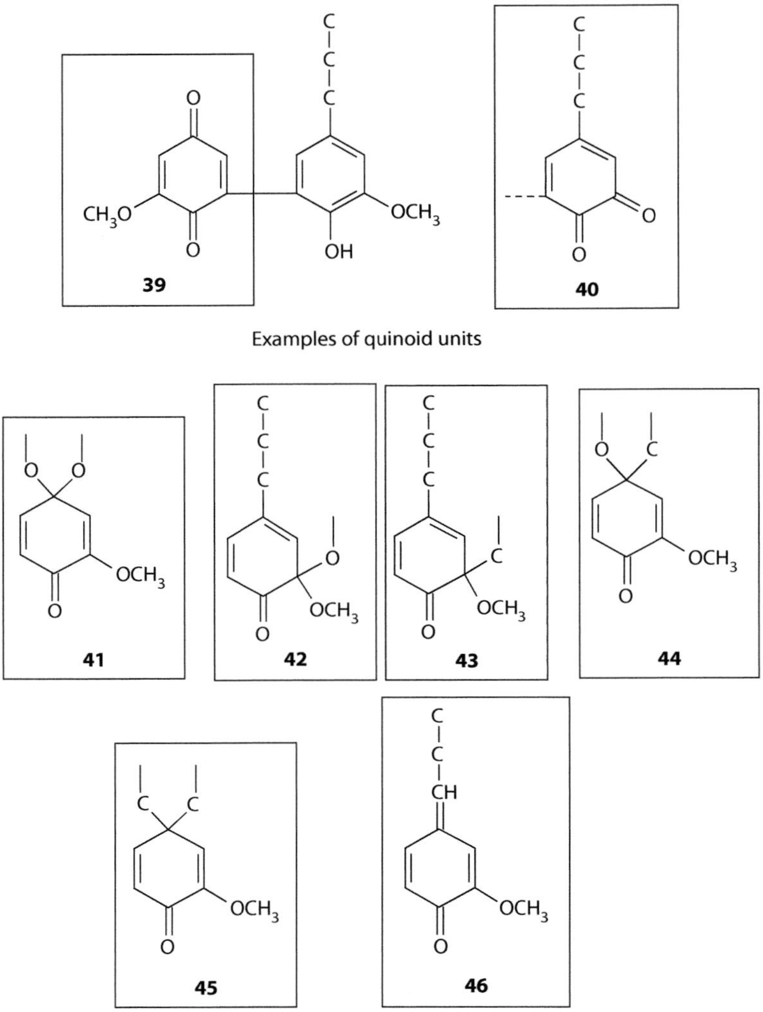

Examples of quinoid units

Examples of cyclohexadienone units

FIGURE 7.6 Examples of quinone and cyclohexadienone units proposed to be present in lignins. Cyclohexadienone units of type **45** are present in the intermediates involved in the biosynthesis of β-1, β-6 and α-6 structures (Fig. 7.5).

of the two types of units. Results from solid-state NMR examinations of ^{13}C–β-labeled lignin in ginkgo wood suggested the presence of 3.5% units of type **47** and 3% units of type **48** [95]. It is tempting to explain the differences between the results by assuming an oxidation of units of type **47** into units of type **48** during the isolation of MWL [96].

Proton NMR spectroscopic examinations of MWL from birch suggest the occurrence of 1% units of type **47** [40] and 1–2% units of type **48** [47]. Degradation

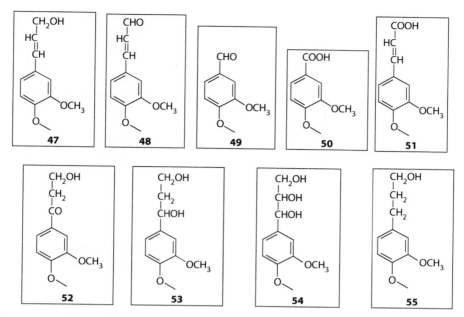

FIGURE 7.7 Different types of end-groups in lignins.

studies [24,97] and/or spectroscopic studies [47,98,99] of MWL reveal the occurrence of units of types **49–51** in lignins. The formation of these units during lignin biosynthesis is conceivable but they may alternatively, or at least in part, arise from oxidation reactions during the preparation of MWL [93,96,100, and 101]. Units of type **52** have recently been detected in lignin [33]. Such units may originate from units of type **7** as a result of reactions caused by the milling during the preparation of MWL [100,101]. An alternative to this explanation, based on the finding that there are small amounts of 1,3-propanediol units (**53**) in lignins, has recently been suggested; such units are assumed to be partly oxidized to type **52** units [13].

Support for the occurrence arylglycerol units (**54**) was obtained in "mild hydrolysis" studies of lignin [102]. According to studies of the formation of formaldehyde on periodate oxidation [103] only a few percent units with glycerol chain could be present in MWLs. The fact that arylglycerols were formed on enzymatic oxidation of p-hydroxycinnamyl alcohols provided further support for the occurrence of units of type **54** in lignins [104]. Final confirmation of this was later obtained in 2D NMR spectroscopic studies [74].

Degradation studies [24] and spectroscopic studies [12,48,105] show that dihydroconiferyl alcohol units (**55**) are present in lignin. Such units cannot be formed by *oxidation* of coniferyl alcohol during lignin biosynthesis (see Introduction). Recent biosynthetic studies show that such units in all probability arise from incorporation of dihydroconiferyl alcohol in the lignin during the oxidative polymerization involved in lignin formation [12].

A variety of degradative [24,106] and spectroscopic studies [45,78,80] have shown that end-groups of glyceraldehyde 2-aryl ether type (35) (Figure 7.5) are present in MWL. Their occurrence in wood lignin has been demonstrated in ozonation studies [107]. The studies point to the presence of 1–2% units of type 35 in lignins. Precursors of the type 32/33 can be expected to be included in most of the estimates of units of type 35; a clear distinction of these two categories of units has only been achieved by Zhang and Gellerstedt [80]. The observation that the glyceraldehyde 2-aryl ether structures accumulated in the low molecular weight fraction of methanol/acid-treated lignin is of interest in this context [45]. The release of a side chain as a glyceraldehyde-2-aryl ether may be a part of the biosynthesis leading to structures with a missing side chain other than β-1, such as units 39 and 41 (see Figure 7.5).

Esterified Units

In some lignins, 5–10% of the units are esterified; examples include p-hydroxybenzoate esters in lignins from *Salix* species (56) [69,108] and p-coumarate esters in grass lignins (57) [109–111]. The γ-positions in lignin are the ones that primarily are esterified [110].

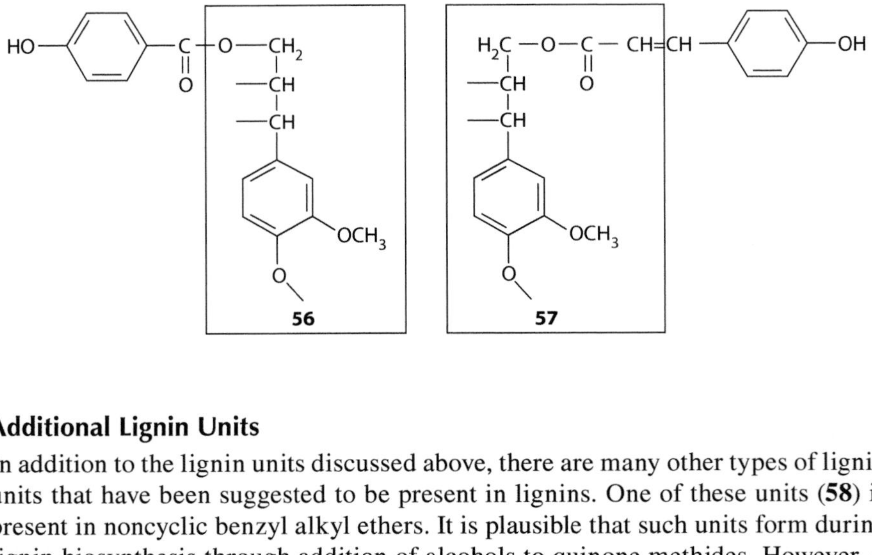

Additional Lignin Units

In addition to the lignin units discussed above, there are many other types of lignin units that have been suggested to be present in lignins. One of these units (58) is present in noncyclic benzyl alkyl ethers. It is plausible that such units form during lignin biosynthesis through addition of alcohols to quinone methides. However, it has been shown in model compound studies that alcohols are not prone to add to quinone methides under conditions similar to those prevailing during lignin biosynthesis [51]. The experimental evidence for the occurrence of units of type 58 in lignin is not conclusive [53,74]. Studies on the etherification of benzyl alcohol groups in lignins [112] indicate that reliable quantitative estimates are lacking, but one cannot exclude the possible occurrence of some 15% units of type 58.

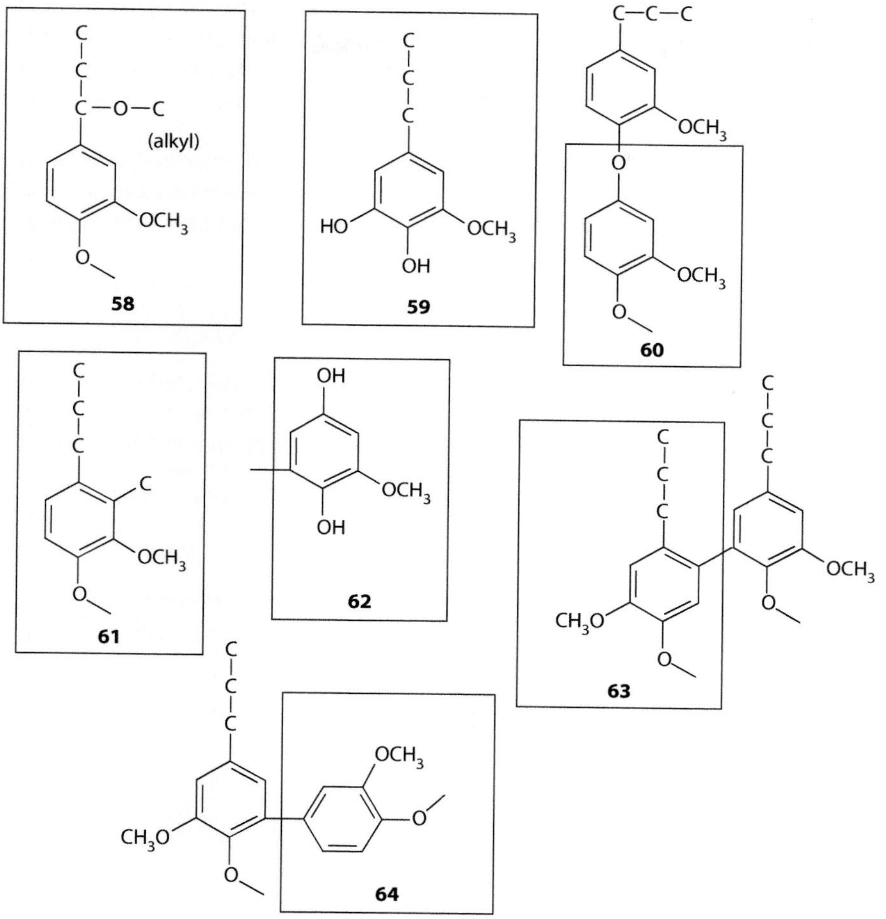

Thioacidolysis studies have shown that large amounts of units of type **59** are present in a lignin from a mutant of maize [9]. Permanganate oxidation studies [113] and hydrogenation studies [53] suggest that small amounts of such units also are present in lignins from normal plants. The formation of such units through artifactitious demethylation of syringyl units may, however, be an explanation of the results from the oxidation and hydrogenation studies.

The occurrence of small amounts of lignin units attached to each other by α–β linkages has been suggested on the basis of thioacetolysis studies [52]. However, α–β linkages may possibly be formed during the alkaline step involved in the thioacetolysis procedure [4]. It has been suggested, based on examinations by fluorescence spectroscopy [114], that small amounts of β–5 structures of the phenylcoumarone type are present in lignins.

Evidence for the occurrence of a series of additional types of lignin units (for instance **60–64**) have been obtained in studies of degradation products obtained on permanganate oxidation of lignins [4]. On the basis of the yields of oxidation products, Adler [4] concluded that only small amounts of units such as **60–64** can be present in lignins.

Other types of units than those discussed above have been suggested to be present in lignins but we have in this presentation focused on lignin units that are well established, and have not given a complete survey of lignin units that have been suggested on the basis of incomplete experimental evidence.

QUANTITATIVE DISTRIBUTION OF DIFFERENT TYPES OF LIGNIN UNITS

The approximate distribution of units with different types of side chains in softwood lignin and a typical hardwood lignin (cf. Table 7.1) are shown in Table 7.2. The distribution of certain types of lignin units in which the aromatic ring is attached

TABLE 7.2
Percentage Distribution of Lignin Units

Type of Unit	Softwood Lignin	Hardwood Lignin
7	35	45
14	1–3	1–5
18	6	1
21	10	4
23	4	8
29, 30	1	<1
31	1	1–2
34	3	5
36, 37	3	2
Units attached to **37**	0–3	0–2
Quinoid units (e.g., **40**)	1–2	2
Cyclohexadienone units (e.g., **42–46**)	3	4
47	2	2
48	3–4	1–2
49	3	1–2
Units with COOH groups (e.g., **50** and **51**) (see Carboxyl Groups section)	4	4
52/53	1–2	1–2
54	2	3
55	1	<1
35	2	2
Units attached to carbohydrates (see The Lignin-Carbohydrate Complex section)	1	2
Additional types of units (see Additional Lignin Units section) (by difference)	13–5	11–0

TABLE 7.3

Distribution of Condensed Lignin Units*

Type of Unit	Softwood Lignin	Hardwood Lignin
15 (including biphenyl units in dibenzodioxocin structures)	23	9
20	4	7
The unit attached to **21**	10	4
37, the unit attached to **36**	3	2
Additional types of condensed units (see Additional Lignin Units section)	5	1

* Lignin units in which the aromatic ring is connected to an adjacent unit. It is assumed that about 45% of the guaiacyl units are condensed (see Introduction).

to an adjacent unit (condensed units) is shown in Table 7.3. The data in the tables are based on the authors' evaluation of different literature reports on the occurrence of individual structural units in lignins. Quantitative lignin data are primarily obtained from studies of isolated lignins. Chemical modification during isolation may give rise to some types of structural units (see the Survey of Different Types of Lignin Units section) due to decomposition of structural elements originally present in the lignin. Studies of lignin *in situ* eliminate this drawback but are insufficient in other respects (see Introduction). Results from *in situ* studies that are of particular interest in this context are the finding that the phenol content in lignins *in situ* is comparatively low [50] and the determination of the distribution of ^{13}C-labeled C–β in phenylpropane units in ginkgo lignin (considered as representative of softwood lignin) [95].

ADDITIONAL COMMENTS ON FUNCTIONAL GROUPS

GENERAL

It is difficult to examine the functional groups in lignins. The main problems when studying lignins *in situ* are: the limitations of the methods that are applicable and the interference of constituents other than lignin. Fractionation and modification during the isolation influences the composition of the functional groups in isolated lignin samples (cf. [24]). As a consequence, results obtained with isolated lignin samples have to be treated with caution, although such samples can be examined by very powerful experimental techniques. Examinations of functional groups in lignins often rely on the use of selective reagents. However, there are many types of structural elements in lignins and the possibility that even rather specific reagents react in more than one way have to be considered (for a discussion of this complication, see [66]).

CARBONYL GROUPS

Carbonyl groups are present in units **31**, **35**, **48**, **49**, and **52** and also in cyclohexa-dienone and quinoid units (Figure 7.6). The total number of units with a carbonyl group in MWL has been determined to be ≈ 20% or more in some studies [77]. Such a large number of carbonyl groups is, however, not compatible with results from NMR spectral examinations. Data given in Table 7.2 are to a large extent based on NMR spectral studies [74,78,80,92].

CARBOXYL GROUPS

Carboxyl groups are present in units **50** and **51**. Several studies suggest that there are only a few percent units carrying a carboxyl group in MWL [41,115,116]. It could be questioned if the observed carboxyl groups are present in lignin *in situ* since they may, at least in part, be formed in connection with the isolation of MWL (see Bonding Patterns and Functional Groups section).

ESTER GROUPINGS

Units esterified with *p*-hydroxybenzoic acid (**56**) are present in certain types of lignin [69,108]. Grass lignins contain units that are esterified with *p*-coumaric acid (**57**) [109–111]. Lactone structures (**28**) have been proposed to be present in lignins [2]; however, evidence for this has only been presented in the case of grass lignins [76].

ETHYLENIC GROUPS

Ethylenic groups are present in units **47, 48,** and **51** and also in cyclohexadienone and quinoid units (Figure 7.6). It is difficult to determine the total number of ethylenic groups. Reported estimates suggest that there are 0.05–0.1 ethylenic groups/phenyl-propane units in softwood lignin [117].

HYDROXYL GROUPS

The total number of hydroxyl groups in MWLs is 1.3–1.5/phenylpropane unit in soft-wood lignins [26,39] and about 1.5/phenylpropane units in hardwood lignins [118,119]. In MWLs from softwoods there are 19–26% phenolic units [26,39,41,116,120] and in MWLs from hardwoods there are 14–18% [118–121]. The proportion of phenolic units in lignin *in situ* has been reported to be rather low [50]. The proportion of free phenolic groups in the *p*-hydroxyphenylpropane (**3**) moiety in lignins is consider-ably larger than the proportion of such groups in the guaiacylpropane (**1**) moiety; guaiacylpropane units in turn carry free phenolic groups more frequently than do syringylpropane units (**2**) [17,18,42,120, and 122].

METHOXYL GROUPS

Methoxyl groups are present in guaiacylpropane units (**1**) and syringylpropane units (**2**). The methoxyl content of a lignin sample reflects the distribution of units **1–3**.

The purity of a lignin sample can, in many cases, be judged based on the methoxyl content. The methoxyl content is usually determined according to modified Zeisel procedures [123]. Alternatively the methoxyl content can be determined by ^{13}C NMR spectroscopy [124]. The methoxyl content can also be roughly estimated based on ^{1}H NMR spectral examinations [79].

THE LIGNIN-CARBOHYDRATE COMPLEX

INTRODUCTION

When isolating lignin from ball-milled wood by extraction with aqueous dioxane, one usually finds that some of the fractions obtained contain lignin and carbohydrate constituents that cannot be separated by solvent extraction or chromatography [20,125,126]. In connection with the preparation of MWL, Björkman [125,126] isolated such fractions and called them lignin-carbohydrate complexes (LCCs). Fractionation experiments suggest that a spectrum of complexes with varying lignin-carbohydrate ratios is present in these products [127]. Both the nature of the bonds and the carbohydrate composition depend on the botanical origin of the sample [128–130] and the particular fractions of the same sample [131,132].

The investigations of the exact nature of the bonds between lignin and carbohydrates have been hampered by the difficulties of preparing well-characterized lignin-carbohydrate complexes in unchanged form. Another difficulty is that the frequency of lignin-carbohydrate bonds is expected to be low in any such sample. It is also difficult to exclude the possibility that lignin-carbohydrate linkages may be formed or broken during the preparation of the samples; the milling can cause rupture of bonds with the formation of radicals causing formation of new linkages [67]. The drastic changes caused by pulping reactions are outside the scope of this chapter.

ISOLATION OF LIGNIN-CARBOHYDRATE COMPLEXES

Most methods for the isolation of lignin-carbohydrate complexes start with solvent extracted wood meal that has been subjected to ball milling in order to increase the surface area of the sample, as in the preparation of MWL. The original procedure for the preparation of MWL prescribes extraction of ball-milled wood with dioxane containing 4% water. The soluble material in such extracts contains lignin, LCCs, and carbohydrates [126,127]. Further extractions of the milled wood residue with solvents, such as dimethyl formamide and dimethyl sulfoxide, result in additional amounts of lignin-carbohydrate complexes [126,133–135]. Lignin-carbohydrate complexes have also been isolated from the residue from the MWL preparation by extraction with hot water [136]. An alternative method [131,132, 137] for the preparation of MWL uses liquid-liquid extraction (pyridine-acetic acid-water and chloroform) instead of a precipitation of the lignin in water from an acetic acid solution. This procedure has a better selectivity for the separation of lignin from carbohydrates, and produces lignin-carbohydrate complexes freed from carbohydrates.

TYPES OF LIGNIN-CARBOHYDRATE BONDS

It has not been possible to isolate degradation products with clearly identifiable bonds between the lignin and the carbohydrate constituents. However, the accumulated evidence from a large number of studies points to the existence of covalent bonds between lignin and the hemicelluloses [131,132,138–145] or pectic substances [135]. For instance, enzymatic hydrolysis with hemicellulases may remove ca. 95% of the carbohydrates from a lignin-carbohydrate complex, leaving residual carbohydrates firmly anchored to the lignin [138]. It has been suggested [139] that the lignin in a lignin-carbohydrate complex from spruce is bonded with ether bonds to arabinose units in arabinoglucuronoxylan and to galactoglucomannan through the primary hydroxyl groups in the carbohydrate moieties. No direct evidence was found for covalent linkages to cellulose [140].

It is usually assumed that the covalent bonds between lignin and carbohydrates in LLCs are formed during the biosynthesis of lignin, through the addition of nucleophiles to quinone methides, formed as intermediates in the oxidation of *p*-hydroxycinnamyl alcohols [67,146,147].

ALKALI-LABILE BONDS

Alkaline treatment of lignin-carbohydrate complexes removes a portion of the carbohydrates [138,140,142,148]. Mild alkaline hydrolysis of isolated lignin samples from birch and spruce with low carbohydrate content (birch, 2.0%; spruce, 0.3%), released the major part of the xylan (the predominant hemicellulose) from birch; from spruce only 50% of the carbohydrates (preferentially xylan) was released [148,149]. This could mean that the birch sample contained mainly alkali-labile lignin carbohydrate linkages, presumably glucuronic acid esters. It has been shown that the 4–O-methylglucuronic acid units in aspen xylan are partly esterified [150]. The spruce sample, on the other hand, evidently contains a large proportion of alkali stable linkages. In model experiments, ester bonds are formed in reactions of model quinone methides with the carboxyl groups of glucuronic acids and pectic substances and the esters have properties in line with results from hydrolysis experiments [151,152].

ALKALI-STABLE BONDS

The bonds that are stable to alkaline hydrolysis are more difficult to characterize, but model compound studies indicate that benzyl ether bonds can be cleaved by acid hydrolysis and are resistant to alkaline hydrolysis [153]. Nonphenolic benzyl alkyl ether model compounds are stable even under alkaline pulping conditions [154]. Watanabe [145] has developed a method for the detection of benzyl ether bonds between lignins and hemicelluloses. The method employs dichlorodicyano-benzoquinone (DDQ) as a specific reagent for the cleavage of such bonds. When acetylated lignin-carbohydrate complexes are treated with DDQ, the carbohydrates bound to the lignin as benzyl ethers are released. The released carbohydrates are methylated and hydrolyzed. The detection of methylated monosaccharides after such treatment is an indication that they have been bound to the lignin as benzyl ethers. Studies with model compounds

[145] have shown that the cleavage reaction is not quantitative; the method, therefore, gives only qualitative data concerning the bonding of the hemicellulose components. The formation of benzyl ether bonds has been demonstrated in model experiments; quinone methide intermediates react, forming benzyl ether bonds with the hydroxyl groups in carbohydrates [146,147,151,152]. It should be pointed out that the formation of ether bonds by this mechanism has been demonstrated in vitro in nonaqueous systems but is suppressed in aqueous systems [152].

LCC SUMMARY

The evidence for ester bonds in lignin-carbohydrate complexes is rather convincing. Both the formation and the cleavage of ester bonds can easily be demonstrated with model compounds. The migration of ester groups along the lignin side chain from the α to the γ position has been observed in model compounds [155]. The evidence for assuming that the alkali resistant bonds are benzyl ether bonds is more circumstantial. The formation of such bonds is difficult to reproduce *in vitro* under aqueous conditions. The DDQ method, depending heavily on the exclusive selectivity of the cleavage reaction, needs confirmation by some other means. Regarding other types of bonding, glycosidic bonds are mentioned from time to time, the mechanisms in plants that lead to the formation of phenol or alcohol glycosides may operate on oligosaccharides or oligolignols [1,130,138,144,156]. There is little experimental evidence for such bonds so far, and there is also a scarcity of reducing end-groups in hemicelluloses that could form such bonds. Other types of bonds are also conceivable. For instance radical reactions during lignification could lead to carbon-carbon linkages [67]. The possibility that certain phenolic acids play a role for the connection of lignin and carbohydrates in grasses has also been discussed [76,99]. Nucleophilic attack on the double bond in hexenuronic acid structures is another speculative proposal [130] that awaits experimental evidence. NMR spectral evidence for the occurrence of lignin carbohydrate linkages have recently been obtained in studies of ^{13}C labeled LCC samples [157]. However, in our opinion, the assignments of the NMR signals require confirmation. Studies of lignin carbohydrate bonds based on ozonation seem to be a promising approach [158].

For reasons apparent from the previous discussion, the nature of the chemical linkage between polysaccharidic and lignin elements in woody tissues is still not well understood. More research is clearly needed. Knowledge of the nature of the lignin carbohydrate connections is of importance for the understanding of the structure and the reactivity of lignin *in situ*.

REFERENCES

1. JM Harkin. Lignin: A natural polymeric product of phenol oxidation. In: WI Taylor, AR Battersby, eds. *Oxidative Coupling of Phenols*. New York: Marcel Dekker, 1967, pp. 243–321.
2. K Freudenberg. The constitution and biosynthesis of lignin. In: K Freudenberg, AC Neish, eds. *Constitution and Biosynthesis of Lignin*. Berlin-Heidelberg: Springer-Verlag, 1968, pp. 47–122.

3. KV Sarkanen, CH Ludwig, eds. *Lignins: Occurrence, Formation, Structure and Reactions.* New York: Wiley-Interscience, 1971.

4. E Adler. Lignin chemistry: Past, present and future. *Wood Sci Technol* 11:169–218, 1977.

5. T Higuchi. Biosynthesis of lignin. In: T Higuchi, ed. *Biosynthesis and Biodegradation of Wood Components.* Orlando: Academic Press, 1985, pp. 141–160.

6. G Brunow, I Kilpeläinen, J Sipilä, K Syrjänen, P Karhunen, H Setälä, P Rummakko. Oxidative coupling of phenols and the biosynthesis of lignin. In: NG Lewis, S Sarkanen, eds. *Lignin and Lignan Biosynthesis.* Washington, DC: American Chemical Society, 1998, pp. 131–147.

7. J Ralph, J Peng, F Lu, RD Hatfield, RF Helm. Are lignins optically active? *J Agric Food Chem* 47:2991–2996, 1999.

8. T Akiyama, K Magara, Y Matsumoto, G Meshitsuka, A Ishizu, K Lundquist. Proof of the presence of racemic forms of arylglycerol-β-aryl ether structure in lignin: Studies on stereo structure of lignin by ozonation. *J Wood Sci* 46:414–415, 2000.

9. C Lapierre, MT Tollier, B Monties. Occurrence of additional monomeric units in the lignins from internodes of a brown midrib mutant of Maize bm3. *C R Acad Sci Ser 3,* 307:723–728, 1988.

10. J Ralph, RD Hatfield, J Piquemal, N Yahiaoui, M Pean, C Lapierre, A-M Boudet. NMR characterization of altered lignins extracted from tobacco plants down-regulated for lignification enzymes cinnamyl-alcohol dehydrogenase and cinnamoyl-CoA reductase. *Proc Natl Acad Sci USA* 95:12803–12808, 1998.

11. J Ralph, RD Hatfield, JH Grabber, HJG Jung, S Quideau, RF Helm. Cell wall cross-linking in grasses by ferulates and diferulates. In: NG Lewis, S Sarkanen, eds. *Lignin and Lignan Biosynthesis.* Washington, DC: American Chemical Society, 1998, pp. 209–236.

12. RR Sederoff, JJ MacKay, J Ralph, RD Hatfield. Unexpected variation in lignin. *Current Opinion in Plant Biology* 2:145–152, 1999.

13. J Ralph, H Kim, J Peng, F Lu. Arylpropane-1,3-diols in lignins from normal and CAD-deficient pines. *Organic Letters* 1:323–326, 1999.

14. Y Tomimura, T Yokoi, N Terashima. Heterogeneity in formation of lignin. V. Degree of condensation in guaiacyl nucleus. *Mokuzai Gakkaishi* 26:37–42, 1980.

15. M Funaoka, I Abe, VI Chiang. Nucleus exchange reaction. In: SY Lin, CW Dence, eds. *Methods in Lignin Chemistry.* Berlin: Springer-Verlag, 1992, pp. 369–386.

16. FD Chan, KL Nguyen, AFA Wallis. Estimation of the aromatic units in lignin by nucleus exchange: A reassessment of the method. *J Wood Chem Technol* 15:473–491, 1995.

17. M Erickson, S Larsson, GE Miksche. Gaschromatographische Analyse von Ligninoxydationsprodukten. VIII. Zur Struktur des Lignins der Fichte. *Acta Chem Scand* 27:903–914, 1973.

18. S Larsson, GE Miksche. Gaschromatographische Analyse von Ligninoxydationsprodukten. IV. Zur Struktur des Lignins der Birke. *Acta Chem Scand* 25:647–662, 1971.

19. ID Suckling, MF Pasco, B Hortling, J Sundquist. Assessment of lignin condensation by GPC analysis of lignin thioacidolysis products. *Holzforschung* 48:501–503, 1994.

20. A Björkman. Studies on finely divided wood. Part I. Extraction of lignin with neutral solvents. *Sven Papperstidn* 59:477–485, 1956.

21. K Lundquist. Wood. In: SY Lin, CW Dence, eds. *Methods in Lignin Chemistry.* Berlin: Springer-Verlag, 1992, pp. 65–70.

22. C Lapierre, B Pollet, B Monties, C Rolando. Thioacidolysis of spruce lignin: GC-MS analysis of the main dimers recovered after Raney nickel desulphuration. *Holzforschung* 45:61–68, 1991.

23. C Rolando, B Monties, C Lapierre. Thioacidolysis. In: SY Lin, CW Dence, eds. *Methods in Lignin Chemistry.* Berlin: Springer-Verlag, 1992, pp. 334–349.

24. C Lapierre, K Lundquist. Investigations of low molecular weight and high molecular weight lignin fractions. *Nord Pulp Pap Res J* 14:158–162, 170, 1999.
25. CH Ludwig, BJ Nist, JL McCarthy. Lignin. XIII. The high resolution nuclear magnetic resonance spectroscopy of protons in acetylated lignins. *J Am Chem Soc* 86:1196–1202, 1964.
26. K Lundquist. NMR studies of lignins. 4. Investigation of spruce lignin by 1H NMR spectroscopy. *Acta Chem Scand* B34:21–26, 1980.
27. G Brunow, K Lundquist, G Gellerstedt. Lignin. In: E Sjöström, R Alén, eds. *Analytical Methods in Wood Chemistry, Pulping and Papermaking.* Berlin: Springer-Verlag, 1998, pp. 77–124.
28. E Adler, JM Pepper, E Eriksoo. Action of mineral acid on lignin and model substances of guaiacylglycerol-beta-aryl ether type. *Ind Eng Chem* 49:1391–1392, 1957.
29. K Lundquist. Acidolysis. In: SY Lin, CW Dence, eds. *Methods in Lignin Chemistry.* Berlin: Springer-Verlag, 1992, pp. 289–300.
30. F Lu, J Ralph. The DFRC method for lignin analysis. 2. Monomers from isolated lignins. *J Agric Food Chem* 46:547–552, 1998.
31. HH Nimz, U Tschirner, M Stähle, R Lehmann, M Schlosser. Carbon-13 NMR spectra of lignins, 10. Comparison of structural units in spruce and beech lignin. *J Wood Chem Technol* 4:265–284, 1984.
32. M Hauteville, K Lundquist, S von Unge. NMR studies of lignins. 7. 1H NMR spectroscopic investigation of the distribution of *erythro* and *threo* forms of β-O-4 structures in lignins. *Acta Chem Scand* B40:31–35, 1986.
33. RM Ede, J Ralph. Assignment of 2D TOCSY spectra of lignins: The role of lignin model compounds. *Magn Reson Chem* 34:261–268, 1996.
34. B Saake, DS Argyropoulos, O Beinhoff, O Faix. A comparison of lignin polymer models (DHPs) and lignins by 31P NMR spectroscopy. *Phytochemistry* 43:499–507, 1996.
35. Y Matsumoto, A Ishizu, J Nakano. Studies on chemical structure of lignin by ozonation. *Holzforschung* 40 (Suppl):81–85, 1986.
36. H Taneda, N Habu, J Nakano. Characterization of the side chain steric structures in the various lignins. *Holzforschung* 43:187–190, 1989.
37. KV Sarkanen, A Islam, CD Anderson. Ozonation. In: SY Lin, C.W. Dence, eds. *Methods in Lignin Chemistry.* Berlin: Springer-Verlag, 1992, pp. 387–406.
38. M Bardet, D Robert, K Lundquist, S von Unge. Distribution of *erythro* and *threo* forms of different types of β-O-4 structures in aspen lignin by 13C NMR using the 2D INADEQUATE experiment. *Magn Reson Chem* 36:597–600, 1998.
39. DR Robert, G Brunow. Quantitative estimation of hydroxyl groups in milled wood lignin from spruce and in a dehydrogenation polymer from coniferyl alcohol using 13C NMR spectroscopy. *Holzforschung* 38:85–90, 1984
40. K Lundquist. 1H NMR spectral studies of lignins. Quantitative estimates of some types of structural elements. *Nord Pulp Pap Res J* 6:140–146, 1991.
41. Z-H Jiang, DS Argyropoulos. The stereoselective degradation of arylglycerol-beta-aryl ethers during kraft pulping. *J Pulp Pap Sci* 20:J183–J188, 1994.
42. DS Argyropoulos. Quantitative phosphorus-31 NMR analysis of six soluble lignins. *J Wood Chem Technol* 14:65–82, 1994.
43. K. Freudenberg, JM Harkin, H-K Werner. Das Vorkommen von Benzylaryläthern im Lignin. *Chem Ber* 97:909–920, 1964.
44. E Adler, GE Miksche, B Johansson. Über die Benzyl-arylätherbindung im Lignin. I. Freilegung von phenolischem Hydroxyl in Ligninpräparaten durch Spaltung leicht hydrolysierbarer Alkyl-arylätherstrukturen. *Holzforschung* 22:171–174, 1968.
45. G Brunow, K Lundquist. On the acid-catalysed alkylation of lignins. *Holzforschung* 45:37–40, 1991.

46. P Karhunen, P Rummakko, A Pajunen, G Brunow. Synthesis and crystal structure determination of model compounds for the dibenzodioxocine structure occurring in wood lignins. *J Chem Soc,* Perkin Trans 1 :2303–2308, 1996.
47. K Lundquist. 1H NMR spectral studies of lignins. Results regarding the occurrence of β-5 structures, β-β structures, non-cyclic benzyl aryl ethers, carbonyl groups and phenolic groups. *Nord Pulp Pap Res J* 7:4–8,16, 1992.
48. RM Ede, G Brunow. Application of two-dimensional homo- and heteronuclear correlation NMR spectroscopy to wood lignin structure determination. *J Org Chem* 57:1477–1480, 1992.
49. I Kilpeläinen, E Ämmälahti, G Brunow, D Robert. Application of three-dimensional HMQC-HOHAHA NMR spectroscopy to wood lignin, a natural polymer. *Tetrahedron Lett* 35:9267–9270, 1994.
50. Y-Z Lai, X-P Guo. Acid-catalyzed hydrolysis of aryl ether linkages in wood. 1. Estimation of non-cyclic α-aryl ether units. *Holzforschung* 46:311–314, 1992.
51. J Sipilä, G Brunow. On the mechanism of formation of non-cyclic benzyl ethers during lignin biosynthesis. Part 2. The effect of pH on the reaction between a β-O-4-type quinone methide and vanillyl alcohol in water-dioxane solutions. The stability of non-cyclic benzyl aryl ethers during lignin biosynthesis. *Holzforschung* 45:275–278, 1991.
52. H Nimz. Beech lignin: Proposal of a constitutional scheme. *Angew Chem Internat Edit* 13:313–321, 1974.
53. A Sakakibara. Hydrogenolysis. In: SY Lin, CW Dence, eds. *Methods in Lignin Chemistry.* Berlin: Springer-Verlag, 1992, pp. 350–368.
54. JC Pew. Evidence of a biphenyl group in lignin. *J Org Chem* 28:1048–1054, 1963.
55. IA Pearl, DL Beyer. Studies on lignin and related products. IX. Cupric oxide oxidation of lignin model substances. *J Am Chem Soc* 76:2224–2226, 1954.
56. SK Bose, KL Wilson, RC Francis, M Aoyama. Lignin analysis by permangate oxidation. I. Native spruce lignin. *Holzforschung* 52:297–303, 1998.
57. M Drumond, M Aoyama, C-L Chen, D Robert. Substituent effects on C-13 chemical shifts of aromatic carbons in biphenyl type lignin model compounds. *J Chem Wood Technol* 9:421–441, 1989.
58. P Karhunen, P Rummakko, J Sipilä, G Brunow. Dibenzodioxocins: A novel type of linkage in softwood lignins. *Tetrahedron Lett* 36:169–170, 1995.
59 DS Argyropoulos, L Jurasek, L Kristofová, Z Xia, Y Sun, E Palus. Abundance and reactivity of dibenzodioxocins in softwood lignin. *J Agric Food Chem* 50:658–666, 2002.
60. P Karhunen, J Mikkola, A Pajunen, G Brunow. The behaviour of dibenzodioxocin structures in lignin during alkaline pulping processes. *Nord Pulp Pap Res J* 14:123–128, 1999.
61. E Adler, K Lundquist. Spectrochemical estimation of phenylcoumaran elements in lignin. *Acta Chem Scand* 17:13–26, 1963.
62. S Li, K Lundquist. Synthesis of lignin models of β-5 type. *Acta Chem Scand* 51:1224–1228, 1997.
63. K Freudenberg, C-L Chen, JM Harkin, H Nimz, H Renner. Observations on lignin. *J Chem Soc, Chem Commun,* 224–225, 1965.
64. N Habu, Y Matsumoto, A Ishizu, J Nakano. The role of the diarylpropane structure as a minor constituent in spruce lignin. *Holzforschung* 44:67–71, 1990.
65. S Li, T Iliefski, K Lundquist, AFA Wallis. Reassignment of relative stereochemistry at C-7 and C-8 in arylcoumaran neolignans. *Phytochemistry* 46:929–934, 1997.
66. K Lundquist, S Li. Structural analysis of lignin and lignin degradation products. Proceedings of 10th International Symposium on Wood and Pulping Chemistry, Vol. 1, Yokohama, 1999, pp. 2–10.

67. YZ Lai, KV Sarkanen. Isolation and structural studies. In: KV Sarkanen, CH Ludwig, eds. *Lignins: Occurrence, Formation, Structure and Reactions.* New York: Wiley-Interscience, 1971, pp. 165–240.
68. S Li, K Lundquist. Acid reactions of lignin models of β-5 type. *Holzforschung* 53:39–42, 1999.
69. LL Landucci, GC Deka, DN Roy. A 13C NMR study of milled wood lignins from hybrid *Salix* clones. *Holzforschung* 46:505–511, 1992.
70. K Ogiyama, T Kondo. On the pinoresinol type of structural units in lignin molecule. IV. The changes of dilactone-yield during enzymic dehydrogenation. *Mokuzai Gakkaishi* 14:416–420, 1968.
71. AFA Wallis. Solvolysis by acids and bases. In: KV Sarkanen, CH Ludwig, eds. *Lignins: Occurrence, Formation, Structure and Reactions.* New York: Wiley-Interscience, 1971, pp. 345–372.
72. S Omori, A Sakakibara. Hydrolysis of lignin with dioxane and water. XI. Isolation of arylglycerol-β-aryl ether and lignan-type compounds from hardwood lignin. *Mokuzai Gakkaishi* 20:388–395, 1974.
73. M Bardet, D Gagnaire, R Nardin, D Robert, M Vincendon. Use of 13C enriched wood for structural NMR investigation of wood and wood components, cellulose and lignin, in solid and in solution. *Holzforschung* 40 (Suppl):17–24, 1986.
74. I Kilpeläinen, J Sipilä, G Brunow, K Lundquist, RM Ede. Application of two-dimensional NMR spectroscopy to wood lignin structure determination and identification of some minor structural units of hard- and softwood lignins. *J Agric Food Chem* 42:2790–2794, 1994.
75. K Lundquist, R Stomberg. On the occurrence of structural elements of the lignan type (β-β structures) in lignins. The crystal structures of (+)-pinoresinol and (±)-*trans*-divanillyltetrahydrofuran. *Holzforschung* 42:375–384, 1988.
76. J Ralph, JH Grabber, RD Hatfield. Lignin-ferulate cross links in grasses: Active incorporation of ferulate polysaccharide esters into ryegrass lignins. *Carbohyd Res* 275:167–178, 1995.
77. E Adler, J Marton. Zur Kenntnis der Carbonylgruppen im Lignin. I. *Acta Chem Scand* 13: 75–96, 1959.
78. K Lundquist, T Olsson. NMR studies of lignins. 1. Signals due to protons in formyl groups. *Acta Chem Scand* B31:788–792, 1977.
79. K Lundquist, S von Unge. NMR studies of lignins. 8. Examination of pyridine-*d5* solutions of acetylated lignins from birch and spruce by 1H NMR spectroscopy. *Acta Chem Scand* B40:791–797, 1986.
80. L Zhang, G Gellerstedt. Detection and determination of carbonyls and quinones by modern NMR techniques. Proceedings of 10th International Symposium on Wood and Pulping Chemistry, Vol. 2, Yokohama, 1999, pp. 164–170.
81. K Lundquist, GE Miksche. Nachweis eines neuen Verknüpfungsprinzips von Guaiacylpropaneinheiten im Fichtenlignin. *Tetrahedron Lett,* 2131–2136, 1965.
82. H Nimz. Über die milde Hydrolyse des Buchenlignins, II. Isolierung eines 1.2-Diaryl-propan-Derivates und seine Überführung in ein Hydroxystilben. *Chem Ber* 98:3160–3164, 1965.
83. K Lundquist. On the occurrence of β-1 structures in lignins. *J Wood Chem Technol* 7:179–185, 1987.
84. C Lapierre, B Pollet, B Monties. Heterogeneous distribution of diarylpropane structures in spruce lignin. *Phytochemistry* 30:659–662, 1991.
85. RM Ede, J Ralph, KM Torr, BSW Dawson. A 2D NMR Investigation of the heterogeneity of distribution of diarylpropane structures in extracted *Pinus radiata* lignins. *Holzforschung* 50:161–164, 1996.

86. H Setälä, A Pajunen, P Rummakko, J Sipilä, G Brunow. A novel type of spiro compound formed by oxidative cross coupling of methyl sinapate with a syringyl lignin model compound. A model for the β-1 pathway in lignin biosynthesis. *J Chem Soc, Perkin Trans* 1:461–464, 1999.

87. J Ralph, J Peng, F Lu. Isochroman structures in lignin: A new β-1 pathway. *Tetrahedron Lett* 39:4963–4964, 1998.

88. H Musso. Phenol coupling. In: WI Taylor, AR Battersby, eds. *Oxidative Coupling of Phenols.* New York: Marcel Dekker, 1967, pp. 1–94.

89. JC Pew, WJ Connors. New structures from the enzymic dehydrogenation of lignin model *p*-hydroxy-α-carbinols. *J Org Chem* 34:580–584, 1969.

90. K Lundquist, L Ericsson. Acid degradation of lignin. VI. Formation of methanol. *Acta Chem Scand* 25:756–758, 1971.

91. L Hemrå, K Lundquist. Acidolytic formation of methanol from quinones and quinoid compounds related to lignin. *Acta Chem Scand* 27:365–366, 1973.

92. DS Argyropoulos, C Heitner. 31P NMR spectroscopy in wood chemistry. Part VI. Solid state 31P NMR of trimethyl phosphite derivatives of chromophores and carboxylic acids present in mechanical pulps; a method for the quantitative determination of *ortho*-quinones. *Holzforschung* 48 (Suppl):112–116, 1994.

93. K Itoh, M Sumimoto, H Tanaka. Comparative studies on the mechanochemistry of guaiacylglycerol- and veratrylglycerol-β-guaiacyl ether. *J Wood Chem Technol* 15:395–411, 1995.

94. BO Lindgren, H Mikawa. The presence of cinnamyl alcohol groups in lignin. *Acta Chem Scand* 11:826–835, 1957.

95. N Terashima, J Hafrén, U Westermark, DL VanderHart. Nondestructive analysis of lignin structure by NMR spectroscopy of specifically 13C-enriched lignins. Part. 1. Solid state study of ginkgo wood. *Holzforschung* 56:43–50, 2002.

96. DY Lee, S Tachibana, M Sumimoto. Mechanochemistry of lignin. II. Mechanochemical reactions of coniferyl alcohol methyl ether. *Cellul Chem Technol* 22:201–210, 1988.

97. K Lundquist. Acid degradation of lignin. II. Separation and identification of low molecular weight phenols. *Acta Chem Scand* 24:889–907, 1970.

98. R Mörck, KP Kringstad. 13C NMR spectra of kraft lignins. II. Kraft lignin acetates. *Holzforschung* 39:109–119, 1985.

99. A Scalbert, B Monties, J-Y Lallemand, E Guittet, C Rolando. Ether linkage between phenolic acids and lignin fractions from wheat straw. *Phytochemistry* 24:1359–1362, 1985.

100. D-Y Lee, M Sumimoto. Mechanochemistry of lignin. III. Mechanochemical reactions of β-O-4 lignin model compounds. *Holzforschung* 44:347–350, 1990.

101. Z-H Wu, M Sumimoto, H Tanaka. Mechanochemistry of lignin. XVII. Factors influencing mechanochemical reactions of veratrylglycerol-β-syringaldehyde ether. *Holzforschung* 48:395–399, 1994.

102. H Nimz. Der Abbau des Lignins durch schonende Hydrolyse. *Holzforschung* 20:105–109, 1966.

103. K Lundquist, R Lundgren. Acid degradation of lignin. Part VII. The cleavage of ether bonds. *Acta Chem Scand* 26:2005–2023, 1970.

104. T Higuchi, F Nakatsubo, Y Ikeda. Enzymic formation of arylglycerols from *p*-hydroxycinnamyl alcohols. *Holzforschung* 28:189–192, 1974.

105. K Lundquist, K Stern. Analysis of lignins by 1H NMR spectroscopy. *Nord Pulp Pap Res J* 4:210–213, 1989.

106. L Berndtson, K Hedlund, L Hemrå, K Lundquist. Synthesis of lignin model compounds for the glyceraldehyde 2-aryl ether type of structure. *Acta Chem Scand* B28:333–338, 1974.

107. Y Matsumoto, A Ishizu, J Nakano. Determination of glyceraldehyde-2-aryl ether type structure in lignin by the use of ozonolysis. *Mokuzai Gakkaishi* 30:74–78, 1984.
108. DCC Smith. *p*-Hydroxybenzoate groups in the lignin of aspen *(Populus tremula)*. *J Chem Soc* 1955:2347–2351, 1955.
109. DCC Smith. Ester groups in lignin. *Nature* 176:267–268, 1955.
110. J Ralph, RD Hatfield, S Quideau, RF Helm, JH Grabber, HJG Jung. Pathway of *p*-coumaric acid incorporation into maize lignin as revealed by NMR. *J Am Chem Soc* 116:9448–9456, 1994.
111. C Crestini, DS Argyropoulos. Structural analysis of wheat straw lignin by quantitative 31P and 2D NMR spectroscopy: The occurrence of ester bonds and α-O-4 substructures. *J Agric Food Chem* 45:1212–1219, 1997.
112. E Adler, G Brunow, K Lundquist. Investigation of the acid-catalysed alkylation of lignins by means of NMR spectroscopic methods. *Holzforschung* 41:199–207, 1987.
113. TK Kirk, K Lundquist. Comparison of sound and white-rotted sapwood of sweetgum with respect to properties of the lignin and composition of extractives. *Sven Papperstidn* 73:294–306, 1970.
114. B Albinsson, S Li, K Lundquist, R Stomberg. The origin of lignin fluorescence. *J Mol Struct* 508:19–27, 1999.
115. J Marton, E Adler. Reactions of lignin with methanolic hydrochloric acid: A discussion of some structural questions. *Tappi* 46[2]:92–98, 1963.
116. E Tiainen, T Drakenberg, T Tamminen, K Kataja, A Hase. Determination of phenolic hydroxyl groups in lignin by combined use of 1H NMR and UV spectroscopy. *Holzforschung* 53:529–533, 1999.
117. CW Dence. Determination of ethylenic groups. In: SY Lin, CW Dence, eds. *Methods in Lignin Chemistry.* Berlin: Springer-Verlag, 1992, pp. 435–445.
118. K Lundquist. NMR studies of lignins. 2. Interpretation of the 1H NMR spectrum of acetylated birch lignin. *Acta Chem Scand* B33:27–30, 1979.
119. O Faix, DS Argyropoulos, D Robert, V Neirinck. Determination of hydroxyl groups in lignins: Evaluation of 1H-, 13C-, 31P -NMR, FTIR and wet chemical methods. *Holzforschung* 48:387–394, 1994.
120. S Li, K Lundquist. A new method for the analysis of phenolic groups in lignins by 1H NMR spectrometry. *Nord Pulp Pap Res J* 9:191–195, 1994.
121. H-M Chang, EB Cowling, W Brown, E Adler, G Miksche. Comparative studies on cellulolytic enzyme lignin and milled wood lignin of sweetgum and spruce. *Holzforschung* 29:153–159, 1975.
122. C Lapierre, C Rolando. Thioacidolysis of pre-methylated lignin samples from pine compression and poplar woods. *Holzforschung* 42:1–4, 1988; C Lapierre, B Monties, C Rolando. Thioacidolysis of diazomethane-methylated pine compression and wheat straw *in situ* lignins. *Holzforschung* 42:409–411, 1988.
123. C-L Chen. Determination of methoxyl groups. In: SY Lin, CW Dence, eds. *Methods in Lignin Chemistry.* Berlin: Springer-Verlag, 1992, pp. 465–472.
124. JR Obst, LL Landucci. Quantitative C-13 NMR of lignins: Methoxyl-aryl ratio. *Holzforschung* 40 (Suppl):87–92, 1986.
125. A Björkman. Lignin and lignin-carbohydrate complexes: Extraction from wood meal with neutral solvents. *Ind Eng Chem* 49:1395–1398, 1957.
126. A Björkman. Studies on finely divided wood. Part 3. Extraction of lignin-carbohydrate complexes with neutral solvents. *Sven Papperstidn* 60:243–251, 1957.
127. BO Lindgren. The lignin-carbohydrate linkage. *Acta Chem Scand* 12:447–452, 1958.
128. TE Timell. Wood hemicelluloses. Part I. *Adv Carbohydrate Chem* 19:247–302, 1964.
129. TE Timell. Wood hemicelluloses. Part II. *Adv Carbohydrate Chem* 20:409–483, 1965.

130. RF Helm. Lignin-polysaccharide interactions in woody plants. In: WG Glasser, RA Northey, TK Schultz, eds. *Lignin: History, Reactions and Materials*. Washington, DC: American Chemical Society, 1999, pp. 161–171.

131. K Lundquist, R Simonson, K Tingsvik. On the composition of dioxane-water extracts of milled spruce wood: Characterization of hydrophilic constituents. *Nord Pulp Pap Res J* 5:107–113, 1990.

132. K Lundquist, R Simonson, K Tingsvik. On the composition of dioxane-water extracts of milled spruce wood: Characterization of hydrophilic constituents. *Nord Pulp Pap Res J* 5:199, 1990.

133. R Tanaka, T Koshijima. Fractionation of Björkman LCC and separation of acetyl glucomannan from the LCC. *Mokuzai Gakkaishi* 18:403–408, 1972.

134. T Koshijima, R Tanaka. Fractionation of the lignin-containing hemicellulose extracted with dimethyl sulfoxide from Japanese red pine. *Cellul Chem Technol* 6:609–620, 1971.

135. G Meshitsuka, ZZ Lee, J Nakano, S Eda. Studies on the nature of lignin-carbohydrate bonding. *J Wood Chem Technol* 2:251–267, 1982.

136. T Watanabe, J Azuma, T Koshijima. A convenient method for preparing lignin-carbohydrate complexes from Pinus densiflora wood. *Mokuzai Gakkaishi* 33:798–803, 1987.

137. K Lundquist, B Ohlsson, R Simonsson. Isolation of lignin by means of liquid-liquid extraction. *Sven Papperstidn* 80:143–144, 1977.

138. JR Obst. Frequency and alkali resistance of lignin-carbohydrate bonds in wood. *Tappi* 65:109–112, 1982.

139. Ö Eriksson, BO Lindgren. About the linkage between lignin and hemicelluloses in wood. *Sven Papperstidn* 80: 59–63, 1977.

140. Ö Eriksson, DAI Goring, BO Lindgren. Structural studies on the chemical bonds between lignins and carbohydrates in spruce wood. *Wood Sci Technol* 14:267–279, 1980.

141. JH Grabber, RD Hatfield, J Ralph, Z Jerzy, N Amrein. Ferulate crosslinking in cell walls isolated from maize cell suspensions. *Phytochemistry* 40:1077–1082, 1995.

142. NN Das, SC Das, AK Mukherjee. On the ester linkage between lignin and 4-O-methyl-D-glucurono-D-xylan in jute fiber (Corchorus capsularis). *Carbohydrate Res* 127:345–348, 1984.

143. JP Joseleau, C Gancet. Selective degradation of the lignin-carbohydrate complex from aspen wood. *Sven Papperstidn* 84:R123–R127, 1981.

144. JP Joseleau, R Kesraoui. Glycosidic bonds between lignin and carbohydrates. *Holzforschung* 40:163–168, 1986.

145. T Watanabe. Structural studies on the covalent bonds between lignin and carbohydrate in lignin-carbohydrate complexes by selective oxidation of the lignin with 2,3-dichloro-5,6-dicyano-1,4-benzoquinone. *Wood Res* 76:59–123, 1989.

146. K Freudenberg, G Grion. Beitrag zum Bildungsmechanismus des Lignins und der Lignin-Kohlenhydrat-Bindung. *Chem Ber* 92:1355–1363, 1959.

147. K Freudenberg, JM Harkin. Modelle für die Bindung des Lignins an die Kohlenhydrate. *Chem Ber* 93:2814–2819, 1960.

148. K Lundquist, R Simonson, K Tingsvik. Studies on lignin carbohydrate linkages in milled wood lignin preparations. *Sven Papperstidn* 83:452–454, 1980.

149. K Lundquist, R Simonson, K Tingsvik. Lignin carbohydrate linkages in milled wood lignin preparations from spruce wood. *Sven Papperstidn* 86:R44–R47, 1983.

150. J Comtat, J-P Joseleau, C Bosso, F Barnoud. Characterization of structurally similar neutral and acidic tetrasaccharides obtained from the enzymic hydrolyzate of a 4-O-methyl-D-glucurono-D-xylan. *Carbohydrate Res.* 38:217–224, 1974.

151. K Tanaka, F Nakatsubo, T Higuchi. Reactions of guaiacylglycerol-β-guaiacyl ether with several sugars II. Reactions of quinonemethide with pyranohexoses. *Mokuzai Gakkaishi* 25:653–659, 1979.

152. J Sipilä, G Brunow. On the mechanism of formation of non-cyclic benzyl ethers during lignin biosynthesis. Part 3. The reactivity of a β-O-4-type quinone methide with methyl-α-D-glucopyranoside in competition with vanillyl alcohol. The formation and the stability of benzyl ethers between lignin and carbohydrates. *Holzforschung* 45 (Suppl):3–7, 1991.

153. B Kosiková, D Joniak, L Kosáková. On the properties of benzyl ether bonds in the lignin-saccharidic complex isolated from spruce. *Holzforschung* 33:11–14, 1979.

154. H Taneda, J Nakano, S Hosoya, H-M Chang. Stability of α-ether type model compounds during chemical pulping processes. *J Wood Chem Technol* 7:485–498, 1987.

155. K Li, RF Helm. Synthesis and rearrangement reactions of ester-linked lignin-carbohydrate model compounds. *J Agric Food Chem* 43:2098–2103, 1995.

156. R Kondo, T Sako, T Iimori, H Imamura. Formation of glycosidic lignin-carbohydrate complex in the enzymatic dehydrogenative polymerization of coniferyl alcohol. *Mokuzai Gakkaishi* 36:332–338, 1990.

157. Y Xie, S Yasuda, H Wu, H Liu. Analysis of the structure of lignin-carbohydrate complexes by the specific 13C tracer method. *J Wood Sci* 46:130–136, 2000.

158. O Karlsson, T Ikeda, T Kishimoto, K Magara, Y Matsumoto, S Hosoya. Ozonation of a lignin-carbohydrate complex model compound of the benzyl ether type. *J Wood Sci* 46:263–265, 2000.

8 Thermal Properties of Isolated and *in situ* Lignin

Hyoe Hatakeyama and Tatsuko Hatakeyama

CONTENTS

INTRODUCTION

Lignin is an amorphous polymer crosslinked with covalent and hydrogen bonds. There are many papers on molecular mass and molecular mass distributions of lignins [1–12]. X-ray diffractograms of lignins indicate that the molecular association of the amorphous chains has a broad distribution compared to synthetic polymers such as polystyrene and quenched polyethyleneterephtalate [13]. The solid-state molecular motion of isolated lignins, such as milled wood lignin (MWL), dioxane lignin (DL), alcoholysis lignin (AL), kraft lignin (KL), blends of these lignins and chemically modified lignins, have been investigated using various experimental techniques.

These methods are dilatometry [14], infrared spectroscopy [15], viscoelastic measurements [16–20], thermal analysis [17,21–34], nuclear magnetic relaxation in both broad-line NMR and pulsed NMR [17,35], and dielectric [36], piezoelectric [37], and acoustic [38] measurements.

In this chapter, polymer properties related to thermal stability and molecular motion of lignin in the solid state are discussed. Several papers on the molecular motion of *in situ* lignin have been published [39–42]. However, it is difficult to measure a single component in a biocomposite such as wood, since it contains three macromolecular species: cellulose, hemicellulose, and lignin. In the past, the sensitivity of available experimental instruments was insufficient to detect the molecular behavior of such a complex material. Therefore, polymer properties of *in situ* lignin are discussed according to results that have recently been obtained using thermal analysis and viscoelastic measurements.

THERMAL STABILITY OF ISOLATED LIGNIN

The thermal stability of various kinds of lignins, such as MWL, DL, AL, and KL has been investigated by thermogravimetry (TG) [22,31,43–45] and simultaneous measurements of TG-Fourier-transform infrared spectroscopy (FTIR) [44,45]. Figure 8.1 shows TG curves and TG derivatograms of KL (Indulin®, $M_n = 1 \times 10^3$, $M_w/M_n = 2.2$) and AL (Alcell®, $M_n = 1 \times 10^3$, $M_w/M_n = 2.1$) at temperatures from 20 to 800°C measured in a flowing nitrogen atmosphere (200 mL/m). TG curves and TG derivatograms of KL, AL, wood powder of Japanese cedar (*Cryptomeria japonica*) and fir (*Abies firma*) are also shown in Figure 8.1 in order to show how the difference in the composition of isolated lignins and wood powder affects their thermal

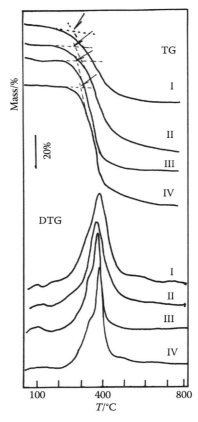

FIGURE 8.1 TG curves and derivative TG (DTG) curves of kraft lignin (KL), alcoholysis lignin (AL) and the wood powder of Japanese cedar and hardwood I; KL, II; AL, III; wood powder of Japanese cedar, IV; wood powder of fir. Decomposition temperature is indicated by arrows.

behavior. Air-dried samples in powder form (about 10 mg) were heated at 20°C/min starting from room temperature. The TG curves indicated that a slight mass decrease occurred at temperatures between 0 and 150°C. This mass decrease is attributed to vaporization of residual water in the sample. The air-dried samples contained a certain amount of bound water whose vaporization was detected at a temperature higher than 100°C. Vaporization was not completed while heating to 100°C, due to the interaction between water molecules and the hydroxyl group of the samples [46,47]. Thermal decomposition of the samples started at around 200°C. The extrapolated decomposition temperature (indicated with the arrow in Figure 8.1) of KL, AL and two kinds of wood powder was observed at around 321, 294, 324, and 331°C, respectively. Derivative TG curves of lignin samples were broader than those of the wood powder. The peak temperature (T_p) appeared in derivative TG curves of lignin at around 390°C. In contrast, T_p of the wood powder was observed at around 370°C and 390°C. Accordingly, in the derivative TG curves of the wood powder, the peak at 390°C was attributed to lignin and a shoulder at 370°C was attributed to cellulose. The residual mass of lignin at 500°C was higher than that of wood powder. This characteristic feature of lignin thermal decomposition in a flowing nitrogen atmosphere indicates that lignin phenolic groups condense during thermal degradation. This condensation during thermal decomposition is considered to be very similar to the process of pyrolysis under nitrogen [48].

Figures 8.2 and 8.3 show stacked curves obtained by TG-Fourier transform infrared spectroscopy (FTIR) of KL measured in air and nitrogen. TG-FTIR measurements were carried out using a Seiko TG 220 themogravimeter equipped with a JASCO FTIR7000 spectrometer [44,45,49,50]. Sample mass was 10 mg and heating rate was 200°C/min. Airflow rate was controlled at 100 mL/min. The gases evolved during the thermal degradation of KL were simultaneously analyzed by FTIR. Spectra were recorded at 30-second intervals; each spectrum is the average of 10 one-second scans. The spectral resolution was 1 cm⁻¹.

The major absorption for KL in air at 2345 cm⁻¹ (Figure 8.2) was attributed to CO_2. The peak appeared at around 320°C, reached maximum height at around

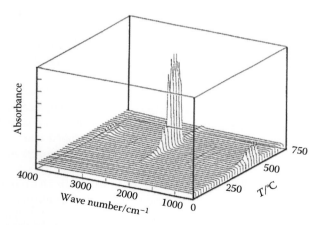

FIGURE 8.2 TG-FTIR of kraft lignin (KL) measured in air.

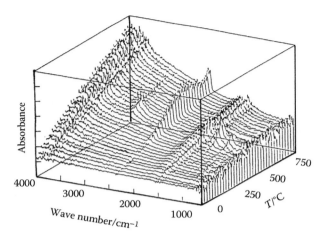

FIGURE 8.3   TG-FTIR of kraft lignin (KL) measured in N_2.

450°C and was complete at around 550°C. Two other small peaks were observed at 3740 cm^{-1} (vH$_2$O) and at 670 cm^{-1} (vCO$_2$). The absorption peaks for KL under nitrogen (Figure 8.3) corresponded to CO$_2$ (2358 cm^{-1}), C–O–C (1128cm^{-1}), C=O (1720 cm^{-1}), CH (2850 and 2920 cm^{-1}) and water (3400 and 3740 cm^{-1}).

Figures 8.1 through 8.3 indicate that lignin was stable in the initial stage of thermal treatment and that the residual mass was large compared with that of polysaccharides. At the high temperatures where gasification occurs, lignin decomposes into CO$_2$ and H$_2$O in air. In nitrogen, lignin decomposes into CO$_2$, H$_2$O and compounds having C–O–C, C=O and CH groups.

MOLECULAR MOTION OF ISOLATED LIGNIN

GLASS TRANSITION BEHAVIOR OF ISOLATED LIGNIN

The glassy state is, thermodynamically, a nonequilibrium state. Thus, the glass transition is a molecular relaxation process having a time dependent nature. The glass transition behavior of various types of lignins has mainly been investigated by dilatometry [14], differential scanning calorimetry (DSC) [21,22,25], viscoelastic measurements [18,19] and broad-line NMR [17,35].

Amorphous polymers like lignin exhibit a glass transition when a hard, brittle, glass-like state is transformed into a rubbery or viscous state by heating. Relaxation phenomena are observed below and above the glass transition temperature (T_g). The glass transition of polymers is observed by DSC as a stepped increase in the heat capacity (C_p) of the sample during heating due to enhancement of molecular motion in the polymer.

Figure 8.4 shows the DSC heating curve of kraft lignin (KL) and alcoholysis lignin (AL). KL, supplied by the Westvaco Co., had a molecular mass (M_n) of 1.1×10^3 g/mol and molecular mass distribution (M_w/M_n) = 2.2. The AL had a molecular mass (M_n) of 1.0×10^3 and molecular mass distribution (M_w/M_n) = 2.1.

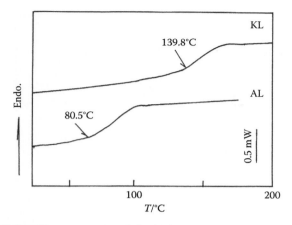

FIGURE 8.4 DSC heating curves of kraft lignin (KL) and alcoholysis lignin (AL).

Measurements were carried out using a Seiko differential scanning calorimeter at a heating rate of 10°C/min, a sample mass of 5 mg and in a N_2 atmosphere. KL showed a base line difference showing glass transition at around 140°C. In contrast, the base line difference of AL was observed at around 83°C. Glass transition temperature (T_g) is defined as the extrapolated onset temperature [51] and is indicated by the arrow in Figure 8.4. As an amorphous polymer, lignin undergoes chain segment motion upon heating. This motion is characteristic of all amorphous polymers, and is indicated by an endothermic shift in the DSC curve (see Figure 8.4). The main chain motion of various types of lignin has been investigated and the values agreed well with each other [13,18,21].

The T_g of lignin is observed at a temperature range that is typical of that for amorphous synthetic polymers [51]. The T_g value of lignin is explained by various molecular factors, such as the presence of rigid phenyl groups in the main chain [17], crosslinking, interchain hydrogen bonding [26] and its molecular mass [22]. The T_g difference between KL and AL may be attributed to the difference in the crosslinking density of these lignins.

The change in heat capacity at T_g (ΔC_p) was 0.23 J g^{-1} deg^{-1} for KL and 0.39 J g^{-1} deg^{-1} for AL. This suggests that the molecular mobility of KL is less than that of AL. This agrees well with the T_g difference between KL and AL, indicating that KL has higher crosslinking density than AL. However, the ΔC_p value of both lignins indicates that the molecular mobility is similar to that of synthetic polymers if differences in T_g between the polymers is taken into account, since it is known that ΔC_p decreases with increasing T_g [52].

X-ray diffractograms indicate that lignin frozen in the glassy state has a broader distribution of structures than synthetic amorphous polymers, such as polystyrene [13]. The maximum point of the broad X-ray peak pattern is an index of the average intermolecular distance d. Figure 8.5 shows d of dioxane lignin (DL) and methylated DL (MDL) determined from such measurements as a function of temperature. DL was extracted from wood powder (cedar, *Hyderia decuurens*) and MDL was prepared by methylation of this DL using diazomethane [26]. A Rigaku X-ray diffractometer

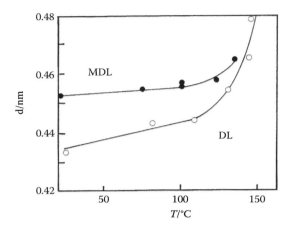

FIGURE 8.5 Intermolecular distance *(d)* of dioxane lignin (DL) and methylated dioxane lignin (MDL) as a function of temperature.

equipped with a temperature controlling system was used for the measurements. Figure 8.5 shows that *d* increased at around 140°C suggesting that the intermolecular distance expands at T_g. The *d* values of MDL were higher than those of DL. It is thought that methylating the OH groups of lignin increases the intermolecular distance. At the same time, the number of intermolecular hydrogen bonds decreases due to the decrease of the number of hydroxyl groups by methylation.

Heat capacities (C_p) of DL [25] and wood cellulose [53] (crystallinity 38%, calculated from X-ray diffractometry [49]), are shown in Figure 8.6. The C_p values of lignin were higher than those of cellulose, which monotonically increased with increasing temperature. The effect of crystallinity on C_p values of cellulose is reported elsewhere [25].

Due to the nonequilibrium nature of the glassy state, its enthalpy is reduced as a function of time and temperature. When lignin is cooled slowly, or annealed at an appropriate temperature less than T_g, the endothermic peak at T_g increases as a function of time. The rate of peak intensity increases depending on temperature and time of annealing. Accordingly, enthalpy relaxation can be observed in lignin in the glassy state.

Figure 8.7 shows DL annealed at 122°C for various time intervals. The endothemic peak of DL increases with increasing time. At the same time, the peak temperature shifts to higher temperature. In general, the relaxation time of the glassy state of polymers can be calculated using DSC data. Usually, the relaxation time of lignin is slower than that of engineering plastics having phenyl groups in the main chain. This slow relaxation of lignin may be attributed to crosslinking of lignin molecules.

GLASS TRANSITION IN THE PRESENCE OF WATER

The molecular motion of hydrophilic polymers is enhanced in the presence of water [47]. Water molecules break intermolecular hydrogen bonds and make

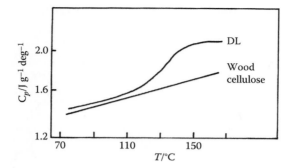

FIGURE 8.6 Heat capacity of dioxane lignin (DL) and wood cellulose.

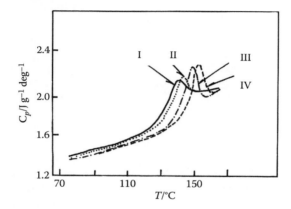

FIGURE 8.7 C_p curves of dioxane lignin (DL) annealed at 122°C for various time intervals I; 60 min, II; 100 min, III; 940 min, IV; 1300 min.

segmental motion occur easily. It is reported that T_g of polymers having hydrophilic groups in the molecular chain, such as OH groups, decreases with increasing water content until the water content reaches the maximum amount of bound water [26,47]. The number of hydroxyl groups influences the amount of water bound to lignin [54]. Figure 8.8 shows a schematic illustration of a lignin molecule associated with water.

Figure 8.9 shows the relationships between T_g and water content for DL and methylated dioxane lignin (MDL). The T_g of DL in the dry state (corresponding to $Wc = 0$ in Figure 8.7) was 152°C and that of MDL was 125°C. It is reasonable to consider that Methylation of hydroxyl groups will decrease T_g. Since the hydroxyl hydrogen atom is replaced with a methyl group, the number of inter and/or intrachain hydrogen bonds decreases. The molecular rotation of the large methoxyl groups enhances the segmental motion of the lignin molecular chain. To determine the effect of water, samples of known water content were hermetically sealed in an aluminium pan and T_g was measured. With increasing W_c, T_g decreased and leveled off at 0.10 g/g for DL and 0.08 g/g for MDL. The T_g at each leveled off W_c was 57°C for DL and 67°C for MDL. This indicates that the increased number of methoxyl groups in MDL increased the hydrophobicity of this lignin.

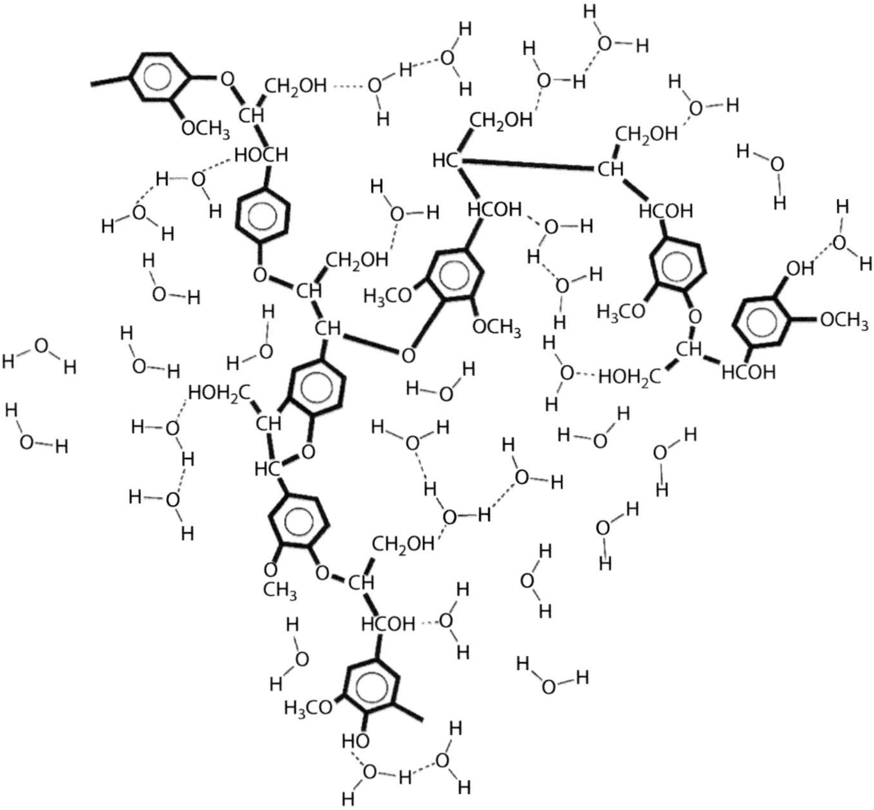

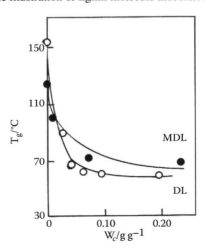

FIGURE 8.8 Schematic illustration of lignin molecule associated with water.

FIGURE 8.9 Relationships between T_g's of dioxane lignin (DL), methylated dioxane lignin (MDL) and water content (W_c).

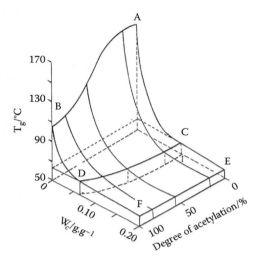

FIGURE 8.10 Three dimensional diagram between T_g, W_c, and degree of acetylation of dioxane lignin (DL).

To investigate the influence of the number of hydroxyl groups on the change of lignin T_g in the presence of water, DL was acetylated to different degrees using acetic anhydride. The optical density of the $C = O$ stretching infrared band of the acetyl groups was used to determine the degree of acetylation [26]. Figure 8.10 shows a three-dimensional diagram of T_g (°C), W_c (g/g) and degree of acetylation (%). The line AB indicates that T_g decreased with increasing acetyl content. The introduction of the bulky acetyl group enhanced the segmental motion of lignin. The curves of T_g vs W_c at constant acetyl content showed that T_g decreased with increasing water content. This was due to the disruption of hydrogen bonds by the addition of water. The line CD indicates the water content where the T_g reached a constant value, approximately $W_c = 0.07$ g/g [47]. The above results agree well with those reported elsewhere [1,41,55,56].

MOLECULAR MOTION OF *IN SITU* LIGNIN

Detection of the segmental motion of lignin in wood is difficult for various reasons. In DSC, a high sensitivity is necessary to detect the limited mass of lignin in wood. In dynamic mechanical analysis using conventional viscoelastic measurements the dynamic modulus (E') of wood exceeds a measurable range (over $E' = 10^{10}$ Pa), especially at a low temperature. However, due to recent advances in thermal analyzers and computer systems, it has been become possible to observe the segmental motion of *in situ* lignin.

GLASS TRANSITION BEHAVIOR OF *IN SITU* SOFTWOOD LIGNIN IN THE DRY STATE

Figure 8.11 shows DSC heating curves of wood powder (fir, *Abies firma*) annealed at various temperatures. For the sample annealed at 131°C, the gap in the baseline can be observed at 131°C (indicated by the arrow in Figure 8.11). As the plots in

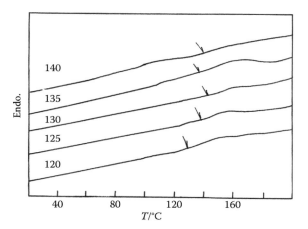

FIGURE 8.11 DSC heating curves of wood powder of fir treated at various temperatures for 60 min.

Figure 8.6 indicate, dry cellulose shows no transition in a temperature range from 0 to 200°. On this account, it is reasonable to consider that the baseline gap observed in Figure 8.11 is attributed to the glass transition of *in situ* lignin. As reported previously [1,7,21,22], T_g's of isolated lignin are usually observed in a temperature range from ca. 90 to 190°C.

The T_g's are affected by various factors such as molecular weight, molecular weight distribution, crosslinking and change of chemical structure, which depend on the isolation process. However, in the case of the T_g of *in situ* lignin, it may be appropriate to consider that the change of T_g depends only on the characteristics of the wood, such as species and growth history. The T_g value of *in situ* lignin shown in Figure 8.11 is reasonable, considering that crosslinking reactions occur during the isolation of lignin. The heat capacity difference at T_g of *in situ* lignin (0.03 J/g deg in wood) is smaller than that of the isolated lignin (0.3 J/g deg). This is due to the coexistence of lignin with polysacchades in wood. The free rotational and translational motion of the lignin is limited because of the intermolecular bonding between lignin and cellulose and hemicellulose.

Figure 8.12 shows the relationship between T_g (T_g is indicated by the arrow in Figure 8.11) and the heat-treatment temperature. T_g increased with increasing heat treatment, suggesting that the molecular arrangement of lignin varied with heat treatment. The T_g increase by heat treatment was about 15°C, a value smaller than that of isolated lignins. This suggests that the motion of *in situ* lignin is restricted because a part of the lignin exists as lignin-carbohydrate complexes. Tsujiyama et al. have reported DSC studies on lignin-carbohydrate complexes [33]. The T_g value (ca. 140°C) is in accord with the present result.

GLASS TRANSITION BEHAVIOR OF *IN SITU* SOFTWOOD AND HARDWOOD LIGNIN

Fresh samples of three kinds of wood, Japanese cedar (*Cryptomeria japonica*), camellia (*Camellia japonica*) and ginkgo (*Ginkgo biloba*) were cut into blocks

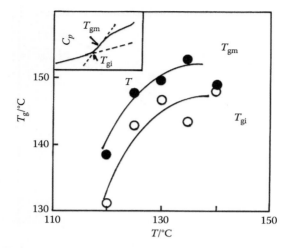

FIGURE 8.12   Relationships between T_g and temperature of heat-treatment of wood powder of fir. Determination of glass transition from C_p curves. T_{gi} = initial temperature of glass transition; T_{gm} = middle temperature of glass transition.

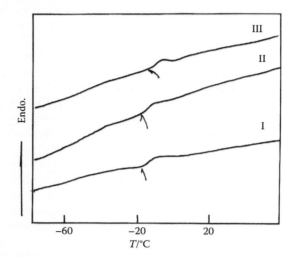

FIGURE 8.13   DSC heating curves of natural woods (g water / g sample) I; Japanese cedar (0.039), II; ginkgo (0.044), III; camellia (0.20).

(width 20 mm, length 100 mm, thickness 20 mm). Each block was sliced and sandwiched between filter papers and maintained for 20 hours at 20°C and RH 70% under a slight pressure. The water content of the conditioned samples was 0.039 g/g for Japanese cedar, 0.044 g/g for camellia and 0.020 g/g for ginkgo. Each sample was hermetically sealed in an aluminum pan and was measured by DSC. Figure 8.13 shows DSC heating curves. A baseline gap due to the glass transition occurred at around −20°C, which can be attributed to the main chain motion of *in situ* lignin. In order to confirm that the baseline gap was attributable to T_g, the heating rate

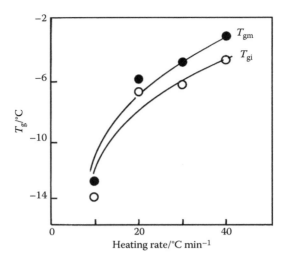

FIGURE 8.14   Heating rate dependency of T_g of natural softwoods (Japanese cedar).

dependency was examined (Figure 8.14). The temperature showing the baseline gap increased with increasing heating rate, indicating that the T_g determination of *in situ* lignin was reliable. Compared with those of isolated lignins, the T_g's of *in situ* lignin decreased markedly because of the association with water. The shift of temperature was about 140°C, a large value compared with the temperature shift of typical hydrophilic synthetic polymers. The temperature decreases observed in the synthetic polymers after the introduction of water were 90–100°C [47,57]. The large shift in T_g of *in situ* lignin is attributed to the simultaneous enhancement of molecular motion of the coexisting polysaccharides and lignin-carbohydrate complexes, caused by the disruption of hydrogen bonding by the association with water.

As we mentioned previously, the ΔC_p value at T_g of wood that corresponded to *in situ* lignin was 0.03 mJ/mg deg, about 10% of ΔC_p of isolated lignin (0.3 J/g deg). The above ΔC_p value of *in situ* lignin is reasonable, if we consider that the lignin content of Japanese cedar is 34.1%, and that lignin molecules in air-dried wood are very rigid due to crosslinking with polysaccharides. The above assumption is supported by viscoelastic measurements [58], which indicated that the viscoelastic properties of lignin in wet conditions are characterized by the properties of wood fiber. If water molecules in wood act as a plasticizer, wood in wet conditions can be treated as a polymer-plasticizer system. The change of T_g values in the polymer-plasticizer system is well described by the modified Gordon-Taylor equation, Equation 8.1 [59].

$$T_g = [(\phi_1 \Delta C_{p1} T_{g1}) + (\phi_2 \Delta C_{p2} T_{g2})] \,/\, [(\phi_1 \Delta C_{p1}) + (\phi_2 \Delta C_{p2})] \qquad (8.1)$$

where ϕ_1 is the weight fraction, ΔC_{p1} the heat capacity and T_{g1} the absolute temperature for water, and ϕ_2 is the weight fraction, ΔC_{p2} the heat capacity and T_{g2} the absolute temperature for wood. According to the reported values, ΔC_{p1} of amorphous ice is 1.94 J/g deg and T_{g1} is 135 K [60].

As we noted above, the T_g of wood (fir) was 130°C (403 K) (131°C for the sample annealed at 120°C, as shown in Figure 8.11). For a water content of 5%, the T_g value calculated according to Equation 8.1 was about −14°C. This T_g value is quite reasonable in comparison with the data shown in Figures 8.13 and 8.14. However, different T_g values have been measured by viscoelastic methods. Hills et al. [61] observed a softening temperature of lignin in radiata pine of 100°C. Olsson and Salmen [58] reported ranges of 64–82°C for aspen, 70–85°C for birch, 92.5–103°C for pine, and 83–100°C for spruce under water-saturated conditions, where the exact value depended on the measuring frequencies. Laborie et al. [62] reported 65–77°C for spruce with ethylene glycol, and 64–75°C for yellow-poplar with ethylene glycol, and Sadoh [63] reported 80°C for birchwood saturated with water. These results suggest that T_g values measured by DSC are very much different from the T_g values measured by viscoelastic methods. This is probably because the influence of absorbed water is different between DSC and viscoelastic measurements, since ice is not a viscoelastic body and the molecular motion is not retarded. The above estimation can be supported by the results of viscoelastic measurements that show T_g of natural Japanese cedar is about 30°C (see Figure 8.15). The samples used in this study were never dried. The above fact also suggests that molecular motion starts at a temperature lower than that of once-dried and then moisturized samples, since intermolecular hydrogen bonding is firmly established in the drying process.

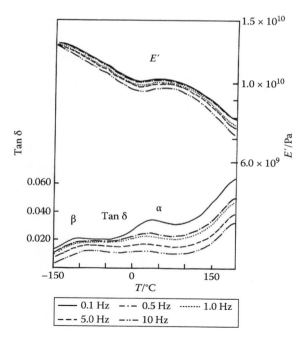

FIGURE 8.15 DMA curves of natural softwood (Japanese cedar) 0.1 Hz, 0.5 Hz, 1.0 Hz, 5.0 Hz and 10 Hz.

VISCOELASTIC PROPERTIES OF SOFTWOOD (JAPANESE CEDAR)

The viscoelastic properties of wood sheets were measured using a dynamic mechanical analyzer (DMA) equipped with a stretching type sample probe. Measurements were carried out at frequencies of 0.1, 0.5, 1.0, 5.0, and 10 Hz at a heating rate of 2°C/min. Data sampling was conducted at 1-second intervals. Wood blocks were sliced into sheets (width 5 mm, length 20 mm, and thickness 0.24 mm). W_c in this stage was 0.120 g/g for Japanese cedar, 0.115 g/g for ginkgo and 0.140 g/g for camellia. The samples containing water were cooled to −150°C from 20°C and heated to 180°C (first heating). The same sample was then cooled from 180°C to −150°C at 50°C/min and heated at 2°C/min from −150 to 180°C (second heating). During the first heating, the sample contained water at temperatures lower than 0°C, which vaporized as the sample was heated at temperatures higher than 0°C. In the second heating, the dry sample was measured.

Figure 8.15 shows the DMA curves of the first heating and Figure 8.16 shows those of the second heating. In the first heating, dynamic modulus (E') decreased in two stages from −150 to 0°C, as shown in Figure 8.15. E' slightly increased from 0°C to 80°C, which corresponded to water vaporization. In the tanδ curve, two dispersions α and β, were observed on going from high to low temperature. The sample became rigid after water vaporization. E' values of the second heating (Figure 8.16) were higher than those of the first heating in a whole range of temperatures. E' decreased smoothly in the second heating. α Dispersion was hardly observed in the second heating curves. However, when tanδ curves of second heating were magnified 10 times, two absorptions were observed, although they are not shown. The

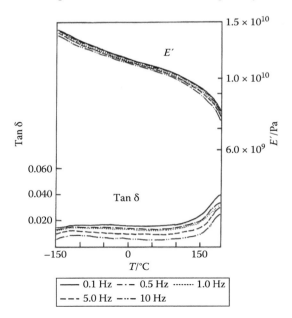

FIGURE 8.16 DMA curves of heat-treated softwood (Japanese cedar) 0.1 Hz, 0.5 Hz, 1.0 Hz, 5.0 Hz and 10 Hz.

water content, W_c, of the orginal samples used for the first heating ranged from 0.11 to 0.13 g/g, which indicated that water in the sample could be categorized as bound water [47,53]. Vaporization of bound water is not complete at 100°C due to the strong intermolecular hydrogen bonding between the hydroxyl group and water molecules [46,47]. Ordinarily, vaporization is complete at around 150°C [46]. In the current case, the sample contained a certain amount of water in the first heating in the temperature region where α dispersion was observed. In hermetically sealed conditions, such as used in DSC, the main chain motion of *in situ* lignin was found at about −20°C. In the DMA, α dispersion was observed at around 20–30°C depending on frequency. This is attributable to the main chain motion. In the second heating curve, water was not present and therefore the molecular motion could not have been enhanced.

The β dispersion of natural softwood (Japanese cedar) was not observed in the second heating. This suggests that the dispersion was related to the relaxation of water restrained in the sample. The dielectric measurement of water-absorbed wood indicated that the relaxation observed in this temperature region was related to the motion of water molecules absorbed on wood although this relaxation is dependent on frequency [36].

The activation energy (E_a) was calculated using the Arrhenius equation. Figure 8.17 shows the relationships between the reciprocal temperature of the tanδ peaks shown in Figure 8.15 and the logarithm of the measured frequencies. From the gradient of lines shown in Figure 8.17, E_a's were calculated: 512 kJ/mol for α dispersion and 59 kJ/mol for β dispersion. The E_a value of the α dispersion was consistent with the main chain motion of polymers [64]. The E_a of the β dispersion was almost the same as that of the relaxation of water sorbed on wood, as estimated by the dielectric measurement [36].

A similar change of dynamic mechanical properties of wood was observed by mild chemical treatments with chemicals such as sodium sulphite. These chemical treatments enhance the molecular motion of lignin by replacing hydroxyl and ether

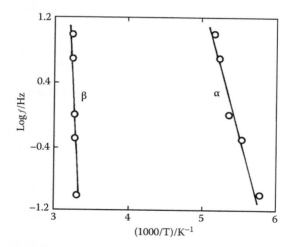

FIGURE 8.17   Relationship between reciprocal peak temperature of tan δ and frequency of natural softwood (Japanese cedar).

groups, which form hydrogen bonds with water, with solvated sulphonate ions that cannot form hydrogen bonds [39,42,65].

CONCLUSIONS

The results obtained by the above measurements suggest that the molecular motion of lignin is similar to that of synthetic polymers having phenyl groups in the main chain. However, because of the presence of hydroxyl groups in the molecular chain, the molecular motion of lignin is markedly influenced by hydrogen bonds to water molecules. It was also suggested that the molecular properties of *in situ* lignin are affected by coexisting polysaccharides through the formation of lignin-carbohydrate complexes.

ACKNOWLEDGMENTS

The authors are grateful to Dr. Shigeo Hirose, of the National Institute of Materals and Chemical Research, for his help in preparaing the manuscript. The helpful assistance of Takanori Yoshida, of Fukui University of Technology, is greatly appreciated.

REFERENCES

1. D A I Goring. Poylmer properties of lignin and lignin derivatives. In: K V Sarkanen, C H Ludwig, eds. *Lignins*. New York: Wiley-Intersicence, 1971, pp. 695–768.
2. R Morck, H Yoshida, P Kringstad, H Hatakeyama. Fractionation of kraft lignin by successive extraction with organic solvents. *Holzforshung* 40: 51–60, 1986.
3. J J Meister, E G Richards. Detemination of a polymer's molecular weight distribution by analytical ultracentrifugation. In: W G Glasser, S Sarkanen, eds. *Lignin, Properties and Materials*. Washington, DC: Am Chem Soc, 1989, pp. 58–81.
4. M E Himmel, L Tatsumoto, K K Oh, K Grohmann, D K Johnson, J L Chum. Detemination of a polymer's molecular weight distribution by analytical ultracentrifugation. In: W G Glasser, S Sarkanen, eds. *Lignin, Properties and Materials*. Washington, DC: Am Chem Soc, 1989, pp. 82–99.
5. E J Siochi, M A Haney, W Mahn, T C Ward. Molecular weight determination of hydroxy-propylated lignins In: W G Glasser, S Sarkanen, eds. *Lignin, Properties and Materials*. Washington, DC: Am Chem Soc, 1989, pp. 100–108.
6. D K Johnson, H L Chum, J A Hyatt. Molecular weight distribution studies using lignin model compounds. In: W G Glasser, S Sarkanen, eds. *Lignin, Properties and Materials*. Washington, DC: Am Chem Soc, 1989, pp. 109–123.
7. K Forss, R Kokkonen, P-E Sagfors. Detemination of molecular mass distribution studies of lignins by gel pemeation chromatography. In: W G Glasser, S Sarkanen, eds. *Lignin, Properties and Materials*. Washington, DC: Am Chem Soc, 1989, pp. 124–133.
8. P Froment, F Pla. Detemination of average molecular weight and molecular weight distribution of lignin. In: W G Glasser, S Sarkanen, eds. *Lignin, Properties and Materials*. Washington, DC: Am Chem Soc, 1989, pp. 134–143.
9. A Lindnder, G Wegener, Characterization of lignins from organosol pulping according to the organocell process. Part 3. Molecular weight detemination and investigation of fractions isolated by GPC. *J Wood Chem Technol* 10: 331–350, 1990.
10. W G Glasser, V Dave, C E Frazier. Molecular weight distribution of semi-commercial lignin derivatives. *J Wood Chem Technol* 13: 543–559, 1993.

11. U Westemark, K Gustafson. Molecular size distribution of wood polymer in birch kraft pulps. *Holtzforshung* 48: 146–150, 1994.
12. D Dong, A L Fricke. Intrinsic viscosity and the molecular weight of kraft lignin. *Polymer* 36: 2075–2078, 1995.
13. T Hatakeyama, H Hatakeyama. Temperature dependence of X-ray of amorphous lignin and polystyrenes. *Polym* 23: 475–499, 1982.
14. M V Ramiah, D A I Goring. The thermal expansion of cellulose, hemicellulose and lignin. *J Polym Sci* Part C11: 27–48, 1965.
15. H Hatakeyama, J Nakano, A Hatano, N Migita. Variation of infrared spectra with temperature for lignin and lignin model compounds. *Tappi* 52: 1724–1728, 1969.
16. M Kimura, H Hatakeyama, J Nakano. Torsional braid analysis of soft wood lignin and its model polymers. *Mokuzai Gakkaishi* 21: 624–628, 1975.
17. H Hatakeyama, K Nakamura, T Hatakeyama. Studies on factors affecting the molecular motion of lignin and lignin-related polystyrene derivatives. *CPPA Trans Tech Sect* 81: 105–110, 1980.
18. S Yano, T Hatakeyama, H Hatakeyama. Primary dispersion region of dioxane lignin. *Rept Prog Polym Phys Japan* 24: 273–274, 1981.
19. T H atakeyema, H Hatakeyama. *Thermal Properties of Green Polymers and Biocomposites*. Dortrecht: Kluwer Academic Publishers, 2004.
20. S Yano, H Hatakeyama, T Hatakeyama. Temperature dependence of the tensile properties of lignin/paper composites. *Polym* 25: 890–893, 1984.
21. H Hatakeyama, K Kubota, J Nakano. Thermal analysis of lignin by differential scanning calorimetry. *Cellulose Chem Technol* 6: 521–526, 1972.
22. H Hatakeyama, K Iwashita, G Meshizuka, J Nakano. Effect of molecular weight on glass transition temperature of lignin. *Mokuzai Gakkaishi* 21: 618–623, 1975.
23. T Nguyen, E Zavadn, E M. Barrall II. Thermal analysis of lignocellulosic materials. Part 1. Unmodified materials. *J Macromol. Sci Rev Macromol Chem* C20: 1–65, 1981. (187 refs are cited.)
24. T Nguyen, E Zavadn, E M. Barrall II. Thermal analysis of lignocellulosic materials. Part 1. Modified Materials. *J Macromol Sci Rev Macromol Chem* C21: 1–60, 1981. (135 refs are cited.)
25. T Hatakeyama, K Nakamura, H Hatakeyama. Studies on heat capacity of cellulose and lignin by differential scanning calorimetry. *Polymer* 23: 1801–1804, 1982.
26. T Hatakeyama, S Hirose, H Hatakeyama. Differential scanning calorimetric studies on bound water in l, 4-dioxane acidolysis lignin. *Makomol* 184: 1265–1274, 1983.
27. S S Kelley, T C Ward, T G Rials, W G Glasser. Engineering plastics from lignin. XVII. Effect of molecular weight on polyurethane film properties. *J Appl Polym Sci* 37: 2961–2971, 1989.
28. D Feldman, D Banu, M. Khoury. Epoxy-lignin polyblends. III. Thermal properties and infrared analysis. *J Appl Polym Sci* 37: 877–887, 1989.
29. M Shigematsu, M Morita, I Sakata. Effect of the addition of lignin-carbohydrate complex on miscibility between hemicellulose and lignin. *Mokuzai Gakkaishi* 37: 50–56, 1991.
30. H Hatakeyama. Thermal analysis in methods. In: S Y Lin, C W Dence, eds. *Lignin Chemistry*. Berlin: Springer-Verlag, 1992, pp. 200–213.
31. K Nakamura, T Hatakeyama, H Hatakeyama. Thermal properties of solvolysis lignin-derived lignocellulose. *Polym Adv Techno* 3: 151–155, 1992.
32. H Hatakeyama, S Hirose, T Hatakeyama, K Nakamura, E Kobashigawa, N Morohoshi. Biodegradable polyurethane from plant components. *J M S-Pure Appl Chem* A32: 743–750, 1995.
33. S Tsujiyama. Glass transition temperature change of lignin-carbohydrate complex degraded by wood-rotting fungi. *Bulletin Kyoto Pref Univ* 47: 43–49, 1995.

34. J Nakano, Y Izuta, T Orita, H Hatakeyama, K Kobashigawa, K Teruya, S Hirose. Thermal and mechanical properties of polyurethanes derived from fractionated kraft lignin. *Sen-I Gakkaishi* 53: 416–422, 1997.

35. H Hatakeyama, J Nakano. Nuclear magnetic resonance studies on lignin in solid state. *Tappi* 53: 472–475, 1970.

36. G Zhao, M Norimoto, T Yamada, T Morooka. Dielectric relaxation of water absorbed on wood. *Mokuzai Gakkaishi* 36: 257–263, 1990.

37. Y Suzuki, N Hirai, M. Ikeda. Piezoelectric relaxation of wood I. Effects of wood species and fine structure on piezoelectric relaxation. *Mokuzai Gakkaishi* 38: 20–28, 1992.

38. T Sasaki, M Norimoto, T Yamada, R M Rowell. Effect of moisture on the acoustical properties of wood. *Mokuzai Gakkaishi* 34: 794–803, 1988.

39. D Atack, C Heitner. Dynamic mechanical properties of sulphonated eastern black spruce. *CPPA Trans Tech Sect* 5: Tr 99–Tr 108, 1979.

40. H Becker, D Noak. Studies on the dynamic torsional viscoelasticity of wood. *Wood Sci Technol* 2: 2: 213–230, 1976.

41. W J Cousins. Elastic modulus of lignin as related to moisture content. *Wood Sci Technol* 10: 9–17, 1976.

42. C Heitner, D Atack. Dynamic mechanical properties of sulphite treated aspen. *Paperi Puu:* 66, 84–89, 1984.

43. S Hirose, H Hatakeyama. A kinetic study on lignin pyrolysis using the integral method. *Mokuzai Gakkaishi* 32: 621–625, 1986.

44. K Nakamura, Y Nishimura, P Zetterlund, T Hatakeyama, H Hatakeyama. TG-FTIR studies on biodegradable polyurethanes. *Themochimica Acta* 282/283: 433–441, 1996.

45. S Hirose, K Kobashigawa, Y Izuta, H Hatakeyama. Thermal degradation of polyurethanes containing lignin studied by TG-FTIR. *Polym International* 47: 247–256, 1998.

46. T Hatakeyama, K Nakamura, H.Hatakeyama. Determination of bound water content in polymers by DTA, DSC and TG. *Thermochimica Acta* 123: 153–161, 1988.

47. H Hatakeyama, T Hatakeyama. Interaction between water and hyrdophilic polymers. *Thermochimica Acta* 308: 3–22, 1998.

48. G G Allan, T. Mattila. High energy degradation. In: K V Sarkanen, C H Ludwig, eds. *Lignins.* New York: Wiley-Interscience, 1971, pp. 575–596.

49. P Zetterlund, S Hirose, T Hatakeyama, H Hatakeyama, A-C Albersson. Thermal and mechanical properties of polyurethanes derived from mono- and disaccharides. *Polymer International* 42: 1–8, 1997.

50. T. Hatakeyama and F. Quinn. *Thermal Analysis.* Chichester: John Wiley, 1994, pp. 107–110.

51. S Nakamura, M Todoki, K Nakamura, H Kanetsuna. Thermal analysis of polymer samples by a round robin method. I. Reproducibility of melting, crystallization and glass transtion temperatures. *Thermochimica Acta* 136: 136–178, 1988.

52. T Hatakeyama, H Hatakeyama. Effect of chemical structure of amorphous polymers on heat capacity difference at glass transition temperature. *Thermochimica Acta* 267: 249–257, 1995.

53. T. Hatakeyama, H Kanetsua, S Ichihara. Thermal analysis of polymer samples by around robin method. III. Heat capacity measurement by DSC. *Thermochimca Acta* 146: 311–316, 1989.

54. K Nakamura, T Hatakeyama, H Hatakeyama. Studies on bound water of cellulose by differential scanning calorimetry. *Tex Res J* 51: 607–613, 1981.

55. D A I Goring. Thermal softening of lignin, hemicellulose and cellulose. *Pulp Paper Mag. Can.* 64: T517, 1963.

56. A H Nissan. Elastic moldulus of lignin as related to moisture content. *Wood Sci Technol.* 11: 147, 1977.

57. K. Nakamura, T Hatakeyama, H Hatakeyama. Differetial scanning calorimetric studies on the glass transition temperature of polyhydroxystyrene derivatives containing sorbed water. *Polymer* 22: 473–476, 1981.
58. A-M Olsson, L Salmen. Viscoelasticity of *in situ* lignin as affected by structure. In: W G Glasser, H. Hatakeyama, eds. *Viscoelasticity of Biomaterials,* ACS Symposium Series, 489, 1990, pp. 133–143.
59. M Gordon, J S Taylor. Ideal copolymers and the second order transitions of synthetic rubber. I. Noncrystalline copolymers. *J Appl Chem* 2: 493–500, 1952.
60. M Sugisaki, H Suga, S. Seki. Calorimetric study on the glassy state. IV. Heat capacities of glassy water and vubic ice. *Bull Chem Soc Jpn* 41: 2591–2599, 1968.
61. W E Hills, A N Rozsa. High temperature and chemical effects on wood stability. Part 2. The effect of heat on the softening of radiata pine. *Wood Sci Technol* 19: 57–66, 1985.
62. M-P G Laborie, L Salmen, C E Frazier. Cooperatively analysis of the *in situ* lignin glass transition. *Holzforschung* 58: 129–133, 2004.
63. T Sadoh. Viscoelastic properties of wood in swelling systems. *Wood Sci Technol* 15: 57–66, 1981.
64. Y Wada. *Physical Properties of Polymers in Solid State.* Tokyo: Baihuukan Publisher, 1981, pp. 381–396.
65. B Vikstrom, P Nelson. Mechanical properties of chemically treated wood and chemimechanical pulps. *Tappi* 63(3): 87–91, 1980.

9 Reactivity of Lignin-Correlation with Molecular Orbital Calculations

Thomas Elder and Raymond C. Fort, Jr.

CONTENTS

INTRODUCTION

To date, and as can be seen from the other chapters of this text, the structure and chemistry of lignin have been described in terms of results from a wide range of chemical or spectroscopic methods to construct a mosaic picture of the polymer. The current chapter continues this process by describing past, present and potential applications of electronic structure calculations to the chemistry of lignin, and correlation with experimental results. Although there may be a tendency to consider experiment and theory as two distinct approaches with little in common, the boundary is becoming increasingly blurred as calculations are extensively integrated with experiment. Improvements in algorithms can provide results with chemical accuracy, such that calculations now represent another "analytical" methodology, and have been described as an "M.O. spectrometer" [1]. Theoretical methods represent another tool that can be used to complement other techniques, a strategy that has found extensive applications in other branches of chemistry. It is in this spirit that this chapter is written. While this chapter is concerned with the specific applications of molecular orbital calculations to lignin, a general and brief introduction to the vocabulary of computational chemistry is also in order.

COMPUTATIONAL CHEMICAL PRELIMINARIES

We are living at a very fortunate and interesting time with respect to computational chemical techniques. Until about the mid-1980s, calculations in chemistry were primarily the domain of relatively few specialists who were largely concerned with improving theoretical predictions on small molecules or even single atoms. The calculations were, to a large extent, studies of the methodology, rather than the molecules. In the recent past, however, due to the rapid improvements in capabilities of computer hardware and software, along with major improvements in accessibility, computational methods have become readily available to practicing chemists. Calculations have become an indispensable tool, similar to spectroscopic resources. A brief perusal of the major chemistry journals will reveal computational applications ranging from small molecules (both organic and inorganic) to large polymer systems and biomolecules.

The general term "molecular modeling" is usually taken to include both classical mechanical and quantum mechanical calculations, along with visualization of results. While the charge of the current chapter is to report on molecular orbital calculations as they have been applied to lignin, a short synopsis of classical mechanical methods and some references to general literature may be of some use.

MOLECULAR MECHANICS

Classical mechanical calculations, also referred to as force-field or molecular mechanics calculations treat molecular structures, as the name implies, as a classical mechanical system, in which the energy of a structure is the sum of a number of subterms that will include, but are not limited to bond stretching, angle bending, torsion, and nonbonded terms. Stretching and bending can be described as harmonic potentials, or by using the somewhat more sophisticated Morse potential. Spring constants describe the stiffness of the bond, and deviations from the specified equilibrium distance or angle, for a given group of atoms, will increase the energy of the system. Torsional energy, associated with twisting or changes in dihedral angles, depends on a parameterized barrier height, multiplicity (i.e., the number of minima encountered in a 360° rotation) and a phase factor describing the geometry at which the minimum energy is found. The simplest nonbonded terms are described by van der Waals interactions using a Lennard-Jones potential and coulombic terms for electrostatics. More sophisticated calculations will begin to include cross-terms (e.g., stretch-bend). It is important to be aware that these calculations do not address the presence of electrons, but rather concentrate only on the behavior of the nuclei.

Owing to the computational simplicity of force-field calculations, very large systems can be examined in short periods of time. These are the methods of choice for studying the structure of synthetic polymers, proteins, nucleic acids, and inorganic networks such as zeolites. Force-fields are also used to provide the energies in molecular dynamics calculations, in which the time evolution of structures can be examined.

The parameter sets, defining the bond lengths, angles, torsion, and nonbonded interactions can be very specific for a given application, such as proteins or nucleic acids. This specificity can lead to surprising failures due to missing parameters even for relatively common atom types and combinations.

Among the major parameter sets used for force-field calculations are those from the University of Georgia, the most recently released of which is MM4 [2], which has very general applicability. Other general force-fields include Sybyl from Tripos Associates [3], and Dreiding [4] and UFF [5], both of which originated at Cal Tech. Among the more specific parameter sets are Gromos [6], AMBER [7], and CHARMm [8] for proteins and nucleic acids, and Momec for inorganic systems [9]. Major references describing the parameterization details and exhaustive examples of force-field calculations include Burkert and Allinger [10], Coomba and Hambley [11], and Rappé and Casewit [12].

MOLECULAR ORBITAL CALCULATIONS

In contrast to force-field calculations in which electrons are not explicitly addressed, molecular orbital calculations, use the methods of quantum mechanics to generate the electronic structure of molecules. Fundamental to the quantum mechanical calculations that are to be performed is the solution of the Schrödinger equation to provide energetic and electronic information on the molecular system. The Schrödinger equation cannot, however, be exactly solved for systems with more than two particles. Since any molecule of interest will have more than one electron, approximations must be used for the solution of the Schrödinger equation. The level of approximation is of critical importance in the quality and time required for the completion of the calculations. Among the most commonly invoked simplifications in molecular orbital theory is the Born-Oppenheimer [13] approximation, by which the motions of atomic nuclei and electrons can be considered separately, since the former are so much heavier and therefore slower moving. Another of the fundamental assumptions made in the performance of electronic structure calculations is that molecular orbitals are composed of a linear combination of atomic orbitals (LCAO).

Molecular orbital calculations can be divided into two large categories, described as *ab initio* and semiempirical, depending on the approximations that are made. The former method is more rigorous, using only mathematical functions and natural constants to describe the atomic orbitals of which the molecular orbitals are composed, via linear combinations. The latter calculations incorporate experimental data and the neglect of the more extensive and time consuming computational details that are explicitly performed in *ab initio* calculations. At the time of this writing, and as a function of computer and basis set, *ab initio* calculations can be feasibly performed on systems with several tens of atoms, while semiempirical calculations can handle several hundreds of atoms.

Ab Initio Calculations

Ab initio calculations are defined in terms of a method or level of theory and the basis set that is used. The methods differ in how interactions between electrons are treated while the basis sets describe the shape of the orbitals. For comprehensive discussions of *ab initio* calculations see the texts by Hehre et al. [14], Leach [15], and Young [16].

The most commonly used method is the Hartree-Fock calculation in which interactions between electrons are treated as the interaction of one electron within an average field of the remaining electrons. Electron interactions are, of course, much more specific than this, and include Pauli repulsions as well as electrostatic ones. Electron correlation can be addressed by various methods, but among the most commonly used are configuration interaction and Moeller-Plesset perturbation theory.

Configuration interaction (commonly abbreviated CI) includes excited states, which, the argument goes, could become populated due to repulsion between electrons. A linear combination of the wavefunctions for the ground state and the excited states should lead to a better (lower) energy for the atom.

In Moeller-Plesset theory, the mixing in of excited states is treated as a series of perturbations with designations MPn (usually MP2, MP3, MP4), where n designates the point at which the series is truncated. Moeller-Plesset theory is less laborious than CI, and thus has displaced the latter method in most *ab initio* calculations, where the computational labor is already high.

Recently, another class of calculations, density functional theory (DFT) has become quite common. This method is faster than *ab initio* calculations for similar levels of performance, and has the advantage that correlation effects are included, at least in part. According to this theory, the properties of a molecular system are functions of the electron density, rather than a wave function (as described by the Schrödinger equation). There is some degree of debate over whether DFT is an *ab initio* calculation or in a class by itself. Parenthetically, one of the recipients of the 1998 Nobel Prize for Chemistry was Walter Kohn, in recognition of his development of density functional theory.

Once a level of theory is selected, a basis set must be chosen. Basis sets are available in a large range of complexity, and the choice exerts considerable influence on the accuracy and time required for the calculation. Basis sets, with their specific and somewhat confusing nomenclature, are incorporated into comprehensive computer programs that will control the calculation and the amount of ancillary information that is reported. Among the standard packages for *ab initio* calculations are Gaussian, Gamess, Jaguar, and HyperChem.

Basis sets will differ in the number of basis functions that are included. The simplest type is a *minimal* basis set, in which each atom is represented by a single orbital of each type, as in sophomore organic chemistry. Thus, a carbon atom would have a 1s orbital, a 2s orbital, and one 2p orbital along each Cartesian axis. An example of a minimal basis set is STO-3G.

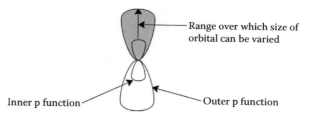

Range over which size of
orbital can be varied

Inner p function

Outer p function

FIGURE 9.1   Split basis set showing inner and outer orbitals.

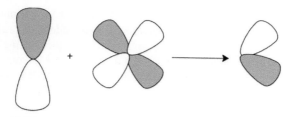

+

FIGURE 9.2   Polarization basis set of a p orbital by mixing with d function.

In the next level of sophistication, a *split valence* basis set, each valence orbital is described by two basis functions representing the inner and outer parts of the orbital. These two terms can be varied independently, and examples are the 3-21G, 6-31G basis sets. Similarly, a triply split basis set is 6-311G. Split valence basis sets will allow the size of the orbitals to change, while the addition of polarization functions, designated as 6-31G(d) or 6-31G*, allow the shape of the orbital to change. This is done by the addition of d-orbitals to all heavy atoms, such that now each nonhydrogen atom will have 15 basis functions (Figure 9.1).

To provide more accurate descriptions of anions, or neutral molecules with unshared pairs, basis sets may be further augmented with so-called diffuse functions. These are intended to improve the basis set at large distances from the nuclei, thus better describing the barely bound electrons of anions. For example, a p-orbital on carbon may be polarized away from the nucleus by mixing into it a d-orbital. Such a basis set will be designated with a "+" as in 6-31+G(d), adding an additional four basis functions to each heavy atom (Figure 9.2).

Increasingly complex basis sets are generated by further splitting of orbitals, and the addition of further polarization and/or diffuse functions. As basis functions are added, the number of orbitals to be evaluated is effectively increased, resulting in a simultaneous and exponential (fourth power) increase in time requirements for a given system. As a consequence, a computational study must take into account the size of the system, and computational resources to establish the feasibility of a particular effort.

Semiempirical Calculations

While *ab initio* calculations have become much more feasible for relatively large molecules, their performance in a reasonable length of time may be predicated on access to high throughput computers. Semiempirical calculations can, however, be readily executed on small workstations or even personal computers. While still based on quantum mechanics, these calculations are more rapid to perform than *ab initio* due to the inclusion of experimental parameter sets and the exclusion of some of the more tedious computational steps.

Historically, there have been two major efforts in the development of semiempirical calculations under the direction of John Pople (Nobel Laureate 1998) and M.J.S. Dewar. The earliest calculations of this nature are, however, the Hückel (HMO) and extended Hückel molecular orbital calculations (EHMO). While not mathematically rigorous, or used extensively at the present time, these calculations can provide considerable insight into the methods and in specific applications can be quite accurate.

True Hückel molecular orbital calculations, among other limitations, only address π-electrons. All atom, semiempirical calculations that include both π- and σ-electrons were developed originally by Pople and coworkers [17] implementing the method of complete neglect of differential overlap (CNDO), followed by the increasingly rigorous method of intermediate neglect of differential overlap (INDO) [18].

For general applications, the methods developed by Pople have largely been supplanted by the calculations from the Dewar group originating at the University of Texas. The first-generation program from this group was MINDO/3 [19], and while successful had systematic errors in heats of formation and bond angles. The subsequent method, the modified neglect of diatomic overlap (MNDO) [20] has been extensively parameterized to cover a larger number of elements, and was the foundation of the more recent AM1 [21] and PM3 [22] methods.

WHAT INFORMATION CAN BE OBTAINED AND HOW IS IT USED?

The output from a molecular orbital calculation, regardless of the specific computational method used, will always report a number of results. These will include the geometry of the system, the total energy for *ab initio* calculations, heat of formation for semiempirical calculations, the energy of each molecular orbital and the coefficients of the atomic orbitals for each atom.

The geometry that is reported at the end of a calculation will depend on how the calculation is performed and what options are initially invoked. The molecular geometry is used as the basic input for all of the calculations that have been discussed thus far, and the geometry can be described in several ways. Although current technology allows the user to sketch an input structure or build it from fragments, the positions of each atom are converted to either Cartesian or internal coordinates for computational purposes. Internal coordinates, or a Z-matrix, describe the molecular structure in terms of bond lengths, bond angles, dihedral angles and connectivity, and for most computational chemistry applications are easier to use than Cartesian coordinates [23]. If diffraction data is available, fractional coordinates that define the dimensions of the unit cell will be reported, and depending on the software, can be input directly. Fractional coordinates are generally converted by the computer program into internal coordinates.

Since the geometry of a molecule will markedly affect the energy, the development of a reasonable geometry is essential. While this might be done with chemical intuition, geometry optimization methods are more consistent and systematic in this regard. Given that energy is a function of geometry, it is theoretically simple to determine by derivative methods the geometry at which the energy is minimized, and therefore representative of the preferred structure. Difficulties arise, however, due to the multidimensional nature of the energy surface. Determining a minimum on such a surface can be problematic, and there are usually several nonlinear optimization methods (e.g., eigenvector following, Davidon-Fletcher-Powell, and Newton–Raphson) incorporated into a computational package [24].

While energy minimizations have become routine, even for relatively sophisticated *ab initio* calculations, it must be borne in mind that the geometry that is identified will only represent a local minimum. Optimization methods are always searching for a minimum and, therefore, will not cross energy barriers, even to reach lower energy structures. As a consequence, conformational searching may be required in order to insure that a suitable geometry is found [25].

Conformational searching involves varying the molecular structure followed by energy calculations, as a means of sampling the potential energy surface and identifying low-energy structures. In theory, changes in bond lengths, angles, and dihedral angles would need to be altered in a rigorous exploration of the surface. In practice, however, changing lengths and angles require large amounts of energy in comparison to changes in dihedral angles that also impart large structural changes. Most conformational search procedures therefore concentrate on modifications to dihedral angles.

There are several methods by which conformational searches can be performed, none of which are guaranteed to identify the global minimum. The most straightforward method is the grid search, in which selected dihedral angles are systematically rotated through a specific number of increments, followed by an energy calculation or minimization. Systematic searches such as these will sample all parts of the surface, but depending on how finely the rotations are divided may miss important conformations. Furthermore, grid searches can be very time consuming, since the number of conformations to be evaluated increases exponentially with the number of bonds considered. For example, for a β-O-4 dimer in which the six rotatable bonds are examined, in 60-degree increments, over 46,000 conformations would be generated. Even at an extremely optimistic calculation rate of one conformation per minute, this calculation would take more than 30 days to complete. Clearly, if a systematic search is to be employed, the researcher must use chemical intuition to limit the scope of the search.

Alternatives to systematic searches include random searches, genetic algorithms, and simulated annealing. A random search, as the name implies, generates random variations in structure, for which energy minimizations are performed and populations of conformers identified. Genetic algorithms begin with a large population of random conformations, the individuals of which have varying levels of "fitness." The more fit members are selected and the procedure is repeated to identify low-energy structures. Simulated annealing is analogous to the annealing process in metallurgy, in that the energy of the structure is increased to a very high temperature, followed by a cooling step. At suitably elevated temperatures, all of conformational space is available (i.e., energy barriers can be crossed). Upon cooling and subsequent energy minimization the low-energy conformers can be selected. In general, conformational searching, particularly the nonsystematic methods, generate so many possible conformations that the calculations must be performed via force-field calculations. Once a smaller set of likely conformations has been selected, molecular orbital calculations can be utilized.

The values reported for the energies of a molecule are in molar quantities and are not directly comparable for nonisomeric structures. Of greater utility are the computed energetics of reactions. The reactions for which the estimation of energetics is possible can be divided into several categories depending on the conservation of electron spin and electron pairs.

Reactions in which the numbers of paired electrons are not conserved are classified as nonisogyric reactions. Examples include homolytic bond dissociations and transition state calculations. In contrast, isogyric reactions conserve the number of paired electrons, such as in heterolytic bond dissociations and some isomerizations. Last, isodesmic reactions are bond conserving and not only maintain the division of electrons into shared and unshared pairs, but also maintain the number and type of bonds between heavy atoms (nonhydrogen).

Of considerable interest and difficulty with respect to calculations is the topic of transition state location. Strictly speaking, the transition state is the structure existing at the highest energy point (a first-order saddle point) on the reaction coordinate between reactants and products.

For reactants and products, experimental data can be developed to define both structures and energies. The energy of the transition state can be defined by an experimental difference from the reactant energy: the activation energy. Kinetic isotope effects can be used to infer structural features for the transition state, but in most cases the description of the molecular arrangement in the transition state is possible only by computation. From the reaction coordinate, the reactant, product, and transition states are all stationary points such that the derivative of the energy with respect to the reaction coordinate is zero. At the minima (reactant, product) the second derivative is greater than zero, while at the maxima (transition states) the second derivative is less than zero. Since the second derivatives of the reaction coordinate are in fact the force constants from which vibrational spectra may be calculated, the minima have all real vibrational frequencies while transition states have one imaginary (negative) frequency, allowing for the verification that a transition state has been identified.

Although defining and characterizing a transition state in this way once it has been found seems straightforward, actually locating it is more complex. For some processes, the reaction coordinate is relatively easy to define, making the transition state structure straightforward to characterize. For some reactions simple bond lengths or dihedral angles will correspond to the reaction coordinate. In many other cases, however, with multiple bonds forming and breaking, no single bond distance or angle adequately describes the process.

The simplest approach for determining the transition state might just be an initial guess at the structure based on chemical intuition or symmetry constraints. Somewhat better is using a transition state that has been obtained at the intended level of calculation for a simpler system, or alternatively saving computation time by locating the specific transition state at a lower level of theory, and then proceeding to the desired level. Other methods are the linear or quadratic synchronous transit (LST and QST) techniques in which the transition state is an average of reactant and product geometries. These are straightforward for unimolecular reactions but more complicated for higher order reactions, where one must create a complex of reactants and/or products. If a useful definition of the reaction coordinate can be made, one can actually follow the reaction from reactants through the transition state to products using the intrinsic reaction coordinate (IRC) approach. Optimizations are performed by sequentially constraining a given geometric term (e.g., bond length or dihedral angle) to fixed values along the reaction coordinate. Alternatively, IRC calculations may begin with a transition state structure and work backward toward reactants or forward toward products.

The atomic orbital coefficients that are printed can be used to determine electron densities, and in most computer programs are employed internally to generate graphical representations of the molecular orbitals. Furthermore, according to frontier molecular orbital theory, bond formation is dependent on overlap between the highest occupied molecular orbital of the nucleophile and lowest unoccupied molecular orbital (LUMO) of the electrophile. The magnitude of the coefficients in the appropriate molecular orbital at a particular atom provides an indication of reactivity.

Another set of values that is almost always printed in the output are the atomic charges. Due to the qualitative familiarity that chemists have with this property it is tempting to use such data as measures of reactivity as well. Charge is not, however, a quantum mechanical term and arises from an arbitrary partitioning of electron density. Among the most common of the partitioning schemes is Mulliken population analysis in which electron density is summed over orbitals centered on atoms, and between nuclei electron density is evenly divided. Due to this somewhat artificial method of assignment, charge must be used with caution. Charge is also sensitive to the basis set that is used, such that the absolute values are not particularly significant and cannot be compared between methods or structures. Relative charges can be use for limited and clearly defined purposes.

A critical part of applying computational methodology to the solution of chemical problems lies in defining the problems in such a way that the "computables" enumerated above can be related to chemical observables.

APPLICATIONS TO LIGNIN CHEMISTRY

The use of molecular orbital calculations as a tool in lignin chemistry has been reviewed in the past by Shevchenko [26], Remko [27] and Elder [28]. It will be seen from these papers that a large proportion of the early work in this area was reported in journals from the former Soviet Union and Eastern Europe. The relative unavailability of these periodicals in the west has made this branch of lignin chemistry somewhat inaccessible to many researchers.

Reports on the application of molecular orbital calculations specifically to the chemistry of lignin dates back at least as far as 1966 and the work of Lindberg and coworkers [29,30,31]. In the paper by Lindberg and Tylli [30], Hückel molecular orbital calculations were used in studies on a range of monomeric model compounds to determine π-electron charge densities, bond orders, and free valences. A linear relationship was detected between these data and the rate of protodeuteration. Similarly, the relative ease of ether cleavage reactions involving the methoxyl group was predicted among the models. Finally, the electrochemical reduction of model aldehydes was generally related to the calculated electronic results.

The later work [31], extended the initial results and revealed an excellent correlation between pK of phenols and the change in π-electron energy between free phenolic and the phenolate anion. Also, a generally linear relationship was reported for the critical oxidation potential (COP) and the energies of the highest occupied molecular orbital for the phenolate anions.

RADICALS

Among the earliest examples of the application of molecular orbital calculations to lignin chemistry is, as might be expected, the formation and reactivity of phenoxy radicals [32]. This work described the use of Pariser-Parr-Pople molecular orbital calculations, comparing various methods to determine the π-electron spin densities for a number of model compounds. The different techniques gave similar results, with the bulk of the unpaired spin density at the phenolic oxygen, followed by the *ortho* and *para* carbons. In a cinnamaldehyde model, the β-position also exhibited considerable unpaired spin density.

In more recent work on coniferyl alcohol, using an all-electron method (MNDO) and geometry optimization [33] it was found that while the phenolic oxygen does indeed have appreciable unpaired spin density, the ortho and para carbons within the aromatic ring are considerably larger in this respect, with the β-position being about equal to the oxygen. Similarly, Russell et al. [34], using AM1 semiempirical calculations on cinnamic acids and alcohols, in concert with electron spin resonance spectroscopy, found that while relatively low, the spin density at the phenolic oxygen increased with methoxyl content. The spin densities at the β-position and the available ortho-carbons were about equal, and interpreted as accounting for the β–β and β–5 linkages that are reported to dominate the polymer produced from coniferyl alcohol.

The coupling of radicals to form σ-complexes and dilignols was examined from a thermodynamic standpoint with PM3 calculations [35]. All combinations of σ-complexes formed through positions with positive spin densities were completed. Among these, the products with significant steric hindrance were energetic, while the β–5, β–β, β–O were the most stable, with the 5-5 couple intermediate in energy. These results are in accord with those reported from experimental results on the initial coupling reactions [36].

PULPING AND BLEACHING

Arguably the most important reactions of lignin from an industrial perspective are those concerned with pulping and bleaching. These have also been the subject of investigation using molecular orbital calculations. Computational research of this type has included work on conventional processes, alternative pulping catalysts and enzymatic bleaching.

Given the importance of the quinone methide in kraft pulping, the nature of this intermediate and its reactivity toward nucleophiles has been the subject of considerable scrutiny. These reactions have been interpreted in terms of hard-soft acid-base (or more generally electrophile/nucleophile) theory. In this theory, a hard acid (or electrophile) will generally have a high-energy LUMO and a positive charge, whereas a hard base (or nucleophile) will have a low-energy highest unoccupied molecular orbital (HOMO) and a negative charge. Interactions between hard reactants occur through coulombic attraction. Conversely, soft bases will have a high-energy HOMO and do not always carry a negative charge, while soft acids will have a low-energy LUMO and may or may not have a positive charge. Soft–soft reactions occur due to interactions between the LUMO of the electrophile and HOMO of the nucleophile, the frontier molecular orbitals (FMO) [37].

These phenomena, with respect to lignin chemistry were summarized by Zarubin and Kiryushina [38] who presented the results of Hückel molecular orbital (presumably EHMO) calculations for a large number of monomeric and dimeric model compounds. From this work it was proposed that the reactive sites in lignin are soft entities, and that the sulfhydryl anion is softer than the hydroxide ion, accounting for the preferential reactions of SH⁻ in kraft pulping.

In a closely related study, Elder et al. [39], using all-atom semiempirical calculations, found that quinone methide intermediates exhibited a small partial negative charge at the α-carbon. This result led to questions over how nucleophilic attack could occur at a negatively charged carbon. Upon examining the frontier molecular orbitals, however, they found that the LUMO of the α-carbon was quite large, indicating the participation of the molecular orbitals rather than coulombic attraction in the initial reactions of kraft pulping.

Hard–soft acid–base arguments were also advanced by Shevchenko and Zarubin [40], using the results of CNDO/S calculations to show that both quinone methides and benzyl cations are soft electrophiles. Furthermore, it was proposed that the addition of a nucleophile to the α-carbon of the quinone methide was stereospecific, due to the nonequivalence of the orbital lobes on either side of the quinone methide plane.

Alternative pulping catalysts have also been addressed through molecular orbital calculations on anthraquinone and its various reactive forms and adducts. Shevchenko and Zarubin [40] interpret the reactivity of the C–10 position and the reduced form of anthraquinone in terms of the atomic orbital contributions to the HOMO. Elder et al. [41] performed MNDO calculations on anthraquinone, its radical anion and dianion, in an attempt to distinguish between a single electron transfer or ionic-adduct mechanism, respectively. The computational results indicated that the reaction of the dianion was considerably more exothermic than the anion-radical. The electronic data were, however, at variance with the proposed adduct mechanism, in that for both of the reduced forms the oxygen atoms had a very large partial negative charge, while the C–10 position carried a partial positive charge. Additionally, the oxygen atoms have generally higher HOMO electron densities than the salient carbons.

Of similar technological importance with pulping are the reactions related to pulp brightness, its generation and reversion. Both of these topics have been the subject of theoretical studies.

Although chlorine bleaching is largely a discontinued practice, the chemistry of lignin chlorination was examined by the application of MNDO calculations to coniferyl and sinapyl alcohols [42]. It has been proposed that the chlorination reaction is an electrophilic aromatic substitution, occurring through the chloronium ion at C–6, with side-chain displacement occurring at C–1, and also attack at the β-carbon in an ethylenic side chain. From calculations, it was reported that the aromatic carbon positions exhibit the largest HOMO coefficients, while the heat of reaction associated with reaction at the β-position is the lowest.

More recently, Garver [43] presented an excellent overview of the reactions of lignin model compounds toward both electrophilic and nucleophilic bleaching agents, comparing computational results with experiment. This MOPAC-93 study was interpreted in terms of charge densities, frontier orbitals and transition state theory. For example, the presence of the γ-carbonyl in coniferaldehyde resulted in a large partial positive charge at the γ-carbon. From these data, questions were raised about the possible mechanism for peroxide attack, and it was suggested that ionized phenolate groups and methoxyl groups would tend to separate the charge rather than diminish the partial positive charge at the carbonyl carbon. Garver [43] also invoked frontier molecular orbital theory in terms of both electron density and the energetics of orbital interaction, describing the relatively small energy differences between the quinone methide LUMO and various attacking nucleophiles. Finally, the reaction between acetoguaiacone (3-methoxy-4-hydroxyacetophenone) and the hydrogen peroxide anion in a totally chlorine free (TCF) bleaching process was considered. It has been proposed in the literature that acetoguaiacone exists as a tautomeric pair, with the hydrogen peroxide anion attacking the α-carbon of the quinone methide form. Calculations indicate that, based on charge density, the reaction alternatively occurs via the acetoguaiacone tautomer.

In other work related to bleaching chemistry, the reactions of chlorine dioxide with monomeric [44] and dimeric [45] lignin model compounds have been studied computationally. These studies closely parallel experimental work in which oxidation mechanisms were proposed [46–51]. In accord with the experimental work, which reports higher reactivity of phenolic compounds, the heats of reaction for these compounds are lower than those for etherified models. The experimentally based mechanisms were generally found to be energetically feasible, but in some cases the electronic results were not consistent with the proposed mechanisms.

The chemistry of enzymatic bleaching has been addressed using calculations [52] in a study of the lignin peroxidase complex II. A relationship was detected between the HOMO energy of the substrates and the association complex. In more recent work, Elder [53] reported on the oxidation reactions of veratryl alcohol itself and its interaction with a dimeric lignin model. Veratryl alcohol is a secondary metabolite produced by *Phanerochaete chrysosporium* and is a putative mediator for lignin peroxidase. Computationally, it was found that the oxidation of veratryl alcohol to veratraldehyde is largely controlled by the initial single-electron transfer, which is quite endothermic. In the addition of water to the veratryl alcohol cation radical, a number of reaction channels may be followed, but the formation of veratraldehyde is favored, in accord with experiment. Interaction of the veratryl alcohol cation radical with a lignin model indicates that when in close proximity, the unpaired electron density is being transferred to the lignin model, whereas at larger separations, the single electron predominates on the veratryl alcohol. In an *ab initio* study directly concerned with the interaction of veratryl alcohol and the lignin peroxidase active site [54], a distinct shift in the unpaired spin density toward the veratryl alcohol was observed, as the interunit distance decreased. Energetically, it was also found that the oxidation of veratryl alcohol is slightly exothermic.

Mediators essential for laccase bleaching of pulp have been the subject of coupled experimental and computational research by Sealy et al. [55]. This study examined the effect of substitution on N-hydroxybenzotriazole, and the activity of phthalimides. N-hydroxybenzotriazole itself was more effective in lowering kappa numbers than any of the derivatives that were tested. Computationally, the bond dissociation energy was generally lower for N-hydroxybenzotriazole than the derivatives. Similarly, the most active of the phthalimide mediators had the lowest bond dissociation energy.

Cole et al. [56] have reported on the brightness reversion reactions of high-yield pulps that have been subjected to lignin retaining bleaching. The reactions of thioglycerols with quinones, which have been implicated in the photoyellowing process, have been examined as a method for brightness stabilization. Computationally, a Michael addition reaction was favored over a redox reaction for the model quinones that were studied. Furthermore, HOMO–LUMO energy differences paralleled the relative reactivities of the quinones. Among the less reactive quinones, the difference in heats of reaction between the Michael addition and redox mechanism were reduced, such that the latter route could become more competitive.

SPECTROSCOPY

Theoretical calculations can be of considerable utility in interpreting, verifying or refuting mechanisms that have been proposed from experimental evidence. Given the complexity of even model compound reactions, however, the mechanisms proposed, while not unreasonable, may not be unequivocal. In contrast, spectroscopic results, while possibly open to interpretation, represent a definite dataset. While a given observation may be explained by a number of conclusions; it is the interpretation and not the spectral data that is open to debate. Given this definition of the results from spectroscopy, the ability of theoretical methods to reproduce or correlate to spectroscopic data is an important measure of their validity.

Among the pioneering work on the application of molecular orbital calculations to lignin in general and spectral data in particular is the large body of research by Milan Remko and coworkers. These efforts, beginning in 1977 [57] were largely concerned with ultraviolet spectra using CNDO/CI and CNDO/s. Beginning with work on monomeric structures [57,58] good agreement with experiment was reported. Subsequent work was performed on quinone methides [58,59]. In later work [60,61], they applied CNDO/S to condensed and β–O–4 dimers. In addition to the work on specific lignin models, flavonoid, coumarone, and stilbene structures were addressed using the π-electron method of Pariser-Parr-Pople, which agreed well with experiment, especially for the determination of absorption maxima [62].

Similar work from the former Soviet Union used Hückel calculations to predict the UV/VIS spectra of lignin substructures [63]. In later papers, all-electron CNDO/S calculations were reported which were in good agreement with experiment [64]. Shevchenko et al. [65] showed the influence of conformation on UV spectra and also reported on the agreement between calculation and experiment. In work related to that of Remko, Burlakov et al. [66] compared theoretical and experimental results for quinone methides.

Working with quinones, implicated in the photoyellowing of high-yield pulps, Cole et al. [33] showed that semiempirical methodology could generate a very good linear relationship between HOMO–LUMO gap and UV absorptions, although absolute values of λ_{max} cannot be calculated at this level of theory. Furthermore, they were capable of reproducing accurately the IR spectrum of methoxybenzoquinone, using *ab initio* calculations and an appropriate scaling factor. Vibrational spectroscopy of lignin models has also been treated using force-field calculations by Jakobsens et al. [67].

While nuclear magnetic resonance spectroscopy has become an invaluable tool in lignin analysis, relatively little effort has been directed toward the calculation of such spectra for lignin model compounds. Liptaj et al. [68] compared conformational data for cinnamaldehyde models derived from NMR to calculation, and similar work on quinone methides at both semiempirical and *ab initio* levels has been reported by Konschin et al. [69].

The calculation of NMR chemical shifts has become feasible recently with improvements in both the speed and capabilities of computational chemical methods [70]. While the bulk of the theoretical work has been directed toward the calculation of ^{13}C chemical shifts, results for ^{17}O [71], ^{31}P [72], and hyperfine coupling constants for nitrogen [73] have also been reported. Calculated ^{13}C chemical shifts for a number of monomeric lignin models have been compared to experiment, with excellent correlations [74]. The sensitivity of the calculations to geometry is highlighted by the poor ability to reproduce the chemical shifts of methoxyl groups and other rotatable structures.

Although there have been few studies of this nature directly concerned with lignin or lignin models, there is a body of work on closely related substituted aromatics. In a study combining NMR and *ab initio* calculations on 1-(p-anisyl) vinyl cations, a 6-31G basis set provided successful information on both chemical shifts and rotational barriers [75]. In subsequent work, Facelli et al. [76] used low-temperature NMR techniques to examine observable differences between the chemical shifts of carbons that are ortho to the methoxyl group in anisole. The ^{13}C chemical shift tensors were calculated with the GIAO method, using self-consistent fields (SCF) and density functional theory and SCF with MP2 electron correlation. It was found that the isotropic shifts were most accurately calculated using density functional theory, while the MP2 method gave somewhat poorer results than SCF. Vanillin and 3-4-dimethoxybenzladehyde were addressed in a coupled computational-experimental study, in which the structures were optimized at the MNDO semiempirical level, followed by GIAO calculations with a 6-311G** basis set [77]. The calculated values are generally high except for the methoxyl carbons. Among the remaining positions, the errors range from 3.4 to 10.9 ppm. This is in contrast to the results of Facelli et al. [76] in which the calculated shifts were generally lower than experiment. The results of Zheng et al. [77] may arise due to the sensitivity of NMR calculations to geometric accuracy [78]. Furthermore, a recent paper [79] compares a number of *ab initio* methods for calculating NMR shielding tensors, and recommends the use of hybrid density functional-Hartee-Fock methods.

In other analytical work, Shevchenko et al. [80] found excellent agreement between theoretical and experimental ionization potentials. Semenov and Khodyreva (81) compare photoelectron spectra with CNDO/S3 calculations and found good results. Semiempirical calculations, on a number of lignin model compounds have also been used as an aid in the interpretation of ESR spectroscopy [34]. In more general research on substituted aromatic systems [82] correlated experimental pK_a values to semiempirical heats of formation and HOMO energies, with correlations in the 0.7–0.9 range.

CONFORMATION

The final area of research on lignin that has been addressed to any large extent by the application of computational methods deals with conformation and the influence of hydrogen bonds. Until the recent appearance of such work by Simon and Eriksson [83,84], this topic was dominated by researchers from the group in Riga, Lativa [85] and Milan Remko and coworkers in Bratislava.

Work on monomeric lignin models was initiated by Remko and Polčin in 1976, using CNDO/2 to examine intramolecular hydrogen bonding in guaiacyl and syringyl models [86], hydroxyphenyl carbonyls [87], and the conformation of *cis*-cinnamaldehyde [88]. The strengths of the hydrogen bonds are reported to be quite variable, with the guaiacyl/syringyl models being very small (0.5–6.3 kJ/mol), while the 2-hydroxyphenyl carbonyls are much larger (33–44 kJ/mole). In both cases a planar conformation was found. Qualitatively similar results were found using peturbation configuration interaction using local orbitals (PCILO) [89].

Based on the results of intramolecular hydrogen bonding, work was begun on intermolecular bonding to elucidate the nature of lignin-carbohydrate complexes. Initially, small polar molecules were used as models for carbohydrates. The complexes formed between the lignin-models and polar molecules were generally strongly hydrogen bonded [90–92]. Analogous work was done with sugar structures representing cellulose and hemicelluloses, and methanol representing the lignin, in which somewhat less energetic hydrogen bond networks were found [93]. As computing power improved, true lignin-carbohydrate complexes were modeled using monosaccharides and benzyl esters and ethers [94,95]. The latter paper indicated a large accessible potential energy surface.

Work on dimeric lignin structures began with the relatively simple biphenyls [96] in which the nonplanarity of the compounds was revealed. A comprehensive study of the conformation of various interunit linkages was published in 1983 [97]. Large, inaccessible areas were found on the potential energy surface of the β-aryl ethers. A solvation study, using PCILO found no relative changes in energies as a function of solvation [98], while within a β-aryl ether, an intramolecular hydrogen bond was detected between the α-OH and β-ether oxygen, along with the confined conformational space previously reported [99].

Using semiempirical calculations, Shevchenko et al. [100] studied the impact of conformation on charge and frontier molecular orbitals in quinone methides and benzyl cations. Charges were sensitive to conformation, while the frontier molecular orbitals were not. These data were used to interpret the stereoselectivity of quinone methides toward nucleophiles. Additional related work addressed the Z–E conformation about the C–1–C–α double bond in quinone methides. Konshcin et al. [69] reported that according to NMR results, the Z-conformation predominates a result that is in accord with their *ab initio* calculations. In similar work [101] it was proposed that the energetic barrier between the conformations was low enough to allow for interconversion at room temperature, taking issue with results from NMR reported by Ralph and Adams [102]. In subsequent work, it was reported that there is a significant barrier to rotation, as evidenced by both calculation and the resolution of both conformers by NMR [103].

In a series of recent studies, Simon and Eriksson have reported on the conformation and hydrogen bonding patterns of lignin model compounds using both force-field [83] and *ab initio* molecular orbital calculations. Among the latter papers, the first work dealt with a phenoxypropane diol, used to model the β–O–4 system in lignin and compare results from the MM3 force-field and HF/6-31G* calculations with respect to geometry, energy and hydrogen bonding [83].

The geometries obtained by the optimization of several conformers using both computational methods did not correspond well, and the force-field calculations resulted in more flexible structures. Based on the *ab initio* calculations, revised MM3 parameters were developed that improved the agreement between the methods. It was also proposed that the presence of hydrogen bonding would alter the rotation of the aromatic ring. This hypothesis was tested by the use of methoxybenzene and water, placed at varying distances from the methoxyl oxygen. As the interatomic distance increased there was a simultaneous decrease in hydrogen bonding, and increases in both the energy of the complex and rotational barrier.

A subsequent study explicitly addressed the conformation of dimeric β–O–4 structures [84], again using MM3 and HF/6-31G* in conjunction. Preliminary work on monomeric methoxylated compounds indicated the presence of an intramolecular hydrogen bond between primary alcoholic groups and the methoxyl group, stabilizing the structure. From these smaller structures, true β–O–4 dimers were constructed, in an effort to minimize computer time requirements. The most stable conformers for both *threo* and *erythro* configurations were found to be extended, rather than folded structures.

The most recent contribution from this group [104] concerned the elasticity of the β–O–4 structures, as related to the macroscopic mechanical properties of wood. Given that both elongation and compression of a molecular system are mainly concerned with torsional changes, rather than the more energetic perturbations in bond lengths and angles, the energies associated with such rotations are the principal focus of the paper. It was reported that when placed under tension, i.e., bond rotation to accommodate extension, the responses were similar between guaiacyl, syringyl, and *p*-hydroxyphenyl structures. Due to the presence of hydrogen bond networks, however, differences can be observed between threo and erythro configurations. As the methoxyl content increases, the rotational barrier and therefore stiffness increases.

ADDENDUM

During the period between the initial preparation of this chapter and its publication, several papers have appeared in the literature that need to be noted. As might be expected, these more recent papers benefit from improvements in computer technology and software such that the methods used are generally the more sophisticated *ab initio* and density functional theory calculations. The latter offer advantages based on wide and successful application in other areas of chemistry and physics, the ability to address relatively large molecular systems, and the use of exchange correlation potentials, accounting for electron correlation effects, making these calculations more rapid to perform than the perturbation calculations that are done in conjunction with Hartree-Fock calculations.

The reactivity and coupling of lignols and the formation of the lignin polymer has been addressed at such higher levels of theory by Durbeej and Eriksson [105–108], Sarkanen and Chen [109], Elder et al. [110], Shigematsu et al. [111], and Martinez et al. [112].

The coniferyl alcohol radical and six dilignols representing the major interunit linkages present in lignin have been examined at the B3LYP/6-31G(d,p) level followed by single-point calculations using B3LYP/6-311++G(2df,p) [105]. Furthermore, the effect of conformation on spin density was determined. In accord with previous studies, polymerization was predicted to occur at positions with large, positive, unpaired spin density. The coniferyl alcohol radical, in particular, and the dimers, in general, had the highest spin density at the phenolic oxygen as would be expected based on the electronegativity of oxygen.

Durbeej and Eriksson [106] have also examined whether the subsequent dimerization reaction is under thermodynamic control by determining if the distribution of interunit linkages is correlated with the stability of the dimers. The structures studied were β–O–4, β–5, β–β, β–1, 5–5, and 5–O–4 linkages. Conformational searching for each structure was done at the PM3 level, followed by optimization with B3LYP/3-21G to identify a small number of low-energy conformers. These resultant structures were then optimized at the B3LYP/6-31G(d,p) level followed by a single-point B3LYP/6-311G(2df,p) calculation with solvation. The energetic results for the dimers indicated that the β–β structures would be dominant at room temperature, with a small amount of β–5 present. In contrast, experimental results have reported that the major coupling product is β–5, followed by β–β and β–O–4 in relatively large proportions and low levels of 5–5. It is gratifying to note that calculation finds that the β–β and β–5 structures have the lowest (albeit reversed) energies. In contraindication of the thermodynamic control theory, however, the 5–5 structure is calculated to be more stable than the β–O–4.

The β–O–4 coupling reaction has been studied using B3LYP/6-31G(d,p), and B3LYP/6-311G(2df,p) both with and without solvation at fixed Cβ–O4 distances from 2.52 to 1.72Å [107]. The mechanisms evaluated were radical-radical coupling to generate a quinone methide and coupling of a phenoxy radical to coniferyl alcohol resulting in a radical dimer. Since a saddle point was not identified, these are not true transition state calculations, but energetic results were reported. Radical-radical coupling was found to be an exothermic reaction (about −23 kcal/mole) with a maximum energy 2–5 kcal/mole greater than the isolated reactants. In contrast, the reaction between coniferyl alcohol and the phenoxy radical was only slightly exothermic with an energy maximum 6–10 kcal/mole above the reactants. While the former reaction would be favored thermodynamically, the radical formed in the latter reaction could be an important participant in lignin polymerization.

Related to this work is a paper by Shigematsu and coworkers [111] in which the PM5 semiempirical parameter set was used to evaluate the transition state for the coupling of p-coumaryl alcohol radicals through β–O–4 bond formation. In contrast to the density functional results [105], it was reported that the phenoxy oxygen had a lower spin density than the salient carbons. It was also found that the energy barrier from the reactants to the transition state was on the order of 10 kcal/mole, while the overall heat of reaction was approximately −18 kcal/mole. These differences are undoubtedly due to the use of semiempirical calculations rather than the more rigorous density functional methods.

In other work using higher levels of theory, Sarkanen and Chen [109] have examined the interaction of monolignols using MP2/6-31G(d) calculations. This work has been undertaken to evaluate the proposal that a preexisting macromolecular template of lignin influences subsequent polymerization reactions. Geometry optimizations indicate that an energetic minimum is obtained with a planar configuration between aromatic rings on neutral monolignols.

The addition of monolignol radicals to dilignol radicals, a reaction analogous to the endwise polymerization process proposed as a mechanism by which lignin chain growth occurs, was studied computationally by comparing the orbital energy gaps as determined by UHF/6-31g(d) optimizations followed by a single-point UB3LYP/6-311G calculation [110]. In general, the energy gaps were the lowest for combinations known from experiment to undergo cross-coupling, while those with larger energy gaps represented reactions that, based on experiment, do not occur.

The reactivity of monolignols has been examined by Martinez and coworkers [112], reporting on the application of Fukui functions to determine the susceptibility of reactive sites to nucleophilic, electrophilic, and free radical attack. Equilibrium conformations were determined at the B3LYP/6-31G(d,p) level, and electron densities at B3LYP/6-311++G(d,p) for coniferyl, sinapyl, and p-coumaryl alcohols. Given the importance of radical chemistry as part of the lignin polymerization process the authors focus on the results related to free radical attack, indicating that the positions with the largest magnitude of the Fukui function represent the softest and therefore most reactive sites toward radicals. The largest values occurred at the phenolic oxygen and β-carbon. Other positions within the aromatic ring, with appreciable Fukui functions were C–1 and C–4 both of which are sterically hindered and therefore not readily amenable to reaction. It is interesting to note that minimal levels of this parameter were found for the C–5 position in coniferyl alcohol and both C–3 and C–5 in p-coumaryl alcohol, arguing against 5–5′ coupling.

Several of these papers [105–107] along with a discussion of archival reports on the application of computational methods to lignin polymerization are summarized in a review article by Durbeej et al. [108].

Calculations have also been recently applied to photochemical phenomena including the role of stilbenes [113] in photoyellowing of paper, interactions of lignin models with singlet oxygen [114,115] and the behavior of dibenzodioxocins [116,117]. In the first paper [113], time-dependent density functional theory calculations were performed on several substituted stilbene structures and the effect on vertical excitation energy was determined. The results compare well with previously reported theoretical data as well as with experimental results. In the work on singlet oxygen reactions [114,115], using MP2/6-31G(d,p) methods and invoking frontier molecular orbital interactions it was found that for the compounds studied, the HOMO of the model would interact with the LUMO of the singlet oxygen. Furthermore, by evaluating the atomic orbital coefficients it was proposed that etherified and phenolic monomethoxyl dimers would react through the B-ring, while a dimethoxyl model would react through the A-ring, both of which are in accord with reaction products observed experimentally.

As part of a study on the photochemistry of dibenzodioxocins, time-dependent density functional and semiempirical calculations were performed on phenolic and nonphenolic model compounds [116]. Based on experimental observations the reaction coordinates for the sequential cleavage the α–O–4 and β–O–4 bonds of the first excited state were examined. It was found that the activation energy of the α–O–4 bond rupture was similar for both phenolic and nonphenolic models, while the subsequent β–O–4 cleavage was more endothermic for the phenolic model. These results are interpreted in terms of the presence of the activating ether group, which is proposed not to be capable of strengthening the β–O–4 bond. Electron densities indicate that upon α–O–4 bond breaking the phenolic model takes on a quinone-methide structure. In addition, an analysis of spin densities shows that the α–O–4 bond undergoes a homolytic cleavage reaction. The time-dependent density functional theory calculations were used to determine excitation energies and oscillator strengths, which were reported to be in accord with experimental results.

In another study on the chemistry of dibenzodioxocins, PM3 semiempirical calculations were used to assess charge distributions and heats of reactions for a phenolic model and the corresponding anion [117]. As might be expected, the largest differences in the charge distribution are centered on the phenolic and phenolate oxygen. Bond cleavage at α–O–4 and β–O–4, shifted the negative charge to the new, free oxygen site resulting from bond breaking. Furthermore, it was found that both α–O–4 and β–O–4 bond cleavage reactions were endothermic, the latter required more energy.

Computational methods coupled with experimental spectroscopy for lignin have been used by Barsberg et al. [118] and Salazar-Valencia et al. [119]. The former paper utilized density functional theory calculations to aid in the interpretation of resonance Raman spectroscopy of lignin radicals. The results from the calculations, which were in good agreement with experiment, were used to assign the vibrational bands to guaiacyl and syringyl moieties. In the latter paper, MNDO and ZINDO/S CI results were reported for dimeric and trimeric lignin models. As the size of the model increases, it is reported that the oscillator strength begins to approach experimental values.

Among the more important reactions of lignins are oxidations, associated mainly with bleaching chemistry. While some experimental oxidation potentials have been reported for lignin model compounds, they can be difficult to perform, necessitating the use of pulse radiolytic methods. Calculated oxidation potentials may, therefore, be of utility in predicting the tendency of specific groups to undergo redox reactions of industrial importance. This has been addressed in two recent papers, the first of which [120] optimized the geometry of eight monolignols using B3LYP/6-31G(d) calculations. These geometries were subjected to single-point calculations at the B3LYP/6-311++G(2df,p) level of theory, with the determination of zero point energies and solvation. Bond dissociation energies, ionization potentials, spin densities, and charges for the phenolic oxygen were reported. Oxidation potentials were determined using a thermodynamic cycle for two possible mechanisms and compared to published results at pH = 7.0.

The first mechanism assumed the formation of a cation radical while the latter, beginning with a neutral phenol, had a phenoxy radical, a proton and an electron as the reaction products. The former mechanism overestimated the oxidation potential while the latter was more accurate when compared to experimental results for four models.

In a closely related paper [121] the calculated oxidation potentials for eight phenolic compounds for which experimental results are known were correlated to develop a calibration curve. From these data the oxidation potentials of coniferyl alcohol, sinapyl alcohol, anisole, guaiacol, and a pinoresinol dimer were predicted. This paper applied B3LYP/6-31G(d) optimizations to both gas phase and solvated models, and compared the results to experimental data at pH = 0. Based on a correlation coefficient of 0.93 for the calibration curve, the oxidation potentials of the unknowns were determined. The relative results from both of these papers are similar, with dimethoxy compounds having lower oxidation potentials than the monomethoxy compounds.

During the recent past there has been an intense renewal of interest in the use of lignocellulosic biomass as an alternative to fossil-based energy resources. As part of this effort, the thermal degradation of lignin to produce fuels or chemicals has also been a resurgent topic of research. In addition to experimental work of this type, computational methods have been employed in the elucidation of pyrolysis mechanisms of lignin [122,123]. Density functional theory calculations on a phenylethyl ether model compound indicated that the proposed homolytic cleavage reaction has a barrier of 57 kcal/mole occurring at a bond distance of 2.1 Å [122]. Additional work on the selectivity of hydrogen abstraction from the α and β carbons, also performed with density functional theory was found to be in close agreement with experiment [123].

The material properties of lignin have been examined by the calculation of Young's modulus of lignin by subjecting a dimeric model compound to strain, coupled with the determination of energy and stress [124]. The computational results, derived from semiempirical, Hartree–Fock and density functional theory calculations are in agreement with available experimental results. Changes in geometry indicate that modifications in dihedral angles occur in response to linear strain. At larger levels of strain, bond rupture is evidenced by abrupt changes in energy, structure, and charge. Based on the current calculations, the bond scission may be occurring through a homolytic reaction between aliphatic carbon atoms. These results may have implications in the reactivity of lignin especially when subjected to processing methods that place large mechanical forces on the structure.

CONCLUDING REMARKS

It can be seen from the text that through dramatic improvements in computer hardware and software, computational chemical methods have become increasingly accessible to nonspecialists and have also become an integral tool in the practice of chemistry. This is, to some extent, reflected in lignin chemistry, but when compared to other analytical tools that have been brought to bear on this enigmatic polymer there are still relatively few applications of calculation. It has been the intention of the current chapter to demonstrate the feasibility, capabilities, and perhaps even more important, the limitations of these methods and how they can and have been applied to lignin.

REFERENCES

1. Clark, T. 1985. *A Handbook of Computational Chemistry.* Wiley-Interscience. New York.
2. Allinger, N. L., K. Chen, and J-H. Lii. 1996. An improved force field (MM4) for saturated hydrocarbons. *J. Comp. Chem.* 17:642–668.
3. Clark, M., R.D. Cramer III, and N. van Opdenhosch. 1989. Validation of the general purpose tripos 5.2 force field. *J. Comp. Chem.* 10:982–1012.
4. Mayo, S. L., B. D. Olafson, and W. A. Goddard. 1990. DREIDING: A generic force field for molecular simulations. *J. Phys. Chem.* 94:8897.
5. Rappé, A. K., C. J. Casewit, K. S. Colwell, W. A. Goddard III, and W. M. Skiff. 1992. UFF, a rule-based full periodic table force field for molecular mechanics and molecular dynamics simulations. *J. Am. Chem. Soc.* 114: 10024.
6. Hermans, J., H.J.C. Berendsen, W.F. van Gunsteren, and J.P.M. Postma. 1984. A consistent empirical potential for water-protein interactions. *Biopolymers* 23:1513–1518.
7. Weiner, S. J., P. A. Kollman, D. T. Nguyen, and D. A. Case. 1986. An all atom force field for simulations of proteins and nucleic acids. *J. Comp. Chem.* 7:230–252.
8. Brooks, B. R., R. E. Bruccoleri, B. D. Olafson, D. J. States, S. Swaminathan, and M. Karplus. 1983. CHARMM: A program for macromolecular energy, minimization, and dynamics calculations. *J. Comp. Chem.* 4:187–217.
9. Bernhardt, P. V., and P. Comba. 1992. Molecular mechanics calculations of transition metal complexes. *Inorg. Chem.* 31: 2638.
10. Burkert, U., and N. Allinger. 1982. *Molecular Mechanics.* ACS Monograph 177. American Chemical Society. Washington, DC.
11. Coomba, P., and T. Hambley. 1995. *Molecular Modeling of Inorganic Compounds.* VCH. Weinheim.
12. Rappé, A., and C. Casewit. 1997. *Molecular Mechanics across Chemistry.* University Science Books. Sausolito, CA.
13. Born, M., and J. R. Oppenheimer. 1927. On the quantum theory of molecules. *Ann. Physik.* 84:457.
14. Hehre, W., L. Radom, P. v.R. Schleyer, and J. Pople. 1986. *Ab Initio Molecular Orbital Theory.* Wiley-Interscience. New York.
15. Leach, A. R. 1996. *Molecular Modelling: Principles and Applications.* Longman. Harlow, England.

16. Young, D. 2001. *Computational Chemistry: A Practical Guide to Applying Techniques to Real World Problems.* Wiley-Interscience. New York.

17. Pople, J. A., and G. A. Segal. 1965. Approximate self-consistent molecular orbital theory. II. Calculations with complete neglect of differential overlap. *J. Chem. Phys.* 43(10, Pt. 2):S136–S149, discussion, S150–S151.

18. Pople, J. A., David, L. B. and Paul A. D. 1967. Approximate self-consistent molecular-orbital theory. V. Intermediate neglect of differential overlap. *J. Chem. Phys.* 47(6):2026–2033.

19. Bingham, R. C., Michael, J. S. Dewar, and Donald H. Lo. 1975. Ground states of molecules. XXVII. MINDO/3 calculations for carbon, hydrogen, oxygen, and nitrogen species. *J. Am. Chem. Soc.* 97(6):1302–6.

20. Dewar, M. J. S., and W. Thiel. 1977. Ground states of molecules. 38. The MNDO method. Approximations and parameters. *J. Am. Chem. Soc.* 99(15):4899–4907.

21. Dewar, M. J. S., Eve G. Zoebisch, Eamonn F. Healy, and James J. P.Stewart. 1985. Development and use of quantum mechanical molecular models. 76. AM1: A new general purpose quantum mechanical molecular model. *J. Am. Chem. Soc.* 107(13):3902–3909.

22. Stewart, J. J. P. 1990. MOPAC: A semiempirical molecular orbital program. *J. Comput.-Aided Mol. Des.* 4(1):1–105.

23. Young, D. 2001. Constructing a Z-matrix. Chapter 9 in *Computational Chemistry: A Practical Guide to Applying Techniques to Real World Problems.* Wiley-Interscience. New York.

24. Leach, A. R. 1996. Energy minimization and related methods for exploring the energy surface. Chapter 4 in *Molecular Modelling: Principles and Applications.* Longman. Harlow, England.

25. Young, D. 2001. Conformational searching. Chapter 21 in *Computational Chemistry: A Practical Guide to Applying Techniques to Real World Problems.* Wiley-Interscience. New York.

26. Shevchenko, S. M. 1994. Theoretical approaches to lignin chemistry. *Croatica Chemica. Acta.* 67(1):95–124.

27. Remko, M. 1986. Application of quantum chemistry to wood chemistry. *Chem. Listy.* 80(6):606–617.

28. Elder, T. 1989. Application of computational methods to the chemistry of lignin. Chapter 19 in *Lignin: Properties and Materials.* Wolfgang G. Glasser and Simo Sarkanen, eds. ACS Symposium Series, 397. American Chemical Society. Washington, DC.

29. Lindberg, J. J., and H. Tylli. 1966. LCAO-MO investigations on lignin model compounds. *Suomen kemistilehti Z* 39:122.

30. Lindberg, J. J., and H. Tylli. 1967. Investigations on the reactivity of lignin and lignin models by using the LCAO-MO method (Hueckel method). *Finska Kemists. Medd.* 76(1):1–18.

31. Lindberg, J. J., and A. Henriksson. 1970. Reactivity and quantum mechanics of lignin. *Suom. Kemisteseuran Tiedonantoja* 79(2):30–36.

32. Mårtensson, O., and G. Karlsson. 1969. β-electron spin densities of lignin constituents. *Arkiv für kemi* 31(2)5–16.

33. Elder, T., and S. D. Worley. 1984. The application of molecular orbital calculations to wood chemistry: The dehydrogenation of coniferyl alcohol. *Wood Sci. Technol.* 18: 307–315.

34. Russell, W. R., Alex R. F, A. Chesson, and M. J. Burkitt. 1998. Oxidative coupling during lignin polymerization is determined by unpaired electron delocalization with parent phenylpropanoid radicals. *Arch. Biochem. Biophys.* 332(2):357–366.

35. Elder, T., and R. M. Ede. 1995. Coupling of coniferyl alcohol in the formation of dilignols. A molecular orbital study. Vol. I., pp. 115–122. Proceedings of Eighth International Symposium on Wood and Pulping Chemistry. Helsinki, Finland.

36. Chioccara, F., S. Poli, B. Rindone, T. Pilati, G. Brunow, P. Pietikåinen, and H. Setålå. 1993. Regio- and diastereo selective syntheses of dimeric lignans using oxidative coupling. *Acta Chem. Scand.* 47:610–616.

37. Fleming, I. 1976. *Frontier Orbitals and Organic Chemical Reactions.* Wiley-Interscience. New York.

38. Zarubin, M. Y., and M. F. Kiryushina. 1987. Ways of the acceleration of the wood delignification according to the modern ideas of the acid-base interaction. Vol. I, pp. 407–413. Proceedings Fourth International Symposium on Wood and Pulping Chemistry. Paris, France.

39. Elder, T. J., M. L. McKee, and S. D. Worley. 1988. The application of molecular orbital calculations to wood chemistry. V. The formation and reactivity of quinone methide intermediates. *Holzforschung* 42(4):233–240.

40. Shevchenko, S. M., and M. Y. Zarubin. 1993. Stereo-electronic structure and chemical reactivity of structural units of lignin. Vol. I, pp. 64–72. Proceedings Seventh International Symposium on Wood and Pulping Chemistry.

41. Elder, T. J., D. J. Gardner, M. L. McKee, and S. D. Worley. 1989. The application of molecular orbital calculations to wood chemistry. VI. The reactions of anthraquinone under pulping conditions. *J. Wood Chem. Technol.* 9(3):277–292.

42. Elder, T. J., and S. D. Worley. 1985. The application of molecular orbital calculations to wood chemistry. III. The chlorination of lignin model compounds. *Holzforschung* 39(3):173–179.

43. Garver, T. M., Jr. 1996. The application of molecular orbitals to bleaching reactions: Computational approach speeds up new development. *Pulp Paper Can.* 97(4):29–33.

44. Elder, T. 1998. Reactions of lignin model compounds with chlorine dioxide molecular orbital calculations. *Holzforschung* 52(4):371–384.

45. Elder, T. 1999. Reactions of lignin model compounds with chlorine dioxide: Molecular orbital calculations on dimers. *J. Pulp Paper Sci.* 25(2):52–59.

46. Gunnarsson, N. P-I., and S. Ljunggren. 1996. The kinetics of lignin reactions during chlorine dioxide bleaching. Part I. Influence of pH and temperature on the reaction of 1-(4, 4-dimethoxyphenyl) ethanol with chlorine dioxide in aqueous solution. *Acta Chem. Scand.* 50:422–431.

47. Ni, Y., X. Shen, and A. R. P. van Heiningen. 1994. Studies on the reactions of phenolic and non-phenolic lignin model compounds with chlorine dioxide. *J. Wood Sci. Technol.* 14:243–262.

48. Brage, C., T. Eriksson, and J. Gierer. 1991. Reactions of chlorine dioxide with lignin in unbleached pulps. Part II. *Holzforschung* 45:47–152.

49. McKague, A. B., D. W. Reeve, and F. Xi. 1995. Reaction of lignin model compounds with chlorine dioxide and sodium hydroxide. *Nordic Pulp Paper Res.* 10(2):114–118.

50. McKague, A. B., G. J. Kang, and D. W. Reeve. 1994. Reactions of lignin model dimers with chlorine dioxide. *Nordic Pulp Paper Res.* 9(2):84–87.

51. McKague, A. B., G. J. Kang, and D. W. Reeve. 1993. Reactions of a lignin model dimer with chlorine dioxide. *Holzforschung* 47(6):497–500.

52. Andersons, B., and J. Gravitis. 1984. Steric and electronic effects on interaction of model substance of the structural unit of lignin with a peroxidase complex II. *Khimaya Drev.* 1984(5):102–103.

53. Elder, T. 1997. Oxidation of a lignin model compound by the veratryl alcohol cation radical. Results from molecular orbital calculations. *Holzforschung* 51(1):47–56.

54. Elder, T., and D. C. Young. 1999. Molecular orbital calculations on the interactions of veratryl alcohol with the lignin peroxidase active site. In *Lignin: Properties and Materials*. W. Glasser, R. Northey and T. Schultz, Eds. American Chemical Society. Washington, DC.

55. Sealy, J., A. Ragauskas, and T. J. Elder. 1999. Investigations into laccase-mediator delignification of kraft pulps. *Holzforschung* 53(5):498–502.

56. Cole, B. J. W., C. Zhou, and R. C. Fort, Jr. 1996. The bleaching and photostabilization of high yield pulp by sulfur compounds. I. Reaction of thioglycerol with model quinines. *J. Wood Chem. Technol.* 16(4):381–403.

57. Remko, M., and J. Polčin 1977a. LCAO MO investigations on lignin model compounds. IV. CNDO/CI calculations of electronic spectra of oxybenzaldehyde type models. *Chem. Zvesti* 31(2):71–79.

58. Remko, M., and J. Polčin. 1977b. LCAO MO investigations on lignin model compounds. V. CNDO/CI calculations of electronic spectra of quinone methide structures. *Monatsh. Chem.* 108(6):1313–1324.

59. Remko, M., and J. Polčin. 1978. MO investigations on lignin model compounds. X. CNDO/CI calculations of electronic spectra of some extended quinone methide structures. *Z. Phys. Chem. Neue Folge* 110:229–235.

60. Remko, M. 1985. MO investigations on lignin model compounds. XIX. CNDO/S study of electronic spectra of biphenyl structure. *Cell. Chem. Technol.* 19(4):423–429.

61. Remko, M., I. Tvaroska, M. Fiserova, and R. Brezny. 1990. MO investigations on lignin model compounds. XXII. Experimental and CNDO/S study of electronic spectra of β-aryl ether structure. *Z. Phys. Chem.* (Leipzig), 271(5):927–930.

62. Remko, M., and J. Polčin. 1980. Studies of model substances of lignin. IX. Experimental and calculated (PPP) electronic spectra of flavanoid, stilbene and coumarone structures. *Collect. Czech. Chem. Comm.* 45(1):201–209.

63. Chupka, E. I., N. V. Taradada, T. A. Gogotova, and T. A. Moskovtseva. 1979. Use of the MO-LCAO method for evaluating spectral characteristics of the model chromophores of lignin. *Khim. Drev.* 1979(1):79–82.

64. Burlakov, V. M., E. I. Chupka, D. Chuvashev, and G. V. Ratovskii. 1986. Characteristics of the electronic structure of ground and excited states of lignin model compounds. I. Molecular shape. *Cell. Chem. Techol.* 20(6):651–62.

65. Shevchenko, S. M., S. G. Semenov, A. G. Anushkinskii, A. V. Pranovich, and E. I. Evstigneev. 1990. *Zh. Org. Khim* 26(9):1839–1848.

66. Burlakov, V. M., E. I. Chupka, D. Chuvashev, and G. V. Ratovskii. 1992. Characterization of electronic structure of ground and excited states of model lignin compounds. III. Quinones and quinone methides. *Cell. Chem. Techol.* 26(4):421–426.

67. Jakobsons, J., J. Gravitis, V. M. Andrianov, J. Dzelme, and V. G. Dashevskii. 1982. Calculation of frequencies of vibrational spectra of model lignin compounds by the method of atom-atom potential functions. I. Calculation of frequencies of the vibrational spectrum of an aromatic ring. *Khim. Drev.* 1982(5):52–57.

68. Liptaj, T., M. Remko, and J. Polčin. 1980. Analysis of proton NMR spectra of cinnamaldehyde type model substances of lignin. *Coll. Czech. Chem. Comm.* 45(2):330–334.

69. Konschin, H., J. Jakobsons, and S. M. Shevchenko. 1980. A theoretical and experimental investigation of Z. E. Stereoisomerism in some simple lignin p-quinone methides. *J. Mol. Struct.* 238:231–244.

70. De Dios, A. C. 1996. Ab initio calculations of the NMR chemical shift. *J. Progr. Magn. Spect.* 29:229–278.

71. Orendt, A. M., R. Biefofsky, A. B. Pomilio, R. Contreras, and J. C. Facelli. 1991. Ab initio and ^{17}O NMR study of aromatic compounds with dicoordinate oxygenations: Intramolecular hydrogen bonding in hydroxy- and methoxybenzene derivatives. *J. Phys. Chem.* 95:6179–6181.

72. Chesnut, D. B., L. D. Quin, and K. D. Moore. 1993. Characterization of NMR shielding in 7-phosphanobornenes. *J. Am. Chem. Soc.* 115:11984–11990.

73. Carmichael, I. 1995. Correlation effects on the hyperfine splitting in HNCN. *J. Phys. Chem.* 99:6832–6835.

74. Elder, T. 1999. Correlation of experimental and ab initio ^{13}C-NMR chemical shifts for monomeric lignin model compounds. *J. Mol. Struct-Theochem.* 505:257–267.

75. Siehl, H-U., F-P. Kaufmann, and K. Hori. 1992. NMR spectroscopic and computational characterization of 1-(p-anisyl) vinyl cations: Methoxy group rotation as a probe of C_β - S_i, C_β - C, and C_β - H hyperconjugation. *J. Am. Chem. Soc.* 114:9343–9349.

76. Facelli, J. C., A. M. Orendt, Y. J. Jiang, R. Pugmire, and D. M. Grant. 1996. Carbon-13 chemical shift tensors and molecular conformation of anisole. *J. Phys. Chem.* 100:8268–8272.

77. Zheng, G., J. Hu, X. Zhang, L. Shen, C. Ye, and G. A. Webb. 1997. Quantum chemical calculation and experimental measurement of the ^{13}C chemical shift tensors of vanillin and 3-4 dimethoxybenzaldehyde. *Chem. Phys. Lett.* 266:533–536.

78. Foresman, J., and A. Frisch. 1993. *Exploring Chemistry with Electronic Structure Calculations,* 2nd Edition. Gausion, Inc. Pittsburgh, PA.

79. Cheesman, J., G. W. Trucks, T. A. Keith, and M. J. Frisch. 1996. A comparison of models for calculating nuclear magnetic resonance shielding tensors. *J. Chem. Phys.* 104:5497–5509.

80. Shevchenko, S. M., and J. Jakobsons. 1990. Vertical ionization potentials of the compounds related to lignin. 6. Calculation of the potentials using CNDO/S method. *Khim. Drev.* 1990(6):33–35.

81. Semenov, S. G., and N. V. Khodureva. 1992. Quantum-chemical estimate of nonspecific solvent effect on the electronic structure and spectra of molecules modeling nucleophilic fragments of lignin. *Opt. Spektrosk.* 73(2):280–290.

82. Grüber, C., and V. Bub. 1989. Quantum mechanically calculated properties for the development of quantitative structure-activity relationships (QSAR'S): Pk$_A$ values of phenols and aromatic and aliphatic carboxylic acids. *Chemosphere* 19:1595–1609.

83. Simon, J. P., and K.-E. L. Eriksson. 1996. The significance of intra-molecular hydrogen bonding in the β-0-4 linkage of lignin. *J. Mol. Struct.* 384:1–7.

84. Simon, J. P., and K.-E. L. Eriksson. 1999. Computational studies of the three-dimensional structural of guaiacyl β-0-4 lignin models. *Holzforschung* 52(3):287–296.

85. Jakobsons, J., J. Gravitis, and V. G. Dashevskii. 1981. Theoretical analysis of conformations of lignin model compounds: Studies of internal rotation, the relative stability of rotamers and an intramolecular hydrogen bond in vanillin and m-methoxybenzaldehyde molecules. *Zh. Strukt. Khim.* 22(2):43–51.

86. Remko, M., and J. Polčin. 1976. LCAO MO investigations of lignin model compounds. I. CNDO/2 calculations of intramolecular hydrogen bond O-H...OCH$_3$ in models of the guaiacol and syringyl type. *Chem. Zvesti* 30(2):170–173.

87. Remko, M., and J. Polčin. 1977. LCAO MO investigations of lignin model compounds. II. CNDO/2 calculations of intramolecular hydrogen bond O-H...O = C in models of the 2-hydroxyphenylcarbonyl type. *Z. Phys. Chemie* (Leipzig), 258(2):219–224.

88. Remko, M., and J. Polčin. 1977. LCAO MO untersuchungen an Lignin model verbindungen. III. Konformationsstruktur des cis-zimtaldehyds. *Z. Phys. Chemie* (Leipzig), 258(6):1187–1189.

89. Remko, M. 1979. MO investigations on lignin model compounds. VIII. A PCILO study of intramolecular hydrogen bond in guaiacol and o-vanillin. *Adv. Mol. Relax. Int. Pr.* 14(4):315–320.

90. Remko, M., and J. Polčin. 1977. LCAO MO investigations of lignin model compounds. VII. CNDO/2 calculations of intermolecular hydrogen bond formed by guaiacol with some strong proton acceptors. *Z. Phys. Chem. Neue Folge* 106:249–257.

91. Remko, M. 1983. MO investigations on lignin model compounds. XV. Effect of intermolecular hydrogen bond on rotational barrier of the intramolecular hydrogen bonded guaiacol hydroxyl group. *Z. Phys. Chem.* (Wiesbaden), 134:129–134.

92. Remko, M. 1983. MO investigations on lignin model compounds. XVII. Model studies of cooperative effects in hydrogen bonded lignoalchols. *Z. Phys. Chem.* (Munich), 138(2):223–228.

93. Remko, M., and J. Polčin. 1981. MO investigations on lignin model compounds. XIII. Model studies of hydrogen bonds formed in the systems lignin-cellulose and lignin-hemicellulose. *Z. Phys. Chem.* (Weisbaden), 125(2):175–181.

94. Remko, M. 1986. Molecular orbital investigations on lignin model compounds. XX. Conformational analysis of benzyl ether and benzyl ester linkages in lignin-carbohydrate complexes. *Holzforschung* 40(4):205–209.

95. Bizik, F., I. Tvaroska, and M. Remko. 1994. Conformational analysis of ester and ether linkages in lignin-arabinoxylan complexes. *Carbohyd. Res.* 261(1):91–102.

96. Remko, M., and J. Polčin. 1980. MO investigations on lignin model compounds. XII. PCILO calculations of conformational structure and hydrogen bonds in 2-2'-dihydroxybihenyl, 2-2'-dimethoxybiphenyl and 2-2'dihydroxy-3'-methoxybiphenyl. *Z. Phys. Chem.* (Wiesbaden), 120(1):1–8.

97. Remko, M., and I. Sekerka. 1983. MO investigations on lignin model compounds. XVI. Conformational analysis of β-aryl ether, benzyl aryl ether, phenyl coumaran, diphenyl ether and biphenyl structures. *Z. Phys. Chem.* (Weisbaden), 134:135–148.

98. Remko, M., and I. Tvaroska. 1987. Molecular orbital investigation on lignin model compounds. XXI. Solvent effect on the stability of β-aryl ether and benzyl aryl ether dimeric units of lignin. *Holzforschung* 41(6):371–377.

99. Remko, M. 1985. MO investigations on lignin model compounds. XVIII. Effect of the intramolecular hydrogen bond on the conformational stability of the β-aryl ether dimeric structural unit of lignin. *Cell. Chem. Technol.* 19(1):47–50.

100. Shevchenko, S. M., T. J. Elder, S. G. Semenov, and M. Ya. Zarubin. 1995. Conformational effects on the electronic structure and chemical reactivity of lignin model p-quinone methides and benzyl cations. *Res. Chem. Inter.* 21(3–5):413–423.

101. Jakobsons, J., and S. M. Shevchenko. 1990. Theoretical conformational analysis of p-quinonemethides as lignin models. *Latv. PSR Zinat. Akad. Vestis, Kim. Ser.* 1990(2):186–191.

102. Ralph, J., and B. R. Adams. 1983. Determination of the conformation and isomeric composition of lignin model quinone methides by NMR. *J. Wood Chem. Technol.* 3(2):183–194.

103. Ralph, J., Thomas J. Elder, and R. M. Ede. 1991. The stereochemistry of guaiacyl lignin model quinone methides. *Holzforschung* 45(3):199–204.

104. Simon, J. Paul, and K-E.L. Eriksson. 1999. Elastic molecular movements in lignin β-O-4 oligomers under external tensile and compressive forces approximated ab initio calculations of selected bond rotations. Personal communication.

105. Durbeej, B., and L. A. Eriksson. 2003. Spin distribution in dehydrogenated coniferyl alcohol and associated dilignol radicals. *Holzforschung* 57(1):59–61.

106. Durbeej, B., and L. A. Eriksson. 2003. A density functional theory study of coniferyl alcohol intermonomeric cross linkages in lignin: Three dimensional structures, stabilities and the thermodynamic control hypothesis. *Holzforschung* 57(2):150–64.

107. Durbeej, B., and L. A. Eriksson. 2003. Formation of β-O-4 lignin models: A theoretical study. *Holzforschung* 57(5):466–478.
108. Durbeej, B., Y.-N. Wang, and L. A. Eriksson. 2003. Lignin biosynthesis and degradation-a major challenge for computational chemistry. High Performance Computing for Computational Science-VECPAR 2002. *Lecture Notes in Comput. Sci.* 2565:137–165.
109. Sarkanen, S., and Yi-Ru Chen. 2005. Towards a mechanism for macromolecular lignin replication. Vol. II, pp. 407–414. Proceedings Thirteenth International Symposium on Wood and Pulping Chemistry. Auckland, New Zealand.
110. Elder, T., G. Brunow, and R.C. Fort, Jr. 2001. Cross-coupling reactions of lignin model compounds: Computational and experimental results. Vol. I, pp. 55–58. Proceedings Eleventh International Symposium on Wood and Pulping Chemistry. Nice, France.
111. Shigematsu, M., T. Kobayahsi, Hiroyasu Taguchi, and Mitsuhiko Tanahashi. 2006. Transition state leading to β-O′ quinonemethide intermediate of p-coumaryl alcohol analyzed by semi-empirical molecular orbital calculation. *J. Wood Sci.* 52:128–133.
112. Martinez, Carmen, José L. Rivera, Rafael Herrera, José L. Rico, Nelly Flores, José G. Rutiaga, and Pablo López. 2008. Evaluation of the chemical reactivity in lignin precursors using the Fukui function. *J. Mol. Model.* 14:77–81.
113. Durbeej, B., and Leif A. Eriksson. 2005. Photodegradation of substituted stilbene compounds: What colors aging paper yellow? *J. Phys. Chem.* A, 109(25):5677–5682.
114. D'Auria, Maurizio, and Rachele Ferri. 2003. Frontier orbital control in the reactivity of singlet oxygen with lignin model compounds: An ab initio study. *J. Photoch. Photobio.* A, 157:1–4.
115. Bonini, C., and M. D'Auria. 2004. Degradation and recovery of fine chemicals through singlet oxygen treatment of lignin. *Ind. Crop. Prod.* 20:243–259.
116. Machado, A. E. da H., R. De Paula, R. Ruggiero, C. Gardrat, and A. Castellan. 2006. Photophysics of dibenzodioxocin. *J. Photoch. Photobio.* A, 180:165–174.
117. Argyropoulos, D. S., L. Jurasek, L. Krištofová, Z. Xia, Y. Sun, and E. Paluš. 2002. Abundance and reactivity of dibenzodioxocins in softwood lignin. *J. Agri. Food Chem.* 50:658–666.
118. Barsberg, S., P. Matusek, M. Towrie, H. Jørgensen, and C. Felby. 2006. Lignin radicals in the plant cell wall probed by Kerr-gated resonance raman spectroscopy. *Biophys. J.* 90:2978–2986.
119. Salazar-Valencia, P. J., S. T. Pérez-Merchancano, and L. E. Bolívar-Marinéz. Optical properties in biopolymers: Lignin fragments. *Braz. J. Phys.* 36:840–843.
120. Wei, K., S.-W. Luo, Y. Fu, L. Lu, and Q.-X. Guo. 2004. A theoretical study on bond dissociation energies and oxidation potentials of monolignols. *J. Mol. Struct.-Theochem.* 712:197–205.
121. Elder, T., and R. C. F., Jr. 2005. Computational electrochemistry of lignin model compounds. Vol. III, pp. 113–115. Proceedings Thirteenth International Symposium on Wood and Pulping Chemistry. Auckland, New Zealand.
122. Britt, P. F., B. G. Sumpter, and A. C. Buchanan III. 2005. Impact of intramolecular hydrogen bonding on C-O homolysis of lignin model compounds: A theoretical study. Preprints—American Chemical Society—Division of Fuel Chemistry 50:169–171.
123. Beste, A., A.C. Buchanan III, P. F. Britt, B. C. Hathorn, and R. J. Harrison. 2007. Kinetic analysis of the pyrolysis of phenethyl ether: Computational prediction of α/β selectivities. *J. Phys. Chem.* A, 111:12118–12126.
124. Elder, T. 2007. Quantum chemical determinations of Young's modulus of lignin. Calculations on a β-O-4′ model compound. *Biomacromolecules* 8:3619–3627.

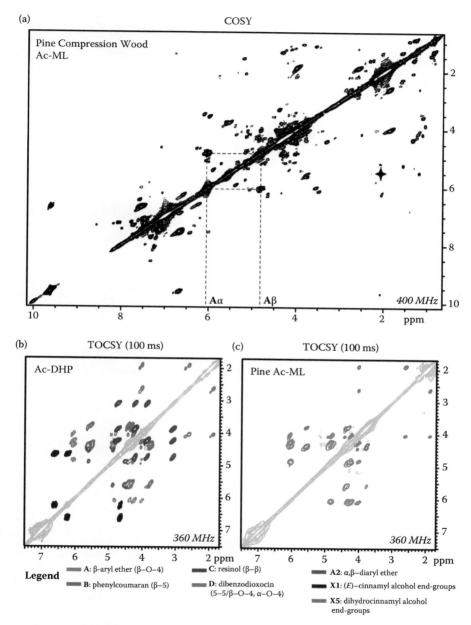

FIGURE 5.2 COSY and TOCSY spectra of acetylated lignins.

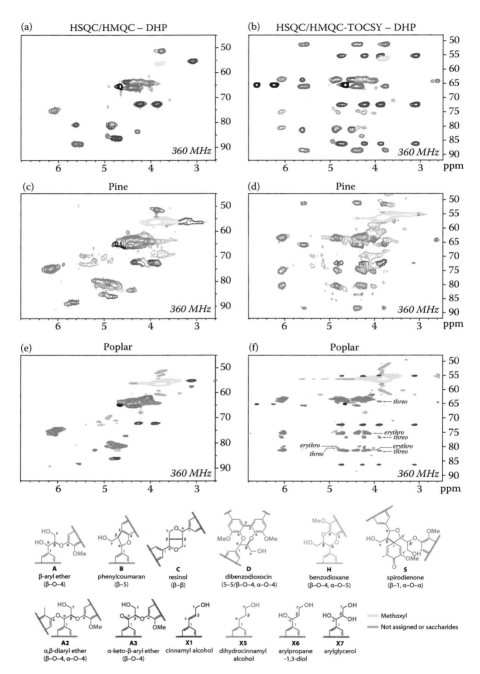

FIGURE 5.3 Heteronuclear correlation spectra of acetylated lignins.

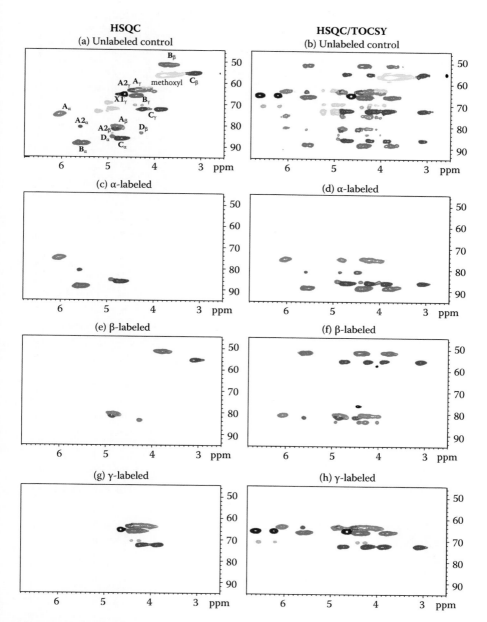

FIGURE 5.4 HSQC and HSQC-TOCSY spectra, and difference-spectra of acetylated synthetic lignins from specifically ^{13}C-enriched coniferyl alcohols.

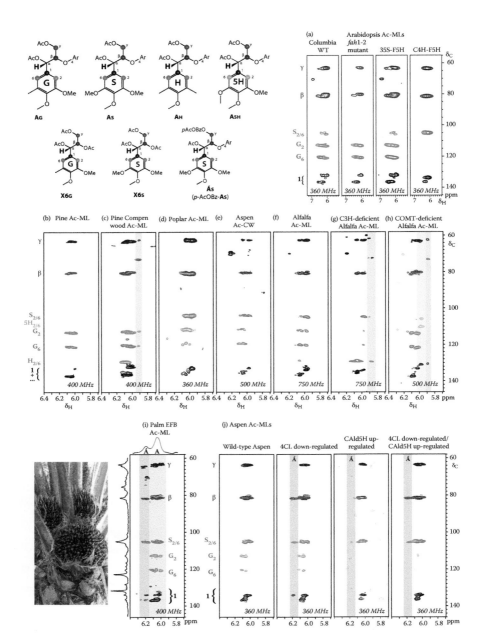

FIGURE 5.5 Heteronuclear long-range correlation (HMBC) spectra of acetylated lignins.

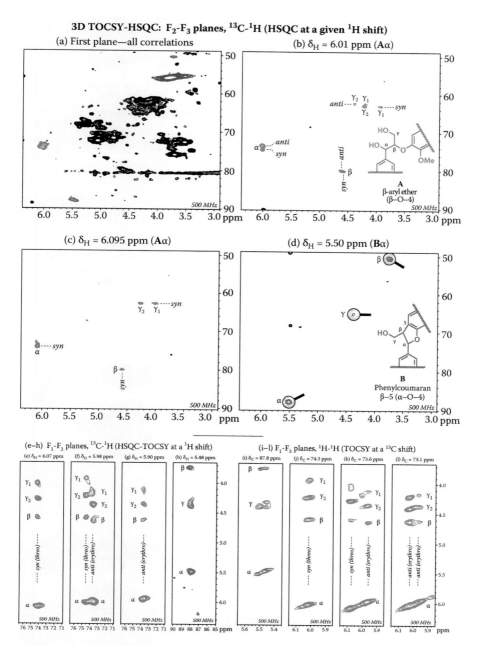

FIGURE 5.6 Slices from a 3D TOCSY-HSQC NMR spectrum of acetylated pine cell walls.

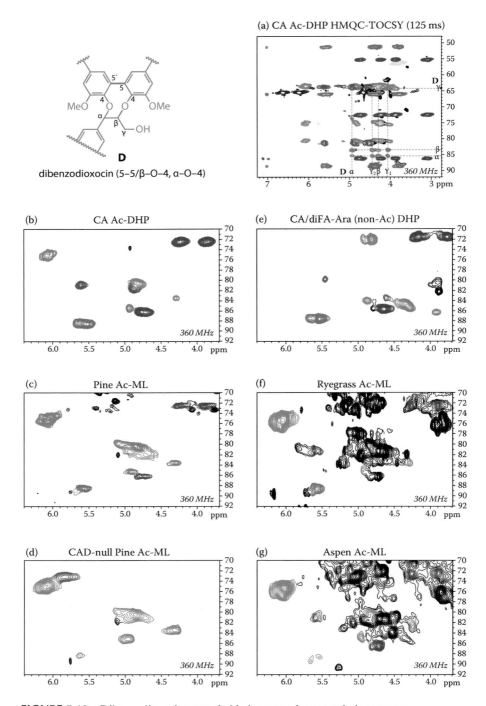

FIGURE 5.10 Dibenzodioxocins revealed in heteronuclear correlation spectra.

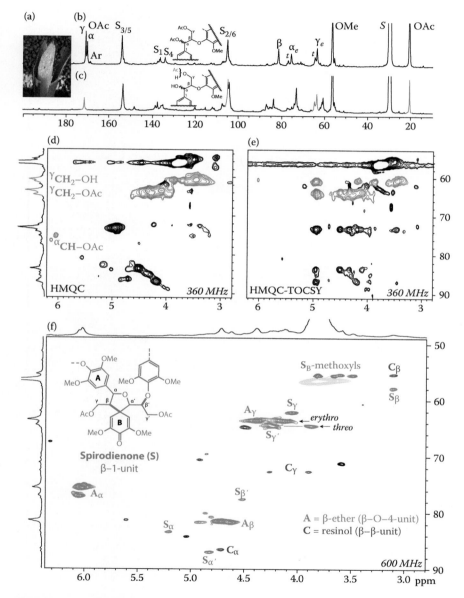

FIGURE 5.11 Spectra from Kenaf isolated lignin revealing natural γ-acetylation and spirodienone units.

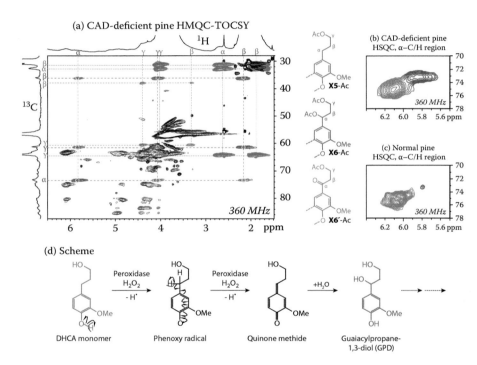

(a) CAD-deficient pine HMQC-TOCSY

^{1}H

^{13}C

360 MHz

X5-Ac

X6-Ac

X6'-Ac

(b) CAD-deficient pine
HSQC, α–C/H region

360 MHz

(c) Normal pine
HSQC, α–C/H region

360 MHz

(d) Scheme

DHCA monomer → Peroxidase H$_2$O$_2$ - H· → Phenoxy radical → Peroxidase H$_2$O$_2$ - H· → Quinone methide → +H$_2$O → Guaiacylpropane-1,3-diol (GPD)

FIGURE 5.12 Dihydroconiferyl alcohol and its derived products in pine lignins.

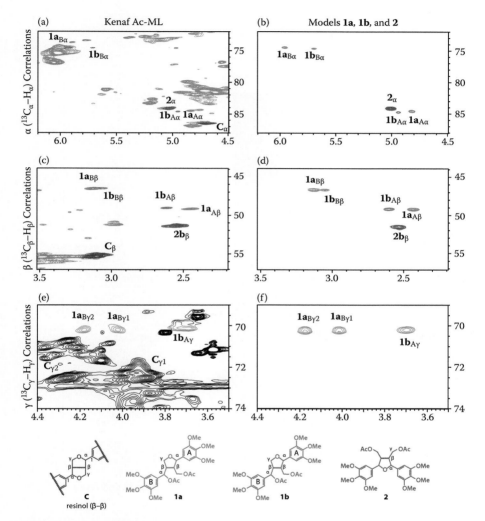

FIGURE 5.15 Novel β–β-coupled structures produced from γ-acylated sinapyl alcohol in Kenaf.

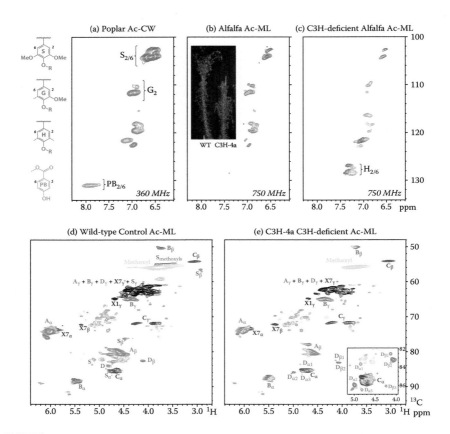

FIGURE 5.16 HSQC spectra of acetylated lignins having different H:G:S ratios.

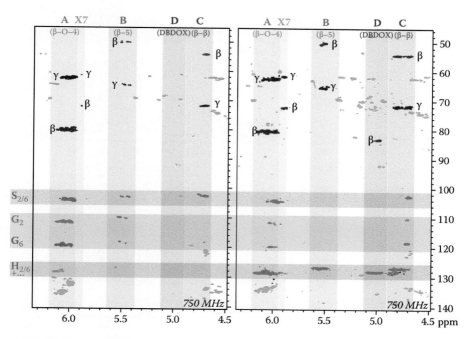

FIGURE 5.17 Partial HMBC spectra of control and C3H-downregulated alfalfa acetylated lignins.

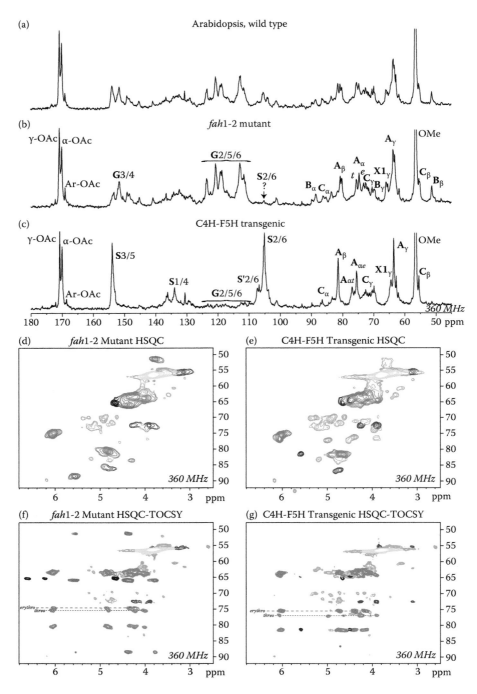

FIGURE 5.18 Spectra from wild-type and F5H-misregulated Arabidopsis lignins.

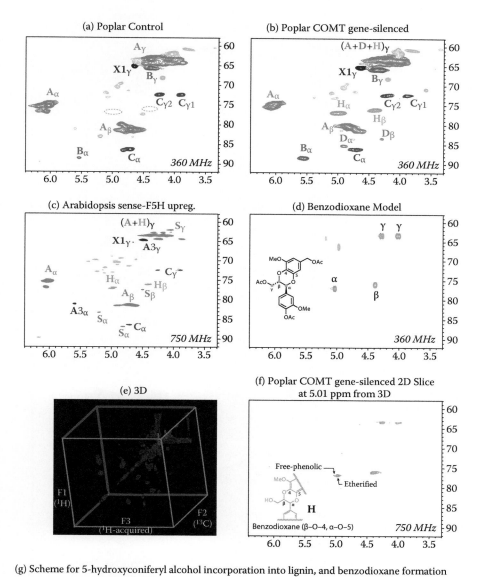

(a) Poplar Control

(b) Poplar COMT gene-silenced

(c) Arabidopsis sense-F5H upreg.

(d) Benzodioxane Model

(e) 3D

(f) Poplar COMT gene-silenced 2D Slice at 5.01 ppm from 3D

(g) Scheme for 5-hydroxyconiferyl alcohol incorporation into lignin, and benzodioxane formation

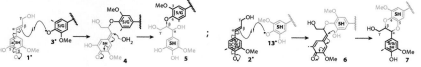

FIGURE 5.19 Partial HMQC spectra highlighting new benzodioxane units in F5H- and COMT-misregulated transgenics.

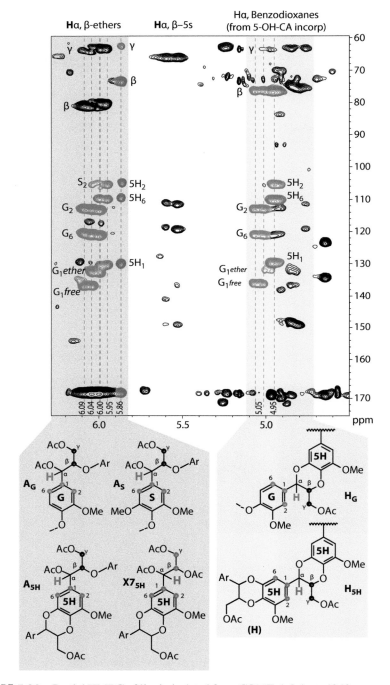

FIGURE 5.20 Partial HMBC of lignin isolated from COMT-deficient alfalfa.

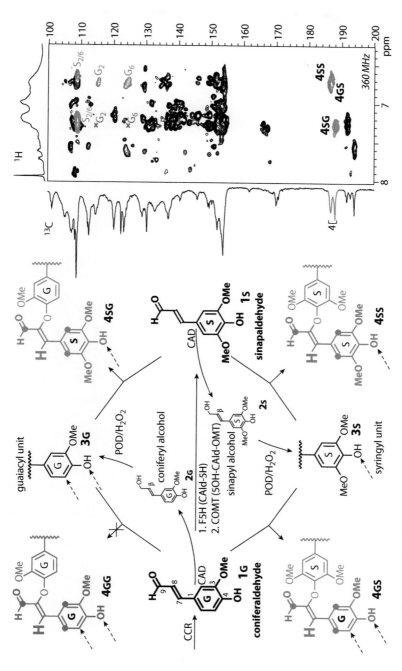

FIGURE 5.21 Cross-coupling of hydroxycinnamaldehydes with lignin units.

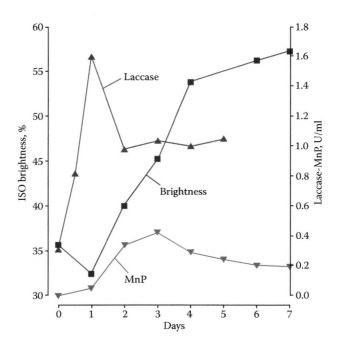

FIGURE 15.7 Production of laccase and MnP by cultures of *Trametes versicolor* during bleaching of hardwood kraft pulp. (Redrawn from Paice, M.G., Reid, I.D., Bourbonnais, R., Archibald, F.S., Jurasek, L., *Appl Environ Microbiol*, 59, 260–265, 1993. With permission.)

10 Chemistry of Alkaline Pulping

Donald Dimmel and Göran Gellerstedt

CONTENTS

INTRODUCTION

By far, the most important chemical pulping process is the kraft process. Its advantages include excellent paper strength and permanence, relatively low energy requirements, chemical recycle and, therefore, low chemical costs, and insensitivity to wood species. The disadvantages are relatively low pulp yields (~40–55% from wood, depending on grade and whether bleached), high capital investment, brown unbleached pulp color, high bleach chemical costs, significant amounts of organic components in the bleaching effluent, and strong odors. An alternative alkaline chemical process is soda/anthraquinone (AQ) pulping. While low in odor, the process requires higher levels of NaOH than the kraft process in order to achieve roughly the same delignification rate and its higher alkali level leads to more carbohydrate damage [1]. Thus, in general, the soda/AQ pulping process leads to lower pulp strength, especially with softwoods.

Using AQ in kraft pulping has several advantages, while still maintaining good pulp strength. The advantages include higher pulp yield and/or lower residual lignin for a given cooking time period. The main disadvantage is related to the cost of AQ, since it is not recoverable. When AQ costs are low, the kraft/AQ process appears particularly attractive; many mills now use such a strategy. Polysulfide pulping is another possibility, although used in fewer mills. Just like the AQ processes, polysulfide pulping gives higher yields; because of the retention of hemicellulose components, the pulps are generally higher in tensile strength, lower in tear strength, and easier to beat or refine [2].

The discussion in this chapter focuses on the lignin reactions that occur during the chemical pulping of wood by soda, kraft, AQ, and polysulfide processes. There are other pulping processes, such as sulfite, organosolv, mechanical, thermo-mechanical and chemical–mechanical, and there are other sources of lignocellulosics. These topics are considered in other chapters. However, worthy of note is the family of sulfite processes, which give high-quality pulps that are more easily bleached than kraft pulps. Sulfite pulping is increasingly out of favor because of difficulties associated with chemical recovery. Another process suffering from the same problem is the alkaline sulfite anthraquinone methanol (ASAM) process. Here, excellent pulp strengths are obtained without the usual kraft odor problems.

Pulping in Alkaline Media

Most chemical pulping processes employee high concentrations of sodium hydroxide (NaOH) and temperatures that are well above the boiling point of the solvent, water. The NaOH swells the wood chips to allow for better chemical penetration. However, its main role is to promote acid/base reactions that facilitate lignin breakdown and dissolution of the lignin fragments. Some aliphatic hydroxyl groups will be ionized by the NaOH and become better nucleophiles in reactions that follow. For many lignin degradation reactions, terminal phenolic groups need to be ionized; the NaOH in the system does this. In addition, smaller phenolic lignin pieces, which are converted to phenolate ions by NaOH, become solubilized.

In general, a high concentration (1M) of NaOH is charged into the pulping digester to compensate for the consumption of NaOH by the acidic wood extractives, by phenolic compounds generated from lignin, and by acidic components associated with carbohydrate degradation reactions. If the pH drops below 11, dissolved lignin will precipitate and be retained in the pulp.

Although NaOH is essential for rapid pulping, the presence of NaOH is quite harmful to the carbohydrates. Reactions with NaOH cause more than 50% of the hemicelluloses and 5–10% of the cellulose to dissolve. Also, reactions with NaOH cause significant drops in the various carbohydrates' polymer length [1], which is related to pulp strength properties. Even so, pulps from alkaline pulping processes are superior to mechanical pulps because lignin removal is highly beneficial to the final pulp strength (and color) properties.

THE KRAFT PROCESS

The kraft process uses a combination of sodium sulfide and sodium hydroxide. Although sodium sulfide (Na_2S) is often mentioned as a key reagent in kraft pulping, it exists in water as NaSH. The role of NaSH, and AQ, in chemical pulping is to facilitate a rapid breakdown of the lignin. Without such pulping additives, long cooking times would be required to remove substantial amounts of lignin; simultaneously, considerable carbohydrate breakdown would occur, leading to a low pulp yield and strength. The NaSH functions primarily as a nucleophile in beneficial lignin degradation reactions, without causing carbohydrate degradation [3]. In standard pulping processes, the levels of NaSH, AQ, and NaOH drop substantially as the cook proceeds. In general, large quantities of NaOH and NaSH are put in the pulping digester at the start so that reasonable levels are still present at the end of the cook.

Pulping is terminated before all of the lignin is removed because the "selectivity" decreases as the cook nears its end. The term selectivity refers to the extent of lignin removal in comparison to carbohydrate degradation, which is reflected in poorer pulp yield and losses in polymeric chain length. Chemical pulping has three characteristic phases: initial, bulk, and residual delignification. At the start of the residual phase, the lignin remaining has structural characteristics that make it resistant to further nucleophilic or electron transfer reactions that aid the breakdown. Consequently, pulping is usually terminated at this point.

OVERVIEW

DIFFERENT PULPING PHASES AND KINETICS

The rate of delignification in the alkaline pulping of wood depends on the presence of either hydrosulfide (Figure 10.1) [4] or AQ [5] (or both). Pulping kinetics have been the subject of numerous investigations; a summary was published in 1989 [6]. A kraft cook can be divided into three distinct phases, each with its own dependency on the concentration of cooking chemicals and temperature (Figure 10.2) [7–9]. The initial delignification phase is characterized by a low activation energy, 40 kJ/mol; as long as a certain NaOH/NaSH concentration is maintained, neither alkali nor

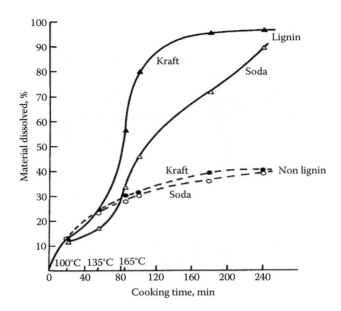

FIGURE 10.1 Removal of lignin and carbohydrates ("non lignin") in kraft and soda pulping of spruce wood. (Reprinted from Stone, J.E. and Clayton, D.W., *Pulp Pap Mag Can*, 61(6), T307–T313, 1960. With permission.)

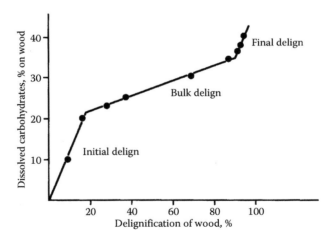

FIGURE 10.2 The three phases—initial, bulk, and final delignification—encountered in kraft pulping of softwood. (Gellerstedt, G. and Lindfors, E.-L., *Holzforschung*, 38, 151–158, 1984.)

hydrosulfide concentration [8] affects the rate. In this part of the cook, approximately 20% of the lignin is dissolved as fragments of rather low apparent molecular weight [9]. The dissolution of lignin is accompanied by a substantial degradation and dissolution of carbohydrates, especially hemicelluloses, predominantly through a peeling reaction [10].

The major portion of the lignin, approximately 70%, is dissolved in the bulk phase, which is characterized by fairly selective lignin removal. Both NaOH and NaSH concentrations affect the rate; the activation energy is in a range that is normal for a chemical reaction, 140–150 kJ/mole [8]. At a degree of delignification of around 90–92%, the selectivity in a kraft cook again changes rather dramatically; prolonged cooking results in very slow delignification, accompanied by the degradation and dissolution of increasing amounts of carbohydrates. The transition point to the residual delignification phase is affected by the hydrosulfide concentration earlier in the cook [11,12]. Thus, a high sulfidity is of particular importance when the cook is about to reach the bulk phase (Figure 10.3). This transition point can, however, also change if the charge of alkali is changed. Thus, a high alkali charge in the cook, or alternatively an addition of alkali late in the cook, will affect the amount of residual lignin in the final delignification phase [13].

The kinetics described above are based on work using traditional differential rate equations; all known characteristics of the kraft process can be conveniently described by using such an approach. It has been argued, however, that this type of kinetics is not valid for polymers, such as lignin, which, when degraded, give rise to fragments of variable molecular weight. Instead, the delignification kinetics should be treated as a degelation process, the reverse of polymerization (gel formation) [14–16]. Although such an approach is attractive from a chemistry point of view, several difficulties arise in the theoretical treatment, since the chemical structure of lignin still is not accurately known and, consequently, assumptions have to be made [17,18].

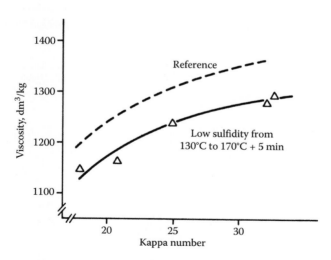

FIGURE 10.3 A laboratory flow-through kraft cook of softwood with a lowered concentration of hydrosulfide ions present in the cooking liquor during the time interval of 60–105 min, after the start at 70°C, corresponding to a temperature interval of 130–170°C + 5 min at 170°C. Pulping selectivity is measured as pulp viscosity as a function of kappa number. (Sjöblom, K., Mjöberg, J., and Hartler, N., *Pap Puu*, 65, 227–240, 1983.)

TOPOCHEMICAL EFFECTS

Lignin reactivity in alkaline pulping is different in different morphological regions of the wood fiber [19,20]. The secondary wall lignin is removed more uniformly than the middle lamella lignin, with the latter starting to react once the cook has reached the bulk phase. This so-called topochemical effect can be conveniently studied using a selective bromination of lignin together with scanning electron microscopy coupled with energy dispersive X-ray analysis (SEM-EDXA) [21]. This technique has been used to demonstrate that AQ in a soda cook greatly facilitates lignin dissolution in all the morphological regions (secondary wall, compound middle lamella, and cell corner middle lamella), and that the delignification power of soda/AQ, in comparison to kraft, is similar in the initial and bulk phases and somewhat better in the residual phase (Figure 10.4) [22].

KAPPA NUMBER SIGNIFICANCE

In chemical pulping, the kappa number is frequently used as a process control parameter for unbleached and partially bleached pulps. It has been found empirically that there is a relationship between the kappa number and the Klason lignin content [23] and, therefore, the kappa number is often used to describe the degree of delignification in (for example) kraft pulping.

The kappa number measurement involves addition of an excess of potassium permanganate to a pulp sample and a determination of the amount of unreacted permanganate after 10 min at a fixed temperature [24]; however, it is not a selective analytical method for lignin. Any functional group present in the pulp that can react with the permanganate ion under the given conditions will contribute to the overall consumption of permanganate. It has been shown that, in addition to the aromatic

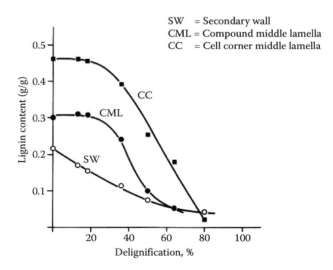

FIGURE 10.4 Topochemical effects during kraft cooking of Douglas-fir. (Saka, S., Thomas, R.J., Gratzl, J.S., and Abson, D., *Wood Sci Technol*, 16, 139–153, 1982.)

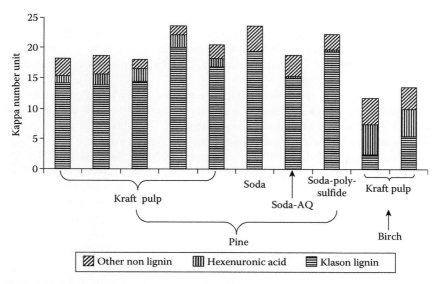

FIGURE 10.5 Contribution from lignin, hexenuronic acid, and "non lignin" structures to the kappa number in various unbleached alkaline pine and birch pulps. (Li, J. and Gellerstedt, G., *Nordic Pulp Pap Res J,* 13(2),153–158, 1998.)

rings in lignin, a number of other structures may react with permanganate during the kappa test [25]. Among these, hexenuronic acid units, formed during alkaline pulping from the uronic acid groups present in xylan hemicelluloses [26,27], will contribute the most. However, several other structures containing double bonds, such as α,β-unsaturated aldehydes and α-ketocarboxylic acids and aldehydes will also react. As a result, the measured pulp kappa number comprises contributions from lignin, hexenuronic acid groups, and other "non lignins," presumably carbohydrate-derived materials (Figure 10.5). The relative amounts of these components can vary widely, depending on wood species and pulping processes/parameters. It has also been shown that alkaline bleaching with oxygen or hydrogen peroxide does not eliminate hexenuronic acids; here, a large portion of the remaining kappa number can be attributable to these acids [27–30].

Recently, a modified kappa number determination method has been developed that more accurately determines the amount of lignin remaining in kraft pulp. The method involves selective removal of all functional groups in the pulp, except the aromatic lignin units, before a kappa number determination [31]. For bleachable grade soft-wood pulps, it has been found that around 80% of the kappa number originates from lignin, whereas in hardwood pulps the corresponding figure is around 50% [31,32].

PULPING CHEMISTRY

NATURE OF REACTIONS AT 170°C

The goal of chemical pulping is to degrade the lignin into (small) fragments that are water-soluble and then separate the aqueous liquor from the fibrous carbohydrate materials. Rapid lignin fragmentation reactions are the key to successful pulping. There are, however, competing reactions that interfere with efficient delignification. This section will consider the beneficial and detrimental reactions of lignin during pulping, relative rates of competing reactions, and the structure of the residual and dissolved lignin after pulping.

Chemical pulping is typically conducted under relatively harsh conditions. Few fundamental studies have addressed the nature of the chemistry of NaOH and NaSH under these conditions. One study showed that the C–H bonds of the methylene hydrogens of 4-hydroxydiphenylmethane underwent partial exchange for C–D when the compound was heated at 170°C for two hours with 1M NaOD in D_2O [33]. This compound should be less acidic than diphenylmethane, which is reported to have a pK_a of 35 at room temperature [34]. Thus, one can estimate that 1M NaOH in water at 170°C has roughly the deprotonating capabilities of $NaNH_2$ (pK_a 35) at room temperature [34]. In the case of lignin, all of the phenolic hydroxyl groups should be ionized under alkaline pulping conditions, along with a substantial percentage of the aliphatic hydroxyl groups and some C–H bonds. Polyanions will likely be the rule, rather than the exception.

The acidity and basicity of compounds should change with temperature. As the temperature increases, reaction rates generally increase due to additional vibrational energies imparted to the bonds. In addition, because of the breakdown of hydrogen bonds, hydroxide ion should be less solvated and, thus, more reactive as the temperature increases [35,36]. The combination of these effects probably accounts for hydroxide ion being an apparent very strong base at 170°C in water. In contrast, exchange reactions with $NaSD/D_2O$ have demonstrated that NaSH is an insignificant base in the presence of NaOH at 170°C [33].

Both NaOH and NaSH could be functioning as nucleophiles during pulping. It is well known that NaSH is an excellent nucleophile at room temperature in reactions with simple substrates, such as CH_3I [37]. Under these conditions, NaSH is approximately 10 times more nucleophilic than NaOH. Some of this difference can probably be attributed to greater solvation of the smaller HO^- ion, in comparison to HS^- [35]. Solvation should be less as the temperature increases [38]. This raises the question of the relative nucleophilicities of HO^- and HS^- at 170°C. One study has shown that HS^- is a stronger nucleophile, by a factor of 11, than HO^- in the cleavage of a methyl glucoside [39]. Another study has shown that HS^- is 20 times stronger than HO^- in the cleavage of a benzyl ether substrate at 195°C [40]. However, HS^- has only half the nucleophilicity of HO^- in reactions at an unsaturated carbon at 195°C [41].

QUINONE METHIDE CHEMISTRY

A key intermediate in many lignin reactions is a quinone methide. This intermediate is formed by loss of an anion from the C_α-position of a phenolate ion (Figure 10.6, step I). The ease of this step is related to the stability of both the C_α-leaving group and the quinone methide. With leaving groups such as Cl^-, $CH_3CO_2^-$, and ArO^-, substituted quinone methides (such as shown in Figure 10.6) readily form at room temperature [42–44]. However, with a poorer leaving group, such as HO^-, temperatures near 110°C have been suggested as necessary to generate the quinone methide [45,46].

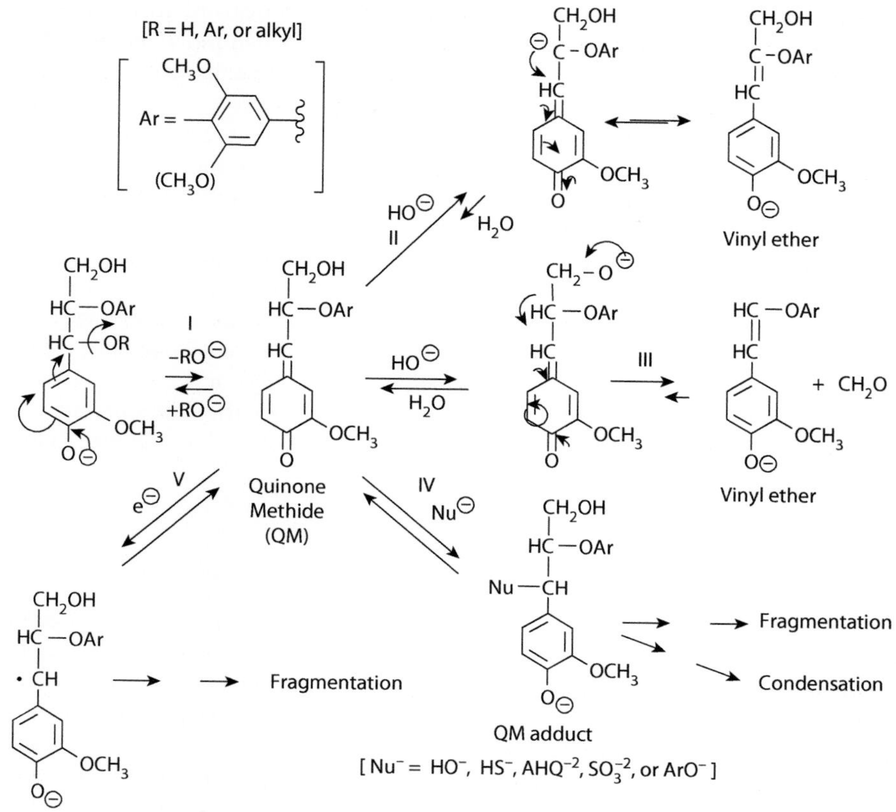

FIGURE 10.6 Quinone formation and reactions for the case of a terminal β-aryl ether phenol group.

A quinone methide is a relatively unstable species; the unsubstituted case has been reported to have a lifetime of 15 seconds in methanol at room temperature [47]. However, more substituted, more hindered quinone methides can be prepared in the presence of water and can have long lifetimes (years) in chloroform solutions [48]. The instability of quinone methides can be attributed to facile reactions that regenerate an aromatic system. An anion (nucleophile) can add to the α-carbon to rearomatize the system via a familiar Michael reaction [49]. The added anion can also be a leaving group (Figure 10.6, step I); in other words, quinone methide formation is a reversible reaction [46,50,51]. Alternatively, the anion can be pulping reagent, such as HO⁻, HS⁻, SO₃⁻², and AHQ⁻², or a phenolate by-product (Figure 10.6, step IV). Addition of the pulping reagent often leads to fragmentation (the Fragmentation via Nucleophilic Chemistry: Soda and Kraft Pulping section). Addition of a phenolate leads to condensation and a build up of size of the molecule (the Condensation Reactions section). Rearomatization can also occur by way of electron transfer to the quinone methide [52].

Two other ways to rearomatize a quinone methide result in the production of enol ethers. One case involves deprotonation of the β-carbon (Figure 10.6, step II), which is simply an extended enolization reaction. The other rearomatization pathway that gives rise to an enol ether is a reverse aldol reaction in which the γ-carbon is lost as formaldehyde (Figure 10.6, step III). Both reactions will be discussed in greater depth later in this chapter.

Evidence for much of the chemistry that is outlined in Figure 10.6 comes from studying lignin model compounds. The models are generally dimer representations of lignin, having a β-aryl ether bond—the most prevalent linkage connecting lignin monomer units. The advantage of models is that the structures of the starting material and products are easy to define and reactions often follow simple kinetics, since the models are either soluble in aqueous alkali or can be made soluble with the addition of a co-organic solvent, such as dioxane. However, these advantages are offset by the models' inability to mimic the heterogeneous environment present when wood is delignified. Even so, the chemistry of models appears to correlate well with structural studies performed on lignins isolated after the pulping and bleaching of wood. Only a few reactions have been studied using polymer lignin models [53–56].

FRAGMENTATION VIA NUCLEOPHILIC CHEMISTRY: SODA AND KRAFT PULPING

The most important fragmentation reaction in alkaline pulping is the cleavage of the β-aryl ether linkage in β–O–4 structures. This reaction fragments the lignin polymer, along with a simultaneous liberation of a new phenolic hydroxyl group. In the alkaline cooking liquor, phenolic groups are ionized and contribute to the solubility of the lignin fragments. Under kraft and soda/AQ pulping conditions, the fragmentation reaction proceeds smoothly; in soda liquor alone, the reaction is much slower. In phenolic units, the reaction of a β–O–4 structure proceeds via the formation of a quinone methide (Figure 10.7). In a kraft cook, this intermediate will add a hydrogen sulfide ion, which in turn attacks the β-carbon atom. As a result, the β-substituent is eliminated and an episulfide is formed [57,58]. In further reaction steps, the episulfide is first degraded with formation of elemental sulfur and a coniferyl alcohol structure [57,59]. Elemental sulfur is rapidly converted into polysulfide by reaction with the hydrogen sulfide ions present in the cooking liquor and, subsequently, the polysulfide disproportionates to form hydrogen sulfide and thiosulfate [60]. The coniferyl alcohol structure can be found in the cooking liquor (as coniferyl alcohol) [61], but is rapidly degraded to structures like vanillin, vinyl-guaiacol, and others [62].

After a rapid ionization step, formation of a quinone methide is the rate-determining step in the set of reactions that follow [63]. Once formed, the quinone methide has several options for its further conversion, including addition of (a) hydroxide ion (e.g., the backward reaction), (b) hydrosulfide ion (the desired reaction), (c) other nucleophiles, or (d) even electrons (electron transfer reactions). Reducing sugars, such as glucose and ascorbic acid, also promote cleavages of β–O–4 linkages [64]. Fullerton and Wilkins propose that the fragmentation reaction involves addition of a sugar enediol to a quinone methide intermediate and subsequent breakdown of the adduct; however, an electron transfer mechanism could also explain the fragmentation

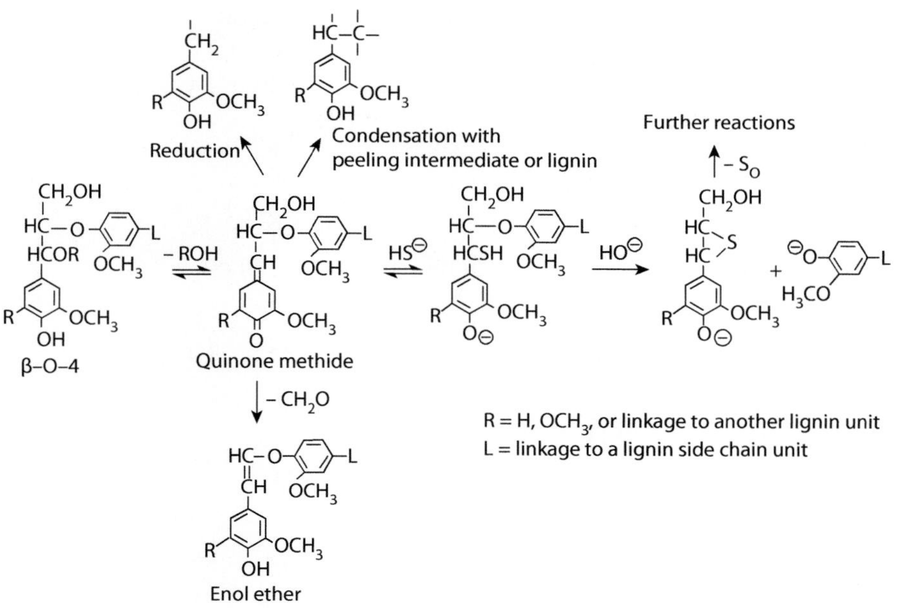

FIGURE 10.7 Reaction scheme for the cleavage of phenolic β–O–4 structures in lignin during kraft cooking conditions. Competing reactions are also indicated; L denotes a lignin residue.

chemistry. Addition to the quinone methide of other types of nucleophiles from lignin or carbohydrates will result in condensation products; such reactions are further discussed in the Condensation Reactions section. Furthermore, the quinone methide intermediate may loose formaldehyde or a β-hydrogen to give enol ether structures (the Competing Reactions: Vinyl Ethers and Stilbenes section). The presence of reduced carbon atoms in the residual lignin after the kraft cook (the Lignin Reduction Reactions section) also indicates that reduction of the quinone methide, probably by an electron transfer process, may play a significant role.

Recently, the fragmentation of phenolic β–O–4 structures in kraft cooking has been reinvestigated using mild reaction conditions [65,66]. The observed rapid degradation of the lignin structure was not, however, accompanied by a corresponding formation of coniferyl alcohol. This result indicated a more complex degradation of the episulfide intermediate; a suggestion is outlined in Figure 10.8.

The presence of polysulfide in a kraft pulping liquor cannot be completely avoided since any elemental sulfur present or formed will rapidly react with the hydrosulfide ions to give polysulfide. In a few pulp mills, polysulfide liquor, prepared by selective oxidation of white liquor, is used to increase pulp yield by decreasing the extent of the carbohydrate peeling reaction. In this reaction, polysulfide converts polysaccharide end-groups (especially of hemicelluloses) into aldonic acid groups, presumably through the interaction with an enediol intermediate. Polysulfide also reacts with unsaturated side chain structures in lignin. Here, the double bond is converted into a carbonyl group, as exemplified in Figure 10.9 [62,67–69]. Polysulfide may, however, also act as a strong nucleophile and compete with hydrosulfide ions for available

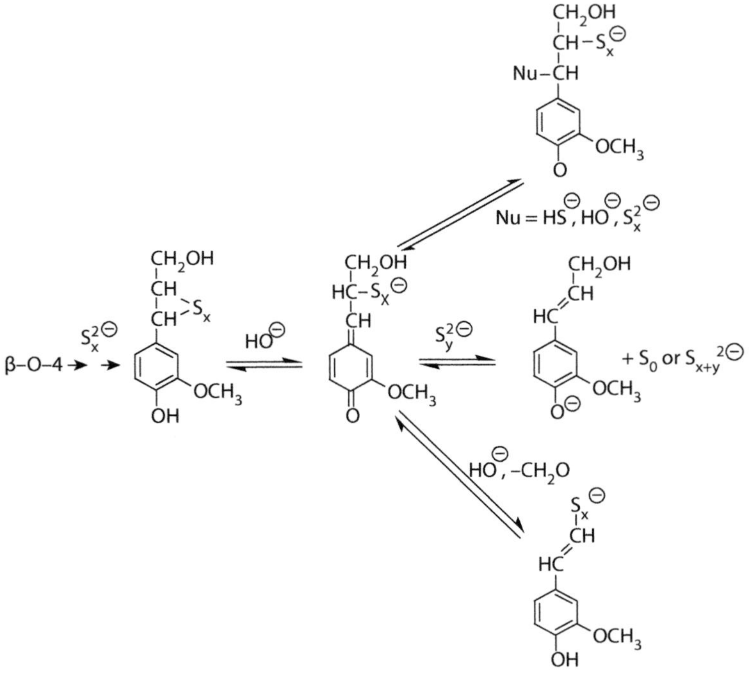

FIGURE 10.8 Alternative routes for the further reactions of episulfide intermediates (cf. Figure 10.7); S_x^{2-} and S_y^{2-} denote polysulfide ions. (Berthold, F., Lindfors, E.-L., and Gellerstedt, G., *Holzforschung,* 52, 481–489, 1998.)

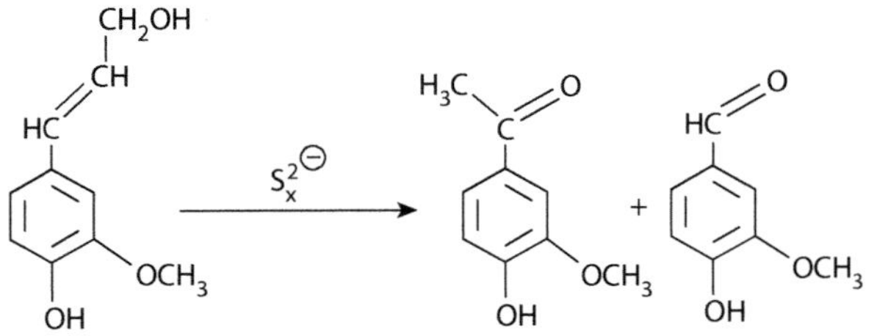

FIGURE 10.9 Products obtained on oxidation of coniferyl alcohol with polysulfide solution at pulping temperature. (Brunow, G. and Miksche, G.E., *Acta Chem Scand,* B29, 349–352, 1975; Berthold, F. and Gellerstedt, G., *Holzforschung,* 52, 490–498, 1998.)

quinone methides (Figure 10.8). In β–O–4 structures, such an intermediate will induce an intramolecular cleavage of the β-aryl ether linkage in complete analogy to the sulfide mechanism shown above (Figure 10.7) [66]. It has been reported that polysulfide is, in fact, better at fragmenting a β-aryl ether linkage in particular lignin model dimers than is NaSH [70].

In contrast to the sulfide/polysulfide mechanism for cleavage of the β–O–4 linkage in phenolic structures, anthrahydroquinone (AHQ) is able to promote the

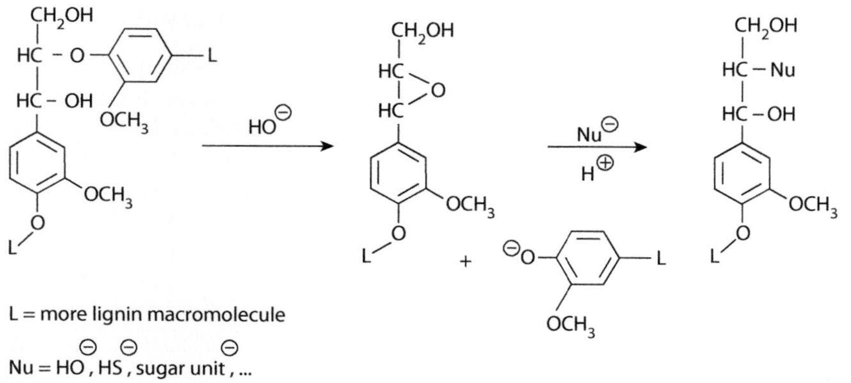

L = more lignin macromolecule

Nu = HO$^{\ominus}$, HS$^{\ominus}$, sugar unit, ...

FIGURE 10.10   Cleavage of a nonphenolic β–O–4 structure and further reaction alternatives of the intermediately (epoxide) structure. (Gierer, J., *Wood Sci Technol*, 14, 241–266, 1980.)

β-aryl ether cleavage with formation of considerable amounts of coniferyl alcohol. This has been verified both by analysis of cooking liquor samples [61,71,72] and by-product analysis in model compound experiments [73].

In nonphenolic structures, the β–O–4 linkage is cleaved via participation of the α- (or γ-) hydroxyl group [74]. As shown in Figure 10.10, the reaction proceeds with formation of an epoxide, which can react further and be opened by attack of nucleophiles, such as NaSH and NaOH, or by nucleophilic sites in lignin or by carbohydrates [75,76]. In the latter cases, the reaction will again result in condensation products (the Condensation Reactions section).

Under kraft pulping conditions, the rate of cleavage of the β–O–4 linkages in phenolic structures, as well as in nonphenolic lignin structures, have been investigated using model compounds. For the phenolic structure, the cleavage reaction is independent of the concentrations of hydroxide and hydrosulfide ions, whereas the nonphenolic structure is strongly influenced by the hydroxide concentration [77]. The activation energies for both types of structures are around 120 kJ/mol. Based on the pseudo first-order reaction rate constants, the calculated kinetic half-lives at 170°C, under realistic kraft cooking conditions, are 1.5 and 44 min for the phenolic and nonphenolic structures, respectively. The latter value indicates that the cleavage of nonphenolic β–O–4 structures plays a crucial role in the overall delignification in kraft pulping.

The rate of β–O–4 cleavage under kraft cooking conditions was much greater for structures having a syringyl structure, in comparison to a guaiacyl structure [78]. Addition of AHQ to soda cooking liquor strongly increases the extent of β–O–4 cleavage; the degree of reaction reaches the same order of magnitude as a corresponding reaction in a kraft or a kraft/AHQ liquor [79].

Kinetic and product investigations of the behavior of lignin in kraft pulping using model compounds have been used frequently, providing detailed chemical information on individual lignin structures. It must be emphasized, however, that a direct "translation" of such results to the behavior of lignin in the pulping of wood has possible flaws [80].

FRAGMENTATION BY AHQ: ADDUCT VS ELECTRON TRANSFER CHEMISTRY

When AQ was first introduced in 1977, there was a flurry of research activity aimed at trying to explain how an organic material used in catalytic amounts (<0.1%) could roughly match the delignification effects of 6% NaSH (kraft system). The initial research studies indicated that AQ was reduced to AHQ by carbohydrates. In the process, carbohydrate aldehyde end-groups were oxidized to acid end-groups. Since the acid end-groups are more stable in alkali, less carbohydrate degradation occurs and pulp yields increase [81]. In an alkaline solution, the AHQ would be in the form of its dianion, AHQ^{-2}. Model compound studies indicate that AHQ^{-2} is oxidized to AQ by lignin, while the latter fragments [52]. Repetition of the AQ/AHQ^{-2} cycle explains the catalytic activity of AQ, higher pulp yields, and faster delignification rates (Figure 10.11).

This simple redox picture has some shortcomings, however. First, products have been observed in pulping liquors and model studied that indicate carbohydrates and lignins are both oxidized and reduced [82,83]. Lignin reduction and oxidation reactions by AQ/AHQ are discussed in the Oxidation and Reduction Reactions and Lignin Reduction Reactions sections. The finding of small amounts of 3,4-dideoxyhexonic acid in AQ-pulping liquors suggests an α,β-dicarbonyl compound was reduced by AHQ to give the observed carbohydrate acid [83]. Are these extra redox reactions important to the observed good impact that low doses of AQ have on pulping or are they just side reactions? Most of the important yield-reducing carbohydrate reactions occur early in pulping (during the warm up), while the lignin reactions occur later (at higher temperatures). Redox reactions involving carbohydrate early and involving lignin late in the cook would help in recycling AQ to AHQ and, therefore, be beneficial.

The second shortcoming of the redox cycle picture is that AQ is relatively insoluble in aqueous alkali at 170°C [84]. Since reaction rates between two insoluble

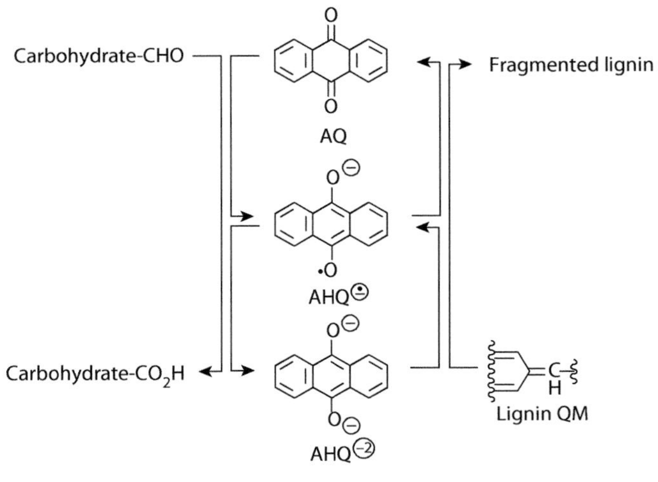

FIGURE 10.11 Redox chemistry for AQ in alkaline; two electron transfers (AQ directly to AHQ^{-2} and back) and/or one electron transfers involving AHQ ion radicals.

species is extremely low, a soluble form of AQ (AHQ radical anion) would be required to oxidize insoluble carbohydrates, which are the ones that account for the increased pulp yields when using AQ. Reaction between insoluble AQ and soluble carbohydrates likely occurs, but provides no pulp yield benefits, since these carbohydrates are not retained. The formation of aldonic acid end units on cellulose when pulping with AQ indicates the occurrence of insoluble carbohydrate oxidation. The efficiency of recycling the catalyst would increase greatly if both the oxidized and reduced forms were water-soluble. This will be the case if the redox cycle involves the partial oxidized AHQ ion radical (to be discussed later and shown in the lower half of Figure 10.11).

Early model studies established that AHQ^{-2} could react with a quinone methide (1) at room temperature to give a C–C bonded adduct (2), and that warming a β-aryl ether dimer adduct in alkali led to fragmentation (Figure 10.12) [52]. This mechanism is both simple and similar to the well-established QM-SH adduct mechanism in the kraft pulping system. The mechanism is in agreement with the observation of coniferyl alcohol (3) as a product in model degradations and in pulping [43,52,61]. More of compound 3 is observed with AQ pulping than with soda or kraft pulping [61,72].

Extensive research by Dimmel and coworkers has shown that another mechanism is also possible between AHQ^{-2} and QMs. They propose that single electron transfer (SET) chemistry promotes the delignification of wood [52]. As is the case for the adduct mechanism, this proposed chemistry is based solely on model compound studies and can account for coniferyl alcohol-type products. Evidence for SET chemistry comes from a combination of studies involving electrochemistry [85], steric effects [86], substituent effects [51], C_α-radical cyclization [87], and kinetics [88,89]. Interested readers should consult these references for specifics. The essence of the chemistry (Figure 10.13) is that AHQ^{-2} delivers an electron to a lignin QM (step 1), leading to lignin fragmentation and oxidization of AHQ^{-2} to AHQ ion radical (steps 2 and 3); the ion radical can then oxidize a carbohydrate aldehyde end unit to an acid end unit (step 4), completing a cycle of redox reactions.

Summing the four steps in Figure 10.13 provides step (5), which indicates that electrons are being transferred from carbohydrates to lignin, leading to stabilized carbohydrate acid end units and lignin fragmentation. One would not expect the

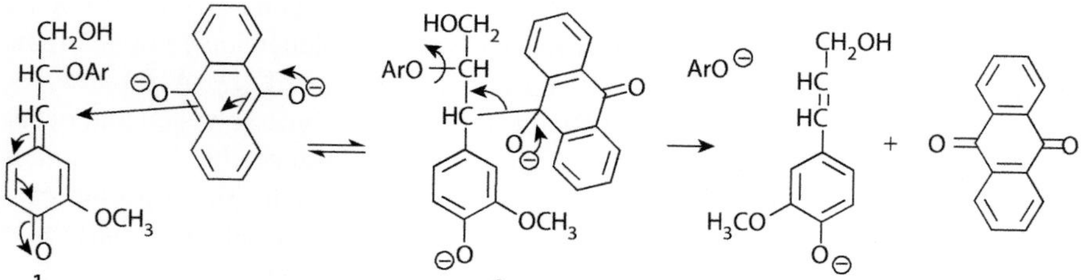

FIGURE 10.12 Proposed adduct mechanism for AHQ^{-2}-induced β-aryl ether dimer fragmentation. (Dimmel, D.R., *J Wood Chem Technol*, 5, 1, 1985.)

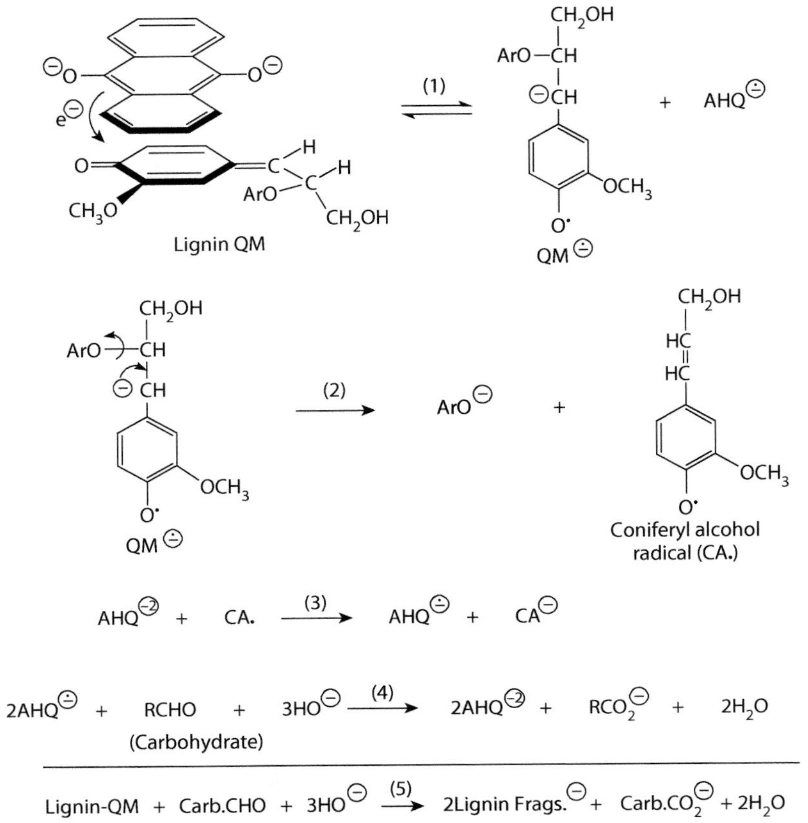

FIGURE 10.13 Redox reactions of AHQ ion radicals and dianions that lead to lignin fragmentation and carbohydrate stabilization. (Dimmel, D.R., *J Wood Chem Technol*, 5, 1, 1985.)

two polymers to collide with the right geometry to pass electrons between them; however, the water-soluble AHQ mediator facilitates the chemistry. Similar reactions are well known in other fields [90].

Adduct and SET reaction pathways could both be occurring during pulping. However, it is likely that one is a lower-energy pathway and dominates. The evidence for SET dominance is substantial. Adduct reactions in general, and QM-AHQ in particular, are reversible [46,50,51]. Lignin QM-AHQ adducts may not be on the reaction pathways of fragmentation. Preparations of hindered QM-AHQ adducts, such as would exist with lignin, are difficult (requiring organic solvents) and occur in low yield [52]. One would expect that fragmentation of β-aryl ether lignin models would be faster for HS⁻ vs AHQ⁻² based on steric effects; yet, just the opposite trend is observed [51,88,89]. Radicals in general [91] and AHQ ion radical in particular [92] have been observed during pulping. Several examples exist of AHQ⁻² participating in SET reactions in aqueous alkali at elevated temperatures [88,89,93,94]. In addition, the SET mechanism can explain the unusual relationship that exists between delignification and reagent concentration. For AQ pulping, kappa number decrease correlates with the square root of the AQ concentration [95]; presumably, AHQ^{-2} can initiate two SET QM fragmentations. A linear-dose relationship is observed [96] for SH^-, where nucleophilic substitution (adduct) mechanisms are suggested.

Some fairly hindered catalysts, such as the rosindones [97] and metal porphyrins [98] are as efficient as AQ at low concentrations in delignifying wood and/or fragmenting models. This observation is contrary to what one would expect with an adduct mechanism; metal ion complexes have been proposed to fragment lignin models by SET pathways. Moreover, carbohydrates have been shown to facilitate β-ether cleavage in lignin models [99] and can also transfer electrons to QMs [87].

CONDENSATION REACTIONS

Under alkaline conditions, phenols are prone to undergo a variety of condensation reactions. Such reactions would result in new alkali-stable carbon–carbon linkages in lignin, resulting in a higher molecular weight polymer with lower solubility in the alkaline pulping liquor. For these reasons, lignin reactions giving rise to condensation have attracted much attention; many experiments have been carried out with lignin model compounds in order to identify the structural prerequisites and the types of reactions. One general condensation type (A) involves addition of a carbanion to a quinone methide, for which several different options are possible (Figure 10.14). A second type (B) is the addition of formaldehyde to two phenolic rings with free five-positions [100,101]. The formaldehyde might originate from eliminated γ-hydroxymethyl groups in lignin or reverse aldol carbohydrate reactions.

Reactions of type A (Figure 10.14) involve a carbanion, originating from an ionized phenol structure in lignin, and a quinone methide. The anion can be located either in the aromatic ring or, in some cases, in the side chain of a conjugated aromatic system. Normally, the resulting product will be a substituted diarylmethane structure, but

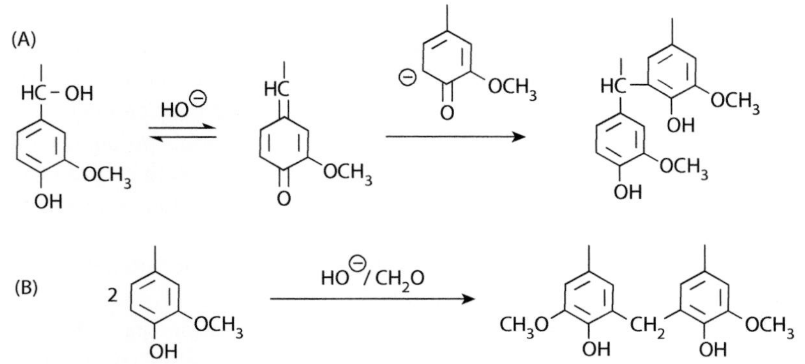

FIGURE 10.14 Suggested condensation reactions in alkaline pulping based on model compound studies. (Gierer, J., Imsgard, F., and Pettersson, I., *Appl Polym Symp,* 28, 1195–1210, 1976.)

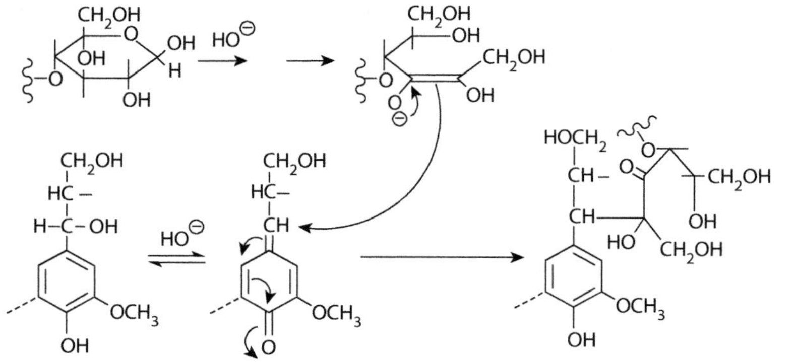

FIGURE 10.15 Model compound reaction between a quinone methide and an enediol intermediate to form a stable lignin-carbohydrate complex. (Fullerton, T.J. and Wilkins, A.L., *J Wood Chem Technol*, 5, 189–201, 1985.)

conjugation will result in products with longer aliphatic chains between the aromatic rings. In reaction type B, the product is a simple diarylmethane structure.

Treatment of a phenolic β–O–4 structure with alkali in the presence of a reactive phenol, such as 2,6-xylenol, has been shown to induce a cleavage of the β-aryl ether linkage through formation of a diarylmethane structure and a subsequent aryl participation reaction [102,103]. The final product, a stilbene, can be isolated in low yield. A different type of condensation reaction has been observed with nonphenolic β–O–4 structures that have been cleaved in alkali to give rise to an epoxide intermediate (the Fragmentation via Nucleophilic Chemistry: Soda and Kraft Pulping section) [75,76]. The epoxide may undergo further ring opening by NaOH, NaSH, NaOR, or a phenolate ion (Figure 10.10). The latter is technically a condensation reaction; so is the reaction with NaOR, where R = a carbohydrate unit (polymer). An alkali stable lignin-carbohydrate complex (LCC) is generated. Stable LCC structures may also be formed by reaction of quinone methides with carbohydrate-derived enedione structures as shown in Figure 10.15 [64,104].

The condensation chemistry presented above is based almost exclusively on the reactions of lignin model compounds. The condensation reactions of a soluble quinone methide with soluble and insoluble phenolate ions have been compared [54,55]. The two soluble reactants gave high amounts of condensation product, while the heterogeneous reaction between the soluble and insoluble reactants produced low yields. The model studies suggest that, once lignin goes into solution, reactions with another soluble piece to give a bigger, perhaps insoluble structure is much more likely than reacting with an insoluble (residual) lignin polymer. Reaction between two insoluble polymers will be highly unlikely.

Studies indicate that AHQ retards lignin-like condensation reactions of vanillyl and syringyl alcohols [94,105,106] and of an isolated lignin [105]. The effect was attributed to AHQ ions transferring electrons to intermediate quinone methide structures that were incapable of fragmenting. The transfer results in an ion radical that can undergo further reactions that lead to destruction of the QM and, thus, limit QM/phenolate ion reactions. This chemistry will be discussed in the Lignin Reduction Reactions section.

COMPETING REACTIONS: VINYL ETHERS AND STILBENES

The formation of a quinone methide usually constitutes the rate-determining step in the reactions of phenolic lignin units in alkaline media [51,63,88,89,107]. The subsequent addition of a nucleophile is assumed to be fast; nevertheless, competing elimination reactions may take place at a comparable rate. In kraft, polysulfide, and soda/AQ pulping, the strongly nucleophilic hydrosulfide, polysulfide, and AHQ anions, respectively, promote reactions that result in comprehensive cleavage of β–O–4 structures. Here, the competing elimination reactions only yield small or trace amounts of the corresponding enol ether structures (Figure 10.6) [65]. On the other hand, lignin intermediate quinone methide structures having β–5, β–β, and β–1 linkages can only react irreversibly by elimination of either a proton or a γ-hydroxymethyl group. The resulting structure in such a case will be an o,p′-dihydroxystilbene [107,108], a 1,4-diphenylbutadiene [109] and a p,p′-dihydroxystilbene [110], respectively (Figure 10.16).

Under soda pulping conditions, the elimination reaction becomes more prominent and, in model experiments, a phenolic β–O–4 structure is to a large extent converted into the corresponding enol ether by elimination of the γ-hydroxymethyl group [111]. The fact that soda pulping can be used to slowly dissolve lignin with

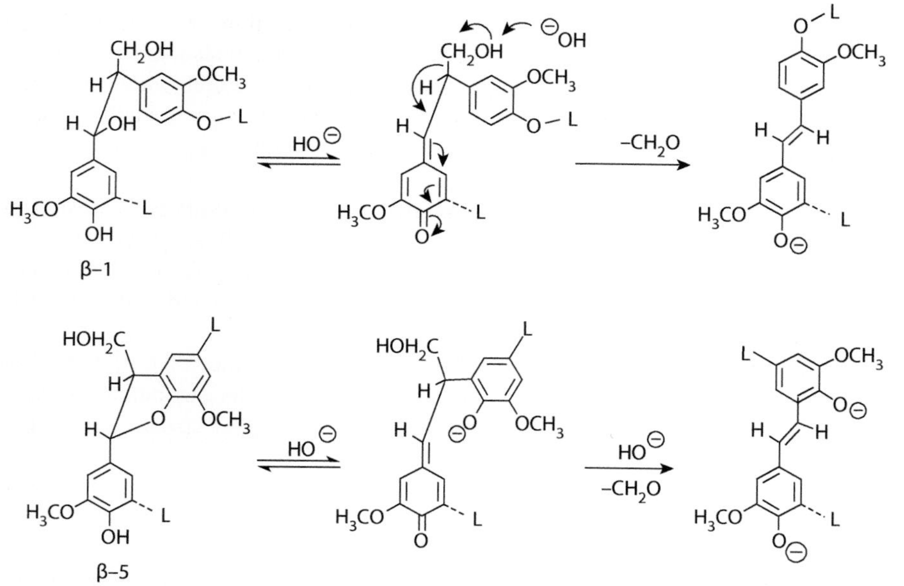

FIGURE 10.16 Formation of stilbenes from β–1 and β–5 lignin sub-structures during kraft (and soda) pulping. (Gierer, J., Pettersson, I., and Smedman, L.-Å., *Acta Chem Scand*, 26, 3366–3376, 1972; Nimz, H., *Chem Ber*, 99, 469–474, 1966.)

a chemical structure similar to that of kraft lignin [9,112] could be a consequence of a slower degradation of vinyl ether structures over the course of the rather long soda cook. Model experiments support a degradation that is the reverse of the vinyl ether formation chemistry, providing a quinone methide that can fragment by a series of different pathways [113]. One of the possibilities is a homolytic cleavage of the allylic β–O–4 linkage in the quinone methide intermediate. This reaction is favored in the case of syringyl (hardwood) structures but is of minor importance when guaiacyl units are linked together [114]. Other β-ether cleavage reactions, such as the involvement of an epoxide, must also be considered in order to explain the observed behavior.

Oxidation and Reduction Reactions

The introduction of carbonyl groups into lignin should facilitate β-ether cleavage of a nonphenolic lignin unit; normally, nonphenolic lignin units are difficult to degrade. It is well known that nucleophilic displacement is activated by having a carbonyl attached to the carbon under attack [115,116]. In addition, an α-carbonyl group will be a better acceptor of electrons than an α-saturated carbon, and will facilitate electron transfer chemistry. Thus, an α- or γ-carbonyl group in lignin will activate cleavage of β-ether units by one of several possible mechanisms, including direct nucleophilic displacement, neighboring nucleophilic displacement, electron-transfer, or elimination-Michael addition-neighboring nucleophilic displacement [117] (Figure 10.17, pathways a–d).

An oxidation is needed to introduce carbonyl groups. In all likelihood, pulping in an oxygen atmosphere will introduce such groups. It is known that oxygen/alkali pulping will delignify wood [118]; however, because there is also substantial oxidation and degradation of carbohydrate components, commercialization of such a process has not occurred. It has been concluded that some lignin oxidation occurs during AQ pulping since increased levels of carbonyl products have been observed (Figure 10.18) [82,119].

Another pulping reagent that could oxidize lignin is polysulfide. The formation of aryl-glyoxylic acids (4), e.g., oxidation products, has been observed in the pulping liquors obtained from polysulfide treatment of softwoods and hardwoods (Figure 10.19) [120]. Both polysulfide and AQ are mild oxidants and, thus, carbohydrate damage is not very extensive [2,121].

The oxidative pretreatment of pine shavings with 2,3-dichloro-5,6-dicyano-1,4-benzoquinone (DDQ) hastens delignification; the effect was attributed to introduction of carbonyl groups into the α-position of β-aryl ether structures, thereby facilitating cleavage of these structures [122].

Lignin Reduction Reactions

Reduction reactions are quite common during pulping; β-ether cleavage by hydrosulfide and AHQ ions are prime examples. In addition, lignin products containing α-CH_2 units have been observed in kraft pulping liquors [123]; such units are not common in native lignin and, therefore, must have been produced by the action of the

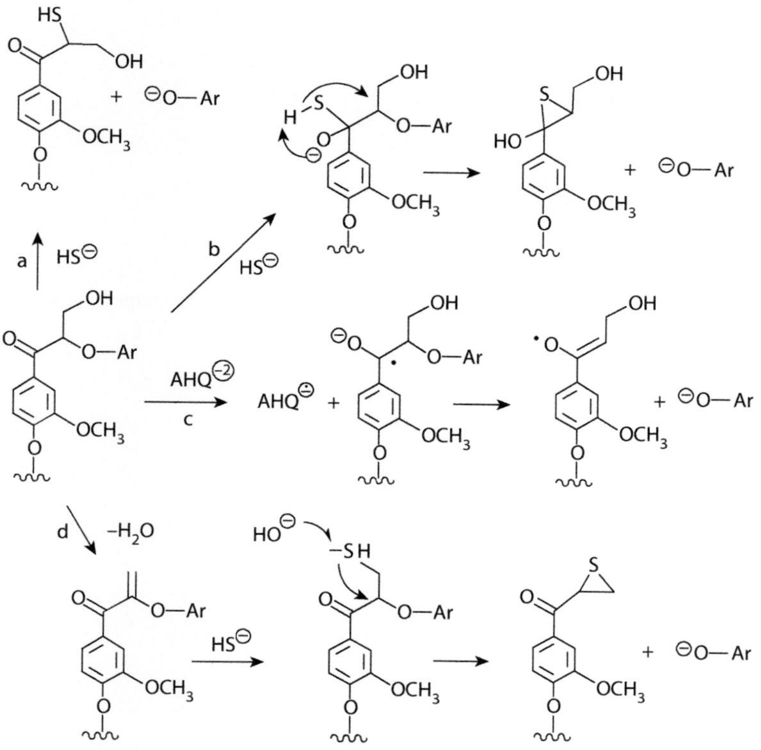

FIGURE 10.17 α-Carbonyl group activation of β-ether cleavage in a nonphenolic lignin unit.

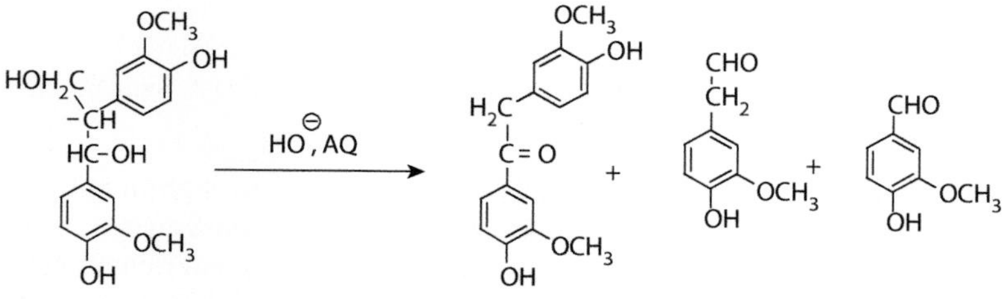

FIGURE 10.18 Oxidation products formed on soda/AQ treatment of a β–1 lignin model compound. (Hise, R.G., Seyler, D.K., Chen, C.-L., and Gratzl, J.S., Oxidative-Hydrolytic Processes in Alkaline Pulping. Proc Int Symp Wood Pulp Chem, Paris, France, 1, 391–398, 1987.)

pulping reagent with the lignin structure. α-Reduced products have been observed from lignin model/AHQ reactions [94,124]. In each reaction, αβ-aryl ether was not available for an easy fragmentation reaction. For example, treatment of syringyl alcohol with AHQ gives, in addition to the expected dimer product 6, two α-reduction products, methylsyringol (8) and a dimer 7 (Figure 10.20).

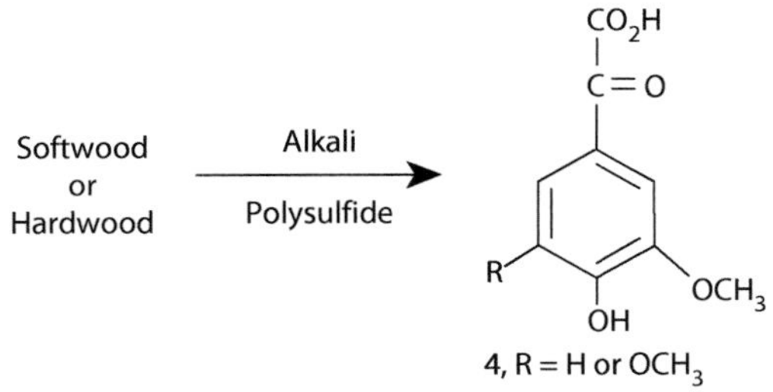

FIGURE 10.19 Aryl-glyoxylic acids observed in polysulfide pulping. (Berthold, F., Lindgren, C.T., and Lindström, M.E., *Holzforschung*, 52, 197–199, 1998.)

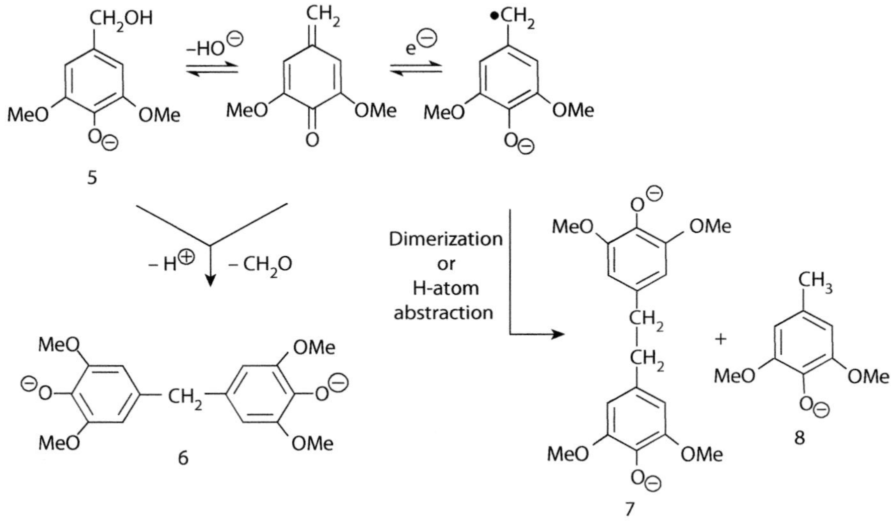

FIGURE 10.20 Syringyl alcohol/AHQ reduction reaction. (Smith, D.A. and Dimmel, D.R., *J Wood Chem Technol*, 14, 297, 1994.)

Reaction of β-methoxy ether models with AHQ led to significant amounts of reduction (9 → 13 + 14, Figure 10.21) [124]; this is a consequence of CH_3O^- being a much poorer leaving group than ArO^-. Electron transfer chemistry most likely explains the observed reduction reactions; such chemistry, shown in Figure 10.20 and Figure 10.21, may become prominent when a fragmentation process is not available, or is high energy, as is the case of radical anion intermediate 12 (Figure 10.21).

How do reduction reactions that lead to α-CH_2-groups affect pulping efficiencies? Several typical (β-aryl ether) cleavage pathways would be blocked, which would have a negative effect on efficient delignification. However, because quinone methide formation would be prohibited, undesirable vinyl ether formation and condensation reactions would also be blocked. Bulk-phase fragmentation reactions, involving a $C_\gamma O^-$ displacement on a C_β-OAr, would still be possible.

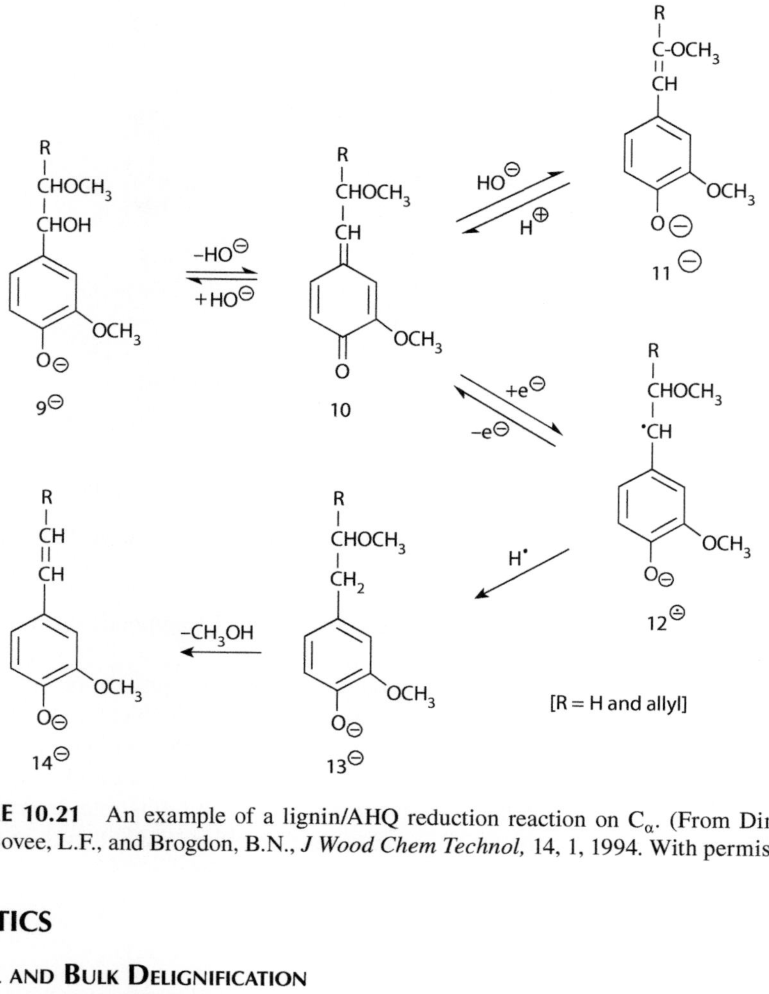

FIGURE 10.21 An example of a lignin/AHQ reduction reaction on C_α. (From Dimmel, D.R., Bovee, L.F., and Brogdon, B.N., *J Wood Chem Technol*, 14, 1, 1994. With permission.)

KINETICS

INITIAL AND BULK DELIGNIFICATION

Several studies point to the fact that the slow step in initial-phase lignin fragmentation is formation of a QM intermediate, while the slow step in bulk-phase fragmentation is cleavage of the β-aryl ether bond. Substituent effects on lignin model fragmentation reactions have proved useful for defining the various rate-determining steps [125]. Phenolic β-aryl ether lignin model compounds (15), which were capable of forming quinone methides at elevated temperatures (Figure 10.22), were studied under soda (NaOH), kraft, and soda/AQ pulping conditions. Under soda conditions, substituent changes on the β-aryl (B) ring had a large effect on fragmentation reactions of the models; changes on ring A showed only small effects. In addition, increasing the NaOH concentration accelerated fragmentation of model 15; similar acceleration effects have been reported [126,127]. This result indicates that the β-ether bond is being broken in the rate-determining step (Figure 10.22, 16 → 18). A similar case exists for non phenolic α-OH substituted units [128]. Vinyl ether

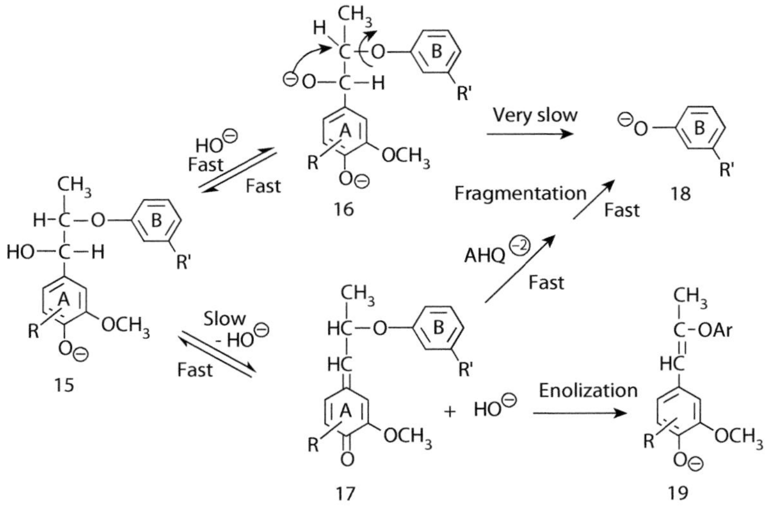

FIGURE 10.22 Reactions of a substituted lignin dimer model. (Dimmel, D.R. and L.F., Schuller, *J Wood Chem Technol*, 6, 535, 1986.)

formation (19), a side reaction that competes with model fragmentation, was more prominent at low alkali concentration.

For reactions of compound 15 under soda/AQ conditions, just the reverse trends were observed [51]. Substituent changes on the β-aryl (B) ring had no effect on the extent of fragmentation, while changes on ring A showed major effects. The explanation is that the lowest-energy fragmentation pathway in this case involves 15 → 17 → 18, for which the first step, quinone methide formation, is rate determining. Substituents on ring B are too far removed to have an effect. Substituents on ring A help stabilize the resulting QM, making its formation easier.

COMPETING REACTIONS

The relative rates between lignin fragmentation reactions and various competing reactions are critical to the effectiveness of a particular pulping process. In this regard, two significant studies have been directed at understanding the relative rates of initial-phase delignification reactions [89,127,129–131]. Each study employed phenolic lignin model compounds capable of forming quinone methides; thus, the studies focused on initial phase chemistry, a phase that is important to achieving low final lignin removal.

Gierer and Ljunggren confirmed that QM formation was the rate-determining step in the overall degradation and condensation reactions of a β-aryl ether phenolic model [127]. A combination of kinetic studies and competition experiments led Gierer and Ljunggren to propose the following rate order for reactions involving a lignin-type QM:

Addition of HS⁻ to a QM >
Addition of a carbanion to a QM (phenol condensation) >

Elimination of formaldehyde from the γ-carbon of a QM to give a vinyl ether
   (Figure 10.6, step III) >>
Formation of a QM (Figure 10.6, step I) >
β-proton abstraction of a QM to give a vinyl ether (Figure 10.6, step II)

One of the difficulties in establishing the detailed chemistry of initial phase pulping is the dominance of the delignification rate by the first step in the process—
formation of a quinone methide intermediate. The rates of subsequent steps are
difficult to determine; however, the rates of these steps are critical to the partitioning
of the QMs toward fragmentation processes (β-aryl ether cleavage) as opposed to
undesirable competing reactions, such as vinyl ether formation and condensation
reactions [57].

Brogdon and Dimmel employed an unusual β-aryl ether phenolic model (20,
Figure 10.23) to study fast QM reactions [89,129–131]. The model (20) generated a
QM (2) at high temperature in alkali and, in the absence of competing reactions, the
terminal hydroxyl group on the long side chain internally reacted with the QM intermediate to give a cyclic product 22. The rates of other QM reactions were "timed"
relative to the cyclization event by determining the yields of different products in
comparison to the yield of 22. The rate of disappearance of model 20 was the same
for soda, alkaline sulfite, soda/AHQ, and kraft cases. This indicates a common slow
step, namely QM formation. In terms of product differences, AHQ gave high levels
of fragmentation (24) and little cyclization (22). Soda and alkaline sulfite were just
the reverse. Under kraft pulping conditions, the cyclization was more rapid then fragmentation, but the rates were somewhat comparable. Small amounts of vinyl ether
products (23) were also observed, but only in the soda case.

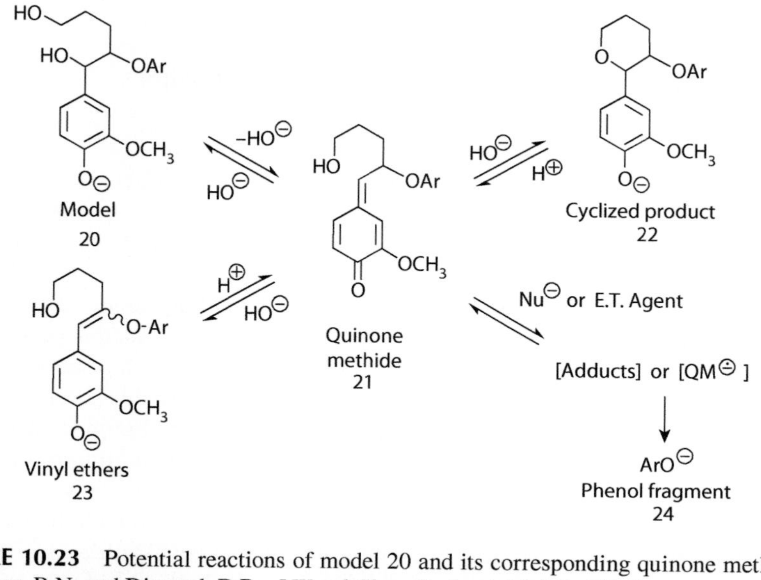

FIGURE 10.23 Potential reactions of model 20 and its corresponding quinone methide 2.
(Brogdon, B.N. and Dimmel, D.R., *J Wood Chem Technol,* 16, 261, 1996.)

The data in the AHQ case indicated that fragmentation pathways from the quinone methide were considerably lower in energy than the cyclization pathway, suggesting that a unique chemistry (electron transfer) was occurring. In the kraft case, the fragmentation pathway (*via* an adduct) is somewhat higher in energy than the cyclization pathway, since the latter dominates initially; however, fairly efficient fragmentation still results because the cyclized product can reform the QM for further reaction with HS⁻. The data indicate that, when equal concentrations of reagents are employed, the fragmentation efficiencies for model 20 (Figure 10.23) were $AHQ^{-2} \gg NaSH \gg NaOH, Na_2SO_3$. In pulping systems, the concentration of AQ/AHQ is much less than that of HS⁻; consequently, the reagent effects are comparable. The study results with model 20 suggested that sulfite ion probably added rapidly and reversibly to QM 21, but at high pH, the α-sulfonated adduct was short lived, and cyclization of the quinone methide dominated. A sulfonated adduct was not observed as a product [89].

Model 20 was also used to examine possible synergistic effects when using mixtures of sulfite/AHQ in water and sulfide/AHQ in solvents other than pure water [129,130]. The addition of alcohol solvents to aqueous soda/AHQ systems produced a large increase in model fragmentation; no increase was observed for methanol addition to soda, kraft, and alkaline sulfite systems. These unusual synergistic effects involving AHQ further point to a unique chemistry operating in this case.

Model 20 was also treated with 2,6-xylenol, 1,5-anhydrocellobiitol, amylose, and amines; none of these substances caused much fragmentation; cyclization reactions dominated [131]. The results indicate that condensation reactions between the QM 21 and phenolates or carbohydrates are much slower than fragmentation reactions of 20 with sulfide or AHD. The addition of amines to soda cooks of 20 provided little additional model fragmentation; instead, vinyl ethers (23) were observed in substantial amounts.

STRUCTURE OF RESIDUAL AND DISSOLVED LIGNIN

COMPONENTS IN THE COOKING LIQUOR

Many lignin model experiments have been carried out to elucidate the features of alkaline pulping processes [57,132]. Today, probably almost all possible reaction modes have been identified, but their relative importance for the behavior of lignin in wood is still not understood in detail. Collections of analytical data on lignins, either residual or dissolved, are difficult due to the heterogeneous nature of the polymer. Nevertheless, the application of analytical techniques, such as permanganate oxidation, acidolysis/thioacidolysis, GPC, elemental and methoxyl analysis, phenol determination, FTIR, 1-H, 13-C, and 31-P NMR, have revealed several features about isolated residual lignin, as well as precipitated, dissolved lignin from alkaline pulping processes. In addition, a variety of low molecular weight lignin fragments have been identified in cooking liquors.

Analysis of the low molecular weight components in a kraft cooking liquor yields a large number of monomeric and dimeric lignin fragments [123a,133,134]. Among these, several dimeric compounds originating from β–1, β–5, β–O–4, and β–β

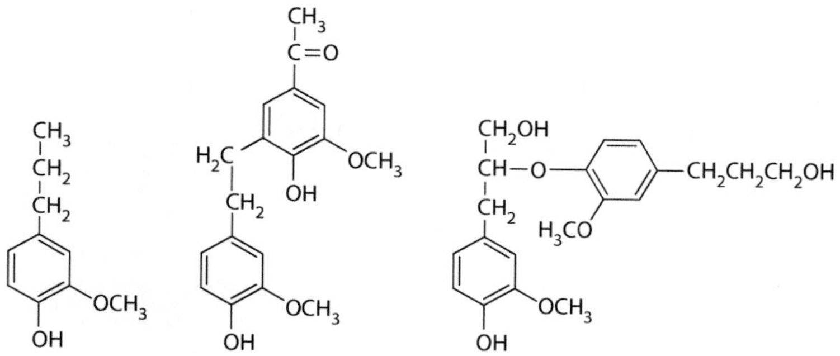

FIGURE 10.24 Some low molecular weight reduction products isolated from kraft black liquor. (Gierer, J. and Lindeberg, O., *Acta Chem Scand,* B34, 161–170, 1980.)

structures have been identified, which support both the structural features of lignin and the reactions taking place in alkaline pulping. Some representative structures are shown in Figure 10.24. In some of the dimers, as well as in several monomer compounds, a saturated side chain is present, indicating that reductive reactions were likely operating during the cook (cf. Figure 10.21). In addition, structures formed by condensation reactions between lignin and carbohydrate-derived fragments have been found [135].

THE β–O–4 STRUCTURE

Both chemical and spectroscopic methods have been employed for analyses of isolated lignin samples precipitated from cooking liquors [9], liberated from fibers by either acid [136,137] or after enzymatic hydrolysis [138,139]. In addition, chemical analyses have been used directly on pulps to obtain information about changes in the lignin structure. The importance of lignin fragmentation by the cleavage of β–O–4 linkages, described in the Fragmentation via Nucleophilic Chemistry: Soda and Kraft Pulping section, has been verified in several investigations using acidolysis [140,141], thioacidolysis [142,143], 31-P [144] and 13-C NMR analysis [145,146]. During kraft pulping of softwood (pine), the β–O–4 content in the fiber lignin decreases almost linearly with the degree of delignification; however, even at low kappa numbers, the residual lignin still contains β–O–4 structures (Figure 10.25). In hardwood (birch) pulps, the cleavage reaction is faster and the remaining amount of β–O–4 structures can be close to zero at kappa numbers below 20 [142].

The total amount of residual β–O–4 structures can vary somewhat with the conditions used in the kraft cook [143,146]; the erythro isomers react faster than the threo isomers, resulting in a predominance of threo isomers towards the end of the cook [144,147]. For the dissolved lignin, the amount of β–O–4 structures is lower, but still substantial at any given kappa number when compared to the residual lignin in the pulp. Toward the end of the cook, however, the values are quite close to each other [140]. These results indicate that the lignin molecules going into solution may occupy a ball-like shape, with limited possibilities for the cooking chemicals to penetrate and further degrade the molecule.

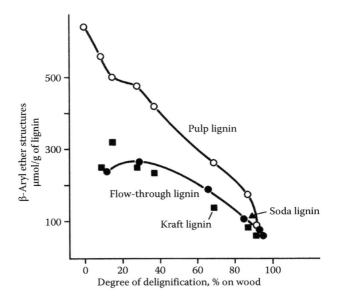

FIGURE 10.25 Changes in β–O–4 structure content as a function of delignification, as determined by acidolysis analysis. Dissolved lignin is from a normal laboratory cook (kraft lignin) and from a flow-through cook. (Reprinted from Gellerstedt, G., Lindfors, E.L., Lapierre, C., and Monties, B., *Svensk Papperstidn,* 87, R61–R67, 1984.)

The cleavage of β–O–4 structures in kraft pulping, due to the presence of hydrosulfide ions, is always accompanied by a minor reaction that gives rise to the formation of enol ether structures [140,141,148,149]. The chemistry takes place even in modern, well-impregnated kraft cooks having a high sulfidity; the elimination of the terminal hydroxymethyl groups in the phenylpropane units to give enol ethers occurs without the involvement of hydrosulfide ions. Obviously, the heterogeneous nature of the pulping process may result in local deficiencies of hydrosulfide ions in the fiber wall during the course of the cook. In soda pulping, the formation of enol ethers from β–O–4 structures should be the predominant reaction type based on the results from model experiments [111]. This is, however, not the case; slightly increased amounts of enol ether structures, as compared to kraft lignins, can be seen by thioacidolysis analysis of both isolated residual lignin and lignin dissolved in the cook. Since the remaining amount of β–O–4 structures is similar in soda and kraft lignins, a comprehensive cleavage of ether linkages also takes place in soda pulping, although at a lower rate (as discussed in the Fragmentation via Nucleophilic Chemistry: Soda and Kraft Pulping section) [149].

PHENOLIC HYDROXYL GROUPS

The cleavage of β–O–4 structures in alkaline pulping is accompanied by the formation of new phenolic hydroxyl groups (Figure 10.7). The amount of such groups in the residual lignin after a kraft cook is of considerable interest since it relates to the reactivity of the pulp in bleaching. Several analytical methods for the direct quantitative determination of ArOH groups in pulp fibers and in isolated lignin samples have been developed [150–155].

For unbleached kraft pulps having kappa numbers in the 30–35 range, phenolic hydroxyl group contents of 27, 31, and 33 phenols per 100 phenylpropane units have been found, which is more than twice the amount in spruce native lignin [150,151]. For a soda pulp with a kappa number around 40, a value of 17% phenols was obtained [150]. These figures should be regarded as approximate mean values since the residual kraft lignin is highly heterogeneous. Residual lignin molecules covering a wide range of structural identities from (almost) dissolved kraft lignin to (almost) native lignin can, thus, be expected to be present in the pulp. Indirect pieces of evidence, such as the amounts of β–O–4 structures and phenolic hydroxyl groups before and after oxygen delignification, support such a conclusion [156,157]. In addition, several factors, such as the concentration of pulping chemicals, cooking temperature, wood species, and washing efficiency may have an influence on the average level of phenolic groups.

Analysis of the dissolved lignin from kraft or soda pulping gives, as expected, much higher values for the content of phenols. Values ranging from 50–70 phenols per 100 phenylpropane units, depending on pulp yield, have been found, with the lower values at the high pulping yields; phenolic contents in soda lignins are somewhat lower than in kraft lignins [132,150,158,159].

POLYDISPERSITY, THE PRESENCE OF LIGNIN-CARBOHYDRATE COMPLEXES

The lignin going into solution in alkaline pulping processes is highly polydisperse [160–162]. As the cook proceeds, the average apparent molecular weight (size) distribution increases successively (Figure 10.26) [9,112]. At the end of the cook, a portion of the (dissolved) lignin molecules seems to be too large to be washed out of the fiber walls; consequently, a "leaching" effect of lignin from kraft pulps can always be observed [163,164]. Dissolved pine lignin has been found to contain trace amounts of carbohydrates, with xylose and arabinose the predominant neutral sugars [9,112]. The isolated residual lignins from birch and pine kraft pulps have been shown to contain chemical linkages to the various polysaccharides present in the pulp [165,166]. In a different experimental approach, Gellerstedt and colleagues achieved the selective breakdown of cellulose by an endoglucanase, followed by dissolution of the pulp in alkali and successive precipitation of the various LCCs, resulting in a separation of glucomannan-lignin, xylan-lignin, and cellulose-lignin complexes [167,168]. Altogether, about 90% of the lignin in a spruce kraft pulp was found to be chemically linked to carbohydrates, with the glucomannan–lignin complex being predominant (Figure 10.27) [167]. The presence of alkali-stable linkages between lignin and carbohydrates can be expected to slow down the dissolution of lignin and, thus, contribute to the existence of "residual" lignin in alkaline pulps.

"CONDENSED" LIGNIN STRUCTURES

Traditionally, the structural units in lignin have been divided into "condensed" and "uncondensed" with the former being structures having a substituent in the C–5

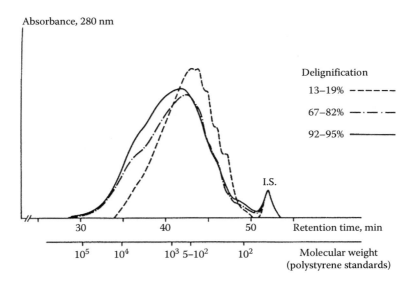

FIGURE 10.26 Gel permeation chromatograms of dissolved kraft lignin taken out at various intervals of delignification from a flow-through kraft cook of softwood. I.S. = internal standard. (Reprinted from Robert, D.R., Bardet, M., Gellerstedt, G., and Lindfors, E.-L., *J Wood Chem Technol*, 4(3), 239–263, 1984. With permission.)

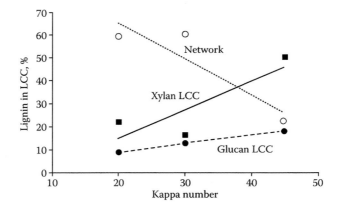

FIGURE 10.27 Effect of the degree of delignification on lignin-carbohydrate complexes during kraft pulping. Approximately 90% of the lignin is linked to carbohydrates with a predominance in the "network" structure at low kappa numbers (the network structure is a glucomannan-lignin complex). (Lawoko, M., Berggren, R., Berthold, F., Henriksson, G., and Gellerstedt, G., *Holzforschung*, 58, 603–610, 2004.)

position, e.g., 5–5, β–5, and 4–O–5 structures [169]. In addition, lignin units being substituted in the aromatic C–6 position should be included in the amount of condensed structures. In spruce native lignin (milled wood lignin, MWL) the condensed structures account for approximately 55%, based on the amount of quaternary aromatic carbons, as revealed by 13-C NMR [147]. After a kraft cook, the frequency of condensed structures is almost unchanged in the dissolved lignin [147], whereas the residual lignin shows a considerable increase in the number of quaternary aromatic carbons [170]. Thus, lignin structures like biphenyl (5–5), phenylcoumarane (β–5) and biphenyl ether (4–O–5) are more abundant in the residual lignin. Analysis by 31-P NMR of isolated kraft pulp lignins, as well as analysis by oxidative degradation of dissolved lignins from a flow-through kraft cook, gives further support for this view, although the analytical data only take into account the phenolic portion of the lignin [171,172].

LIGNIN CONDENSATION

Lignin condensation reactions during alkaline pulping have been suggested to be the most important reaction type for decreasing or preventing final delignification from taking place [98,173–176]. From a chemistry point of view, such reactions should occur and, based on model experiments, several modes of reaction have been demonstrated (the Condensation Reactions section) [98,173,176]. Analysis of kraft lignin samples using quantitative 13-C or 31-P NMR has also led to the conclusion that condensation reactions may take place during the cook [148,176]. By applying a "nucleus exchange" analytical technique directly on pulps, Chiang and Funaoka concluded that the residual lignin in kraft pulps undergoes successive rearrangements, eventually resulting in comprehensive formation of structures of the diarylmethane type as the cook proceeds [174,175]. However, the validity of this method has been questioned [177].

The presence of condensed lignin structures of the diarylmethane type (Figure 10.14) has been supported by analysis of isolated kraft lignins with 13-C [147,148] and 31-P NMR [154] and by the presence of diguaiacyl- and disyringylmethane in kraft pulping liquors [133]. Recently, it was shown by two-dimensional heteronuclear single quantum coherence (HSQC) NMR that the sharp 13-C NMR signal located around 29 ppm, normally attributed to 5,5'-diguaiacylmethane units [147,148], does not belong to a diarylmethane structure [156]. It has also been reported that a lignin model attached to a polymeric support, capable of forming a quinone methide, did not yield any condensation product with added 2,6-dimethoxyphenol [55].

A laboratory kraft cook of a softwood in the presence of 2,6-xylenol, a phenolic marker, produced chemistry indicative of radical coupling reactions between phenolic units in lignin [178]. If it is assumed that such reactions also take place between the residual fiber lignin and lignin fragments present in the pulping liquor, this chemistry could explain the increase in the number of biphenyl structures towards the end of a kraft cook [172]. The reactivity of a biphenyl structure should be [HO−]-dependent, since the pK_a-values for the two ArOH groups are quite different from those of simple phenols. Due to the formation of a relatively stable biphenyl H-bonded ring structure (Figure 10.28), ionization of the first phenolic-OH is rather easy (pK_a ~7), while the second ArOH is rather difficult (pK_a ~11) [179]. Toward the end of the cook, when the concentration of alkali has been reduced, a biphenyl structure will exist in the H-bonded ring structure and possibly be less prone to undergo cleavage of any attached β–O_4 structures.

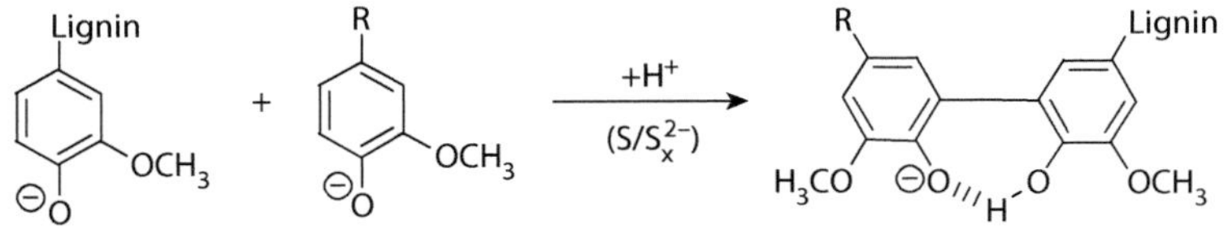

FIGURE 10.28 Suggested mode of coupling between a dissolved lignin fragment and an end unit in the residual fiber lignin. The sulfur/polysulfide system is assumed to promote one-electron transfer reactions, leading to the loss of two C_5-hydrogens and coupling. At modest alkali levels, a stabilized ring system can form through hydrogen bonding. (Majtnerova, A. and Gellerstedt, G., *Nordic Pulp Pap Res J*, 21, 129–134, 2006.)

IMPROVING PULPING EFFICIENCY

CHANGING THE PROCESS

Improving the efficiency of delignification is a difficult task. Much of the chemistry that takes place in the initial phase of delignification appears to greatly influence the overall efficiency of the whole delignification process. This phase is dominated by the reactions of quinone methides; the formation of these intermediates is the slow step in the process. In order to speed up delignification, one needs to find a way to lower the activation energy associated with producing quinone methides. This can be done by having (a) extra substituents on the phenolic ring, which results in a more stable quinone methide, or (b) a better leaving group on the α-carbon. With regard to the first way, hardwoods delignify more easily than softwoods; in part, this can be explained by the fact that hardwoods also contain syringyl units that would give rise to more stable quinone methide intermediates.

A second approach is to try to force quinone methide intermediates to follow fragmentation pathways, rather than undesirable pathways. The competition between the various pathways is shown in Figure 10.29. The various differences in energies depicted in Figure 10.29 are based upon the research of Gierer and Ljunggren and of Dimmel and coworkers [51,89,125,127,129–131]. For the sake of simplicity, only one curve is shown for the production of vinyl ether products; there are at least two pathways. There are two obvious ways to encourage fragmentation over other types of reactions. First, use chemistry that has a low activation energy associated with fragmentation, e.g., electron transfer AHQ chemistry. The second is to increase the concentration of the reagent that promotes fragmentation. Increasing the concentration of NaSH in kraft pulping would not be expected to improve the rate, since quinone methide formation is the rate-determined step in reactions with NaSH; however, higher concentrations of NaSH mean more frequent collisions with quinone methides and, thus, greater levels of fragmentation versus, for example, condensation.

Another obvious way is to decrease the NaOH concentration during the initial phase, where quinone methide chemistry dominates. This should decrease vinyl ether formation, since the rate of such a reaction is related to the NaOH concentration (Figure 10.29). Having a high NaOH concentration in the bulk phase will benefit delignification. In addition, replacing the liquor during the cook will remove the dissolved lignin and carbohydrate components, thereby minimizing condensation

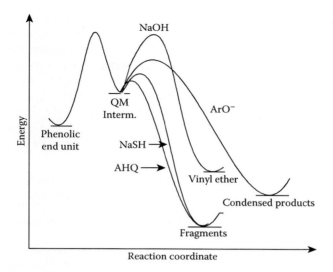

FIGURE 10.29 Depiction of the energetics for initial-phase reactions of phenolic lignin end units in the presence of NaOH, NaSH, and AHQ. (Dimmel, D.R. and Schuller, L.F., *J Wood Chem Technol,* 6, 565, 1986.)

reactions. All of these operations are used in newer, more selective pulping processes, such as EMCC™, SuperBatch™, RDH™, and others, which start with a high NaSH concentration and a lower NaOH concentration, followed by NaOH addition to give a more even distribution of NaOH during the cook and, in some cases, displacement of the liquors [180]. Some of these same principles have been applied to kraft/AQ cooking processes [181] and to a limited extent to soda/AQ [182].

CHANGING THE TREE

Changing the composition of the raw material supply would have a very large impact on pulping efficiencies. Several research groups are examining this approach. For example, by using genetic engineering techniques, Chiang has developed aspen trees that have 45% less lignin and 15% more cellulose [183]. The cellulose/lignin ratio has been improved from 2.0 to 4.3. These altered aspen trees display rapid growth, an enhanced root system, and large leaves. Pulping of these trees has not yet been reported; however, one would expect the pulping to be easier than that of normal aspen, just because there is less lignin to be removed. In another genetic engineering approach, Chapple has produced plants that overexpress the enzyme that regulates the production of syringyl monomer units and is producing species with high amounts of syringyl lignin [184]. Here again, such plants should be very easy to pulp.

Still another approach is to search for trees that contain lignin that is easily fragmented and dissolved during pulping. Mutant loblolly pine trees that are deficient in the enzyme cinnamyl alcohol dehydrogenase (CAD) have been obtained through directed breeding [185]. The lignin in the wood of CAD-deficient trees has a different pool of precursors [186]. In comparison to a normal 12-year-old loblolly pine, wood from a 12-year-old CAD-deficient tree was much more easily delignified [187]. The high reactivity of CAD-deficient wood appears to be related to a lower molecular weight and a higher phenolic content lignin [187]. Young, partially CAD-deficient trees pulped about 25% easier than the corresponding same-age normal tree [188]. However, the delignification rates for 14-year-old partially CAD-deficient and normal trees were quite similar [188]. Even so, partially CAD-deficient loblolly pine trees may be a preferred raw material source since they grow at a significantly faster rate than normal loblolly pine trees.

REFERENCES

1. (a) CR Mitchell, JH Ross. The Effect of Variables in the Soda Process. *Pulp Pap Mag Can* 33(3):35, 1932. (b) PJ Nelson, GM Irvine. Soda-AQ Pulping and the Tear Strength Problem. Proc Int Symp Wood Pulp Chem, Raleigh, NC, May 22–25, 1989, p. 515. (c) PJ Nelson, GM Irvine. Tearing Resistance in Soda-AQ and Kraft Pulps. *Tappi* J 75(1):163, 1992.
2. (a) N Sanyer, JF Laundrie. Factors Affecting the Yield Increase and Fiber Quality in Polysulfide Pulping of Loblolly Pine, Other Softwoods, and Red Oak. *Tappi* 47(10):640, 1964. (b) RP Kibblewhite, AD Bawden. Structural Organizations and Papermaking Qualities of Kraft, Soda-AQ, Neutral Sulfite-AQ, Polysulfide, and Polysulfide-AQ Pulps. *Appita* 42:275, 1989.
3. DA Blythe, LR Schroeder. Degradation of a Nonreducing Cellulose Model,1,5-Anhydro-4-O-β–D-Glucopyranosyl-D-Glucitol, under Kraft Pulping Conditions. *J Wood Chem Technol* 5:313, 1985.
4. JE Stone, DW Clayton. The Role of Sulfide in the Kraft Process. *Pulp Pap Mag Can* 61(6):T307–T313, 1960.
5. H Holton. Soda Additive Softwood Pulping: A Major New Process. *Pulp Pap Can* 78(10):T218–T223, 1977.
6. TM Grace, EW Malcolm. *Pulp and Paper Manufacture. Volume 5. Alkaline Pulping.* Montreal: The Joint Textbook Committee of the Paper Industry, TAPPI, CPPA, 1989, pp. 45–73.
7. PJ Kleppe. Kraft Pulping. *Tappi* 53(1):35–47, 1970.
8. A Teder, L Olm. Extended Delignification by Combination of Modified Kraft Pulping and Oxygen Bleaching. *Pap Puu* 63:315–326, 1981.
9. G Gellerstedt, E-L Lindfors. Structural Changes in Lignin during Kraft Pulping. *Holzforschung* 38:151–158, 1984.
10. R Kondo, KV Sarkanen. Kinetics of Lignin and Hemicellulose Dissolution during the Initial Stage of Alkaline Pulping. *Holzforschung* 38:31–36, 1984.
11. K Sjöblom, J Mjöberg, N Hartler. Extended Delignification in Kraft Cooking through Improved Selectivity. Part 1. The Effects of the Inorganic Composition of the Cooking Liquor. *Pap Puu* 65:227–240, 1983.
12. P Axegård, J-E Wiken. Delignification Studies: Factors Affecting the Amount of "Residual Lignin." *Svensk Papperstidn* 86(15):R178–R184, 1983.
13. CT Lindgren, ME Lindström. The Kinetics of Residual Delignification and Factors Affecting the Amount of Residual Lignin during Kraft Pulping. *J Pulp Pap Sci* 22:J290–J295, 1996.
14. JF Yan, DC Johnson. Delignification and Degelation: Analogy in Chemical Kinetics. *J Appl Pol Sci* 26:1623–1635, 1981.
15. JF Yan. Molecular Theory of Delignification. *Macromolecules* 14:1438–1445, 1981.
16. JF Yan. Kinetics of Delignification: A Molecular Approach. *Science* 215:1390–1392, 1982.

17. JF Yan, F Pla, R Kondo, M Dolk, JL McCarthy. Lignin 21: Depolymerization by Bond Cleavage Reactions and Degelation. *Macromolecules* 17:2137–2142, 1984.
18. M Dolk, JF Yan, JL McCarthy. Lignin 25: Kinetics of Delignification of Western Hemlock in Flow-through Reactors under Alkaline Conditions. *Holzforschung* 43:91–98, 1989.
19. BJ Fergus, DAI Goring. The Topochemistry of Delignification in Kraft and Neutral Sulphite Pulping of Birch Wood. *Pulp Pap Mag Can* 69(9):T314–T322, 1969.
20. JR Wood, DAI Goring. The Distribution of Lignin in Fibres Produced by Kraft and Acid Sulphite Pulping of Spruce Wood. *Pulp Pap Mag Can* 74(9):T309–T313, 1973.
21. S Saka, RJ Thomas, JS Gratzl. Lignin Distribution: Determination by Energy-Dispersive Analysis of X-rays. *Tappi* 61(1):73–76, 1978.
22. S Saka, RJ Thomas, JS Gratzl, D Abson. Topochemistry of Delignification in Douglas-Fir Wood with Soda, Soda-Anthraquinone and Kraft Pulping as Determined by SEM-EDXA. *Wood Sci Technol* 16:139–153, 1982.
23. TM Grace, EW Malcolm. *Pulp and Paper Manufacture. Volume 5. Alkaline Pulping.* Montreal: The Joint Textbook Committee of the Paper Industry, TAPPI, CPPA, 1989, p. 405.
24. (a) SCAN-C1:77, 1977 Scandinavian Pulp. Paper and Board Testing Committee, Stockholm, Sweder; (b) TAPPI Test Method T 236 cm-85, 1996. TAPPI Press, Atlanta, GA, USA.
25. J Li, G Gellerstedt. On the Structural Significance of the Kappa Number Measurement. *Nordic Pulp Pap Res J* 13(2):153–158, 1998.
26. MH Johansson, O Samuelson. Epimerization and Degradation of 2-O-(4-O-methyl-α-D-glucopyranosyluronic acid)-D-xylitol in Alkaline Medium. *Carbohydr Res* 54:295–299, 1977.
27. J Buchert, A Teleman, V Harjunpää, M Tenkanen, L Viikari, T Vuorinen. Effect of Cooking and Bleaching on the Structure of Xylan in Conventional Pine Kraft Pulp. *Tappi J* 78(11):125–130, 1995.
28. G Gellerstedt, J Li. An HPLC Method for the Quantitative Determination of Hexeneuronic Acid Groups in Chemical Pulps. *Carbohydr Res* 294:41–51, 1996.
29. M Tenkanen, G Gellerstedt, T Vuorinen, A Teleman, M Perttula, J Li, J Buchert. Determination of Hexenuronic Acid in Softwood Kraft Pulps by Three Different Methods. *J Pulp Pap Sci* 25:306–311, 1999.
30. CA-S Gustavsson, W Wafa Al-Dajani. The Influence of Cooking Conditions on the Degradation of Hexenuronic Acid, Xylan, Glucomannan and Cellulose during Kraft Pulping of Softwood. *Nordic Pulp Pap Res J* 15:160–167, 2000.
31. J Li, G Gellerstedt. Oxymercuration-Demercuration-Kappa Number: An Accurate Estimation of the Lignin Content in Chemical Pulps. *Nordic Pulp Pap Res J* 17:410–414, 2002.
32. G Gellerstedt, J Li, O Sevastyanova. The Relationship between Kappa Number and Oxidizable Structures in Bleached Kraft Pulps. *J Pulp Pap Sci* 28:262–266, 2002.
33. DA Smith, DR Dimmel. High Temperature Proton Exchange Reactions by Hydroxide in Water. *J Wood Chem Technol* 4:75, 1984.
34. J March. *Advanced Organic Chemistry: Reactions, Mechanisms, and Structure,* 3rd Ed., McGraw-Hill, New York, 1985, p. 222.
35. J March. *Advanced Organic Chemistry: Reactions, Mechanisms, and Structure,* 3rd Ed., McGraw-Hill, New York, 1985, pp. 197–201.
36. H Kwart. Temperature Dependence of the Primary Kinetic Hydrogen Isotope Effect as a Mechanistic Criterion. *Acc Chem Res* 15:401–408, 1982.
37. J March. *Advanced Organic Chemistry: Reactions, Mechanisms, and Structure,* 3rd Ed., McGraw-Hill, New York, 1985, Chapter 10.
38. DK Bohme, GI Mackay. Bridging the Gap Between Gas Phase and Solution: Transition in the Kinetics of Nucleophilic Displacement Reactions. *J Am Chem Soc* 103:978, 1981.

39. DA Blythe, LR Schroeder. Degradation of a Nonreducing Cellulose Model, 1,5-Anydro-4-O-β–D-glucopyranosyl-D-glucitol, under Kraft Pulping Conditions. *J Wood Chem Technol* 5:313–334, 1985.

40. GA Reed, DR Dimmel, EW Malcolm. Nucleophilicities of Selected Ions in Water at 195°C. *J Org Chem* 58:6372, 1993.

41. GA Reed, DR Dimmel, EW Malcolm. The Influence of Nucleophiles on the High Temperature Aqueous Isomerization of cis-Cinnamic Acid. *J Org Chem* 58:6364, 1993.

42. DR Dimmel, D Shepard. Regioselective Alkylations of Anthrahydroquinone and Anthrone in Water with Quinonemethides and Other Alkylating Agents. *J Org Chem* 47:22, 1982.

43. LL Landucci. Formation of Carbon-Linked Anthrone-Lignin and Anthrahydroquinone-Lignin Adducts. *J Wood Chem Technol* 1:61, 1981.

44. J March. *Advanced Organic Chemistry: Reactions, Mechanisms, and Structure,* 3rd Ed., McGraw-Hill, New York, 1985, pp. 310–316.

45. GE Miksche. Lignin Reactions in Alkaline Pulping Processes (Rate Processes in Soda Pulping). In: *Chemistry of Delignification with Oxygen, Ozone, and Peroxides,* Ed. RP Singh, J Nakano, JS Gratzl. Uni Publishers Co., Tokyo, 1980, pp. 107–120.

46. J Ralph. The Reactivity of Lignin (Model) Quinone Methides. Proc Int Symp Wood Pulp Chem, Raleigh, NC, May 22–25, 1989, p. 61.

47. LK Dyall, S Winstein. Nuclear Magnetic Resonance Spectra and Characterization of Some Quinone Methides. *J Am Chem Soc* 94:2196, 1972.

48. J Ralph, BR Adams. Determination of the Conformation and Isomeric Composition of Lignin Model Quinone Methides by NMR. *J Wood Chem Technol* 3:183, 1983.

49. J March. *Advanced Organic Chemistry: Reactions, Mechanisms, and Structure,* 3rd Ed., McGraw-Hill, New York, 1985, p. 711.

50. ID Suckling. Enhanced Cleavage of β-Aryl Ether Bonds in Lignin Model Compounds during Sulfite-Anthraquinone Pulping. *J Wood Chem Technol* 8:43, 1988.

51. DR Dimmel, LF Schuller. Structural/Reactivity Studies (II): Reactions of Lignin Model Compounds with Pulping Additives. *J Wood Chem Technol* 6:565, 1986.

52. DR Dimmel. Electron Transfer Reactions in Pulping Systems (I): Theory and Applicability to Anthraquinone Pulping. *J Wood Chem Technol* 5:1, 1985, and references therein.

53. PB Apfeld, LF Bovee, RA Barkhau, DR Dimmel. Insoluble Lignin Model (2): Preparation, Characterization and Reactions of a Polymer-Bound β-Aryl Ether Model. *J Wood Chem Technol* 8:483, 1988.

54. RA Barkhau, EW Malcolm, DR Dimmel. Insoluble Lignin Models (4): Condensation Reactions of a Polymer-Bound Guaiacylpropanol Model. *J Wood Chem Technol* 10:269, 1990.

55. RA Barkhau, EW Malcolm, DR Dimmel. Insoluble Lignin Models (5): Preparation and Alkaline Condensation Reactions of a Polymer-Bound Quinone Methide. *J Wood Chem Technol* 14:17, 1994.

56. G Brunow, AI Hatakka, I Kilpelainen, TK Lundell. Use of Novel Polystyrene-Bound Lignin Models as Substrates for Lignin Peroxidases from Phlebia Radiata. Proc Int Symp Wood Pulp Chem, Melbourne, Australia, Vol. 1, 165 (1991).

57. J Gierer. The Reactions of Lignin during Pulping. *Svensk Papperstidn* 73(18):571–596, 1970.

58. J Gierer. Chemistry of Delignification. *Wood Sci Technol* 19:289–312, 1985.

59. G Brunow, T Ilus, GE Miksche. Reactions of Sulphur during Sulphate Pulping. *Acta Chem Scand* 26:1117–1122, 1972.

60. L Gustafsson, A Teder. The Thermal Decomposition of Aqueous Polysulfide Solutions. *Svensk Papperstidn* 72(8):249–260, 1969.

61. RD Mortimer. The Formation of Coniferyl Alcohol during Alkaline Delignification with Anthraquinone. *J Wood Chem Technol* 2:383–415, 1982.
62. J Gierer, O Lindeberg. Reactions of Lignin during Sulfate Pulping. Part XV. The Behaviour of Intermediary Coniferyl Alcohol Structures. *Acta Chem Scand* B32:577–587, 1978.
63. GE Miksche. Über das Verhalten des Lignins bei der Alkalikochung. VIII. Isomerisierung der Phenolatanionen von erythro- und threo-Isoeugenol-β-(2-methoxyphenyl)-äther uber ein Chinonmethid. *Acta Chem Scand* 26:4137–4142, 1972.
64. TJ Fullerton, AL Wilkins. The Mechanism of Cleavage of β-ether Bonds in Lignin Model Compounds by Reducing Sugars. *J Wood Chem Technol* 5:189–201, 1985.
65. F Berthold, E-L Lindfors, G Gellerstedt. Degradation of Guaiacylglycerol-β-guaiacyl Ether in the Presence of HS⁻ or Polysulfide at Various Alkalinities. Part I. Degradation Rate and the Formation of Enol Ether. *Holzforschung* 52:398–404, 1998.
66. F Berthold, E-L Lindfors, G Gellerstedt. Degradation of Guaiacylglycerol-β-guaiacyl Ether in the Presence of NaHS or Polysulfide at Various Alkalinities. Part II. Liberation of Coniferyl Alcohol and Sulphur. *Holzforschung* 52:481–489, 1998.
67. G Brunow, GE Miksche. Oxidation of p-hydroxystyrenes by Aqueous Sodium Polysulfide. *Acta Chem Scand* B29:349–352, 1975.
68. G Brunow, GE Miksche. Some Reactions of Lignin in Kraft and Polysulfide Pulping. *Appl Polym Symp* 28:1155–1168, 1976.
69. F Berthold, G Gellerstedt. Influence of Polysulfide on the Rate of Degradation of Six p-OH Styrene Structures at Two OH-concentrations. *Holzforschung* 52:490–498, 1998.
70. DR Dimmel, S Anderson, P Izsak. A Study Aimed at Understanding the AQ/Polysulfide Synergistic Effect in Alkaline Pulping. *J Wood Chem Technol* 23:141–159, 2003.
71. RG Hise, C-L Chen, JS Gratzl. Chemistry of Phenolic β-aryl Lignin Structures in Alkaline Pulping. Proc Int Symp Wood Pulp Chem, Tsukuba, Japan, Vol. 2, pp. 166–171, 1983.
72. R Kondo, KV Sarkanen. Formation and Reaction of Coniferyl Alcohol during Alkaline Pulping. *J Wood Chem Technol* 4:301–311, 1984.
73. J Gierer, O Lindeberg, I Noren. Alkaline Delignification in the Presence of Anthraquinone/ Anthrahydroquinone. *Holzforschung* 33:213–214, 1979.
74. J Gierer. Chemical Aspects of Kraft Pulping. *Wood Sci Technol* 14:241–266, 1980.
75. J Gierer, S Wännström. Formation of Ether Bonds between Lignins and Carbohydrates during Alkaline Pulping Processes. *Holzforschung* 40:347–352, 1986.
76. J Gierer, I Noren, S Wännström. Formation of Condensation Products on Treatment of Non-phenolic Lignin Units of the β-aryl Ether Type with Alkali: Model Studies on a Novel Mode of Alkaline Lignin Condensation. *Holzforschung* 41:79–82, 1987.
77. S Ljunggren. The Significance of Aryl Ether Cleavage in Kraft Delignification of Softwood. *Svensk Papperstidn* 83(13):363–369, 1980.
78. R Kondo, Y Tsutsumi, H Imamura. Kinetics of β-aryl Ether Cleavage of Phenolic Syringyl Type Lignin Model Compounds in Soda and Kraft Systems. *Holzforschung* 41:83–88, 1987.
79. JR Obst, LL Landucci, N Sanyer. Quinones in Alkaline Pulping: β-ether Cleavage of Free Phenolic Units in Lignin. *Tappi* 62(1):55–59, 1979.
80. JR Obst. Kinetics of Alkaline Cleavage of β-aryl Ether Bonds in Lignin Models: Significance to Delignification. *Holzforschung* 37:23–28, 1983.
81. (a) BI Fleming, GJ Kubes, JM MacLeod, HI Bolker. Soda Pulping with Anthraquinone: A Mechanism. *Tappi* 61(6):43 (1978). (b) O Samuelson. Carbohydrates Reactions during Alkaline Cooking with Addition of Quinones. *Pulp Pap Can* 81(8):68, 1980, and references therein. (c) K Ruoho, E Sjöstrom. Improved Stabilization of Carbohydrates by the Oxygen-Quinone System. *Tappi* 61(7):87, 1978. (d) A Satohi, J Nakano, A Ishizi, Y Nomura, M Nakamura. Studies on Reaction Mechanism of Quinone in Alkaline Cooking: Acceleration of Delignification and Stabilization of Carbohydrates. *Japan Tappi* 33(6):410, 1979.

82. JS Gratzl. The Reaction Mechanisms of Anthraquinone in Alkaline Pulping. Proc EUCEPA Symp, Helsinki, Finland, 1980, Paper 12.

83. O Samuelson, LA Sjöberg. Spent Liquors from Sodium Hydroxide Cooking with Addition of Anthraquinone. *Cell Chem Technol* 12:463, 1978.

84. C Storgard-Envall, DR Dimmel. Dissolving Reactions of Anthraquinone at High Temperatures. *J Wood Chem Technol* 6:367, 1986.

85. DR Dimmel, LF Perry, HL Chum, PD Palasz, P. D. Electron Transfer Reactions in Pulping Systems (II): Electrochemistry of Anthraquinone/Lignin Model Quinonemethides. *J Wood Chem Technol* 5:15, 1985.

86. DR Dimmel, LF Schuller. Electron Transfer Reactions in Pulping Systems (III): A Study of Steric Effects in Lignin Model/AQ Reactions. *J Wood Chem Technol* 6:345, 1986.

87. DA Smith, DR Dimmel. Electron-Transfer Reactions in Pulping Systems (V): The Application of an Intramolecular Cyclization Rx as a Detector of Electron-Transfer to Quinonemethides. *J Org Chem* 53:5428, 1988.

88. DR Dimmel, LF Schuller, PB Apfeld. Electron Transfer Reactions in Pulping Systems IV: An Example of Dramatic Reactivity Differences for Fragmentation of a β-Aryl Ether Bond by AHQ^{-2} and HS$^-$. *J Wood Chem Technol* 7:97, 1987.

89. BN Brogdon, DR Dimmel. Fundamental Study of Relative Delignification Efficiencies (I). Conventional Pulping Systems. *J Wood Chem Technol* 16:261, 1996.

90. N Tsubokawa, T Endo, M Okawara. Electron Transfer between Insoluble Dihydronicotinamide Polymer and Insoluble Benzoquinone Polymer. *J Polym Sci* 20:2205, 1982.

91. TN Kleinert. Mechanisms in Alkaline Delignification. II. Free Radical Reactions. *Tappi* 49(3):126, 1966.

92. (a) M Hocking, HI Bolker, BI Fleming. Investigation of Anthraquinone-Catalyzed Alkaline Pulping via Component Modeling and Electron Spin Resonance Experiments. *Can J Chem* 58:1983, 1980. (b) SM Mattar, BI Fleming. AQ Radicals in Alkaline Pulping. *Tappi* 64(4):136, 1981. (c) JE Doyle, FO Looney. Electron Spin Resonance Investigation of Quinone-Assisted Soda Pulping. *Appita* 36:219, 1982.

93. K Kuroda, DR Dimmel. Electron Transfer Reactions in Pulping Systems (VI): Alcohol Bisulfite Pulping. *J Wood Chem Technol* 12:313, 1992.

94. DA Smith, DR Dimmel. Electron Transfer Reactions in Pulping Systems (IX): Reactions in Syringyl Alcohol with Typical Pulping Reagents. *J Wood Chem Technol* 14:297, 1994.

95. DP Werthemann. Why Redox Pulping Catalysts Fit the Square Root Relationship. *J Wood Chem Technol* 1:169, 1981.

96. DP Werthemann. Sulfide and Anthraquinone-like Catalysts Delignify Wood via Different Chemical Mechanisms in Alkaline Pulping. *Tappi* 65(7):98, 1982.

97. DP Werthemann, H Huber-Emden, PM Bersier, J Kelemen. High Catalytic Activity of Rosindone and Related Compounds in Alkaline Pulping. *J Wood Chem Technol* 1:185, 1981.

98. (a) PA Watson, LJ Wright, TJ Fullerton. Reactions of Metal Ion Complexes with Lignin Model Compounds. I. Co(TSPP) as a Single Electron Transfer Catalyst and Implications for the Mechanism of AQ pulping. *J Wood Chem Technol* 13:371, 1993. (b) PA Watson, LJ Wright, TJ Fullerton. Reactions of Metal Ion Complexes with Lignin Model Compounds. II. Fe(TSPC) Catalyzed Formation of Oxidized Products in the Absence of Oxygen. *J Wood Chem Technol* 13:391, 1993. (c) PA Watson, LJ Wright, TJ Fullerton. Reactions of Metal Ion Complexes with Lignin Model Compounds. III. Rh(TSPP) Catalyzed Formation of Guaiacol from β-Aryl Ethers in Exceptionally High Yield. *J Wood Chem Technol* 13:411, 1993.

99. (a) TJ Fullerton, LJ Wright. Enhanced Beta-Ether Cleavage of Lignin Model Compounds by Reducing Sugars. *Tappi J* 67(3):78, 1984. (b) TJ Fullerton, AL Wilkins. Mechanism of Cleavage of Beta-Ether Bonds in Lignin Model Compounds by Reducing Sugars. *J Wood Chem Technol* 5:189, 1985. (c) J Jansen, TJ Fullerton. Influence of Carbohydrates and Related Compounds on the Alkaline Cleavage of the beta-Aryl Ether Linkage in a Phenolic Lignin Model Compound. *Holzforschung* 41:359, 1987.

100. J Gierer, F Imsgard, I Pettersson. Possible Condensation and Polymerization Reactions of Lignin Fragments during Alkaline Pulping Processes. *Appl Polym Symp* 28:1195–1210, 1976.

101. S Yasuda, B-H Yoon, N Terashima. Chromophoric Structures of Alkali Lignin IV. Chromophoric Structures of Condensation Products from Guaiacylglycerol-β-aryl Ether [γ-^{13}C]. *Mokuzai Gakkaishi* 26(6):421–425, 1980.

102. J Gierer, I Pettersson. Studies on the Condensation of Lignins in Alkaline Media. Part II. The Formation of Stilbene and Arylcoumaran Structures through Neighbouring Group Participation Reactions. *Can J Chem* 55(4):593–599, 1977.

103. J Gierer, O Lindeberg. Studies of the Condensation of Lignins in Alkaline Media. Part III. The Formation of Stilbenes, Arylcoumarans and Diarylmethanes on Treatment of Spruce Wood Meal with Alkali and White Liquor in the Presence of Xylenols. *Acta Chem Scand* B33:580–582, 1979.

104. F Letumier, E Ämmälahti, J Sipilä, T Vuorinen. A Facile Route of Formation of Lignin-Xylan Complexes during Alkaline Pulping. Proc Int Pulp Bleach Conf, Halifax, Canada, Vol. 2, p 11–15, 2000.

105. DR Dimmel, D Shepard, TA Brown. The Influence of Anthrahydroquinone and Other Additivies on the Condensation Reactions of Vanillyl Alcohol. *J Wood Chem Technol* 1:123, 1981.

106. DA Smith, DR Dimmel. Electron Transfer Reactions in Pulping Systems (VIII): Reaction of Syringyl Alcohol in Aqueous Alkali. *J Wood Chem Technol* 14:279, 1994.

107. GE Miksche. Uber das Verhalten des Lignins bei der Alkalikochung VI. Zum Abbau von p-Hydroxy-phenylcumaranstrukturen durch Alkali. *Acta Chem Scand* 26:3269–3274, 1972.

108. J Gierer, I Pettersson, L-Å Smedman. The Reactions of Lignin during Sulfate Pulping. Part XII. Reactions of Intermediary o,p'-dihydroxystilbene Structures. *Acta Chem Scand* 26:3366–3376, 1972.

109. J Gierer, L-Å Smedman. The Reactions of Lignin during Sulfate Pulping. Part XI. Reactions of Pinoresinol with Alkali and with White Liquor. *Acta Chem Scand* 25:1461–1467, 1971.

110. H Nimz. Isolierung von zwei weiteren Abbauphenolen mit einer 1,2-Diaryl-propan-Struktur. *Chem Ber* 99:469–474, 1966.

111. J Gierer, I Noren. Uber die Reaktionen des Lignins bei der Sulfatkochung II. Modellversuche zur Spaltung von Aryl-alkylätherbindungen durch Alkali. *Acta Chem Scand* 16:1713–1729, 1962.

112. DR Robert, M Bardet, G Gellerstedt, E-L Lindfors. Structural Changes in Lignin during Kraft Cooking. Part 3. On the Structure of Dissolved Lignins. *J Wood Chem Technol* 4(3):239–263, 1984.

113. DR Dimmel, LF Bovee. Pulping Reactions of Vinyl Ethers. *J Wood Chem Technol* 13:583, 1993.

114. GE Miksche. Zum alkalischen Abbau von Arylglycerin-β-(2,6-dimethoxy-4-alkylaryl)-ätherstrukturen. *Acta Chem Scand* 27:1355–1368, 1973.

115. J March. *Advanced Organic Chemistry: Reactions, Mechanisms, and Structure,* 3rd Ed., McGraw-Hill, New York, 1985, pp. 301–302.

116. J Gierer, S Ljunggren, P Ljunquist, I Noren. The Reactions of Lignin during Sulfate Pulping. Part 18. The Significance of α-carbonyl Groups for the Cleavage of β-aryl Ether Structures. *Svensk Papperstidn* 84:75–82, 1980.

117. J Gierer and S Ljunggren. The Reactions of Lignin during Sulfate Pulping. Part 16. The Kinetics of the Cleavage of β-aryl Ether Linkages in Structures Containing Carbonyl Groups. *Svensk Papperstidn* 83:71–81, 1979.

118. PJ Kleppe. Future of Oxygen in Pulp Manufacture. *Papier* 39(10A):V8, 1985.

119. (a) RG Hise, C-L Chen, JS Gratzl. Chemistry of Phenolic beta-Aryl Lignin Structures in Alkaline Pulping with Quinone Additives. Proc Can Wood Chem Symp, Niagara Falls, Canada, pp. 71–72, 1982. (b) RG Hise, DK Seyler, C-L Chen, JS Gratzl. Oxidative-Hydrolytic Processes in Alkaline Pulping. Proc Int Symp Wood Pulp Chem, Paris, France, Vol. 1, pp. 391–398, 1987.

120. F Berthold, CT Lindgren, ME Lindström. Formation of (4-hydroxy-3-methoxyphenyl)-glyoxylic Acid and (4-hydroxy-3,5-dimethoxyphenyl)-glyoxylic Acid during Polysulfide Treatment of Softwood and Hardwood. *Holzforschung* 52:197–199, 1998.

121. TJ Blain. Anthraquinone Pulping: Fifteen Years Later. *Tappi J* 76(3):137, 1993.

122. J Gierer, I Noren. Oxidative Pretreatment of Pine Wood to Facilitate Delignification during Kraft Pulping. *Holzforschung* 36:123, 1982.

123. (a) J Gierer, O Lindeberg. Reactions of Lignin during Sulfate Pulping. Part XIX. Isolation and Identification of New Dimers from a Spent Sulfate Liquor. *Acta Chem Scand* B34:161–170, 1980. (b) L Lowendahl, G Petersson, O Samuelson. Phenolic Compounds in Kraft Black Liquor. *Svensk Papperstidn* 81:392, 1978. (c) K Niemelä. GLC-MS Studies on Pine Kraft Black Liquors. *Holzforschung* 42:169, 1988.

124. DR Dimmel, LF Bovee, BN Brogdon. Electron Transfer Reactions in Pulping Systems (VII): Degradation Reactions of β-Methoxy Lignin Models. *J Wood Chem Technol* 14:1, 1994.

125. DR Dimmel, LF Schuller. Structural/Reactivity Studies (I): Soda Reactions of Lignin Model Compounds. *J Wood Chem Technol* 6:535, 1986.

126. JR Obst, N Sanyer. Effect of Quinones and Amines on Cleavage Rate of Beta-O-4 Ethers in Lignin During Alkaline Pulping. *Tappi* 63(7):111, 1980.

127. J Gierer, S Ljunggren. The Reactants of Lignin during the Sulfate Pulping Part 17. Kinetic Treatment of the Formation and Competing Reactions of Quinone Methide Intermediates. *Svensk Papperstidn* 83:503, 1979.

128. WE Collier, TH Fisher, LL Ingram, AL Harris, TP Schultz. Alkaline Hydrolysis of Nonphenolic β-O-4 Lignin Model Dimers: Further Studies of the Substituent Effect on the Leaving Phenoxide. *Holzforschung* 50:420, 1996.

129. BN Brogdon, DR Dimmel. Fundamental Study of Relative Delignification Efficiencies (II). Combinations of Pulping Additives. *J Wood Chem Technol* 16:285, 1996.

130. BN Brogdon, DR Dimmel. Fundamental Study of Relative Delignification Efficiences (III). Organosolv Pulping Systems. *J Wood Chem Technol* 16:297, 1996.

131. BN Brogdon, DR Dimmel. Competing Reactions Affecting Delignification in Pulping Systems. *J Wood Chem Technol* 16:405, 1996.

132. J Gierer. Chemistry of Delignification. Part 2: Reactions of Lignin during Bleaching. *Wood Sci Technol* 20:1–33, 1986.

133. K Niemelä. Low-Molecular Weight Organic Compounds in Birch Kraft Black Liquor. Annales Academiae Scietiarum Fennicae, Series A, No. 229. Helsinki University of Technology, 1990.

134. F Berthold, G Gellerstedt. Reactive Structures Formed during the Initial Phase of a Kraft Cook. Proc Int Symp Wood Pulp Chem, Beijing, China, Vol. 3, pp. 160–163, 1993.

135. J Gierer, S Wännström. Formation of Alkali-Stable C-C-bonds between Lignin and Carbohydrate Fragments during Kraft Pulping. *Holzforschung* 38:181–184, 1984.

136. G Gellerstedt, J Pranda, E-L Lindfors. Structural and Molecular Properties of Residual Birch Kraft Lignins. *J Wood Chem Technol* 14:467, 1994.
137. W Wafa Al-Dajani, G Gellerstedt. On the Isolation and Structure of Softwood Residual Lignins. *Nordic Pulp Pap Res J* 17:193–198, 2002.
138. B Hortling, M Ranua, J Sundquist. Investigation of the Residual Lignin in Chemical Pulps. Part 1. Enzymatic Hydrolysis of the Pulps and Fractionation of the Products. *Nordic Pulp Pap Res J* 5:33–37, 1990.
139. S Wu, DS Argyropoulos. An Improved Method for Isolating Lignin in High Yield and Purity. *J Pulp Pap Sci* 29:235–240, 2003.
140. G Gellerstedt, EL Lindfors, C Lapierre, B Monties. Structural Changes in Lignin during Kraft Cooking. Part 2. Characterization by Acidolysis. *Svensk Papperstidn* 87:R61–R67, 1984.
141. G Gellerstedt, EL Lindfors. On the Formation of Enol Ether Structures in Lignin during Kraft Cooking. *Nordic Pulp Pap Res J* 2:71–75, 1987.
142. G Gellerstedt, EL Lindfors, C Lapierre, D Robert. The Reactivity of Lignin in Birch Kraft Cooking. Proc European Workshop Lignocell Pulp, Hamburg, Germany, pp. 224–227, 1990.
143. C Gustavsson, K Sjöström, W Wafa Al-Dajani. The Influence of Cooking Conditions on the Bleachability and Chemical Structure of Kraft Pulps. *Nordic Pulp Pap Res J* 14:71–81, 1999.
144. BC Ahvazi, DS Argyropoulos. Thermodynamic Parameters Governing the Stereoselective Degradation of Arylglycerol-β-aryl Ether Bonds in Milled Wood Lignin under Kraft Pulping Conditions. *Nordic Pulp Pap Res J* 12:282–288, 1997.
145. PM Froass, AJ Ragauskas, J Jiang. Chemical Structure of Residual Lignin from Kraft Pulp. *J Wood Chem Technol* 16:347–365, 1996.
146. PM Froass, AJ Ragauskas, J Jiang. Nuclear Magnetic Resonance Studies. 4. Analysis of Residual Lignin after Kraft Pulping. *Ind Eng Chem Res* 37:3388–3394, 1998.
147. KP Kringstad, R Mörck. ^{13}C-NMR Spectra of Kraft Lignins. *Holzforschung* 37:237–244, 1983.
148. G Gellerstedt, D Robert. Quantitative ^{13}C NMR Analysis of Kraft Lignins. *Acta Chem Scand* B41:541–546, 1987.
149. G Gellerstedt, W Wafa Al-Dajani. Bleachability of Alkaline Pulps. Part 1. The Importance of β-aryl Ether Linkages in Lignin. *Holzforschung* 54:609–617, 2000.
150. G Gellerstedt, E-L Lindfors. Structural Changes in Lignin during Kraft Cooking. Part 4. Phenolic Hydroxyl Groups in Wood and Kraft Pulps. *Svensk Papperstidn* 87:R115–R118, 1984.
151. RC Francis, Y-Z Lai, CW Dence, TC Alexander. Estimating the Concentration of Phenolic Hydroxyl Groups in Wood Pulps. *Tappi J* 74(9):219–224, 1991.
152. DR Robert, G Brunow. Quantitative Estimation of Hydroxyl Groups in Milled Wood Lignin from Spruce and in a Dehydrogenation Polymer from Coniferyl Alcohol Using ^{13}C NMR Spectroscopy. *Holzforschung* 38:85–90, 1984.
153. M Barelle. A New Method for the Quantitative ^{19}F NMR Spectroscopic Analysis of Hydroxyl Groups in Lignins. *Holzforschung* 47:261–267, 1993.
154. DS Argyropoulos. Quantitative Phosphouros-^{31} NMR Analysis of Lignins: A New Tool for the Lignin Chemist. *J Wood Chem Technol* 14:45–63, 1994.
155. A Gärtner, G Gellerstedt, T Tamminen. Determination of Phenolic Hydroxyl Groups in Residual Lignin Using a Modified UV-method. *Nordic Pulp Pap Res J* 14:163–170, 1999.
156. G Gellerstedt, L Zhang. Chemistry of TCF-bleaching with Oxygen and Hydrogen Peroxide. *Am Chem Soc Symp Series* 785:61–72, 2001.
157. G Gellerstedt, E-L Lindfors. Hydrophilic Groups in Lignin after Oxygen Bleaching. *Tappi J* 70(6):119–122, 1987.

158. P Månsson. Quantitative Determination of Phenolic and Total Hydroxyl Groups in Lignins. *Holzforschung* 37:143–146, 1983.

159. E Evstigneyev, H Maiyorova, A Platonov. Lignin Functionalization and the Alkaline Delignification Rate. *Tappi J* 75(5):177–182, 1992.

160. Y Liu, S Carriero, K Pye, DS Argyropoulos. A Comparison of the Structural Changes Occurring in Lignin during Alcell and Kraft Pulping of Hardwoods and Softwoods. *Am Chem Soc Symp Series* 742:447–464, 2000.

161. R Mörck, H Yoshida, KP Kringstad. Fractionation of Kraft Lignin by Successive Extraction with Organic Solvents I. Functional Groups, ^{13}C NMR-spectra and Molecular Weight Distributions. *Holzforschung* 40(suppl):51–60, 1986.

162. R Mörck, A Reimann, KP Kringstad. Fractionation of Kraft Lignin by Successive Extraction with Organic Solvents III. Fractionation of Kraft Lignin from Birch. *Holzforschung* 42:111–116, 1988.

163. BD Favis, JM Willis, DAI Goring. High Temperature Leaching of Lignin from Unbleached Kraft Pulp Fibers. *J Wood Chem Technol* 3:1–7, 1983.

164. (a) BD Favis, JM Willis, DAI Goring. The Effect of Electrolytes on the Leaching of Lignin from Unbleached Kraft Pulp Fibers. *J Wood Chem Technol* 3:9–16, 1983. (b) J Li, A Phenix, M MacLeod, Diffusion Lignin Macromolecules within the Fibre Walls of Kraft Pulp. Part I: Determination of the restricted diffusion coefficient under alkaline conditions, *Can J Chem Eng* 75(1):16, 1997. (c) J Li, C Mui. Effects of Lignin Diffusion on Kraft Delignification Kinetics as Determined by Liquor Analysis. Part I. An elemental study. *J Pulp Paper Sci* 25(11):373, 1999.

165. M Tenkanen, T Tamminen, B Hortling. Investigation of Lignin-Carbohydrate Complexes in Kraft Pulps by Selective Enzymatic Treatments. *Appl Microbiol Biotechnol* 51:241–248, 1999.

166. T Iversen, S Wännström. Lignin-Carbohydrate Bonds in a Residual Lignin Isolated from Pine Kraft Pulp. *Holzforschung* 40:19–22, 1986.

167. M Lawoko, R Berggren, F Berthold, G Henriksson, G Gellerstedt. Changes in the Lignin-Carbohydrate Complex in Softwood Kraft Pulp during Kraft and Oxygen Delignification: Lignin-Polysaccharide Networks II. *Holzforschung* 58:603–610, 2004.

168. M Lawoko, G Henriksson, G Gellerstedt. Structural Differences Between the Lignin-Carbohydrate Complexes Present in Wood and in Chemical Pulps. *Biomacromolecules* 6:3467–3473, 2005.

169. A Sakakibara, Y Sano. Chemistry of Lignin. In *Wood and Cellulosic Chemistry,* 2nd Ed. Ed. D N-S Hon, N Shiraishi. Marcel Dekker Inc, New York, 2001, pp. 128–129.

170. G Gellerstedt, E-L Lindfors, M Pettersson, D Robert. Reactions of Lignin in Chlorine Dioxide Bleaching of Kraft Pulps. *Res Chem Intermed* 21:441–456, 1995.

171. Z-H Jiang, DS Argyropoulos. Isolation and Characterization of Residual Lignins in Kraft Pulps. *J Pulp Pap Sci* 25:25–29, 1999.

172. G Gellerstedt, K Gustafsson. Structural Changes in Lignin during Kraft Cooking. Part 5. Analysis of Dissolved Lignin by Oxidative Degradation. *J Wood Chem Technol* 7:65–80, 1987.

173. S Yasuda, B-H Yoon, N Terashima. Chromophoric Structures of Alkali Lignin IV. Chromophoric Structures of Condensation Products from Guaiacylglycerol-β-aryl Ether [γ-^{13}C]. *Mokuzai Gakkaishi* 26:421–425, 1980.

174. VL Chang, M Funaoka. The Formation and Quantity of Diphenylmethane Type Structures in Residual Lignin during Kraft Delignification of Douglas-Fir. *Holzforschung* 42:385–391, 1988.

175. VL Chang, M Funaoka. The Dissolution and Condensation Reactions of Guaiacyl and Syringyl Units in Residual Lignin during Kraft Delignification of Sweetgum. *Holzforschung* 44:147–155, 1990.

176. BC Ahvazi, G Pageau, DS Argyropoulos. On the Formation of Diphenylmethane Structures in Lignin under Kraft, EMCC, and Soda Pulping Conditions. *Can J Chem* 76:506–512, 1998.

177. FD Chan, KL Nguyen, AFA Wallis. Estimation of the Aromatic Units in Lignin by Nucleus Exchange: A Reassessment of the Method. *J Wood Chem Technol* 15:473–491, 1995.

178. A Majtnerova, G Gellerstedt. Radical Coupling: A Major Obstacle to Delignification in Kraft Pulping. *Nordic Pulp Pap Res J* 21:129–134, 2006.

179. M Ragnar, CT Lindgren, N-O Nilvebrant. pKa-values of Guaiacyl and Syringyl Phenols Related to Lignin. *J Wood Chem Technol* 20:277–305, 2000.

180. (a) PO Tikka. Conditions to Extend Kraft Cooking Successfully. Proc Tappi Pulp Conf, Boston, p. 699, 1992. (b) S Nordén, R Reeves, D Dahl. Bleaching of Extremely Low Kappa Southern Pine, Cooked by the Superbatch Process. Proc Tappi Pulp Conf, Boston, p. 159, 1992. (c) EK Andrews. RDH [Rapid Displacement Heating] Kraft Pulping to Extend Delignification, Decrease Effluent, and Improve Productivity and Pulp Properties. *Tappi J* 72(11):55, 1989.

181. (a) JE Jiang. Extended Delignification of Southern Pine with Polysulfide and Anthraquinone. Proc Tappi Pulp Conf, Atlanta, p. 313, 1993. (b) G Katz. Anthraquinone and Anthraquinone Polysulfide Pulping for Extending Digester Delignification. Proc Tappi Pulp Conf, Atlanta, p. 323, 1993. (c) N Hartler. Extended Delignification in Kraft Cooking: A New Concept. *Svensk Papperstidn* 81:483, 1978.

182. NH Shin. A Modified Soda-AQ/Oxygen Pulping Process as an Alternative to Kraft Process for the Production of Bleached Softwood Pulp. PhD Thesis, North Carolina State University, Raleigh, NC, 1988.

183. VL Chiang. Genetic Engineering of Lignin and Cellulose Biosynthesis in Trees for Wood Quality Improvement. Proc Brazilian Symp Chemistry of Lignins and Other Wood Components, Belo Horizonte, Brazil, Oral Presentations, pp. 9–18, 2001.

184. K Meyer, AM Shirley, JC Cusumano, DA Bell-Lelong, C Chapple. Lignin Monomer Composition Is Determined by the Expression of a Cytochrome P450-dependent Monooxygenase in *Arabidopsis*. *Proc Natl Acad Sci USA* 95:6619, 1998.

185. JJ MacKay, DM O'Malley, T Presnell, FL Booker, MM Campbell, RW Whetten, RR Sederoff. Inheritance, Gene Expression and Lignin Characterization in a Mutant Pine Deficient in Cinnamyl Alcohol Dehydrogenase. *Proc Natl Acad Sci USA* 94:8255, 1997.

186. (a) C Lapierre, B Pollet, J MacKay, R Sederoff. Lignin Structure in a Mutant Pine Deficient in Cinnamyl Alcohol Dehydrogenase. *J Agr Food Chem* 48:2326, 2000. (b) J Ralph, JJ MacKay, RD Hatfield, DM O'Malley, RW Whetten, RR Sederoff. An Abnormal Lignin in a Mutant Loblolly Pine (*Pinus taeda* L.). *Science* 277:235, 1997. (c) J MacKay, DR Dimmel, JJ Boon. Pyrolysis MS Characterization of Wood from CAD-Deficient Pine. *J Wood Chem Technol* 21:19, 2001.

187. DR Dimmel, JJ MacKay, EM Althen, C Parks. Pulping and Bleaching of CAD-Deficient Wood. *J Wood Chem Technol* 21:1, 2001.

188. DR Dimmel, JJ MacKay, C Courchene, J Kadla, JT Scott, DM O'Malley, SE McKeand. Pulping and Bleaching of Partially Cad-Deficient Wood. *J Wood Chem Technol* 22:235–248, 2002.

11 Chemistry of Pulp Bleaching

Göran Gellerstedt

CONTENTS

GENERAL OVERVIEW

The main purpose of bleaching is to increase the whiteness and the cleanliness of the fibers to obtain better contrast and color reproduction in printing together with improved paper machine runnability. Other important fiber properties such as water absorbancy can also be changed dramatically. The World production of fully bleached

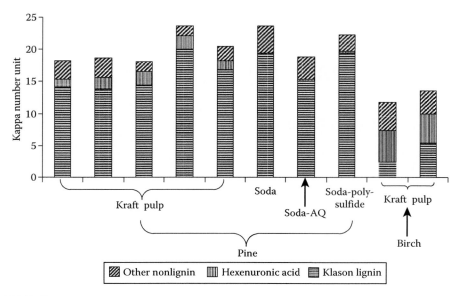

FIGURE 11.1 Contribution to the kappa number in unbleached alkaline pulps from lignin, hexenuronic acid, and "non lignin" structures.

chemical pulp is in the order of 83 million tons/y, with the US, Canada, Sweden, and Brazil as major producers of market pulp. A rather steady average production increase of approximately 3% per year takes place worldwide, although most of the new capacity is added in South America and in Southeast Asia. Bleached mechanical pulps are usually used directly in integrated mills, predominantly, for the production of a large variety of printing papers.

Bleachable grades of chemical pulp (kraft or sulfite) usually contain about 1.5–4.5% residual lignin in the fiber. In addition, various amounts of hexenuronic acid is present in kraft pulps with hardwood pulps usually having much higher levels than softwood due to the larger amount of xylan in the former. Other carbohydrate related "contaminants," containing double bonds and/or carbonyl groups, also seem to be present in kraft pulps although the exact structure is unknown. In addition, certain chemical pulps, notably hardwood pulps, may contain considerable amounts of extractives [1,2]. Altogether, these constituents in the unbleached fiber contribute to color and reactivity. The total amount of these types of structures is usually measured as the kappa number of the pulp. In Figure 11.1, some representative values are given to illustrate the relative contribution of lignin, hexenuronic acid and other "non-lignin" structures to the kappa number in various unbleached kraft and kraft-like pulps [3].

In mechanical and chemimechanical pulping, virtually all wood components are present in the unbleached fiber with the pulp yield usually being around 90–97%. Lignin is the predominant component contributing to the color of these pulps. The small amounts of coniferaldehyde and quinone structures present in lignin both have light absorption within the visible light range (Figure 11.2) [4,5]. Some chemical

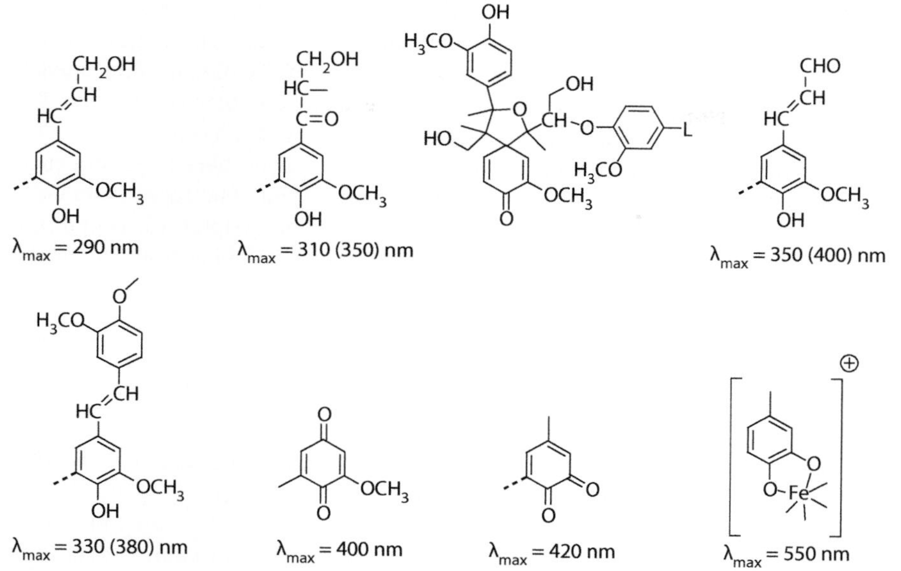

FIGURE 11.2 Conjugated structures present in wood and/or unbleached mechanical pulps. Light absorption maxima are given for solution and, when available, for the solid state (within brackets).

changes occur in the lignin during the chip refining process cause the lignin to darken. Other changes include a cleavage of a small portion of the β–O–4 linkages thus giving rise to an increase of the amount of free phenolic hydroxyl groups together with a formation of carbonyl groups [6]. Chemimechanical pulping that uses the addition of sodium sulfite to the wood results in less discoloration of the fibers but with sulfonation of certain lignin structures taking place.

In most studies to date, the chemical reactions encountered with various bleaching agents have been investigated using isolated lignin or lignin model compounds as substrate. Based on such work, most of the reaction principles and mechanisms involved have been elucidated. For some bleaching agents, the detailed reaction kinetics has also been identified. In addition, much work has been devoted to the identification of individual components found in bleaching effluents from chemical pulp bleaching. Here, environmental issues have been the driving force with one important ambition being the identification of compounds with toxic and/or mutagenic properties.

Despite the fact that elemental chlorine is an efficient delignifying agent for chemical pulps, its use results in the formation of large amounts of chlorinated organic matter of both high and low molecular weight. Many such compounds are known to be toxic and, as a consequence, the use of elemental chlorine for pulp bleaching has, in recent years, gradually decreased. Despite the fact that different countries have different legislative requirements, a further rapid decrease in the use of chlorine can therefore be foreseen. In some countries, e.g., Sweden, this shift of bleaching technology was already initiated in the 1980s due to the strong pressure exerted by different environmental groups. The major replacement bleaching chemical is chlorine dioxide (Elemental Chlorine Free bleaching, ECF) although a parallel development toward a completely chlorine free bleaching technology based on oxygen containing chemicals has also taken place (Totally Chlorine- Free bleaching, TCF). To a minor extent, bleaching sequences involving enzyme treatment of pulp fibers are being used. The rapid increase in general knowledge about enzyme systems described in Chapter 15, Biopulping and Biobleaching, may result in the development of new "biobleaching" technologies in the future.

CHEMICAL OVERVIEW

GENERAL ASPECTS

Mechanical pulps are bleached with as little dissolved organic matter as possible either reductively with dithionite (Y) (hydrosulfite) or oxidatively with alkaline hydrogen peroxide (P). This so-called lignin retaining bleaching cannot produce a completely white fiber since trace amounts of chromophoric structures seem to be either inaccessible or unreactive to the bleaching agent.

In chemical pulping, the cook has to be interrupted before all the lignin has been removed to prevent the deterioration of pulp quality. When the (kraft) cook reaches the final delignification phase, the rate of lignin dissolution slows down considerably and prolonged cooking will result in an increased degradation and dissolution of carbohydrates. At this point, the hydrolytic reactions resulting in depolymerization and dissolution of lignin cease and, in practice, further delignification and bleaching is done using oxidation chemistry. The small amount of residual lignin present in the fiber is removed in the bleach plant in a series of alternating oxidative and extractive stages with the ultimate goal being a clean and white cellulosic fiber.

Originally, this was accomplished with chlorine (C) as the first oxidative stage followed by an alkaline extraction (E). The final brightness increase was done with chlorine dioxide (D), e.g., in the sequence D E D. In some mills, sodium hypochlorite (H) was also used. The growing awareness that chlorine and hypochlorite cause severe environmental problems, forced a development toward less harmful bleaching chemicals and around 1970, oxygen (O) was introduced on a technical scale to partially replace the first C-stage. Further use of oxygen and small amounts of hydrogen peroxide (P) were effective as additives in the E-stages resulting in an improved extraction efficiency or, alternatively, in a reduced charge of chlorine dioxide. Around 1990, efficient hydrogen peroxide as well as ozone stages were installed in totally chlorine-free sequences (TCF-bleaching) to meet the ever increasing environmental requirements. At present, the worldwide use of TCF-bleaching in various sequences contributes a minor market share of bleached chemical pulp, however, and elemental chlorine-free (ECF) bleaching with chlorine dioxide is predominant in all major pulp producing countries.

In the bleaching of chemical pulp, oxidation reactions result in the formation of carboxyl groups in the lignin that under alkaline conditions will ionize and, thus,

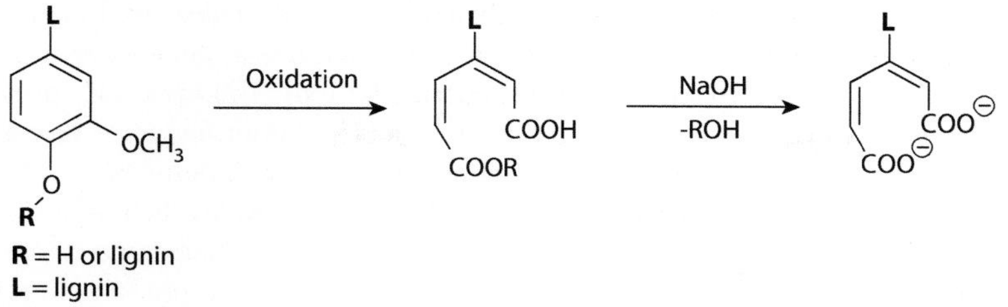

R = H or lignin
L = lignin

FIGURE 11.3 General reaction scheme for the oxidation of an aromatic unit in lignin and its alkaline hydrolysis.

provide increased solubility (Figure 11.3). This simple reaction principle is not enough, however, to completely eliminate the lignin from the fibers and fragmentation reactions are also required to produce molecules that are small enough to penetrate out from the swollen network structure of the fiber wall. The initial bleaching kinetics, which causes a very rapid decrease in lignin content, is usually followed by a much slower delignification rate and, consequently, one oxidative/extractive stage is not enough to completely eliminate the lignin. Another important but much less investigated lignin dissolution mechanism is the diffusion through the fiber wall [7]. For other reactive components present in the fibers such as hexenuronic acid and "non lignin" structures, the behavior toward various bleaching agents is variable [8,9]. The chemistry and technology of bleaching as well as the reactions between lignin structures and various bleaching agents have been reviewed [10–12].

THE STRUCTURE OF RESIDUAL LIGNIN

The structure of the residual lignin, present in unbleached chemical (kraft) pulps, has been the subject of numerous studies but, however, our knowledge is far from complete (see Chapter 10). The most apparent feature of the unbleached pulp lignin is, however, its heterogeneity with structural units that range from virtually native to highly degraded that resembles the dissolved lignin. Thus, all the original coupling modes, except the dibenzodioxocin structure, between the phenylpropane units can still be found in the residual lignin, albeit in different relative proportions in comparison to the native lignin [13,14]. Certain structures such as those of the β–5 and β–1 types are present in part as the corresponding stilbene [15]. A small amount of the original β–O–4 structures can be found as the corresponding enol ether structure in addition to intact β–O–4 structures [16,17]. In comparison to wood lignin, the content of free phenolic hydroxyl groups in the residual kraft pulp lignin is much higher, [18,19] but, due to the structural heterogeneity, the average value may include a large range of different values [20–23]. During the kraft cook, many of the terminal hydroxymethyl groups are split off as formaldehyde and, although a portion of these may incorporate into the lignin through condensation reactions, the major part is found in the cooking liquor [24]. The residual lignin also contains saturated carbons of the methylene type. The predominant origin of these is the native lignin, which contains small amounts of certain reduced structures

such as dihydroconiferyl alcohol and secoisolariciresinol. These will survive the
kraft cook. In addition, extractives of the fatty acid type are incorporated into the
residual lignin to a small extent, either by chemical or ionic linkages, thus providing
further amounts of unreactive saturated structures [25]. Other features contributing
to the difficulty in removing the last portion of lignin in kraft pulps is the presence
of chemical linkages between lignin and carbohydrates [26]. All the major polysaccharides in the fiber, i.e., cellulose, xylan, and glucomannan, have been found to be
linked to lignin, probably with benzyl ether linkages being the predominant mode
of coupling.

NUCLEOPHILIC REACTIONS

The Peroxide Stage

In oxidative bleaching of mechanical pulps with alkaline hydrogen peroxide (P),
the major mode of reaction is nucleophilic attack by the peroxide anion on electron
deficient carbon atoms in the lignin. The predominant source of such carbons is in
various types of conjugated carbonyl structures with the primary reaction resulting
in addition of the peroxy group to the lignin (Figure 11.4). In subsequent reaction
steps, the conjugated structure is broken down with formation of carboxyl groups
in the fibers as well as of low molecular weight organic acids. The conditions chosen
in the bleaching operation are such that a selective elimination of chromophoric
groups in lignin is achieved without much delignification. The metal ion catalysed
decomposition of hydrogen peroxide into oxygen and water via the formation of radical intermediates must be minimized, however, both from economic and chemistry
points of view. This is usually done by pretreatment of the unbleached pulp with a
chelating agent (Q-stage).

Nucleophilic reactions encountered in chemical pulp bleaching, again involve alkaline hydrogen peroxide, but also the extraction stage(s). In contrast to mechanical pulp
bleaching, the conjugated (colored) structures formed during a kraft cook are rather
resistant to alkaline hydrogen peroxide since structures, like those of the coniferaldehyde and quinone types, have been virtually eliminated. Therefore, the mild reaction
conditions used for mechanical pulps can no longer be applied and the peroxide stage
must be performed such that the lignin is eliminated from the fibers [27]. This can be
achieved by raising the temperature to a level where new conjugated carbonyl groups
can be created in the lignin. One such structure, which starts to form at a noticeable
rate around 100°C, is the quinone methide [28] and, consequently, a preservation of
intact phenolic benzyl alcoholic structures throughout the cook is beneficial when a
P-stage is applied [17].

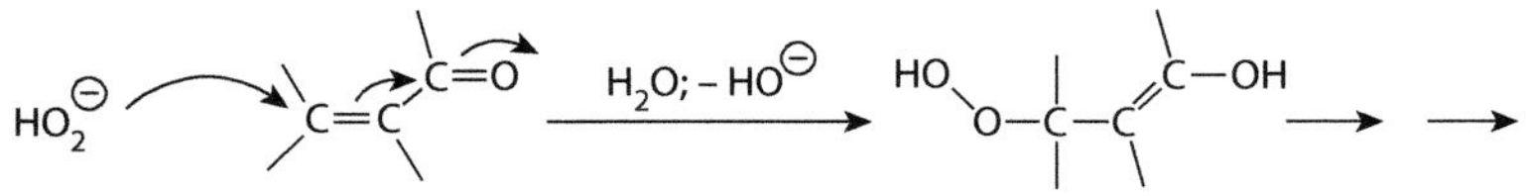

FIGURE 11.4 General scheme for addition of hydroperoxy ion to a conjugated carbonyl
structure.

The Alkaline Extraction Stage

An alkaline extraction stage (E) is used to neutralize any carboxyl groups formed in a preceding acidic oxidation stage, usually a chlorine dioxide or a chlorine stage. Thereby, the solubility of the oxidized lignin is greatly increased and the lignin can be separated from the fibers. The nucleophilic hydroxyl ion, sometimes in combination with a small amount of hydrogen peroxide and/or oxygen, also acts as an efficient dechlorination agent for chlorinated lignin structures formed in the acidic stage. This mechanism can remove up to around 70% of all organically bound chlorine with a corresponding formation of inorganic chloride (Figure 11.5) [29].

ELECTROPHILIC REACTIONS

Most of the major bleaching agents for chemical pulps are electrophilic in nature as compared to chemical pulping, where delignification is dominated by nucleophilic reactions. Once, the extent of the latter have ceased, the remaining lignin can only be attacked by a different type of general chemical principle. Among the various bleaching agents, large differences in reactivity are encountered. Whereas chlorine dioxide reacts with a high rate of reaction with phenolic lignin structures [30], chlorine as well as ozone show very high rates of reactivity with all types of aromatic structures present in pulps [31,32]. In contrast to these bleaching agents, peracetic acid and oxygen both have rather low reaction rates [33,34]. Peracetic acid is able to oxidize both phenolic and nonphenolic aromatic rings; oxygen, on the other hand, reacts exclusively with phenols. These differences in reactivity and the fact that the oxygen-based bleaching agents (i.e., oxygen, hydrogen peroxide, ozone, and peracetic acid) all require certain precautions in order not to decompose into radical species have resulted in a rapid

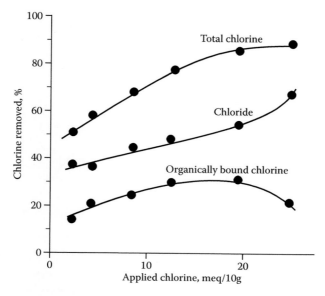

FIGURE 11.5 Formation of chloride on treatment of chlorinated kraft pulp with alkali.

and comprehensive change of bleaching technologies. In particular, the increased use of hydrogen peroxide in especially designed P-stages have forced the development of pulp pretreatment with a chelating agent (Q-stage) prior to the peroxide stage in order to avoid the metal catalyzed decomposition of hydrogen peroxide.

The Oxygen/Hydrogen Peroxide System

The reaction between oxygen and lignin requires a high temperature, a high oxygen pressure and alkaline conditions in order to proceed at a measurable rate. Under such conditions, the created phenolate ions can transfer one electron to oxygen thus giving rise to a resonance stabilized phenoxy radical and a superoxide ion (Figure 11.6). The latter, being the anion of the hydroperoxy radical (pK_a-value = 4.9), can react with itself to give oxygen and hydrogen peroxide according to Equation 11.1.

$$O_2^{\ominus} + O_2^{\ominus} + 2H^{\oplus} \longrightarrow H_2O_2 + O_2 \qquad (11.1)$$

Under alkaline conditions, the rate of this reaction is low (Figure 11.7) [35], but superoxide ion, being a strong Brönsted base, may abstract a proton from an organic

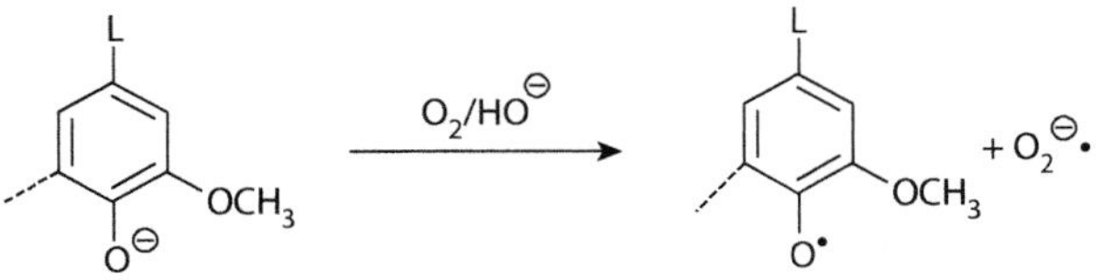

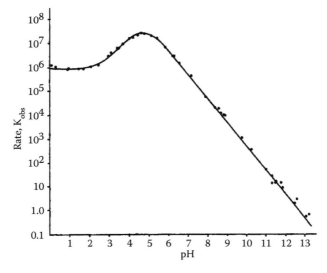

FIGURE 11.6   Initial reaction on treatment of a phenolic lignin unit with oxygen and alkali.

FIGURE 11.7   The influence of pH on the dismutation reaction of superoxide. (Reprinted with permission of Wiley-Blackwell from Bielski, B.H.J., *Photochem Photobiol*, 28, 645–649, 1978.)

substrate and thus give a fast dismutation into oxygen and hydrogen peroxide [36]. The presence of active transition metal ions can catalyze the dismutation reaction, shown in Equation 11.1, thus increasing the rate of formation of hydrogen peroxide [37]. In practice, the activity of the catalytically powerful manganese and iron ions in the dismutation reaction can be efficiently decreased by the addition of magnesium salt to the oxygen bleaching stage [38]. However, despite this addition, a complete deactivation of transition metal ions cannot be achieved. In addition to an enhanced rate of formation of hydrogen peroxide, these metal ions will also increase the rate of decomposition of hydrogen peroxide with the intermediate formation of superoxide ions and hydroxyl radicals according to the reactions shown in Equation 11.2 [39,40].

$$H_2O_2 + HO_2^{\ominus} \longrightarrow HO^{\bullet} + O_2^{\ominus \bullet} + H_2O$$

$$H_2O_2 + Fe^{2+}(Mn^{2+}) \longrightarrow HO^{\ominus} + HO^{\bullet} + Fe^{3+}(Mn^{3+}) \tag{11.2}$$

$$H_2O_2 + Fe^{3+}(Mn^{3+}) + 2HO^{\ominus} \longrightarrow O_2^{\ominus \bullet} + Fe^{2+}(Mn^{2+}) + 2H_2O$$

In the bleaching of either mechanical or chemical pulp with alkaline hydrogen peroxide, the bleaching stage must always be preceded by a treatment of the pulp with a chelating agent like diethylenetriaminepentaacetic acid (DTPA) or EDTA in order to reduce the content of transition metal ions to acceptable levels.

The hydroxyl radical (pK_a-value = 11.9) is a powerful oxidant with an oxidation potential of 2.0 volt vs SHE. It reacts with organic substrates either by addition or hydrogen abstraction, whereas with inorganic substrates, electron transfer reactions and, in some cases, radical radical interactions take place [41]. With aromatic and olefinic structures like in lignin, the hydroxyl radical reacts by addition to the unsaturated system. Further reaction steps give rise to a variety of degradation products. In carbohydrates and polysaccharides, radical centers will be created through hydrogen abstraction. In subsequent reaction steps, various oxidized sugar moieties are formed that in polymeric structures like cellulose will give rise to chain scission if exposed to alkaline conditions.

The Ozone System

Ozone bleaching is carried out under weakly acidic conditions with the pH-value being around 3. At this pH-range, the self decomposition of ozone, shown in Equation 11.3, is low since the concentration of hydroxyl ions is low [42,43].

$$O_3 \rightleftharpoons O + O_2$$

$$O + H_2O \longrightarrow 2HO^{\bullet}$$

$$HO^{\bullet} + O_3 \longrightarrow HO_2^{\bullet} + O_2 \tag{11.3}$$

$$O_3 + HO_2^{\bullet} \longrightarrow HO^{\bullet} + 2O_2$$

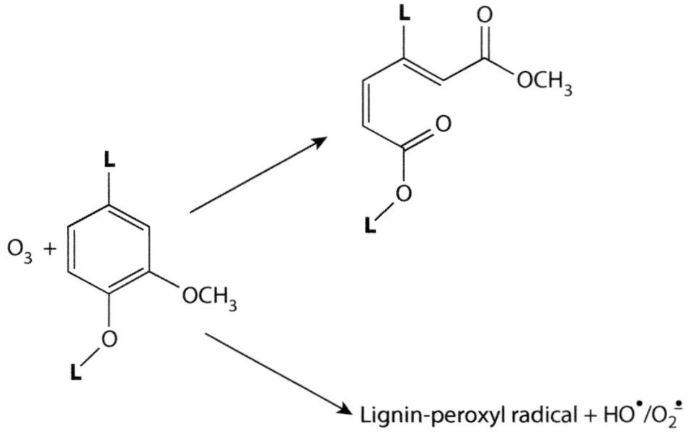

FIGURE 11.8 Initial reactions between an aromatic unit in lignin and ozone. In addition to oxidative cleavage of the aromatic ring (ozonolysis), a direct formation of radical intermediates takes place.

Certain metal ions, like ferrous ions, catalyze the decomposition to oxygen and hydroxyl radicals [44]. However, the direct reaction between lignin and ozone is of greater importance for the formation of hydroxyl radicals in ozone bleaching is in addition to the desirable ozonolysis reaction also gives rise to a direct formation of hydroxyl radicals (Figure 11.8) [45]. At pH values approaching the pK_a for the hydroperoxy radical or higher, the rate of reaction with ozone increases at least 10^5 times and, under such conditions, the flux of hydroxyl radicals makes the bleaching with ozone impractical.

The Chlorine Dioxide/Chlorine System

Chlorine dioxide itself has an unpaired electron and its major reaction with pulp lignin is through an electron transfer reaction from the aromatic rings with formation of chlorite ion (Figure 11.9). Phenols as well as phenol ethers react with about the same rate of reaction whereas phenolate ions shows a reactivity that is around 10^6 times higher [30,46,47]. Therefore, despite the fact that the bleaching with chlorine dioxide is carried out under acidic conditions, the free phenolic units in lignin (in their anionic form) react much faster than the etherified. In this reaction, a phenoxy radical is formed that can react with a new molecule of chlorine dioxide in a radical-radical coupling reaction with formation of a chlorite ester. In an analogous way, aliphatic double bonds if present in the residual lignin may also react with chlorine dioxide with formation of chlorite esters [48]. In a subsequent hydrolysis step, hypochlorous acid and chlorite can be formed together with a variety of oxidized organic products [49]. Being a strong oxidant, hypochlorous acid may oxidize the chlorite present with the formation of chlorate as shown in Equation 11.4 [50]; however, the hypochlorous acid will also give oxidation and chlorination of the lignin. The latter reaction may proceed directly or through its equilibrium product elemental chlorine (Equation 11.5). The latter equilibrium is pH-dependant and an increase will result in a progressive reduction of the concentration of chlorine according to Equations 11.5 and 11.6. Based on the corresponding equilibrium constants, K_1 and K_2, the relationship between the various chlorine species and pH can be calculated (Figure 11.10).

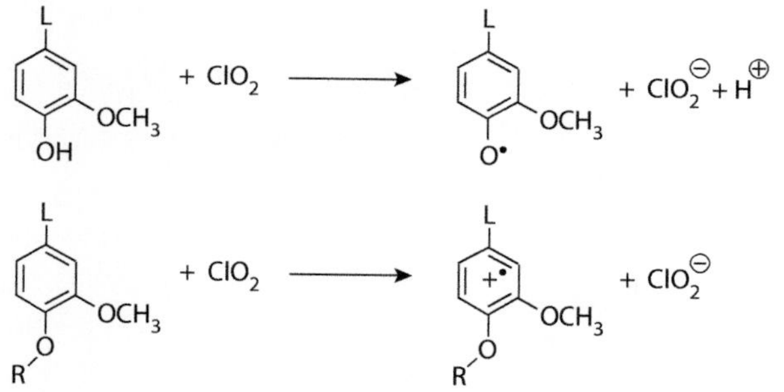

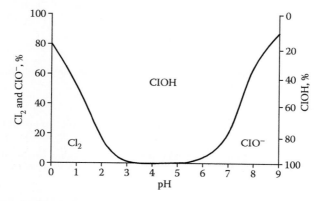

FIGURE 11.9 Initial reaction between an aromatic lignin unit and chlorine dioxide. Formation of chlorite.

FIGURE 11.10 Amount of chlorine, hypochlorous acid and hypochlorite present in aqueous solution at different pH-values.

$$ClOH + ClO_2^{\ominus} \longrightarrow Cl^{\ominus} + ClO_3^{\ominus} + H^{\oplus} \qquad (11.4)$$

$$ClOH + Cl^{\ominus} + H^{\oplus} \rightleftharpoons Cl_2 + H_2O \qquad (11.5)$$

$$K_1 = \frac{[ClOH]\,[Cl^{\ominus}]\,[H^{\oplus}]}{[Cl_2]} = 3.9 \times 10^{-4} \text{ at } 25°C$$

$$ClOH \rightleftharpoons ClO^{\ominus} + H^{\oplus} \qquad (11.6)$$

$$K_2 = \frac{[ClO^{\ominus}]\,[H^{\oplus}]}{[ClOH]} = 5.6 \times 10^{-8} \text{ at } 25°C$$

An aqueous solution of chlorine constituted the major bleaching agent for cellulosic fibers for almost 200 years. The reactions between lignin structures and

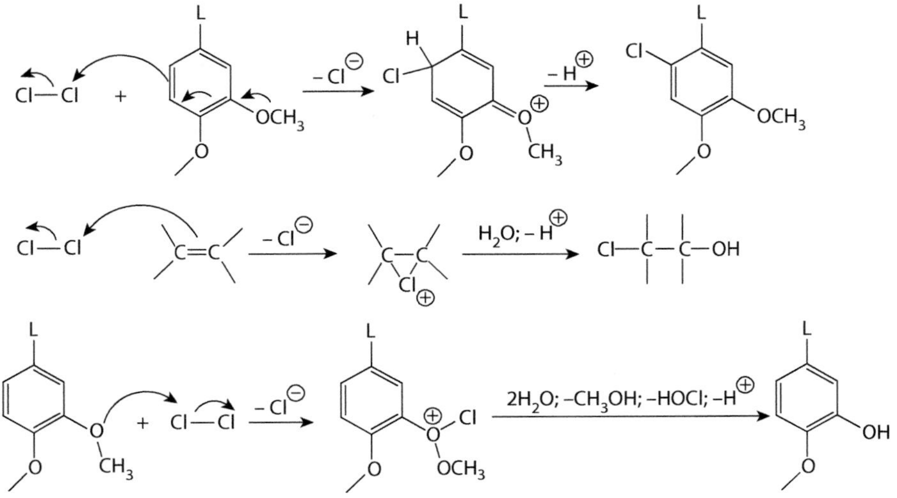

FIGURE 11.11 Initial reaction types, i.e., substitution, addition, and hydrolysis, when lignin is reacted with aqueous chlorine.

chlorine involve substitution of aromatic rings, addition to olefinic double bonds and hydrolysis of alkyl aryl ether linkages as summarized in Figure 11.11 [10–12]. In addition, a variety of oxidation reactions may take place due to the simultaneous presence of hypochlorous acid.

DETAILED CHEMISTRY

OXYGEN DELIGNIFICATION CHEMISTRY

After kraft (or sulfite) pulping, the residual fiber lignin is structurally heterogeneous. Among the various structural units present, the free phenolic end-groups are, however, by far the most reactive toward electrophilic reagents. This is taken advantage of in the oxygen delignification stage that, when applied, proceeds directly after the kraft (sulfite) digestion. In practice, typical reaction conditions can be 100°C for 60 min and with an oxygen pressure of around 0.6 MPa corresponding to an oxygen charge of ~22 kg/tonne of pulp. The degree of delignification, measured as change of kappa number, is in the order of 35–65% depending on wood species and exact process conditions. The carbohydrate-derived contributions to the kappa number are, however, not decreased during an oxygen stage and, consequently, for both softwood and hardwood, the dissolution of lignin is much greater than indicated by the kappa number reduction (Figure 11.12) [51]. It has been shown that the rate of delignification in the oxygen stage proceeds in two phases, with the first being approximately 20 times faster than the second [52]. Both phases are, however, dependent on alkali concentration and oxygen pressure.

The reactions between lignin phenolic structures and oxygen have been the subject of numerous studies. In this work, the oxidation mechanisms, the detailed

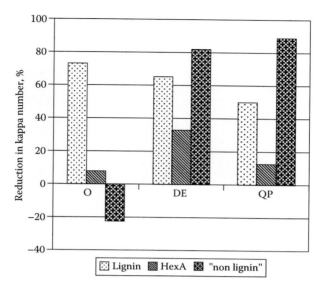

FIGURE 11.12 Changes in the amounts of lignin, hexenuronic acid, and "non lignin" structures when a softwood kraft pulp is successively bleached with oxygen-alkali (O), chlorine dioxide followed by alkaline extraction (DE) and treatment with a chelating agent followed by alkaline hydrogen peroxide (QP). (Reprinted with permission from Li, J., Sevastyanova, O., and Gellerstedt, G., *Nordic Pulp Pap Res J*, 17, 415–419, 2002.)

degradation pathways [53–68] as well as the reaction kinetics has been elucidated [69–74]. The reactions of polymeric lignin have also been studied both with isolated lignin preparations [75–77] and through chemical and spectroscopic analyses of dissolved and residual lignin after an oxygen stage [78–84]. The primary reaction step in the oxidation of a phenolic structure is the slow electron transfer from the phenolate anion to oxygen with the formation of a superoxide ion and a phenoxy radical. The phenoxy radical, stabilized by resonance, reacts with a second molecule of oxygen to form an organic peroxy radical that is rapidly reduced to a hydroperoxide or, alternatively, reacts directly with a superoxide ion [85]. In subsequent steps, the organic hydroperoxide isomers will react further and the aromatic ring can be either degraded, or eliminated from the rest of the lignin macromolecule (Figure 11.13) [53]. Demethoxylation is also possible and approximately 1.5 kg of methanol per tonne of pulp has been found in the effluent from an oxygen stage [86].

At least for low molecular weight phenols, the degradative reactions are always accompanied by radical-radical coupling reactions to give dimers. With creosol (2-methoxy-4-methylphenol) as the substrate, virtually all reaction pathways have been identified, by which an aromatic ring related to the guaiacyl structure is degraded. These are shown in Figure 11.14 [57]. Similar work has also been done with various t-butyl-substituted phenols [63]. For more ligninlike model compounds, i.e., structures with an α-hydroxy (or α-ether) group in the side chain, the general reaction scheme shown in Figure 11.14, is further expanded to encompass side chain elimination and formation of an aldehyde [62]. In addition, on oxidation with oxygen/alkali,

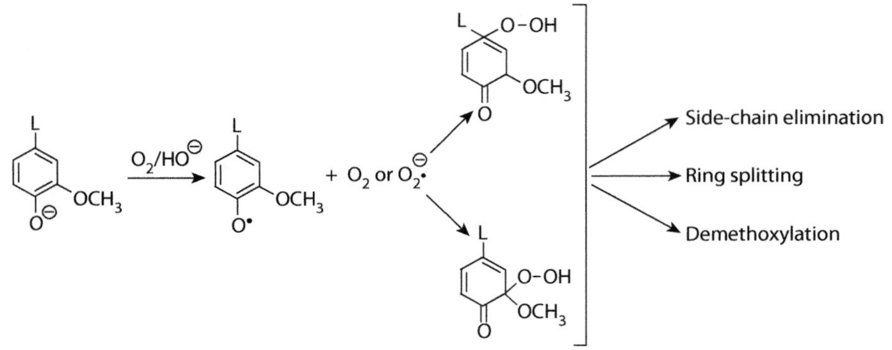

FIGURE 11.13   The various modes of reaction of a phenolic lignin unit with oxygen-alkali.

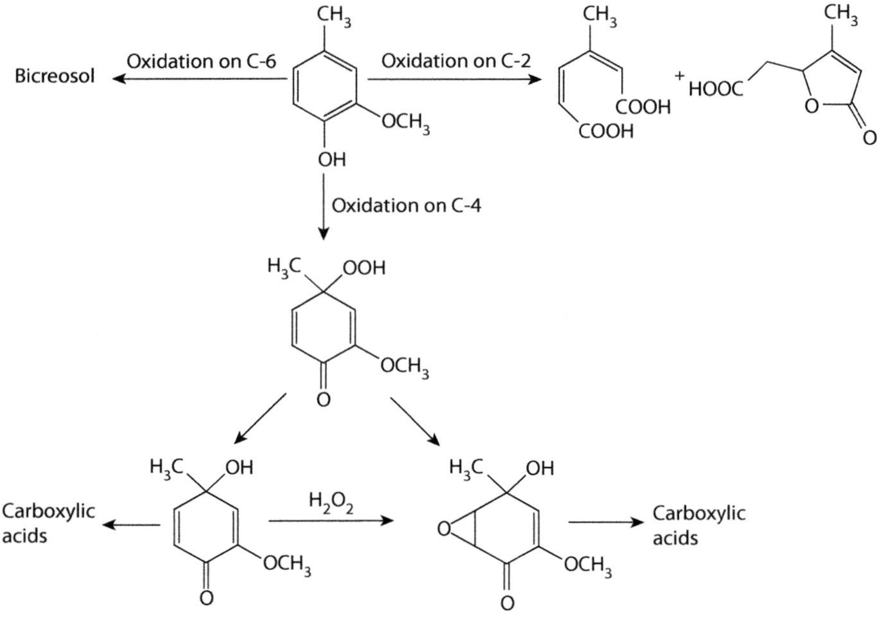

FIGURE 11.14   Reaction products obtained on oxidation of creosol with oxygen-alkali.

lignin structures having a styrene or a stilbene structure may give rise to extended resonance structures that involve the olefinic carbons [62,64].

During the oxidation of organic substrates (autoxidation), hydrogen peroxide is always formed and may participate in the overall reaction. The chemistry of peroxide bleaching is further described in the Peroxide Bleaching Chemistry section. Hydrogen peroxide may, however, decompose with the formation of hydroxyl radicals and super-oxide ions, which in further reaction steps either can recombine to give oxygen and water or react with the substrate [87–92]. The decomposition reaction is catalyzed by transition metal ions (Fenton chemistry). In oxygen delignification (as in peroxide, ozone and peracetic acid bleaching), the presence of hydroxyl radicals is detrimental since the high reactivity of this radical can result in indiscriminate oxidation reactions with the cellulose. Since the rate of reaction between hydroxyl radicals and lignin structures is somewhat higher than that with carbohydrates, the presence of lignin in the pulp fibers will to some extent lower the occurrence of such reactions. The difference is small, however [93]. In addition, hydrogen peroxide, formed or added, can act as a scavenger for hydroxyl radicals (Equation 11.7).

$$HO^\bullet + H_2O_2 \longrightarrow HO_2^\bullet + H_2O \quad k = 3 \times 10^7 \, M^{-1}s^{-1}$$

$$HO^\bullet + HO_2^\ominus \longrightarrow O_2^{\ominus\bullet} + H_2O \quad k = 8 \times 10^8 \, M^{-1}s^{-1}$$

(11.7)

In model experiments, it has been shown that the major reaction between lignin structures and the hydroxyl radical under alkaline conditions seems to be (i) addition to the aromatic ring in nonphenolic units, (ii) electron abstraction by phenolate ions, and (iii) hydrogen abstraction from the side chain. The last reaction can be accompanied by side chain cleavage whereas the first will give rise to the formation of a phenolic hydroxyl group [88,92]. All these reaction types have been thoroughly investigated using gamma irradiation for the generation of hydroxyl radicals. Based on the reaction pattern and the products that are formed, this type of oxidation of lignin seems to promote delignification.

The rate of oxidation with oxygen of various aromatic structures related to lignin has been studied in well controlled experiments. The data obtained demonstrate that most phenolic structures are degraded at moderate rates under technically relevant conditions. Thus, the kinetic half-life for a series of phenolic structures was found to be in the order of 15–90 min. In special cases, such as for catechol, styrene and stilbene structures, the rates were much higher, however [72,73,94]. These data indicate that the conditions encountered in an oxygen delignification stage, only result in a partial degradation of the predominant types of phenolic structures.

The residual lignin present in kraft pulp after an oxygen stage as well as dissolved lignin has been analyzed using various chemical and spectroscopic methods. As expected, the content of phenolic hydroxyl groups is strongly reduced in the residual lignin after oxidation [76,79] but, still, all types of phenolic end-groups are present albeit in different relative amounts. Thus, the differences in reaction rates, found in model experiments, are such that the 5-5' structures survive to a greater extent than guaiacyl end-groups and p-hydroxyphenyl structures react at a lower rate than guaiacyl structures [78,84]. The content of phenolic hydroxyl groups in the residual lignin is in the order of 1 unit per 10 phenylpropane units, which can be compared to a value of around 3 in the unbleached pulp. The dissolved lignin in the oxygen stage, on the other hand, contains 3–4 phenolic units per 10 phenylpropane units, i.e., an apparent higher value than the original. This discrepancy can only be explained by the assumption that the residual lignin in the unbleached pulp has a heterogeneous structure ranging from "native" to severely degraded. The latter portion may resemble the dissolved kraft lignin, which has a content of phenolic groups of around 7 per 10 phenylpropane units. In agreement with this, recent NMR data show that the apparent content of β–O–4 structures in the residual lignin is higher after, as compared to before the oxygen stage [13]. Since the most degraded portion of the residual unbleached pulp lignin should resemble the dissolved kraft lignin, a low content of remaining β–O–4 structures will accompany the high content of phenolic units. Consequently, a preferential degradation and dissolution of this material in the oxygen stage should leave behind a residual lignin with a less degraded structure.

Alkaline oxygen oxidation of phenols results in the formation of a variety of acidic products as indicated in Figure 11.14. In guaiacyl structures, the major route seems to be oxidation of the carbon atoms 3 and 4 in the ring and formation of muconic acid structures. Detailed analysis reveals, however, that further degradation may give rise to various low molecular mass aliphatic mono- and di-carboxylic acids such as lactic acid, oxalic acid, and others [75]. In the oxidation of pulp lignin, the formation of (ionized) carboxyl groups is essential since these will render the lignin molecules more water soluble. In addition, a certain depolymerization of lignin fragments can be anticipated to occur since the molecular mass distribution of the dissolved lignin from the oxygen stage is similar to that from the kraft cook [95]. The number of carboxyl groups in the dissolved lignin from the oxygen stage is high, 6–7 units per 10 phenylpropane units [79].

PEROXIDE BLEACHING CHEMISTRY

Mechanical Pulp Bleaching

Alkaline hydrogen peroxide in the presence of sodium silicate bleach mechanical pulps (SGW, TMP, CTMP). In order to further stabilize the bleaching liquor against metal ion induced decomposition, the pulp must be pretreated with a chelating agent, usually EDTA or DTPA. Typically, the bleaching is carried out at a temperature of 60–70°C during 60–120 min. Under these conditions, the yield loss is low, 1–4%, and the major reactions are lignin brightening together with a deacetylation of hemicelluloses and demethoxylation/dissolution of pectin [96]. Native lignin-derived chromophores such as coniferaldehyde [97] and quinone structures [98] as well as other types of conjugated carbonyl structures [97,99] all react with the strongly nucleophilic hydroperoxy anions albeit, with largely different reaction rates. At higher temperatures (~100°C), the equilibrium between phenolic benzyl alcohol structures and the corresponding quinone methides starts to form at a noticable rate [100]. Being a conjugated carbonyl structure, the quinone methide can react further with a hydroperoxy anion resulting in lignin degradation [101–103] that is utilized in the lignin-degrading peroxide bleaching of chemical pulps.

The reaction between conjugated carbonyl groups and alkaline hydrogen peroxide causes the destruction of the chromophore and the formation of acidic products. In Figures 11.15 through 11.17, such reactions are exemplified for structures of the aryl-α-carbonyl [97,99,104], coniferaldehyde [97] and quinone [98] types. The stoichiometry of these reactions is, however, much dependent on the absence of any decomposition of hydrogen peroxide since, otherwise, an increased consumption of hydrogen peroxide and a less favorable stoichiometry will result. The latter reaction

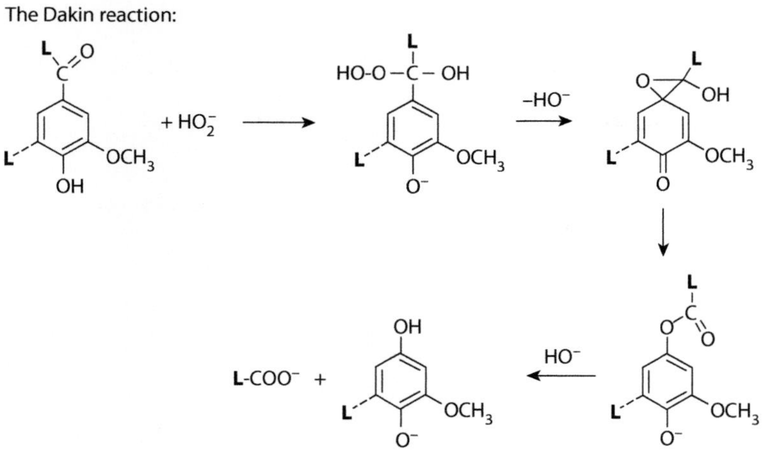

The Dakin reaction:

FIGURE 11.15 Reaction mechanism for oxidation of a phenolic aryl-α-carbonyl structure with alkaline hydrogen peroxide.

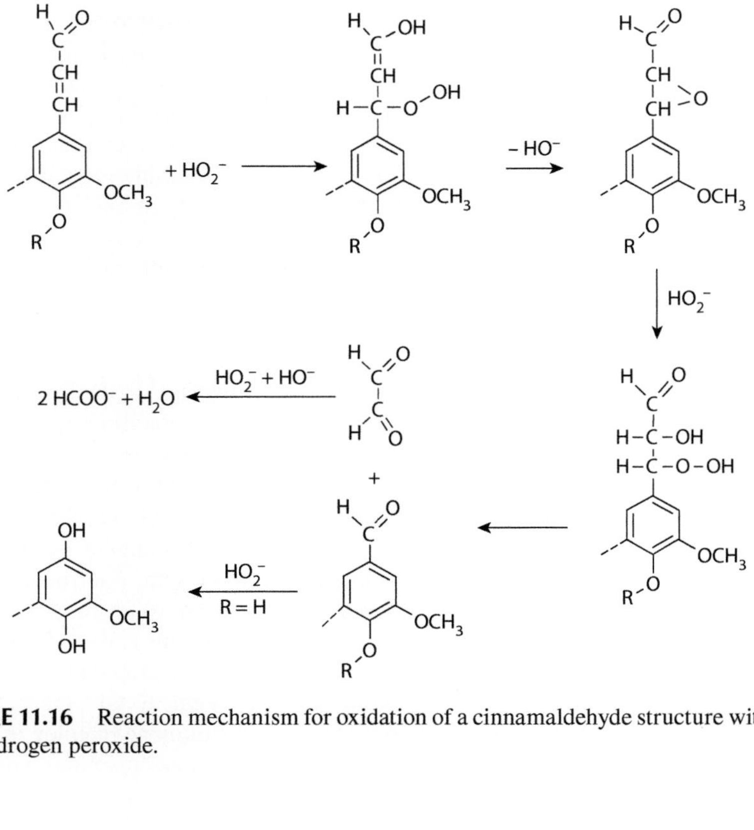

FIGURE 11.16 Reaction mechanism for oxidation of a cinnamaldehyde structure with alkaline hydrogen peroxide.

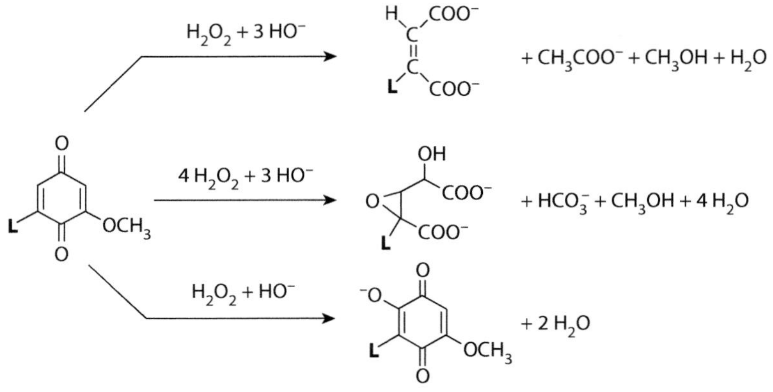

FIGURE 11.17 Reaction products on oxidation of a quinone structure with alkaline hydrogen peroxide.

has been demonstrated using acetoguaiacone as a model compound. In the absence of any transition metal ions, the stoichiometry and products are those shown in Figure 11.15, whereas the addition of trace amounts of Mn(III)-DTPA or Fe(III)-DTPA results in a strongly increased consumption of hydrogen peroxide and a higher reaction rate. At the same time, the product mixture becomes much more complex and, in addition to the expected methoxyhydroquinone, further oxidation to methoxy-p-benzoquinone-derived products takes place [97,105,106]. For phenolic lignin structures substituted with a good leaving group, such as an ether function in the α-position of the side chain, the formation of a quinone methide is facilitated and may occur at very mild conditions [107]. Thus, under the conditions of mechanical pulp bleaching, such methylene quinone structures can form to some extent. Using a β–1 structure as model compound, it has been shown that the corresponding stilbene can be formed provided that the pH is high [108].

Chemical Pulp Bleaching

The use of hydrogen peroxide in chemical pulp bleaching has gradually developed from only being an additive in the alkaline extraction stage (sometimes together with oxygen) to a true peroxide stage pressurized with oxygen and performed at around 100°C. The finding that manganese ions can be efficiently removed from the pulp in a separate chelating stage (Q-stage) preceding the P-stage paved the way for this development [27,109]. It was recognized early that chemical pulps cannot be efficiently bleached with alkaline hydrogen peroxide unless the lignin is removed from the fibers [27]. The desirable reactions that take place have been shown to result in a comprehensive oxidation of lignin and formation of hydrophilic groups [110–115]. Thus, in addition to the removal of chromophores [116], an oxidative degradation of phenolic lignin structures via the formation of quinone methides seems to take place. In that reaction pathway, an addition of hydroperoxy anion to the quinone methide followed by epoxide formation and subsequent cleavage (the "Dakin-like" reaction) will result in a lignin fragmentation (Figure 11.18) [101–103].

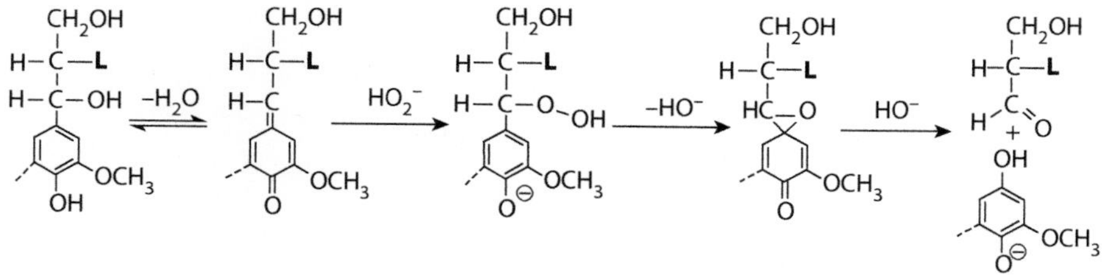

FIGURE 11.18 Reaction mechanism for oxidation of a phenolic benzylalcohol structure with alkaline hydrogen peroxide.

CHLORINE DIOXIDE BLEACHING CHEMISTRY

At present, chlorine dioxide constitutes the most important bleaching agent for chemical pulps. The gradual replacement of elemental chlorine with chlorine dioxide in the industry has been driven by the concern that bleaching effluents originating from chlorine bleaching may contain harmful chlorinated substances. This topic is further treated in the Chlorine Bleaching Chemistry section. With chlorine dioxide, the residual lignin in chemical pulps is efficiently oxidized and degraded although chlorination also occurs to a limited extent. In a combined bleach plant effluent based on chlorine dioxide bleaching, the efficient dechlorination of lignin taking place in alkaline conditions produces very low amounts of residual organically bound chlorine.

The reactions of lignin with chlorine dioxide have been thoroughly investigated both with lignin model compounds and by analyses of the structure of residual lignin using chemical and spectroscopic methods. Furthermore, a large number of low molecular weight compounds have been identified in bleaching liquors with special emphasis put on the presence of chlorinated compounds (the Dissolved Chlorinated Material section). On oxidation of a phenolic or a nonphenolic lignin structure, the aromatic ring is rapidly degraded with formation of muconic acid and quinone structures as predominant end products (Figures 11.19 through 11.20) [117–121]. Kinetic data have shown that phenols are much more reactive than the corresponding nonphenolic structure due to the presence of phenolate anion. Despite the low concentration of the latter at technical bleaching conditions, the overall bleaching rate will be determined by the amount of phenols present in the lignin. Once these have been consumed, an oxidation of olefinic groups may take place [122–125]. During lignin oxidation, a simultaneous step-wise reduction of chlorine dioxide takes place and chlorite, hypochlorous acid, chlorine and chloride are all present in the bleaching liquor, either as transient intermediates or as stable end products [118,126,127]. In addition, substantial amounts of chlorate are formed resulting in a loss of bleaching agent. The presence of hypochlorous acid (and chlorine) in the bleaching liquor causes further oxidation of the lignin but also causes generation of organically bound chlorine. The oxidation of lignin is accompanied by the formation of methanol from both phenolic and nonphenolic structures through demethylation of aromatic methoxyl groups. A close relationship has been found between the demethylation reaction and the degree of delignification [128].

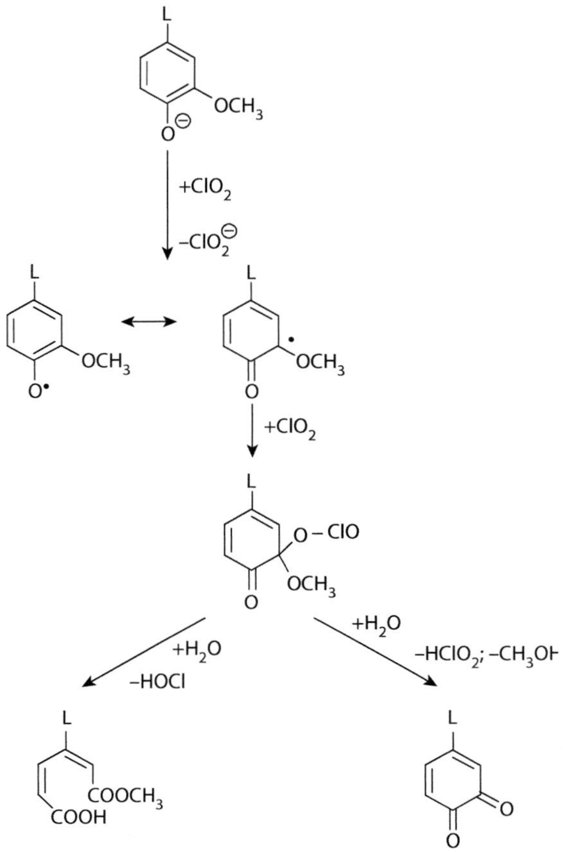

FIGURE 11.19 Reaction mechanism for oxidation of a phenolic lignin unit with chlorine dioxide.

The further reactions between partially degraded lignin structures and chlorine dioxide, such as those encountered in the final bleaching stages, have not been investigated in detail [cf. 129]. Based on studies on the reactions between isolated and conjugated double bonds and chlorine dioxide under mild acidic conditions, it can be concluded, however, that hydroxylation, oxidation and chlorination reactions may all take place [125,126]. The reactions are shown in Figure 11.21. Again, the chlorination reaction is caused by the presence of hypochlorous acid/chlorine in the system rather than by chlorine dioxide itself.

On bleaching of kraft pulp with chlorine dioxide, most of the phenolic units present in the pulp lignin after pulping and subsequent oxygen delignification are degraded. In the fully bleached pulp, only trace amounts of such structures remain. This is illustrated in Figure 11.22, which shows the successive decrease of some phenolic endgroups in the pulp lignin in a chlorine dioxide based bleaching sequence of softwood kraft pulp [130]. Isolation of the residual lignins from the same series of pulp reveals

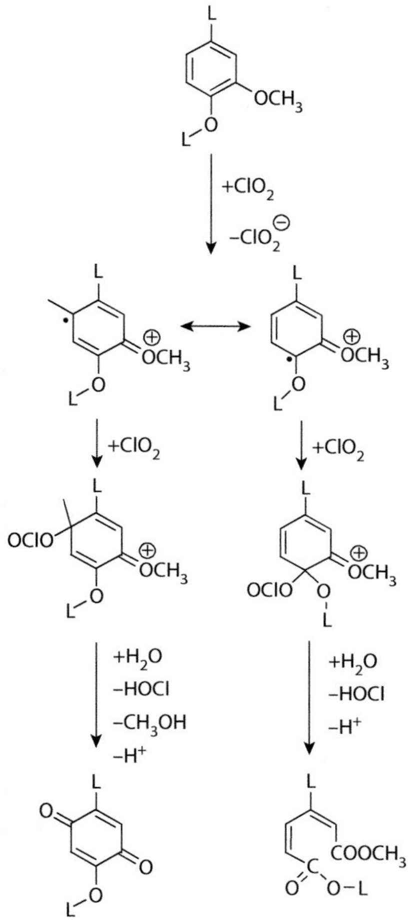

FIGURE 11.20 Reaction mechanism for oxidation of a non-phenolic lignin unit with chlorine dioxide.

that as the lignin becomes successively oxidized, a gradual decrease of both methoxyl groups and organically bound chlorine takes place (Figure 11.23). On ^{13}C NMR analysis of the isolated lignin after prebleaching employing the sequence OD (EOP) (kappa number of pulp 3.1), the spectrum still shows all the features of a lignin structure albeit with an additional large and broad signal at around 170 ppm originating from carboxyl carbons [130,131]. Obviously, the delignification of the fibers in the bleaching plant seems to take place by a stepwise oxidation and solubilisation of lignin surfaces leaving a rather intact lignin structure behind and exposed for a new oxidation reaction.

In comparison to chlorine (see the next section, Chlorine Bleaching Chemistry), chlorine dioxide has a much higher selectivity in its reactions with lignin, and in analogy to model experiments, oxidation is the major mode of reaction. This has been demonstrated by bleaching experiments of kraft pulp in which the prebleaching was

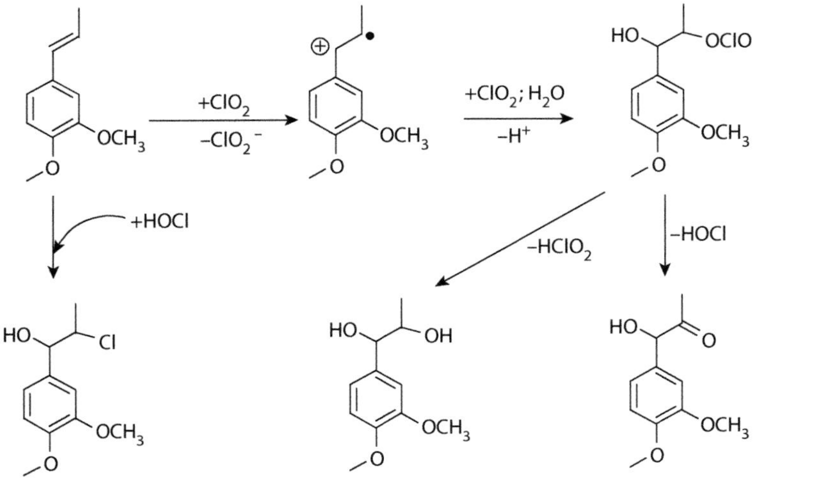

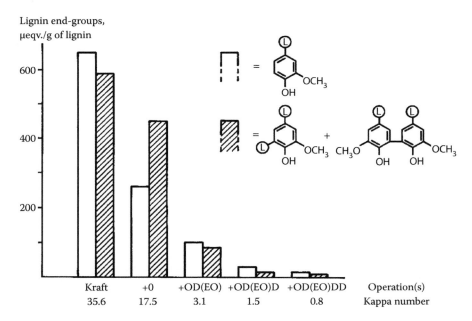

FIGURE 11.21   Reaction mechanism for oxidation of an olefin structure with chlorine dioxide.

FIGURE 11.22   Remaining amount of specific phenolic end-groups in lignin when a soft-wood kraft pulp is bleached according to the sequence OD(EO)DD.

carried out with different degrees of substitution of chlorine with chlorine dioxide. By ultrafiltration of the total bleaching effluent from each experiment and analysis of the high molecular weight material ($M_W > 1000$ Da) for the content of methoxyl groups and organically bound chlorine, a gradual decrease in chlorine was accompanied by a higher remaining amount of methoxyl groups (Figure 11.24) [132].

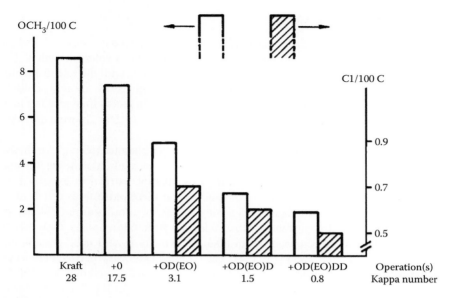

FIGURE 11.23 Successive decrease of aromatic methoxyl groups and organically bound chlorine in the residual fiber lignin when a softwood kraft pulp is bleached according to the sequence OD(EO)DD.

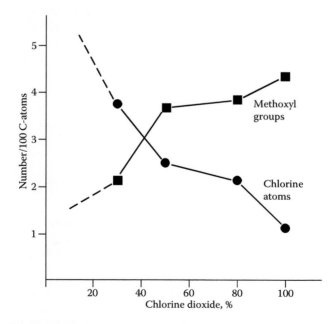

FIGURE 11.24 Comparison of remaining amount of methoxyl groups and presence of organically bound chlorine in the dissolved lignin after prebleaching (CE → DE) of softwood kraft pulp using different ratios of chlorine to chlorine dioxide.

CHLORINE BLEACHING CHEMISTRY

The high bleaching power of elemental chlorine when applied on cellulosic fibers has been known since the late eighteenth century. The efficiency of chlorine made it the major delignification and bleaching agent for chemical pulps until the late twentieth century. The growing knowledge that chlorine bleaching gives rise to highly chlorinated organic matter in bleaching effluents has, however, in recent time resulted in a gradual replacement of chlorine for other bleaching agents such as chlorine dioxide, oxygen, and hydrogen peroxide.

The reaction between chlorine and residual lignin in a chemical pulp is extremely fast but far from complete. This has been shown by repeated chlorine treatments of a kraft pulp, each treatment being followed by alkaline extraction, as shown in Figure 11.25 [133]. A reasonable explanation to the observed effect, which also has been noticed for other oxidative bleaching agents, is a formation of "blocking groups" in the lignin that prevents further oxidation from taking place. Only when these groups have been removed by an alkaline extraction, will it be possible to carry out a new reaction cycle [133,134]. The detailed chemistry between chlorine and lignin structures has been studied using lignin model compounds as well as pulp and much of this work has been summarized before [135,136]. The importance of demethylation reactions for the overall lignin reactions was, however, recently investigated by the use of veratryl alcohol, a simple model for the aromatic units in lignin [137,cf.138]. On treatment with chlorine water, a large number of reaction products could be identified as shown in Figure 11.26. Obviously, a comprehensive chlorination takes place together with demethylation and oxidation reactions. To a minor extent, the side chain is also eliminated. From these and other published data, a

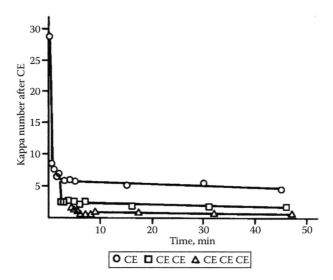

FIGURE 11.25 Successive dissolution of lignin from a softwood kraft pulp after repeated treatment with chlorine followed by alkaline extraction (CE). (Reprinted with permission of de Gruyter from Berry, R.M., and Fleming, B.I., *Holzforschung*, 41, 177–183, 1987.)

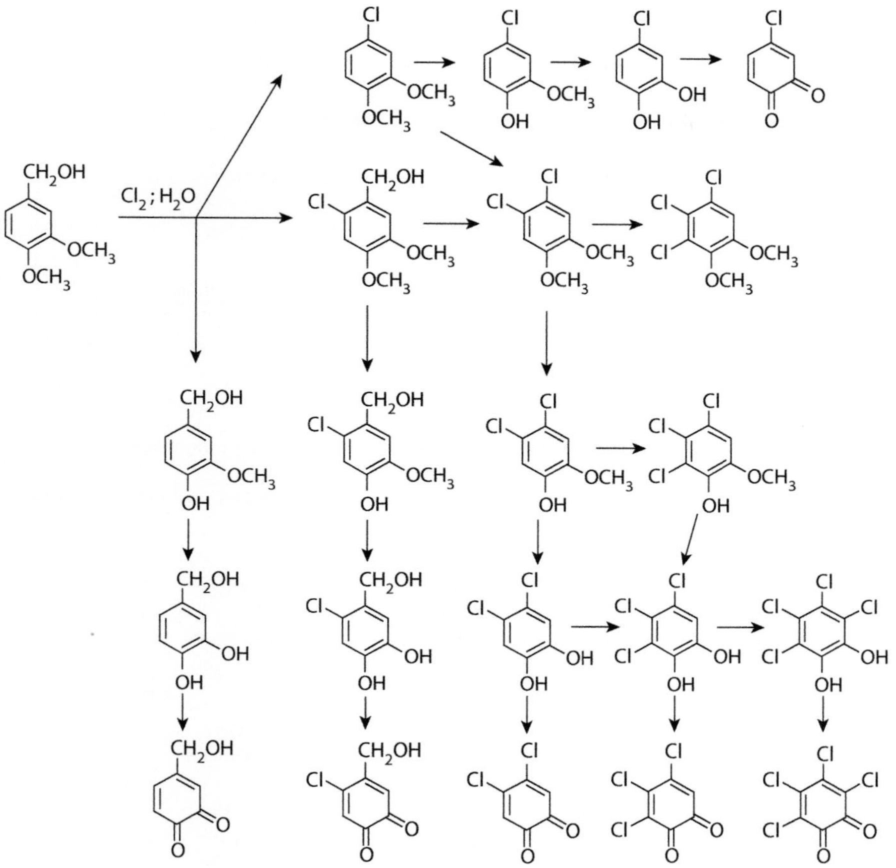

FIGURE 11.26 Model experiment demonstrating the large variety of reaction products when a simple lignin-like structure (veratryl alcohol) is treated with aqueous chlorine.

general scheme summarizing the reactions between chlorine and lignin can be made (Figure 11.27). Further support for this scheme has been obtained by analysis of low and high molecular weight material isolated from bleaching liquors [139–142].

Dissolved Chlorinated Material

The suspicion that bleaching effluents from bleaching of chemical pulp with elemental chlorine might contain chlorinated phenols was raised early [143] and could be confirmed only a few years later [144]. Today, a large number (>300) of low molecular weight compounds originating from lignin as well as from carbohydrates and extractives are known to be present in bleaching liquors using chlorine in the sequence. Many of these are chlorinated and include a variety of phenols originating from lignin as shown in Figure 11.28 [142]. In analogy to the high degree of chlorination of lignin taking place as a result of bleaching with elemental chlorine

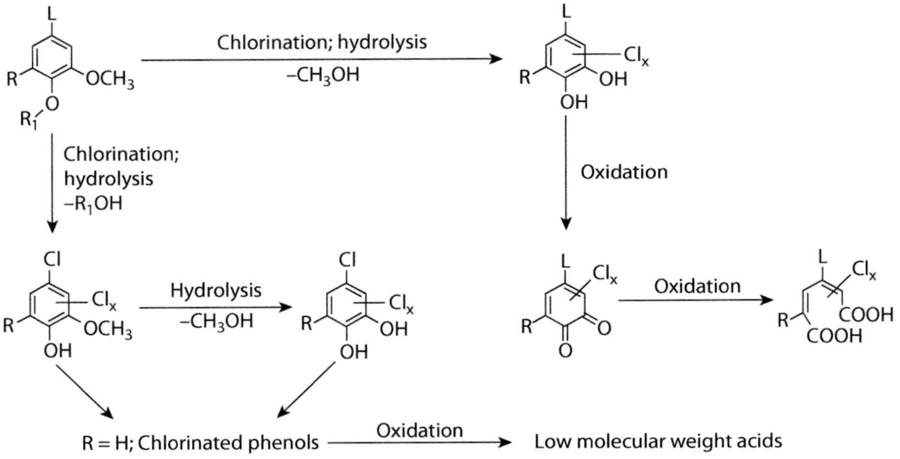

FIGURE 11.27 General reaction scheme for the reactions between aromatic lignin units and aqueous chlorine.

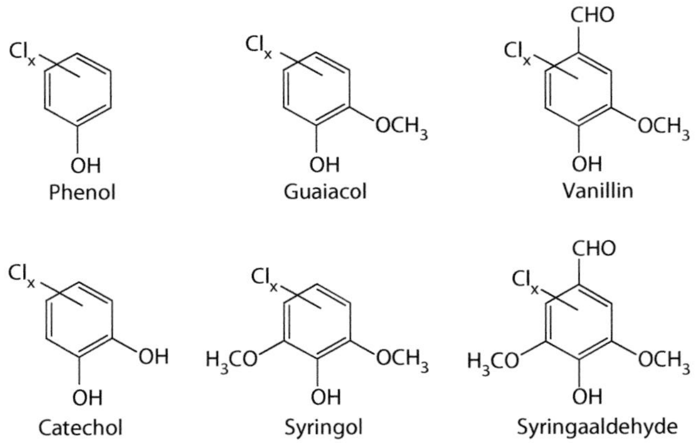

FIGURE 11.28 Major types of chlorinated phenols present in bleaching liquors after chlorine bleaching.

(cf. Figure 11.24), highly chlorinated phenols were found to be prevalent in such bleaching liquors. On successive substitution of chlorine with chlorine dioxide, a rapid decrease chlorinated phenols could be observed, however (Figure 11.29a through c) [145], and in bleaching sequences using chlorine dioxide as the sole chlorine-containing bleaching agent, only a minor formation of mono- and di-chlorinated phenols was detected [146]. Analogous results were also found on analysis of the dissolved high molecular weight lignin material from various bleaching sequences with carbon/chlorine ratios ranging from 15 (chlorine bleaching) to 260 (chlorine dioxide bleaching) [146,147]. The finding that 1,1-dichlorodimethyl sulfone is present in bleaching liquors from either chlorine or chlorine dioxide based bleaching

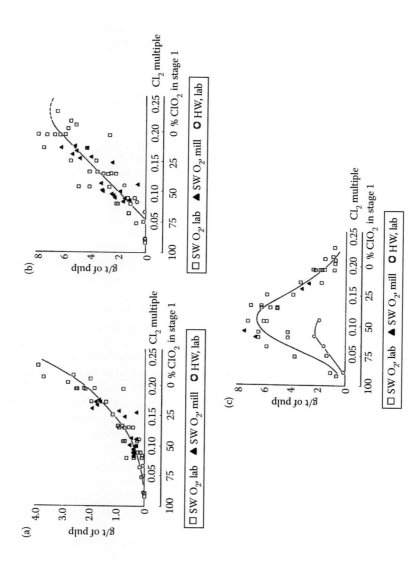

FIGURE 11.29 Formation of tetrachloroguaiacol (a), 3,4,5-trichloroguaiacol (b), and dichloroguaiacol (c) as a function of the amount of chlorine dioxide charged in the prebleaching of softwood kraft pulp.

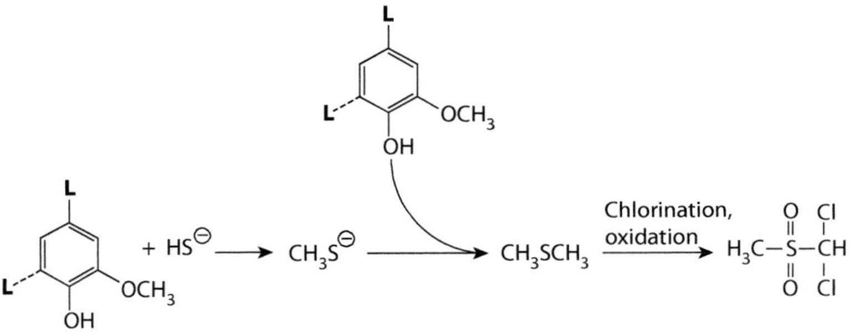

FIGURE 11.30 Reaction mechanism for the formation of 1,1-dichlorodimethyl-sulfone found in bleaching liquors.

sequences is somewhat surprising. Obviously, a portion of the dimethyl sulfide formed in the kraft digester, despite being a volatile compound, is "dissolved" in the pulp and subjected to both chlorination and oxidation when reaching the bleach plant as outlined in Figure 11.30 [148].

OZONE BLEACHING CHEMISTRY

An ozone (Z) bleaching stage can be used to partially or completely replace a chlorine dioxide stage, i.e., rather early in a bleaching sequence. One important characteristic of ozone is its extremely high reactivity and, in technical operation, the efficient mixing of ozone with the pulp suspension is a difficult task. Nevertheless, medium consistency bleaching can be done by the use of high shear mixing usually with two mixers in series with a short retention time in between. As an alternative, high consistency bleaching is also available. In both cases, the total reaction time is short and in the order of seconds to a few minutes. The redox potential of ozone is 2.08 volt vs SHE and, consequently, almost all types of substrates can be oxidized. This has been utilized in a variety of technical applications and water purification with ozone is of widespread importance. For the bleaching of chemical pulps, ozone was introduced in the early 1990s, however, the number of mills employing an ozone stage is very limited.

The reactions between ozone and a variety of organic substrates have been thoroughly investigated from a kinetic, as well as from a mechanistic point of view. Thus, it has been found that the absolute rates of reaction may differ by several orders of magnitude between typical representatives for polysaccharides and lignin, i.e., glucose and phenol. Large differences may, however, also be encountered between individual ligninlike structures as shown in Table 11.1 [32,149,150]. Based on these data, a high degree of selectivity in pulp bleaching should be expected. This is, however, not the case and, in addition to the desirable Criegee reaction, a formation of radical intermediates take place when lignin is oxidized (Figure 11.31, cf. Figure 11.8) [151]. The latter may in subsequent reaction steps attack both lignin and polysaccharides [88,89,152], thus giving rise to further lignin oxidation but also to oxidation of carbohydrate chains.

TABLE 11.1

Rate Constants for the Ozone Oxidation of Various Aromatic Compounds and Comparison with Simple Sugars

Lignin Model Compound [149]	Relative Rate Constant, $k_i/k_{creosol}$
3,4-Dimethoxycinnamyl alcohol	5
4-Methylsyringol	2.4
Creosol	1
Vanillin	0.32
p-Cresol	0.23
Vanillic acid	0.12
Veratrylglycol	0.051
Methyl-β-D-glucopyranoside	0.005
Misc. Compounds [32,150]	**Absolute Rate Constant, $M^{-1}s^{-1}$**
Styrene	30,000^^
Phenol	1,300^^
Anisole	290
Glucose	0.45

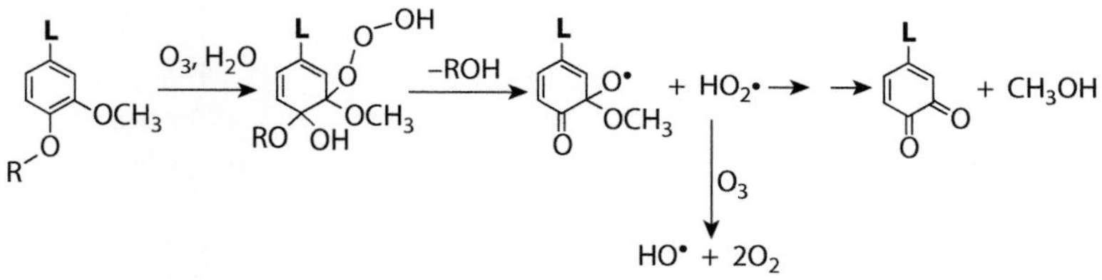

FIGURE 11.31 Reaction mechanism for the formation of hydroxyl radical on ozone treatment of an aromatic lignin unit.

The classical Craigee reaction involves addition of ozone across a double bond and subsequent fragmentation of the intermediate ozonide into carbonyl containing reaction products [153,154]. A nice example is shown in Figure 11.32 where a lignin model compound of the biaryl type is shown to be successively degraded with formation of ester and carbonyl groups [154]. In addition to the formation of products of the muconic acid type, the step-wise ozone oxidation of aromatic rings gives rise to a variety of low molecular weight degradation products including formic acid and carbon dioxide. By subjecting catechol and other simple phenols to ozonation in water, these degradation path-ways have been clarified rather completely (Figure 11.33) [155].

On ozonation of polymeric lignin, the major reaction seems to be a formation of muconic acid structures by oxidation of phenolic end-groups. After the initial rapid reaction, however, only a slow further oxidation seems to take place, as shown in Figure 11.34 [156], again illustrating the shortcomings of making far-going interpretations of reactions in technical systems based only on experiments with low molecular weight model compounds. An additional complication in the interpretation of

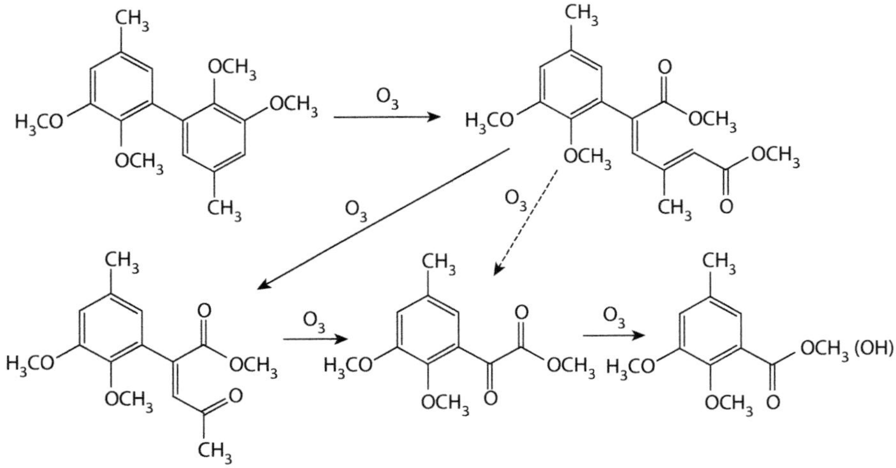

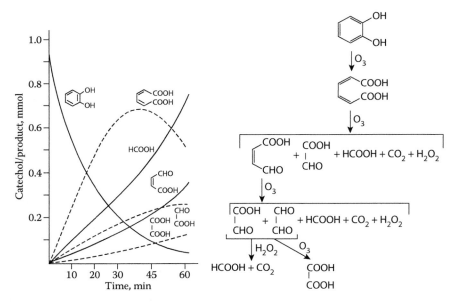

FIGURE 11.32   Successive ozonolysis of a non phenolic lignin model compound of the biaryl type.

FIGURE 11.33   Reaction scheme showing the complete aqueous ozone oxidation of catechol to low molecular weight fragments.

laboratory based ozone oxidation experiments is the fact that the reaction usually is carried out by continuous addition of ozone gas during a long time whereas the technical system involves mixing of gas and reaction with a pulp suspension during a very short time. Although all bleaching agents are added in such a way, the high reactivity of ozone adds to the uncertainty of laboratory data.

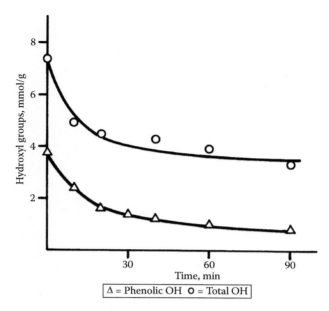

FIGURE 11.34   Aqueous ozone treatment of kraft lignin. Changes in phenolic and total content of hydroxyl groups as a function of ozonation time.

PERACETIC ACID BLEACHING CHEMISTRY

Peracetic acid has been known since a long time as a selective delignifying agent and it can be used for the preparation of holocellulose from wood. It has also been used as a bleaching agent for chemical pulps and in a few mills worldwide; a peracetic acid stage is currently employed. The reaction between lignin structures and peracetic acid is rather slow and reaction rates, measured as kinetic half lives at 25°C, are in the order of 10–300 min [157]. Under mild acidic conditions, peracetic acid acts as an electrophile and it can oxidize electron-rich aromatic structures with formation of quinones, which ultimately are further oxidized to muconic acid structures. The reaction sequence is illustrated in Figure 11.35 [157,158]. In lignin structures having a benzylalcohol function, a degradation of the side-chain structure may take place with the most likely reaction being either an oxidation of the α-carbon followed by a Baeyer–Villiger rearrangement or formation of an epoxide and hydrolysis as shown in Figure 11.36 [158–160].

Other Peracids

In addition to peracetic acid, other peracids such as peroxysulfuric acid [161,162] and peroxyphosphoric acid [159] have been investigated as oxidants for lignin structures. Although certain differences in reactivity exist, the general mode of reaction, viz. aromatic hydroxylation, ring opening via quinone formation, side chain oxidation and Baeyer–Villiger rearrangement seem to be common features. The fact that lower pH-values generally increase the reaction rate [159] is, however, not advantageous from a pulp bleaching point of view.

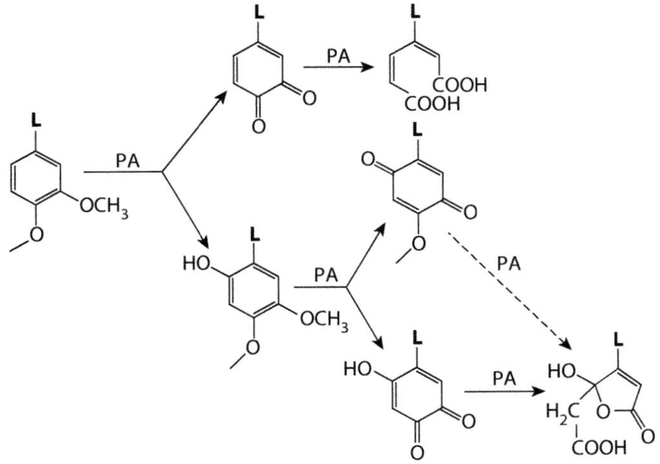

FIGURE 11.35 Reaction scheme for the oxidation of aromatic lignin units with peracetic acid (PA).

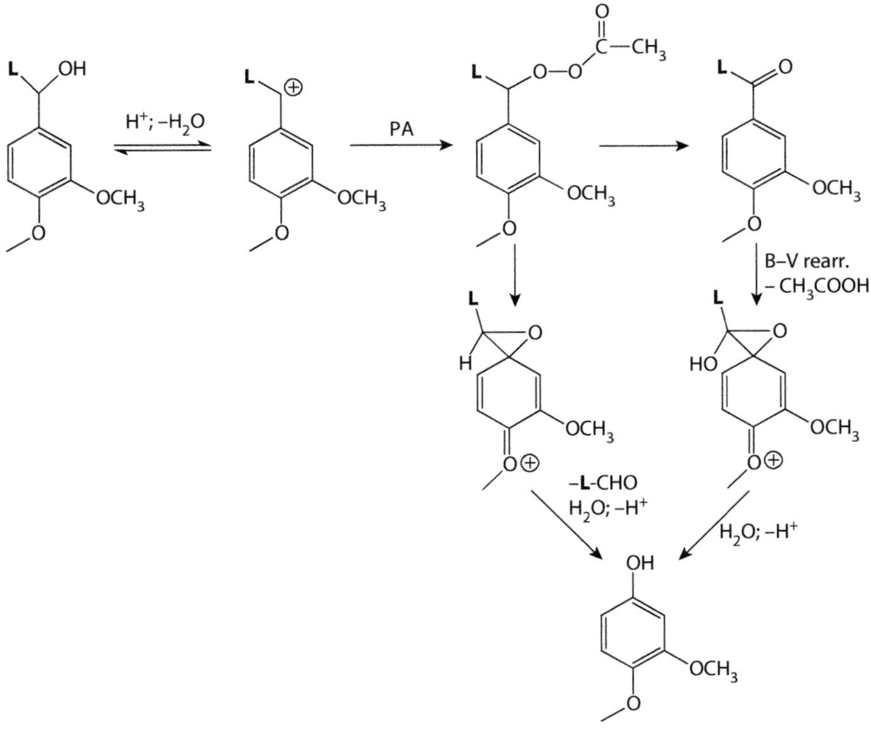

FIGURE 11.36 Reaction mechanism for side chain cleavage in benzylalcohol structures on oxidation with peracetic acid (PA).

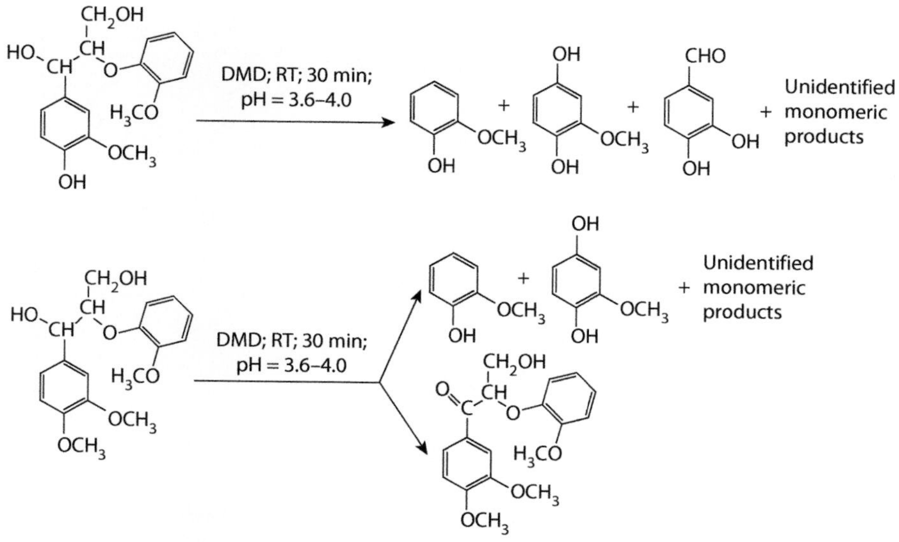

FIGURE 11.37 Reaction products obtained on oxidation of phenolic and non-phenolic lignin model compounds with dimethyldioxirane (DMD).

As a further extension of the bleaching with peracids, the use of "activated oxygen" has been suggested [163,164]. The activated oxygen compound, a dioxirane, can be conveniently prepared by mixing monoperoxysulfate (Oxone®) with, e.g., acetone in an aqueous bicarbonate solution. The resulting dimethyldioxirane can either be isolated by distillation or prepared *in situ* and used to achieve a substantial degree of delignification of kraft pulp. The amount of oxone required is high, however, making the cost of this bleaching stage prohibitive. The reactions of dimethyldioxirane with lignin structures have been investigated with selected model compounds [165]. Oxidative side chain cleavage, demethylation, aromatic hydroxylation, quinone formation, and oxidation of benzylalcohol groups were among the identified reaction types. Some of these are shown in Figure 11.37. These products are consistent with a mechanism proceeding via electron and proton transfer reactions.

THE EXTRACTION STAGE

In bleaching, any acidic bleaching stage must be followed by an alkaline extraction stage in order to dissolve and remove the oxidized lignin from the fibers. In practice, the alkaline stage is often reinforced with oxygen and/or hydrogen peroxide in order to further improve the lignin dissolution and to reduce the amount of bleaching agent that is required, usually chlorine dioxide. The alkaline conditions result in a neutralization of carboxylic acid and phenolic groups and, in addition, any ester groups formed in the oxidative treatment will become hydrolyzed. As a result, the water solubility increases and the oxidized lignin can be washed out from the fibers thus reactivating the residual fiber lignin for new oxidation (cf. Figure 11.25). A further increase of water solubility is obtained by nucleophilic attack of hydroxyl

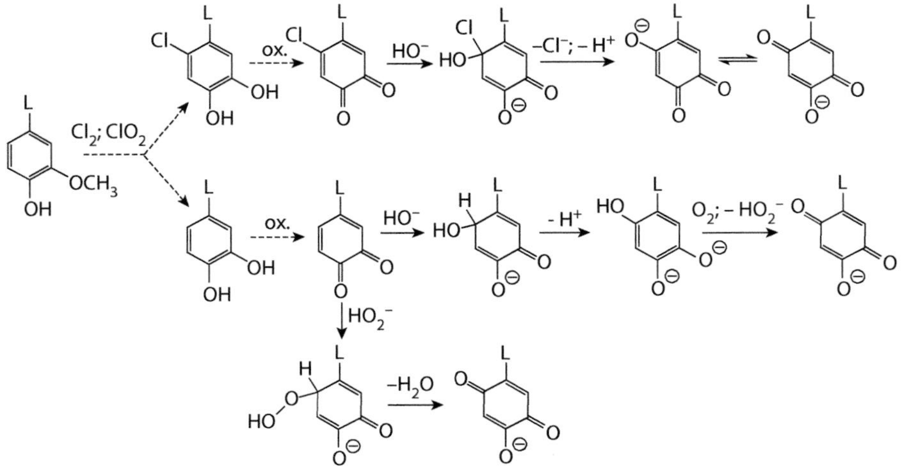

FIGURE 11.38 Reaction scheme showing the formation of hydroxyquinone structures on treatment of quinones with alkali and with alkali in the presence of oxygen and hydrogen peroxide.

ions and hydroperoxy ions (when present) on quinone structures formed as a result of oxidation in the preceding chlorine or chlorine dioxide stage. Such reactions give rise to the formation of highly water soluble hydroxyquinone structures as outlined in Figure 11.38 [166–169] as well as direct oxidation of quinones by hydroperoxy ions [98]. At the same time, an efficient elimination of organically bound chlorine takes place (cf. Figure 11.5) [29,166].

OTHER BLEACHING AGENTS

Oxidation with Polyoxometalate (POM)

The possibility of utilizing oxygen at a neutral or slightly acidic pH-value for a direct or indirect oxidation of lignin has been studied both on kraft pulp, lignin and lignin model compounds. So far, no commercial application as a bleaching stage has been developed. The direct reaction of lignin with oxygen in acidic media was found to give an oxidative cleavage of side chains as well as products originating from acidolysis of lignin [170] but the reaction conditions were rather harsh. By employing a polyoxometalate (POM), having a lower energetic barrier to lignin oxidation than oxygen, the reaction can, however, be performed using much milder conditions. Oxygen can then be used to reoxidize the POM, either in a separate reaction step ("the stoichiometric approach") [171,172] or in a simultaneous operation ("the catalytic approach") [173,174]. Both alternatives have been extensively elaborated. In the former case, monosubstituted Keggin-types of POM $[X^{n+} VW_{11}O_{40}]^{(8-n)-}$ have been used where the central tetrahedron is an XO_4 unit ($X^{n+} = Al^{3+}$, Si^{4+} or P^{5+}). With the use of three equivalents of the POM, a virtually quantitative cleavage of a phenolic lignin model compound was achieved using mild reaction conditions as shown in Figure 11.39 [175,176]. The drawback with this approach is, however, the very large quantity of

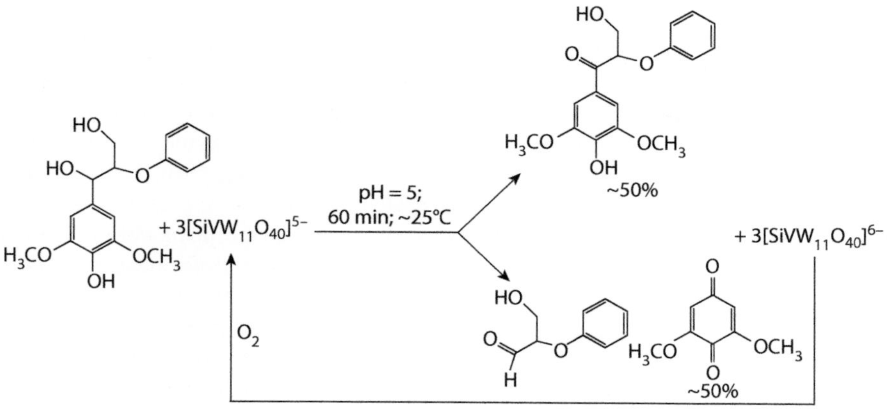

FIGURE 11.39 The stoichiometric oxidation of a phenolic lignin model compound with a POM structure.

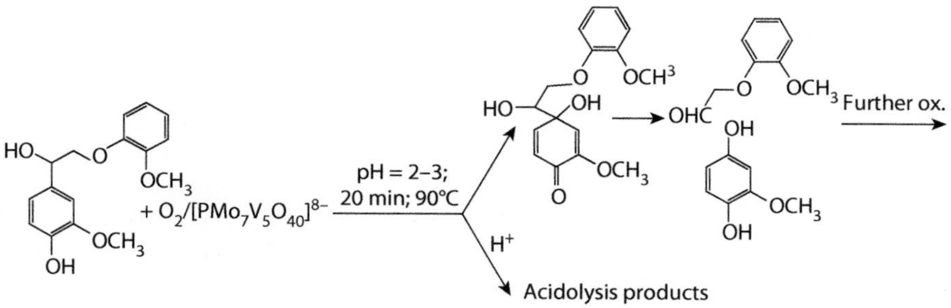

FIGURE 11.40 Catalytic oxidation of a phenolic lignin model compound with acidic oxygen in the presence of a POM structure.

POM that is required and which has to be quantitatively reoxidized by oxygen before the next run, making the technology highly demanding. The reoxidation step will, however, also give rise to a complete wet combustion of the dissolved organic material into carbon dioxide and water.

In the catalytic approach, a small quantity of the POM is present in the reaction medium together with oxygen thus achieving a continous reoxidation of the catalyst and a simultaneous complete oxidation of the organic material. The most frequently used Keggin structure, $[PMo_7V_5O_{40}]^{8-}$, has been shown to oxidize a phenolic lignin structure under mild but rather acidic reaction conditions. Due to the acidic conditions, a simultaneous acidolysis reaction also took place (Figure 11.40) [177,178]. A similar type of oxidation can also be performed with acidic hydrogen peroxide in the presence of a catalytic amount of molybdate [179]. Whether a stoichiometric or a catalytic approach is used in a technical operation, environmental as well as economic reasons will require a technology allowing a virtually quantitative recovery of the POM. The possibility for development of a rather closed pulp mill based on POM technology is, however, attractive and has received much interest.

Laccase/Mediator Oxidation

The increased knowledge about lignin biodegradation involving oxidative enzymes such as lignin peroxidases, manganese peroxidases and laccases has resulted in an attention in such enzymes as potential bleaching agents for chemical pulp (see also Chapter 15, Biopulping and Biobleaching). The fact that the enzyme itself is unable to penetrate the fiber cell wall because of its size has, however, prevented any technological advancement until it was demonstrated that laccase together with a low molecular weight "mediator" and oxygen was able to give a significant delignification effect on kraft pulp under mild conditions [180,181]. These findings have initiated a comprehensive development work aiming at elucidating the chemistry and bleaching efficiency as well as finding efficient and nondegradable mediators. So far, a major drawback in this type of bleaching seems, however, to be the irreversible loss of a substantial portion of the mediator in each run [182], since, in a technical operation, a quantitative recycling of the mediator is a necessary prerequisite [183].

The initial work on laccase oxidation of lignin in the presence of 2,2′-azino-bis (3-ethylbenzothiazoline-6-sulfonic acid), ABTS, as mediator demonstrated that simple nonphenolic model compounds could be oxidized as shown in Figure 11.41 [184]. Thereby, the function of laccases was given a broader importance than just as oxidant of phenolic substrates. Subsequent work has demonstrated that in the

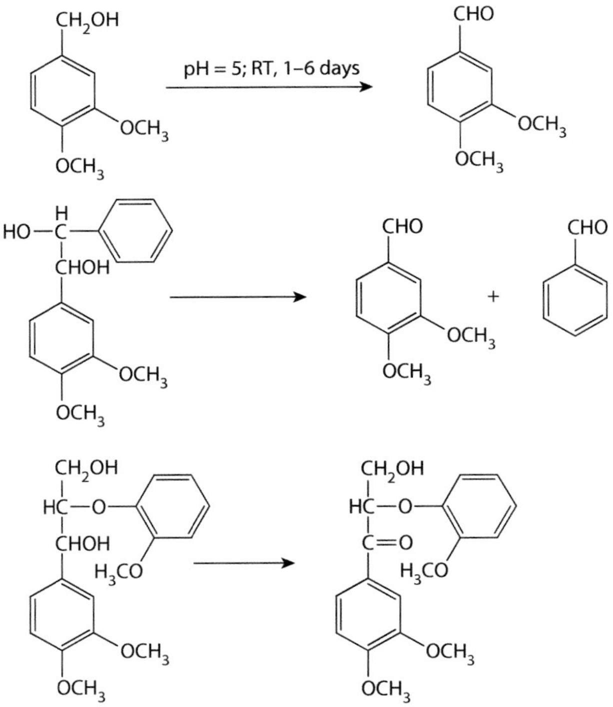

FIGURE 11.41 Oxidation of non phenolic lignin model compounds with laccase in the presence of ABTS as a mediator.

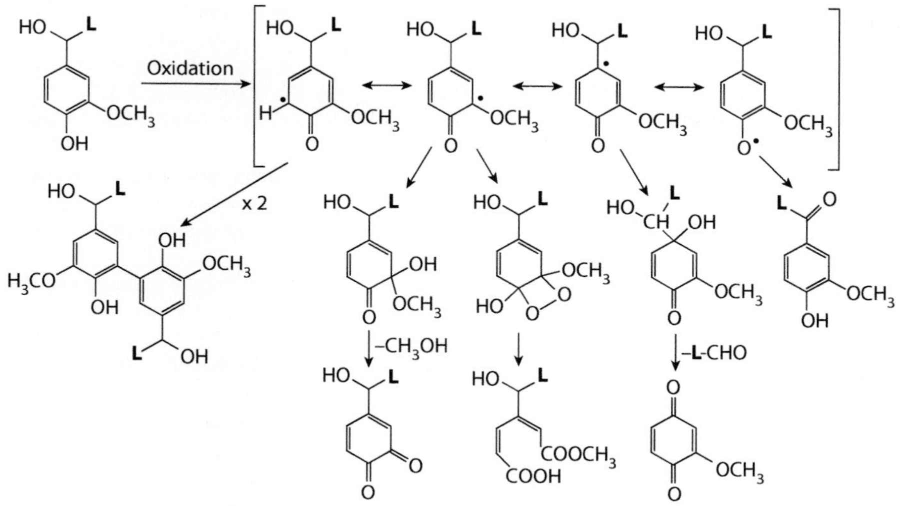

FIGURE 11.42 Oxidation of a phenolic lignin model compound with laccase and oxygen in the presence of HBT as a mediator.

presence of the mediator, 1-hydroxybenzotriazole (HBT), and oxygen, a simple phenolic lignin model compound can be oxidatively degraded whereas in the absence of HBT, oxidative coupling products are produced (Figure 11.42) [185]. Furthermore, nonphenolic structures are oxidized, predominantly by introduction of side chain carbonyl groups [186]. Analysis of isolated residual lignin from kraft pulp and oxygen delignified kraft pulp after treatment with laccase in the presence of oxygen and N-hydroxy-N-phenylacetamide, NHA, as mediator gave results in support of the model experiments, viz a lower amount of phenolic end-groups, a decrease in methoxyl groups and an increase in carbonyl and carboxyl groups [187].

REFERENCES

1. S-J Shin, LR Schroeder, Y-Z Lai. Impact of residual extractives on lignin determination in kraft pulps. *J Wood Chem Technol* 24:139–152, 2004.
2. MM Costa, JL Colodette. The impact of kappa number composition on Eucalyptus kraft pulp bleachability. *Braz J Chem Eng* 24:61–71, 2007.
3. J Li, G Gellerstedt. On the structural significance of the kappa number measurement. *Nordic Pulp Pap Res J* 13:153–158, 1998.
4. L Zhang, G Gellerstedt. Quantitative 2D HSQC NMR determination of polymer structures by selecting suitable internal standard references. *Magn Reson Chem* 45:37–45, 2007.
5. F Imsgard, I Falkehag, KP Kringstad. On possible chromophoric structures in spruce wood. *Tappi* 54(10):1680–1684, 1971.
6. M Johansson, G Gellerstedt. On chromophores and leucochromophores formed during the refining of wood. *Nordic Pulp Pap Res J* 17:5–8, 2002.
7. K Ala-Kaila, J Li, O Sevastyanova, G Gellerstedt. Apparent and actual delignification response in industrial oxygen-alkali delignification of birch kraft pulp. *Tappi J* 2(10):23–27, 2003.

8. J Li, O Sevastyanova, G Gellerstedt. The distribution of oxidizable structures in ECF- and TCF-bleached kraft pulps. *Nordic Pulp Pap Res J* 17:415–419, 2002.
9. M Ragnar. On the importance of the structural composition of pulp for the selectivity of ozone and chlorine dioxide bleaching. *Nordic Pulp Pap Res J* 16:72–79, 2001.
10. CW Dence, DW Reeve. *Pulp Bleaching: Principles and Practice.* TAPPI Press, Atlanta, 1996.
11. J Gierer. The chemistry of delignification: A general concept, Part II. *Holzforschung* 36:55–64, 1982.
12. J Gierer. Chemistry of delignification. Part 2: Reactions of lignins during bleaching. *Wood Sci Technol* 20:1–33, 1986.
13. G Gellerstedt, L Zhang. Chemistry of TCF-bleaching with oxygen and hydrogen peroxide. In: DS Argyropoulos, ed. *Oxidative Delignification Chemistry: Fundamentals and Catalysis.* ACS Symp Series, 785, 2001, pp. 61–72.
14. Z-H Jiang, DS Argyropoulos. Isolation and characterization of residual lignins in kraft pulps. *J Pulp Pap Sci* 25:25–29, 1999.
15. J Gierer. The reactions of lignin during pulping. *Svensk Papperstidn* 73:571–596, 1970.
16. G Gellerstedt, E-L Lindfors. On the formation of enol ether structures in lignin during kraft cooking. *Nordic Pulp Pap Res J* 2:71–75, 1987.
17. G Gellerstedt, W Wafa Al-Dajani. Bleachability of alkaline pulps. Part 1. The importance of β-aryl ether linkages in lignin. *Holzforschung* 54:609–617, 2000.
18. G Gellerstedt, E-L Lindfors. Structural changes in lignin during kraft cooking. Part 4. Phenolic hydroxyl groups in wood and kraft pulps. *Svensk Papperstidn* 87:R115–R118, 1984.
19. RC Francis, Y-Z Lai, CW Dence, TC Alexander. Estimating the concentration of phenolic hydroxyl groups in wood pulps. *Tappi J* 74(9):219–224, 1991.
20. J Wang, Z-H Jiang, DS Argyropoulos. Isolation and characterization of lignin extracted from softwood kraft pulp after xylanase treatment. *J Pulp Pap Sci* 23:J47–J51, 1997.
21. PM Froass, AJ Ragauskas, JE Jiang. NMR studies. Part 3. Analysis of lignins from modern kraft pulping technologies. *Holzforschung* 52:385–390, 1998.
22. ST Moe, AJ Ragauskas. Oxygen delignification of high-yield kraft pulp. Part 1. Structural properties of residual lignins. *Holzforschung* 53:416–422, 1999.
23. FS Chakar, AJ Ragauskas. The effects of oxidative alkaline extraction stages after laccase HBT and laccase NHAA treatments: An NMR study of residual lignins. *J Wood Chem Technol* 20:169–184, 2000.
24. H Araki, Y Tomimura, N Terashima. Radiotracer experiments on lignin reactions. V. The behaviour of lignin in kraft pulping process. *Mokuzai Gakkaishi* 23:378–382, 1977.
25. G Gellerstedt. Condensation in kraft pulping: A reality? Proceedings of 12th International Symposium on Wood and Pulping Chemistry, Madison, WI, 2003, Vol. 1, pp. 1–8.
26. M Lawoko, G Henriksson, G Gellerstedt. New method for quantitative preparation of lignin-carbohydrate complex from unbleached softwood kraft pulp: Lignin polysaccharide networks I. *Holzforschung* 57:69–74, 2003.
27. G Gellerstedt, I Pettersson. Chemical aspects of hydrogen peroxide bleaching. Part II. The bleaching of kraft pulps. *J Wood Chem Technol* 2:231–250, 1982.
28. J Gierer, S Ljunggren. The reactions of lignin during sulfate pulping. Part 17. Kinetic treatment of the formation and competing reactions of quinone methide intermediates. *Svensk Papperstidn* 82:503–512, 1979.
29. AW Kempf, CW Dence. Structure and reactivity of chlorolignin II. Alkaline hydrolysis of chlorinated kraft pulp. *Tappi* 53(5):864–873, 1970.
30. J Hoigne, H Bader. Kinetics of reactions of chlorine dioxide (OClO) in water I. Rate constants for inorganic and organic compounds. *Water Res* 28:45–55, 1994.
31. KV Sarkanen, RW Strauss. Demethylation of lignin and lignin models by aqueous chlorine solutions I. Softwood lignins. *Tappi* 44:459–464, 1961.

32. J Hoigne, H Bader. Rate constants of reactions of ozone with organic and inorganic compounds in water I. Non-dissociating organic compounds. *Water Res* 17:173–183, 1983.
33. DC Johnson. Lignin reactions in delignification with peroxyacetic acid. In: RP Singh, J Nakano, JS Gratzl, eds. *Chemistry of Delignification with Oxygen, Ozone and Peroxides.* Uni Publishers Co., Ltd., Tokyo, 1980, pp. 217–228.
34. E Johansson, S Ljunggren. The kinetics of lignin reactions during oxygen bleaching, Part 4. The reactivities of different lignin model compounds and the influence of metal ions on the rate of degradation. *J Wood Chem Technol* 14:507–525, 1994.
35. BHJ Bielski. Reevaluation of the spectral and kinetic properties of HO_2 and O_2^- free radicals. *Photochem Photobiol* 28:645–649, 1978.
36. DT Sawyer, MJ Gibian. The chemistry of superoxide ion. *Tetrahedron* 35:1471–1481, 1979.
37. C Bull, GJ McClune, JA Fee. The mechanism of Fe-EDTA catalyzed superoxide dismutation. *J Am Chem Soc* 105:5290–5300, 1983.
38. J Liden, LO Öhman. Redox stabilization of iron and manganese in the +II oxidation state by magnesium precipitates and some anionic polymers. Implications for the use of oxygen-based bleaching chemicals. *J Pulp Pap Sci* 23:J193–J199, 1997.
39. C Walling. Fenton's reagent revisited. *Acc Chem Res* 8:125–131, 1975.
40. R Agnemo, G Gellerstedt. The reactions of lignin with alkaline hydrogen peroxide. Part II. Factors influencing the decomposition of phenolic structures. *Acta Chem Scand* B33:337–342, 1979.
41. LM Dorfman, GE Adams. Reactivity of the hydroxyl radical in aqueous solutions. National Bureau of Standards, Department of Commerce, Washington DC, 1973, Report No NSRDS-NBS-46.
42. J Staehelin, RE Buhler, J Hoigne. Ozone decomposition in water studied by pulse radiolysis. 2. OH and HO_4 as chain intermediates. *J Phys Chem* 88:5999–6004, 1984.
43. K Sehested, H Corfitzen, J Holcman, CH Fischer, EJ Hart. The primary reaction in the decomposition of ozone in acidic aqueous solutions. *Environ Sci Technol* 25:1589–1596, 1991.
44. J Hoigne, H Bader, WR Haag, J Staehelin. Rate constants of reactions of ozone with organic and inorganic compounds in water: III. *Water Res* 19:993–1004, 1985.
45. M Ragnar, T Eriksson, T Reitberger, P Brandt. A new mechanism in the ozone reaction with lignin like structures. *Holzforschung* 53:423–428, 1999.
46. JE Wajon, DH Rosenblatt, EP Burrows. Oxidation of phenol and hydroquinone by chlorine dioxide. *Environ Sci Technol* 16:396–402, 1982.
47. PI Gunnarsson, SCH Ljunggren. The kinetics of lignin reactions during chlorine dioxide bleaching. Part 1. Influence of pH and temperature on the reaction of 1-(3,4-dimethoxyphenyl)ethanol with chlorine dioxide in aqueous solution. *Acta Chem Scand* 50:422–431, 1996.
48. C Rav-Acha, E Choshen. Aqueous reactions of chlorine dioxide with hydrocarbons. *Environ Sci Technol* 21:1069–1074, 1987.
49. Y Ni, GJ Kubes, ARP van Heiningen. Rate processes of chlorine species distribution during chlorine dioxide prebleaching of kraft pulp. *Nordic Pulp Pap Res J* 7:200–204, 1992.
50. Y Ni, GJ Kubes, ARP van Heiningen. Mechanism of chlorate formation during bleaching of kraft pulp with chlorine dioxide. *J Pulp Pap Sci* 19:J1–J5, 1993.
51. J Li, O Sevastyanova, G Gellerstedt. The distribution of oxidizable structures in ECF- and TCF-bleached kraft pulps. *Nordic Pulp Pap Res J* 17:415–419, 2002.
52. L Olm, A Teder. The kinetics of oxygen bleaching. *Tappi* 62(12):43–46, 1979.
53. K Kratzl, P Claus, W Lonsky, JS Gratzl. Model studies on reactions occurring in oxidations of lignin with molecular oxygen in alkaline media. *Wood Sci Technol* 8:35–49, 1974.

54. FW Vierhapper, E Tengler, K Kratzl. Zur Sauerstoffoxidation von Kreosolderivaten in alkalisch-wässriger Lösung. *Monatsh Chem* 106:1191–1201, 1975.
55. K Kratzl, PK Claus, A Hruschka, FW Vierhapper. Theoretical fundamentals on oxygen bleaching and pulping. *Cell Chem Technol* 12:445–462, 1978.
56. P Fricko, M Holocher-Ertl, K Kratzl. Zur Oxidation von Kreosol mit Sauerstoff in alkalischer Lösung. *Monatsh Chem* 111:1025–1041, 1980.
57. M Holocher-Ertl, K Kratzl. Modellversuche zur Sauerstoffbleiche von Zellstoff. 4 Mitt. Zum oxidativen Abbau von Lignin: Abbau von Kreosol (2-Methoxy-4-methylphenol) mit molekularem Sauerstoff, Wasserstoffperoxid und Ozon. *Holzforschung* 36:11–16, 1982.
58. RC Eckert, H-m Chang, WP Tucker. Oxidative degradation of phenolic lignin model compounds with oxygen and alkali. *Tappi* 56(6):134–138, 1973.
59. H-m Chang, JS Gratzl. Ring cleavage reactions of lignin models with oxygen and alkali. In: RP Singh, J Nakano, JS Gratzl, eds. *Chemistry of Delignification with Oxygen, Ozone and Peroxides*. Uni Publishers Co., Ltd., Tokyo, 1980, pp. 151–163.
60. J Gierer, F Imsgard. Studies on the autoxidation of t-butyl-substituted phenols in alkaline media. 1. Reactions of 4-t-butylguaiacol. *Acta Chem Scand* B31:537–545, 1977.
61. J Gierer, F Imsgard. Studies on the autoxidation of t-butyl-substituted phenols in alkaline media. 2. Reactions of 4,6-di-t-butylguaiacol and related compounds. *Acta Chem Scand* B31:546–560, 1977.
62. J Gierer, F Imsgard, I Noren. Studies on the degradation of phenolic lignin units of the β-aryl ether type with oxygen in alkaline media. *Acta Chem Scand* B31:561–572, 1977.
63. J Gierer, F Imsgard. The reactions of lignins with oxygen and hydrogen peroxide in alkaline media. *Svensk Papperstidn* 80:510–518, 1977.
64. J Gierer, N-O Nilvebrant. Studies on the degradation of residual lignin structures by oxygen. Part I. Mechanism of autoxidation of 4,4'-dihydroxy-3,3'dimethoxystilbene in alkaline media. *Holzforschung* 40(suppl):107–113, 1986.
65. T Aoyagi, S Hosoya, J Nakano. A new reaction site in lignin during O_2-alkali treatment. *Mokuzai Gakkaishi* 25:783–788, 1979.
66. LL Landucci. Metal-catalyzed phenoxy radical generation during lignin oxidation. *Trans Techn Sect CPPA* 4(1):25–29, 1978.
67. LL Landucci. Electrochemical behavior of catalysts for phenoxy radical generation. *Tappi* 62(4):71–74, 1979.
68. H Xu, Y-Z Lai. Reactivity of lignin diphenylmethane model dimers under mild alkali-O_2 conditions. *J Wood Chem Technol* 17:223–234, 1997.
69. SA Wallick, KV Sarkanen. Effect of pH on the autoxidation kinetics of vanillin. *Wood Sci Technol* 17:107–116, 1983.
70. S Ljunggren. Kinetic aspects of some lignin reactions in oxygen bleaching. *J Pulp Pap Sci* 12(2):J54–J57, 1986.
71. S Ljunggren. The kinetics of lignin reactions during oxygen bleaching. Part 1. The reactivity of p,p'-dihydroxystilbene. *Nordic Pulp Pap Res J* 5:38–43, 1990.
72. S Ljunggren, E Johansson. The kinetics of lignin reactions during oxygen bleaching. Part 2. The reactivity of 4,4'-dihydroxy-3,3'-dimethoxy-stilbene and β-aryl ether structures. *Nordic Pulp Pap Res J* 5:148–154, 1990.
73. S Ljunggren, E Johansson. The kinetics of lignin reactions during oxygen bleaching. Part 3. The reactivity of 4-n-propylguaiacol and 4,4'-di-n-propyl-6,6'-biguaiacol. *Holzforschung* 44:291–296, 1990.
74. S Ljunggren, E Johansson. The kinetics of lignin reactions during oxygen bleaching. Part 4. The reactivities of different lignin model compounds and the influence of metal ions on the rate of degradation. *J Wood Chem Technol* 14:507–525, 1994.
75. RA Young, J Gierer. Degradation of native lignin under oxygen-alkali conditions. *Appl Polym Symp* 28:1213–1223, 1976.

76. TE Crozier, DC Johnson, NS Thompson. Changes in a southern pine dioxane lignin on oxidation with oxygen in sodium carbonate media. *Tappi* 62(9):107–111, 1979.

77. B Ericsson, KV Sarkanen, T Tiedeman. Degradation of hardwood lignin by oxygen pulping at pH 9. In: RP Singh, J Nakano, JS Gratzl, eds. *Chemistry of Delignification with Oxygen, Ozone and Peroxides.* Uni Publishers Co., Ltd., Tokyo, 1980, pp. 173–187.

78. G Gellerstedt, K Gustafsson, E-L Lindfors. Structural changes in lignin during oxygen bleaching. *Nordic Pulp Pap Res J* 1(3):14–17, 1986.

79. G Gellerstedt, E-L Lindfors. Hydrophilic groups in lignin after oxygen bleaching. *Tappi J* 70(6):119–122, 1987.

80. G Gellerstedt, L Heuts. Changes in the lignin structure during a totally chlorine free bleaching sequence. *J Pulp Pap Sci* 23(7):J335–J339, 1997.

81. G Gellerstedt, L Heuts, D Robert. Structural changes in lignin during a totally chlorine free bleaching sequence. Part II. An NMR study. *J Pulp Pap Sci* 25(4):111–117, 1999.

82. Y Sun, D Argyropoulos. Fundamentals of high-pressure oxygen and low-pressure oxygen-peroxide (EOP) delignification of softwood and hardwood kraft pulps: A comparison. *J Pulp Pap Sci* 21(6):J185–J190, 1995.

83. F Asgari, DS Argyropoulos. Fundamentals of oxygen delignification. Part II. Functional group formation/elimination in residual kraft lignin. *Can J Chem* 76:1606–1615, 1998.

84. DS Argyropoulos, Y Liu. The role and fate of lignin's condensed structures during oxygen delignification. *J Pulp Pap Sci* 26(3):107–113, 2000.

85. S Marklund, G Marklund. Involvement of the superoxide anion radical in the autoxidation of pyrogallol and a convenient assay for superoxide dismutase. *Eur J Biochem* 47:469–474, 1974.

86. K Pfister, E Sjöström. Characterization of spent bleaching liquors. Part 5. Composition of material dissolved during oxygen-alkali delignification. *Pap Puu* 61:525–528, 1979.

87. R Agnemo, G Gellerstedt. The reactions of lignin with alkaline hydrogen peroxide. Part II. Factors influencing the decomposition of phenolic structures. *Acta Chem Scand* B33:337–342, 1979.

88. J Gierer, E Yang, T Reitberger. The reactions of hydroxyl radicals with aromatic rings in lignins studied with creosol and 4-methylveratrol. *Holzforschung* 46:495–504, 1992.

89. J Gierer, E Yang, T Reitberger. On the significance of the superoxide radical (O_2^-/ $HO_2\cdot$) in oxidative delignification studied with 4-t-butylsyringol and 4-t-butylguaiacol. *Holzforschung* 48:405–414, 1994.

90. J Gierer, E Yang, T Reitberger. The reactions of chromophores of the stilbene type with the hydroxyl radical (HO·) and the superoxide radical (O_2^-/$HO_2\cdot$). Part 1. The cleavage of the conjugated double bond. *Holzforschung* 50:342–352, 1996.

91. J Gierer, E Yang, T Reitberger. The reactions of chromophores of the stilbene type with the hydroxyl radical (HO·) and the superoxide radical (O_2^-/$HO_2\cdot$). Part 2. reactions other than cleavage of the conjugated double bond. *Holzforschung* 50:353–359, 1996.

92. J Gierer, T Reitberger, E Yang, B-H Yoon. Formation and involvement of radicals in oxygen delignification studied by the autoxidation of lignin and carbohydrate model compounds. *J Wood Chem Technol* 21:313–341, 2001.

93. M Ek, J Gierer, K Jansbo, T Reitberger. Study on the selectivity of bleaching with oxygen-containing species. *Holzforschung* 43:391–396, 1989.

94. S Ljunggren, G Gellerstedt, M Pettersson. Chemical aspects on the degradation of lignin during oxygen bleaching. Proceedings of 6th International Symposium on Wood and Pulping Chemistry, Melbourne, 1981, Vol. 1, pp. 229–236.

95. G Gellerstedt. Gel permeation chromatography. In: SY Lin, CW Dence, eds. *Methods in Lignin Chemistry.* Berlin: Springer-Verlag, 1992, pp. 487–497.

96. B Holmbom. Molecular interactions in wood fibre suspensions. 9th International Symposium on Wood and Pulping Chemistry, Montreal, Canada. Proceedings, Vol. 1, pp. PL3-1–PL3-6, 1997.

97. G Gellerstedt, R Agnemo. The reactions of lignin with alkaline hydrogen peroxide. Part III. The oxidation of conjugated carbonyl structures. *Acta Chem Scand* B34:275–280, 1980.

98. G Gellerstedt, H-L Hardell, E-L Lindfors. The reactions of lignin with alkaline hydrogen peroxide. Part IV. Products from the oxidation of quinone model compounds. *Acta Chem Scand* B34:669–673, 1980.

99. MB Hocking, K Bhandari, B Shell, TA Smyth. Steric and pH effects on the rate of Dakin oxidation of acylphenols. *J Org Chem* 47:4208–4215, 1982.

100. GE Miksche. Uber das Verhalten des Lignins bei der Alkalikochung. VIII. Isomerisierung der Phenolatanionen von erythro- und threo-Isoeugenol-β-(2-methoxyphenyl)-äther uber ein Chinonmethid. *Acta Chem Scand* 26:4137–4142, 1972.

101. S Omori, CW Dence. The reactions of alkaline hydrogen peroxide with lignin model dimers. Part 2. Guaiacylglycerol-β-guaiacyl ether. *Wood Sci Technol* 15:113–123, 1981.

102. JF Kadla, H-m Chang, H Jameel. The reactions of lignins with hydrogen peroxide at high temperature. Part 1. The oxidation of lignin model compounds. *Holzforschung* 51:428–434, 1997.

103. L Heuts, G Gellerstedt. Oxidation of guaiacylglycerol-β-guaiacyl ether with alkaline hydrogen peroxide in the presence of kraft pulp. *Nordic Pulp Pap Res J* 13:107–111, 1998.

104. S Omori, CW Dence. The reactions of alkaline hydrogen peroxide with lignin model dimers. Part 1. Phenacyl α-aryl ethers. *Wood Sci Technol* 15:67–79, 1981.

105. G Gellerstedt, I Pettersson, S Sundin. Chemical aspects of hydrogen peroxide bleaching. Proceedings of 1st International Symposium on Wood and Pulping Chemistry, Stockholm, 1981, Vol. 2, pp. 120–124.

106. RW Pero, CW Dence. The role of transition metals on the formation of color from methoxyhydroquinone: An intermediate in the peroxide bleaching of TMP. *J Pulp Pap Sci* 12:192–197, 1986.

107. B Johansson, GE Miksche. Uber die Benzyl-arylätherbindung im Lignin. II. Versuche mit Modellen. *Acta Chem Scand* 26:289–308, 1972.

108. G Gellerstedt, R Agnemo. The reactions of lignin with alkaline hydrogen peroxide. Part V. The formation of stilbenes. *Acta Chem Scand* B34:461–462, 1980.

109. J Basta, L Holtinger, J Höök. Controlling the profile of metals in the pulp before hydrogen peroxide treatment. Proceedings of 6th International Symposium on Wood and Pulping Chemistry, Melbourne, 1991, Vol. 1, pp. 237–244.

110. G Gellerstedt, L Heuts. Changes in the lignin structure during a totally chlorine free bleaching sequence. *J Pulp Pap Sci* 23:J335–J340, 1997.

111. G Gellerstedt, L Heuts, D Robert. Structural changes in lignin during a totally chlorine free bleaching sequence. Part II: An NMR study. *J Pulp Pap Sci* 25:111–117, 1999.

112. A Gärtner, G Gellerstedt. Oxidation of residual lignin with alkaline hydrogen peroxide. Part I: Changes in hydrophilic groups. *J Pulp Pap Sci* 26:448–454, 2000.

113. JF Kadla, H-m Chang, H Jameel. The reactions of lignins with high temperature hydrogen peroxide. Part 2. The oxidation of kraft lignin. *Holzforschung* 53:277–284, 1999.

114. Y Sun, M Fenster, A Yu, RM Berry, DS Argyropoulos. The effect of metal ions on the reaction of hydrogen peroxide with kraft lignin model compounds. *Can J Chem* 77:667–675, 1999.

115. L Jurasek, L Kristofova, Y Sun, DS Argyropoulos. Alkaline oxidative degradation of diphenylmethane structures: Activation energy and computational analysis of the reaction mechanism. *Can J Chem* 79:1394–1401, 2001.

116. A Gärtner, G Gellerstedt. Oxidation of residual lignin with alkaline hydrogen peroxide. Part II. Elimination of chromophoric groups. *J Pulp Pap Sci* 27:244–248, 2001.

117. B Lindgren. Chlorine dioxide and chlorite oxidations of phenols related to lignin. *Svensk Papperstidn* 74:57–63, 1971.

118. JJ Kolar, BO Lindgren, B Pettersson. Chemical reactions in chlorine dioxide stages of pulp bleaching. *Wood Sci Technol* 17:117–128, 1983.

119. C Brage, T Eriksson, J Gierer. Reactions of chlorine dioxide with lignins in unbleached pulps. Part I. *Holzforschung* 45:23–30, 1991.

120. C Brage, T Eriksson, J Gierer. Reactions of chlorine dioxide with lignins in unbleached pulps. Part I. *Holzforschung* 45:147–152, 1991.

121. M Zawadzki, T Runge, A Ragauskas. Facile detection of *ortho-* and *para*-quinone structures in residual kraft lignin by ^{31}P NMR spectroscopy. *J Pulp Pap Sci* 26:102–106, 2000.

122. S Ljunggren, P-I Gunnarsson. Some kinetic aspects of lignin reactions in chlorine dioxide bleaching. 8th Int Symp Wood Pulping Chem, Helsinki, Finland. Proceedings, Vol. 2, pp. 303–308, 1995.

123. P-I Gunnarsson, S Ljunggren. The kinetics of lignin reactions during chlorine dioxide bleaching. Part 1. Influence of pH and temperature on the reaction of 1-(3,4-dimethoxy-phenyl)ethanol with chlorine dioxide in aqueous solution. *Acta Chem Scand* 50:422–431, 1996.

124. E Chosen, R Elits, C Rav-Asha. The formation of cation-radicals by the action of chlorine dioxide on p-substituted styrenes and other alkenes. *Tetrahedron Lett* 27:5989, 1986.

125. C Rav-Asha, E Chosen. Aqueous reactions of chlorine dioxide with hydrocarbons. *Environ Sci Technol* 21:1069–1074, 1987.

126. JJ Kolar, BO Lindgren. Oxidation of styrene by chlorine dioxide and by chlorite in aqueous solution. *Acta Chem Scand* B36:599–605, 1982.

127. Y Ni, GJ Kubes, ARP van Heiningen. Rate processes of AOX formation and chlorine species distribution during ClO$_2$ prebleaching of kraft pulp. Int Pulp Bleaching Conf, Stockholm, Sweden. *Proc* 2:195–218, 1991.

128. Y Ni, GJ Kubes, ARP van Heiningen. Characterization of pulp demethylation during chlorine dioxide bleaching. *Wood Sci Technol* 29:87–94, 1995.

129. C Mateo, C Chirat, D Lachenal. The chromophores remaining after bleaching to moderate brightness. *J Wood Chem Technol* 24:279–288, 2004.

130. G Gellerstedt, E-L Lindfors, M Pettersson, D Robert. Reactions of lignin in chlorine dioxide bleaching of kraft pulps. *Res Chem Intermed* 21:441–456, 1995.

131. TM Runge, AJ Ragauskas. NMR analysis of oxidative alkaline extraction stage lignins. *Holzforschung* 53:623–631, 1999.

132. G Gellerstedt, E-L Lindfors, M Pettersson, E Sjöholm, D Robert. Chemical aspects on chlorine dioxide as a bleaching agent for chemical pulps. 6th Int Symp Wood Pulping Chem, Melbourne, Australia. Proceedings, Vol. 1, pp. 331–336, 1991.

133. RM Berry, BI Fleming. Why do chlorination and extraction fail to delignify unbleached kraft pulp completely? *Holzforschung* 41:177–183, 1987.

134. Y Ni, GJ Kubes, ARP van Heiningen. A new mechanism for pulp delignification during chlorination. *J Pulp Pap Sci* 16(1): J13–J19, 1990.

135. CW Dence. Halogenation and nitration. In: KV Sarkanen, CH Ludwig, eds. *Lignins: Occurrence, Formation and Reactions*. Wiley-Interscience, New York, pp. 373–422, 1971.

136. CW Dence. Chemistry of chemical pulp bleaching. In: CW Dence, DW Reeve, eds. *Pulp Bleaching. Principles and Practice*. TAPPI Ppress, Atlanta, pp. 125–160, 1996.

137. X Shen, A van Heiningen. Delignification mechanism during chlorine bleaching of kraft pulp; the formation of chlorinated phenolic compounds from the cleavage of alkyl aryl ether linkages. 6th International Symposium on Wood and Pulping Chemistry, Melbourne, Australia. Proceedings, Vol. 1, pp. 557–561, 1991.

138. Y Ni, GJ Kubes, ARP van Heiningen. Methanol number: A fast method to determine lignin content of pulp. *J Pulp Pap Sci* 16(3):J83–J86, 1990.

139. K Lindström, F Österberg. Characterization of the high molecular mass chlorinated matter in spent bleach liquors (SBL). Part 1. Alkaline SBL. *Holzforschung* 38:201–212, 1984.

140. F Österberg, K Lindström. Characterization of the high molecular mass chlorinated matter in spent bleach liquors (SBL). Part 2. Acidic SBL. *Holzforschung* 39:149–158, 1985.

141. F Österberg, K Lindström. Characterization of the high molecular mass chlorinated matter in spent bleach liquors (SBL). 3-mass spectrometric interpretation of aromatic degradation products in SBL. *Org Mass Spectrometry* 20:515–524, 1985.

142. KP Kringstad, K Lindström. Spent liquors from pulp bleaching. *Environ Sci Technol* 18(8): 236A–248A, 1984.

143. W Sandermann. Polychlorierte aromatische Verbindungen als Umweltgifte. *Naturwissenschaften* 61:207–213, 1974.

144. K Lindström, J Nordin. Gas chromatography-mass spectrometry of chlorophenols in spent bleach liquors. *J Chromatogr* 128:13–26, 1976.

145. P Axegård. Improvement of bleach plant effluent by cutting back on Cl_2. *Pulp Pap Can* 90(5):T183–T187, 1989.

146. L Strömberg, R Mörck, F de Sousa, O Dahlman. Effects of internal process changes and external treatment on effluent chemistry. In: MR Servos, KR Munkittrick, JH Carey, GJ van der Kraak, eds. *Environmental Fate and Effects of Pulp and Paper Mill Effluents,* St Lucie Press, Delray Beach, Florida, pp. 3–19, 1996.

147. OB Dahlman, AK Reimann, LM Strömberg, RE Mörck. High-molecular-weight effluent materials from modern ECF and TCF bleaching. *Tappi J* 78(12):99–109, 1995.

148. K Lindström, R Schubert. Determination by MS-MS of 1,1-dichlorodimethyl sulfone from pulp mill bleach plant effluents in aquatic organisms. *J High Res Chrom & Chrom Commun* 7:68–73, 1984.

149. T Eriksson, J Gierer. Relative rates of ozonation of lignin model compounds. 3rd International Symposium on Wood and Pulping Chemistry, Vancouver, Canada. Proceedings, Vol. 2, pp. 29–30, 1985.

150. J Hoigne, H Bader. Rate constants of reactions of ozone with organic and inorganic compounds in water. 2. Dissociating organic compounds. *Water Res* 17:185–194, 1983.

151. M Ragnar, T Eriksson, T Reitberger, P Brandt. A new mechanism in the ozone reaction with lignin like structures. *Holzforschung* 53:423–428, 1999.

152. MN Schuchmann, C von Sonntag. Reactions of ozone with D-glucose in oxygenated aqueous solution: Direct action and hydroxyl radical pathway. *Aqua* 38:311–317, 1989.

153. H Kaneko, S Hosoya, J Nakano. Degradation of lignin with ozone: Reactions of biphenyl and a-carbonyl type model compounds with ozone. *Mokuzai Gakkaishi* 27:678–683, 1981.

154. T Eriksson, J Gierer. Studies on the ozonation of structural elements in residual kraft lignins. *J Wood Chem Technol* 5:53–84, 1985.

155. Y Yamamoto, E Niki, H Shiokawa, Y Kamiya. Ozonation of organic compounds. 2. Ozonation of phenol in water. *J Org Chem* 44:2137–2142, 1979.

156. P Månsson, R Öster. Ozonation of kraft lignin. *Nordic Pulp Pap Res J* 3:75–81, 1988.

157. DC Johnson. Lignin reactions in delignification with peracetic acid. In: RP Singh, J Nakano, JS Gratzl, eds. *Chemistry of Delignification with Oxygen, Ozone and Peroxides.* Uni Publishers Co., Ltd. Tokyo, Japan, pp. 217–228, 1980.

158. JF Kadla, H-m Chang. The reactions of peroxides with lignin and lignin model compounds. In: DS Argyropoulos, ed. *Oxidative Delignification Chemistry: Fundamentals and Catalysis.* ACS Symposium Series, 785, pp. 108–129, 2001.

159. T Zhu, JF Kadla, H-m Chang, H Jameel. Reactions of lignin with peroxymonophosphoric acid: The degradation of lignin model compounds. *Holzforschung* 57:44–51, 2003.

160. M Zawadzki, A Ragauskas. Pulp properties influencing oxygen delignification bleachability. Tappi Pulping Conference, Orlando, Florida. Proceedings, Vol. 1, pp. 323–333, 1999.

161. H Kawamoto, H-m Chang, H Jameel. Reaction of peroxyacids with lignin and lignin model compounds. 8th International Symposium on Wood and Pulping Chemistry, Helsinki, Finland. Proceedings, Vol. 1, pp. 383–390, 1995.

162. T Kishimoto, JF Kadla, H-m Chang, H Jameel. The reactions of lignin model compounds with hydrogen peroxide at low pH. *Holzforschung* 57:52–58, 2003.

163. C-L Lee, K Hunt, RW Murray. Activated oxygen, a selective bleaching agent for chemical pulps. Part I. Laboratory bleaching with isolated and in-situ-generated activated oxygen. *J Pulp Pap Sci* 20:J125–J130, 1994.

164. AJ Ragauskas. Investigation of dimethyldioxirane as a bleaching reagent for kraft pulp. *Tappi J* 76(7):87–90, 1993.

165. DS Argyropoulos, Y Sun, RM Berry, J Bouchard. Reactions of dimethyldioxirane with lignin model compounds. *J Pulp Pap Sci* 22:J84–J90, 1996.

166. SA Braddon, CW Dence. Structure and reactivity of chlorolignin. I. Alkaline hydrolysis of chlorine-substituted lignin model compounds. *Tappi* 51(6):249–256, 1968.

167. G Gellerstedt, B Pettersson. Autoxidation of lignin. *Svensk Papperstidn* 83:314–318, 1980.

168. BN Brogdon, LA Lucia. New insights into lignin modification during chlorine dioxide bleaching sequences (III): The impact of modifications in the (EO) versus E stage on the D_1 stage. *J Wood Chem Technol* 25:133–147, 2005.

169. BN Brogdon, LA Lucia. New insights into lignin modification during chlorine dioxide bleaching sequences (IV): The impact of modifications in the (EP) and (EOP) stages on the D_1 stage. *J Wood Chem Technol* 25:149–170, 2005.

170. J Gierer, N-O Nilvebrant. Studies on the degradation of lignin by oxygen in acidic media. 2nd International Symposium on Wood and Pulping Chemistry, Tsukuba, Japan. Proceedings, Vol. 4, pp. 109–112, 1983.

171. IA Weinstock, RH Atalla, RS Reiner, CJ Houtman, CL Hill. Selective transition-metal catalysis of oxygen delignification using water-soluble salts of polyoxometalate (POM) anions. Part I. Chemical principles and process concepts. *Holzforschung* 52:304–310, 1998.

172. RH Atalla, IA Weinstock, JS Bond, RS Reiner, DM Sonnen, CJ Houtman, RA Heintz, CG Hill, CL Hill, MW Wemple, YuV Geletii, EMG Barbuzzi. Polyoxometalate-based closed systems for oxidative delignification of wood pulp fibers. In: DS Argyropoulos, ed. *Oxidative Delignification Chemistry: Fundamentals and Catalysis*. ACS Symposium Series, 785, pp. 313–326, 2001.

173. DV Evtuguin, C Pascoal Neto. New polyoxometalate promoted method of oxygen delignification. *Holzforschung* 51:338–342, 1997.

174. DV Evtuguin, C Pascoal Neto. Catalytic oxidative delignification with Keggin-type molybdovanadophosphate heteropolyanions. In: DS Argyropoulos, ed. *Oxidative Delignification Chemistry: Fundamentals and Catalysis*. ACS Symposium Series, 785, pp. 342–355, 2001.

175. IA Weinstock, KE Hammel, MA Moen, LL Landucci, S Ralph, CE Sullivan, RS Reiner. Selective transition-metal catalysis of oxygen delignification using water-soluble salts of polyoxometalate (POM) anions. Part II. Reactions of α-$[SiVW_{11}O_{40}]^{5-}$ with phenolic lignin-model compounds. *Holzforschung* 52:311–318, 1998.

176. VA Grigoriev, CL Hill, IA Weinstock. Polyoxometalate oxidation of phenolic lignin models. In: DS Argyropoulos, ed. *Oxidative Delignification Chemistry: Fundamentals and Catalysis*. ACS Symposium Series, 785, pp. 297–312, 2001.

177. DV Evtuguin, C Pascoal Neto, H Carapuca, J Soares. Lignin degradation in oxygen delignification catalysed by $[PMo_7V_5O_{40}]^{8-}$ polyanion. Part II. Study on lignin monomeric model compounds. *Holzforschung* 54:511–518, 2000.

178. DV Evtuguin, C Pascoal Neto. Lignin degradation reactions in aerobic delignification catalyzed by heteropolyanion $[PMo_7V_5O_{40}]^{8-}$. In: DS Argyropoulos, ed. *Oxidative Delignification Chemistry: Fundamentals and Catalysis*. ACS Symposium Series, 785, pp. 327–341, 2001.

179. V Kubelka, RC Francis, CW Dence. Delignification with acidic hydrogen peroxide activated by molybdate. *J Pulp Pap Sci* 18:J108–J114, 1992.

180. R Bourbonnais, MG Paice. Demethylation and delignification of kraft pulp by *Trametes versicolor* laccase in the presence of 2,2′-azinobis-(3-ethylbenzthiazoline-6-sulphonate). *Appl Microbiol Biotechnol* 36:823–827, 1992.

181. HP Call, I Mucke. History, overview and applications of mediated lignolytic systems especially laccase-mediator-systems (Lignozym®-process). *J Biotechnol* 53:163–202, 1997.

182. J Sealey, A Ragauskas. Fundamental investigations into the chemical mechanisms involved in laccase mediator biobleaching. 9th International Symposium on Wood and Pulping Chemistry, Montreal, Canada. Proceedings, Vol. 1, pp. F1-1–F1-4, 1997.

183. R Bourbonnais, D Rochefort, MG Paice, D Leech. Development of stable redox complexes to mediate delignification of kraft pulp by laccase. In: DS Argyropoulos, ed. *Oxidative Delignification Chemistry: Fundamentals and Catalysis*. ACS Symposium Series, 785, pp. 391–399, 2001.

184. R Bourbonnais, MG Paice. Oxidation of non-phenolic substrates: An expanded role for laccase in lignin biodegradation. *FEBS Lett* 267:99–102, 1990.

185. C Crestini, DS Argyropoulos. On the role of 1-hydroxybenzotriazole as mediator in laccase oxidation of residual kraft lignin. In: DS Argyropoulos, ed. *Oxidative Delignification Chemistry: Fundamentals and Catalysis*. ACS Symposium Series, 785, pp. 373–390, 2001.

186. J Freudenreich, M Amann, E Fritz-Langhals, J Stohrer. Understanding the Lignozym®-process. International Pulp Bleaching Conference, Helsinki, Finland. Proceedings, Vol. 1, pp. 71–76, 1998.

187. K Poppius-Levlin, T Tamminen, A Kalliola, T Ohra-aho. Characterization of residual lignins in pulps delignified by laccase/N-hydroxyacetanilide. In: DS Argyropoulos, ed. *Oxidative Delignification Chemistry: Fundamentals and Catalysis*. ACS Symposium Series, 785, pp. 358–372, 2001.

12 The Chemistry of Lignin-Retaining Bleaching: Oxidative Bleaching Agents

Gordon Leary and John A. Schmidt

CONTENTS

INTRODUCTION

The chromophores that reduce the brightness of mechanical pulps are predominately carbonyl structures in the lignin—ketones, aldehydes, and quinones. These structures are native to the lignin, but they can also be formed during storage and processing of mechanical and high-yield pulps. Bleaching agents used for mechanical pulps must brighten the pulp while maintaining their characteristic high yield. Unlike chemical pulp bleaching, where the objective is to remove residual lignin, mechanical pulp bleaching must alter the structure of chromophores to noncolored functional groups without removing a significant amount of lignin. This can be achieved by using reductive or oxidative bleaching agents. Reductive bleaching is discussed in Chapter 13. This chapter deals with oxidative bleaching agents.

The oxidative agents most commonly used for chemical pulps, chlorine dioxide and (now rarely) chlorine, are unsuitable for mechanical pulps because they will darken the lignin. This darkening is immaterial in chemical pulp bleaching, since the lignin is ultimately removed from the fibers. The only oxidative agents known to brighten mechanical pulps are peroxides. Alkaline hydrogen peroxide is the only one that is used commercially, but there has been significant research on the use of various peracids, ozone and the active-oxygen species formed *in situ* from these agents. This chapter is primarily about the chemistry of lignin-retaining bleaching, but it will also discuss the combination of peroxide with bleaching agents that may degrade and remove some of the lignin from pulps. The structures of some lignin model compounds (II–XI) that are frequently referred to throughout the text are collected in Figure 12.1.

HYDROGEN PEROXIDE

CHEMISTRY RELATED TO BLEACHING

Several aspects of aqueous hydrogen peroxide chemistry are relevant to chemical and mechanical pulp bleaching.

Hydrogen Peroxide as a Source of Hydroxyl ($\bullet$OH) Radicals

Hydrogen peroxide readily decomposes to form hydroxyl radicals. These radicals show little selectivity and attack both lignin and cellulose, helping to solubilize the residual lignin in chemical pulps but having the negative effect of reducing the cellulose degree of polymerization (DP) [1,2].

The reactions of hydroxyl radicals, formed in mechanical pulp bleaching from alkaline peroxide, are entirely undesirable. They represent an unprofitable decomposition of the bleaching agent.

Gierer [3] has reviewed the reactions of the hydroxyl radical with lignin. It attacks both the lignin side chains and the aromatic rings, but the ionized form (O^{-}) is less reactive and at pHs above the pK_a (11.9) only side chain cleavage was observed.

The hydroxyl radicals may be formed by homolytic cleavage of hydrogen peroxide (Equation 12.1) or by metal ion catalyzed decomposition (Equations 12.2 and 12.3) [4]. Oxygen gas is one of the decomposition products (Equations 12.4 and 12.5).

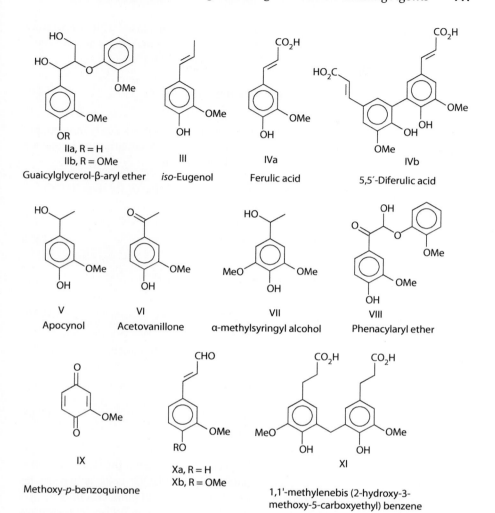

FIGURE 12.1   Structures of various lignin model compounds cited throughout the text.

$$H_2O_2 \longrightarrow 2 \, {}^{\bullet}OH \tag{12.1}$$

$$H_2O_2 + M^{n+} \longrightarrow {}^{-}OH + {}^{\bullet}OH + M^{(n+1)+} \tag{12.2}$$

$${}^{-}OH + {}^{-}OOH + M^{(n+1)+} \longrightarrow O_2^{\bullet-} + H_2O + M^{n+} \tag{12.3}$$

$$O_2^{\bullet-} + M^{(n+1)+} \longrightarrow O_2 + M^{n+} \tag{12.4}$$

$$O_2^{\bullet-} + {}^{\bullet}OH \longrightarrow O_2 + {}^{-}OH \tag{12.5}$$

The rate of homolytic cleavage of hydrogen peroxide increases with temperature and is also accelerated by the presence of alkali, reaching a maximum at a pH equal to the pK_a of hydrogen peroxide, according to Equation 12.6 [5]:

$$H_2O_2 + {}^-OOH \longrightarrow {}^\bullet OH + O_2^{\bullet -} + H_2O \qquad (12.6)$$

To limit the decomposition induced by transition metals (particularly manganese, iron, chromium, and copper) alkaline peroxide bleaching formulations may include chelating agents, silicates, phosphates, or earth metal salts.

Hydrogen Peroxide as a Weak Acid; The Nucleophilicity of Its Base HOO⁻

Hydrogen peroxide is a weak acid (Equation 12.7):

$$H_2O + H_2O_2 \rightleftharpoons {}^-OOH + H_3O^+ \qquad (12.7)$$

The equilibrium can also be expressed as in Equation 12.8:

$$H_2O + {}^-OOH \rightleftharpoons H_2O_2 + {}^-OH \qquad (12.8)$$

Consequently, two equilibrium constants can be written (Equations 12.9 and 12.10):

$$K_a = [H^+]\,[HOO^-]/[H_2O_2] \qquad (12.9)$$

$$K_b = [H_2O_2]\,[HO^-]/[HOO^-] \qquad (12.10)$$

The value of pK_a decreases (K_a increases) with temperature. At 15°C the pK_a is 11.9, at 25°C it is 11.6, and at 35°C 11.45 [6,7]. This means the equilibrium concentration of HOO⁻ ions increases with temperature and, since HOO⁻ is the active species in lignin-retaining peroxide bleaching, the rate of mechanical pulp bleaching by peroxide should also increase with temperature. Unfortunately, at higher temperatures the benefits of more rapid bleaching are usually more than offset by an increase in the rate of peroxide decomposition and by an increase in the rates of lignin darkening reactions (reactions of lignin with alkali and oxygen). As the temperature is increased the dissociation constant of water (and the concentration of HO⁻ ions) increases to an even greater degree than K_a for hydrogen peroxide. It can be calculated [8–10] that the ratio of [HO⁻]/[HOO⁻] in a half neutralized solution of hydrogen peroxide in water rises from less than 0.1 at 10°C to about 0.25 at 120°C. Teder and Tormund [11] expressed this temperature dependence in terms of K_b, Equation 12.11, where T is in degrees absolute:

$$pK_b = (1330/T) - 2.13 \qquad (12.11)$$

Consequently reactions dependent on the [HO⁻] concentration, such as the darkening of mechanical pulps in alkaline peroxide, may increase with temperature more rapidly than the rate of bleaching. The alkaline peroxide brightness ceiling may fall with increasing reaction temperatures. The optimum $NaOH/H_2O_2$ ratios for mechanical pulp bleaching also fall with rising temperatures [12,13].

Alkaline Hydrogen Peroxide as a Lignin-Retaining Bleaching Agent

Under alkaline conditions hydrogen peroxide attacks carbonyl groups in lignin and can brighten mechanical pulp without substantial lignin removal. The mechanism involves attack by the HOO^- ion, which is a strong nucleophile [14–16]. The principal lignin carbonyl structures attacked are cinnamaldehydes, quinones, aryl aldehydes, and aryl ketones. The chemistry of these reactions has been reviewed by Dence [17]. Nucleophilic attack by HOO^- cleaves the enone side chains of cinnamaldehydes with the formation of benzaldehyde and benzoic acid end-groups [18,19]. The benzaldehydes, as well as aryl ketone groups, may be degraded to hydroquinones such as 2-methoxy hydroquinone (I) by displacement of the carbonyl side chain, the Dakin reaction. It is specific for o- and p-hydroxy substituted aryl carbonyl compounds (Figure 12.2). Hocking et al. [20] have isolated hydroquinone monoacetate, as an intermediate product of the Dakin oxidation of p-hydroxy acetophenone. In alkali, the acetate rapidly hydrolyses to the hydroquinone. Hydroquinones seem to be stable to peroxide but quinones, which are formed by the reaction of hydroquinones with oxygen, react with hydroperoxide ion in ring-opening reactions that leave carboxylic groups attached to the residual lignin framework [21]. The carboxylic acid groups, through hydrogen bonding, contribute to the gain in interfiber bonding strength [22] observed when mechanical pulps are bleached with alkaline peroxide.

Carbonyl groups in nonphenolic lignin units are much more resistant to alkaline peroxide, but they are oxidized to some extent via cleavage of the bond between the α and β side chain carbon atoms, again forming carboxylic acids.

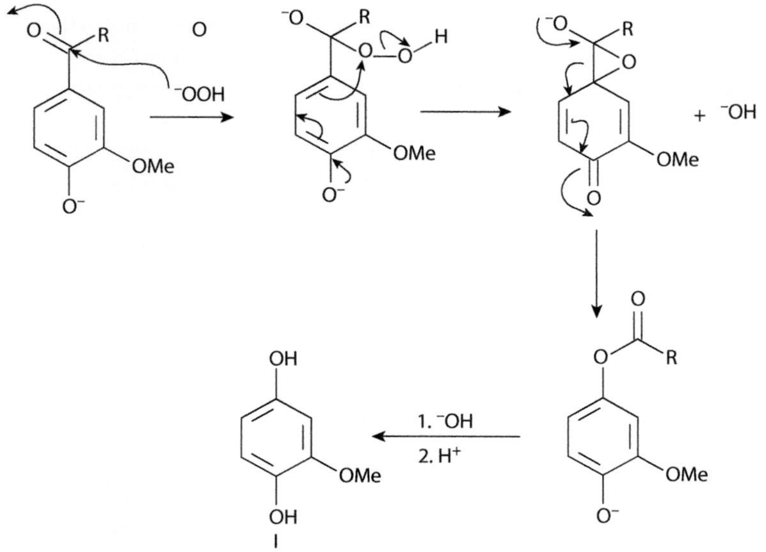

FIGURE 12.2 Dakin reaction of o-and p- hydroxyl, aryl-α-carbonyl compounds.

When it is used to bleach mechanical pulps, alkaline hydrogen peroxide is usually considered to be a lignin-retaining bleaching agent. However, some lignin is dissolved and this appears to help the brightness without necessarily degrading the lignin chromophores. Garver et al. [23] have shown that the brightness of stone groundwood bleached with alkaline peroxide correlates with the amount of lignin (and hemicellulose) that dissolves. The dissolved material is very accessible to the peroxide and is itself degraded, consuming the bleaching chemical and reducing the overall efficiency in bleaching.

Unionized Hydrogen Peroxide as a Weak Nucleophile

Unionized hydrogen peroxide (HOOH) is a much weaker nucleophile than hydroperoxide ion (HOO⁻) but, from model compound studies [65], it may be able to react with benzyl carbonium ions in competition with lignin condensation reactions.

REACTIONS OF HYDROGEN PEROXIDE WITH LIGNIN CHROMOPHORES

Reactions with Lignin Model Compounds

Reactions of Unstabilized Hydrogen Peroxide Involving Hydroxyl Radicals

Various studies [24–27] have shown that lignin phenolic model compounds undergo ring cleavage when treated with unstabilized peroxide. This is due to attack by hydroxyl radicals formed by hydrogen peroxide decomposition. The phenolic compounds studied, e.g., creosol or α-methyl syringyl alcohol (VII, Figure 12.1), represented typical lignin structural units with alkyl or hydroxy-alkyl (benzyl alcohol) side chains. Cleavage of lignin side chains may also occur. β–O–4 lignin interunit linkages (e.g., IIa, IIb, Figure 12.1), which are stable to attack by hydroperoxide ions, can be cleaved by hydroxyl radicals [28,29].

The role of hydroxyl radicals in these ring and interunit cleavage reactions has been confirmed by adding peroxide stabilizers. When sodium silicate was present there was no reaction with either simple phenol model compounds or guaiacylglycerol-β-guaiacyl ether (Figure 12.1, IIa) [28,30,31].

Kojima and Kayama [32] observed ring coupling reactions and hydroxylation. Hydroxyl radicals from unstabilized hydrogen peroxide caused radical coupling and formed hydroxylated *iso*-eugenol (Figure 12.1, III).

The production of hydroxyl radicals may be enhanced by addition of benzoyl peroxide [33] or by subjecting the hydrogen peroxide to UV irradiation [34,35]. Ferulic acid (Figure 12.1, IV), which is resistant to hydroperoxide anion attack, was degraded by hydrogen peroxide subjected to UV irradiation [36].

Reactions of Stabilized Alkaline Hydrogen Peroxide
(Attack by Hydroperoxide Ions)

Dakin and Dakin-like Reactions   Cinnamaldehydes or aromatic structures containing carbonyl substituents *para* to a free phenolic hydroxyl group (e.g., *p*-hydroxyacetophenones) react with stabilized alkaline peroxide to form benzaldehydes and hydroquinones, respectively [37,38]. Structures containing aryl-α-carbonyl substituents *para* to an etherified phenolic group are degraded more slowly [39].

Conditions that promote the formation of quinone methides will enable the reaction of lignin units that do not contain electron-withdrawing substituents such as carbonyl groups. *p*-Hydroxy benzyl alcohols do not react with stabilized alkaline hydrogen peroxide at traditional bleaching temperatures (below 70°C), but at 90°C quinone methide formation is enhanced and they will undergo Dakin-like reactions, forming hydroquinones [40]. Guaiacylglycerol-β-guaiacyl ether (IIa in Figure 12.1) was degraded at 90°C over three hours [41] but structures that facilitate the formation of quinone methides may react at lower temperatures. The unexpected reaction of the phenolic lignin model compound apocynol (V in Figure 12.1) with peroxide at 50°C was attributed to the formation of a somewhat stabilized quinone methide [33].

Hydroquinone and Quinone Formation   Hydroquinone formation from lignin model compounds containing an aryl carbonyl group *para* to a phenolic hydroxyl, by the Dakin reaction, has been well documented. Phenacyl α-aryl ethers (VIII in Figure 12.1) react principally by displacement of the side chain in a Dakin reaction to form a hydroquinone and a carboxylic acid (Figure 12.3) [38]. Hosoya and Nakano [39] showed that other lignin model dimers containing *p*-hydroxy aryl carbonyl groups produced hydroquinones when treated with stabilized alkaline peroxide at 50°C.

Nonni et al. [42] subsequently confirmed the validity of these model compound experiments when they found methoxy hydroquinone (I) in the spent peroxide bleaching liquors of a thermomechanical pulp (TMP) mill. Holah and Heitner [43] showed, at long reaction times and with a large excess of peroxide, that an absorption band with $\lambda_{max} = 320$ nm was removed. They attributed this to slow removal of α-carbonyl structures, which would presumably have occurred by a Dakin reaction.

Using neutral (pH 7) hydrogen peroxide Oki et al. [44] isolated methoxy hydroquinone and methoxy benzoquinone from the reaction of vanillin and acetovanillone with hydrogen peroxide. Other *p*-hydroxybenzaldehydes were degraded in a similar way. Hirashima and Sumimoto [45] obtained evidence for the presence of quinones in mechanical pulps, using a photochemical reaction with acetaldehyde.

There is no doubt that hydroquinones and hence their oxidation products, quinones, are formed by the Dakin oxidation when mechanical pulps are treated with

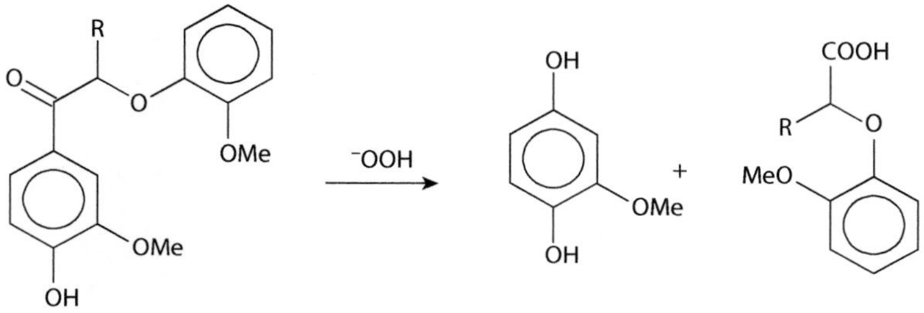

FIGURE 12.3   Dakin reaction of phenacyl α-aryl ethers. (After Omori, S. and Dence, C.W., *Wood Sci. Technol.*, 15, 67–79, 1981; Hoyosa, S. and Nakano, J., *Mokuzai Gakkaishi*, 26, 97–101, 1980.)

alkaline peroxide. It is surprising, however, that they are isolated as the final products. Hydroquinones are very susceptible to oxidation to quinones, which are readily degraded by alkaline peroxide. However, in the absence of oxygen, or if the peroxide is stabilized against decomposition, there may be insufficient oxygen present to oxidize all the hydroquinones to quinones. Supporting this, Lachenal et al. [46] found that adding oxygen gas during peroxide bleaching increased the final pulp brightness.

The quinones and ring cleavage products found in the reaction of unstabilized peroxide with lignin model compounds like α-methyl syringyl alcohol (Figure 12.1 VII) [25] are a result of hydroxyl radical attack [28,30].

Lee and Sumimoto [47] suggested that peroxide-bleached mechanical pulps yellow readily because the hydroquinones they contain are "leucochromophores," which may be subsequently oxidized further to colored compounds. In particular they showed that peroxide treatment of phenolic stilbene (XIII), which was formed from phenylcoumaran (XII) by mechanochemical action, gave rise to a labile hydroquinone that was easily oxidized by air to an intensely colored stilbenequinone (XIV) (Figure 12.4). It should be noted that these conclusions were based on experiments with model compounds. Conversely, Zhu et al. [48] found that mechanochemical action had no significant effect on the photosensitivity of mechanical pulps.

In summary, there is abundant evidence that alkaline peroxide reacts with lignin to form hydroquinones, some of which are not oxidized further under the bleaching conditions. It seems likely that the p-quinones detected in bleached TMP have been formed by subsequent oxidation by air.

Quinone Cleavage and Resistance to Peroxide Treatment   Hydrogen peroxide readily oxidizes o- and p-quinones via ring cleavage, forming colorless, open-chain carboxylic acids [21]. This is the predominant reaction, however, colored products may also be formed, particularly at high alkali concentrations. There is firm evidence for the formation of hydroxy-substituted quinones. In addition, the formation of quinone polymers has also been proposed. The hydroxy-quinones are probably formed

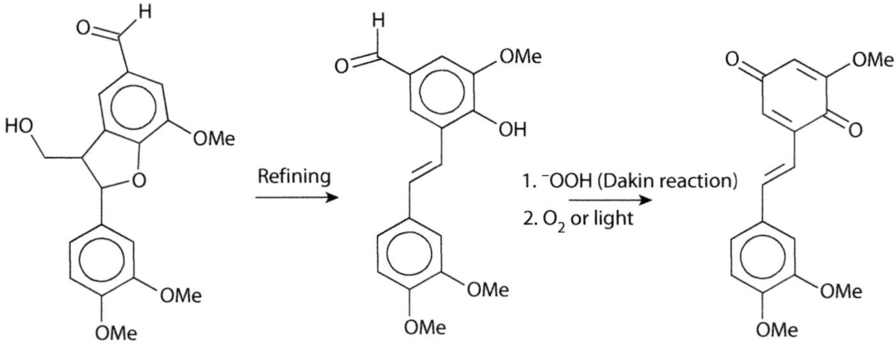

FIGURE 12.4   Mechanochemical action converts phenyl coumaran XII to phenolic stilbene XIII, which is subsequently oxidized to stilbene quinine XIV. (After Lee, D.Y. and Sumimoto, M., *Holzforschung,* 45 (suppl), 15–20, 1991.)

via attack of a hydroxide ion in competition with the more nucleophilic hydroperoxide ion [21,25]. Hydroxy-quinones were detected following alkaline peroxide treatment of model o- and p-quinones. In particular, the hydroxy-quinone (XVI) was obtained as a minor product on treatment of 4-t-butyl-1,2-benzoquinone (XV) with peroxide at pH 9 and 25°C (Figure 12.5) [21]. The hydroxy-quinone was resistant to further peroxide attack, as might be expected due to the higher electron density that results from the hydroxyl substitution. However, XVI could be degraded under more aggressive conditions: pH 11 and 40°C.

Earlier, Hosoya et al. [49] also found that although quinones reacted with alkaline peroxide, they were hard to decolorize completely. The resistance to bleaching was most marked with 2-methoxy-1,4-benzoquinone, which remained colored even when treated with as much as a 10-molar excess of alkaline peroxide. By comparison, 2,6-dimethoxy-1,4-benzoquinone was bleached by a five-molar excess of peroxide. This suggests that the vacant five position in 2-methoxy-1,4-benzoquinone is involved in the formation of persistent color. Hydroxylation at this position is possible but Hosoya et al. [49] suggested that the color was due to polymerization. Alkali-induced quinone polymerization through the five-position has been reported by Erdtman and Granath [50] and in neutral or acidic conditions by Forsskahl et al. [51]. They also obtained the colored dibenzofuran-1,4-quinone (XVII), a derivative of a 5-5' linked dimer, by irradiating methoxy-1,4-benzoquinone (Figure 12.6). Hosoya, Seike and Nakano [49] isolated colored material after treating methoxy-1,4-benzoquinone with alkali alone but they did not fully characterize the product.

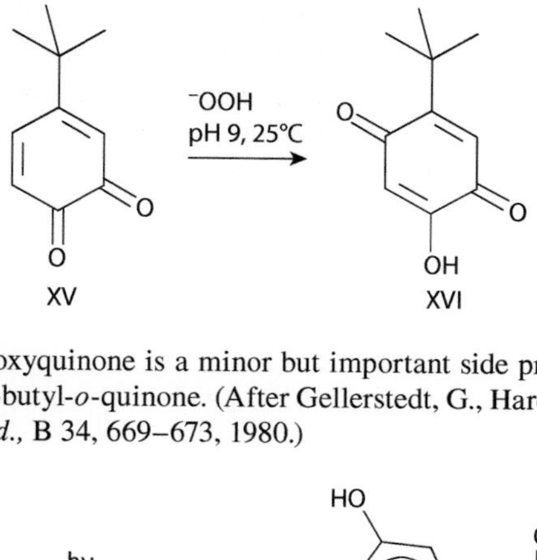

FIGURE 12.5 A hydroxyquinone is a minor but important side product in the reaction of alkaline peroxide with t-butyl-o-quinone. (After Gellerstedt, G., Hardell, H-L., and Lindfors, E-L., *Acta. Chem. Scand.*, B 34, 669–673, 1980.)

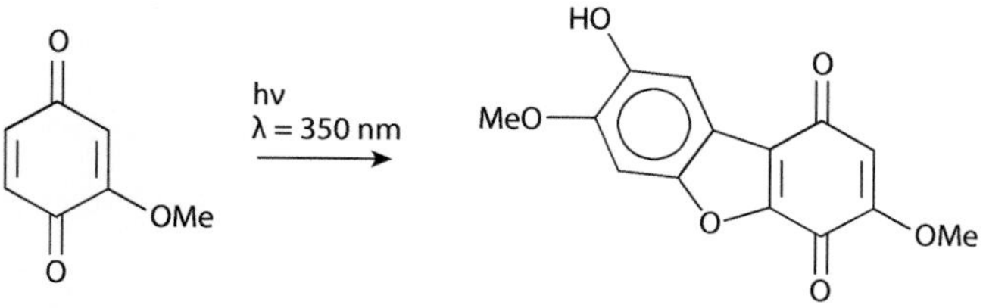

FIGURE 12.6 Dibenzofuran XVII was obtained by UV irradiation of methoxy-p-benzoquinone. (After Forsskahl, I., Gustafsson, J., and Nybergh, A., *Acta. Chem. Scand.*, B 35, 389–394, 1981.)

In a separate study, Pero and Dence [52] also found methoxy-1,4-benzoquinone difficult to bleach with peroxide. They postulated that both hydroxy-quinone and quinone condensation products were formed.

The partial resistance of quinones to bleaching by alkaline peroxide suggests that their formation and incomplete removal is a cause of alkali darkening of mechanical pulps, and a reason there is a brightness ceiling in alkaline peroxide mechanical pulp bleaching. It has not been proven if this is due to the formation of hydroxy-quinones, quinone polymers or both. Quinones in TMP may be insufficiently mobile to form polymers by reacting with each other. Leary and Giampaolo [53] used UV spectral data to attribute alkali darkening of TMP to hydroxy-quinone formation. He et al. [54], on the other hand, found that darkening by alkali could be reversed by subsequent alkaline peroxide bleaching.

α-β *Unsaturated Compounds*   Cinnamaldehydes (e.g., coniferaldehyde, Xa) are degraded by alkaline peroxide with cleavage of the α-β double bond to form benzaldehydes [37]. Cinnamic acids, which occur in association with lignin and hemicelluloses in grasses, are much more stable. Ferulic acid (IVa) and its esterified or etherified derivatives, ethyl ferulate and 3,4-dimethoxy cinnamic acid, reacted very slowly with alkaline peroxide at 50°C [55].

The α-β double bond in hydroxy- or methoxy-substituted stilbenes should also be stable to hydroperoxide ion. It is the electron withdrawing carbonyl group in cinnamaldehydes that makes their double bond susceptible to nucleophilic attack.

Stilbenes are unlikely to occur in native lignin, but they may be formed from phenolic β-1 structures [56,57]. Nonni and Dence (Figure 12.7) [56] found that reacting 1,2-di-guaiacyl-1,3-propanediol (XVIII) with alkali under vacuum gave *p*-hydroxystilbenes corresponding both to loss of formaldehyde (XIX) and to simple dehydration (XX). In the presence of peroxide, or even trace amounts of oxygen, the

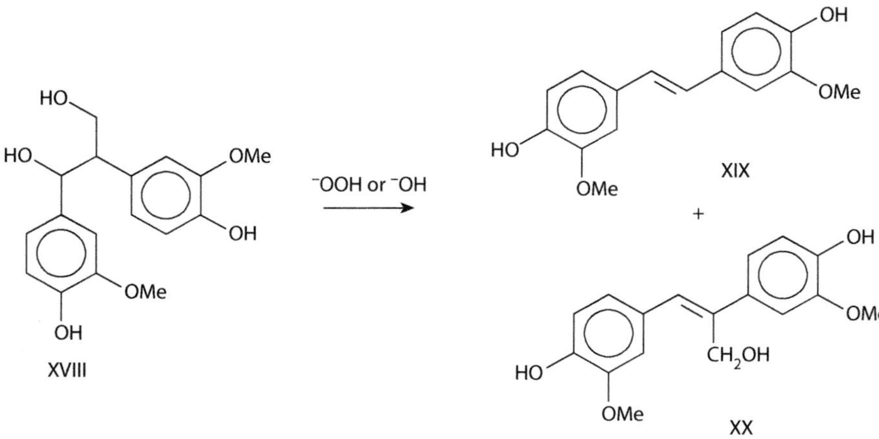

FIGURE 12.7   Formation of phenolic stilbenes from the degradation of β-1 structures. (After He, Z., Ni, Y., and Zhang, E., *J. Wood. Chem. Technol.*, 24, 1–13, 2004; Nonni, A.J. and Dence, C. W., *Holzforschung*, 42, 37–46, 1988.)

phenolic stilbenes were further degraded to various substituted benzaldehydes and benzoic acids. This suggests that such stilbenes should not survive typical peroxide bleaching conditions. However, Gellerstedt and Agnemo [57] were able to isolate phenolic stilbenes as products of the reaction of the α-ethyl ether of XVIII with alkaline peroxide. Also, di-guaiacylstilbene (XIX) has been found in the acid hydrolyzate of bleached mechanical pulps [58].

Although they are not susceptible to nucleophilic attack, guaiacyl or syringyl stilbenes react in alkaline peroxide with cleavage of the exocyclic double bond to form the corresponding benzoic acids, benzaldehydes and benzaldehyde degradation products. Pan et al. [59] found that diguaiacylstilbene (XIX) formed vanillic acid, vanillin and methoxy hydroquinone when treated with alkaline hydrogen peroxide. 3,5-Dimethoxy-4-hydroxy stilbene also reacted although the less activated 4-hydroxystilbene was stable.

Nonni and Dence [60] suggested that the hydroxy-stilbenes might react in alkaline peroxide by the addition of undissociated hydrogen peroxide, acting as an electrophile. An alternative is that hydroxy-stilbenes are reactive with alkali and oxygen. Johansson and Ljungren [61] and Wong et al. [62] observed that 4-hydroxy-substituted stilbenes disappeared rapidly when treated with oxygen in alkali. Pan et al. [59] were unable to confirm this. They found that 4-hydroxy stilbene was stable in alkaline peroxide regardless of the presence or absence of oxygen.

A paper by Gierer et al. [63] probably resolves the mechanism. These workers found that a number of hydroxy- or methoxy-substituted stilbenes would react with hydroxyl radicals, or with a combination of superoxide and hydroxyl radicals. The formation of superoxide and hydroxyl radicals by hydrogen peroxide decomposition is hard to prevent and incomplete chelation of metal ions might explain the inconsistent results reported by the earlier workers.

Models for Residual Lignins   Smith and McDonough [64] studied the degradation of a diphenylmethane model compound for residual lignin, 1,1′-methylenebis[2-hydroxy-3-methoxy-5-(carboxyethyl)benzene (XI, Figure 12.1), by hydrogen peroxide in the presence of transition metals. Hydroxy radicals would be produced. Cu (II), potassium ferricyanide and Mn and Fe salts increased the rate of degradation, but the Mn and Fe salts decomposed the peroxide too rapidly to achieve effective model compound degradation.

Reactions of Acidic Hydrogen Peroxide (Involving Attack by HOOH)
Kishimoto et al. [65] treated 4-substituted guaiacyl and veratryl lignin model compounds with acidic hydrogen peroxide stabilized by the addition of DTPA. The reactions were carried out with a three molar excess of peroxide at 70°C in aqueous or aqueous dioxane solution and pHs in the range 1–3. The compounds were more susceptible to attack by peroxide at the lower pHs. Guaiacyl compounds reacted more rapidly than their methylated analogues.

The α-hydroxy compounds reacted more readily than the corresponding α-carbonyl compounds. A mechanism involving nucleophilic attack on benzyl carbonium ions by HOOH was proposed to account for this. A mechanism involving perhydronium ion ($H_3O_2^+$) was discounted. There was no evidence of the direct ring hydroxylation by $H_3O_2^+$ that is observed with peracetic or persulfuric acids.

Reactions with Chromophores in Lignin Preparations

Alkaline Peroxide

Sulfate lignin was extensively degraded by treatment with unstabilized peroxide at pH 11 and 100°C [66].

Several workers have studied the reaction of alkaline peroxide with softwood milled wood lignin (MWL). Kano et al. [67] observed a decrease in the number of α-carbonyl groups and the formation of methanol when MWL was treated with stabilized peroxide. However, some α-carbonyl groups in lignin appear to be resistant to alkaline peroxide. Spittler and Dence [68] compared the changes in the absorbance at 350 nm of MWL solutions brought about by different bleaching treatments, including alkaline peroxide. They confirmed the important contribution of carbonyl groups to the color of high-yield pulps, and found evidence for some "relatively electron-rich" carbonyls that were resistant to alkaline peroxide and $NaBH_4$, but that were reduced by BH_3 in tetrahydrofuran. Kanitskaya et al. [69] reported that when dioxane lignin was treated with alkaline peroxide the number of α-carbonyl groups actually increased. The increase in α-carbonyl group content was accompanied by the expected decrease in coniferaldehyde structures.

Pan et al. [70] used enzymes to isolate lignins from TMP and TMP treated with alkaline peroxide. Thioacidolysis and ^{13}C NMR spectroscopy revealed a loss of coniferaldehyde structures, but little other change.

Alkaline peroxide treatment of an isolated residual kraft lignin resulted in small decreases in the concentration of carboxylic, phenolic and aliphatic OH groups, according to the ^{31}P NMR spectrum of its trimethyl phosphite derivative [71].

Alkaline (and acidic) peroxide was used to bleach hydroxypropyl lignin [72].

The yellow color of wheat straw, attributed at least in part to ferulic acid structures associated with the lignin, is not bleached by alkaline peroxide treatment [73].

Neutral and Acidic Hydrogen Peroxide

Hydrogen peroxide in acidic or neutral conditions is used by some pulp mills, often in conjunction with oxygen delignification, to bleach chemical pulps with the intention of degrading rather than retaining the lignin. High temperatures are often employed although Oki et al. [44] found that soluble dioxane lignin or MWL was degraded by neutral hydrogen peroxide at 20°C. They noted a decrease in methoxyl and phenolic hydroxyl group contents and a drop in syringaldehyde and vanillin yields following nitrobenzene oxidation. There was an increase in the number of carboxylic acids per C_9 unit.

PERACETIC ACID

Peracetic acid is prepared by mixing hydrogen peroxide and acetic acid in the presence of an acid catalyst. A commercial solution (e.g., of 35–40%) of peracetic acid in aqueous acetic acid contains some free hydrogen peroxide.

CHEMISTRY RELATED TO BLEACHING

In Acid Solutions

Peracetic acid is an effective delignifying agent because it degrades and solubilizes lignin and is slow to react with carbohydrates. In its unionized form it is a strongly electrophilic reagent that will hydroxylate aromatic rings and convert olefinic double bonds to epoxides or 1,2-glycols (e.g., styrene to 1-phenyl-1,2-ethanediol). The reactions with aromatic compounds are consistent with the addition of OH+ to the aromatic ring [74,75]. Norman and Smith [76] have summarized the evidence that the hydroxylation occurs via OH+ (although probably not as a discrete intermediate) rather than via hydroxyl radical. Lignin ring substituents such as OH or OMe direct the OH+ addition to *ortho* or *para* ring positions, and the hydroxylation is followed by oxidation and/or demethylation to form *o*- or *p*-quinones. The quinones react further with excess peracetic acid to give cyclic and alicyclic muconic acids (Figure 12.8). *o*-Quinones are more reactive than the *p*-quinones [77].

Lignin model compounds containing aryl carbonyl groups undergo Baeyer–Villiger oxidation to esters and their hydrolysis products (e.g., hydroquinones) [78,79].

In Neutral or Alkaline Solutions

In neutral or moderately alkaline solutions, peracetic acid can be used to brighten mechanical pulps with little dissolution of the lignin. The peracetate ion (CH_3COOO^-), like the hydroperoxide ion (HOO^-), is a strong nucleophile that selectively reacts with quinones and aryl carbonyl groups but does not attack electron-rich aromatic rings. Since the pK_a of peracetic acid is about 8.2, the high pHs necessary to form the HOO^- ion for bleaching with hydrogen peroxide can be avoided. Thus McDonough [80] found that the rate of reaction of peracetic acid with the lignin model compound, acetovanillone (VI), increases with pH to a maximum (~8.5), close to the pK_a. A decline in the rate of reaction at higher pHs was attributed to the ionization of the acetovanillone phenoxide group.

REACTIONS OF PERACETIC ACID WITH LIGNIN CHROMOPHORES

Reactions with Lignin Model Compounds

Acidic Solutions

The guaiacyl and syringyl aromatic rings are activated for electrophilic attack and, judging by the reaction of lignin model compounds, most structures in lignin are vulnerable to peracetic acid oxidation.

Simple Model Compounds   The reaction with 4-alkyl phenols (e.g., creosol) was slow [81] but phenols with α-hydroxy groups, such as vanillyl alcohol and apocynol (V), formed hydroquinones and hydroquinone acetatates. Presumably, the α-hydroxy groups were first oxidized to the corresponding α-carbonyl compounds.

Carbonyl Compounds   Peracetic acid attacks α-carbonyl groups in lignin model compounds [75,78,82–86]. The reaction occurs readily according to a Baeyer–Villiger oxidation. Nimz and Schwind [78] obtained significant yields of hydroquinone

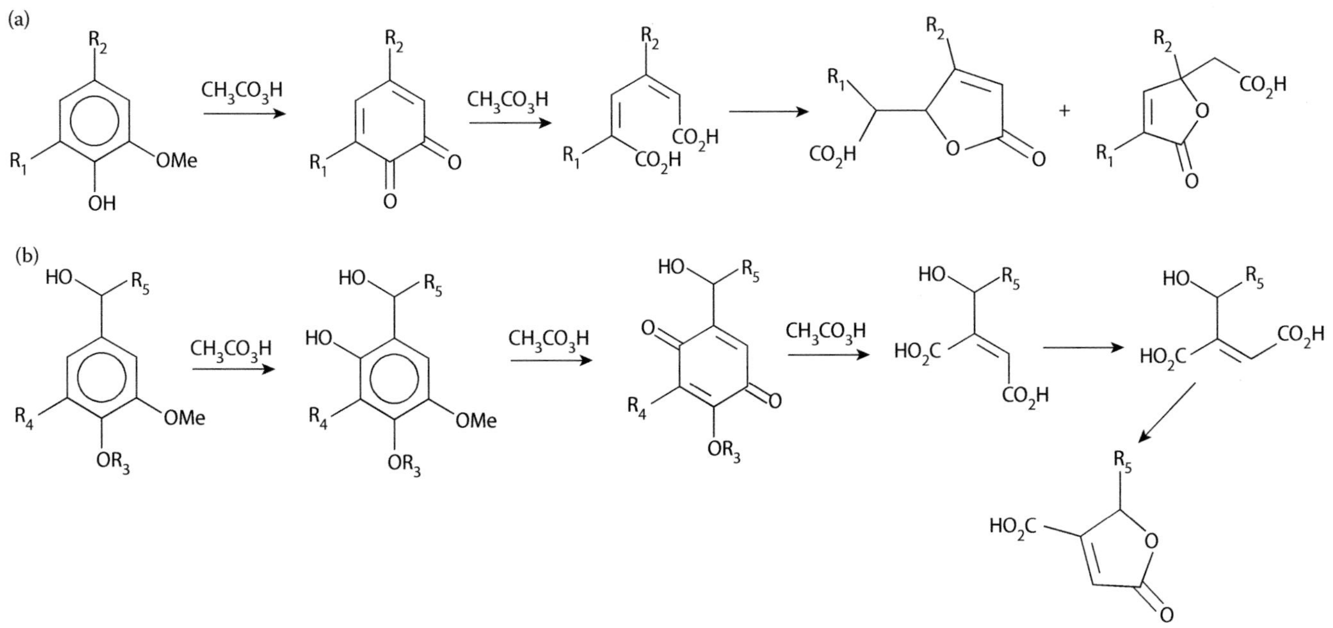

FIGURE 12.8 Reaction of lignin aromatic units with peracetic acid after Lai and Sarkanen. (After Lai, Y.-Z. and Sarkanen, K.V., *Tappi*, 51, 449–453, 1968.) (a) Oxidative ring opening of free phenolic groups to muconic acids via an *o*-quinone, followed by lactonization. (b) oxidative ring opening of etherified lignin phenolic groups via *p*-quinones, followed by lactonization.

esters, hydroquinones, and *p*-quinones from vanillin, syringaldehyde, veratraldehyde and acetovanillone. The aldehydes reacted faster than the ketone acetovanillone. Methylation of the phenolic hydroxyl made little difference to the rate; vanillin and veratraldehyde reacted at similar rates.

α-β *Unsaturated Compounds* Model compounds with unsaturated side chains react with peracetic acid via epoxidation and cleavage of the olefinic double bond. Treatment of solutions of model stilbenes with peracetic acid at pH 5 and 50°C resulted in extensive degradation. Less than 20% of diguaiacyl stilbene (XIX in Figure 12.7) and mono-syringyl stilbene were recovered after one hour; the less reactive 4-hydroxy stilbene was 60% destroyed [59]. The aldehydes formed were further degraded to quinones and probably muconic acids, although no degradation products were isolated.

Iso-eugenol (III) and dehydro-di-*iso*eugenol (XXI) also form epoxides and hence diols, which are further oxidized to carbonyl compounds and carbonyl degradation products (Figure 12.9) [87].

Lignin Model Dimers Peracetic acid also degrades lignin by cleavage of β-aryl ether linkages, oxidative elimination of side chains and, to a lesser degree, demethylation to catechols and *o*-benzoquinones. Muconic acids may be formed through oxidative cleavage of the aromatic rings.

The β–O–4 guaiacylglycol lignin model compounds XXIII and XXII were somewhat resistant to peracetic acid (Figure 12.10) [85]. After treatment with 3% peracetic acid for 48 hours at 30°C, 28% of XXIII and 75% of XXII were recovered. Degradation proceeded via demethylation of the phenolic ring to give 2-(2methoxy phenoxy)-1-(3,4-dihydroxyphenyl) ethane (9.2%) (XXIV) and cleavage of carbon–carbon bonds in the side chain. The methyl ether XXII degraded more slowly because it was first demethylated to XXIII. The range of simple aromatic products isolated included: acetovanillone (VI, 1.5%), 2-methoxyphenoxy acetic acid (XXV, 5.9%), catechol (4.3%), 2-methoxyhydroquinone (I), 2-methoxy-1,4-benzoquinone (0.4%) and traces of vanillin, guaiacol, and protocatechuic acids. Muconic, malonic, and oxalic acids were also formed.

Degradation of lignin β–O–4-linked model dimers with C_3 side chains—1-(4-hydroxy-3-methoxyphenyl)-2-(2-methoxyphenoxy)propanol (XXVIa), 1-(3,4-dimethoxyphenyl)-2-(2-methoxy-4-methylphenoxy)propanol (XXVIb) and guaiacyl- and veratryl- glycerol-β-guaiacyl ether (IIa and IIb) followed several distinct pathways. These are summarized in Figures 12.10 (glycerol side chain) and 12.11 (methyl group in the γ-position) and tabulated below [75,84,85]:

(i) β-guaiacyl ether cleavage; in one route (path A) forming guaiacol or *p*-creosol and introducing a carbonyl group into the C_3 side chain, in another (path B) forming *o*-quinones from the B-ring. The *o*-quinones are rapidly oxidized to muconic acids.

(ii) Cleavage of the alkyl-aryl carbon–carbon bond to form guaiacoxy aldehydes (path C).

(iii) Benzyl hydroxyl oxidation to an α-carbonyl group (path D).

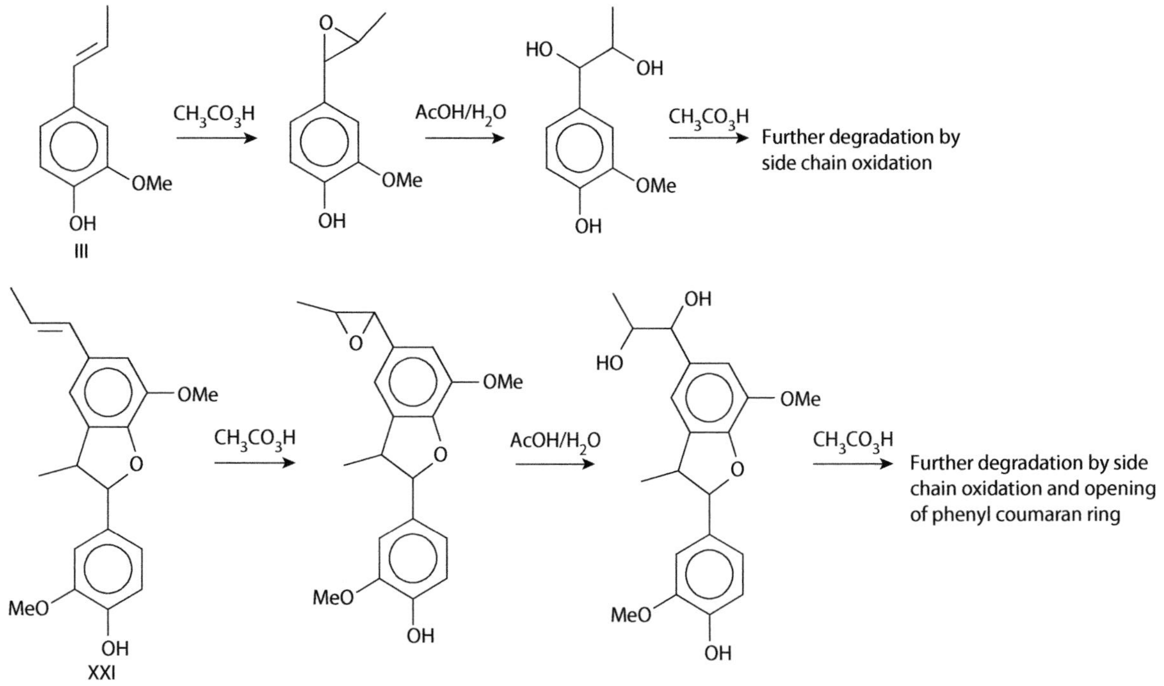

FIGURE 12.9 Reaction of isoeugenol and dehydrodiisoeugenol with peracetic acid. (After Oki, T., Okubo, K., and Ishikawa, H., *Mokuzai Gakkaishi*, 18, 601–610, 1972.)

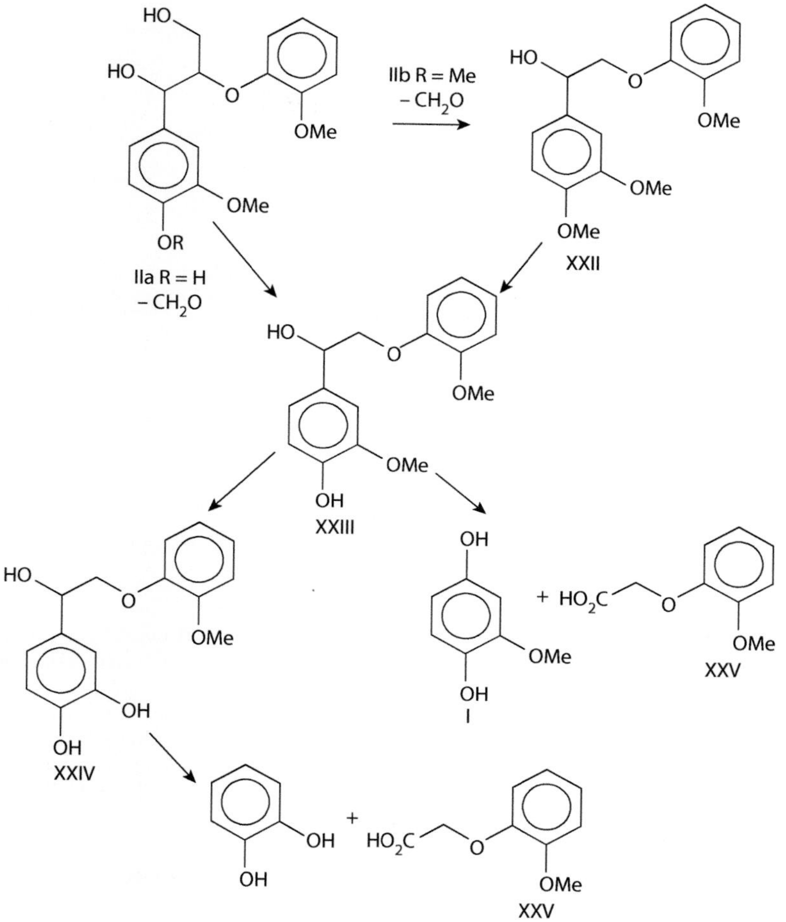

FIGURE 12.10 One pathway of the peracetic acid degradation of guaiacyl- and veratrylglycerol-β-aryl ethers proceeds via the guaiacylglycol-β-aryl ether XXIII. (After Sakai, K., and Kondo, T., *Tappi,* 55, 1702–1706, 1972.)

(iv) Demethylation of ring methoxyl groups.

(v) Aromatic ring hydroxylation.

(vi) In the case of the glycerol derivatives, β-γ carbon–carbon cleavage with loss of the terminal side chain carbon as formaldehyde to form guaiacylglycol-β-guaiacyl ether (XXIII), which is then degraded by demethylation to the corresponding catechol and 2-methoxyphenoxyacetic acid (Figure 12.10).

The phenyl coumaran, dihydro-dehydro-di-*iso*eugenol (XXVIIa) and its methyl ether (XVIIb) reacted via cleavage of the B-ring, forming a dilactone (Figure 12.12) [79].

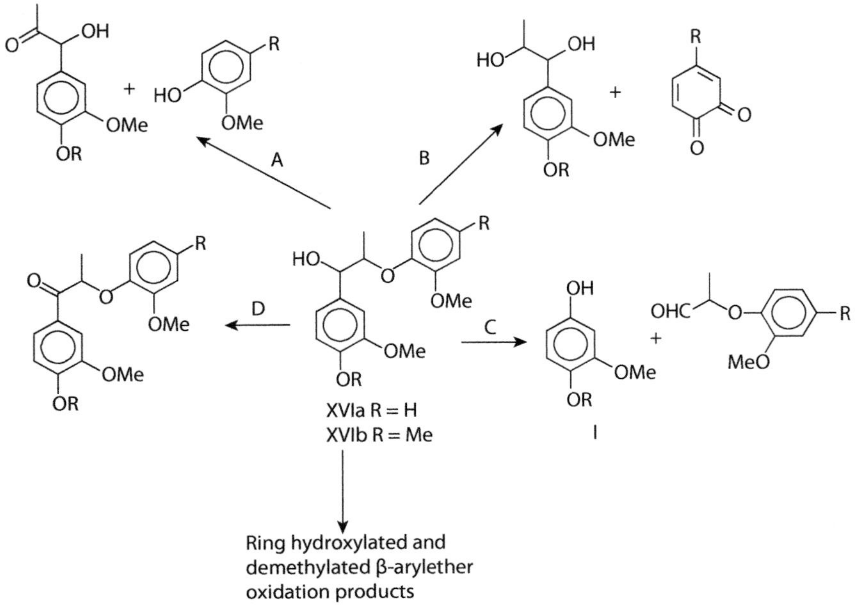

FIGURE 12.11 General reaction pathways for the peracetic acid degradation of β–O–4 lignin model dimers. (After Lawrence, W., McKelvey, R.D., and Johnson, D.C., *Svensk. Papperstidn.*, 83, 11–18, 1980.)

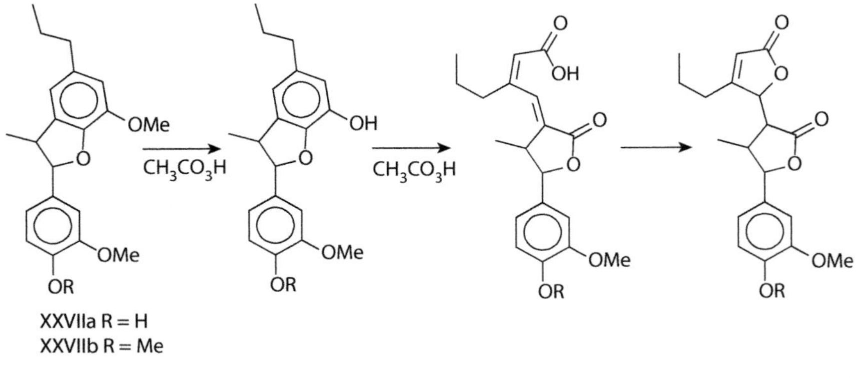

FIGURE 12.12 Oxidation of dehydro-dihydrodiisoeugenol by peracetic acid at pH 3 after Nimz. (After Nimz, H.R. and Schwind, H., *Cellulose Chem. Technol.*, 13, 35–46, 1979.)

Neutral or Alkaline Solutions

p-alkyl and p-*hydroxyalkyl Phenols*   Several groups [78,82,88] have investigated the reaction of lignin model phenols with peracetic acid in the pH range 7–12. In these neutral or alkaline solutions phenols exist as a mixture of phenoxide ions and unionized phenols; peracetic acid may react either as electrophilic peracetic acid (attack by ^+OH) or nucleophilic peracetate (CH_3COO^-).

The reactions of peracetic acid with *p*-alkyl and *p*-hydroxyalkyl phenols can be attributed to electrophilic reaction of unionized peracetic acid. Phenol was unreactive but degradation occurred when methoxyl or other electron donating substituents were added. Creosol, guaiacol, vanillyl alcohol, or apocynol (V) are more readily oxidized at pH near 9, confirming that ionized phenols are more susceptible than unionized phenols to electrophilic attack. An observed decline in the rate of reaction above pH 9 can be attributed to the decreased concentration of peracetic acid due to ionization to peracetate ions.

Creosol was not very reactive. At pH 8, according to Nimz and Schwind [78], over 70% of unreacted creosol was recovered after treatment for one hour at 60°C. However, McDonough and Rapson [88] reported that malonic and methylmaleic acids were found when creosol was treated with peracetic acid at pH 9. They suggested that nucleophilic attack by peracetate anions occurred at the ring carbon, C6. But this mechanism is doubtful since hydroperoxide anion, a stronger nucleophile than peracetate [15], does not react with creosol.

Vanillyl alcohol was oxidized to vanillic acid [82]. Apocynol (V in Figure 12.1) acetovanillone (VI in Figure 12.1) gave 2-methoxy-4-acetoxyphenol, the product of Baeyer–Villiger oxidation of acetovanillone, as the major product. This suggests that the oxidation of apocynol by neutral peracetic acid occurs by initial oxidation to acetovanillone and then follows the same mechanism. The same collection of organic acids were observed in the reactions of both apocynol and acetovanillone (oxalic, malonic, hydroxymalonic maleic, myristic) as well as methoxyhydroquinone and methoxybenzoquinone [82].

Pan et al. [89] have compared the chemistry of peracetate and alkaline hydrogen peroxide bleaching of coniferaldehyde (Xa) and its methylether, 3,4-dimethoxy cinnamaldehyde (Xb) (Figure 12.13). The products formed from coniferaldehyde by the two reagents were similar, α-β bond cleavage to give vanillin, vanillic acid, and the Dakin degradation product methoxy hydroquinone. In both cases, the β-γ cleavage product homovanillic acid (XXXa) was also isolated.

In contrast to coniferaldehyde, 3,4-dimethoxycinnamaldehyde (Xb) gave quite different yields of products with the two bleaching treatments. The alkaline peroxide treatment gave predominately veratraldehyde and veratric acid, products of α-β bond cleavage (route A, Figure 12.13). Some unreacted starting material was recovered. The peracetate treatment was a faster and more quantitative reaction and it gave a high yield of homoveratric acid (XXXb), a product of β-γ cleavage. It was suggested that in both cases the first step was a similar epoxidation of the α-β bond. Peracetic acid, owing to its higher oxidizing power, was able to oxidize the epoxy aldehyde (XXVIII) to the epoxy acid (XXIX), which then either rearranged or decarboxylated, cleaving the β-γ bond (route B, Figure 12.13). This would imply that peracetic acid bleaching, compared to alkaline peroxide bleaching, should give more stable products (saving bleaching reagent) that are less likely to degrade to quinone chromophores.

Dimeric Lignin Model Compounds   The β–O–4 model compound, guaiacylglycerol-β-guaiacyl ether (IIa), was relatively stable to peracetic acid at pH 7 [79]. The saturated phenyl coumarans, dihydro-dehydro-di-*iso*eugenol and its methyl

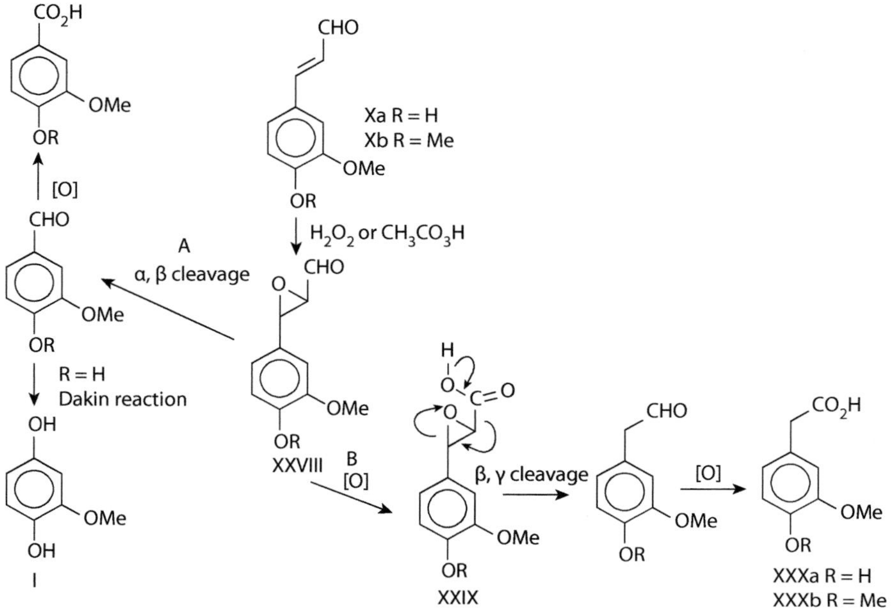

FIGURE 12.13 Reaction of coniferaldehyde and its methyl ether with alkaline hydrogen peroxide or peracetate. (Redrawn from Pan, G.X., Spencer, L. and Leary, G. J., A comparative study on reactions of hydrogen peroxide and peracetic acid with lignin chromophores. Part 1. The reaction of coniferaldehyde model compounds. *Holzforschung*, 54, Berlin: Walter de Gruyter, 2000, pp. 144–152, Scheme 2. With permission.)

ether (XXVIIa, XXVIIb, Figure 12.12) reacted very slowly, most being recovered unchanged after 2 hours at 60°C. These results indicate that the main structure of the lignin polymeric chain should not react with neutral or alkaline peracetic acid.

Unsaturated Lignin Model Compounds   Double bonds in lignin model compounds are attacked by peracetate ions. Dehydro-di-*iso*eugenol (XXI, Figure 12.9) reacted with epoxidation of the aliphatic double bond and formation of the diol. The double bonds in stilbenes [59] and coniferaldehyde [90] are also cleaved. Ferulic acid (IVa) and its ethyl ester reacted slowly at 50°C; the methyl ether, 3,4-dimethoxy cinnamic acid, was much less reactive and was almost quantitatively recovered [55]. The reactions of ferulic acid and its ethyl ester (both in the trans form) were accompanied by trans-cis isomerization, perhaps an indication of reversible phenoxy radical formation. Homovanillic acid (XXXa) was also formed; the proposed mechanism involved epoxidation of the α-β double bonds followed by decarboxylation.

Phenols with p-*carbonyl Substituents*   Acetovanillone (VI) reacted more rapidly than phenols that had no *p*-carbonyl substituent. The maximum rate of reaction occurred near pH 8 [88]. Strumila and Rapson [82] obtained Baeyer–Villiger oxidation products, methoxyhydroquinone and its monoacetate, from neutral peracetic acid treatment of acetovanillone. From vanillin they obtained only vanillic acid, although Nimz and Schwind [78] reported the formation of methoxyhydroquinone and methoxy-*p*-benzoquinone.

β–O–4 model compounds with α-carbonyl groups underwent the normal Baeyer–Villiger oxidations to hydroquinones and quinones [79].

Reactions of Peracetic Acid with Chromophores in Lignin Preparations
Acidic Solutions
Oki et al. [91] and Sakai and Kondo [90] found that peracetic acid degraded and depolymerized dioxane lignins. In hardwood lignins the electron-rich syringyl nuclei were attacked in preference to the guaiacyl units. Sakai and Kondo [90] confirmed that the β–O–4 ether cleavage observed in peracetic acid treatment of model compounds occurred on treatment of milled wood lignins from pine and birch. β–O–4 ether cleavage seemed to occur more readily than oxidative elimination of the MWL side chains.

In Neutral or Alkaline Solutions
Nimz and Schwind [79] used ¹³C NMR to confirm that, as expected from the model compound experiments, peracetic acid at pH 7 is selective in its degradation of milled wood and DHP lignins. Olefinic side chains and α-carbonyl groups were attacked. There was also some degradation of phenyl coumarans and a decrease in the number of phenolic hydroxyl groups. The degradation was sufficient for 20–30% of the MWL to become water soluble after treatment with 100–150% of its weight of peracetic acid. In contrast to treatment with peracetic acid at pH 3, which turned the lignins yellow, the insoluble lignin was bleached.

Jaaskelelainen and Poppius-Levlin [92] found that the molecular weight, methoxyl content and phenolic hydroxyl content of residual lignin in pulps treated with neutral peracetic acid was unchanged, while the carbonyl and carboxylic acid contents increased.

HYDROGEN PEROXIDE ACTIVATION

Hydrogen peroxide or sodium perborate in commercial detergents may be activated to bleach at lower temperatures and pHs by the use of acetylated compounds such as tetra acetyl ethylene diamine (TAED) [93,94]. Under mildly alkaline conditions, TAED reacts rapidly with the anion of hydrogen peroxide forming peroxyacetic acid. Such *in situ* generation has the advantage of avoiding the need to ship and store peracetic acid. Mainly applied to chemical pulp bleaching, TAED-peroxide has been tested with TMP [95] where the lower pHs lead to retention of bulk though lower brightness. Hu et al. [96] also discussed the use of TAED/peroxide for bleaching TMP. They found it oxidized acetovanillone two orders of magnitude faster than traditional alkaline peroxide.

Peracetic acid is also formed during conventional alkaline peroxide bleaching by the reaction of HOO⁻ with the acetyl groups in the hemicelluloses. Kang and Ni [97] found that peracetic acid was formed *in situ* in this way, although it did not give a higher brightness and led to a lower level of residual peroxide.

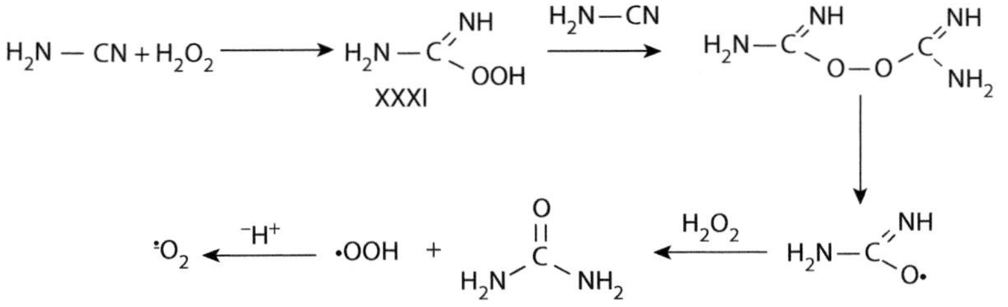

FIGURE 12.14 Generation of superoxide anion from cyanimide and alkaline peroxide. (Kadla, J.F., Chang, H-M., and Gratzl, J.S., *Holzforschung,* 52, 506–512, 1998.)

Alkaline hydrogen peroxide may be activated more aggressively by adding cyanamide or dicyandiamide. These systems have enhanced reactivities toward lignin. Most proposed applications of the cyanamide-peroxide system have been to chemical pulp bleaching, but it is interesting to compare cyanamide-peroxide with a conventional alkaline peroxide lignin-retaining treatment. The cyanamide-peroxide has the advantage of being a fast reaction operating at lower pHs (~9.5) and temperatures (60°C). It is thought to operate through electrophilic attack, compared to the nucleophilic mechanism for alkaline peroxide.

Sturm and Kuchler [98] proposed that in cyanamide-peroxide bleaching a peroxyimidic acid intermediate (XXXI) was formed (Figure 12.14). Kadla et al. [99,100] have shown that this intermediate probably decomposes to form superoxide, which is the active delignification agent [100]. In model compound studies, these authors found that the most significant reaction is with free phenolic OH groups. Evidence was obtained from EPR and product analysis of a predominately free radical mechanism. The model compounds creosol, apocynol (V in Figure 12.1) and acetovanillone (VI in Figure 12.1) reacted via aromatic hydroxylation, demethoxylation, radical coupling, and oxidative ring opening.

Reaction of cyanamide-peroxide with a soluble kraft lignin was more rapid than the corresponding alkaline peroxide oxidation [101]. When cyanamide-peroxide was used in eight-molar excess at 60°C, after one hour only 30% of the kraft lignin was recovered by acid precipitation. This compared with an 80% recovery for the alkaline peroxide treatment. The peroxide in the cyanamide-peroxide system was totally consumed, whereas more than 70% residual peroxide was obtained in the alkaline peroxide experiments. Most of the peroxide in the cyanamide-peroxide system was consumed within 10 minutes.

The recovered cyanamide-peroxide lignin had greatly increased carboxylic acid content and had undergone more demethoxylation than the lignin obtained from the traditional alkaline peroxide treatment. Both the cyanamide-peroxide and the alkaline-peroxide preparations had lost about 30% of their initial phenolic hydroxyl groups. Vanillic acid was the compound isolated in the highest yield but 4-hydroxy 5-methoxyphthalic acid (XXXII), obtained from β-6 linked lignin units, and substantial amounts of 4-hydroxy 5-methoxy isophthalic acid (XXXIII) were also formed (Figure 12.15). 4-Hydroxy 5-methoxy isophthalic acid would be derived from phenyl coumaran and 5-5' biphenyl units in the lignin.

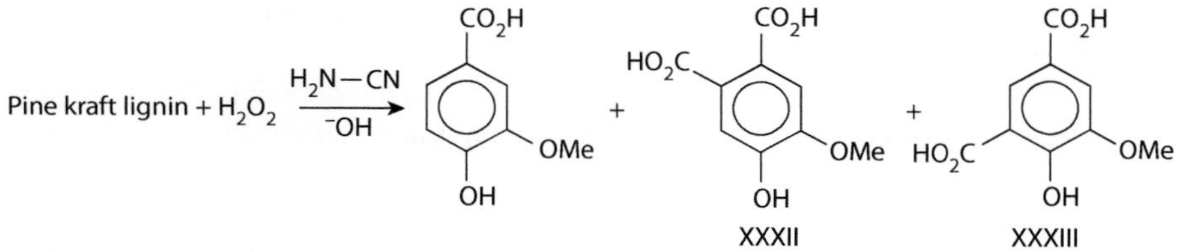

FIGURE 12.15 Major products isolated in the reaction of pine kraft lignin with cyanamide-activated peroxide. (After Kadla, J.F., Chang, H-M., Chen, C-L., and Gratzl, J.S., *Holzforschung*, 52, 513–519, 1998.)

AKALINE PEROXIDE COMBINED WITH OTHER REAGENTS: OZONE, PERSULFURIC ACID, AND DIMETHYLDIOXIRANE (DMD)

Oxygen + Peroxide

The use of oxygen and alkaline peroxide in the same bleaching stage to bleach mechanical pulps was patented by Lachenal et al. [46]. But their claim of an improved brightness is apparently contrary to work by Ni et al. [102], who obtained a higher brightness for maple bleached chemithermomechanical pulp (BCTMP) when they excluded oxygen by bleaching in a pressurized nitrogen atmosphere.

Excluding all traces of oxygen gas from an alkaline peroxide bleaching stage is very difficult to achieve, and oxygen is easily generated by metal-catalyzed decomposition of peroxide. Ni and coworkers regularly vented their pressurized bleaching reactor to release the oxygen formed from the peroxide. Thus, one might consider whether the enhanced brightness was a result of improved penetration of bleaching liquors into the pulp brought about by repeated pressure cycling.

There are plausible mechanisms to explain both positive and negative impacts of oxygen. A positive effect by oxygen could result from oxidation of hydroquinones to quinones that the peroxide could then cleave, or from the removal of stilbenes, perhaps via superoxide chemistry [63]. A negative oxygen effect could fit with suppression of alkali darkening, which involves oxygen, though He et al. [54] found that the chromophores formed in alkali darkening were fully bleachable by hydrogen peroxide.

Ozone and Ozone + Peroxide

Ozone was added to the eye of a refiner to enhance the TMP quality and to reduce energy consumption by facilitating fiber separation [103]. The ozone mainly oxidized the wood extractives but, as well as saving energy, a 1–2% charge of ozone on the TMP rejects resulted in a higher pulp strength and improved subsequent peroxide bleachability.

Three decades earlier Liebergott [104] and Soteland [105] reported separately that ozone, known on its own to yellow mechanical pulps, would improve groundwood pulp brightness and brightness stability if combined with peroxide. Both workers claimed that the brightness obtained was greater than the brightness reached without ozone pretreatment. The doses of ozone and peroxide used by Liebergott were quite small (1% and 0.6% on pulp, respectively) and pulp yields were over 99%. Liebergott's treatment also enhanced the pulp strength; increases in breaking length of 50–100% were obtained [104].

The alkaline peroxide and the ozone were applied to the pulp at high consistency in a rapid sequence that amounted to a single stage of treatment. The mechanism has never been elaborated, but a presumed effect is that the ozone rapidly oxidized the pulp with the formation of carboxylic (muconic) acids that improved the pulp strength. The yellow color—perhaps quinones—produced by the ozone treatment was removed by the peroxide. A similar rationale has been proposed for adding hydrogen peroxide to an ozone bleaching stage for chemical pulp with, in that case, the possibility that hexeneuronic acids were also removed by the ozone [106].

Alternatively, there is a very wide range of reactions that can occur between peroxide and ozone, encompassing a great variety of reactive radical species (see, for example Lesko et al. [107].

A more recent investigation into the use of ozone, at different pHs, as a pre-treatment to peroxide bleaching of mechanical pulps concluded that ozone was not an effective agent. Although the pulp's mechanical properties were improved by the ozone treatment, a 6% ozone treatment was no better than a 1% peroxide treatment [108].

Ozone enhanced the bleachability of an alkaline peroxide wheat straw mechanical pulp but caused significant losses in yield, particularly when it was followed by a peroxide bleaching stage [109].

Persulfuric Acid and Peroxide

Peroxymonosulfuric acid (PMS) is a stronger oxidizing agent than hydrogen peroxide or peracetic acid [110] and can act as either an electrophilic or a nucleophilic reagent [111]. It has generally been studied as an alternative to hydrogen peroxide or peracetic acid in chemical pulp bleaching.

Pan et al. [112] showed that PMS (as the buffered oxone salt) was active in destroying ferulic acid (IV), which is resistant to peroxide and reacts only slowly with peracetic acid. They then used PMS alone and in two-stage sequences (PMS/P and P/PMS) with alkaline peroxide (P), to brighten wheat straw mechanical pulp. Wheat straw pulp contains a significant number of ferulic acid (IVa in Figure 12.1) and diferulic acid (IVb, in Figure 12.1) structures. With charges equivalent to 2% of active oxygen on pulp, both PMS and PMS/P achieved brightness gains of 7 points. When the charges were increased to 6% active oxygen, further brightness gains of 7.5, 8 and 9.2 points were made with P, P/PMS and PMS/P sequences. These trends roughly followed the loss of ferulic acid units from the straw. It was concluded that PMS could degrade some of the ferulic acid chromophores and/or make others more susceptible to second-stage peroxide bleaching.

Dimethyl Dioxirane (DMD) and DMD + Peroxide

Dioxiranes, of which DMD is the simplest member, are powerful electrophilic oxidizing agents, able to transfer oxygen to aromatic rings and double bonds. DMD was identified by Ragauskas [113] and Lee and coworkers [114] as a very effective bleaching agent for chemical pulps. It can be added directly to pulps or prepared *in situ* from acetone and PMS.

On its own, DMD is too powerful an electrophilic reagent to be considered for lignin-retaining bleaching. But, in careful conjunction with alkaline peroxide, it has been found useful in bleaching the ferulic acid-linked lignins in wheat straw mechanical pulps. Pan et al. [112] evaluated DMD for its ability to oxidize ferulic acid (IVa), ferulic acid ethyl ester and diferulic acid (IVb). All these model compounds were completely oxidized at room temperature.

When the wheat straw mechanical pulps were treated with DMD alone, the brightness gain was similar to that achieved by PMS. A two stage DMD/P or P/DMD treatment with peroxide (P) showed enhancement over treatment PMS/P or with P alone [112].

CONCLUSION

In the 1990s, the Canadian government mounted a significant research effort to improve the properties of mechanical pulps and thereby expand their markets [115], for which one of the authors (Gordon Leary) served as executive director from 1994 to 1996. The ambitious goal of the network was to achieve "90, 90 and 9"—mechanical pulp produced in over 90% yield, with an ISO brightness exceeding 90% and with a breaking length of more than 9 km. Of course, only an economic amount of bleaching agent should be used.

Targets of 90, 90 and 9 will require improvements in the entire chain of processes leading from wood chips to paper: wood handling, pulping, as well as bleaching and wet-end practices. But perhaps oxidative bleaching can contribute most. It can, in principle, do the following:

- Maintain a high yield by retaining the lignin and hemicelluloses
- Give a high brightness by eliminating the remaining chromophores
- Increase strength as a result of carboxylic acid formation

Optical brightening agents (OBAs) are commonly used to increase brightness of chemical pulps by as much as 10 points, as well as increasing whiteness. Conventionally, the use of mechanical pulps has been discouraged in furnishes where an OBA is used, since it is known that the lignin will decrease the effectiveness of the OBA. Nonetheless, mechanical pulps at the upper end of their brightness range (80% for softwoods, 85% for hardwoods) do respond to OBAs, and suppliers have been recently been promoting this application. This may represent one way to achieve 90% brightness in a wood-containing paper [116], if not in the pulp exiting a bleaching tower.

Higher strength could be achieved through a first stage treatment that enhances interfiber bonding without producing chromophores resistant to a second bleaching stage. Ozonolysis is an example.

Progress has been made toward the 90, 90, 9 objective, and each of the individual targets can probably be achieved. However, we are still well short of the combined achievement. A high yield militates against a high brightness, and stronger pulps are usually not as bright, or have lower yields. Thus, 90, 90, and 9 remains a goal for future research.

REFERENCES

1. C Chirat and D Lachenal. Effect of hydroxyl radicals on cellulose and pulp and their occurrence during ozone bleaching. *Holzforschung* 51:147–154, 1997.
2. D Lachenal. Hydrogen peroxide as a delignifying agent. In: CW Dence and DW Reeve, eds. *Pulp Bleaching: Principles and Practice*. Atlanta: Tappi Press, 1996, pp. 347–362.
3. J Gierer. Formation and involvement of superoxide and hydroxyl radicals in TCF bleaching processes: A review. *Holzforschung* 51:34–46, 1997.
4. HS Isbell, E Parks and RG Naves. Reactions of carbohydrates with hydroperoxides VII. Degradation of reducing sugars and related compounds by alkaline hydrogen peroxide in the presence and absence of iron and magnesium salts. *Carbohydr. Res.* 45:197–204, 1975.
5. JL Roberts and DT Sawyer. Base-induced generation of superoxide ion and hydroxyl radical from hydrogen peroxide. *J. Am. Chem. Soc.* 100:329–330, 1978.
6. O Legrini, E Oliveros, and AN Braun. Photochemical processes for water treatment. *Chem. Rev.* 93:671–698, 1993.
7. MG Evans and N Uri. The dissociation constant of hydrogen peroxide and the electron affinty of the HO_2 radical. *Trans. Faraday Soc.* 45:224–230, 1949.
8. KY Salnis, KP Mishchenko, and EI Flis. Thermodynamics for the dissociation fo hydrogen peroxide in aqueous solutions. *Zhur. Neorg. Khim.* 2:1985–1989, 1957.
9. A Teder. The equilibrium between hydrogen sulfite and sulfite ions. *Svensk. Papperstidn.* 17:704–705, 1975.
10. LG Sillen and AE Martell. Stability constants of metal ion complexes. Specialist Periodical Reports 17, London: Royal Society of Chemistry, 1964.
11. A Teder and D Tormund. The equilibrium between hydrogen peroxide and the peroxide ion: A matter of importance in peroxide bleaching. *Svensk. Papperstidn.* 83:106–109, 1980.
12. J Hook, S Wallin, and G Akerlund. Optimization and control of two-stage peroxide bleaching. Tappi Pulping Conference, Seattle, WA, 1989, pp. 267–275.
13. J Kappel and J Sbaschnigg. Bleaching of groundwood pulp at temperatures up to 95°C. *Pulp Paper Can.* 92:T229–T234, 1991.
14. E Koubek, ML Haggett, CJ Battalgia, MI-r Kairat, HY Pyun, and JO Edwards. Kinetics and mechanism of the spontaneous decomposition of some peroxoacids, hydrogen peroxide and t-butyl hydroperoxide. *J. Am. Chem. Soc.* 85:2263–2268, 1963.
15. WP Jencks and J Carriuolo. Reactivity of nucleophilic reagents towards esters. *J. Am. Chem. Soc.* 82:1778–1786, 1960.
16. JE McIsaac, LR Subbaraman, HA Mulahusen, and EJ Behrman. The nucleophilic reactivity of peroxy anions. *J. Org. Chem.* 37:1037–1041, 1972.
17. CW Dence. Chemistry of mechanical pulp bleaching. In: CW Dence and DW Reeve, eds. *Pulp Bleaching: Principles and Practise*. Atlanta: Tappi Press, 1996, pp. 160–181.
18. J Gierer. The chemistry of delignification. A general concept. Part II. *Holzforschung* 3:55–64, 1982.
19. J Gierer. Chemistry of delignification. Part 2. Reactions of lignins during bleaching. *Wood Sci. Technol.* 20:1–33, 1986.
20. MB Hocking, M Ko, and TA Smith. Detection of intermediates and isolation of hydroquinone monoacetate in the Dakin oxidation of *p*-hydroxyacetophenone. *Can. J. Chem.* 56:2646–2649, 1978.
21. G Gellerstedt, H-L Hardell, and E-L Lindfors. The reactions of lignin with alkaline hydrogen peroxide. Part IV. Products from he oxidation of quinone model compounds. *Acta. Chem. Scand.* B 34:669–673, 1980.
22. S Katz, N Liebergott, and AM Scallan. A mechanism for the alkali strengthening of mechanical pulp. *Tappi* 64:97–100, 1981.

23. TM Garver, KJ Maa, EC Xu, and DG Holah. Size-exclusion chromatography of the soluble lignin products from hydrogen peroxide brightening of mechanical pulps. *Res. Chem. Intermed.* 21:503–519, 1995.

24. M Holocher-Ertl and K Kratzl. Model experiments on oxygen bleaching of pulp. IV. On the oxidative degradation of lignin: Degradation of Creosol (2-methoxy-4-methylphenol) with oxygen, hydrogen peroxide and ozone. *Holzforschung* 36:11–16, 1982.

25. AW Kempf and CW Dence. The reactions of hardwood model compounds with alkaline hydrogen peroxide. *Tappi* 58:104–108, 1975.

26. CW Bailey and CW Dence. Reactions of alkaline hydrogen peroxide with softwood lignin model compounds, spruce milled-groundwood lignin and spruce groundwood. *Tappi* 52:491–500, 1969.

27. IV Sen'ko, TS Anikeenko, AD Alekseev, and VM Reznikov. Mechanism of hydrogen peroxide oxidation of wood and its components. (7) Oxidation of lignin model compounds with hydrogen peroxide in an alkaline solution. *Khim. Drev.* 30–34, 1981.

28. S Omori and CW Dence. The reactions of alkaline hydrogen peroxide with lignin model dimers. Part 2. Guaiacylglycerol-β-guaiacyl ether. *Wood Sci. Technol.* 15:113–123, 1981.

29. K Tatsumi and N Terashima. Oxidative degradation of lignin. (7) Cleavage of β-O-4 linkage of guaiaclglycerol-β-ether by hydroxyl radicals [from hydrogen peroxide]. *Mokuzai Gakkaishi* 31:316–317, 1985.

30. R Agnemo and G Gellerstedt. The reactions of lignin with alkaline hydrogen peroxide. Part II. Factors influencing the decomposition of phenolic structures. *Acta. Chem. Scand.* B 33:337–342, 1979.

31. J Gierer and F Imsgard. The reactions of lignin with oxygen and hydrogen peroxide in alkaline media. *Svensk. Papperstidn.* 80:510–518, 1977.

32. Y Kojima and T Kayama. Reactions of a lignin model compound containing a ring-conjugated double bond with hydrogen peroxide. *Res. Bull. Coll. Expt. Forest Hokkaido University* 40:783–794, 1983.

33. S Hosoya, T Kondo and J Nakano. Reactivity of lignin towards alkaline hydrogen peroxide. *Mokuzai Gakkaishi* 25:777–782, 1979.

34. K Tatsumi and N Terashima. Oxidative degradation of lignin. (4) Reactivities of monomeric lignin model compounds with hydrogen peroxide. *Mokuzai Gakkaishi* 27:873–878, 1981.

35. Y-P Sun, B Yates, J Abbot, and C-L Chen. Oxidation of lignin model compounds containing an α-carbonyl group and a ring-conjugated double bond by hydrogen peroxide-uv photolysis. *Holzforschung* 50:226–232, 1996.

36. Y-P Sun, B Yates, J Abbot, and C-L Chen. ESR study of lignin model compounds irradiated by uv (254 nm photons) in the presence of hydrogen peroxide. *Holzforschung* 50:233–236, 1996.

37. G Gellerstedt and R Agnemo. The reactions of lignin with alkaline hydrogen peroxide. Part III. The oxidation of conjugated carbonyl structures. *Acta. Chem. Scand.* B 34:275–280, 1980.

38. S Omori and CW Dence. The reactions of alkaline hydrogen peroxide with lignin model dimers. Part 1. Phenacyl α-aryl ethers. *Wood Sci. Technol.* 15:67–79, 1981.

39. S Hoyosa and J Nakano. Reaction of α-carbonyl group in lignin during alkaline hydrogen peroxide bleaching. *Mokuzai Gakkaishi* 26:97–101, 1980.

40. JF Kadla, H-M Chang, and H Jameel. The reactions of lignins with hydrogen peroxide at high temperature. Part 1. The oxidation of lignin model compounds. *Holzforschung* 51:428–434, 1997.

41. L Heuts and G Gellerstedt. Oxidation of guiacylglycerol-β-guaiacyl ether with alkaline hydrogen peroxide in the presence of kraft pulp. *Nord. Pulp Paper Res. J.* 13:107–111, 1998.

42. AJ Nonni, LE Falk, and CW Dence. Methoxy-*p*-benzoquinone and methoxyhydroquinone as models for chromophore changes in the bleaching of softwood mechancial pulps. I. The detection of methoxyhydroquinone in the spent liquor from the peroxide bleaching of thermomechanical pulp. *J. Wood. Chem. Technol.* 2:223–229, 1982.

43. DG Holah and C Heitner. Alkaline peroxide bleaching of mechanical and ultra high-yield pulps. *J. Pulp Paper Sci.* 18:J161–J165, 1992.

44. T Oki, H Ishikawa, and K Okubo. On the brightening of lignin by neutral hydrogen peroxide. *Japan Tappi* 32:368–376, 1978.

45. H Hirashima and M Sumimoto. Basic chromophore and leucochromophore in mechanical pulps: Possible repetition of photochemical reduction-oxidation-reduction. *Tappi* 77:146–154, 1994.

46. D Lachenal, C Bourne, and C de Choudens. Process for bleaching a mechanical pulp with hydrogen peroxide. US Patent 4756798, 1987.

47. DY Lee and M Sumimoto. Mechanochemistry of lignin V. An intensive chromophore in mechanical pulps bleached with alkaline hydrogen peroxide. *Holzforschung* 45 (suppl):15–20, 1991.

48. JH Zhu, C Archer, F MacNab, G Andrews, G Kubes, and DG Gray. The influence of mechanochemical action on the photosensitivity of refiner pulps. *J. Pulp Paper Sci.* 23:J305–J310, 1997.

49. S Hosoya, K Seike, and J Nakano. Bleaching of high yield pulp: Reactions of quinones and quinone polymers with some reducing agents. *Mokuzai Gakkaishi* 22:314–319, 1976.

50. H Erdtman and M Granath. Studies on humic acids. V. The reaction of *p*-benzoquinone with alkali. *Acta. Chem. Scand.* 13:811–816, 1954.

51. I Forsskahl, J Gustafsson, and A Nybergh. The photochemical discolouration of methoxy-*p*-benzoquinone in solution. *Acta. Chem. Scand.* B 35:389–394, 1981.

52. RW Pero and CW Dence. Methoxy-*p*-benzoquinone and methoxyhydroquinone as models for chromophore changes in the bleaching of softwood mechanical pulps. II. The effect of peroxide charge, reaction pH, and transition metal ions. *J. Wood. Chem. Technol.* 3:195–222, 1983.

53. G Leary and D Giampaolo. The darkening reactions of TMP and BTMP during alkaline peroxide bleaching. *J. Pulp Paper Sci.* 25:141–147, 1999.

54. Z He, Y Ni, and E Zhang. Alkaline darkening and its relationship to peroxide bleaching of mechanical pulp. *J. Wood. Chem. Technol.* 24:1–13, 2004.

55. GX Pan, L Spencer, and GJ Leary. Reactivity of ferulic acid and its derivatives toward hydrogen peroxide and peracetic acid. *J. Agric. Food Chem.* 47:3325–3331, 1999.

56. AJ Nonni and CW Dence. The reactions of alkaline hydrogen peroxide with lignin model dimers. Part 3. 1,2-Diaryl-1,3-propanediols. *Holzforschung* 42:37–46, 1988.

57. G Gellerstedt and R Agnemo. The reactions of lignin with alkaline hydrogen peroxide. Part V. The formation of stilbenes. *Acta. Chem. Scand.* B 34:461–480, 1980.

58. G Gellerstedt and L Zhang. Formation and Reactions of Leucochromophoric structures in high yield pulping. *J. Wood. Chem. Technol.* 12:387–412, 1992.

59. GX Pan, L Spencer, and GJ Leary. A comparative study on reactions of hydrogen peroxide and peracetic acid with lignin chromophores. Part 2. The reaction of stilbene-type model compounds. *Holzforschung* 54:153–158, 2000.

60. AJ Nonni and CW Dence. The reactions of alkaline hydrogen peroxide with lignin model dimers. Part 4. Substituted stilbenes and undissociated hydrogen peroxide. *Holzforschung* 42:99–104, 1988.

61. S Ljungren and E Johansson. The kinetics of lignin reactions during oxygen bleaching. Part 2. The reactivity of 4,4'-dihydroxy-3,3'-dimethoxystilbene and β-aryl structures. *Nord. Pulp Paper Res. J.* 5:148–154, 1990.

62. DF Wong, G Leary, and G Arct. The role of stilbenes in bleaching and colour stability of mechanical pulps. I. The reaction of lignin model stilbenes with alkali and oxygen. *Res. Chem. Intermed.* 21:329–342, 1995.

63. J Gierer, E Yang, and T Reitberger. The reactions of chromophores of the stilbene type with the hydroxy radical and the superoxide radical. *Holzforschung* 50:342–253, 1996.

64. PK Smith and TJ McDonough. Transition metal ion catalysis of the reaction of a residual lignin-related compound with alkaline hydrogen peroxide. *Svensk. Papperstidn.* 88:R106–R112, 1985.

65. T Kishimoto, JF Kadla, H-M Chang, and H Jameel. The reactions of lignin model compounds with hydrogen peroxide at low pH. *Holzforschung* 57:52–58, 2003.

66. M Lukavoca, J Klanduch, and S Kovac. Oxidative degradation of sulphate lignin using hydrogen peroxide. *Holzforschung* 31:13–18, 1977.

67. T Kano, S Hosoya, and J Nakano. On the brightening of lignin in cold soda pulp during peroxide bleaching. *Mokuzai Gakkaishi* 19:331–334, 1973.

68. TD Spittler and CW Dence. Destruction and creation of chromophores in softwood lignin by alkaline peroxide. *Svensk. Papperstidn.* 80:275–284, 1977.

69. LV Kanitskaya, AN Zakazov, OA Rossinskii, AV Rokhin, and VA Babkin. Structural changes in spruce [picea] dioxane lignin under the action of hydrogen peroxide and sodium borohydride. *Izv. VUZ. Lesn. Zh.* 147–154, 1993.

70. X Pan, D Lachenal, C Lapierre, B Monties, V Neirinck, and D Robert. On the behaviour of spruce [Picea Abies] thermomechanical pulp lignin during hydrogen peroxide bleaching. *Holzforschung* 48:429–435, 1994.

71. Y Sun and DS Argyropoulos. A comparison of the reactivity and efficiency of ozone, chlorine dioxide, dimethyldioxirane and hydrogen peroxide with residual kraft lignin. *Holzforschung* 50:429–435, 1996.

72. CA Barnett and WG Glasser. Bleaching of hydroxypropyl lignin with hydrogen peroxide. In: WG Glasser and S Sarkanen, eds. *Lignin: Properties and Materials.* 347, Washington, DC: American Chemical Society, 1988, pp. 436–451.

73. GX Pan, JL Bolton, and GJ Leary. Determination of ferulic and *p*-coumaric acids in wheat straw and the amounts released by mild acid and alkaline peroxide treatment. *J. Agric. Food Chem.* 46:5283–5288, 1998.

74. Y-Z Lai and KV Sarkanen. Delignification by peracetic acid II. Comparative study on softwood and hardwood lignins. *Tappi* 51:449–453, 1968.

75. W Lawrence, RD McKelvey, and DC Johnson. The peroxyacetic acid oxidation of a lignin-related β-aryl ether. *Svensk. Papperstidn.* 83:11–18, 1980.

76. ROC Norman and JRL Smith. Mechanisms of aromatic hydroxylation and ring-opening reactions. In: TE King, HS Mason, and M Morrison, eds. *Oxidases and Related Redox Systems, 1.* New York: Wiley, 1965, pp. 131–155.

77. JC Farrand and DC Johnson. Peroxy acetic acid oxidation of 4-methylphenols and their methyl ethers. *J. Org. Chem.* 36:3606–3612, 1971.

78. HR Nimz and H Schwind. Oxidation of monomeric lignin model compounds with peracetic acid. *Cellulose Chem. Technol.* 13:35–46, 1979.

79. HR Nimz and H Schwind. Oxidation of lignin and lignin model compounds with peracetic acid. International Symposium on Wood and Pulping Chemistry, Stockholm, 1981, 2, pp. 105–112

80. TJ McDonough. Peracetic acid decomposition and oxidation of lignin model compounds in alkaline solutions. PhD thesis, University of Toronto, 1972.

81. DC Johnson. Lignin reactions in delignification with peroxyacetic acid. In: JS Gratzl, J Nakano, and RP Singh, eds. *Chemistry of Delignification with Oxygen, Ozone and Peroxides.* Tokyo: Uni Publishers, 1980, pp. 217–228.

82. G Strumila and H Rapson. Reaction products of neutral peracetic acid oxidation of model lignin phenols. *Pulp Paper Can.* 76:T276–T280, 1975.

83. K Sakai and T Kondo. Delignification in peracid bleaching. VII. Kinetic aspects on reactivities of guaiacyl and syringyl units in lignin. *Mokuzai Gakkaishi* 21:39–42, 1975.

84. K Sakai and T Kondo. Delignification in peracid bleaching VI. Reaction of the β-aryl ether type model compounds with peracetic acid. *Tappi* 55:1702–1706, 1972.

85. T Oki, K Okubo, and H Ishikawa. The oxidative degradation of guaiacylglycerol-β-guaiacyl ether by peracetic acid. *Mokuzai Gakkaishi* 20:549–557, 1974.

86. T Oki, K Okubo and H Ishikawa. The oxidative degradation by peracetic acid of 2-(2-methoxyphenoxy)-1-(4-hydroxy-3-methoxyphenyl)ethanol and 2-(2-methoxyphenoxy)-1-(3,4-dimethoxyphenyl) ethanol. *Mokuzai Gakkaishi* 18:601–610, 1972.

87. T Oki, K Okubo, and H Ishikawa. The peracetic acid oxidation of isoeugenol and dehydrodiisoeugenol. *Mokuzai Gakkaishi* 20:89–97, 1974.

88. TJ McDonough and H Rapson. Reactions of peracetic acid with lignin model compounds in alkaline solution. *Trans. Tech. Sec. CPPA* 1:12–17, 1975.

89. GX Pan, L Spencer, and GJ Leary. A comparative study on reactions of hydrogen peroxide and peracetic acid with lignin chromophores. Part 1. The reaction of coniferaldehyde model compounds. *Holzforschung* 54:144–152, 2000.

90. K Sakai and T Kondo. Delignification in peracid bleaching. VIII. The reaction of the β–aryl ether structure in milled wood lignin. *Mokuzai Gakkaishi* 21:87–92, 1975.

91. T Oki, H Ishikawa and K Okubo. Characteristics of oxidative degradation of lignin by peroxide and oxygen-alkali methods. *Mokuzai Gakkaishi* 24:406–414, 1978.

92. A-S Jaaskelainen and K Poppius-Levlin. Chemical changes in residual lignin structure during peroxyacetic acid delignification of softwood kraft pulp. *Nord. Pulp Paper Res. J.* 14:116–122, 1999.

93. HG Hauthal, H Schmidt, HJ Scholz, J Hofmann, and W Pritzow. Studies concerning the mechanism of bleaching activation. *Tenside Surfact. Det.* 27:187–193, 1990.

94. NA Turner and AJ Matthews. Enhanced delignification and bleaching using tetraacetylethylenediamine (TAED)-activated peroxide. Tappi Pulping Conference, Montreal, 1998, 3, pp. 1269–1276.

95. S Hsieh, C Agarwal, RW Maurer, and J Mathews. The effectiveness of TAED on the peroxide bleaching of mechanical, chemical and recycled pulp. *Tappi* 5:27–31, 2006.

96. QY Hu, C Daneault, and S Robert. Use of a nitrogen-centered peroxide activator to increase the chromophore removal potential of peroxide: Lignin model compound study. *J. Wood. Chem. Technol.* 26:165–174, 2006.

97. G Kang and Y Ni. Formation of peracetic acid during peroxide bleaching of mechanical pulps. *Appita* 60:70–73, 2007.

98. WGJ Sturm and JG Kuchler. The nitrilamine reinforced hydrogen peroxide bleaching of kraft pulps. Non-Chlorine Bleaching Conference, Hilton Head, SC, 1993, pp. 31–41

99. JF Kadla, H-M Chang, and JS Gratzl. Reactions of lignin with cyanamide activated hydrogen peroxide. Part 1. The degradation of lignin model compounds. *Holzforschung* 52:506–512, 1998.

100. JF Kadla, H-M Chang, C-L Chen, and JS Gratzl. Reactions of lignin with cyanamide activated hydrogen peroxide. Part 2. The degradation mechanism of phenolic lignin model compounds. *Holzforschung* 52:513–519, 1998.

101. JF Kadla and H-M Chang. Reactions of lignin with cyanamide activated activated hydrogen peroxide. Part 3. The degradation of pine kraft lignin. *Holzforschung* 56:76–84, 2002.

102. Y Ni, Q Jiang and Z Li. Bleaching of Maple CTMP pulp to high brightness using the P_{N2} process. *Appita* 53:404–406, 2000.

103. M Petit-Conil. Use of ozone in mechanical pulping processes. *Revue ATIP* 57:17–26, 2003.

104. N Liebergott. Paprizone process for brightening and strengthening groundwood. *Paper Trade Journal* 155:28–29, 1971.

105. N Soteland. Effect of ozone on some properties of groundwoods of four species, I. *Norsk. Skogindustri* 25:61–66, 1971.

106. L Cesar, U Suess, R da Silva, L Peixoto, and M Antonio. Addition of hydrogen peroxide in the ozonation stage to improve pulp bleachability. *Papel* 65:45–49, 2002.

107. TM Lesko, AJ Colussi, and MR Hoffman. Hydrogen isotope effects and mechanism of aqueous ozone and peroxone decompositions. *J. Am. Chem. Soc.* 126:2004.

108. S Robert, JP Lamothe, and C Daneault. Treatment of thermomechanical pulp from tamarack (Larix laricinia) by ozone and ozone-peroxide bleaching process: Paper mechanical properties. *J. Pulp Paper Sci.* 17:J68–J72, 1991.

109. GX Pan and GJ Leary. The bleachability of wheat straw alkaline peroxide mechanical pulp. *Cellulose Chem. Technol.* 34:537–547, 2001.

110. CJ Biermann and JD Kronis. Bleaching chemistry: Oxidation potentials of bleaching agents. *Prog. Paper Recycling* 6:65–70, 1997.

111. T Delagoutte, D Lachenal, and H Ledon. Delignification and bleaching with peracids. Part 1: Comparison with hydrogen peroxide. *Pap. ja Puu* 81:506–510, 1999.

112. GX Pan, CI Thomson, and GJ Leary. The role of ferulic acid removal during bleaching of wheat straw mechanical pulp with peroxygen reagents. *Holzforschung* 57:282–283, 2003.

113. AJ Ragauskas. Investigation of dimethyldioxirane as a bleaching agent for kraft pulp. *Tappi* 76:87–90, 1993.

114. C-L Lee, K Hunt, and RW McMurray. Dimethyldioxirane as a non-chlorine agent for chemical pulp bleaching. *Tappi* 76:137–140, 1993.

115. HI Bolker. Focussing scientific excellence on mechanical pulping: The mechanical and chemimechanical pulps network. *Pulp Paper Can.* 94:T133–T138, 1993.

116. M Scheringer, K Halder, and K Hungerbuhler. Comparing the environmental performance of fluorescent whitening agents with peroxide bleaching of mechanical pulp. *J. Ind. Ecol.* 3:77–95, 2000.

13 The Chemistry of Lignin-Retaining Bleaching: Reductive Bleaching Agents

Sylvain Robert

CONTENTS

Mechanical pulps may be bleached either by oxidative or reductive processes that remove colored chromophores [1,2]. In water, the reductive processes are limited by the strength of the chemical used, because highly reductive chemicals like metal hydrides (NaH and LiAlH$_4$) also react with water. Other reducing agents have been extensively studied in the past, for example some metal cation like Cr^{2+} and U^{3+} [3,4], that can easily reduce chromophores and bleach lignin model compounds, are themselves colored and are not suited for the bleaching of pulp.

HYDROSULFITE, Y

In today's pulp and paper industry, the most widely used and inexpensive reducing bleaching chemical for high-yield pulps is sodium dithionite, also known as sodium hydrosulfite [5]. Hydrosulfite reduces aldehyde and keto groups of lignin to alcohols, and quinones to hydroquinones [4]. However, by reacting with oxygen in solution, hydrosulfite also produces byproducts known to cause corrosion of steel [6–9]. In addition to hydrosulfites, this chapter focuses on the investigation of the bleaching potential of new reducing chemicals that can replace hydrosulfite without its detriments.

It has been shown initially [10] that the brightening of lignin with sodium hydrosulfite is based on the addition of bisulfite to the conjugated double bonds and the reduction of quinoid structures of lignosulfonates, lignin model compounds (vanillin, acetoguaiacone, isoeugenol, a chalcone derivative and *ortho*-quinone) and spruce groundwood [10]. However, this work has also shown that conjugated carbonyl groups are not easily reduced (if at all) to alcoholic groups. Bisulfite anions are generated by the dissociation of hydrosulfite during the reduction of chromophores in pulp as shown in Equation 13.1 [11].

$$S_2O_4^{2-} + 2\ H_2O \rightarrow 2\ HSO_3^- + 2\ H^+ + 2\ e^- \qquad (13.1)$$

However, experimental results obtained, clearly indicated the superior bleaching effects of $Na_2S_2O_4$ over $NaHSO_3$, but without any explanation concerning the reaction mechanism involved [10].

Polčin and Rapson [12] have shown that hydrosulfite predominantly attacks simple quinoid, α,β-unsaturated aldehyde and anthocyanidine structures found in groundwood pulps prepared from western hemlock (*Tsuga heterophylla*) and eastern spruce (*Picea glauca*). A few years later, de Vries et al. [13,14] have demonstrated that hydrosulfite can reduce many types of aldehydes and ketones in solution according to the mechanism shown in Figure 13.1. Ketones were found to react sluggishly in water, esters are hydrolyzed, while carboxylic acids and amines are not reduced.

Electron spin resonance spectroscopy (ESR) has determined that the unusually long sulfur–sulfur bond in $S_2O_4^{2-}$ dissociates to $SO_2^{-\bullet}$ radicals in aqueous solutions [15,16].

This work has led to more studies on the reaction mechanism of reduction by hydrosulfite, Chung [17] proposed that the mechanism proceeds by stepwise electron transfer from the reducing agent to the carbonyl group to form a ketyl radical intermediate that can then abstract a hydrogen atom from the medium, dimerize to pinacol, or undergo further reduction (Figure 13.2).

Wan et al. [18] have established the major role played by the radical anion $SO_2^{-\bullet}$ originating from many different parent ions and chemical environments. Their time-resolved CIDEP observations showed that the radical species can be generated either by photolysis, or by the thermal dissociation as shown in Equation 13.2:

$$S_2O_4^{2-} \rightleftharpoons 2SO_2^{-\bullet} \qquad (13.2)$$

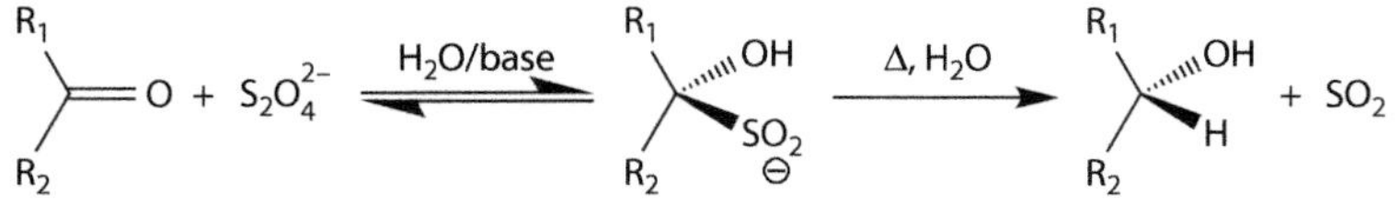

FIGURE 13.1 Reduction of carbonyl group by dithionite under aqueous basic conditions. (After de Vries, J.G., van Bergen, T.J., and R.M., Kellogg, *Synthesis*, 4, 246–247, 1977; de Vries, J.G. and Kellogg, R.M., *J Org Chem*, 45, 4126–4129, 1980.)

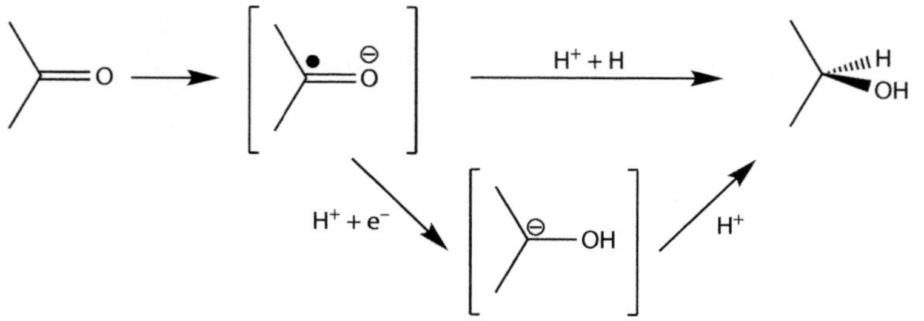

FIGURE 13.2   Reduction of carbonyl group in a stepwise electron transfer reaction. (Redrawn from Chung, S.K., *J Org Chem,* 46(20), 5457–5458, 1981.)

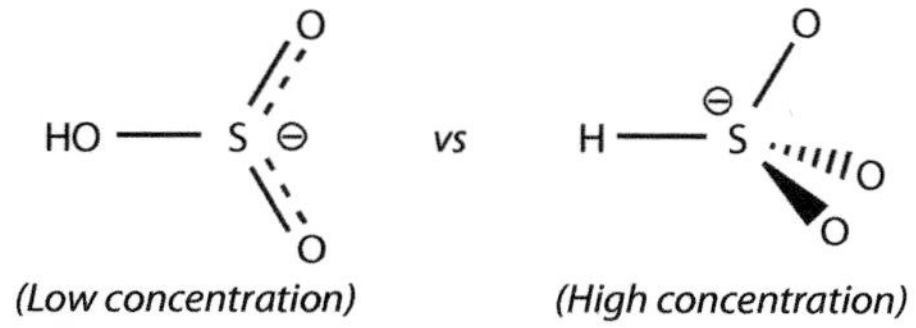

(Low concentration)   (High concentration)

FIGURE 13.3   Bisulfite anion structure. (After Golding, R.M., *J Chem Soc,* 3711–3716, 1960.)

The literature [11] has shown that, in water, hydrosulfite is converted to bisulfite (Equation 13.1). On the other hand, Wan [18] has clearly shown that the following thermal dissociation reaction does *not* occur (Equation 13.3) in aqueous solutions of sodium bisulfite:

$$HSO_3^- \not\leftrightarrow SO_2^{-\bullet} + {}^{\bullet}OH \qquad (13.3)$$

Molecular modeling on these species using semi empirical calculations with PM3 in HyperChem [19] suggested a solvated bisulfite anion is more stable than the $SO_2^{-\bullet}$ anion by about 400 kJ/mol. Wan has also elaborated on the structure of the bisulfite anion as proposed by Golding [20] as a function of its concentration in aqueous media (Figure 13.3).

Molecular calculations of the solvated anions in water indicate that they are both stable with heats of formation of −900 kJ/mol and −944 kJ/mol for the $H\text{-}SO_3^-$ and $HO\text{-}SO_2^-$ structures, respectively [18]. It is thus interesting to note that at low concentration, $HO\text{-}SO_2^-$ can easily yield the $SO_2^{-\bullet}$ radical anion and an $^{\bullet}OH$ radical.

It is also possible that electron exchange between sulfur dioxide radical anions create sulfur dioxide and sulfur dioxide dianions (sulfoxylates), as shown in Equation 13.4. These species are also known for their reducing properties.

$$2\,SO_2^{-\bullet} \rightarrow SO_2 + SO_2^{2-} \qquad (13.4)$$

In the presence of oxygen in solution, complex mixtures of sulfite, sulfate, thiosulfate and polythionates ions are created [11]. The primary influence on the thiosulfate level created comes from the acidic conditions used in bleaching (pH <5) through acid-catalyzed disproportionation reactions (Equation 13.5) [7,18]:

$$2 \, S_2O_4{}^{2-} + H_2O \rightarrow 2 \, HSO_3{}^- + S_2O_3{}^{2-} \tag{13.5}$$

Hydrosulfite can also be oxidized by air into bisulfite and bisulfate anions (Equation 13.6) [7]:

$$S_2O_4{}^{2-} + O_2 + H_2O \rightarrow HSO_3{}^- + HSO_4{}^- \tag{13.6}$$

Wan also outlined that the bleaching process may involve a number of secondary complicated equilibria. For example, they stressed that when $SO_2{}^{-\bullet}$ radical anion transfers the unpaired electron to some organic substrate, it becomes a neutral SO_2 molecule that be rapidly hydrated according to Equation 13.7 [21]:

$$SO_2 + H_2O \rightleftharpoons HSO_3{}^- + H^+ \tag{13.7}$$

Dence [11] presented various reaction mechanisms present during hydrosulfite bleaching for *ortho*-benzoquinoid structure by a sulfur dioxide radical anion (top of Figure 13.4), and for the reduction of coniferaldehyde type structures (bottom of Figure 13.4).

Wan et al. [18] have also presented evidence of the primary electron transfer reactions occurring between the $SO_2{}^{-\bullet}$ radical anion and quinones, Q (Equations 13.8 and 13.9):

$$SO_2{}^{-\bullet} + Q \leftrightarrow SO_2 + Q^{-\bullet} \tag{13.8}$$

$$SO_2{}^{-\bullet} + Q^{-\bullet} \rightarrow SO_2 + Q^{2-} \tag{13.9}$$

They were not able to directly detect the $Q^{-\bullet}$ radical anions in their ESR experiments. However, their experiments with a model compound representative of β–O–4 units type found in lignin (α-guaiacoxyacetoveratrone, Figure 13.5 [22], gives rise to *ortho*-quinones after irradiation with UV light [18]. These *ortho*-quinones were bleached afterward with hydrosulfite to yield the colorless Q^{2-} dianions to support their proposed mechanism (Equations 13.8 and 13.9).

As a side note, this same set of experiments also explained the effect of the addition of calcium carbonate on the increased efficiency of hydrosulfite bleaching [18]. Using phenanthrenequinone as a model compound, they observed the complexation of Ca^{2+} with $Q^{-\bullet}$, resulting in a yellow colloid subsequently easily bleached by subsequent addition of hydrosulfite according to Equations 13.10 and 13.11:

$$2 \, Q^{-\bullet} + Ca^{2+} \rightarrow CaQ_2 \downarrow \tag{13.10}$$

$$CaQ_2 + 2 \, SO_2{}^{-\bullet} \rightarrow Ca^{2+} + 2 \, Q^{2-} + 2 \, SO_2 \tag{13.11}$$

This indicates the importance of the $SO_2{}^{-\bullet}$ radical anion as the key species active in hydrosulfite reductive chemistry.

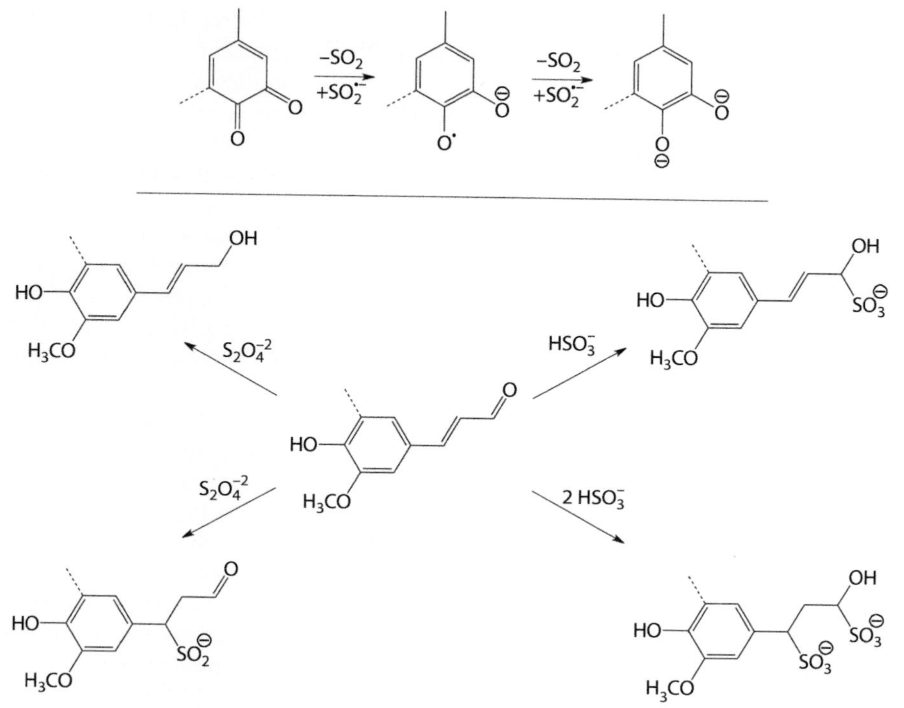

FIGURE 13.4 Reduction of *ortho*-benzoquinoid and coniferaldehyde-type structures during hydrosulfite bleaching. (Adapted from Dence, C.W., *In Chemistry of Mechanical Pulp Bleaching in Pulp Bleaching: Principles and Practice,* Tappi Press, Atlanta, Georgia, 161–182, 1996.)

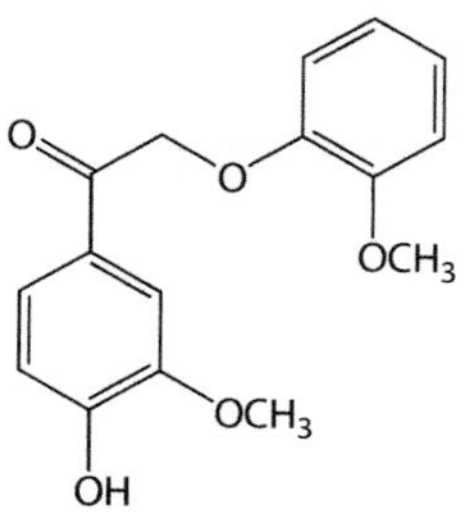

FIGURE 13.5 α-guaiacoxyacetoveratrone. (From Wan, J.K.S., Shkrob, I.A., and Depew, M.C., *Photochemistry of Lignocellulosic Materials*, Am. Chem Soc, Washington, DC, 99–110, 1993.)

FORMAMIDINE SULFINIC ACID, FAS

Also known as thiourea dioxide (TUDO) or amino(imino)methanesulfinic acid (FAS), formamidine sulfinic acid is a reducing agent mainly used in the bleaching of deinked pulps to remove colored dyes [23–28].

Research, mainly by the team of Daneault, has shown that FAS can be efficiently used as a bleaching chemical for softwood TMP and that the reaction of FAS with pulp is extremely fast, with maximum brightness obtained after 15 minutes. The brightness achieved is temperature dependent, and is also proportional to the applied chemical charge. Within the range of FAS used in these studies, from 0 to 2%, the highest brightness was obtained at pH 10. However, it is expected that higher brightness values may be achieved at pH 12 for higher FAS charges, because the ISO Brightness ceiling is not yet obtained. It is also demonstrated that pH 12 is also suitable in order to decrease in the yellowness of the pulp. Since the floor value of b* (yellow, blue coordinates in CIE La*b* system) is not obtained within the charge range studied, lower values of b* may probably be expected at higher FAS charges.

It is also shown that FAS-treated pulps at either pH 8 or 10 yellow more rapidly than unbleached pulps under light-induced yellowing during the first 21 hours. However, after a long period of time of exposure to light, FAS-treated pulps are much more stable. In all cases, FAS bleaching of the pulps at pH 12 causes slower aging than either untreated or bleached with FAS at pH 8 or 10. The intensity of light-induced yellowing is of the same order for all FAS-treated pulps.

The reductive power of FAS has been studied in literature. Early works by Nakagawa and Minami [29] on many ketones have led to the proposed reduction mechanism presented in Figure 13.6, where the reduction proceeds by a one electron transfer followed by radical formation that leads either to a secondary alcohol or a dimer.

Works by Gupta et al. [30] and by Song et al. [31] have established, by X-ray single crystal diffraction techniques and *ab initio* calculations with Gaussian (see ref. 5 in 31) and GAMESS (see ref. 6 in 31), the structure of the FAS/TUDO molecule (Figure 13.7). The thiourea portion of the molecule has a planar conformation and the sulfur atom in the dioxide part is at the apex of a trigonal pyramid formed with the two oxygen and the carbon atoms at the base. The valence structure is suggested to be a combination of zwitterions forms.

This led Makarov [32], in his extensive survey of the chemistry of sulfur-containing reducing agents to report interesting facts. First, quantum chemical calculations have shown that solvation of FAS in water decreases the total energy of the system, leading to enhanced tautomerisation. Also, hydrosulfite contains abnormally long S–S bond and FAS, as well as sodium hydroxymethanesulfinate (or Rongalite), contain abnormally long C–S bonds, highly prone to rupture. It is this structural feature that accounts

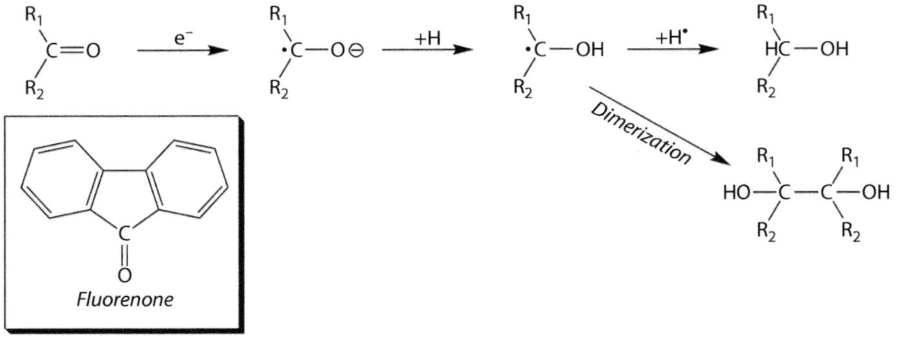

FIGURE 13.6 Fluorenone reduction reaction mechanism. (Adapted from Nakagawa, K. and Minami, K., *Tetrahedron Lett.*, 5, 343–346, 1972.)

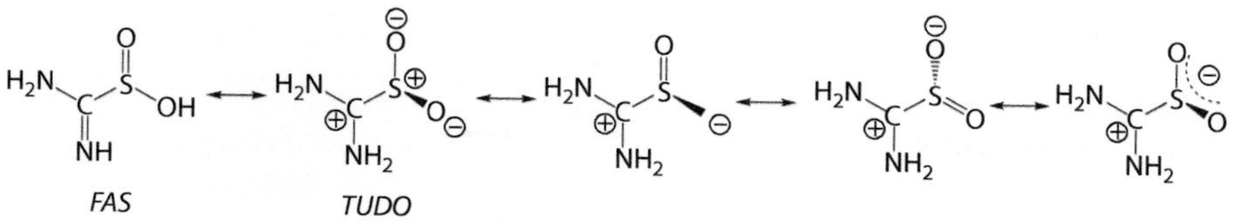

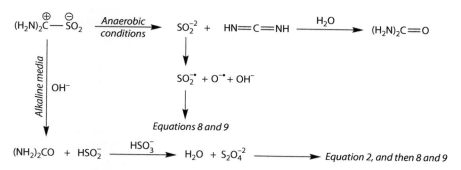

FIGURE 13.7 Formamidine sulfinic acid structure. (Adapted from Gupta, D., Soman, R., and Dev, S., *Tetrahedron*, 38(20), 3013–3018, 1982; Song, J.S., Kim, E.H., Kang, S.K., Yun, S.S., Suh, I.H., Choi, S.S., Lee, S., and Jensen, W.P., *Bull Korean Chem. Soc.*, 17(2), 201–205, 1996.)

FIGURE 13.8 Formamidine sulfinic acid decomposition and reducing reaction mechanism. (Adapted from Makarov, S.V., *Russ. Chem. Rev.*, 70(10), 885–895, 2001.)

for the high reducing activities of these compounds. He has also shown that the chemical reactivity of hydrosulfites, HMS, and FAS and their analogues are governed by transformations of the same intermediates, namely sulfoxylic acid, H_2SO_2 or its anions and the $SO_2^{-\bullet}$ radical anion.

This led to the following proposed reaction mechanism (Figure 13.8) by which FAS/TUDO can be used as a reducing bleaching agent on pulp, i.e., through the $SO_2^{-\bullet}$ radical anion intermediate like hydrosulfite.

PHOSPHOROUS DERIVATIVES, PHOSPHINATES AND PHOSPHINES

Reducing phosphorous chemicals such as [33] sodium hydroxymethylphosphinate (NaHMPP), phosphorous pentachloride, phenylphosphonous dichloride, some phenylphosphoranes, phenylphosphinic and phenylphosphonic acids as well as spirophosphorane are able to bleach a thermomechanical pulp (TMP) made from balsam fir (*Abies balsamea*) and black spruce (*Picea mariana*) at acidic pH [33]. This effect has been attributed to the presence of a reducing hydrogen atom of the phosphorous atom. The efficiency in decreasing order is shown in Figure 13.9.

It should also be noted that the presence of acidic hydrogen on the oxygen atom that is linked to the phosphorous atom is detrimental to the bleaching efficiency

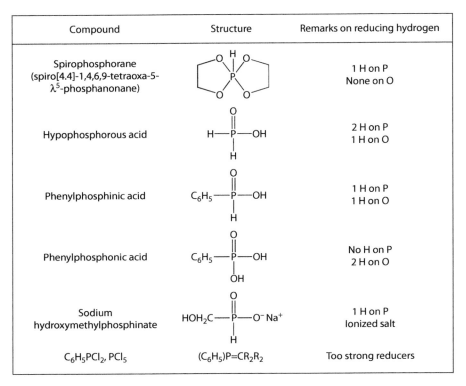

Compound	Structure	Remarks on reducing hydrogen
Spirophosphorane (spiro[4.4]-1,4,6,9-tetraoxa-5-λ^5-phosphanonane)		1 H on P None on O
Hypophosphorous acid	H—P—OH	2 H on P 1 H on O
Phenylphosphinic acid	C_6H_5—P—OH	1 H on P 1 H on O
Phenylphosphonic acid	C_6H_5—P—OH	No H on P 2 H on O
Sodium hydroxymethylphosphinate	HOH_2C—P—O$^-$ Na$^+$	1 H on P Ionized salt
$C_6H_5PCl_2$, PCl_5	$(C_6H_5)P{=}CR_2R_2$	Too strong reducers

FIGURE 13.9 Some reducing phosphorous compounds. (From Djerdjouri, N.E. and Robert, S., 9th Intl Symp Wood Pulp. Chem Montréal, June 9–12, 23-1–23-3, 1997.)

of the molecule. Also shown in this work [33], very powerful reducing chemicals like phosphorous pentachloride or phenylphosphonous dichloride, react rapidly with water and are not even considered a viable alternative as bleaching chemicals.

Recently, there has been very interesting research by Hu and James on phosphine bleaching of mechanical pulps [34–38]. The bleaching effect of phosphorous chemicals were partly based on their reducing potential, as well as the great attention they were receiving as inhibitors of light-induced color reversion of high-yield pulps [39–45]. Hu and James have shown that, among the phosphorus compounds, tris(hydroxymethyl)phosphine (THP), $P(CH_2OH)_3$, and tetrakis (hydroxymethyl) phosphonium chloride/sulfate (THPC/THPS), $[P(CH_2OH)_4]Cl/[P(CH_2OH)_4]_2SO_4$, are capable of bleaching mechanical pulps such as spruce TMP to levels similar to that obtained using sodium hydrosulfite (Y) [34]. Interestingly, 1,2-bis [bis(hydroxymethyl)-phosphino]ethane (BBHPE), $(HOCH_2)_2PCH_2CH_2P(CH_2OH)_2$, shows bleaching power higher than that of THP or Y at similar charges [35]. These authors also explored the reaction mechanism of THP and BBHPE [37], as shown in Figure 13.10.

Their conclusion is that THP and BBHPE are both capable of saturating/removing the C=O and/or C=C double bonds in the lignin model chromophores 2-methoxy-

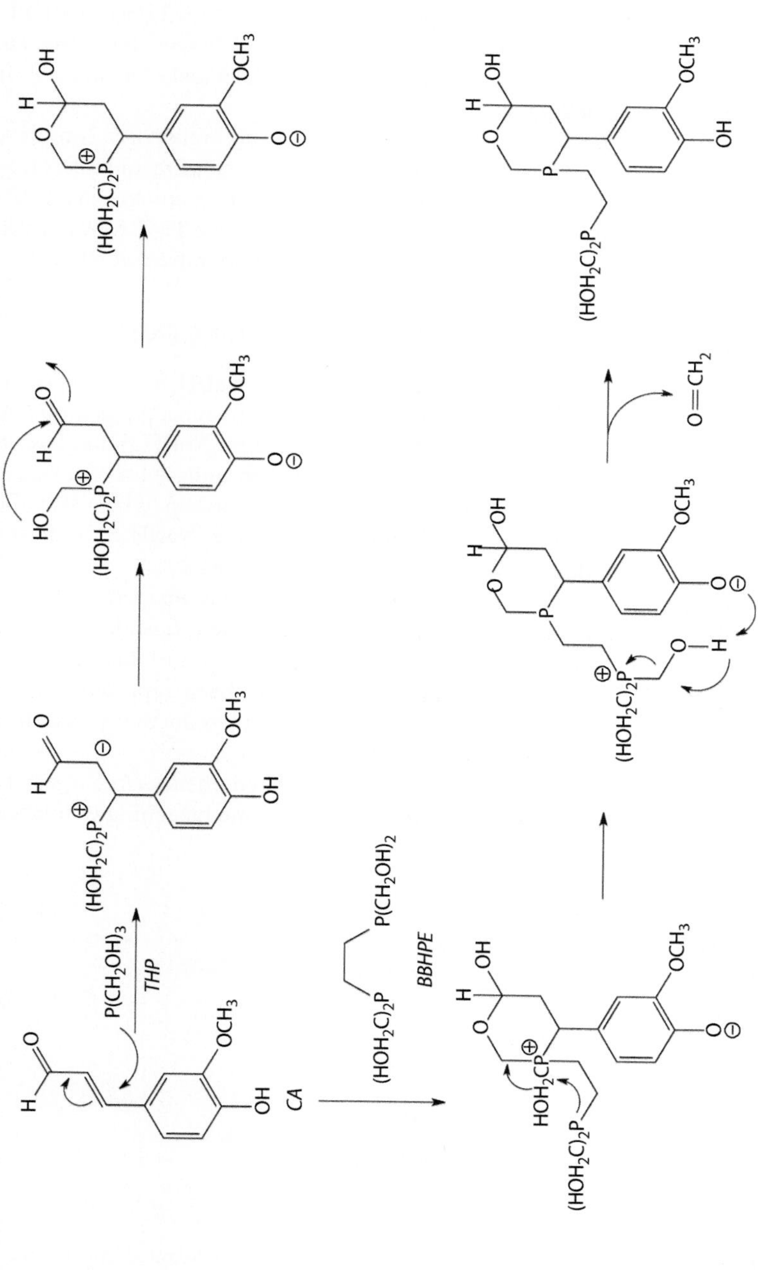

FIGURE 13.10 Proposed reactions of coniferaldehyde (CA) with THP and BBHPE. (Adapted from Chandra, R.P., Hu, T.Q., James, B.R., Ezhova, M.B., and Moiseev, D.V., *J. Pulp Paper Sci.*, 33(1), 15–22, 2007.)

para-quinone and coniferaldehyde. The phosphines decreased the etherified conifer-aldehyde structures in a spruce TMP MWL (milled wood lignin from TMP) and removed the *ortho*- and *para*-quinones introduced via Fremy's salt oxidation to the spruce TMP. The major reaction of the model chromophores with THP or BBHPE is the nucleophilic addition of the tertiary phosphine to a C=O or C=C bond, and such a nucleophilic addition may be responsible for the higher heat and moisture stability of THP- or BBHPE-bleached pulps.

THP and BBHPE are equally reactive toward the model chromophores 2-methoxy-*para*-quinone and coniferaldehyde, and toward the *ortho*- and *para*-quinones intro-duced to the spruce TMP. However, BBHPE shows a higher reactivity than THP or Y toward the etherified coniferaldehyde structures in the spruce TMP MWL. Such a higher reactivity is likely responsible for the higher bleaching power of BBHPE.

SULFURATED BOROHYDRIDE, LALANCETTE'S REAGENT

The strong reducing potential of sulfurated borohydride, $NaBH_2S_3$, also known as the Lalancette's Reagent [46] prompted investigation of this reactant as a potential reductive bleaching agent of mechanical and high-yield pulps. $NaBH_2S_3$ has interme-diate reducing potential between $LiAlH_4$ and $NaBH_4$. When sodium borohydride and sulfur are allowed to react at room temperature in tetrahydrofuran (THF) [47], there is a rapid evolution of hydrogen, and sulfurated borohydride, $NaBH_2S_3$, is formed. The reaction stops before the formation of thioperborates ($NaBS_3$).

Lalancette's reagent reduces aldehydes to alcohols at low temperature [48,49] and forms sulfides and thiols at about 60°C [50]. The reagent has also been used to prepare thioacetals in quantitative yields [51]. The reduction of ketones [52,53], oximes [54], epoxides [55] end episulfides [56] has also been reported. The two most interesting reactions related to pulp bleaching are the reduction of aldehydes (Figure 13.11) and ketones (Figure 13.12).

The results obtained [46] showed that at equal applied chemical charges (1%), Lalancette's reagent and sodium hydrosulfite have the same bleaching efficiencies.

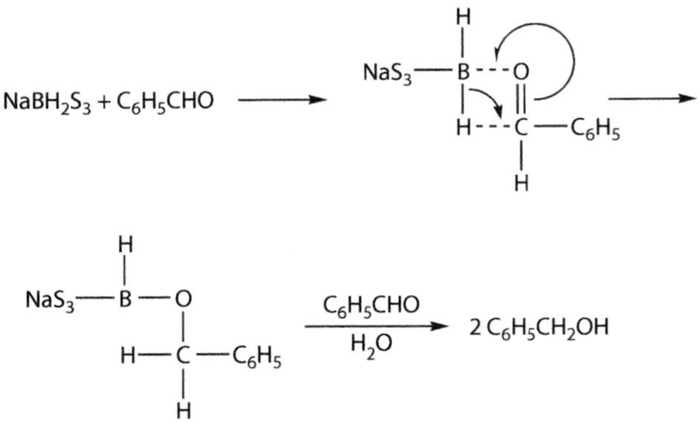

FIGURE 13.11 Reduction of aldehydes by sulfurated borohydride. (Adapted from Lalancette, J.M. and Frêche, A., *Can. J. Chem.*, 48(15), 2366–2371, 1970; Ramanujam, V.M.S. and Trieff, N.M., *J. Chem. Soc. Perkins 2: Phys. Org. Chem.*, 15, 1811–1815, 1976.)

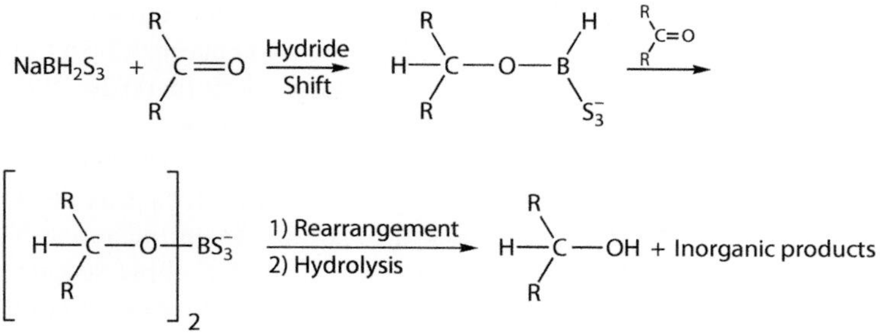

FIGURE 13.12   Reduction of ketones by sulfurated borohydride. (Adapted from Lalancette, J.M. and Frêche, A., *Can. J. Chem.*, 48(15), 2366–2371, 1970; Ramanujam, V.M.S. and Trieff, N.M., *J. Chem. Soc. Perkins 2: Phys. Org. Chem.*, 15, 1811–1815, 1976.)

However, because this chemical is synthesized from the (costly) sodium boro-hydride, and because its own hydrolysis reaction rate is too high, sulfurated borohydride is not a suitable alternative to hydrosulfite. But interestingly, this study suggests the need for milder and more selective reductive chemicals: chemicals able to selectively and rapidly react with carbonyl groups, and not with water to any extent.

AMINEBORANES

Research on sulfurated borohydride gave rise to the study of amineboranes [57–60]. Amineboranes are known to be highly selective reducers of carbonyl groups, like aldehydes and ketones, while amineboranes are inert toward carboxylic acids and esters [61]. It is also shown [62] that the reduction of benzaldehyde produces benzyl alcohol, and not phenol. This is a very interesting result for the wood chemist as phenolic structures, and not primary aromatic alcohols, are involved in the photoinduced yellowing of paper [63]. Furthermore, this reduction of the carbonyl groups can easily, and efficiently, be done in water, in which amineboranes are stable. Amineboranes are remarkably stable (thermally and hydrolytically), water soluble, easily handled and totally non reactive towards atmospheric or dissolved oxygen. Many are also available on an industrial scale.

tert-butylamineborane complex (TBAB), $(CH_3)_3CNH_2:BH_3$, and borane-ammonia complex (BA), $BH_3:NH_3$, were first used to bleach a balsam fir–black spruce TMP. These results proved that both amineboranes can successfully increase the brightness of the pulp. In fact, the brightness obtained is higher than those found with hydrosulfite under the same standard conditions (1% charge, 60°C). Amineboranes can be used at any pH from 4 to 10 (lower pH gives better results), while the use of hydrosulfite is restricted to a pH around 5.5 ± 0.2 [5]. It should be noted that the final pH in both amineboranes bleaching was about 8.

The reduction in the CIE b* color coordinate is even more important for amineboranes at the same % w/w applied. When compared with hydrosulfite at the same conditions and mole ratio applied on pulp, the results show that the decrease in CIE b* color coordinate is more important than the one obtained with hydrosulfite. Note however that to obtain the same molar ratio (5.7 mmol) for amineboranes as for hydrosulfite, the applied charge is 0.5% for *tert*-butylamineborane (TBAB) and only 0.175% for borane-ammonia! Even at an applied charge lower that 0.2%, the increase in Brightness is 3.6% ISO, and the decrease in CIE b* is 1 unit!

The amineborane bleaching was also compared to a 1% hydrogen peroxide bleaching (P). In this latter case, the brightness is slightly higher for the peroxide bleaching, while the decrease in the CIE b* value is still better for the amineborane bleaching stage. Both amineboranes were then used in two-stages bleaching sequences with hydrogen peroxide. The P/TBAB sequence results in an increase of 5.6 points over a single peroxide stage. Brightness in excess of 76% ISO can thus easily be achieved using a P/TBAB sequence. All this time, it was shown that in all bleaching using amineboranes, no adverse effect on mechanical properties was observed. These results show that amineboranes does not affect the cellulosic matrix.

FUNDAMENTAL CHEMISTRY

Hutchins et al. [61,64] showed that in alkaline media, the reduction of carbonyl groups proceeds by an initial, slow formation of a ketone-amineborane complex, followed by a fast intramolecular hydride transfer. In acidic media, the rate-determining step involves the attack of amineborane on a protonated carbonyl, which is formed in a rapid preequilibrium, through the creation of a four-center transition state [65–67]. Since alkali darkening, this mechanism explains why, for the same reaction time, it is possible to reach a higher brightness at low pH than at high pH (Table 1 in ref. 60). In lignin, the main targets of bleaching agents are ketone and aldehyde groups. Hydrogen peroxide is well known to easily attack ketones, while slowly reacting with aldehydes [68]. Therefore, the use of amineboranes is complementary to the alkaline peroxide bleaching stage.

The selectivity of the amineboranes attack on lignin was verified with the variation of the UV-visible light absorption coefficient, k_{abs}, of very thin sheets (10 g.m^{-2}) of paper using a diffuse reflectance accessory (integrating sphere). This procedure is well described in literature [23,63,69,70]. UV-visible spectra indicate that the light absorption at 350 nm is strongly decreased upon bleaching. In the literature [63], the 350 nm absorption region is assigned to the coniferaldehyde type structures of lignin. For comparison using experimental data from NIST [71], aromatic ketones in solution show maximum absorption between 240 and 330 nm (240 nm for acetophenone, 305 nm for α-acetylacetophenone, 310 nm for *o*-methoxyacetophenone), *para*-quinones between 330 and 430 nm (330 nm for benzophenone, 365 nm for *p*-benzoquinone and 430 nm for 2-methyl-*p*-benzoquinone) and *ortho*-quinones around 400 to 405 nm (400 nm for *o*-benzoquinone, 405 nm for β-naphtoquinone and 3,5-di-*tert*-butyl-*o*-benzoquinone). These spectra thus support the selectivity of the reducing reaction involved. The importance of the decrease in the light absorption coefficient by 80 m^2·kg^{-1} at 350 nm, confirms that the yield of the reaction with amineboranes is very high. Schmidt and Heitner [72] have also shown that the extensive reduction of a black spruce pulp with sodium borohydride, a very strong reducer, results in a maximum decrease of −65 m^2·kg^{-1} at 350 nm. For further comparison, hydrosulfite decreases the light absorption coefficient at 350 nm by about −20 m^2·kg^{-1}

at best. Since no increase in the light absorption coefficient spectra was found between 300 and 700 nm, the authors concluded that amineboranes modify the coniferaldehyde types chromophores to chemical species that absorb light below 300 nm. Andrews [62] had also shown that TBAB is highly selective and favors the attack on benzaldehyde instead of acetophenone in water at a ratio of 98:2; a very strong indication of the selectivity of TBAB toward aldehyde over ketones (325 nm).

Pedneault et al. [60] also used TBAB as the second stage of a two stage bleach with hydrogen peroxide. They showed that that the chemical structures oxidized by hydrogen peroxide are not reduced by TBAB stage and that the two chemicals are complementary to one another.

These studies have thus shown that amineboranes (*tert*-butylamineborane and borane-ammonia complexes) are bleaching chemicals more efficient than hydrosulfite. They are noncorrosive, and their bleaching degradation products are boric acid and an amine. However, their main disadvantage is their high cost, even if their optimal bleaching conditions can be obtained at about 10°C lower than what is used for hydrosulfite.

MODEL COMPOUNDS AND SELECTIVITY

The above studies raise the question of the selectivity of the bleaching reaction toward other lignin chromophores and leads to the work on the bleaching chemistry of *tert*-butylamineborane complex with lignin model compounds (Figure 13.13) in aqueous solutions [73].

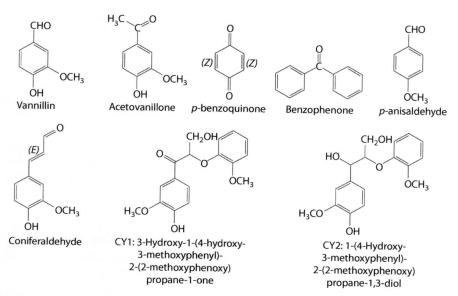

FIGURE 13.13 Lignin model compounds used in the selectivity reaction study of tert-butylamineborane complex. (From Pellerin, C., Pedneault, C., and Robert, S., *J Pulp Paper Sci*, 26(12), 436–440, 2000.)

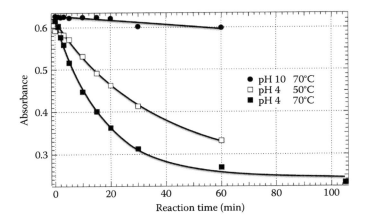

FIGURE 13.14 Reduction kinetics of vanillin reduction by TBAB at various pH and temperature. (From Pellerin, C., Pedneault, C., and Robert, S., *J Pulp Paper Sci,* 26(12), 436–440, 2000.)

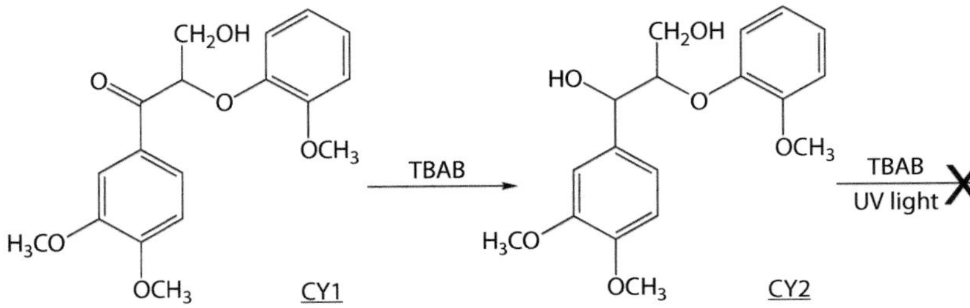

FIGURE 13.15 Reductive reaction pathway of CY1 by TBAB. (From Pellerin, C., Pedneault, C., and Robert, S., *J. Pulp Paper Sci.,* 26(12), 436–440, 2000.)

Pellerin et al. [73] showed, for instance, that the reduction of the vanillin carbonyl with TBAB proceeds rapidly at 70°C at pH 4, and is negligible in alkaline media (Figure 13.14). This decrease in reactivity is explained by the involvement of the phenolate ion at alkaline pH to the resonance structure, resulting in an increase in electron density on the carbonyl carbon, thus making the carbon atom unfavorable to nucleophilic attack, and thus to reduction by TBAB. This result, with this model compound [73], confirms the fact that pulp bleaching with TBAB is faster in acidic media at high temperature [58].

The difference in reactivity TBAB with vanillin in acidic versus alkaline conditions corresponds to what is observed when using wood pulp. With acetovanillone, there is no significant reduction of the carbonyl group under any of the experimental conditions tested. However, the reaction scheme of *para*-benzoquinone presents a fast reduction reaction, even at room temperature, to produce hydroquinone.

The CY1/CY2 reaction scheme is very interesting (Figure 13.15) because β–O–4 units type represents about 50% of the known structural skeleton of softwood lignin [74]. The slow but quantitative reduction of this model ketone compound (CY1)

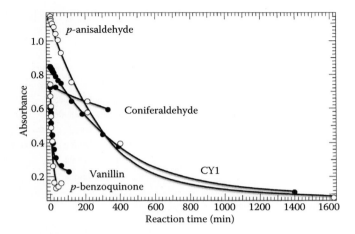

FIGURE 13.16   Kinetics of reduction of some model compounds. (From Pellerin, C., Pedneault, C., and Robert, S., *J. Pulp Paper Sci.*, 26(12), 436–440, 2000.)

results in an alcohol (CY2) that does not react with UV light or with TBAB. This proves that amineborane does not attack alcoholic β–O–4 structure types on lignin, keeping the structural integrity of lignin intact. The brightness gain obtained by the reduction of the CY1 type structure can thus be considered as stable.

From the study of model compounds, kinetics of the reduction reaction with TBAB can be obtained. Some of the results are shown in Figure 13.16.

All the kinetics obtained can easily be fitted by a simple exponential function of the following type (Equation 13.12):

$$Abs = k_0 + k_1 e^{-k_2 t} \qquad (13.12)$$

The results obtained from these curve fittings are presented in Table 13.1. Since the correlation coefficients are very high (>0.99), a simple exponential is required to fit the data. It means that only one single order chemical reaction is involved.

From the results presented in Table 13.1, one can see that TBAB attacks quinones more rapidly than any other chromophores. TBAB also selectively attacks aldehydes over ketones. The selectivity is based on the reaction rates, and the yield is based on the observed final absorbance at the end of the reaction time. More so, aromatic ketones must be activated to react with TBAB by the presence of some electron attracting functional group near the carbonyl group. Finally, alcohol groups are totally inert toward the action of TBAB. Since hydrogen peroxide is known to attack ketones faster than aldehydes [68], the two bleaching chemicals do not compete for the same chromophores and are complementary to one another.

This study on the bleaching chemistry of amineboranes with lignin model compounds has resulted in a bleaching efficiency ranking for those commercially available amineboranes [75]. The most efficient is *tert*-butylamineborane, closely followed by borane-ammonia. The third most efficient is dimethylamineborane, followed by the tertiary amines (trimethylamineborane and triethylamineborane).

TABLE 13.1
Kinetic Data—Reduction Rate Constants

Product	k_2 (min^{-1})	Structure
p-benzoquinone	0.1023	Quinone
Vanillin	0.0578	Aromatic aldehyde
p-anisaldehyde	0.0034	Aromatic aldehyde
CY1	0.0025	Aromatic ketone e$^-$ attractor close by
Coniferaldehyde	0.0018	Aromatic ketone α-β unsaturated
Acetovanillone	-----	Aromatic ketone
Benzophenone	-----	Aromatic ketone
Alcohol derivatives of the above model compounds	----	Alcohol

Source: From Pellerin, C., Pedneault, C., and Robert, S., *J. Pulp Paper Sci.*, 26(12), 436–440, 2000.

TABLE 13.2
Bleaching Efficiency (ISO Brightness Gain) on Softwood TMP and pK_a of Amineboranes

Amine	pK_a	ISO Gain ($\pm 0,5$)
Ammonia	9.26	7,0
Trimethylamine	9.74	4,0
Iso-propylamine	10.60	n/a
Methylamine	10.64	n/a
Ethylamine	10.64	n/a
Dimethylamine	10.72	5,0
Triethylamine	10.76	4,0
tert-butylamine	10.83	7,5
Diethylamine	10.99	n/a

Source: From L'Ecuyer, N., Lobit, C., Geldes, A., and Robert, S., 12th Intl Symp Wood Pulp Chem, Madison, WI, June 9–12, 143–146, 2003.

Work published by Robert [76–78] have tried to fully establish a structure-reactivity relationship of amineboranes toward bleaching of mechanical pulps using molecular simulation. The literature suggests that the reductive strength of the amineboranes is proportional to the basicity of the uncomplexed amine moiety [61]. According to this hypothesis, tertiary amines should have been the most efficient, but the experiments have shown otherwise (see Table 13.2).

MOLECULAR SIMULATION

Molecular mechanics calculations are routinely used by chemists to calculate molecular equilibrium geometries, conformations and properties, as well as UV and IR spectra. Quantum chemical calculations, both *ab initio* and semiempirical, as well as DFT methods [79], while significantly more costly in terms of computer power and CPU time usage, are also becoming common for these tasks, and are also able to supply quantitative thermochemical and kinetic data. Most important, the theories underlying calculations have now evolved to a stage where a variety of important quantities, among them molecular equilibrium geometry and reaction energies, may be obtained with sufficient accuracy to be useful.

Molecules like amineboranes can usually be modeled in vacuum easily using quantum mechanics methods. However, to perform a simulation of those molecules in an aqueous media, one must sometimes, if not often, put *ab initio* quantum mechanics calculations aside and use semiempirical quantum mechanics models because water molecules rapidly add up to a tremendous number of atoms that goes beyond the limits of quantum mechanics calculations on personal computers.

Robert's team [76–78] performed *ab initio* and DFT calculations using Gaussian 03W v6.0 [80] on various amineboranes (Table 13.3). In Gaussian 03W, the aqueous media was modeled using a continuum of constant dielectric constant: the Polarizable Continuum Model (PCM) solvation method in its IEFPCM (Integral Equation Formalism) version [81]. The latest version of HyperChem (v7.51) [19] was also used to perform PM3 calculations since this software does contain the parameters for the boron atom, as opposed to the original PM3 model. *Ab initio* and DFT calculations were also performed directly in HyperChem without any add-on. In HyperChem,

---

TABLE 13.3
Amineboranes

Commercially Available

$BH_3:NH_3$	BA
$BH_3:NH_2tBu$	TBAB
$BH_3:NHMe_2$	DMAB
$BH_3:NMe_3$	TMAB
$BH_3:NH(CH_2CH_2)_2O$	MPB
$BH_3:NEt_3$	TEAB

Synthesized

$BH_3:NH_2iPr$	iPAB
$BH_3:NHEt_2$	DEAB

Not yet synthesized, but calculated

$BH_3:NH_2Me$	MAB
$BH_3:NH_2Et$	EAB

Source: From L'Ecuyer, N., Lobit, C., Geldes, A., and Robert, S., 12th Intl Symp Wood Pulp Chem, Madison, WI, June 9–12, 143–146, 2003; L'Ecuyer, N., Laplante, M.C., and Robert, S., 13th Intl Symp Wood Fibre Pulp Chem, Auckland, New Zealand, May 16–19, 2, 265–271, 2005.

the aqueous media was modeled using explicit solvent molecules. This method uses a lot more computer power than the continuum method, but it has the advantage of determining specific hydrogen bond interactions between the solute and the solvent.

Experimental bleaching of a softwood thermomechanical pulp using aminebo-ranes were presented in the literature [58,60,77]. The usual comparison between bleaching agents is generally done in % w/w (for a same chemical) or in active equivalent (either electron for redox reaction or active chlorine or active hydrogen). However, since the amineboranes reactive part is the hydrogen located on the boron atom [76], and since the amineboranes studied vary greatly in molecular weight, the authors compared their bleaching efficiency either as an ISO brightness gain or post color (PC) number per mmol of amineborane per 100 g of pulp.

The PC number can easily be calculated from the following Equation 13.13 [63]:

$$PC = 100\left(\frac{k_1}{s_1} - \frac{k_0}{s_0}\right) \tag{13.13}$$

where k_1 is the light absorption coefficient of the sample (the treated handsheets) after reaction and s_1 is the light scattering coefficient of the same sample after reaction. k_0 and s_0 refer to the unreacted sample, in this case, the unbleached pulp. k and s are related to the reflectance of the handsheets R by the following Equation 13.14:

$$\frac{k}{s} = \frac{(1 - 0.01R)^2}{0.02R} \tag{13.14}$$

The experimental results on bleaching are presented in Table 13.4. The best bleaching results are expressed either as the highest Brightness Gain, or as the lowest

TABLE 13.4
Brightness Gain and PC Numbers of Bleached Handsheets

	Brightness Gain (%ISO) / mmol Amineborane / 100 g Pulp	PC Number / mmol Amineborane / 100 g Pulp
MPB	.559 ± .007	−.42 ± .03
TBAB	.435 ± .006	−.33 ± .02
DEAB	.348 ± .006	−.28 ± .02
TEAB	.307 ± .008	−.26 ± .01
iPAB	.302 ± .005	−.24 ± .02
DMAB	.196 ± .004	−.161 ± .009
TMAB	.195 ± .005	−.164 ± .009
BA	.144 ± .002	−.112 ± .007

Source: From L'Ecuyer, N., Laplante, M.C., and Robert, S., 13th Intl Symp Wood Fibre Pulp Chem, Auckland, New Zealand, May 16–19, 2, 265–271, 2005.

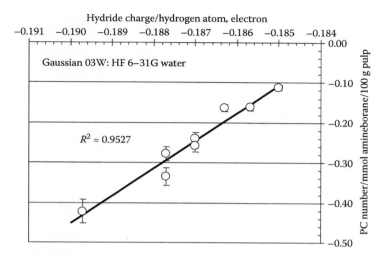

FIGURE 13.17   PC Number per mmol of amineborane per 100 g of pulp as a function of the hydride charge calculated by HF 6-31G using Gaussian 03W with IEFPCM. (From Hivon, D. and Robert, S., 11th Intl Symp Wood Pulp Chem, Nice, France, June 11–14, 375–381, 2001.)

PC Number. Both have the same meaning. The PC Number for all the amineborane tested is then plotted against the hydride charge obtained by molecular simulation using the Mulliken population analysis.

These results are presented in Figure 13.17 for an *ab initio* Hartree–Fock, 6-31G in water using IEFPCM. Error bars in the experimental evaluation of PC Numbers vary between 5.5 and 7% and are often smaller than the size of the symbols used.

These experimental results and calculations show that all the primary (iPAB, TBAB), secondary (DMAB, DEAB, MPB), tertiary (TEAB, TMAB) and quaternary (BA) amineboranes share a common structure reactivity relationship: the bleaching efficiency toward high-yield softwood TMP is greater for the amineborane where the charge delocalization induced by the amine moiety [82,83] increases the negative charge on the hydrogen located on the boron atom.

CONCLUSIONS

There have been some notable advances in the elucidation of the reaction mechanisms of hydrosulfite itself. The importance of the radical anion species has been demonstrated in detail, and the implication of these discoveries has opened the way to new vistas to bleach wood pulps using reductive chemistry.

The last decade have thus shown that some new interesting chemicals, either phosphorous-based of boron based, can be used as new reductive bleaching chemicals for high-yield softwood pulps. These chemicals can be used in multistage bleaching sequences along with hydrogen peroxide. Phosphines and amineboranes can both be used to reduce lignin chromophores.

Phosphines like THP and BBHPE are both capable of saturating/removing the C=O and/or C=C double bonds in lignin model compounds. They are also capable of decreasing the etherified coniferaldehyde structures in a spruce TMP milled wood lignin (MWL) and removing the *ortho-* and *para*-quinones artificially introduced on a spruce TMP. The main reaction observed is the nucleophilic addition of the tertiary phosphine to a C=O or C=C bond.

For amineboranes, it was shown that the reaction is highly selective, and the yields are, in some cases very high. Amineboranes also react rapidly with quinones, then with aromatic aldehydes, and finally slowly with aromatic ketones. Amineboranes are sensitive to the electron density on the carbonyl group. Since TBAB and hydrogen peroxide do not attack ketones and aldehydes with the same reaction rate and the same extent, reduction bleaching with TBAB is complementary to oxidative bleaching using hydrogen peroxide.

The link between the structure of amineboranes and their bleaching efficiency of unbleached softwood mechanical pulps is the partial charge localized on the boron's hydride hydrogen in water. Predictions based on the pK_a of the amine part, or of the generic substitution degree of the amine do not explain the experimental bleaching efficiency of laboratory tested amineboranes. The prediction obtained with computation, based on the partial charges of the hydride, does fit the experimental data. Delocalization of the charges over the amineborane molecule caused by the presence of water molecules all around may explain this behavior, as well as the asymmetric steric hindrance of the alkyl groups on the nitrogen atom. However, since the alkyl part of the amineborane also governs some of the water solubility of the amineborane, the size of the alkyl group cannot be too large if these molecules must be used in aqueous solutions.

Exploration of these new phosphorous-based and boron based reducing chemicals, as well as new insights on hydrosulfite reductive bleaching that will be developed from these results, is very promising for the future.

REFERENCES

1. R Barton, C Tredway, M Ellis, E Sullivan. Chapter 19 in *Pulp and Paper Manufacture,* 3rd ed. Vol. 2, Joint Textbook Committee of the Paper Industry, Montréal, Canada, 1987, pp. 227–237.
2. DW Reeve. *Introduction to the Principles and Practice of Pulp Bleaching in Pulp Bleaching: Principles and Practice.* Edited by CW Dence, DW Reeve. Atlanta, Georgia: Tappi Press, 1996, pp. 1–24.
3. RA Fleury, WH Rapson. Characterization of chromophoric groups in groundwood and lignin model compounds by reaction with specific reducing agents. *Pulp Paper Mag Can.* 68(6):62–18, 1968.
4. WH Rapson. Mechanisms of groundwood bleaching. *APPITA* 23(2):102–114, 1969.
5. C Daneault, S Robert, PY Dionne. Groundwood post-refining hydrosulfite bleaching. *Pulp and Paper Can* 95(7):31–35, 1994.
6. AM Devaney, RG Guess. Sodium thiosulfate in hydrosulfite bleaching. *J Pulp Paper Sci* 8(3):60–64, 1982.
7. A Garner. Sources of thiosulfate in paper machine white water. Part II: Thiosulfate formation during sodium hydrosulfite brightening. *J Pulp Paper Sci* 10(3):51–57, 1984.
8. LD Bond, N Sweeney, W Giust, A Fluet, MG Fairbank, P Whiting. Controlling thiosulfate resulting from hydrosulfite brightening of mechanical pulps. *J Pulp Paper Sci* 17(2):30–33, 1991.

9. JP Casey. Chapter 5, Bleaching. In *Pulp and Paper: Chemistry and Chemical Technology,* Vol. 1, 3rd ed. New York: John Wiley & Sons, 1980, pp. 633–764.

10. S Hosoya, H Hatakeyama, J Nakano. Brightening of Lignin in pulp with sodium hydrosulfite. *J Jap Wood Res Soc* 16(3):140–144, 1970.

11. CW Dence. Section 3.2, Reactions of sodium dithionite with chromophores systems in lignin. In *Chemistry of Mechanical Pulp Bleaching in Pulp Bleaching: Principles and Practice.* Edited by CW Dence, DW Reeve. Atlanta, Georgia: Tappi Press, 1996, pp. 161–182.

12. J Polčin, WH Rapson. Effects of bleaching agents on the absorption spectra of lignin in groundwood pulps: Part I. Reductive bleaching. *Pulp Paper Mag Can* 72(3):69–90, 1971.

13. JG de Vries, TJ van Bergen, RM Kellogg. Sodium dithionite as a reductant for aldehydes and ketones. *Synthesis* (4):246–247, 1977.

14. JG de Vries, RM Kellogg. Reduction of aldehydes and ketones by sodium dithionite. *J Org Chem* 45:4126–4129, 1980.

15. PW Atkins, A Horsfield, MCR Symons. Oxides and oxyions of the non-metals. Part VII. SO_2^- and ClO_2. *J Chem Soc* 5220–5225, 1964.

16. RG Rinker, TP Gordon, WH Corcoran. Electron spin resonance studies of sodium dithionite and sodium formaldehyde sulfoxylate. *Inor Chem* 3(10):1467–1469, 1964.

17. SK Chung. Mechanism of sodium dithionite reduction of aldehydes and ketones. *J Org Chem* 46(20):5457–5458, 1981.

18. RS Pemberton, MC Depew, C Heitner, JKS Wan. Some mechanistic insights onto a model bleaching process of quinones by bisulfite and dithionite: An ESR-CIDEP study. *J Wood Chem Technol* 15(1):65–83, 1995.

19. HyperChem(TM) Professional 8.0.6, Hypercube, Inc., 1115 NW 4th Street, Gainesville, Florida 32601, USA.

20. RM Golding, Ultraviolet absorption studies of the bisulfite-pyrosulfite equilibrium. *J Chem Soc* 3711–3716, 1960.

21. FA Cotton, G Wilkinson, CA Murillo, M Bochmann. Chapter 12 in *Advanced Inorganic Chemistry,* 6th ed. Wiley, Chicester, 1999, pp. 496–546.

22. JKS Wan, IA Shkrob, MC Depew. Primary photophysical and photochemical processes of α-guaiacoxyacetoveratrone. ACS Symp Series 531 (Photochemistry of Lignocellulosic Materials), C. Heitner and J.C. Scaiano, Eds. Washington, DC, Am. Chem Soc, 1993, pp. 99–110.

23. C Daneault, S Robert, C Leduc. Bleaching of mechanical pulp with formamidine sulfinic acid. *Cell Chem Technol* 28(2):205–217, 1994.

24. C Daneault, S Robert, C Leduc. Formamidine sulfinic acid used as a bleaching chemical on softwood TMP. *Res Chem Intermed* 21(3–5):521–533, 1995.

25. C Daneault, C Leduc. Bleaching efficiency of formamidine sulfinic acid (FAS) in comparison to hydrosulfite, borohydride, and peroxide in one and two stages. *Tappi J* 78(7):153–160, 1995.

26. G Galland, L Magnin, B Carre, P Larnicol. Best use of bleaching chemicals in deinking lines. *Revue ATIP* 56(1):10–18, 23–25, 2002.

27. H Song, S Wang, Z Chen. Study on FAS in pilot deinking of mixed color waste paper. Emerging Technologies of Pulping & Papermaking, Proceedings of the International Symposium on Emerging Technologies of Pulping & Papermaking, 2nd, Guangzhou, China, Oct. 9–11, 333–338, 2002.

28. E Poppel, Z Lado. Bleachability of alkaline and neutral medium deinked waste paper stocks. *Cell Chem Technol* 36(3–4):303–316, 2003.

29. K Nakagawa, K Minami. Reduction of organic compounds with thiourea dioxide. I. Reduction of ketones to secondary alcohols. *Tetrahedron Lett* 5:343–346, 1972.

30. D Gupta, R Soman, S Dev. Thiourea: A convenient reagent for the reductive cleavage of olefin in ozonolysis products. *Tetrahedron* 38(20):3013–3018, 1982.

31. JS Song, EH Kim, SK Kang, SS Yun, IH Suh, SS Choi, S Lee, WP Jensen. The structure and ab initio studies of thiourea dioxide. *Bull Korean Chem Soc* 17(2):201–205, 1996.
32. SV Makarov. Recent trends in the chemistry of sulfur-containing reducing agents. *Russ Chem Rev* 70(10):885–895, 2001.
33. NE Djerdjouri, S Robert. Reductive bleaching of high yield pulps using phosphorus chemicals. 9th Intl Symp Wood Pulp. Chem Montréal, June 9–12, 23-1–23-3, 1997.
34. TQ Hu, BR James, D Yawalata, MB Ezhova. A new class of bleaching and brightness stabilizing agents. Part I: Bleaching of mechanical pulps. *J Pulp Paper Sci* 30(8):233–240, 2004.
35. TQ Hu, BR James, D Yawalata, BM Ezhova, RP Chandra. A new class of bleaching and brightness stabilizing agents. Part II: Bleaching power of a bisphosphine. *J Pulp Paper Sci* 31(2):69–75, 2005.
36. TQ Hu, BR James, D Yawalata, BM Ezhova. A new class of bleaching and brightness stabilizing agents. Part III: Brightness stabilization of mechanical pulps. *J Pulp Paper Sci* 32(3):131–136, 2006.
37. RP Chandra, TQ Hu, BR James, MB Ezhova, DV Moiseev. A new class of bleaching and brightness stabilizing agents. Part IV: Probing the bleaching chemistry of THP and BBHPE. *J Pulp Paper Sci* 33(1):15–22, 2007.
38. TQ Hu, E Yu, BR James, P Marcazzan. The fate of phosphorous in the bleaching of spruce TMP with the new bleaching agent: THPS. *Holzforschung* 62:389–396, 2008.
39. P Fornier de Violet, A Nourmamode, N Colombo, J Zhu, A Castellan. Photochemical brightness reversion of peroxide bleached mechanical pulps in the presence of various additives. *Cell Chem Technol* 24(2):225–235, 1990.
40. A Castellan, JH Zhu, N Colombo, A Nourmamode. An approach to understanding the mechanism of protection of bleached high-yield pulps against photoyellowing by reducing agents using the lignin model dimer: 3,4-dimethoxy-α-(2′-methoxyphenoxy)-acetophenone. *J Photochem Photobiol. A: Chem* 58(2):263–271, 1991.
41. JX Guo, DG Gray. Inhibition of light-induced yellowing of the high-yield pulp by sodium hydroxymethylphosphinate. *J Pulp Paper Sci* 22(2):64–70, 1996.
42. TQ Hu, BR James, CL Lee, Towards inhibition of yellowing of mechanical pulps. Part I. Catalytic hydrogenation of lignin model compounds under mild conditions. *J Pulp Paper Sci* 23(4):153–156, 1997.
43. TQ Hu, BR James, CL Lee. Towards inhibition of yellowing of mechanical pulps. Part II. Water-soluble catalysts for the hydrogenation of lignin model compounds. *J Pulp Paper Sci* 23(5):200–205, 1997.
44. TQ Hu, BR James, Y Wang. Towards inhibition of yellowing of mechanical pulps. Part III. Hydrogenation of milled wood lignin. *J Pulp Paper Sci* 25(9):312–317, 1999.
45. TQ Hu, BR James. Towards inhibition of yellowing of mechanical pulps. Part IV. Photostability of hydrogenated lignin model compounds. *J Pulp Paper Sci* 26(5):173–175, 2000.
46. S Robert, C Daneault, M Foesser. Sulfurated borohydride used as bleaching chemical on a softwood TMP. *Rev Chem Interm* 21(3–5):563–576, 1995.
47. JM Lalancette, A Frêche, R Monteux. Reductions with sulfurated borohydrides. I. Preparation of sulfurated borohydrides. *Can J Chem* 46(16):2754–2757, 1968.
48. JM Lalancette, A Frêche, JR Brindle, M Laliberté. Reductions of functional groups with sulfurated borohydrides: Application to steroidal ketones. *Synthesis* 10:526–532, 1972.
49. JM Lalancette, A Frêche. Reductions with sulfurated borohydrides. II. Reduction of typical aldehydes. *Can J Chem* 47(5):739–742, 1969.
50. JR Brindle, JL Liard. Facile synthesis of thiols and sulfides. *Can J Chem* 53(10):1480–1483, 1975.
51. JM Lalancette, A Lachance. Facile preparation of thioacetals in neutral medium starting from sodium borohydride. *Can J Chem* 47(5):859–860, 1969.

52. JM Lalancette, A Frêche. Reductions with sulfurated borohydrides. V. Reductions of ketones. *Can J Chem* 48(15):2366–2371, 1970.
53. VMS Ramanujam, NM Trieff. Reduction of some ketones using sulfurated sodium borohydride (NaBH$_2$S$_3$): Kinetics and mechanism. *J Chem Soc Perkins 2: Phys Org Chem* 15:1811–1815, 1976.
54. JM Lalancette, JR Brindle. Reductions with sulfurated borohydrides. IV. Reduction of oximes. *Can J Chem* 48(5):735–737, 1970.
55. JM Lalancette, A Frêche. Reductions with sulfurated borohydrides. VII. Reactions with epoxides. *Can J Chem* 49(24):4047–4053, 1971.
56. JM Lalancette. M Laliberté. Preparation of 1,2-dithiols from episulfides, *Tetrahedron Lett* 16:1401–1404, 1973.
57. C Pedneault, S Robert, C Pellerin. Blanchiment par de nouveaux agents réducteurs: Les boranamines sont plus efficaces en milieu acide. *Les Papetières du Québec* 7(3):24–28, 1996.
58. C Pedneault, S Robert, C Pellerin. Bleaching with new reductive chemicals: Replacement of hydrosulfite. *Pulp Paper Can* 98(3):51–55, 1997.
59. C Pellerin, C Pedneault, S Robert. Amineboranes Used as New Selective Reducing Chemicals on High Yield Pulps. 9th Intl Symp Wood Pulp. Chem Montréal, Canada, pp. 86-1–86-4, 1997.
60. C Pedneault, C Pellerin, S Robert. Amineboranes as new reductive bleaching chemicals on softwood pulp single stage and multistage processes. *Tappi J* 82(2):110–114, 1999.
61. RO Hutchins, K Learn, B Nazer, D Pytlewski. Amineboranes as selective reducing and hydroborating agents: A review. *Organic Preparations and Procedure Int* 16(5):335–372, 1984.
62. GC Andrews, TC Crawford. Chemoselectivity in the reduction of aldehydes and ketones with amineborane. *Tetrahedron Lett* 21(8):693–696, 1980.
63. JA Schmidt, C Heitner. Use of UV-visible diffuse reflectance spectroscopy for chromophore research on wood fibers: A review. *Tappi J* 76(2):117–123, 1993.
64. CF Lane. The Borane-amine Complexes. *Aldrichimica Acta* 6(3):51–58, 1973.
65. SS Jr White, HC Kelly. Kinetics and mechanism of the morpholine-borane reduction of methyl alkyl ketones. *J Amer Chem Soc* 92(14):4203–4209, 1970.
66. TC Wolfe, HC Kelly. Kinetics and mechanism of the morpholine-borane reduction of substituted acetophenones and benzaldehydes. *J Chem Soc Perkin Trans. 2: Phys Org Chem* (1972–1999) 14:1948–1950, 1973.
67. HC Kelly, MB Giusto, FR Marchelli. Amineborane reductions in aqueous acid media. *J Amer Chem Soc* 86(18):3882–3884, 1964.
68. DG Holah, C Heitner. The color and UV-visible absorption spectra of mechanical and ultra-high yield pulps treated with alkaline hydrogen peroxide. *J Pulp Paper Sci* 18(5):161–165, 1992.
69. JA Schmidt, C Heitner. Light-induced yellowing of mechanical and ultrahigh yield pulps. Part 1. Effect of methylation, sodium borohydride reduction and ascorbic acid on chromophore formation. *J Wood Chem Technol* 11(4):397–418, 1991.
70. C Heitner. Light-induced yellowing of wood-containing papers: An evolution of the mechanism. ACS Symp Series 531 (Photochemistry of Lignocellulosic Materials), C Heitner, JC Scaiano, Eds. Washington, DC, Amer Chem Soc., 1993, pp. 2–25.
71. V Talrose, EB Stern, AA Goncharova, NA Messineva, NV Trusova, MV Efimkina. UV/ Visible Spectra in NIST Chemistry WebBook no. 69. PJ Linstrom, WG Mallard, eds. National Institute of Standards and Technology, Gaithersburg MD, 20899, http://webbook. nist.gov (November 25, 2008).
72. JA Schmidt, C Heitner. Use of UV-visible diffuse reflectance spectroscopy for chromophore research on wood fibers: A review. *Tappi J* 76(2):117–123, 1993.

73. C Pellerin, C Pedneault, S Robert. Bleaching chemistry of amine boranes with lignin model compounds. *J Pulp Paper Sci* 26(12):436–440, 2000.
74. A Sakakibara, Y Sano. *Chemistry of Lignin in Wood and Cellulosic Chemistry,* 2nd ed. Ed. DNS Hon and N Shiraishi. Marcel Decker Inc., New York, 1991, pp. 109–174.
75. N Roy. Étude de blanchiments réducteurs à l'aide de dérivés borés et sulfurés et leur influence sur la photoréversion. Master thesis, Université du Québec à Trois-Rivières (UQTR), 2000.
76. D Hivon, S Robert. Reductive bleaching of mechanical pulps: A molecular simulation study of reaction mechanisms of amineboranes. 11th Intl Symp Wood Pulp Chem, Nice, France, June 11–14, 375–381, 2001.
77. N L'Ecuyer, C Lobit, A Geldes, S Robert. Reductive bleaching of mechanical pulps by amineboranes: Molecular simulation and reaction mechanisms. 12th Intl Symp Wood Pulp Chem, Madison, WI, June 9–12, 143–146, 2003.
78. N L'Ecuyer, MC Laplante, S Robert. Molecular simulation and reaction mechanisms of the reductive bleaching of mechanical pulps by amineboranes. 13th Intl Symp Wood Fibre Pulp Chem, Auckland, New Zealand, May 16–19, Volume 2, 265–271, 2005.
79. PW Atkins, RS Friedman. *Molecular Quantum Mechanics,* 4th ed. Oxford: Oxford Univ. Press, 2004, 588 pages.
80. Gaussian 03, Revision D.01, MJ Frisch, GW Trucks, HB Schlegel, GE Scuseria, MA Robb, JR Cheeseman, JA Montgomery, Jr, TVreven, KN Kudin, JC Burant, JM Millam, SS Iyengar, J Tomasi, V Barone, B Mennucci, M Cossi, G Scalmani, N Rega, GA Petersson, H Nakatsuji, M Hada, M Ehara, K Toyota, R Fukuda, J Hasegawa, M Ishida, T Nakajima, Y Honda, O Kitao, H Nakai, M Klene, X Li, JE Knox, HP Hratchian, JB Cross, V Bakken, C Adamo, J Jaramillo, R Gomperts, RE Stratmann, O Yazyev, AJ Austin, R Cammi, C Pomelli, JW Ochterski, PY Ayala, K Morokuma, GA Voth, P Salvador, JJ Dannenberg, VG Zakrzewski, S Dapprich, AD Daniels, M C Strain, O Farkas, DK Malick, AD Rabuck, K Raghavachari, JB Foresman, JV Ortiz, Q Cui, AG Baboul, S Clifford, J Cioslowski, BB Stefanov, G Liu, A Liashenko, P Piskorz, I Komaromi, RL Martin, DJ Fox, T Keith, MA Al-Laham, CY Peng, A Nanayakkara, M Challacombe, PMW Gill, B Johnson, W Chen, MW Wong, C Gonzalez, JA Pople, Gaussian, Inc., Wallingford, CT, 2004.
81. E Cancès, B Mennucci, J Tomasi. A new integral equation formalism for the polarizable continuum model: Theoretical background and applications to isotropic and anisotropic dielectrics. *J Chem Phys* 107(8):3032–3041, 1997.
82. LG Wade. *Organic Chemistry*. Englewood Cliffs, NJ: Prentice-Hall, 1987, 923 pages.
83. CR Noller. *Chemistry of Organic Compounds,* 3rd ed. Philadelphia: W.B. Saunders Company, 1965, 989 pages.

14 Lignin Biodegradation

Karl-Erik L. Eriksson

CONTENTS

GROWTH OF WHITE-ROT FUNGI

GROWTH AND DEGRADATION OF WOOD

In North America, 1600–1700 species of wood-rotting fungi have been identified, 94% of which are white-rotters [1]. Most of these white-rot fungi colonize hardwood trees (angiosperms) with their lower lignin content and higher hemicellulose content, but many also degrade softwoods (gymnosperms).

Schacht, in 1863, was the first to describe the decomposition phenomena caused by fungi decaying wood. However Hartig, was the first to help generate a more common understanding that wood degradation was caused by biological processes [2]. He also contributed significantly to the understanding of wood decomposition and identified different types of decay. The three main types of wood-rotting fungi are classified as white-rotters, brown-rotters and soft-rotters. White-rot fungi are the only ones that can degrade lignin to any extent, while brown-rot and soft-rot fungi mainly degrade wood polysaccharides [3]. In the forest it is easy to distinguish between white- and brown-rotters by the color of the rotted wood. The ability to degrade lignin and the bleached white color of advanced decay caused by white-rot fungi suggest that different enzyme systems are employed by this type of wood-rotting fungi compared to the brown-rotters. Bavendamm [4] and Davidson et al. [5] used color formation around mycelia, on galic and tannic acid media to distinguish between white-rot and brown-rot fungi. This is a sensitive and accurate technique since only white-rot fungi produce extracellular phenoloxidases giving rise to quinone formation from phenols. The intensity and appearance of the color reaction varies considerably between different species of white-rot fungi. Due to differences in phenoloxidase reactions on kraft lignin agar plates, Ander and Eriksson [6] could divide 25 studied white-rot fungi into two groups that also differed in other properties.

White-rot fungi decay wood in several different ways that lead to both macroscopic and microscopic differences in the remaining wood. The most common way of attacking wood is probably to remove all the cell wall components simultaneously and in proportion to their occurrence in the cell wall while others preferentially degrade the lignin component [7–11] The term white-rot was mostly used for fungi that preferentially attacked lignin. Hubert [12] categorized decay by white-rot fungi as white-pocket, white-mottled, white-stringy, etc., depending upon the macroscopic characteristics. Blanchette [13–15] stated that no matter which method of classification was used, the decay caused by these fungi should be first classified as white-rot. White-rot fungi colonize wood quickly and the ray parenchyma cells are often the first to be colonized [3]. Fungal hypha penetrate from cell to cell via pit structures or bore holes through the cell walls. The type of wood being decayed influences the order in which lignin, cellulose, and hemicelluloses are degraded and this also differs among species of white-rot fungi [16–19]. The decay gives rise to either a simultaneous rot of lignin, cellulose and the hemicelluloses, or a preferential attack of one or more components. One prominent pattern of cell wall degradation in gymnosperms is an erosion of the cell wall from the cell lumen toward the middle lamella. Enzymes diffusing from the fungal hypha attack the wood immediately surrounding the hypha. In this process, all cell wall components are degraded [3]. After removal of the secondary wall layers, erosion continues into the middle lamella. Holes develop in certain areas and these holes become progressively larger until the entire cell wall is destroyed. The cell corners, known to have a high lignin content, are the most resistant to degradation and are sometimes the only remaining part of the cell in advanced stages of degradation. The attack of the cell wall from the lumen toward the middle lamella is a common trait and has been frequently observed [20–24].

The most common pattern of attack is for the white-rot fungi to attack all the wood components simultaneously (Figure 14.1). This normally happens through the progressive pattern of degradation described above. However, the most interesting type of white-rot is when the lignin is preferentially degraded within all cell wall layers. Some species, such as *Phellinus pini, Ganoderma tsugae,* and *Ceriporiopsis subvermispora* seem to always cause selective delignification at least of certain types of wood [25]. The picture of specific lignin degradation by *Phellinus pini* is presented in Figure 14.2. When lignin is preferentially removed from the entire cell wall, it causes a separation of the cells, particularly when the middle lamella is extensively attacked. However, lignin is also degraded in the secondary wall in this specific attack but there are no lysis zones, erosions troughs, or thinned areas. In softwoods the fibers may be totally depleted of middle lamella, but cell wall integrity is maintained. In the selective attack on the lignin, the crystalline nature of cellulose is not destroyed. It is believed that degradation of crystalline cellulose takes the concerted action of endo- and exo-glucanases. Studies of the plant cell wall degrading enzymes produced by *C. subvermispora* showed that this fungus did not produce cellobiohydrolase [26]. Blanchette et al. [27] used ultrastructural, immunocytochemical, and UV absorption spectroscopy techniques to elucidate the progressive changes that occur within woody cell walls during decay by *C. subvermispora*. Marker proteins of varying molecular weights were infiltrated into sound and decayed wood to evaluate changes in cell wall

FIGURE 14.1 Simultaneous attack on all the wood components by the white-rot fungus *Phanerochaete chrysosporium.*

FIGURE 14.2 Selective attack on the lignin by the white-rot fungus *Phellinus pine.* (Courtesy of R. A. Blanchette.)

porosity. Insulin (MW 5730) readily penetrated the outer regions of secondary walls of wood after two weeks of decay. Myoglobin (molecular weight 17,600) could not penetrate the cell walls until after four weeks and ovalbumin (mol. weight 44,287) could penetrate only after eight weeks when decay was in an advanced stage. None of the proteins used could defuse into cell walls of untreated control wood samples.

A third type of white-rot is a combination of selective delignification and simultaneous decay within the same type of wood. This is sometimes called a mottled type of white-rot with both forms of decay present [18,28,29]. In wood decayed by *Ganoderma applanatum*, both white zones, which are specifically delignified and tan areas, where a simultaneous attack has taken place, can be observed. The varieties of selectively delignified and simultaneously decayed wood are tremendous both among strains of the same fungus and among different fungal species. The pattern of degradation also has a lot to do with the wood itself. Other factors that influence the degradation pattern are wood nitrogen content, temperature, and humidity. Low nitrogen content has been found to play a crucial role in wood decay [30–32]. Low levels of nitrogen have been shown to stimulate lignin degradation, and to trigger the ligninolytic system, i.e., production of lignin depolymerizing enzymes such as lignin peroxidase (LiP) and manganese peroxidase (MnP) [33,34].

Biotechnical applications of white-rot fungi and their isolated ligninolytic enzymes have a potential for industrial use. Delignification of wood chips to reduce refining energy in mechanical pulping and consumption of wood chemicals in kraft cooking has been proposed [35]. Delignification of wood, straw, and bagasse to increase digestibility by ruminants, has also been demonstrated [36]. The use of ligninolytic enzymes, particularly laccase and MnP produced by white-rot fungi, are also of great potential for bleaching of wood pulp [37]. The use of white-rot fungi for treatment of waste bleach liquors to reduce color, toxicity, and mutagenicity has also attracted interest as well as treatment of soils or waste waters to remove environmental pollutants such as PCB, DDT, and dioxins [3].

PHYSIOLOGICAL DEMANDS FOR EXPRESSION OF LIGNINOLYTIC ENZYMES IN SUBMERGED CULTURES

Among the microorganisms that degrade lignin, the white-rot *Basidiomycetes* are the most efficient; they degrade lignin more rapidly and extensively than other groups of microorganisms. Although lignin is an energy-rich material, it has not been possible to demonstrate that it can serve as the sole carbon and energy source for any known microorganism [38]. For lignin degradation to take place, white-rot fungi require an additional, more easily utilizable carbon source [39–41].

The organism of choice for studies of lignin biodegradation has been the white-rot fungus *Phanerochaete chrysosporium* [3]. The ligninolytic system of *P. chrysosporium* is in effect only during secondary metabolism triggered mainly by nitrogen starvation [34]. However, secondary metabolism can also be triggered by carbon or sulphur starvation [42–44]. These findings are true for *P. chrysosporium* and many other white-rot fungi. However, there are also examples of white-rot fungi that are not so strongly regulated by nitrogen starvation. Addition of nitrogen to certain fungal cultures has even been found to increase lignin degradation [3]. Such fungi may be found in nitrogen rich environments such as cattle dung, whereas in fungi growing in wood, where they encounter low nitrogen concentrations [30] lignin degradation would be repressed. High nitrogen content represses lignin degradation (^{14}C-lignin → [14]CO$_2$), as first discovered by Kirk and coworkers [33,34] with *P. chrysosporium* and *Coriolus versicolor.* This was a major discovery that greatly stimulated and benefitted lignin biodegradation research. It was later found that many amino acids, especially glutamate, glutamine, and histidine, strongly suppressed lignin degradation [45,46].

Veratryl alcohol is an important secondary metabolite associated with the ligninolytic system. Many white-rot fungi, particularly those producing LiP, have been demonstrated to produce veratryl alcohol *de novo* [47,48]. Veratryl alcohol is synthesized using the phenylalanine pathway [49]. Other culture parameters that affect lignin degradation also have an impact on veratryl alcohol synthesis. Veratryl alcohol has been shown to induce LiP activity and so does lignin. However, the increased LiP activity does not seem to give rise to a significant increase in total lignin metabolism by *P. chrysosporium,* i.e., lignin → [14]CO$_2$, which may indicate that LiP is not the sole rate-limiting component.

Since lignin degradation is almost entirely an oxidative process increased oxygen levels enhance lignin degradation considerably in various white-rot fungi [50]. It was found by Kirk et al. [33,51] that cultures of *P. chrysosporium,* maintained under an atmosphere of 5% O$_2$, released only 1% of total available ^{14}C-ring-DHP as [14]CO$_2$ after 35 days of incubation while approximately 47 and 57% of the ^{14}C-label was converted to [14]CO$_2$ if the oxygen percentage was 21 and 100%, respectively. It was also found by the same workers, that the maximum rate of [14]CO$_2$ evolution was approximately three times higher at 100% O$_2$-atmosphere compared to in air. Similar findings in other fungi suggest that the beneficial effect of O$_2$ on lignin biodegradation might be applicable to white-rot fungi in general.

Agitation of *P. chrysosporium* cultures was originally found to strongly suppress ligninolytic activity (^{14}C-lignin → [14]CO$_2$) and it was reported that LiP activity was undetectable in agitated cultures of this fungus [33,50–55]. There are, however, conflicting reports in the literature concerning agitation and lignin degradation. Reid et al. [56] reported that agitated cultures of *P. chrysosporium,* in which the mycelium had formed single large pellets, readily released [14]CO$_2$ from ^{14}C-ring-DHP. Under these conditions, lignin in aspen wood was also degraded to water soluble products and CO$_2$. Production of both LiP and oxidation of lignin to CO$_2$ in agitated cultures of wild-type and mutant strains of *P. chrysosporium* have later been demonstrated in several laboratories [57–59]

ENZYMOLOGY OF LIGNIN DEGRADATION

In *P. chrysosporium,* LiP and MnP and the H$_2$O$_2$-generating system seem to be the major components of the extracellular lignin degrading system [57,60–64]. While MnP expression is regulated at the level of gene transcription similarly to LiP, i.e., by the depletion of nutrient nitrogen [65], MnP activity is also dependent upon the presence of Mn^{2+} in the culture medium [66,67]. The *mnp* gene transcription is regulated by Mn^{2+} [67–70]. MnP is also regulated at the gene transcription level by heat shock

[71] H_2O_2, and other chemical stress [72]. Gold and coworkers [73] have demonstrated in stationary cultures supplemented with 180 µM Mn, that the transcription levels of *mnp1* and *mnp2* increased approximately 100 and 1700 fold respectively over basal levels. Also, the two genes were expressed differently in stationary and agitated cultures, i.e., in stationary cultures, *mnp2* was the major expressed *mnp* gene whereas in agitated cultures it was the *mnp1* gene. The enzymology of MnP will now be discussed in more detail.

Manganese Peroxidases

Besides LiP, one of the most studied lignin degrading phenoloxidases is MnP, another heme containing extracellular fungal peroxidase. MnP was first identified in ligninolytic cultures of *Phanerochaete chrysosporium* as a Mn(II) and H_2O_2 dependent oxidase [55]. Later it was purified and characterized from both *P. chrysosporium* and *Trametes versicolor* [74–77].

MnP is ubiquitous among white-rotters, and is capable of oxidizing phenolic lignin units, various phenols and lignin model compounds. Recently MnP was even shown to degrade nylon [78]. MnP has a ferric, high spin pentacoordinate iron pr\otoporphyrin IX heme cofactor. MnP from *P. chrysosporium* contains about 15% carbohydrates.

The mechanism of MnP catalysis has been studied in detail [79]. It was shown that the ferric MnP native enzyme donates two electrons to H_2O_2, and its heme is oxidized to a ferrolyl state while a porphyrin radical is formed (Figure 14.3). Thus, the enzyme acquires two oxidizing equivalents, one located on the iron, and another on the porphyrin ring. This form of the enzyme is named Compound I. In the next step, Compound I reacts with a Mn^{2+} ion or a terminal phenolic substrate, oxidizing either one, Mn^{2+} to Mn^{3+}, or the phenolic substrate to its phenoxy radical, and is converted to Compound II in this process. If neither of these substrates is encountered, Compound I can be reduced to Compound II by a second molecule of H_2O_2, which in turn is

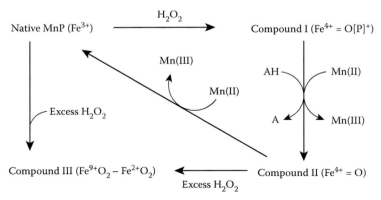

FIGURE 14.3 The catalytic cycle and the five oxidation states of Mn-peroxidase. Reaction paths 3 → 5 → 4 → 3 indicate one catalytic cycle of the enzyme. ROOH = H_2O_2; AH_2 = phenolic substrate. (Based on Wariishi, H., Akileswaran, L., and Gold, M.H., *Biochemistry*, 27, 5365–5370, 1988. With permission.)

probably oxidized to $HO_2^{\cdot}/O_2^{-}$. Compound II can be reduced to the native enzyme by oxidizing another molecule of Mn^{2+}. Mn^{3+} ions, formed in each of these reactions, then oxidize the terminal phenolic substrates. Mn^{3+} ions have to be properly chelated by organic acids to be able to oxidize these substrates. Even in the absence of the enzyme, suitably chelated Mn^{3+} ions were shown to act as a ligninolytic agent and oxidize veratryl alcohol (VA), lignin and lignin model compounds [80].

In the presence of excess H_2O_2, both native enzyme and Compound II can be converted to the inactive state of the enzyme, "Compound III." Therefore, the concentration of H_2O_2 must be controlled very strictly in MnP reactions.

As explained above, MnP is a Mn^{2+} and H_2O_2-dependent enzyme, and it oxidizes Mn^{2+} to Mn^{3+}, which, in turn oxidizes lignin and other substrates such as aromatic phenols, phenolic lignin model compounds and high molecular weight chlorolignins [81,82]. In order for Mn^{3+} to diffuse away from the enzyme and oxidize other compounds, it has to be sufficiently stable, and also able to dissociate from the active site of the enzyme. These requirements are satisfied when Mn^{3+}, formed by the peroxidase, complexes with organic acids that are metabolic products of the fungus. Most Mn^{3+} chelators are carboxylic acids; malonate, oxalate, and lactate are most commonly and effectively employed for this purpose [83]. As mentioned, these chelators facilitate the dissociation of enzyme-manganese complex by chelating Mn^{3+} [84].

Another important function of H_2O_2 is to induce MnP gene transcription. Studies with *P. chrysosporium* showed that MnP is induced 1.6-fold upon addition of Mn(II) and H_2O_2 [72]. Chemical stress caused by compounds like ethanol, sodium arsenite, and 2,4-dichlorophenol, as well as heat shock and molecular oxygen (in the presence of Mn) were also found to increase the levels of MnP transcripts. Although Mn^{2+} alone was shown to function as a MnP inducer in *P. chrysosporium* [66,85], in studies with another fungus, *Phlebia radiata*, contradictory results were reported [86]. This group demonstrated that high concentrations of Mn(II) had no influence on the levels of LiP, MnP, or laccase. Moreover, high Mn containing cultures exhibited less efficient DHP mineralization. Malonate itself was also inhibitory to DHP mineralization. However, when Mn(II) and malonate were added together to the cultures, production of all three above mentioned enzymes were elevated.

From the kinetic and model studies it was postulated that the catalytic site of the *P. chrysosporium* MnP is located at or near the heme edge of δ-meso-carbon [87]. When the crystal structure of this MnP was elucidated, the active site was found to consist of a His ligand hydrogen bonded to an Asn residue, and a distal side peroxide binding pocket formed by a catalytic His and Arg [88]. Mn(II) binding site was shown to be at the propionate ends of the heme, Mn(II) being hexacoordinated by an Asp, two Glu residues, a heme propionate and two water molecules (Figure 14.4).

Each molecule of MnP was first reported to contain two Ca^{2+} ions, which are believed to have a structural role. It has been shown that the thermal stability of the enzyme depends on the presence of these Ca^{2+} ions. Thermal inactivation appears to be a two-step process; loss of a Ca^{2+} decreases the enzyme's stability [89]. If excess Ca^{2+} is added to the medium at this stage, the enzyme can be reactivated. However, further inactivation caused by the loss of heme, cannot be reversed. In another study

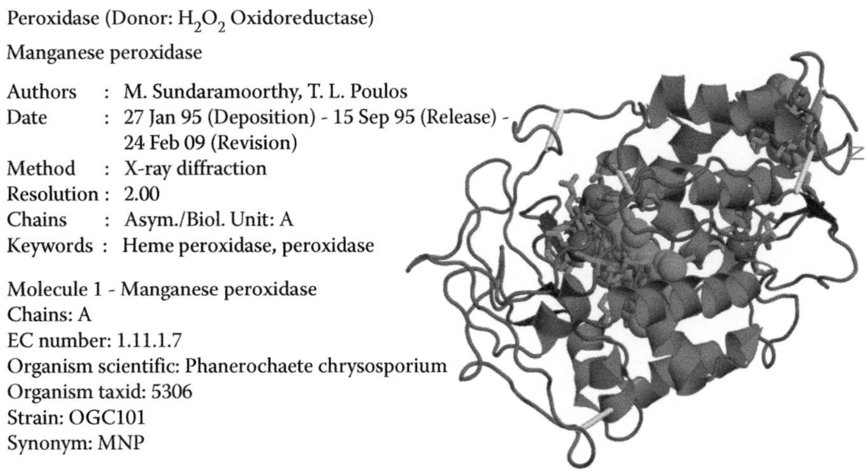

Peroxidase (Donor: H$_2$O$_2$ Oxidoreductase)
Manganese peroxidase

Authors : M. Sundaramoorthy, T. L. Poulos
Date : 27 Jan 95 (Deposition) - 15 Sep 95 (Release) -
24 Feb 09 (Revision)

Method : X-ray diffraction
Resolution : 2.00
Chains : Asym./Biol. Unit: A
Keywords : Heme peroxidase, peroxidase

Molecule 1 - Manganese peroxidase
Chains: A
EC number: 1.11.1.7
Organism scientific: Phanerochaete chrysosporium
Organism taxid: 5306
Strain: OGC101
Synonym: MNP

FIGURE 14.4 X-ray diffraction of manganese peroxidase, resolution 2.06 Angstroms. (From Jena Library of Biological Macromolecules – JenaLib; www.fli-leibniz.de/IMAGE.html; Hühne, R., Koch, F.T., and Sühnel, J., *Brief Funct Genomic Proteomic*, 6, 220–239, 2007. With permission.)

Asymmetric/Biological Unit (10, 10)

No.	Name	Evidence	Residues	Description
01	AC1	Software	ASN A:98, THR A:99, ILE A:100, ASN A:131, NAG A:362, HOH A:498	Binding site for residue NAG A 361
02	AC2	Software	MET A:94, GLN A:95, ASN A:98, NAG A:361	Binding site for residue NAG A 362
03	AC3	Software	SER A:174, ASP A:191, THR A:193, THR A:196, ASP A:198	Binding site for residue CA A 371
04	AC4	Software	ASP A:47, GLY A:62, ASP A:64, SER A:66, HOH A:493, HOH A:545	Binding site for residue CA A 372
05	AC5	Software	GLU A:35, GLU A:39, ASP A:179, HEM A:396, HOH A:441, HOH A:520	Binding site for residue MN A 381
06	AC6	Software	HIS A:38, GLU A:39, ARG A:42, PHE A:45, GLU A:143, PRO A:144, LEU A:169, SER A:172, HIS A:173, ALA A:176, ARG A:177, ALA A:178, ASP A:179, LYS A:180, VAL A:181, MN A:381, HOH A:478, HOH A:520, HOH A:556, HOH A:650	Binding site for residue HEM A 396
07	CA1	Author	CA A:371, THR A:196, ASP A:191, SER A:174, THR A:193, ASP A:198	Proximal calcium binding site
08	CA2	Author	CA A:372, ASP A:47, GLY A:62, SER A:66, ASP A:64, HOH A:493, HOH A:545	Distal calcium binding site
09	GL1	Author	NAG A:361, NAG A:362, ASN A:131	Carbohydrate binding site
10	MN	Author	MN A:381, HEM A:396, GLU A:35, GLU A:39, ASP A:179, HOH A:441, HOH A:520	Manganese binding site

Ligands, Modified Residues, Ions (4, 6)

No.	Name	Count	Type	Full Name
1	CA	2	Ligand/Ion	Calcium ion
2	HEM	1	Ligand/Ion	Protoporphyrin IX containing FE
3	MN	1	Ligand/Ion	Manganese (II) ion
4	NAG	2	Ligand/Ion	N-acetyl-D-glucosamine

FIGURE 14.4 (Continued)

Timofeevski et al. [90] showed that a molecule of MnP contained 4 Ca^{2+} ions, two of which are released upon thermal inactivation. In fact, at higher temperatures Mn^{2+} stabilized the enzyme more efficiently than did Ca^{2+}.

From the results of kinetic studies with *P. chrysosporium* MnP, it was suggested that since enzyme activity was not stimulated by the size of the chelators, enzyme must bind free Mn(II) rather than the enzyme-chelator complex, and compete with the chelators for free Mn(II). It was also speculated that the optimal chelators should have relatively low Mn(II) binding constants [91]. However, Kuan et al. [92] opposed this suggestion. They argued that good chelators such as oxalate, malonate and lactate facilitate the reaction between Comp II and Mn(II). However, chelators that do not readily complex with Mn(II) such as succinate and polyglutamate did not stimulate the same reaction. Therefore, Compound II of MnP preferentially reacts with chelated Mn(II).

Some of the other extracellular fungal enzymes produced simultaneously with MnP appear to work in accord with MnP. Studies of the simultaneous action between laccase and MnP, which coexist in the cultures of various fungi, showed that laccase worked in concert with MnP in the degradation of ligninsulfonates and solubilization of Hevea lignins, the highest degradation was obtained in both cases when the enzymes were applied together [93,94]. Similarly, an interaction model between MnP and another extracellular enzyme produced by *T. versicolor*; cellobiose dehydrogenase (CDH) was also proposed [95]. According to this model, CDH supports MnP in different ways: 1: CDH oxidizes cellobiose to cellobionic acid, an efficient Mn(III) chelator. 2: Insoluble MnO_2, formed from the disproportionation of Mn(III), is returned into the soluble Mn pool in the form of Mn(II) and Mn(III) by CDH. This reaction not only provides extra Mn(II) for MnP catalysis, but also stimulates MnP production. 3: Quinones are converted to the corresponding phenolics by CDH, and these phenolics are substrates for MnP.

As already mentioned, MnP can oxidize phenolic lignin substructures and various phenols, but is unable to oxidize nonphenolic lignins. However, in the presence of certain compounds, which could act as mediators, degradation of nonphenolic lignin substructures was also reported. Wariishi et al. [96] found that in the presence of glutathione (GSH), MnP could efficiently oxidize veratryl alcohol (VA), anisyl alcohol, and benzyl alcohol. They demonstrated that the formed Mn^{3+} oxidizes thiol to a thiyl radical, which in turn abstracts a hydrogen from the substrate and forms the corresponding aldehyde. Substrate oxidation was at least two magnitudes higher when the reactions were performed under anaerobic conditions.

In another study by Bao et al. [97], it was shown that in the presence of Tween 80, MnP and Mn^{2+} could oxidize a β–O–4 lignin model compound without the need for addition of H_2O_2. They suggested a mechanism called *MnP mediated lipid peroxidation,* by which lipid peroxy-radicals are generated. These radicals easily abstract hydrogens from substrates. The yields of the two polar oxidation products detected by HPLC were 20 and 26%, respectively. The dimer was not oxidized when MnP or Mn^{2+} were omitted, when a free radical scavenger was added, or when the reaction was conducted under argon. The same mechanism of nonphenolic lignin degradation was demonstrated in *Ceriporiopsis subvermispora* cultures in which MnP but not LiP is produced [98].

D'Annibale and his colleagues [99] performed experiments on VA oxidation using MnP from *Lentinus edodes*. They demonstrated two different types of reactions occurring in the presence and absence of GSH. When GSH was not included in the reaction mixture of MnP, Mn^{2+}, VA, H_2O_2 and chelator, aromatic ring cleavage, side chain oxidization and dimerization of VA were detected. (The main metabolite was γ-muconolactone (58%), and three dimerization products were also observed [31] 2.3 and 1.8%, respectively). However, with the addition of GSH, veratraldehyde was the only metabolite formed. Veratryl alcohol oxidation was 32 and 14% in nonthiol and thiol mediated reactions, respectively. Under both conditions, oxidation depended strictly on the presence of both Mn^{2+} and H_2O_2.

Various fungi, including *P. chrysosporium,* and *T. versicolor* were shown capable of degrading lignin and bleaching kraft pulp. These abilities were mostly correlated with the MnP activity found in the culture fluids. Reid [100] demonstrated that in a 14-day culture of *T. versicolor* on unbleached softwood kraft pulp followed by a DED bleaching sequence increased brightness from 33 to 61% and lowered kappa number from 24 to 8.5.

Various papers have been published that establish that MnP can increase brightness and decrease pulp kappa number. For instance, MnP purified from *T. versicolor,* caused most of the demethylation and delignification of the kraft pulp when compared to the effect of the complete, cellfree culture solution [101]. In that study, it was also demonstrated that MnP bleached kraft pulp brownstock and caused release of methanol. The highest level of bleaching effect was obtained in cultures of *T. versicolor* when MnP production and activity were at their peak values. It was also found that MnP showed preference for oxidation of phenolic lignin substructures. Various dikaryotic and monokaryotic strains of *T. versicolor* were tested for their ability to bleach softwood or hardwood kraft pulp [102]. It was found that a monokaryotic strain named 52J caused more than two magnitudes higher brightness gain compared to its parent dikaryotic strain probably due to that biomass and dark pigment production by strain 52J was much less than that by the dikaryon. In 1995, Addleman and others produced MnP deficient mutants of the parent *T. versicolor* 52J, and those mutants were not only unable to bleach or delignify hardwood kraft pulp, but also released less ethylene from 2-keto methiolbutyric acid and produced less $^{14}CO_2$ from ^{14}C-ring labeled DHP. When purified MnP was added to pulp cultures, pulp bleaching and delignification as well as methanol release were restored to a certain extent [103].

Katagiri et al. [104] investigated the effects of MnP on pulp bleaching in a solid culture system in which *P. chrysosporium* and *T. versicolor* were incubated with unbleached hardwood kraft pulp. In five days cultures, the brightness increase was 30 and 15% ISO-points, respectively.

A series of studies on the role of MnP in pulp bleaching and lignin degradation was performed by Kondo et al. [105]. The pulp bleaching studies were performed with *Phanerochaete sordida* YK-624, and demonstrated a 21.4% brightness gain in seven days by utilizing a membrane system that prevented a direct contact between the hyphae and the kraft pulp. They concluded that enzymes such as peroxidases and phenoloxidases, which can diffuse through the membrane, were responsible for the bleaching effect, and demonstrated a positive correlation between the level of MnP and brightening of the pulp. The same group also worked with partially purified MnP from YK-624, and found that the enzyme, in the presence of Mn(II), Tween 80, sodium malonate, with continuous addition of H_2O_2, could decrease the kappa number by 6 points and increase brightness by 10 points in hardwood kraft pulp in only 24 hrs. Brightness increased 43% ISO after six separate treatments with MnP combined with alkaline extraction. This was the first report about the utilization of a MnP/Tween 80 system for pulp bleaching [105]. The same researchers also found that $MnSO_4$ was in fact dispensable for pulp bleaching by MnP [106]. Thus, 13 points of brightness increase was obtained in the presence of MnP, continuous addition of H_2O_2, and 2 mM of oxalate. It was concluded that Mn(II) already present in the pulp was sufficient for the function of MnP catalysis. These researchers also studied another fungus; a *Ganoderma sp.* strain YK-505. It was found to bleach pulp under unfavorable conditions such as at high temperatures and hydrogen peroxide concentrations although other MnPs could not maintain their activities at such conditions [107]. MnP isozyme G-1 from YK-505 exhibited optimum activity at 55°C and both isozymes G-1 and G-2 seemed to be very stable to thermal denaturation. Continuous addition of 5–100 mM H_2O_2 at a rate of 1.5 ml/h increased brightness of pulp by both of these isozymes although MnPs from most other sources would be rapidly inactivated at such excessive H_2O_2 concentrations. Even an initial addition of 1 or 2 mM H_2O_2 to the MnP preparation from YK-505 did not inhibit the enzyme's pulp brightening ability. Mn(II) was also found not to be required for biobleaching of oxygen delignified kraft pulp by *Bjerkandera sp.* strain BOS55. MnP was produced by this strain, grown on pulp, even without addition of Mn(II) [108]. External addition of $MnSO_4$ did not cause a further increase in brightness, although the bleaching ability was explained by the presence of $MnSO_4$ in the pulp according to Harazono et al. [107].

In another study, when MnP from the fungus IZU-154 was incubated with softwood or hardwood kraft pulp in the presence of Mn^{2+}, glucose oxidase and glucose, the amount of effective chlorine needed to increase brightness was substantially lowered. After the enzyme treatment, chlorine consumption required to increase the brightness to 85% ISO decreased 51, 66, and 69% for normal-lignin-content softwood kraft pulp, low-lignin-content softwood kraft pulp and hardwood kraft pulp respectively [109].

More recent investigations with the *Bjerkandera sp* strain BOS55 revealed the importance of organic acids for pulp bleaching in the absence of Mn(II) [110]. The advantages with organic acids were summarized as: Better pH-buffering effects, stimulation of phenoloxidase production, induction of useful secondary metabolites such as oxalate and veratryl alcohol, and reduction of oxygen radicals. They also suggested that MnP from this strain might possibly function independently of Mn(II). Reid and Paice [111], investigated the effects of MnP from *T. versicolor* on residual lignin in softwood kraft pulp. They demonstrated that although MnP partially oxidized the lignin in the pulp it did not degrade it to soluble fragments. Finally, Reid [112] reports that the activities of the known enzymes of *T. versicolor*, i.e., laccase and MnP, do not account for all the lignin degradation by this organism, and suggested that additional enzymes might play a role as well.

LACCASE AND REDOX-MEDIATORS

Laccase was first identified by Yoshida [113], as a proteinaceous substance that catalyzed the lacquer curing process and it was first referred to as diastase. Bertrand [114], who had identified one of the defining reactions catalyzed by the enzyme, i.e., the ability to oxidize hydroquinone, implemented the name laccase [115].

Laccases are so widely distributed in the fungi that it is very possible that this enzyme may be ubiquitous in these organisms [3,116]. Laccases have also been found in a large variety of plant species [117], in about a dozen studied insects [118], and has also recently been found in the bacterium *Azospirillum lipoferum* [119]. Laccase has a variety of functions including participation in lignin biosynthesis [117], degradation of plant cell-walls [120,121], plant pathogenicity [122], and insect sclerotization [123].

Laccases (p-diphenol: O_2 oxidoreductase; EC 1.10.3.2) catalyze the oxidation of *p*-diphenols with the concurrent reduction of dioxygen to water. However, the actual substrate specificities of laccases are often quite broad and vary with the source of the enzyme [116,117]. Laccases are members of the blue copper oxidase enzyme family. Members of this family have four cupric (Cu^{2+}) ions where each of the known magnetic species (type 1, type 2, and type 3) is associated with a single polypeptide chain. In the blue copper oxidases the Cu^{2+} domain is highly conserved and, for some time, the crystallographic structure of ascorbate oxidase, another member of this class of enzymes, has provided a good model for the structure of the laccase active site [124,125]. The crystal structure of the Type-2 Cu depleted laccase from *Coprinus cinereus* at 2.2. A resolution has also been elucidated [126].

To differentiate between the enzyme activities of laccase and other phenoloxidases is not a trivial matter. The reason is that laccases, unlike most enzymes, are relatively nonspecific in terms of their substrates. Not only *p*-diphenols, but also *o*-diphenols, aryldiamines, aminophenols, and hydroxyindols may be oxidized by laccases [116,127–129]. Laccases are also capable of oxidizing monophenols but are rapidly inactivated by the products' phenoxyradicals. It is therefore not surprising that a great deal of confusion exists in the laccase literature due to the lack of a direct definition of what constitutes a laccase.

Recent studies suggest that in white-rot fungi, the combination of laccase with either LiP and/or MnP is a more common combination of phenoloxidases than the LiP/MnP pattern found in *P. chrysosporium*. While laccases from different fungal species serve different purposes, laccases produced by white-rot fungi are believed to mainly participate in the degradation of lignin. Although used for the same purpose, these fungal laccases could have very different characteristics such as carbohydrate content, redox potential, substrate specificity, thermal stability, etc., depending on the fungal species. One problem in efforts to assign to laccase a substitute role for either LiP or MnP in lignin degradation has been that the redox potentials of the laccases studied so far have not been high enough to remove electrons from nonphenolic aromatic substrates. Such structures make up 90% of lignin structures and laccases alone cannot oxidize these predominately nonphenolic substructures of lignin. However, Sariaslani et al. [130] demonstrated that laccase could oxidize a nonphenolic aromatic compound, rotenone, in the presence of chlorpromazine. The oxidation of methoxylated benzyl alcohols by laccase in the presence of syringaldehyde was demonstrated by Kawai et al. [131]. Bourbonnais and Paice [132] showed that two artificial laccase substrates ABTS (2,2′-azinobis-[3-ethylbenzoline-6-sulfonate]) and remazol blue, could act as redox-mediators, which enable laccase to oxidize nonphenolic lignin model compounds. The importance of these findings were extended when the same authors later demonstrated that kraft pulp could be partially delignified and demethylated by laccase from *Trametes versicolor* in the presence of ABTS [133]. Call and Mücke in Germany [134] were the first to apply the laccase mediator concept to pulp bleaching in pilot plant scale. They used 1-hydroxybenzotriazole (1-HBT) as redox mediator. With the rapidly developing interest for laccase-based bleaching, investigations were started at the University of Georgia, regarding the role played by laccases in lignin degradation by various white-rot fungi. An extensive screening was undertaken to identify white-rot fungi that produce large amounts of laccase. *Pycnoporus cinnabarinus*, an isolate obtained from decaying pine wood in Queensland, Australia, was found and turned out to be an ideal candidate for in-depth studies to examine the significance of laccase in lignin degradation. Crucial for the choice of *P. cinnabarinus* as a model organism was that it produces only one isolectric form of laccase and small amounts of an as yet unidentified peroxidase with substrate specificities that do not fit those of either LiP or MnP [135]. Despite the lack of LiP and MnP production, the rate of lignin degradation by *P. cinnabarinus* is comparable to that of *P. chrysosporium* [136]. *P. cinnabarinus* laccase has the same traits as practically all other laccases from white-rot fungi. The molecular mass, 76,500 Da, as determined by SDS-PAGE, is comparable to other fungal laccases. Spectroscopic characterization showed a typical laccase spectrum both in the UV/visible region as with EPR technique. The results confirmed the presence of 4 Cu(II) ions typical for an intact laccase active center. The level of glycosylation was about 9% with mannose as the predominant glycosyl residue (>70%). Comparison of the N-terminal protein sequence of the *P. cinnabarinus* laccase with those of other fungal laccases, showed the closest similarity to laccase II from *T. versicolor* (86%). High similarity was also found with other laccases from white-rot fungi while in contrast the N-terminal sequences of laccases isolated from non-wood-rotting fungi such as *Neurospora crassa* and the yeastlike fungus *Cryptococcus neoformans* were significantly different with similarities of 18 and 0%, respectively. We took the results of these investigations as a sign that the lack of LiP and MnP did not seem to require compensation by a laccase with an unusually high redox potential for lignin degradation. The results were also taken as support for the possible production by the fungus of a physiological enzyme-mediator system allowing for the oxidation of nonphenolic lignin structures by the *P. cinnabarinus* laccase.

When *P. cinnabarinus* was grown at 30°C in shake flask cultures with glucose as the sole carbon source, a red pigment and laccase activity developed and accumulated in parallel in the culture solutions (Figure 14.5) [137]. The red color turned out to be in good agreement with the spectrum reported previously for cinnabarinic acid (CA) [138,139]. However, CA turned out not to act as a redox mediator for the laccase while the o-aminophenol, 3-hydroxyantranilic acid, demonstrated to be a precursor for CA in other biological systems, did. We demonstrated in *in vitro* studies with purified laccase that it catalyzed the formation of CA by oxidative coupling

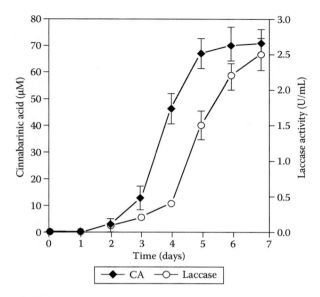

FIGURE 14.5 Laccase and cinnabarinic acid production during growth of *P. cinnabarinus*. (From Eggert, C., Temp, U., Dean, J.F.D., and Eriksson, K-E.L., *FEBS Lett*, 376, 202–206, 1995. With permission.)

of 3-HAA [137]. The K_m value obtained for 3-HAA was in the same range as for other common laccase substrates such as guaiacol.

Synthetic substrates, representing lignin substructures, are commonly used to assess ligninolytic activities [3]. Three such model compounds representing both phenolic and nonphenolic structures commonly found in lignin, were used to test the ability of *P. cinnabarinus* laccase to oxidize these substrates with and without 3-HAA. With laccase only, the model compounds containing phenolic hydroxyl groups were oxidized while in the presence of laccase and 3-HAA the nonphenolic model was also oxidized. To investigate if the laccase/3-HAA couple could depolymerize lignin a synthetic polymer (DHP) prepared from ^{14}C-ring-labeled coniferyl alcohol was used. Repolymerization reactions are a common feature of *in vitro* systems for lignin degradation, which typically present a problem for demonstrating ligninolytic activity toward high molecular weight substrates by LiP and other phenoloxidases. To avoid this problem we used in our experiments, a cassette to contain the high molecular weight substrates while degradation fragments can rapidly diffuse through a dialysis membrane at both sides of the cassette with a nominal cut-off of 10 kDa, (Figure 14.6) [140]. We found that with a high molecular weight laccase outside the membrane and the high molecular weight substrate inside the membrane, nothing happened. However, with addition of 3-HAA, outside the cassette + laccase, radioactive material from the labeled DHP diffused into the outer reaction chamber. Gel permeation chromatography of the reaction products resulted in a broad peak of material ranging from molecular weights approximately around 4000 to monomeric phenolic compounds. This experiment clearly showed that treatment with the laccase/3-HAA couple led to fragmentation of the synthetic lignin. Since

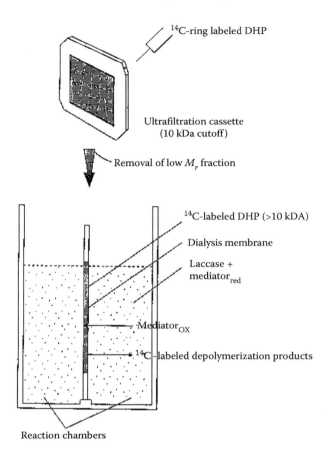

FIGURE 14.6 Picture detailing an apparatus for testing the ability of ligninolytic systems based on redox mediators to fragment high M_r polymeric substrates. Radio-labeled polymere is injected into a commercially available dialysis membrane cassette that is sandwiched between two reaction chambers. Various combinations of enzymes, mediators, substrates, or reductants may be added to each chamber, and fragmentation of the polymere can be followed by quantitating the low M_r material passing through the membrane. (From Eggert, C., Temp, U., Dean, J.F.D., and Eriksson, K-E.L., *FEBS Lett*, 391, 144–148, 1996. With permission.)

the enzymes involved in lignin degradation are too large to penetrate the unaltered wood-cell wall [141] the use of low molecular weight diffusable compounds to oxidize the polymer makes sense.

To further demonstrate the importance of laccase in lignin degradation by the white-rot fungus *P. cinnabarinus* we produced laccase-less mutants of the fungus. We could show that these laccaseless mutants were greatly reduced in their ability to metabolize ^{14}C-ring-labeled DHP. However, [14]CO$_2$ evolution in cultures of these mutants could be restored to levels comparable to those of the wild type cultures by addition of purified *P. cinnabarinus* laccase. This clearly indicates that laccase is essential for lignin degradation by this white-rot fungus [142].

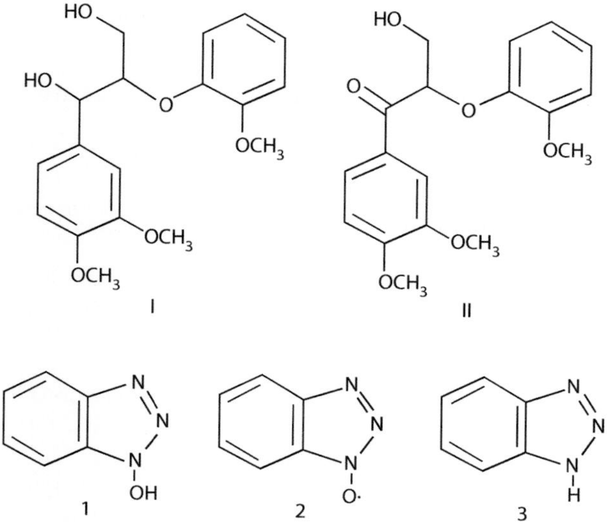

FIGURE 14.7 Structures of mediator and lignin model compounds. (From Li, K., Helm, R.F., and Eriksson, K-E.L., *Biotechnol Appl Biochem*, 27, 239–243, 1998. With permission.)

Although a laccase-mediator system seems to be both an interesting and a promising method for environmentally benign pulp bleaching, there are several hurdles before such systems can be applied in pulp mills. The laccase mediators identified so far are still too expensive, the effect of laccase mediator systems in pulp bleaching are still not satisfactory and the mechanisms underlying these laccase/mediator reactions are yet unclear. To screen for more efficient laccase mediators, a fast screening system was developed by Li et al. [143] (Figure 14.7).

Monitoring the oxidation of compound I by HPLC was found to be useful for a fast screening of many potentially effective laccase mediators. A lignin structure such as the ketone II, can easily be degraded by hydrogen peroxide under alkaline condition via a Dakin type reaction [143]. Thus, if the lignin interunit linkage I was oxidized to form a ketone structure II in lignin, lignin macromolecules would be depolymerized when treated with an alkaline solution of hydrogen peroxide. While oxidation of dimer I is carried out in a homogeneous solution, removal of lignin in pulp is a reaction in a heterogeneous system. Therefore a laccase mediator system unable to oxidize dimer I is unlikely to efficiently remove lignin in pulp. Using dimer I, a large number of putative laccase mediators were screened. It was found in this screening work that phenols were not effective laccase mediators, nor were compounds containing both a hydroxyl and an amino group on an aromatic ring such as 3-HAA. In general, aromatic amines are not ideal as laccase mediators since most of them are toxic to the environment. ABTS has been intensively investigated and used as a laccase mediator [133,144,145]. However, laccase plus ABTS did not result in any oxidation of dimer I. Although dyes are not very desirable laccase mediators because of their strong color, several dyes were investigated as well. Neither dyes nor heterocyclic compounds containing S and N atoms were effective as mediators. The last category of laccase mediators

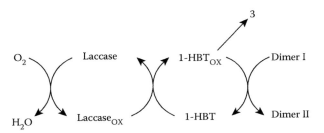

SCHEME 14.1 Proposed mechanism for the laccase-mediator oxidation of non-phenolic lignins. The label 3 in the scheme refers to compound 3 in Figure 14.7. (From Li, K., Helm, R.F., and Eriksson, K-E.L., *Biotechnol Appl Biochem*, 27, 239–243, 1998. With permission.)

we investigated was compounds containing hydroxylamino groups (HO-NR$_2$). These categorizes of compounds such as 1-HBT, used by Call and Mücke [134], turned out to be the most effective laccase mediators we have found so far.

To better understand the mechanisms for the oxidation of lignin by a laccase mediator system, a laccase from *Polyporous* sp, kindly provided by Novozymes, was used in combination with 1-HBT. The redox mediator was found to be partly regenerated during the oxidation of lignin dimer I in the presence of laccase. A free radical of 1-HBT generated by laccase was probably responsible for the oxidation of I [146]. The free radical of 1-HBT was, however, transformed to benzotriazole, which could not mediate the oxidation of I. A proposed mechanism for the laccase mediator oxidation of nonphenolic lignins is given in Scheme 14.1.

To investigate the importance, not only of laccase mediators, but also of laccases per se, several laccases were studied for the oxidation of the nonphenolic lignin dimer I. In the presence of the redox mediators 1-HBT or violuric acid, it was found that the oxidation rates of dimer I by the laccases differed considerably. In oxidation of dimer I, both 1-HBT and violuric acid were to some extent, consumed. The consumption rate followed the same order of laccases as the oxidation rates of dimer I. The oxidation rate of dimer I was found to be dependent on both k_{cat} and the stability of the laccase in question. Both 1-HBT and violuric acid inactivated the laccases—violuric acid to a greater extent then 1-HBT. The presence of dimer I in the reaction mixture slowed down this inactivation. Inactivation seems to be mainly due to the reaction of the redox mediator free-radical with the laccases. No relationship between the carbohydrate content of the laccases and their inactivation was found. When the redox potential of the laccases is in the range of 750–800 mV, i.e., about that of the redox mediator, a further increase in redox potential does not affect k_{cat} and the oxidation rate of dimer I [147].

REFERENCES

1. RL Gilbertson. Wood-rotting fungi of North America. *Mycologia* 72:1–49, 1980.
2. R Hartig. *Die Zersetzungserscheinungen des Holzes der Nadelholzbaume und der Eiche in forstlicher, botanischer und chemischer Richtung.* Berlin/New York: Springer, 1878,151 pp.

3. K-EL Eriksson, RA Blanchette, P Ander. *Microbial and Enzymatic Degradation of Wood and Wood Components.* Berlin: Springer Verlag, 1990, 416 pp.
4. W Bavendamm. Über das Vorkommen und den Nachweis von Oxydasen dei holzzerstörenden Pilzen. *Z Pflanzenkr Pflanzenschultz* 38:257–276, 1928.
5. RW Davidson, WA Campbell, DJ Blaisdell. Differentiation of wood-decaying fungi by their reactions on gallic or tannic acid medium. *J Agr Res* 57:683–695, 1938.
6. P Ander, K-E Eriksson. Selective degradation of wood components by white-rot fungi. *Physiol. Plant* 41:238–248, 1977.
7. E Björkman, O Samuelson, E Ringström, T Bergek, E Malm. Decay injuries in spruce forests and their importance for the production of chemical paper pulp and rayon pulp. Kgl Skogshögskolan Skr No. 4, Stockholm, 1949, 73 pp.
8. L Jurásek. Zm_ny v mikrostrukture zdrevnatelé bunecné blány pri rozkladu drevokaznými houbami. (Changes in the microstructure of the lignified cell wall during attack by wood-destroying fungi). *Biologia* 10:569–579, 1955.
9. H Meier. Über den Zellwandabbau durch Holzvermorschungspilze und die submikroskopische Struktur von Fichtentracheiden und Birkenholzfasern. *Holz Roh-Werkst* 13:323–338, 1955.
10. R Schmid, W Liese. Über die mikromorphologischen Veränderungen der Zellwandstrukturen von Buchen- und Fichtenholz beim Abbau durch Polyporus versicolor (L.) Fr. *Arch Mikrobiol* 47:260–276, 1964.
11. H Von Aufsess, H von Pechmann, H Graessle. Fluoreszenzmikroskopische Beobachtungen an pilzbefallenem Holz. *Holz Roh-Werkst* 26:50–61, 1968.
12. EE Hubert. *An Outline of Forest Pathology.* New York: Wiley, 1931, 543 pp.
13. RA Blanchette. Wood decomposition by Phellinus (Fomes) pini: A scanning electron microscopy study. *Can J Bot* 58:1496–1503, 1980.
14. T Nilsson. Defining fungal decay types: A proposal for discussion. Intern Res. Group on Wood Preserve Document IRG/WP/1264, 1985.
15. L Otjen, RA Blanchette. A discussion of microstructural changes in wood during decomposition by white rot basidiomycetes. *Can J Bot* 64:905–911, 1986b.
16. WG Campbell. The Chemistry of wood. III. The effect on wood substance of Ganoderma applanatum (Pers.) Pat., Fomes fomentarius (Linn.) Fr., Polyporus adustus (Willd.) Fr., Pleurotus ostreatus (Jacq.) Fr., Armillaria mellea (Vahl.) Fr., Trametes pini (Brot.) Fr., and Polystictus abietinus (Dicks.) Fr. *Biochem J.* 26:1829–1838, 1932.
17. TK Kirk, TL Highley. Quantitative changes in structural components of conifer woods during decay by white- and brown-rot fungi. *Phytopathology* 55:739–745, 1973.
18. RA Blanchette. Selective delignification of eastern hemlock by Ganoderma tsugae. *Phytopathology* 74:153–160, 1984.
19. RA Blanchette. Manganese accumulation in wood decayed by white-rot fungi. *Phytopathology* 74:725–730, 1984.
20. H Greaves, JF Levy. Comparative degradation of the sapwood of scotch pine, beech and birch by Lenzites trabea, Polystictus versicolor, Chaetomium globosum and Bacillus polymyxa. *J Inst Wood Sci* 15:55–63, 1965.
21. WW Wilcox. Changes in wood microstructure through progressive stages of decay. US For Serv. Res. Papa FPL-70, 1968, 46 pp.
22. W Liese. Ultrastructural aspects of wood tissue disintegration. *Ann Rev Phytopathol* 8:231–258, 1970.
23. D Dirol. Etude in vitro de la colonisation et de la dégradation structurale du bois de Hêtre par *Coriolus versicolor* (L.) *Rev Mycol* 40:295–317, 1976.
24. K Messner, H Stachelberger. Transmission electron microscope observations of white rot caused by Trametes hirsuta with respect to osmiophilic particles. *Trans Br Mycol Soc* 83:209–216. 1984.

25. RA Blanchette. Delignification by wood-decay fungi. *Ann Rev Phytopathol* 29:381–398, 1991.
26. A Sethuraman, DE Akin, K-EL Eriksson. Plant cell wall degrading enzymes produced by the white-rot fungus *Ceriporiopsis subvermispora*. *Biotechnol Appl Biochem* 27(1):37–47, 1998.
27. RA Blanchette, EW Krueger, JE Haight, M Akhtar, DE Akin. Cell wall alterations in loblolly pine wood decayed by the white-rot fungus, Ceriporiopsis subvermispora. *J Biotechnol* 53:203–213, 1997.
28. JE Adaskaveg, RL Gilbertson. In vitro decay studies of selective delignification and simultaneous decay by the white rot fungi *Ganoderma lucidum* and *G. tsugae*. *Can J Bot* 64:1611–1619, 1986.
29. I Dill, G Kraepelin. Palo Podrido: Model for extensive delignification of wood by Ganoderma applanatum. *Appl Environ Microbiol* 52:1305–1312, 1986.
30. EB Cowling, W Merrill. Nitrogen in wood and its role in wood deterioration. *Can J Bot* 44:1539–1554, 1966.
31. MP Levi, EB Cowling. Role of nitrogen in wood deterioration. VII Physiological adaption of wood-destroying and other fungi to substrates deficient in nitrogen. *Phytopathology* 59:460–468, 1969.
32. W Merrill, EB Cowling. Role of nitrogen in wood deterioration. IV Relationship of natural variation in nitrogen content of wood to its susceptibility to decay. *Phytopathology* 56:1324–1325, 1966.
33. TK Kirk, E Schultz, WJ Connors, LF Loren, JG Zeikus. Influence of cultural parameters lignin metabolism by *Phanerochaete chrysosporium*. *Arch Microbiol* 117:227–285, 1978.
34. P Keyser, TK Kirk, JG Zeikus. Ligninolytic enzyme system of *Phanerochaete chrysosporium*: synthesized in the absence of lignin in response to nitrogen starvation. *J Bacteriol* 135: 790–797, 1978.
35. M Akhtar, RA Blanchette, TK Kirk. Fungal delignification and biomechanical pulping of wood. In: Special Issue: Biotechnology in the Pulp and Paper Industry. K.-E.L Eriksson, Ed. Adv. Biochem. Eng./Biotechnol., Springer Verlag, 1997, pp. 159–196.
36. DE Akin, WH Morrison III, LL Rigsby, GR Gamble, A Sethuraman, K-EL Eriksson. Biological delignification of plant components by the white-rot fungi *Ceriporiopsis subvermispora* and *Cyathus stercoreus*. *Anim Feed Sci Technol* 63:305–321, 1996.
37. P Bajpai, PK Bajpai. Reduction of organochlorine compounds in bleach plant effluents. In: Special Issue: Biotechnology in the Pulp and Paper Industry. Eriksson, K.-E.L., Ed. Adv. Biochem. Eng./Biotechnol., Springer Verlag, 1997, pp. 213–260.
38. TK Kirk, RL Farrell. Enzymatic "Combustion": The microbial degradation of lignin. *Ann Rev Microbiol* 41:465–515, 1987.
39. P Ander, K-E Eriksson. Influence of carbohydrates on lignin degradation by the white-rot fungus *Sporotrichum pulverulentum*. *Svensk Papperstidn* 78:643–652, 1975.
40. TK Kirk, WJ Connors, JG Zeikus. Requirement for a growth substrate during lignin decomposition by two wood-rotting fungi. *Appl Environ Microbiol* 32:192, 1976.
41. SA Drew, KL Kadam. Lignin metabolism by *Aspergillus fumigatus* and white-rot fungi. *Dev Ind Microbiol* 20:153, 1979.
42. TW Jeffries, S Choi, TK Kirk. Nutritional regulation of lignin degradation by *Phanerochaete chrysosporium*. *Appl Environ Microbiol* 42:290–296, 1981.
43. JA Buswell, B Mollet, E Odier. Ligninolytic enzyme production by *Phanerochaete chrysosporium* under conditions of nitrogen sufficiency. *FEMS Microbiol Lett* 25:295–299, 1984.
44. TK Kirk. Physiology of lignin metabolism by white-rot fungi. In *Lignin Biodegradation: Microbiology, Chemistry and Potential Applications,* Vol. 2. Kirk TK, Higuchi T, and Chang H-m, Eds. Boca Raton, FL: CRC Press, 1980, p. 51.

45. P Fenn, S Choi, TK Kirk. Ligninolytic activity of *Phanerochaete chrysosporium*: physiology of suppression by NH_4^+ and L-glutamate. *Arch Microbiol* 130:66–71, 1981.

46. P Fenn, TK Kirk. Relationship of nitrogen to the onset and suppression of ligninolytic activity and secondary metabolism in *Phanerochaete chrysosporium*. *Arch Microbiol* 130:59–65, 1981.

47. K Lundquist, TK Kirk. *De novo* synthesis and decomposition of veratryl alcohol by a lignin-degrading basidiomycete. *Phytochemistry* 17:1676, 1978.

48. E De Jong, JA Field, JAM Bont. Ary alcohols in the physiology of white-rot fungi. *FEMS Microbiol Rev* 13:153–188, 1994.

49. M Shimada, F Nakatsubo, TK Kirk, T Higuchi. Biosynthesis of the secondary metabolite veratryl alcohol in relation to lignin degradation in *Phanerochaete chrysosporium*. *Arch Microbiol* 129:321–324, 1981.

50. JA Buswell, E Odier. Lignin biodegradation. *CRC Crit Rev Biotechnol* 6:1–60, 1987.

51. TK Kirk, HHJ Yang, P Keyser. The chemistry and physiology of the fungal degradation of lignin. *Dev Ind Microbiol* 19:51–61, 1978.

52. BD Faison, TK Kirk. Factors involved in the regulation of a ligninase activity in *Phanerochaete chrysosporium*. *Appl Environ Microbiol* 42:299, 1985.

53. S Dagley. Microbial Catabolism the carbon cycle and environmental pollution. *Naturwissenschaften* 65:85, 1978.

54. M Tien, TK Kirk. Lignin degrading enzyme from the hymenomycete *Phanerochaete chrysosporium* burds. *Science* 221:661–663, 1983.

55. M Kuwahara, JK Glenn, MA Morgan, MH Gold. Separation and characterization of two extracellular H_2O_2-dependent oxidases from ligninolytic cultures of *Phanerochaete chrysosporium*. *FEBS Lett* 169(2):247–250, 1984.

56. ID Reid, EE Chao, SS Dawson. Lignin degradation by *Phanerochaete chrysosporium* in agitated cultures. *Can J Microbiol* 31:88–90, 1985.

57. MH Gold, M Kuwahara, AA Chiu, JK Glenn. Purification and characterization of an extracellular hydrogen peroxide requiring diarylpropane oxygenase from the white-rot basidiomycete, *Phanerochaete*. *Arch Biochem Biophys* 34:353–362. 1984.

58. M Leisola, A Fiechter. Ligninase production in agitated conditions by *Phanerochaete chrysosporium*. *FEMS Microbiol Lett* 29:33–36, 1985.

59. A Jäger, S Croan, TK Kirk. Production of ligninases and degradation of lignin in agitated submerged cultures of *Phanerochaete chrysosporium*. *Appl Environ Microbiol* 50:1274–1278, 1985.

60. MH Gold, M Alic. Molecular biology of the lignin-degrading basidiomycete *Phanerochaete chrysosporium*. *Microbiol Rev* 57:605–622, 1993.

61. MH Gold, TM Cheng, MB Mayfield. Isolation and complementation studies of auxotrophic mutants of the lignin-degrading basidiomycete *Phanerochaete chrysosporium*. *Appl Environ Microbiol* 44:996–1000, 1982.

62. MH Gold, H Wariishi, K Valli. Extracellular peroxidases involved in lignin degradation by the white-rot basidiomycete *Phanerochaete chrysosporium* ACS Symp Ser 389:127–140, 1989.

63. RL Farrell. A new key enzyme for lignin degradation. *Ann. N.Y. Acad. Sci.* 501:150–158, 1987.

64. D Cullen, P.J. Kersten. Enzymology and molecular biology of lignin biodegradation. In *The Mycota III, Biochemistry and Molecular Biology,* 2nd Edition. R. Brambl and G.A, Marzluf, eds. Springer Verlag, 2004, pp. 249–273.

65. DG Pribnow, MB Mayfield, VJ Nipper, JA Brown, MH Gold. Characterization of a cDNA encoding a manganese peroxidase, from the lignin-degrading basidiomycete *Phanerochaete chrysosporium*. *J Biol Chem* 264:5036–5040, 1989.

66. P Bonnarme, TW Jeffries. Mn(II) regulation of lignin peroxidases and manganese-dependent peroxidases from lignin-degrading white-rot fungi. *Appl Environ Microbiol* 56:210–217, 1990.

67. JA Brown, JK Glenn, MH Gold. Manganese regulates expression of manganese peroxidase by *Phanerochaete chrysosporium. J Bacteriol* 172:3125–3130, 1990.

68. JA Brown, M Alic, MH Gold. Manganese peroxidase gene transcription in *Phanerochaete chrysosporium*: activation by manganese. *J Bacteriol* 173:4101–4106, 1991.

69. JA Brown, D Li, M Alic, MH Gold. Heat shock induction of manganese peroxidase gene transcription in *Phanerochaete chrysosporium. Appl Environ Microbiol* 59:4295–4299, 1993.

70. JM Gettemy, B Ma, M Alic, MH Gold. Reverse transcription-PCR analysis of the regulation of the manganese peroxidase gene family. *Appl Environ Microbiol* 64(2):569–574, 1997.

71. BJ Godfrey, L Akileswaran, MH Gold. A reporter gene construct for studying the regulation of manganese peroxidase gene expression. *Appl Environ Microbiol* 60:1353–1358, 1994.

72. Li D, M Alic, JA Brown, MH Gold. Regulation of manganese peroxidase gene transcription by hydrogen peroxide, chemical stress, and molecular oxygen. *Appl Environ Microbiol* 61:341–345, 1995.

73. JM Gettemy, D Li, M Alic, MH Gold. Truncated gene reported system for studying the regulation of manganese peroxidase. *Curr Genet* 31:519–524, 1998.

74. A Paszczynski, V-A Huynh, R Crawford. Enzymatic activities of an extracellular, manganese-dependent peroxidase from *Phanerochaete chrysosporium. FEMS Microbiol Lett* 29:37–41, 1985.

75. JK Glenn, MH Gold. Purification and characterization of an extracellular Mn(II)-dependent peroxidase from the lignin-degrading basidiomycete, *Phanerochaete chrysosporium. Arch Biochem Biophys* 242:329–341, 1985.

76. A Paszczynski, V-A Huynh, R Crawford. Comparison of ligninase-1 and peroxidase-M2 from the white-rot fungus *Phanerochaete chrysosporium. Arch Biochem Biophys* 244(2):750–765, 1986.

77. T Johansson, PO Nyman. A manganese(II)-dependent extracellular peroxidase from the white-rot fungus *Trametes versicolor. Acta Chemica Scandinavica* B41, 762–765, 1987.

78. T Deguchi, Y Kitaoka, M Kakezawa, T Nishida. Purification and characterization of a nylon-degrading enzyme. *Appl Environ Microbiol* 64(4):1366–1371, 1998.

79. H Wariishi, L Akileswaran, MH Gold. Manganese peroxidase from *Phanerochaete chrysosporium*: spectral characterization of the oxidized states and the catalytic cycle. *Biochemistry* 27, 5365–5370, 1988.

80. IT Forrester, AC Grabski, RR Burgess, GF Leatham. Manganese, Mn-dependent peroxidases, and the biodegradation of lignin. *Biochem Biophys Res Com* 157(3):992–999, 1988.

81. H Wariishi, K Valli, MH Gold. Oxidative cleavage of a phenolic diarylpropane lignin model dimer by manganese peroxidase from *Phanerochaete chrysosporium. Biochemistry* 28:6017–6023, 1989

82. R Lackner, E Srebotnik, K Messner. Oxidative degradation of high molecular weight chlorolignin by manganese peroxidase of *Phanerochaete chrysosporium. Biochem Biophys Res Com* 178(3): 1092–1098, 1991.

83. K Kishi, H Wariishi, L Marquez, HB Dunford, MH Gold. Mechanism of manganese peroxidase compound II reduction: Effect of organic acid chelators and pH. *Biochemistry* 33:8694–8701, 1994.

84. H Wariishi, HB Dunford, ID MacDonald, MH Gold. Manganese peroxidase from the lignin-degrading basidiomycete *Phanerochaete chrysosporium*. Transient state kinetics and reaction mechanism. *J Biol Chem* 246(6):3335–3340, 1989.

85. FH Perie, MH Gold. Manganese regulation of manganese peroxidase expression and lignin degradation by the white rot fungus *Dichomitus squalens*. *Appl Environ Microbiol* 57(8):2240–2245, 1991.

86. AM Moilanen, T Lundell, T Vares, A Hatakka. Manganese and malonate are individual regulators for the production of lignin and manganese peroxidase isozymes and in the degradation of lignin by *Phlebia radiata*. *Appl Microbiol Biotechnol* 45:792–799, 1996.

87. RZ Harris, H Wariishi, MH Gold, PR Ortiz de Montellano. The catalytic site of manganese peroxidase. Regiospecific addition of sodium azide and alkylhydrazines to the heme group. *J Biol Chem* 266(14): 8751–8758, 1991.

88. M Sundaramoorthy, K Kishi, MH Gold, TL Poulos. The crystal structure of manganese peroxidase from *Phanerochaete chrysosporium* at 2.06-A resolution. *J Biol Chem* 269(52):32759–32767, 1994.

89. GRJ Sutherland, SD Aust. The effects of calcium on the thermal stability and activity of manganese peroxidase. *Arch Biochem Biophys* 332(1):128–134, 1996.

90. SL Timofeevski, SD Aust. Kinetics of calcium release from manganese peroxidase during thermal inactivation. *Arch Biochem Biophys* 342(1):169–175, 1997.

91. H Wariishi, K Valli, MH Gold. Manganese(II) oxidation by manganese peroxidase from the basidiomycete *Phanerochaete chrysosporium*. *J Biol Chem* 267(33):23688–23695, 1992.

92. I-C Kuan, KA Johnson, M Tien. Kinetic analysis of manganese peroxidase. *J Biol Chem* 268(27): 20064–20070, 1993.

93. HJ Bae, YS Kim. Degradation of ligninsulfonates by simultaneous action of laccase and Mn-peroxidase. Proceedings of the 6th international conference on biotechnology in the pulp and paper industry: Advances in applied and fundamental research. Facultas-Universitatsverlag, Vienna, Austria. 1995, pp. 393–396.

94. H Galliano, G Gas, JL Seris, AM Boudet. Lignin degradation by *Rigidoporus lignosus* involves synergistic action of two oxidizing enzymes: Mn peroxidase and laccase. *Enzyme Microbiol Technol* 13, 478–482, 1991.

95. BP Roy, MG Paice, FS Archibald, SK Misra, LE Misiak. Creation of metal-complexing agents, reduction of manganese dioxide, and promotion of manganese peroxidase-mediated Mn(III) production by cellobiose:quinone oxidoreductase from *Trametes versicolor*. *J Biol Chem* 269(31), 19745–19750, 1994.

96. H Wariishi, K Valli, V Renganathan, MH Gold. Thiol-mediated oxidation of nonphenolic lignin model compounds by manganese peroxidase of *Phanerochaete chrysosporium*. *J. Biol. Chem.* 264(24), 14185–14191, 1989.

97. W Bao, Y Fukushima, KA Jensen Jr, MA Moen, KE Hammel. Oxidative degradation of non-phenolic lignin during lipid peroxidation by fungal manganese peroxidase. *FEBS Lett.* 354: 297–300, 1994.

98. KA Jensen Jr, W Bao, S Kawai, E Srebotnik, KE Hammel. Manganese-dependent cleavage of nonphenolic lignin structures by *Ceriporiopsis subvermispora* in the absence of lignin peroxidase. *Appl Environ Microbiol* 62(10):3679–3686, 1996.

99. A D'Annibale, C Crestini, E De Mattia, GG Sermanni. Veratryl alcohol oxidation by manganese-dependent peroxidase from *Lentinus edodes*. *J Biotechnol* 48:231–239, 1996.

100. ID Reid. Biological bleaching of softwood kraft pulp with the fungus *Trametes (Coriolus) versicolor*. *Tappi J* 73(8):149–153, 1990.

101. MG Paice, ID Reid, R Bourbonnais, FS Archibald, L Jurasek. Manganese peroxidase, produced by *Trametes versicolor* during pulp bleaching, demethylates and delignifies kraft pulp. *Appl Environ Microbiol* 59(1):260–265, 1993.

102. K Addleman, F Archibald. Kraft pulp bleaching and delignification by dikaryons and monokaryons of *Trametes versicolor*. *Appl Environ Microbiol* 59(1):266–273, 1993.

103. K Addleman, T Dumonceaux, MG Paice, R Bourbonnais, F Archibald. Production and characterization of *Trametes versicolor* mutants unable to bleach hardwood kraft pulp. *Appl Environ Microbiol* 61(10):3687–3694, 1995.

104. N Katagiri, Y Tsutsumi, T Nishida. Correlation of brightening with cumulative enzyme activity related to lignin biodegradation during biobleaching of kraft pulp by white rot fungi in the solid-state fermentation system. *Appl Environ Microbiol* 61(2):617–622, 1995.

105. R Kondo, K Harazono, K Sakai. Bleaching of hardwood kraft pulp and manganese peroxidase secreted from *Phanerochaete sordida* YK-624. *Appl Environ Microbiol* 60(12):4359–4363, 1994.

106. K Harazono, R Kondo, K Sakai. Bleaching of hardwood kraft pulp with manganese peroxidase from *Phanerochaete sordida* YK-624 without addition of $MnSO_4$. *Appl Environ Microbiol* 62(3):913–917, 1996.

107. K Harazono, R Kondo, K Sakai. Bleaching of kraft pulp with manganese peroxidase secreted from *Ganoderma* sp. YK-505: Improvement of bleaching system by using the tolerant enzyme. *ISWPC* G4, 1–4, 1997.

108. MT Moreira, G Feijoo, R Sierra-Alvarez, J Lema, JA Field. Manganese is not required for biobleaching of oxygen-delignified kraft pulp by the white rot fungus *Bjerkandera* sp. Strain BOS55. *Appl Environ Microbiol* 63:1749–1755, 1997.

109. K Ehara, Y Tsutsumi, T Nishida. Biobleaching of softwood kraft pulp with manganese peroxidase. *ISWPC* 26:1–4, 1997.

110. MT Moreira, G Feijoo, T Mester, P Mayorga, R Sierra-Alvarez, JA Field. Role of organic acids in the manganese-independent biobleaching system of *Bjerkandera* sp. strain BOS55. *Appl. Environ. Microbiol.* 64(7), 2409–2417, 1998.

111. ID Reid, MG Paice. Effects of manganese peroxidase on residual lignin of softwood kraft pulp. *Appl Environ Microbiol* 64(6):2273–2274, 1998.

112. ID Reid. Fate of residual lignin during delignification of kraft pulp by *Trametes versicolor*. *Appl Environ Microbiol* 64(6):2117–2125, 1998.

113. H Yoshida. Chemistry of lacquer (urushi). Part I. *J Chem Soc* 43:472–486, 1883.

114. MG Bertrand. Sur la laccase et sur le pouvoir oxydant de cette diastase. *CRAS Paris* 118:1215–1218, 1894.

115. MG Bertrand. Sur la laccase et sur le pouvoir oxydant de cette diastase. *CRAS Paris* 120:266–269, 1895.

116. AM Mayer, E Harel. Polyphenol oxidases in plant. *Phytochemistry* 18:193–215, 1979.

117. JFD Dean, K-EL Eriksson. Laccase and the deposition of lignin in vascular plants. *Holzforschung* 48:21–24, 1994.

118. BR Thomas, M Yonekura, TD Morgan, TH Czapla, TL Hopkins, KJ Kramer. A trypsin-solubilized laccase from pharate pupal integument of the tobacco hornworm, Manduca sexta. *Insect Biochem* 19:611–622, 1989.

119. D Faure, ML Boullant, R Bally. Isolation of Azospirillum lipoferum 4T Tn5 mutants affected in melanization and laccase activity. *Appl Environ Microbiol* 60:3412–3415, 1994.

120. J-L Yang, B Pettersson, K-E Eriksson. Development of bioassays for the characterization of pulp fiber surfaces. I. Characterization of various mechanical pulp fiber surfaces by specific cellulolytic enzymes. *Nordic Pulp and Paper Res J* 3:19–25, 1988.

121. (a) RC Kuhad, A Singh, K-EL Eriksson. Microorganisms and enzymes involved in the degradation of plant fiber cell wall components. In: Special Issue: Biotechnology in the Pulp and Paper Industry, K-EL Eriksson, ed. Adv. Biochem. Eng./Biotechnol., Springer Verlag, 1997, pp. 45–126. (b) RC Kuhad, AJ Singh, KK Tripathi, RK Saxena, K-EL Eriksson. Microorganisms as an Alternative Source of Protein. *Nutrition Reviews* 55(3):65–75, 1997.

122. M Spaghi, P Jendet, R Bessis, P Leroux. Degradation of stilbene-type phytoalexins in relation to the pathogenicity of *Botrytis cinerea* to grapevines. *Plant Pathology* 45:139–144, 1996.

123. SO Anderson. Sclerotization and tanning of the cuticle. In: *Comprehensive Insect Physiology, Biochemistry and Pharmacology,* Vol. 3. GA Kerkut, LI Gilbert, eds. New York: Pergammon Press Ltd., 1985, pp. 59–74.

124. A Messerschmidt, A Rossi, R Ladenstein, R Huber, M Bolognesi, G Gatti, A Marchesini, R Petruzelli, A Finazzi-Agro. X-ray crystal structure of the blue oxidase ascorbate oxidase from zucchini. Analysis of the polypeptide fold and a model of the copper sites and ligands. *J Mol Biol* 206:513–529, 1989.

125. A Messerschmidt, R Huber. The blue oxidases, ascorbate oxidase, laccase and ceruloplasmin: Modeling and structural relationships. *Eur J Biochem* 187:341–352, 1990.

126. V Ducros, AM Brzozowski, KS Wilson, SH Brown, P Østergaard, P Schneider, DS Yaver, AH Pedersen, GJ Davies. Crystal structure of the type-2 Cu depleted laccase from *Coprinus cinereus* at 2.2 Å resolution. *Nature Struct. Biol.* 5(4):310–316, 1998.

127. BG Malmström, L-E Andréasson, B Reinhammar. Copper-containing oxidases and superoxide dismutase. In: *The Enzymes,* Vol.12. PD Boyer, ed. New York: Academic Press, 1975, pp. 507–579.

128. B Reinhammar. Laccase. In: *Copper Proteins and Copper Enzymes,* Vol. III. R Lontie, ed. Boca Raton, FL: CRC Press, 1984, pp. 1–35.

129. W Cai, R Martin, B Lemaure, J-L Leuba, V Pétiard. Hydroxyindols: A new class of laccase substrates. *Plant Physiol Biochem* 31:441–445, 1993.

130. FS Sariaslani, JM Beale Jr, P Rosazza. Oxidation of rotenone by Polyporus anceps laccase. *J Nat Prod* 47(4):692–697, 1984.

131. S Kawai, T Umezawa, T Higuchi. Oxidation of methoxylated benzyl alcohols by laccase of Coriolus versicolor in the presence of syringaldehyde. *Wood Research* 76:10–16, 1989.

132. R Bourbonnais, MG Paice. Oxidation of non-phenolic substrates, an expanded role for laccase in lignin biodegradation. *FEBS Lett* 267(1):99–102, 1990.

133. R Bourbonnais, MG Paice. Demethylation and delignification of kraft pulp by Trametes versicolor laccase in the presence of 2,2'azinobis-(3-ethylbenzthiazoline-6-sulphonate). *Appl Microbiol Biotechnol* 36:823–827, 1992.

134. HP Call, I Mücke. State-of-the-art enzyme bleaching and disclosure of a breakthrough process. Conference Proceedings; in TAPPI Pulping Conference Proceedings, San Diego, CA, TAPPI, 1994, pp. 83–101.

135. C Eggert, U Temp, JFD Dean, K-EL Eriksson. A fungal metabolite mediates laccase degradation of nonphenolic lignin structures and synthetic lignin by laccase. *FEBS Lett* 391:144–148, 1996.

136. A Hatakka, AK Uusi-Rauva. Degradation of ^{14}C-labeled poplar wood lignin by selected white-rot fungi. *Eur J Appl Microbiol Biotechnol* 17:235–242, 1983.

137. C Eggert, U Temp, JFD Dean, K-EL Eriksson. Laccase-mediated formation of the phenoxazinone derivative, cinnabarinic acid. *FEBS Lett* 376:202–206, 1995.

138. S Christen, P Southwell-Keely, R Stocker. Oxidation of 3-hydroxyanthranilic acid to the *Phenoxazinone cinnabarinic* acid by peroxyl radicals and by compound I of peroxidases or catalase. *Biochemistry* 31:8090–8097, 1992.

139. O Toussaint, K Lerch. Catalytic oxidation of 2-amino phenols and ortho hydroxylation of aromatic amines by tyrosinase. *Biochemistry* 26:8567–8571, 1987.

140. C Eggert, U Temp, K-EL Eriksson. Laccase-producing white-rot fungus lacking lignin peroxidase and manganese peroxidase. In: ACS Symposium Series 655, *Enzymes for Pulp and Paper Processing.* TW Jeffries, L Viikari, eds., Washington, DC: American Chem Soc., 1996, pp. 130–150.

141. E Srebotnik, K Messner, R Foisner. Penetrability of white-rot degraded pine wood by the lignin peroxidase of *Phanerochaete chrysosporium*. *Appl Environ Microbiol* 54:2608–2614, 1988.

142. C Eggert, U Temp, K-EL Eriksson. Laccase is essential for lignin degradation by the white-rot fungus *Pycnoporus cinnabarinus*. *FEBS Lett* 407:89–92, 1997.

143. K Li, H Bermek, K-EL Eriksson. A new technique for screening of laccase/mediators. TAPPI Biol. Sci. Symp. San Francisco, 1997, pp. 349–353.

144. R Bourbonnais, MG Paice, ID Reid, P Lanthier, M Yaguchi. Lignin oxidation by laccase isozymes from Trametes versicolor and role of the mediator 2,2'-azinobis-(3-ethylbenzthiazoline-6-sulphonate) in kraft lignin depolymerization. *Appl Environ Microbiol* 61(5):1876–1880, 1995.

145. R Bourbonnais, MG Paice MG. Enzymatic delignification of kraft pulp using laccase and a mediator. *Tappi J* 79(6):199–204, 1996.

146. K Li, RF Helm, K-EL Eriksson. Mechanistic studies of the oxidation of a non-phenolic lignin model compound by the laccase/i-hbt redox system. *Biotechnol Appl Biochem* 27:239–243, 1998.

147. Li, K.C., F. Xu, and K.E.L. Eriksson. Comparison of fungal laccases and redox mediators in oxidation of a nonphenolic lignin model compound. *Appl Environ Microbiol* 65(6):2654–2660, 1999.

15 Biopulping and Biobleaching

I. D. Reid, R. Bourbonnais, and M. G. Paice

CONTENTS

INTRODUCTION

This review covers the application of microorganisms and the enzymes that they produce in the manufacturing of chemical and mechanical pulp. Biopulping research, mainly aimed at mechanical pulping, has focussed on application of fungi as a pretreatment prior to refining. Biobleaching, on the other hand, usually employs microbial enzymes, especially in production of bleached kraft pulp. Although the primary target of biopulping and biobleaching is lignin modification or removal, in fact other effects are also evident, such as increased cell wall porosity in biopulping, and hemicellulose hydrolysis in biobleaching. Thus the review covers mechanistic aspects of both lignin modification and other cell wall changes associated with biological pulping and bleaching. Currently biopulping with fungal cultures and biobleaching with laccase and mediators have only been demonstrated at a pilot scale, while biobleaching with xylanase is in use in several kraft mills.

BIOPULPING

Biopulping refers to the treatment of wood chips or other lignocellulosic raw materials with microorganisms or the enzymes that they produce, in order to improve subsequent mechanical or chemical pulp manufacturing processes. This usually involves lignin biodegradation, but may also include other chemical or structural changes, such as the destruction of pit membranes to improve cooking liquor penetration. Most biopulping research has focussed on white-rot fungi. In "biomechanical" pulping, the aim of biological treatment is to replace pretreatment chemicals for mechanical pulping, reduce energy demand and increase paper strength. For chemical pulping, biopulping is intended to reduce the amount of cooking chemicals, to increase the cooking capacity, or to enable extended cooking, resulting in lower consumption of bleaching chemicals. Increased delignification efficiency results in an indirect energy saving for pulping, and reduces pollution [1].

BIOPULPING MECHANISM

The original idea for biopulping was that biological delignification would replace chemical delignification. Screening for biopulping fungi emphasized extensive and selective lignin removal [2–5]. This approach succeeded in identifying several fungi that were effective in biomechanical pulping [6–8]. It quickly became apparent, however, that the benefits of biomechanical pulping—lower refining energy requirement, higher strength—were not highly correlated with the extent of lignin removal from the wood [6–8]. In the two-week fungal treatments that are being suggested for commercial biopulping, total wood weight loss is only 2% [9]. Data for lignin removal under these conditions have not been reported, but it must be less than 7% of the total lignin even if all the weight lost is lignin. It appears that some kind of lignin modification, rather than lignin removal, may be responsible for the biopulping effect.

Ceriporiopsis subvermispora is capable of lignin degradation all the way to CO_2, and it produces the enzymes laccase and Mn-peroxidase [10–16]. It does not produce lignin peroxidase, although its DNA contains sequences homologous to the *lip* gene from *Phanerochaete chrysosporium* [17]. Despite the absence of lignin peroxidase (LiP), *C. subvermispora* can degrade nonphenolic lignin model compounds by α, β-cleavage as LiP does [18]. Mn-peroxidase may catalyze this reaction in the presence of unsaturated fatty acids [19].

C. subvermispora and other selectively delignifying fungi usually colonize wood via the ray cells. From there, the fungal hyphae grow along the lumina of the fibers [20]. After two weeks treatment of pine wood with *C. subvermispora*, the fiber walls did not show obvious structural disruption. Staining with uranyl acetate revealed a band of increased electron density within the secondary wall adjacent to the lumen (Figure 15.1) [21]. With longer exposure to the fungus, this zone of increased staining extended farther and farther into the secondary wall. The increased staining might have been caused by formation of acidic groups that would bind uranyl ions, or by easier penetration of the ions into the fiber wall [21]. Both the secondary walls and the middle lamellae of colonized cells had decreased UV absorbance, showing some changes to their lignin.

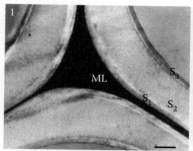
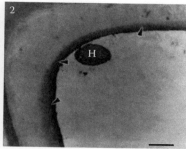

FIGURE 15.1 Modification of secondary wall of *Pinus taeda* by two-week treatment with *Ceriporiopsis subvermispora*, revealed by staining with uranyl acetate. (1) Untreated wood, (2) Treated for two weeks. Bar = 1 μm. Arrows show increased staining. (Reproduced from Blanchette, R.A., Krueger, E.W., Haight, J.E., Akhtar, M., and Akin, D.E., *J Biotechnol*, 53(2–3), 203–213, 1997. With permission.)

Infiltration of the cell walls with proteins of defined size showed that the porosity of the wall on a molecular scale increased as the fungal treatment progressed. Insulin (5,700 Da) could not penetrate the secondary walls of untreated wood, but could penetrate the same areas as were stained by uranyl acetate in wood treated with *C. subvermispora*. Myoglobin (17,600 Da) was excluded from sound cell walls or walls treated for two weeks with *C. subvermispora*, but could penetrate the secondary walls after four weeks treatment. Ovalbumin (44,287 Da) could not penetrate cell walls treated for less than eight weeks with the fungus [21].

Since laccase and Mn-peroxidase have about the same molecular size as ovalbumin, they must also be unable to penetrate the cell walls before decay is well advanced. However, chelated Mn(III), the direct product of Mn-peroxidase action, should be able to diffuse into the cell walls, and may be responsible for the initial modification of the lignin in the secondary walls and the middle lamellae.

Biodegradation of wood polysaccharides may also contribute to the effect, especially for biochemical pulping. Treatment of wood with a nonligninolytic fungus, *Ophiostoma piliferum*, known commercially as the Cartapip process, can increase delignification during kraft pulping, apparently by removing extractives and rupturing pit membranes, thus increasing the permeability of the wood to pulping chemicals [20,22,23]. Wood treatment with the enzyme cellulase increases the rate of diffusion of NaOH into the wood [24] and aids kraft pulping [25–27]. White-rot fungi such as *C. subvermispora* increase wood permeability by the same mechanisms as they colonize it.

BIOPULPING TECHNOLOGY

Biomechanical Pulping

The historical development of biomechanical pulping has been frequently reviewed [28–55]. Although the idea of using delignifying fungi to aid pulping was proposed more than 40 years ago [28], and a patent was granted in 1976 [56], systematic development of the technology awaited the establishment in 1987 of the Biopulping Consortium centered at the USDA Forest Products Laboratory in Madison, Wisconsin [57]. The key finding is that incubation of wood chips with certain lignin-degrading fungi for a few weeks lowers the energy required to refine the chips into mechanical pulps, and improves the strength properties of the resulting pulps.

Economic analysis predicts that biological pretreatment could save money in mechanical pulp production [9], and some attempts have been made to commercialize the process [54].

Energy Savings

Treatment with white-rot fungi has been claimed to lower the amount of refining energy required to produce mechanical pulp of a desired freeness from hardwoods [6,58–66], softwoods [7,58,60,61,63,67–70], straw [71,72], kenaf [73], jute [74], or bagasse [75], with reported energy savings ranging from 10 to 48%. A large fraction of the total energy savings happens during the early stages of mechanical pulping, when the wood or other raw material is broken down into individual fibers and development of fiber flexibility begins [6,65]. Production of coarse mechanical pulp from fungally treated aspen chips can take as little as $^1/_3$ of the energy required by untreated chips [6]. The biomechanical pulps contain more long fibers and fewer fines than corresponding control pulps, suggesting that the fungal treatment favors fiber separation and decreases fiber damage during defibration [7,76]. Subsequent refining of the separated fibers, to increase their flexibility and fibrillate their surfaces, as measured by decreases in pulp freeness, appeared to consume as much energy in the treated aspen samples as in the controls [6]. A similar concentration of energy savings in the early stages of mechanical pulping was shown in the relationship of tensile index to consumed refining energy for Norway spruce and aspen treated with *C. subvermispora* [61]. Treatment of loblolly pine chips with several white-rot fungi [60], however, and treatment of aspen with *C. subvermispora* [65] resulted in energy savings during both early and later stages of refining (Figure 15.2). Several studies have shown that treatment of coarse mechanical pulps with white-rot fungi can decrease the energy required to refine them [77–79].

Strength Improvements

At the same freeness level, mechanical pulps made from biologically pretreated chips gave stronger handsheets than the untreated control pulps. The increased strength appears to result from improved interfiber bonding: the zero-span tensile strength of the fibers was little changed, but the tensile, tear, and burst strengths went up by as much as 80%, 160%, and 250% [8,65,76]. The small (~20%) increases in zero-span tensile strength that have been reported may result from the higher content of long fibers in the biomechanical pulps. Hardwoods tend to show larger tensile strength increases (17–70%) than softwoods (6–24%) (Figure 15.3). Increases in tensile strength are loosely correlated with increases in tear strength (correlation coefficient = +0.84) and burst strength (correlation coefficient = +0.94) for most samples; the correlation is more apparent for hardwoods than for softwoods. Aspen treated with *C. subvermispora* [65] showed exceptionally high tear strength, and aspen treated with *Dichomitus squalens* [76] showed exceptionally high burst strength (Figure 15.3). Burst and tear strength were more responsive to pretreatment

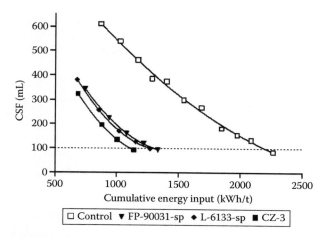

FIGURE 15.2 Development of freeness during refining of mechanical pulps from aspen chips either untreated (□) or pretreated with three strains of *Ceriporiopsis subvermispora* (FP-90031-sp, L-6133-sp, and CZ-3). (Redrawn from Akhtar,. M., *Holzforschung*, 48, 199– 202, 1994. With permission.)

with white-rot fungi than tensile strength was; on average, a unit increase in tensile strength was associated with a 1.5 times larger increase in tear strength and a 2.3 times larger increase in burst strength (Figure 15.3). Thermomechanical pulping of poplar chips also gave higher tensile and tear strength after pretreatment with selected white-rot fungi [59]. Treatment of coarse mechanical pulps with white-rot fungi also led to increased strength properties [61,77–80].

The publications of the Biopulping Consortium present their results on energy savings and strength enhancement separately, based on pulp freeness. Replotting the data for strength (tensile index) versus refining energy input shows more clearly that the fungal pretreatments allow simultaneous increases in pulp strength and decreases in energy consumption (Figure 15.4).

Brightness Loss

Treatment of wood chips [8,61,67,81,82] or coarse mechanical pulps [61,77–80,83] with white-rot fungi generally leads to lower brightness and scattering coefficients. Decreased scattering probably results from both lower fines content and increased interfiber bonding in the biomechanical pulps, and contributes to the decreased brightness. In addition to the decreased scattering, light absorption by the fibers increases, probably as a result of formation of quinones and other chromophores during lignin oxidation. The brightness of aspen biomechanical pulps can be increased by conventional bleaching with peroxide [76,84] or hydrosulfite [84], although higher than normal charges of bleaching chemicals are needed [9,61,84]. As a result of their low scattering coefficients, bleached biomechanical pulps have lower opacity than control refiner mechanical pulps, but higher opacity than bleached CTMP [84]. The propensity of bleached aspen biomechanical pulps to photochemical reversion was similar to that of other mechanical pulps [84]. No data on bleaching of softwood biomechanical pulps has been published.

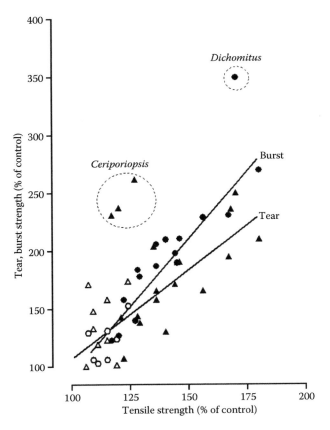

FIGURE 15.3 Correlation between changes in burst, tear and tensile strength of refiner mechanical pulps resulting from biological pretreatment. Triangles, tear strength; circles, burst strength; open symbols, softwood; filled symbols, hardwood. Circled points were excluded from the regression analysis. (Data extracted from Akhtar, M., Attridge, M.C., Myers, G.C., Blanchette, R.A., *Holzforschung*, 47, 36–40, 1993. With permission.)

Biochemical Pulping

Biochemical pulping has received much less attention than biomechanical pulping. The available information has been comprehensively summarized in two book chapters [20,85].

Kraft

Pretreatment of wood chips with white-rot fungi has been reported to accelerate delignification during subsequent kraft pulping. This leads to lower kappa numbers at a given pulp yield, based on the weight of the treated chips. When the weight losses during fungal treatment are factored in, however, the pulp yield based on the original wood weight is lower with the fungal treatment than without it. Fungal

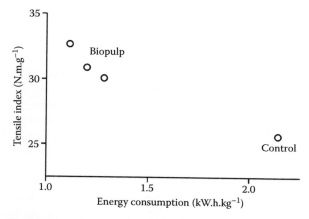

FIGURE 15.4 Pretreatment of aspen wood chips with three strains of *Ceriporiopsis subvermispora* lowered the energy required to refine to 100 mL CSF and increased the tensile strength of the pulp. (Data redrawn from Akhtar, M., *Holzforschung*, 48, 199–202, 1994. With permission.)

pretreatments have also been reported to increase the tensile and burst strengths of kraft pulps, and to decrease their brightness [20].

Oriaran and coworkers intensively studied the effects of pretreatment with *P. chrysosporium* on kraft pulping of aspen and red oak [86,87]. Their experiments covered a range of fungal pretreatment times, kraft cooking times, and effective alkali levels. Unfortunately the kraft cooking conditions they used gave pulps with high kappa numbers [75–104 for aspen, 64–128 for red oak], and their experiments don't give any information about the effects of fungal pretreatment on kraft cooking to normal commercial levels (kappa numbers near 15 for hardwood kraft pulps). Rocheleau et al. found that pretreatment of aspen chips for three to six weeks with *Phlebia tremellosa* gave slightly lower kappa number and yield after kraft pulping [22]. In contrast to other studies, they found that the fungal treatment slightly increased unbleached pulp brightness. A weakness of this study is the use of unsterilized aspen chips, a situation in which *Phlebia tremellosa* does not compete well with indigenous "weed" fungi [88]. Growth of contaminating fungi on 80% of the chips inoculated with *P. tremellosa* was noted [22], bringing into question the extent of colonization of the wood by *P. tremellosa* in these experiments.

Wolfaardt et al. screened wood-inhabiting fungi for ability to lower the kappa number after sequential fungal treatment and kraft pulping of pine chips [89]. They found that *Stereum hirsutum* was especially effective, giving a 30% decrease in kappa number after nine weeks treatment. The *Stereum* treatment did not change the relationship between kappa number and pulp yield, viscosity, or alkali consumption. Wood terpenes (α-pinene) inhibited colonization of the chips by *S. hirsutum*. Wolfaardt et al. concluded that treatment with *S. hirsutum* could be used to shorten the cooking time required for kraft pulping of pine, but that the benefits were unlikely to surpass the costs of the treatment [89]. They did not report weight losses resulting from the fungal treatment or the bleachability of the bio-kraft pulps.

Sulfite

Pretreatment of hardwood and softwood chips with *C. subvermispora* and some other white-rot fungi increased the rate of delignification during magnesium-based sulfite cooking, leading to a reduction in kappa number under standard cooking conditions [90]. The increased delignification was offset by losses of tear and tensile strength, and a decrease in brightness. Shorter cooking times reduced the brightness loss, and led to a 3.5% gain in brightness after bleaching [20]. Pretreatment of loblolly pine chips with *C. subvermispora* for two weeks increased the rates of both delignification and carbohydrate dissolution in sodium bisulfite cooking, so that the relationship between kappa number and yield was the same for treated and control chips [20]. In calcium-acid sulfite cooking, however, the biotreated chips showed a selective increase in delignification, leading to lower kappa numbers without yield loss (Figure 15.5). The pulps produced from the biotreated chips had lower shives content than the control pulps [20,91]. Their lower kappa numbers did not make the biosulfite pulps easier to bleach; because of their lower initial brightness, the biosulfite pulps required the same charges of either peroxide or formamidine sulfinic acid (FAS) as the control pulps to reach a brightness of 80% [20,91]. The FAS-bleached biosulfite pulps were less yellow (b*) than the control pulps. The Microtox™ toxicity of the spent liquor from cooking the biotreated chips was substantially less than that of the control [20,91], apparently because of biodegradation of toxic extractives by the fungus [92]. Screening of 10 cultures of white-rot fungi for treatment of eucalyptus chips before sulfite pulping for dissolving pulp manufacture indicated that *C. subvermispora* was the most effective species [93]. In this

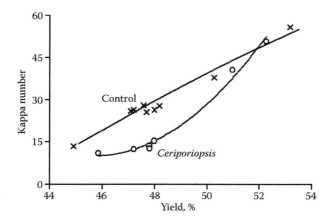

FIGURE 15.5 Kappa number-yield relationships for calcium-acid sulfite pulping of loblolly pine chips treated for two weeks with *Ceriporiopsis subvermispora* CZ-3 (○) or untreated (X). (Reprinted from Messner, K., Koller, K., Wall, M.B., Akhtar, M., Scott G.M., In: RA Young, M Akhtar, eds. *Environmentally Friendly Technologies for the Pulp and Paper Industry*, New York: John Wiley & Sons, Inc., 385–419, 1998. With permission.)

system, the fungal pretreatment increased rather than decreased the brightness of the unbleached pulp.

The potential benefits of fungal treatment before sulfite pulping are increased digester throughput and reduced effluent toxicity. Possible increases in pulp bleachability need to be further investigated.

Organosolv

Pretreatment of eucalyptus chips with the white-rot fungus *Trametes versicolor* for a month has been reported to accelerate their delignification with methanol/water [85]. Longer treatment with *T. versicolor*, or treatment with *P. chrysosporium*, caused nonselective decay of the wood, leading to organosolv pulps with higher residual lignin content. Treatment of pine chips with *C. subvermispora* has also been reported to increase the rate and extent of delignification during cooking with formic acid/acetone [85].

Pulping Fungi

C. subvermispora has attracted the most attention as a biopulping fungus for both biomechanical and biochemical pulping [20,52]. It is a white-rot fungus, probably related to *Phlebia* and *Phanerochaete* [94]. Another species, *Phlebia subserialis*, has better characteristics for solid-state fermentation [55], but it has not yet received as much study as *C. subvermispora*. *P. chrysosporium*, which has also been called *Sporotrichum pulverulentum*, *Peniophora* "G," *Chrysosporium pruinosum*, and *Sporotrichum pruinosum* [94] was studied intensively because of its status as the best-known lignin-degrading fungus, and it is effective for biomechanical pulping of hardwood [6,8].

BIOBLEACHING

Biobleaching of kraft pulps has been investigated using white rot fungi, ligninolytic enzymes, and hemicellulase enzymes. The latter target xylan or lignin carbohydrate linkages and have been applied extensively at full scale in many kraft mills. Here we will focus on biological bleaching that targets lignin, namely with white-rot fungi or ligninolytic enzymes. For xylanase bleaching mechanisms the reader is referred to articles published elsewhere [95,96]

PRODUCTION OF ENZYMES DURING PULP BLEACHING BY FUNGI

White-rot fungi such as *P. chrysosporium* [39,97] and *Trametes versicolor* [98,99] can delignify and bleach kraft pulp when cultured for periods of between several days and weeks (Figure 15.6). The rate of bleaching depends on the kraft pulp lignin content [100], but is too slow for use in a conventional bleach plant. Thus most research effort has focussed on identifying the underlying biochemical mechanisms so that the enzymes responsible for bleaching can be used directly on pulp. A direct physical contact between fungal mycelium and pulp fibers is not required for bleaching, as evidenced by experiments with immobilized fungus [101], and with fungal mycelia separated from pulp by filters [102,103]. However, sustained diffusion of certain secreted

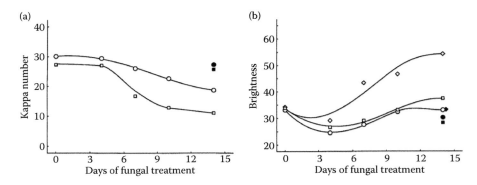

FIGURE 15.6   Time course of (a) delignification (kappa number decrease) and (b) bleaching (brightness increase) of softwood kraft pulp by the fungus *Trametes versicolor*. Symbols: circles—pulp before extraction; squares—pulp after extraction; triangles—pulp after chlorine dioxide bleaching. Filled circles are controls where no fungal treatment was performed. (From Reid, I.D., Paice, M.G., Ho, C., and Jurasek, L., *Tappi J.*, 73(8), 149–153, 1990. With permission.)

components is required, otherwise cell free cultures do not cause bleaching [102]. Such components could be hydrogen peroxide, required by peroxidases, or unstable mediators, required by laccase (see the Laccase/Mediators Bleaching section).

White-rot fungi often produce combinations of lignin peroxidase, manganese peroxidase, and/or laccase when performing extensive lignin degradation [for reviews, see 104,105]. Initial research with *P. chrysosporium* identified lignin peroxidase (LiP) as responsible for lignin depolymerization [106]. However, for the fungus *Trametes versicolor*, LiP activity was inhibited by metavanadate while fungal bleaching was not, indicating that LiP is not necessary for bleaching [102]. Furthermore, Katagiri et al. [107] found that in solid state cultures, *P.chrysosporium* bleached pulp but did not produce detectable levels of lignin peroxidase.

Several lines of evidence indicate that manganese peroxidase (MnP) is a key enzyme in pulp bleaching by *T. versicolor*. The enzyme is induced by the presence of pulp (Figure 15.7) [108], as are organic acids that are required for chelation of Mn(III) [109], and mutants of the fungus lacking MnP are unable to bleach pulp [110]. The bleaching activity of the mutant is partially restored by addition of exogenous MnP. In softwood pulp bleaching experiments with *P. chrysosporium*, Katagari et al. [107] concluded that MnP may be involved in delignification provided Mn(II) concentrations in pulp are sufficient. Another white-rot fungus, *Bjerkandera* sp., has been found to produce manganese-independent pulp bleaching [111,112]. This fungus produces LiP, MnP and a third peroxidase, namely manganese-independent peroxidase (MIP) or "versatile" peroxidase (VP). In spite of the appearance of MIP during pulp bleaching by *Bjerkandera,* there was a strong correlation between brightness gains and the appearance of MnP activity, leading the authors to conclude that an MnP is probably responsible for bleaching while functioning in some Mn-independent manner [113].

Laccase is also produced by *T.versicolor* during bleaching of kraft pulp (Figure 15.7) [114]. As discussed below, laccase will polymerize and darken lignin unless a

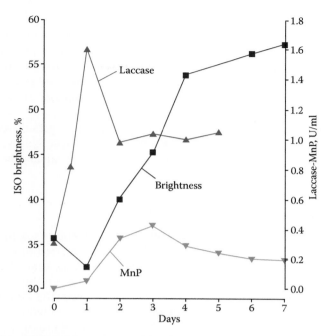

FIGURE 15.7 **(See color insert following page 336.)** Production of laccase and MnP by cultures of *Trametes versicolor* during bleaching of hardwood kraft pulp. (Redrawn from Paice, M.G., Reid, I.D., Bourbonnais, R., Archibald, F.S., Jurasek, L., *Appl Environ Microbiol*, 59, 260–265, 1993. With permission.)

mediator is present. In spite of extensive searching, we have found no evidence for the production of a natural mediator during pulp bleaching by *T.versicolor*. A pigment precursor, 3-hydroxyanthranilic acid (3-HAA), was found to be produced by the white-rot fungus, *Pycnoporus cinnabarinus*, and was proposed to act as a natural fungal laccase mediator [115,116]. In the presence of this fungal metabolite, laccase was shown to oxidize and cleave nonphenolic lignin compounds and a synthetic lignin polymer (DHP), although no evidence of pulp delignification by this mediator has been shown to date.

As well as peroxidases and laccases, several other redox enzymes have been implicated in residual lignin bleaching. Many fungi produce cellobiose dehydrogenase (CDH), which oxidizes cellobiose and cellooligomers and concurrently reduces phenols, metals and, to a limited extent, oxygen [117]. Potential interactions and synergies between MnP and CDH include reduction of MnO_2 to $Mn(II)$ [118], generation of sugar acids that complex with $Mn(II)$ [118], conversion of nonphenolic lignin substructures to phenolic structures by hydroxyl radicals generated through Fenton chemistry, with $Fe(II)$ being cycled by CDH [119], and reduction of quinones and cation radicals to phenols to provide new substrates for MnP [120]. However, in spite of sustained efforts there has been no conclusive demonstration of a significant role for CDH in pulp bleaching. In fact, CDH-deficient strains of

T.versicolor, produced by transforming protoplasts with a plasmid carrying a cdh gene, were still able to bleach kraft pulp at a rate comparable with the wild-type [121]. Another enzyme that may facilitate the action of MnP in pulp delignification by *P. chrysosporium* is an NADH-dependent ferrireductase, which was found to reduce manganese dioxide to Mn(II) in unbleached hardwood kraft pulp [122]. However, oxalate is also commonly produced by such fungi and can also act as a reducing agent for MnO_2.

Finally, several enzymes from white-rot fungi are able to generate hydrogen peroxide, which is required for peroxidatic activity. These include glucose-2-oxidase [123] and veratryl alcohol oxidase [124]. Hydrogen peroxide can also be generated nonenzymically by several routes.

MANGANESE PEROXIDASE AND OTHER PEROXIDASES

There are at least three families of fungal peroxidases implicated in lignin degradation, namely manganese peroxidase, MIP (sometimes referred to as VP [125]), and lignin peroxidase, although, as discussed above, the latter family has not been directly linked to kraft pulp bleaching. The molecular biology of lignin-degrading fungal peroxidases has been the subject of several reviews [125,126]. Lignolytic peroxidases are usually secreted as multiple isozymes due to multiple duplicated or unique genes, often on different alleles, and due to posttranslational modifications such as glycosylation. Expression of the peroxidase genes is often triggered by nutrient limitation (carbon or nitrogen), but not always. Transcription of MnP is usually regulated by Mn(II), and sometimes by heat shock or chemical stress [127]. Based on homology analysis, the peroxidases can be placed mainly in three family groups, one containing LiP enzymes, a second containing MnP, and a third containing both LiP and MnP enzymes from *Trametes versicolor* and *Pleurotus* and also the VPs [126]. We will now review the chemistry of pulp bleaching by peroxidases.

Manganese Peroxidase

Manganese peroxidase (MnP) was discovered in 1983 [128], and its role in oxidation of lignin model compounds and depolymerization of DHP lignin has been well documented [129–131]. However, it was only in 1993 that the enzyme was found to be important in delignification and bleaching of kraft pulp by white-rot fungi [109]. MnP oxidizes Mn(II) to Mn(III), which in complexed form is the proximal oxidant for lignin (Figure 15.8). Mn(III) malonate was used in the first report of pulp bleaching by MnP [108], but other complexing agents such as oxalate and glycolate, which are produced by *T.versicolor* under fungal bleaching culture conditions, are also effective [109], as are sugar acids [118]. The addition of nonionic surfactants such as Tween 80 to bleaching solutions of MnP enhances the bleaching effect at low pulp consistency [132,133]. Unsaturated fatty acids have a similar effect, which may indicate a role for peroxy radicals in delignification [134].

The chemistry of pulp bleaching by MnP is complex and not fully understood. As with laccase-catalyzed bleaching, methanol is formed during delignification by MnP [108], and may result from oxidative demethylation of phenolic subunits (Figure 15.9).

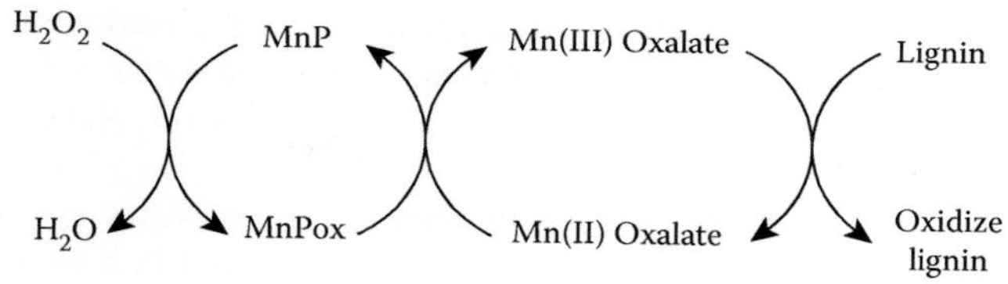

FIGURE 15.8 Oxidation of residual lignin by MnP and complexed manganese ions.

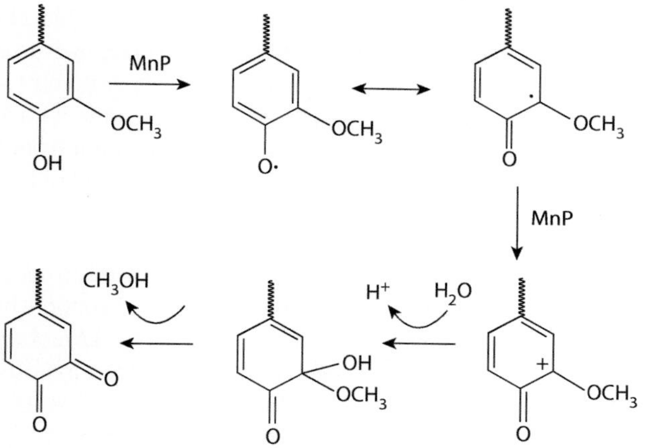

FIGURE 15.9 Oxidation of guaiacyl groups to o-quinonoid structures by manganese peroxidase.

Quinone formation has been identified by the hypochromic effect of sodium borohydride or trimethyl phosphite treatment of alkaline solubilized lignin from MnP bleaching [135]. MnP is known to oxidize phenolic subunits at the C-alpha position to give C-alpha carbonyls [136], but somewhat surprisingly, these were not observed during MnP catalyzed delignification [135]. This may be because such structures are unstable under the oxidative alkaline extraction conditions that are required to remove lignin after MnP bleaching [137].

The initial oxidation of lignin by MnP, as measured by kappa number change, does not result in solubilization of lignin, because the klason lignin content remains constant [138]. It is only during subsequent alkaline extraction that the klason lignin content of the pulp decreases. A much improved delignification is observed [139] if the subsequent extraction stage includes hydrogen peroxide (either Ep or P stages). This may be because of an increased content of o-quinone structure in MnP-treated residual lignin; such structures (Figure 15.9) are known to be susceptible to alkaline peroxide oxidation [140].

The MnP catalytic cycle resembles that of the classical heme peroxidase horseradish peroxidase [141], and also LiP [142]. A two-electron transfer from hydrogen peroxide to native MnP results in MnP Compound I, which can then perform a one electron oxidation of chelated Mn(II). The resulting MnP Compound II performs another oxidation of Mn(II) and returns the enzyme to the native form [142]. MnP Compound I, but not Compound II, is also reducible by some phenolic subtrates, so both direct and Mn(II) mediated oxidation of phenols by MnP are possible. Importantly for pulp bleaching, the native enzyme can be converted to Compound III by an excess of hydrogen peroxide, as might be encountered in a bleaching tower. Compound III is often regarded as a dead-end form of the enzyme, although it can in fact decay to native enzyme with the release of superoxide [141]. A further complication is that the stable, high-redox potential Mn(III) complexes produced by MnP are also subject to dissociation to Mn(II) and Mn(IV) by excess peroxide [143]. Thus it is necessary to add peroxide at a controlled rate in order to drive the bleaching reactions of MnP. In conventional bleaching and brightening reactions, hydrogen peroxide is added to pulp at medium to high consistency with a high shear mixer [144], allowing the maximum kinetic effect to occur. This would not be possible with MnP. Instead, *in situ* peroxide formation has been proposed, such as generation from oxygen by an oxidase [145], or by generation of organic peroxides from lipids [132].

Other Peroxidases

As mentioned above, there is little evidence from the literature to support a role for lignin peroxidase in kraft pulp bleaching. However, several reports indicate that a manganese-independent bleaching of kraft pulp is possible by a *Bjerkandera* species. This species and *Pleurotus eryngii* have been found to produce a hybrid enzyme (VP) with activity towards Mn (II) and nonphenolic substrates such as veratryl alcohol [146,147]. A gene for a similar enzyme was also found in *T. versicolor* [125,148]. The oxidation of the nonphenolic substrates occurs in the absence of Mn(II). Currently, there is no firm evidence that this VP can oxidize residual lignin in kraft pulp. However, the enzyme is capable of oxidizing azo dyes and phthalocyanin complexes in an Mn(II) independent manner [149]. The reports of Mn-independent bleaching by *Bjerkandera* species may thus be ascribed to VP with both MnP and nonphenolic substrate range, rather than to the previously reported MnP.

LACCASE/MEDIATOR BLEACHING

Laccase is a blue copper-containing oxidase that catalyses the four electron reduction of oxygen to water with simultaneous one-electron oxidation of various aromatic substrates [150]. The enzyme is widely distributed in plants and fungi and particularly is abundant in many lignin-degrading white-rot fungi [151,152]. Laccase has a broad substrate range, including phenols, aromatic amines and other easily oxidized aromatic compounds [153]. Oxidation of lignin by fungal laccase has been studied quite intensively since the early 1970's. In lignin, only phenolic subunits are attacked by laccase, giving oxygen-centered radicals which usually react further though nonenzymatic routes. Ishihara [154] showed that oxidation of phenolic compounds by laccase often results in polymerization and quinone formation. Extensive studies on model compounds by Higuchi's group showed that oxidation of phenolic lignin dimers by laccase can also lead to C–α oxidation, Cα–Cβ cleavage and alkyl-aryl cleavage [155,156].

There is no direct evidence that laccase alone can degrade lignin in wood and pulp. In fact, wood and pulp lignins are poor substrates for laccase because lignin has a low content of free phenolic hydroxyl groups. Also, laccase is a large molecule (molecular weight around 70,000). Consequently, the enzyme cannot diffuse easily within the fiber wall to contact lignin directly [157]. However, it was found that the substrate range of laccase can be extended to nonphenolic subunits of lignin when an appropriate cosubstrate is present [158]. Thus, as shown in Figure 15.10, laccase can efficiently oxidize veratryl alcohol 1 to veratraldehyde 2, oxidize the nonphenolic ß–O–4 dimer, veratrylglycerol-ß-guaiacyl ether 3 to its C–α keto derivative 4 and cleave the nonphenolic ß-1dimer, dimethoxyhydrobenzoin 5 to give veratraldehyde 2 and benzaldehyde 6 only when a mediator such as 2,2′-azinobis (3-ethylbenzthiazoline-5-sulphonate), (ABTS) is present [158]. Identical reactions were also shown with syringaldehyde as cosubstrate [159] but the efficiency of the mediated reactions were almost 100 times lower than with ABTS [160]. Muheim et al. [160] explained these reactions by abstraction of a hydrogen atom from the C–α position to give hydroxy-substituted ketyl radicals, which are subsequently oxidized to the corresponding aldehyde or ketone functionalities.

The application of laccase and mediator for kraft pulp delignification was first demonstrated using ABTS as redox mediator [161]. The delignification reaction was shown to be preceded by demethylation of most free phenolic methoxyl groups in lignin [162]. Following these studies, the concept of lignin degradation by laccase and mediator was developed and is illustrated in Figure 15.11. The current hypothesis is that the redox mediator acts as a diffusible electron carrier between residual lignin in the fiber wall and the large laccase molecule in solution. The overall reaction is a

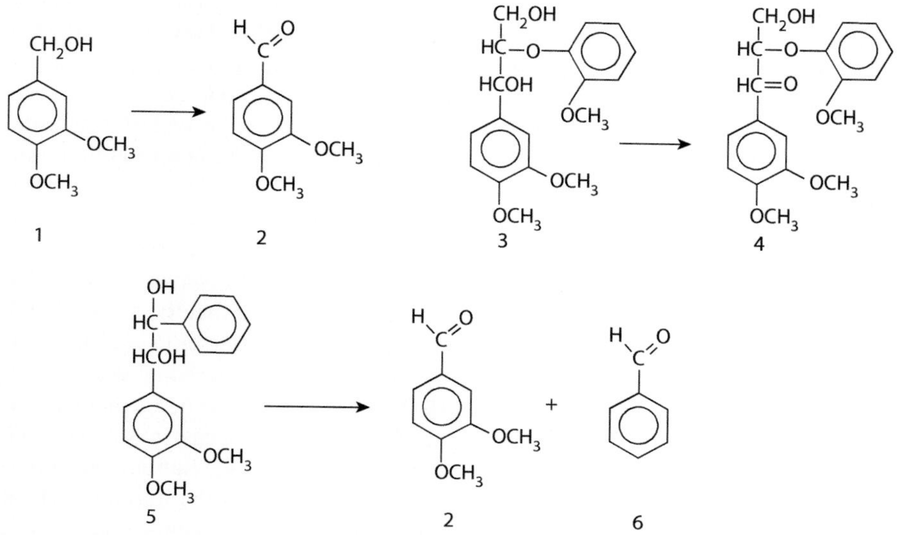

FIGURE 15.10 Oxidation of lignin model compounds by laccase and ABTS. (From Bourbonnais, R., Paice, M.G., *FEBS Letters*, 267, 99–102, 1990. With permission.)

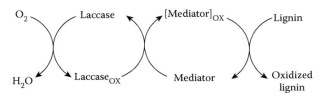

FIGURE 15.11 Schematic representation of lignin degradation by laccase and mediator system. The structure of various mediators, including ABTS and HBT are shown in Table 15.1.

continuous flow of electrons from lignin to oxygen leading to the formation of water and oxidized lignin.

Using this concept, over 50% delignification of kraft pulp was reported with laccase and another mediator, 1-hydroxybenzotriazole (HBT) [163], or by repeated treatment of laccase/ABTS under industrial process conditions [164]. Delignification with laccase and HBT was applied to pilot plant trials of totally chlorine free (TCF) kraft pulp bleaching [165]. Sequential treatment with laccase/ABTS or HBT followed by dimethyldioxirane (DMD) [166] or alkaline hydrogen peroxide [167–169] was also shown to improve kraft pulp bleaching efficiency. The ability of ABTS or HBT to delignify and bleach various kraft pulps was reported in several papers [170–179]. Application of laccase and mediators has also been described for degradation of polycyclic aromatic hydrocarbons (PAHs) [180–184], selective oxidation of aromatic methyl groups [185] and benzyl alcohols [186–189], and bleaching of textile dyes [190–192]. A pigment precursor, 3-HAA, was found to be produced by the white-rot fungus, *P. cinnabarinus*, and was proposed to act as a natural fungal laccase mediator [193]. In the presence of this fungal metabolite, laccase was shown to oxidize and cleave nonphenolic lignin compounds and a synthetic lignin polymer dehydrogenatively polymerized (DHP), although no evidence of pulp delignification has been shown to date. In a subsequent paper Li et al. [194] showed that 3-HAA does not play an important role in the fungal degradation of lignin.

To better understand the mechanism of chemical interactions between laccase, mediator and lignin, numerous model compounds studies were undertaken using ABTS and HBT as mediator. As previously shown with ABTS, nonphenolic lignin model compounds were oxidized with laccase and HBT [177,195–200]. However, the rate of veratryl alcohol and ß–O–4 dimer 3 oxidation was higher with HBT than with ABTS, and the reaction products with the ß–1 dimer 5 were distinct for the two mediators, as shown by formation of the C–α carbonyl derivative 7 only in the presence of HBT (Figure 15.12) [177]. Srebotnik [199] showed that the laccase/HBT couple was able to oxidize and cleave a more reactive nonphenolic β–O–4 dimer containing a *para*-alkoxyl substituent 8 (Figure 15.12). The C–α carbonyl derivative 9 produced during the reaction with laccase/HBT was also shown to be further cleaved to compounds 10 and 11. In the initial phase of the reaction, laccase in the presence of ABTS or HBT was shown to cause polymerization of phenolic lignin model compounds or soluble kraft lignin preparations, but then degrades the initially formed polymers [201,202]. Condensed phenolic substructures of kraft lignin such as 5-5′, α–5 diphenylmethane and stilbene were found to be degraded mainly via side-chain oxidation reactions by laccase mediator [203,204].

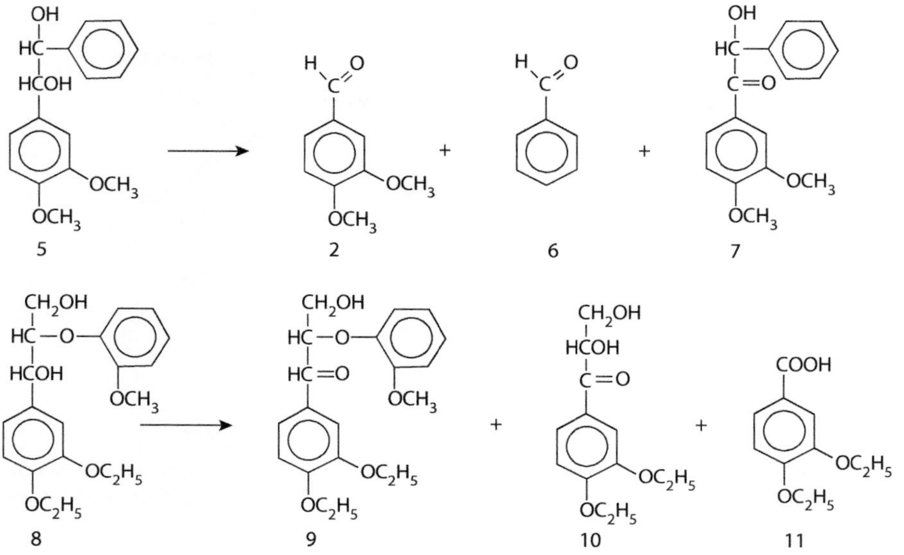

FIGURE 15.12 Lignin model dimers oxidation catalyzed by laccase and HBT. (From Bourbonnais, R., Paice, M.G., Freiermuth, B., Bodie, E., Borneman, S., *Applied and Environmental Microbiology*, 63(12), 4627–4632, 1997. With permission.)

ABTS and HBT are known to form radical intermediates following chemical, electrochemical and enzymatic (laccase) oxidation [205,206]. Electroanalytical studies showed that the ABTS$^{+\bullet}$ cation radical, which is easily generated by laccase, can be further oxidized to ABTS^{2+} dication by the enzyme at a very slow rate [207]. These two radical intermediates are stable and electrochemically reversible [206,207]. The high redox potential intermediate ABTS^{2+} (E$^{o'}$ = 1.08V vs standard hydrogen electrode (SHE)) was shown to be responsible for the oxidation of nonphenolic lignin compounds while the lower redox potential radical ABTS$^{+\bullet}$ (E$^{o'}$ = 0.67V vs SHE) reacted only with phenolic structures in lignin (Figure 15.13). Unlike ABTS, the radical intermediate of HBT (E$_p$ = 1.08V vs SHE), which is presumably the benzotriazolyl-1-oxide radical [205], is unstable and decays rapidly to unreactive breakdown products, including benzotriazole [173,208,209]. As with ABTS^{2+} dication, due to its high redox potential, the radical intermediate of HBT is only generated slowly by laccase, but can then efficiently oxidize nonphenolic lignin subunits. Coupling reactions involving ABTS or HBT radical intermediates with lignin phenolic compounds have also been reported [201,207,208,210]. The instability of HBT radical and the coupling behavior of ABTS radicals prevent their use as true catalysts for pulp delignification.

As a result of extensive screening, a growing number of nitrogen containing compounds were also found to mediate pulp delignification with laccase. Apart from ABTS and HBT, they include 1-nitroso-2-naphthol-3,6-disulfonic acid (NNDS) and 2-nitroso-1-naphthol-4-sulfonic acid (HNNS) [176,177,211,212], phenothiazines [190], violuric acid, and N-hydroxyacetanilide (NHA) [176,213–217].

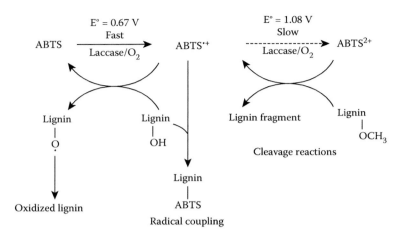

FIGURE 15.13 Schematic representation of pulp delignification with laccase and ABTS. (Adapted from Bourbonnais, R., Leech, D., Paice, M.G., *Biochimica et Biophysica Acta*, 1379(3), 381–390, 1998. With permission.)

Recently, transition metal complexes, such as potassium octacyanomolybdate ($K_4Mo(CN)_8$), were found to efficiently catalyze pulp delignification with laccase [218,219]. Due to their redox properties and stability, these metal complexes can be recycled and reused with the same efficiency. Efficient pulp delignification was also reported by using the electrochemically oxidized mediators, violuric acid [220] and potassium octacyanomolybdate [221] without using laccase. The latter report showed that molecular oxygen was not involved during electrochemical pulp delignification with octacyanomolybdate, which acts most likely as an electron carrier and not as an oxygen-activated agent [221]. Polyoxometalates were also shown to catalyze pulp delignification in the presence of laccase, but the rate of delignification was relatively low [222]. Table 15.1 summarizes the redox and enzymatic oxidation rate characteristics of these mediators in relation to their capacity to delignify kraft pulp. It seems likely that the efficiency of a mediator to delignify pulp is much more related to its redox potential then to its oxidation rate by laccase. Thus, HBT with a high redox potential and a very slow rate of enzymatic oxidation, is one of the best mediators in this series. On the other hand, a high redox potential mediator does not guarantee an efficient delignification, as shown by the results obtained with promazine, which has radical intermediate redox potentials similar to those of ABTS, but with a much lower rate of delignification. Other factors such as radical stability, diffusion rate, and kinetics of electron transfer between mediator and lignin are likely involved in the overall efficiency of the delignification reaction.

The efficiency of enzymatic delignification is also dependent on the redox and stability characteristics of the laccase used during the treatment. As reported in various studies [214,217,223–226] high redox potential laccases are more efficient in oxidizing high redox potential mediators. Thus, fungal laccases with potential around 0.75 to 0.8 V are much more efficient than the plant laccases. However, fungal laccases with similar redox potential were shown to have large differences in their kinetic

TABLE 15.1

Properties of Laccase Mediators

Mediator	Redox Potential vs NHE[1] V	Relative O_2 Uptake Rate[2]	Pulp[3] Delignification, %
HBT	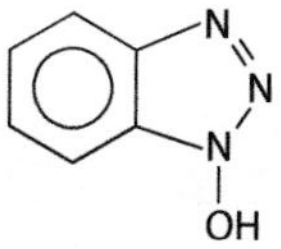 1.04*	1	45
$K_4Mo(CN)_8$	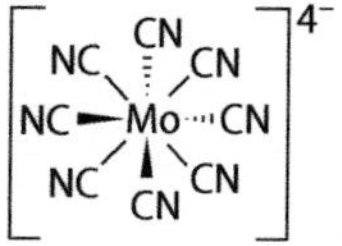 0.75	37	43
NHA	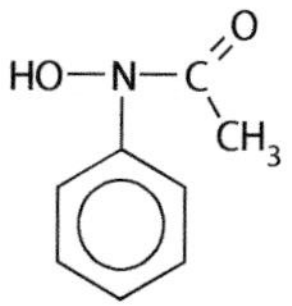 0.83*; 1.01*	42	42

(Continued)

TABLE 15.1
Properties of Laccase Mediators (Continued)

Mediator	Redox Potential vs NHE[1] V	Relative O_2 Uptake Rate[2]	Pulp[3] Delignification, %
Violuric acid	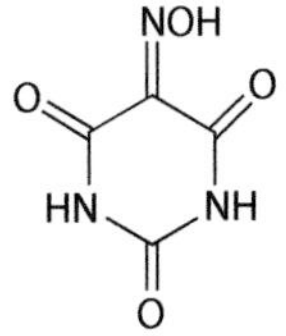 0.91	14	42
ABTS	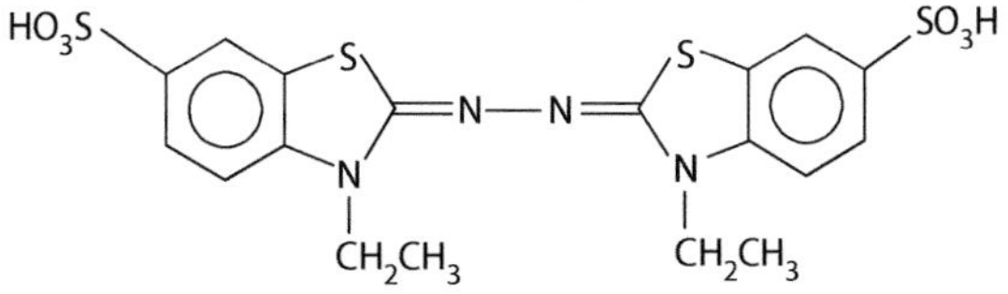 0.67; 1.09	100	37
HNNS	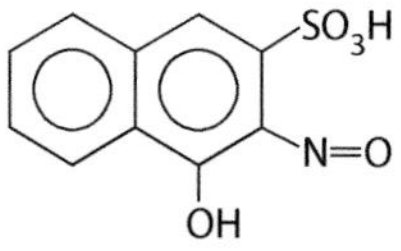 0.91*; 1.10*	46	34

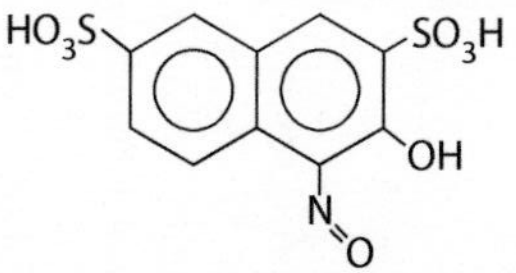 NNDS HO₃S (SO₃H, OH, N=O structure)	0.82	46	29
Promazine CH₂CH₂CH₂N(CH₃)₂ (phenothiazine structure)	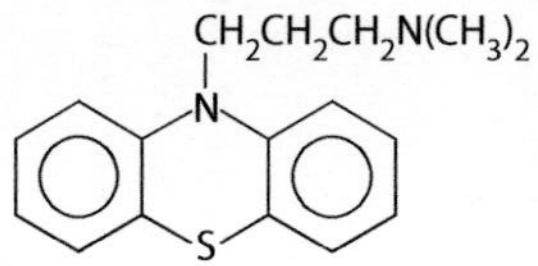 0.77; 1.06*	59	18

Source: R Bourbonnais, D Rochefort, MG Paice, S Renaud, D Leech. Development of stable redox complexes to mediate delignification of kraft pulp by laccase. ACS Symposium Series 785: 391–399, 2001. With permission.

[1] NHE: Standard Hydrogen Electrode. Where two values are shown, this indicates that two electron transfers were observed within the measured range.

[2] In the absence of pulp.

[3] Softwood oxygen delignified kraft pulp, kappa no = 15. Mediator applied at 1% on pulp followed by a peroxide reinforced alkaline extraction (Ep).

*Electrochemically irreversible, redox potentials are lower than the values shown.

parameters (k_m and k_{cat}) with selected mediators [214]. The stability of the enzyme to free radicals produced by mediator oxidation is also an important factor in the overall delignification process [214,217]. To be easily and economically employed in an industrial bleaching sequence, the laccase will likely have to be thermostable, active over a wide pH range, and resistant to deactivation by radical intermediates. Genetic engineering of laccase has produced high-yield recombinant enzymes [224], but further research is needed to produce higher potential and more stable laccases.

CONCLUDING REMARKS

The complex and variable structure of lignin in wood and pulp, combined with the morphology of wood-derived fibers that makes contact between microorganisms, enzymes, and lignin difficult, have made research in the areas of biopulping and biobleaching a challenging subject. Nevertheless, we have seen a marked increase in our understanding of the mechanisms of fungal and enzymic pulping and bleaching in the last 25 years. Furthermore, in biobleaching, there has been a significant application of xylanases in kraft mills. This has occurred in part due to the rapid advances in biotechnology and molecular biology, resulting in more effective, cheaper enzymes. Protein engineering is used routinely to produce superior xylanases and more stable peroxidases, and metabolic pathway engineering is talked about as a possibility for fungal systems. There is also a rapidly growing database of fungal genomes; the total genome of *P. chrysosporium* has been sequenced and is publicly available [227]. In spite of these advances, the fundamental barrier to competing with conventional chemical and mechanical pulping and bleaching remains, namely the inaccessibility of enzymes to lignin in a cell wall structure. This barrier probably means that there will be only a limited niche for these techniques in pulping and bleaching.

ACKNOWLEDGMENTS

We wish to thank Paul Bicho of FPinnovations–Paprican Division, Vancouver, for his careful review of the manuscript.

REFERENCES

1. TK Kirk, M Akhtar, RA Blanchette. Biopulping: Seven years of consortia research. Proceedings of TAPPI Biological Sciences Symposium, Atlanta, 1994, pp. 57–66.
2. EC Setliff, WW Eudy. Screening white-rot fungi for their capacity to delignify wood. In: TK Kirk, T Higuchi, H-m Chang, eds. *Lignin Biodegradation: Microbiology, Chemistry and Potential Applications*. Boca Raton: CRC Press, 1980, pp. 135–149.
3. L Otjen, R Blanchette, M Effland, G Leatham. Assessment of 30 white rot basidiomycetes for selective lignin degradation. *Holzforschung* 41:343–349, 1987.
4. RA Blanchette, TA Burnes, GF Leatham, MJ Effland. Selection of white-rot fungi for biopulping. *Biomass* 15:93–101, 1988.
5. T Nishida, Y Kashino, A Mimura, Y Takahara. Lignin biodegradation by wood-rotting fungi. 1. Screening of lignin-degrading fungi. *J Jpn Wood Res Soc* 34(6):530–536, 1988.
6. GF Leatham, GC Myers, TH Wegner. Biomechanical pulping of aspen chips: Energy savings resulting from different fungal treatments. *Tappi J* 73(5):197–200, 1990.

7. M Akhtar, MC Attridge, GC Myers, RA Blanchette. Biomechanical pulping of loblolly pine chips with selected white-rot fungi. *Holzforschung* 47:36–40, 1993.

8. GF Leatham, GC Myers, TH Wegner, RA Blanchette. Biomechanical pulping of aspen chips: paper strength and optical properties resulting from different fungal treatments. *Tappi J* 73(3):249–255, 1990.

9. GM Scott, R Swaney. New technology for papermaking: Biopulping economics. *Tappi J* 81(12):153–157, 1998.

10. C Rüttimann-Johnson, L Salas, R Vicuña, TK Kirk. Extracellular enzyme production and synthetic lignin mineralization by *Ceriporiopsis subvermispora*. *Appl Environ Microbiol* 59:1792–1797, 1993.

11. S Lobos, J Larraín, L Salas, D Cullen, R Vicuña. Isoenzymes of manganese-dependent peroxidase and laccase produced by the lignin-degrading basidiomycete *Ceriporiopsis subvermispora*. *Microbiology–UK* 140:2691–2698, 1994.

12. Y Fukushima, TK Kirk. Laccase component of the *Ceriporiopsis subvermispora* lignin-degrading system. *Appl Environ Microbiol* 61:872–876, 1995.

13. C Salas, S Lobos, J Larraín, L Salas, D Cullen, R Vicuña. Properties of laccase isoenzymes produced by the basidiomycete *Ceriporiopsis subvermispora*. *Biotechnol Appl Biochem* 21:323–333, 1995.

14. E Karahanian, G Corsini, S Lobos, R Vicuña. Structure and expression of a laccase gene from the ligninolytic basidiomycete *Ceriporiopsis subvermispora*. *Biochim Biophys Acta–Gene Struct Expr* 1443(1–2):65–74, 1998.

15. A Sethuraman, DE Akin, KEL Eriksson. Plant-cell-wall-degrading enzymes produced by the white-rot fungus *Ceriporiopsis subvermispora*. *Biotechnol Appl Biochem* 27(Pt. 1):37–47, 1998.

16. S Lobos, L Larrondo, L Salas, E Karahanian, R Vicuña. Cloning and molecular analysis of a cDNA and the Cs-mnp1 gene encoding a manganese peroxidase isoenzyme from the lignin-degrading basidiomycete *Ceriporiopsis subvermispora*. *Gene* 206(2):185–193, 1998.

17. S Rajakumar, J Gaskell, D Cullen, S Lobos, E Karahanian, R Vicuña. *Lip*-like genes in *Phanerochaete sordida* and *Ceriporiopsis subvermispora*, white rot fungi with no detectable lignin peroxidase activity. *Appl Environ Microbiol* 62:2660–2663, 1996.

18. E Srebotnik, KAJ Jensen, S Kawai, KE Hammel. Evidence that *Ceriporiopsis subvermispora* degrades nonphenolic lignin structures by a one-electron-oxidation mechanism. *Appl Environ Microbiol* 63(11):4435–4440, 1997.

19. KAJ Jensen, WL Bao, S Kawai, E Srebotnik, KE Hammel. Manganese-dependent cleavage of nonphenolic lignin structures by *Ceriporiopsis subvermispora* in the absence of lignin peroxidase. *Appl Environ Microbiol* 62:3679–3686, 1996.

20. K Messner, K Koller, MB Wall, M Akhtar, GM Scott. Fungal treatment of wood chips for chemical pulping. In: RA Young, M Akhtar, eds. *Environmentally Friendly Technologies for the Pulp and Paper Industry*. New York: John Wiley & Sons, Inc., 1998, pp. 385–419.

21. RA Blanchette, EW Krueger, JE Haight, M Akhtar, DE Akin. Cell wall alterations in loblolly pine wood decayed by the white-rot fungus, *Ceriporiopsis subvermispora*. *J Biotechnol* 53(2–3):203–213, 1997.

22. MJ Rocheleau, BB Sithole, LH Allen, S Iverson, R Farrell, Y Noel. Fungal treatment of aspen chips for wood resin reduction: A laboratory evaluation. *J Pulp Pap Sci* 24(2):37–42, 1998.

23. MB Wall, G Stafford, Y Noel, A Fritz, S Iverson, RL Farrell. Treatment with *Ophiostoma piliferum* improves chemical pulping efficiency. In: E Srebotnik, K Messner, eds. *Biotechnology in the Pulp and Paper Industry: Recent Advances in Applied and Fundamental Research*. Vienna: Facultas-Universitatsverlag, 1996, pp. 205–210.

24. CJ Jacobs-Young, RA Venditti, TW Joyce. Effect of enzymatic pretreatment on the diffusion of sodium hydroxide in wood. *Tappi J* 81(1):260–266, 1998.

25. CJ Jacobs, RA Vendetti, TW Joyce. Effect of enzyme pretreatments on conventional kraft pulping. *Tappi J* 81(2):143–147, 1998.
26. C Jacobs-Young, JA Heitmann, RA Venditti. Conventional kraft pulping using enzyme pretreatment technology: Role of chip thickness, specie and enzyme combinations. In: BN Brogdon, PW Hart, JC Ransdell, BL Scheller, eds. *Innovative Advances in the Forest Products Industries*. New York: Amer Inst Chemical Engineers, 1998, pp. 1–15.
27. CJ Jacobs-Young, RR Gustafson, JA Heitmann. Conventional kraft pulping using enzyme pretreatment technology: Role of diffusivity in enhancing pulp uniformity. *Paperi Ja Puu-Paper and Timber* 82(2):114–119, 2000.
28. LR Lawson, Jr., CN Still. The biological decomposition of lignin: Literature survey. *Tappi J* 40(9):56A–80A, 1957.
29. P Ander, KE Eriksson. Lignin degradation and utilization by microorganisms. *Prog Ind Microbiol* 14:1–58, 1978.
30. TK Kirk, Hm Chang. Potential applications of bio-ligninolytic systems. *Enz Microb Technol* 3:189–195, 1981.
31. JM Bowman. Biodelignification. *Pap Technol Ind* 24:89–94, 1983.
32. KE Eriksson. White-rot fungus in the service of the pulp and paper industry. *Kemisk Tidskrift* 95:A32, 1983.
33. Kirk, TW Jeffries, GF Leatham. Biotechnology - applications and implications for the pulp and paper industry. *Tappi J* 66:45–51, 1983.
34. KE Eriksson. Advances in microbial delignification. *Biotechnol Adv* 2:149–160, 1984.
35. L Jurasek, MG Paice. *Biotechnology in the Pulp and Paper Industry*. Ottawa: Science Council of Canada, 1984.
36. TK Kirk, TW Jeffries. Microbial technology in wood utilization. Proceedings of TAPPI Research and Development Conference, 1984, pp. 9.
37. KE Eriksson. Swedish developments in biotechnology related to the pulp and paper industry. *Tappi J* 68(7):46–55, 1985.
38. KE Eriksson. Microbial delignification of lignocellulosic materials. *Forestry Chronicle* 61:459–463, 1985.
39. KE Eriksson, TK Kirk. Biopulping, biobleaching and treatment of Kraft bleaching effluents with white-rot fungi. In: M Moo-Young, ed. *Comprehensive Biotechnology*. Oxford: Pergamon Press, 1985, pp. 271–294.
40. EA Griffin. Can we use biotechnology to reduce industrial energy consumption? *Tappi J* 68:56–59, 1985.
41. KE Eriksson, SC Johnsrud. Microbial delignification of lignocellulosic materials. *Papier* 40:33–37, 1986.
42. L Jurasek, M Paice. Pulp, paper, and biotechnology. *Chemtech* 16:360–365, 1986.
43. KE Eriksson. Microbial delignification: Basics, potentials and applications. *FEMS Symp* 43:285–302, 1988.
44. KEL Eriksson, RA Blanchette, P Ander. *Microbial and Enzymatic Degradation of Wood and Wood Components*. Berlin: Springer-Verlag, 1990.
45. L Flandroy. Industrie papetière: Une page à tourner. *Biofutur Juin,* 21–35, 1991.
46. ID Reid. Biological pulping in paper manufacture. *Tibtech* 9:262–265, 1991.
47. TK Kirk, RR Burgess, JW Koning, Jr. The use of fungi in pulping wood: an overview of biopulping research. In: GF Leatham, ed. *Frontiers in Industrial Mycology*. New York: Chapman & Hall, 1992, pp. 99–111.
48. P Ander. Biopulping, biobleaching and the use of enzymes in the pulp and paper industry. *J Kor TAPPI* 25(2):70–76, 1993.
49. K Sakai, R Kondo. Application of ligninolytic fungi to pulping processes. *J Jpn TAPPI* 47:933–943, 1993.
50. R Grant. Enzymes help to increase pulp & paper production. *Pulp and Paper International* 37(8):26–27, 1995.

51. M Akhtar, RA Blanchette, TK Kirk. Fungal delignification and biomechanical pulping of wood. *Adv Biochem Eng/Biotechnol* 57:159–195, 1997.
52. M Akhtar, RA Blanchette, G Myers, TK Kirk. An overview of biomechanical pulping research. In: RA Young, M Akhtar, eds. *Environmentally Friendly Technologies for the Pulp and Paper Industry.* New York: John Wiley & Sons, Inc., 1998, pp. 309–340.
53. HS Sabharwal. Biological approaches for pulping and bleaching of nonwoody plants. In: RA Young, M Akhtar, eds. *Environmentally Friendly Technologies for the Pulp and Paper Industry.* New York: John Wiley & Sons, Inc., 1998, pp. 449–479.
54. GM Scott, M Akhtar, MJ Lentz, TK Kirk, R Swaney. New technology for papermaking: commercialized biopulping. *Tappi J* 81(11):220–225, 1998.
55. GM Scott, M Akhtar, MJ Lentz, RE Swaney. Engineering, scale-up, and economic aspects of fungal pretreatment of wood chips. In: RA Young, M Akhtar, eds. *Environmentally Friendly Technologies for the Pulp and Paper Industry.* New York: John Wiley & Sons, Inc., 1998, pp. 341–383.
56. KE Eriksson, P Ander, B Henningsson, T Nilsson, B Goodell. Method for producing cellulose pulp. US Patent 3,962,033, 1976.
57. TK Kirk, JW Koning, Jr., RR Burgess, M Akhtar, RA Blanchette, DC Cameron, D Cullen, PJ Kersten, EN Lightfoot, GC Myers, M Sykes, MB Wall. Biopulping: A glimpse of the future? Forest Products Laboratory Research Report FPL-RP523, 1993, 74 pp.
58. P Ander, KE Eriksson. Mekanisk massa från förrötad flis: En inledande undersökning. *Sven papperstid* 78:641–642, 1975.
59. I Akamatsu, K Yoshihara, H Kamishima, T Fujii. Influence of white-rot fungi on poplar chips and thermo-mechanical pulping of fungi-treated chips. *J Jpn Wood Res Soc* 30:697–702, 1984.
60. GF Leatham, GC Myers, TH Wegner, RA Blanchette. Energy savings in biomechanical pulping. In: TK Kirk, H-m Chang, eds. *Biotechnology in Pulp and Paper Manufacture Applications and Fundamental Investigations.* Boston: Butterworth-Heinemann, 1990, pp. 17–25.
61. EC Setliff, R Marton, SG Granzow, KL Eriksson. Biomechanical pulping with white-rot fungi. *Tappi J* 73(8):141–147, 1990.
62. RN Patel, KK Rao. Potential use of the white-rot fungus *Antrodiella* sp. (RK1) in the process of biomechanical pulping and bioremediation. Proceedings of Fifth International Conference on Biotechnology in the Pulp and Paper Industry, Kyoto, 1992, p. 65.
63. M Akhtar, MC Attridge, RA Blanchette, GC Myers, MB Wall, MS Sykes, JW Koning, Jr., RR Burgess, TH Wegner, T Kirk. The white-rot fungus *Ceriporiopsis subvermispora* saves electrical energy and improves strength properties during biomechanical pulping of both hardwood and softwood chips. In: M Kuwahara, M Shimada, eds. *Biotechnology in Pulp and Paper Industry.* Tokyo: UNI Publishers, 1992, pp. 3–8.
64. RN Patel, GD Thakker, KR Rao. Potential use of a white-rot fungus *Antrodiella* sp. RK1 for biopulping. *J Biotechnol* 36:19–23, 1994.
65. M Akhtar. Biomechanical pulping of aspen wood chips with three strains of *Ceriporiopsis subvermispora. Holzforschung* 48:199–202, 1994.
66. L Lu, JL Yu, JX Chen. Biological chemimechanical pulping of aspen. In: L Huanbin, Z Huaiyu, X Yimin, eds. *Emerging Technologies of Pulping & Papermaking of Fast-Growing Wood.* Guangzhou: South China University Technology Press, 1998, pp. 424–430.
67. KE Eriksson, L Vallander. Properties of pulps from thermomechanical pulping of chips pretreated with fungi. *Sven papperstid* 85:R33–R38, 1982.
68. RA Blanchette, GF Leatham, M Attridge, M Akhtar, GC Myers. Biomechanical pulping with *C. subvermispora.* US Patent 5,055,159, 1991.
69. S Iverson, R Blanchette, C Behrendt, MB Wall, D Williams. *Phlebiopsis gigantea,* a white rot fungus for biomechanical pulping. Proceedings of TAPPI Biological Sciences Symposium, San Francisco, 1997, pp. 25–28.

70. M Akhtar, MJ Lentz, RA Blanchette, TK Kirk. Corn steep liquor lowers the amount of inoculum for biopulping. *Tappi J* 80(6):161–164, 1997.
71. G Giovannozzi-Sermanni, A D'Annibale, C Perani, A Porri, F Pastina, V Minelli, N Vitale, A Gelsomino. Characteristics of paper handsheets after combined biological pretreatments and conventional pulping of wheat straw. *Tappi J* 77(6):151–157, 1994.
72. S Camarero, JM Barrasa, M Pelayo, AT Martinez. Evaluation of *Pleurotus* species for wheat-straw biopulping. *J Pulp Pap Sci* 24(7):197–203, 1998.
73. HS Sabharwal, M Akhtar, RA Blanchette, RA Young. Biomechanical pulping of kenaf. *Tappi J* 77(12):105–112, 1994.
74. HS Sabharwal, M Akhtar, RA Blanchette, RA Young. Refiner mechanical and biomechanical pulping of jute. *Holzforschung* 49:537–544, 1995.
75. P Bustamante, J Ramos, V Zuniga, HS Sabharwal, RA Young. Biomechanical pulping of bagasse with the white rot fungi *Ceriporiopsis subvermispora* and *Pleurotus ostreatus*. *Tappi J* 82(6):123–128, 1999.
76. GC Myers, GF Leatham, TH Wegner, RA Blanchette. Fungal pretreatment of aspen chips improves strength of refiner mechanical pulp. *Tappi J* 71(5):105–108, 1988.
77. SS Bar-Lev, TK Kirk, H-m Chang. Fungal treatment can reduce energy requirements for secondary refining of TMP. *Tappi J* 65(10):111–113, 1982.
78. GF Leatham, GC Myers. A PFI mill can be used to predict biomechanical pulp strength properties. *Tappi J* 73(4):192–197, 1990.
79. Y Kashino, T Nishida, Y Takahara, K Fujita, R Kondo, K Sakai. Biomechanical pulping using white-rot fungus IZU-154. *Tappi J* 76(12):167–171, 1993.
80. L Pilon, MC Barbe, M Desrochers, L Jurasek, PJ Neumann. Fungal treatment of mechanical pulps: Its effect on paper properties. *Biotech Bioeng* 24:2063–2076, 1982.
81. L Samuelsson, PJ Mjöberg, N Hartler, L Vallander, KE Eriksson. Influence of fungal treatment on the strength versus energy relationship in mechanical pulping. *Sven papperstid* 83:221–225, 1980.
82. TH Wegner, GC Myers, GF Leatham. Biological treatments as an alternative to chemical pretreatment in high-yield wood pulping. *Tappi J* 74(3):189–193, 1991.
83. J Pellinen, J Abuhasan, TW Joyce, Hm Chang. Biological delignification of pulp by *Phanerochaete chrysosporium*. *J Biotechnol* 10:161–170, 1989.
84. M Sykes. Bleaching and brightness stability of aspen biomechanical pulps. *Tappi J* 76(11):121–126, 1993.
85. A Ferraz, L Christov, M Akhtar. Fungal pretreatment for organosolv pulping and dissolving pulp production. In: RA Young, M Akhtar, eds. *Environmentally Friendly Technologies for the Pulp and Paper Industry*. New York: John Wiley & Sons, Inc., 1997, pp. 421–447.
86. TP Oriaran, P Labosky, Jr., PR Blankenhorn. Kraft pulp and papermaking properties of *Phanerochaete chrysosporium*-degraded aspen. *Tappi J* 73(7):147–152, 1990.
87. TP Oriaran, P Labosky, Jr., PR Blankenhorn. Kraft pulp and papermaking properties of *Phanerochaete chrysosporium* degraded red oak. *Wood and Fiber Science* 23:316–327, 1991.
88. ID Reid. Optimisation of solid-state fermentation for selective delignification of aspen wood with *Phlebia tremellosa*. *Enz Microb Technol* 11:804–809, 1989.
89. JF Wolfaardt, JL Bosman, A Jacobs, JR Male, CJ Rabie. Bio-kraft pulping of softwood. In: E Srebotnik, K Messner, eds. *Biotechnology in the Pulp and Paper Industry: Recent Advances in Applied and Fundamental Research*. Vienna: Facultas-Universitatsverlag, 1996, pp. 211–216.
90. K Messner, S Masek, E Srebotnik, G Techt. Fungal pretreatment of wood chips for chemical pulping. In: M Kuwahara, M Shimada, eds. *Biotechnology in Pulp and Paper Industry*. Tokyo: UNI Publishers, 1992, pp. 9–13.

91. GM Scott, M Akhtar, MJ Lentz, M Sykes, S Abubakr. Environmental aspects of bio-sulfite pulping. Proceedings of International Environmental Conference, Atlanta, GA, 1995, pp. 1155–1161.

92. K Fischer, M Akhtar, RA Blanchette, TA Burnes, K Messner, TK Kirk. Reduction of resin content in wood chips during experimental biological pulping processes. *Holzforschung* 48(4):285–290, 1994.

93. S Mosai, JF Wolfaardt, BA Prior, LP Christov. Evaluation of selected white-rot fungi for biosulfite pulping. *Bioresource Technol* 68(1):89–93, 1999.

94. HH Burdsall. Taxonomy of industrially important white-rot fungi. In: RA Young, M Akhtar, eds. *Environmentally Friendly Technologies for the Pulp and Paper Industry.* New York: John Wiley & Sons, Inc., 1998, pp. 259–272.

95. MG Paice, N Gurnagul, DH Page, L Jurasek. Mechanism of hemicellulose-directed prebleaching of kraft pulps. *Enz Microb Technol* 14:272–276, 1992.

96. A Kantelinen, B Hortling, J Sundquist, M Linko, L Viikari. Proposed mechanism of the enzymatic bleaching of kraft pulp with xylanases. *Holzforschung* 47:318–324, 1993.

97. TK Kirk, HH Yang. Partial delignification of unbleached kraft pulp with ligninolytic fungi. *Biotechnol Lett* 1:347–352, 1979.

98. FS Archibald, R Bourbonnais, L Jurasek, MG Paice, ID Reid. Kraft pulp bleaching and delignification by *Trametes versicolor. J Biotechnol* 53(2–3):215–236, 1997.

99. ID Reid. Bleaching kraft pulps with white-rot fungi. In: RA Young, M Akhtar, eds. *Environmentally Friendly Technologies for the Pulp and Paper Industry.* New York: John Wiley & Sons, Inc., 1998, pp. 505–514.

100. ID Reid, MG Paice, C. Ho, L. Jurasek. Biological bleaching of softwood kraft pulp with the fungus *Trametes (Coriolus) versicolor.* Tappi J. 73(8), 149–153, 1990.

101. N Kirkpatrick, I Reid, E Ziomek, MG Paice. Biological bleaching of hardwood kraft pulp using *Trametes (Coriolus) versicolor* immobilized in polyurethane foam. *Appl Microbiol Biotechnol* 33:105–109, 1990.

102. FS Archibald. The role of fungus-fiber contact in the biobleaching of kraft brownstock by *Trametes (Coriolus) versicolor. Holzforschung* 46:305–310, 1992.

103. R Kondo, K Kurashiki, K Sakai. In vitro bleaching of hardwood kraft pulp by extracel-lular enzymes excreted from white rot fungi in a cultivation system using a membrane filter. *Appl Environ Microbiol* 60:921–926, 1994.

104. R ten Have, PJM Teunissen. Oxidative mechanisms involved in lignin degradation by white-rot fungi. *Chemical Reviews* 101:3397–3413, 2001.

105. A Hatakka. Lignin-modifying enzymes from selected white-rot fungi: Production and role in lignin degradation. *FEMS Microbiol Rev* 13:125–135, 1994.

106. M Tien, TK Kirk. Lignin degrading enzyme from the hymenomycete *Phanerochaete chrysosporium. Science* 221:661–663, 1983.

107. N Katagiri, Y Tsutsumi, T Nishida. Biobleaching of softwood kraft pulp by white-rot fungi and its related enzymes. *J Jpn Wood Res Soc* 43(8):678–685, 1997.

108. MG Paice, ID Reid, R Bourbonnais, FS Archibald, L Jurasek. Manganese peroxidase produced by *Trametes versicolor* during pulp bleaching, demethylates and delignifies kraft pulp. *Appl Environ Microbiol* 59:260–265, 1993.

109. B Roy, F Archibald. Effects of kraft pulp and lignin on *Trametes versicolor* carbon metabolism. *Appl Environ Microbiol* 59:1855–1863, 1993.

110. K Addleman, T Dumonceaux, MG Paice, R Bourbonnais, FS Archibald. Production and characterization of *Trametes versicolor* mutants unable to bleach hardwood kraft pulp. *Appl Environ Microbiol* 61:3687–3694, 1995.

111. MT Moreira, G Feijoo, R Sierra-Alvarez, J Lema, JA Field. Manganese is not required for biobleaching of oxygen-delignified kraft pulp by the white rot fungus Bjerkandera sp. strain BOS55. *Appl Environ Microbiol* 63(5):1749–1755, 1997.

112. MT Moreira, G Feijoo, R Sierra-Alvarez, J Lema, JA Field. Biobleaching of oxygen delignified kraft pulp by several white rot fungal strains. *J Biotechnol* 53(2–3):237–251, 1997.

113. MT Moreira, G Feijoo, R Sierra-Alvarez, JA Field. Reevaluation of the manganese requirement for the biobleaching of kraft pulp by white rot fungi. *Bioresour Technol* 70(3):255–260, 1999.

114. MG Paice, FS Archibald, R Bourbonnais, L Jurasek, ID Reid, T Charles, T Dumonceaux. Enzymology of kraft pulp bleaching by *Trametes versicolor*. In: TW Jeffries, L Viikari, eds. *Enzymes for Pulp and Paper Processing*, ACS Symposium Series, Vol. 655. Washington: American Chemical Society, 1996, pp. 151–164.

115. C Eggert, U Temp, KEL Eriksson. Laccase is essential for lignin degradation by the white-rot fungus *Pycnoporus cinnabarinus*. *FEBS Lett* 407(1):89–92, 1997.

116. C Eggert, U Temp, JFD Dean, K-EL Eriksson. A fungal metabolite mediates degradation of non-phenolic lignin structures and synthetic lignin by laccase. *FEBS Lett* 391:144–148, 1996.

117. P Ander, C Misra, RL Farrell, KE Eriksson. Redox reactions in lignin degradation: interactions between laccase, different peroxidases, and cellobiose: Quinone oxidoreductase. *J Biotechnol* 13:189–198, 1990.

118. BP Roy, MG Paice, FS Archibald, SK Misra, LE Misiak. Creation of metal-complexing agents, reduction of manganese dioxide, and promotion of manganese peroxidase-mediated Mn(III) production by cellobiose quinone oxidoreductase from *Trametes versicolor*. *J Biol Chem* 269:19745–19750, 1994.

119. L Hilden, G Johansson, G Pettersson, JB Li, P Ljungquist, G Henriksson. Do the extracellular enzymes cellobiose dehydrogenase and manganese peroxidase form a pathway in lignin biodegradation? *FEBS Lett* 477(1–2):79–83, 2000.

120. P Ander, L Marzullo. Sugar oxidoreductases and veratryl alcohol oxidase as related to lignin degradation. *J Biotechnol* 53(2–3):115–131, 1997.

121. T Dumonceaux, K Bartholomew, L Valeanu, T Charles, F Archibald. Cellobiose dehydrogenase is essential for wood invasion and nonessential for kraft pulp delignification by *Trametes versicolor*. *Enz Microb Technol* 29(8–9):478–489, 2001.

122. H Hirai, T Oniki, R Kondo, K Sakai. Change in oxidation state of manganese atoms in kraft pulp during biological bleaching with white-rot fungus *Phanerochaete sordida* YK- 624. *J Wood Sci* 46(2):164–166, 2000.

123. KE Eriksson, B Pettersson, J Volc, V Musilek. Formation and partial characterization of glucose-2-oxidase, a H_2O_2 producing enzyme in *Phanerochaete chrysosporium*. *Appl Microbiol Biotechnol* 23:257–262, 1986.

124. R Bourbonnais, MG Paice. Veratryl alcohol oxidases from the lignin-degrading basidiomycete *Pleurotus sajor-caju*. *Biochem J* 255:445–450, 1988.

125. AT Martinez. Molecular biology and structure-function of lignin-degrading heme peroxidases. *Enz Microb Technol* 30:425–444, 2002.

126. A Conesa, PJ Punt, CAMJJ van den Hondel. Fungal peroxidases: Molecular aspects and applications. *J Biotechnol* 93(2):143–158, 2002.

127. JA Brown, D Li, M Alic, MH Gold. Heat shock induction of manganese peroxidase gene transcription in *Phanerochaete chrysosporium*. *Appl Environ Microbiol* 59:4295–4299, 1993.

128. JK Glenn, MH Gold. Purification and characterization of an extracellular Mn(II)-dependent peroxidase from the lignin-degrading Basidiomycete *Phanerochaete chrysosporium*. *Arch Biochem Biophys* 242:329–341, 1985.

129. H Wariishi, K Valli, MH Gold. Oxidative cleavage of a phenolic diarylpropane lignin model by manganese peroxidase from *Phanerochaete chrysosporium*. *Biochemistry* 28:6017–6023, 1989.

130. H Wariishi, K Valli, MH Gold. *In vitro* depolymerization of lignin by manganese peroxidase of *Phanerochaete chrysosporium*. *Biochem Biophys Res Comm* 176:269–276, 1991.

131. M Hofrichter. Review: Lignin conversion by manganese peroxidase (MnP). *Enz Microb Technol* 30:454–466, 2002.

132. R Kondo, K Harazona, K Sakai. Bleaching of hardwood kraft pulp with manganese peroxidase secreted from *Phanerochaete chrysosporium*. *Appl Environ Microbiol* 60:4359–4363, 1994.

133. K Ehara, Y Tsutsumi, T Nishida. Role of Tween 80 in biobleaching of unbleached hardwood kraft pulp with manganese peroxidase. *J Wood Sci* 46(2):137–142, 2000.

134. T Watanabe, S Katayama, M Enoki, YH Honda, M Kuwahara. Formation of acyl radical in lipid peroxidation of linoleic acid by manganese-dependent peroxidase from *Ceriporiopsis subvermispora* and *Bjerkandera adusta*. *Eur J Biochem* 267(13):4222–4231, 2000.

135. K Ehara, Y Tsutsumi, T Nishida. Structural changes of residual lignin in softwood kraft pulp treated with manganese peroxidase. *J Wood Sci* 44(4):327–331, 1998.

136. U Tuor, H Wariishi, HE Schoemaker, MH Gold. Oxidation of phenolic arylglycerol beta-aryl ether lignin model compounds by manganaese peroxidase from *Phanerochaete chrysosporium*: Oxidative cleavage of an alpha-carbonyl model compound. *Biochemistry* 31:4986–4995, 1992.

137. DM Li, HL Youngs, MH Gold. Heterologous expression of a thermostable manganese peroxidase from *Dichomitus squalens* in *Phanerochaete chrysosporium*. *Arch Biochem Biophys* 385(2):348–356, 2001.

138. ID Reid, MG Paice. Effects of manganese peroxidase on residual lignin of softwood kraft pulp. *Appl Environ Microbiol* 64:2273–2274, 1998.

139. MG Paice, R Bourbonnais, ID Reid. Sequential bleaching of kraft pulp with oxidative enzymes and alkaline hydrogen peroxide. *Tappi J* 78(9):161–170, 1995.

140. S. Hosoya, J. Nakano, Reactions of alpha-carbonyl group in lignin during alkaline hydrogen peroxide bleaching. *Mokuzai Gakkaishi*, 26 (2): 97–101, 1980.

141. HB Dunford. One-electron oxidations by peroxidases. *Xenobiotica* 25(7):725–733, 1995.

142. H Wariishi, L Akileswaran, MH Gold. Manganese peroxidase from the basidiomycete *Phanerochaete chrysosoporium*: Spectral characterization of the oxidized states and the catalytic cycle. *Biochemistry* 27:5365–5370, 1988.

143. FS Archibald, I Fridovich. The scavenging of superoxide radical by manganous complexes: *In vitro*. *Arch Biochem Biophys* 214:452–463, 1982.

144. JR Anderson, B Amini. Hydrogen peroxide bleaching. In: CW Dence, DW Reeve, eds. *Pulp Bleaching: Principles and Practice*. Atlanta: Tappi Press, 1996, pp. 347–362.

145. MG Paice, FS Archibald, R Bourbonnais, ID Reid, S Renaud. Manganese peroxidase catalyzed bleaching of kraft pulps. Proceedings of TAPPI Biological Sciences Symposium, San Francisco, 1997, pp. 343–345.

146. S Camarero, S Sarkar, FJ Ruiz-Duenas, MJ Martínez, AT Martínez. Description of a versatile peroxidase involved in the natural degradation of lignin that has both manganese peroxidase and lignin peroxidase substrate interaction sites. *J Biol Chem* 274(15):10324–10330, 1999.

147. T Mester, JA Field. Characterization of a novel manganese peroxidase-lignin peroxidase hybrid isozyme produced by *Bjerkandera* species strain BOS55 in the absence of manganese. *J Biol Chem* 273(25):15412–15417, 1998.

148. PJ Collins, MM O'Brien, ADW Dobson. Cloning and characterization of a cDNA encoding a novel extracellular peroxidase from *Trametes versicolor*. *Appl Environ Microbiol* 65(3):1343–1347, 1999.

149. A Heinfling, MJ Martinez, AT Martinez, M Bergbauer, U Szewzyk. Purification and characterization of peroxidases from the dye-decolorizing fungus *Bjerkandera adusta*. *FEMS Microbiol Lett* 165(1):43–50, 1998.

150. B Reinhammar. Laccase. In: R Lontie, ed. *Copper Proteins and Copper Enzymes*. Boca Raton, FL: CRC Press, 1984, pp. 1–35.

151. L Gianfreda, F Xu, J-M Bollag. Laccases: A useful group of oxidoreductive enzymes. *Bioremediation Journal* 3(1):1–26, 1999.

152. C Eggert, U Temp, K-EL Eriksson. Laccase-producing white rot fungus lacking lignin peroxidase and manganese peroxidase: Role of laccase in lignin biodegradation In: TW Jefferies, L Viikari. Eds. *Enzymes for Pulp and Paper Processing*. ACS Symposium Series, 1996, pp. 130–150.

153. CF Thurston. The structure and function of fungal laccases. *Microbiology* 140:19–26, 1994.

154. T Ishihara. The role of laccase in lignin biodegradation. In: HT Kirk T.K., Chang H.M., eds. *Lignin Biodegradation: Microbiology, Chemistry and Potential Applications*. Boca Raton, FL: CRC Press, 1980, pp. 17–31.

155. S Kawai, T Umezawa, T Higuchi. Degradation mechanisms of phenolic beta-1 lignin substructure model compounds by laccase of *Coriolus versicolor*. *Arch. Biochem. Biophys.* 262(1):99–110, 1988.

156. T Higuchi. Mechanisms of lignin degradation by lignin peroxidase and laccase of white-rot fungi. In: NG Lewis, Paice, M.G., eds. *Plant Cell Wall Polymers: Biogenesis and Biodegradation*. ACS Symposium Series, 1989, pp. 482–502.

157. MG Paice, FS Archibald, R Bourbonnais, L Jurasek, ID Reid, T Charles, T Dumonceaux. Enzymology of kraft pulp bleaching by *Trametes versicolor*. In: TW Jeffries, L Viikari, eds. *Enzymes for Pulp and Paper Processing*. Washington, DC: American Chemical Society, 1996, pp. 151–164.

158. R Bourbonnais, MG Paice. Oxidation of non-phenolic substrates: An expanded role for laccase in lignin biodegradation. *FEBS Letters* 267:99–102, 1990.

159. S Kawai, T Umezawa, T Higuchi. Oxidation of methoxylated benzyl alcohols by laccase of *Coriolus versicolor* in the presence of syringaldehyde. *Wood Res.* 76:10–16, 1989.

160. A Muheim, A Fiechter, PJ Harvey, HE Schoemaker. On the mechanism of oxidation of non-phenolic lignin model compounds by the laccase-abts couple. *Holzforschung* 46:121–126, 1992.

161. R Bourbonnais, MG Paice. Demethylation and delignification of kraft pulp by *Trametes versicolor* laccase in the presence of 2,2'-azinobis-(3-ethylbenzthiazoline-6-sulpho-nate). *Appl. Microbiol. Biot.* 36:823–827, 1992.

162. R Bourbonnais, MG Paice, ID Reid. Lignin oxidation and pulp delignification by lac-case of *Trametes versicolor* in the presence of ABTS. In: M Kuwahara, M Shimada, eds. *Biotechnology in Pulp and Paper Industry*. Tokyo: UNI Publishers, 1992, pp. 181–186.

163. HP Call. Process for modifying, breaking down or bleaching lignin, materials contain-ing lignin or like substances. PCT World Patent Application (WO 94/29510) 1994.

164. R Bourbonnais, MG Paice. Enzymatic delignification of kraft pulp using laccase and a mediator. *Tappi J* 79(6):199–204, 1996.

165. HP Call. Further improvements of the Laccase-Mediator-System (LMS) for enzymatic delignification (bleaching) and results from large scale trials. Proceedings of Int Non-Chlorine Bleach Conf, Amelia Island, Florida, 1995.

166. K Hunt, CL Lee, R Bourbonnais, MG Paice. Pulp bleaching with dimethyldioxirane and lignin-oxidizing enzymes. *JPPS* 24(2):55–59, 1998.

167. MG Paice, R Bourbonnais, ID Reid. Bleaching kraft pulps with oxidative enzymes and alkaline hydrogen peroxide. *Tappi J* 78(9):161–169, 1995.

168. K Poppius-Levlin, Wang, W., Ranua, M. TCF bleaching of laccase/mediator-treated kraft pulps. Proceedings of Int pulp Bleaching conf, Helsinki, Finland, 1998, pp. 77–85.

169. FS Chakar, AJ Ragauskas. The effects of oxidative alkaline extraction stages after laccase(HBT) and laccase(NHAA) treatments: An NMR study of residual lignins. *J. Wood Chem. Tech.* 20(2):169–184, 2000.

170. K Poppius-Levlin, W Wang, M Ranua, ML Niku-Paavola, L Viikari. Biobleaching of chemical pulps by laccase/mediator systems. Proceedings of TAPPI Biological Sciences Symposium, San Francisco, 1997, pp. 329–333.

171. K Poppius-Levlin, M Wang, T Tamminen, B Hortling, L Viikari, ML Niku-Paavola. Effects of laccase/HBT treatment on pulp and lignin structures. *J. Pulp Pap. Sci.* 25(3):90–94, 1999.

172. JE Sealey, Runge, T.M., Ragauskas, A.J. Biobleaching of kraft pulps with laccase and hydroxybenzotriazole. Proceedings of Tappi Biol Sci Symp, San Fransisco, CA, 1997, pp. 339–342.

173. J Sealey, AJ Ragauskas. Investigation of laccase/N-hydroxybenzotriazole delignification of kraft pulp. *J. Wood Chem. Technol.* 18(4):403–416, 1998.

174. J Sealey, AJ Ragauskas. Residual lignin studies of laccase-delignified kraft pulps. *Enzyme Microb. Technol.* 23(7–8):422–426, 1998.

175. FS Chakar, AJ Ragauskas. Biobleaching of high lignin content kraft pulps via laccase-mediator systems. Proceedings of Tappi Pulping Conference, Atlanta, 1998, pp. 109–118.

176. M Amann. The Lignozym® process: Coming closer to the mill. Proceedings of 9th International Symposium on Wood and Pulping Chemistry, Montreal, 1997, pp. F4-1–F4-5.

177. R Bourbonnais, MG Paice, B Freiermuth, E Bodie, S Borneman. Reactivities of various mediators and laccases with kraft pulp and lignin model compounds. *Appl. Environ. Microb.* 63(12):4627–4632, 1997.

178. R Bourbonnais, MG Paice, D Leech, B Freiermuth. Reactivity and mechanism of laccase mediators for pulp delignification. Proceedings of TAPPI Biological Sciences Symposium, San Francisco, 1997, pp. 335–338.

179. M Balakshin, CL Chen, JS Gratzl, AG Kirkman, H Jakob. Biobleaching of pulp with dioxygen in the laccase-mediator system. Part 1. Kinetics of delignification. *Holzforschung* 54:390–396, 2000.

180. C Johannes, A Majcherczyk, A Huttermann. Degradation of anthracene by laccase of *Trametes versicolor* in the presence of different mediator compounds. *Appl. Microbiology and Biotechnology* 46(3):313–317, 1996.

181. C Johannes, A Majcherczyk, A Huttermann. Oxidation of acenaphthene and acenaphthylene by laccase of *Trametes versicolor* in a laccase-mediator system. *J. Biotech.* 61(2):151–156, 1998.

182. PJ Collins, MJJ Kotterman, JA Field, ADW Dobson. Oxidation of anthracene and benzo[*a*]pyrene by laccases from *Trametes versicolor*. *Appl. Environ. Microbiol.* 62(12):4563–4567, 1996.

183. S Bohmer, K Messner, E Srebotnik. Oxidation of phenanthrene by a fungal laccase in the presence of 1-hydroxybenzotriazole and unsaturated lipids. *Biochem. Biophys. Res. Commun.* 244(1):233–238, 1998.

184. A Majcherczyk, C Johannes. Radical mediated indirect oxidation of a PEG-coupled polycyclic aromatic hydrocarbon (PAH) model compound by fungal laccase. *Biochim. Biophys. Acta-General Subjects* 1474(2):157–162, 2000.

185. A Potthast, T Rosenau, C-L Chen, JS Gratzl. Selective enzymatic oxidation of aromatic methyl groups to aldehydes. *J Org Chem* 60:4320–4321, 1995.

186. A Potthast, T Rosenau, CL Chen, JS Gratzl. A novel method for the conversion of benzyl alcohols to benzaldehydes by laccase-catalyzed oxidation. *J. Mol. Catal. A* 108:5–9, 1996.

187. A Potthast, T Rosenau, K Fischer. Oxidation of benzyl alcohols by the laccase-mediator system (LMS) - a comprehensive kinetic description. *Holzforschung* 55(1):47–56, 2001.

188. A Majcherczyk, C Johannes, A Hüttermann. Oxidation of aromatic alcohols by laccase from *Trametes versicolor* mediated by the 2,2′ -azino-bis-(3-thylbenzothiazoline-6-sulphonic acid) cation radical and dication. *Applied Microbiology and Biotechnology*, 51(2), 267–276, 1999.

189. M Fabbrini, C Galli, P Gentili, D Macchitella. An oxidation of alcohols by oxygen with the enzyme laccase and mediation by TEMPO. *Tetrahedron Lett.* 42(43):7551–7553, 2001.

190. P Schneider, AH Pedersen. Enhancement of laccase reactions. PCT World Patent Application:WO 95/01426, 1995.

191. P Reyes, MA Pickard, R Vazquez-Duhalt. Hydroxybenzotriazole increases the range of textile dyes decolorized by immobilized laccase. *Biotechnol. Lett.* 21(10):875–880, 1999.

192. Y Wong, J Yu. Laccase-catalyzed decolorization of synthetic dyes. *Water Res.* 33(16):3512–3520, 1999.

193. C Eggert, U Temp, JFD Dean, K-EL Eriksson. A fungal metabolite mediates degradation of non-phenolic lignin structures and synthetic lignin by laccase. *FEBS Lett.* 391(1–2):144–148, 1996.

194. K Li, PS Horanyi, R Collins, RS Phillips, K-EL Eriksson. Investigation of the role of 3-hydroxyanthranilic acid in the degradation of lignin by white-rot fungus *Pycnoporus cinnabarinus*. *Enzyme Microb. Technol.* 28(4–5):301–307, 2001.

195. H Xu, YZ Lai, D Slomczynski, JP Nakas, SW Tanenbaum. Mediator-assisted selective oxidation of lignin model compounds by laccase from *Botrytis cinerea*. *Biotechnol. Lett.* 19(10):957–960, 1997.

196. E Srebotnik, KA Jensen, KE Hammel. Cleavage of nonphenolic lignin structures by laccase in the presence of 1-hydroxybenzotriazole. Proceedings of 7th Int Conf Biotechnol Pulp Paper Ind, Vancouver, 1998, pp. B195–B197.

197. S Kawai, M Asukai, N Ohya, K Okita, T Ito, H Ohashi. Degradation of a non-phenolic beta-O-4 substructure and of polymeric lignin model compounds by laccase of *Coriolus versicolor* in the presence of 1-hydroxybenzotriazole. *FEMS Microbiol. Lett.* 170(1):51–57, 1999.

198. S Kawai, M Nakagawa, H Ohashi. Aromatic ring cleavage of a non-phenolic β-O-4 lignin model dimer by laccase of *Trametes versicolor* in the presence of 1-hydroxybenzotriazole. *FEBS Lett.* 446(2–3):355–358, 1999.

199. E Srebotnik, KE Hammel. Degradation of nonphenolic lignin by the laccase/1- hydroxybenzotriazole system. *J. Biotechnol.* 81(2–3):179–188, 2000.

200. MY Balakshin, CL Chen, JS Gratzl, AG Kirkman, H Jakob. Kinetic studies on oxidation of veratryl alcohol by laccase-mediator system. Part 1. Effects of mediator concentration. *Holzforschung* 54(2):165–170, 2000.

201. A Potthast, T Rosenau, H Koch, K Fischer. The reaction of phenolic model compounds in the laccase-mediator system (LMS): Investigations by matrix assisted laser desorption ionization time-of-flight mass spectrometry (MALDI-TOF-MS). *Holzforschung* 53(2):175–180, 1999.

202. R Bourbonnais, MG Paice, ID Reid, P Lanthier, M Yaguchi. Lignin oxidation by laccase isozymes from *Trametes versicolor* and role of the mediator 2,2'-azinobis(3-ethylbenzthiazoline-5-sulfonate) in kraft lignin depolymerization. *Appl. Environ. Microbiol.* 61:1876–1880, 1995.

203. C Crestini. On the role of 1-hydroxybenzotriazole as mediator in laccase oxidation of residual kraft lignin. In: DS Argyropoulos, ed. *Oxidative Delignification Chemistry.* ACS Symposium Series, 785, 2000, pp. 373–390.

204. C Crestini, DS Argyropoulos. The early oxidative biodegradation steps of residual kraft lignin models with laccase. *Bioorg. Med. Chem.* 6(11):2161–2169, 1998.
205. HG Aurich, W Weiss. Nachweis und reaktionen des benzotriazolyl-1-oxid-radikals. *Chem Ber* 106:2408–2414, 1973.
206. S Hünig, H Balli, H Conrad, A Schott. Polarographie von 2,2'- azinen aromatischer heterocyclen. *Justus Liebigs Ann Chem* 676:52–65, 1964.
207. R Bourbonnais, D Leech, MG Paice. Electrochemical analysis of the interactions of laccase mediators with lignin model compounds. *Biochim. Biophys. Acta (BBA): General Subjects* 1379(3):381–390, 1998.
208. A Potthast, H Koch, K Fischer. The laccase-mediator-system: Reaction with model compounds and pulp. Proceedings of 9th International Symposium on Wood and Pulping Chemistry, Montreal, 1997, pp. F2-1–4.
209. K Kristopaitis, J Kulys, A Palaima. Fungal peroxidase- and laccase-catalyzed oxidation of 1-hydroxybenzotriazole. *Biologija* 4:33–38, 1996.
210. E E Matsumura, E Yamamoto, A Numata, T Kawano, T Shin, S Murao. Structures of the laccase-catalyzed oxidation products of hydroxybenzoic acids in the presence of ABTS. *Agric Biol Chem* 50:1355–1357, 1986.
211. MG Paice, R Bourbonnais, ID Reid, FS Archibald. Kraft pulp bleaching by redox enzymes. Proceedings of 9th International Symposium on Wood and Pulping Chemistry, Montreal, QC, 1997, pp. PL1.1–4.
212. R Zing, J Fletcher, B Freiermuth, K Huber. Method for enhancing the activity of an enzyme, compounds showing enzyme activity enhancement, and detergents containing such compounds. PCT WO 97/06244 1997.
213. J Freudenreich, M Amann, E Fritz-Langhals, J Stohrer. Understanding the Lignozym^R-process. Proceedings of Int Pulp Bleaching Conf, Helsinki, 1998, pp. 71–76.
214. KC Li, F Xu, KEL Eriksson. Comparison of fungal laccases and redox mediators in oxidation of a nonphenolic lignin model compound. *Appl. Environ. Microbiol.* 65(6):2654–2660, 1999.
215. HP Call. A multicomponent system for changing, degrading or bleaching lignin, lignin-containing materials or similar substances and processes for its use. PCT WO 97/36039 1997.
216. HP Call. Multicomponent system for changing, reducing or bleaching lignin, lignin-containing materials or similar substances as well as processes for its application. PCT WO 97/36041 1997.
217. Xu, JJ Kulys, K Duke, K Li, K Krikstopaitis, H-JW Deussen, E Abbate, V Galinyte, P Schneider. Redox chemistry in laccase-catalyzed oxidation of N-hydroxy compounds. *Appl. Environ. Microbiol.* 66(5):2052–2056, 2000.
218. R Bourbonnais, D Rochefort, MG Paice, S Renaud, D Leech. Transition metal complexes: A new class of laccase mediators for pulp bleaching. *Tappi J* 83(10):68, 2000.
219. R Bourbonnais, D Rochefort, MG Paice, S Renaud, D Leech. Development of stable redox complexes to mediate delignification of kraft pulp by laccase. ACS Symposium Series 785: 391–399, 2000.
220. C Padtberg, H-C Kim, M Mickel, S Bartling, N Hampp. Electrochemical delignification of softwood pulp with violuric acid. *Tappi J* 84(4):68, 2001.
221. D Rochefort, R Bourbonnais, D Leech, S Renaud, MG Paice, M.G. Electrochemical oxidation of transition metal-based mediators for pulp delignification. *J Electrochem Soc* 149(1):D15–D20, 2002.
222. MY Balakshin, DV Evtuguin, C Pascoal Neto, A Cavaco-Paulo. Polyoxometalates as mediators in the laccase catalyzed delignification. *J. Mole. Catal. B: Enzymatic* 16(3–4):131–140, 2001.

223. F Xu. Oxidation of phenols, anilines, and benzenethiols by fungal laccases: Correlation between activity and redox potentials as well as halide inhibition. *Biochemistry* 35:7608–7615, 1996.

224. F Xu, W Shin, SH Brown, JA Wahleithner, UM Sundaram, EI Solomon. A study of a series of recombinant fungal laccases and bilirubin oxidase that exhibit significant differences in redox potential, substrate specificity, and stability. *Biochim. Biophys. Acta* 1292(2):303–311, 1996.

225. F Xu, RM Berka, JA Wahleithner, BA Nelson, JR Shuster, SH Brown, AE Palmer, EI Solomon. Site-directed mutations in fungal laccase: Effect on redox potential, activity and pH profile. *Biochem. J.* 334(Pt. 1):63–70, 1998.

226. F Xu. Effects of redox potential and hydroxide inhibition on the pH activity profile of fungal laccases. *J. Biol. Chem.* 272(2):924–928, 1997.

227. D Martinez, LF Larrondo, N Putnam, MDS Gelpke, K Huang, J Chapman, KG Helfenbein, P Ramaiya, JC Detter, F Larimer, PM Coutinho, B Henrissat, R Berka, D Cullen, D Rakhsar. Genome sequence of the lignocellulose degrading fungus *Phanerochaete chrysoporium* strain RP78. *Nat. Biol.* 22(6):695–700, 2004.

16 The Photochemistry of Lignin

Cyril Heitner

CONTENTS

INTRODUCTION

In contrast to cellulose and hemicellulose, lignin is the only polymeric component of the plant fiber that absorbs both visible and near-UV light. In Chapter 2, the absorption of light in the ultraviolet and visible regions of the spectrum was described as characteristic of di- and tri-substituted benzene conjugated to a small population of aromatic carbonyls, α,β-unsaturated carbonyls, quinones and catechols. The excited electronic states formed after light absorption can generate reactive radical species, which in turn react with oxygen to form chromophores. Alternatively, the excited states can sensitize the conversion of triplet oxygen to the very reactive singlet oxygen. The photooxidation of lignin causes depolymerization through cleavage of interunit bonds and yellowing through the oxidation of the aromatic groups. This chapter describes the results of research elucidating the detailed reaction pathways leading to photodegradation and chromophore formation (yellowing).

PHOTODEGRADATION OF LIGNIN

DEPOLYMERIZATION

Irradiation of unbleached kraft pulps in the presence of oxygen followed by alkaline extraction decreased the lignin content as measured by the kappa number [1–3]. As seen in Figure 16.1, UV irradiation at ambient temperature and pressure accelerated the delignification of unbleached kraft pulp. Up to 85% of the residual lignin was removed. However, delignification was accompanied by cellulose degradation. In contrast, oxygen treatment in the dark did not cause delignification or cellulose degradation. The rate of delignification was increased both at acidic and alkaline pH, indicating both acid and alkaline catalyzed photo-solvolysis. Decreasing the oxygen partial pressure from 1 to 0.25 atmospheres had very little effect on the rate of delignification. However, total removal of oxygen did slow delignification [2,3]. In contrast, irradiation of *Eucalyptis grandis* acetic acid organosolv pulp in the presence of oxygen under alkaline conditions did not lead to any significant increase in delignification [4]. The differences in the response to alkaline photooxidative delignification of the kraft and organosolv pulp may be due to the differences in the nature of the residual lignin in the pulp. Lignin in acetic acid organosolv pulp may not have the photoactive groups that lead to photo-oxidative degradation. However, lignin in pulps produced by the alkaline kraft process do undergo facile photo-oxidative degradation. Also, the amount of residual lignin in the kraft pulp at a kappa number in the range of 30 to 50 was generally lower than that of the organosolv pulp at 65 [2,4].

Irradiation of milled wood lignin with UV light in the presence of oxygen causes degradation of the macromolecular structure as indicated by the decrease in

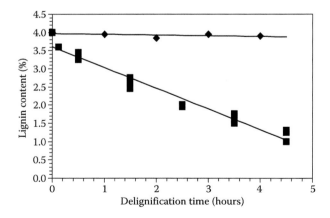

FIGURE 16.1 Irradiation of unbleached kraft pulp in the presence of oxygen decreased lignin content by up to 85%. (Adapted from Marcoccia, B., Goring, D.A.I., and Reeve, D.W., *J Pulp Paper Sci*, 17, J34–J38, 1991. With permission.)

molecular weight [5–9] shown in Figure 16.2. Irradiation of MWL in the absence of oxygen did not decrease the molecular weight [6,8]. Based on evidence from the size exclusion chromatography, shown in Figure 16.3, irradiation under nitrogen atmosphere causes the condensation of MWL to higher molecular weights [6]. These results were confirmed by the higher rate of decrease in benzylic hydroxyl content of

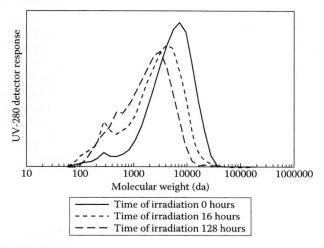

FIGURE 16.2 Irradiation of MWL in the presence of oxygen decreased the molecular weight. (Redrawn from Destiné, J.-N., Wang, J., Heitner, C., and St. John Manley, R., *J Pulp Paper Sci*, 22, J24–J30, 1996. With permission.)

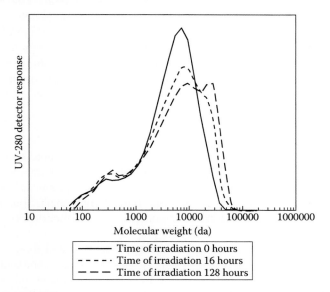

FIGURE 16.3 Irradiation of MWL with UV light in the absence of oxygen (nitrogen atmosphere) increased the molecular weight. (Redrawn from Destiné, J.-N., Wang, J., Heitner, C., and St. John Manley, R., *J Pulp Paper Sci*, 22, J24–J30, 1996. With permission.)

lignin during irradiation in oxygen compared to irradiation in nitrogen. In addition, there was a higher rate of formation of condensed phenolic groups during the irradiation under nitrogen compared to that of oxygen [10].

The addition of a known singlet oxygen quencher, 1-diazabicyclo[2.2.2]octane (DABCO), to MWL during UV irradiation in the presence of oxygen decreases the molecular weight to about 72% of it's original value. Without DABCO, photo-oxidative degradation decreased the molecular to 28% of the original value [11]. Also, peroxyl radicals produced by the thermal decomposition of 2,2′-azobis(amidinopropane)-dihydrochloride (ABAP) in oxygen degrade MWL in a similar fashion as did UV irradiation in the presence of oxygen [9].

Since DABCO is stable to oxidation and does not absorb near UV light ($\lambda > 300$ nm), it is widely used to test for the participation of singlet oxygen ($^1O_2$) in the oxidation of organic compounds [12]. The inhibition of photo-oxidative degradation of MWL by DABCO indicates a major role of singlet oxygen. However, the effect of peroxyl radicals on decreasing molecular weight indicates that oxygen-based free radicals also play a role in photo-oxidative degradation.

LIGNIN DEGRADATION

Cleavage of the β–O–4 Aryl Ether Bond

The most common linkage between lignin monomers (C9 units) is the β–O–4 aryl ether bond. The reactivity of this bond is influenced by the functionality of the α carbon atom, which is most frequently a hydroxyl group. In addition, there is a small but significant amount of phenacyl aryl ether groups (carbonyl at the α carbon), especially in lignin from alkaline peroxide bleached mechanical pulps. Thioacidolysis of lignin from bleached thermomechanical pulp before and after 24 hours UV irradiation showed a decrease in β–O–4 interunit bonds from about 1000 to 150 μmoles/g [13]. This result suggests that photo-oxidative degradation of lignin proceeds in part through cleavage of β–O–4 aryl ether interunit bonds, which is consistent with the observed decrease in molecular weight [8,9]. The magnitude of the decrease in β–O–4 aryl ether interunit bonds indicates that phenacyl aryl ether and guaiacyl glycerol β–O–4 aryl ether groups are both degraded.

Phenacyl Aryl Ether

Gierer and Lin [14] demonstrated the first example of the photoreactivity of the phenacyl aryl ether group. Irradiation of α-guaiacoxyacetoveratrone, **1** shown in Scheme 16.1, with near-UV light ($400 > \lambda > 300$ nm) produced the hydrogen abstraction products derived from the initial radical components: guaiacol, **2** and acetoveratrone, **3**. However, in addition, products of radical dimerization **4**, ring coupling **5** and **6**, oxidation **7** and oxidative degradation **8** have been isolated as shown in Scheme 16.1 [15–21]. The two ring coupling products, **5** and **6** are produced by radical recombination of the phenacyl radical with resonance stabilized mesomers of the guaiacoxy radical, as seen in Scheme 16.2. The formation of radicals from the photodegradation of phenacyl aryl ethers is indicated by the five line electron spin resonance (ESR) signal assigned to the phenacyl radical superimposed on the singlet attributed to the phenoxy radical [22].

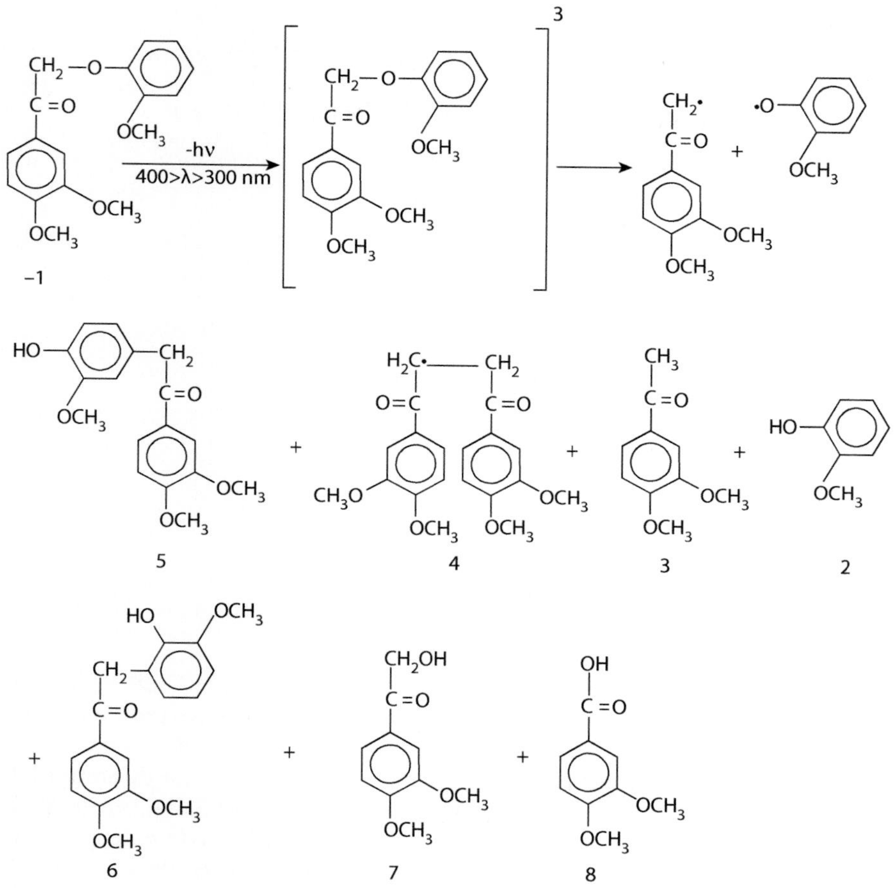

SCHEME 16.1 Photochemistry of α-guaiacoxyacetoveratrone. (Adapted from Scaiano, J.C. and Netto-Ferreira, J.C., *J. Photochem.*, 32, 253–259, 1986; Vanucci, C., Fornier de Violet, P., Bouas-Laurent, H. and Castellan A. *J Photochem Photobiol A: Chemistry*, 4, 1251–265, 1988; Netto-Ferreira, J.C. and Scaiano, J.C., *Tetrahedran Letters*, 30, 443–446, 1980; Castellan, A., Colombo, N., Cucuphat, C., Fornier, de Violet, P., *Holzforschung*, 43, 179–185, 1989; Castellan, A., Colombo N., Fornier, de Violet, P., Nourmamode A. and Bouas-Laurent H., Proceedings International Symposium on Wood and Pulping Chemistry, Raleigh, 421–430, 1989; Netto-Ferreira J.C., Avellar I.G.J. and Scaiano J.C., *J Org Chem*, 55, 89–92, 1990; Fukagawa N. and Ishizu A., *J Wood Chem Technol*, 11, 263–289, 1991.)

It was initially assumed that the degradation of the phenacyl aryl ether bond occurred through the n-π* triplet excited state [14–18]. However, quenching experiments and time-resolved chemically induced dynamic nuclear polarization (CIDNP) spectra have indicated that there is significant photodegradation through the singlet excited state [23–27].

The multiplicities of the excited states leading to the photo-degradation of various phenacyl arylethers were determined using the application of Kapstein's rule to the

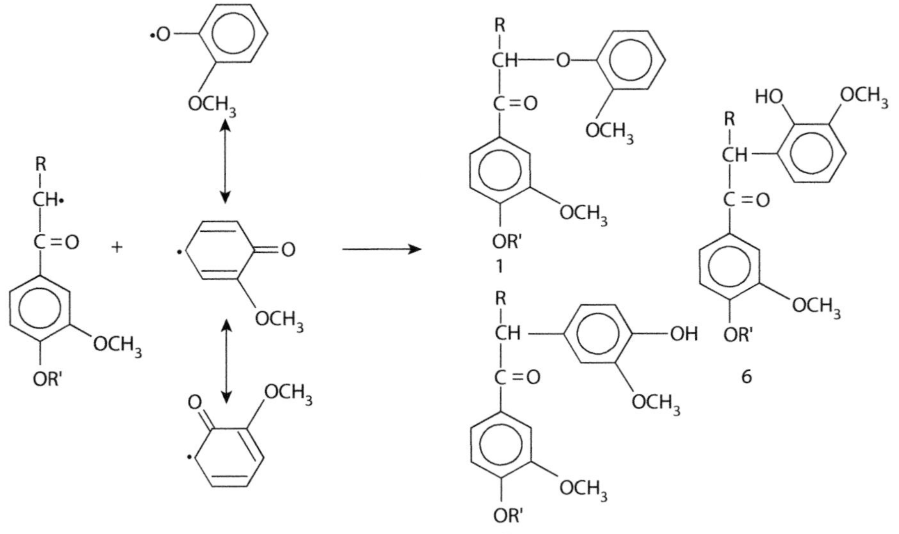

SCHEME 16.2   Recombination of phenacyl radical with resonance stabilizede mesomers of the guaiacoxy radical. (Redrawn from Fukagawa. N. and Ishizu, A., *J Wood Chem Technol*, 11, 263–289, 1991.)

polarization directions of transient and stable products observed by CIDNP ^{1}H-NMR spectra [26,27]. After the Irradiation of the lignin model, α-guaiacoxyacetoveratrone (**1**, α-GAV), both the dark and CIDNP steady state spectra indicated the formation of degradation products such as guaiacol, **2**, acetoveratrone, **3**, and 1,4-diveratryl-1,4-butadione, **4**, α–(3-methoxy-4-hydroxy-phenyl)-acetoveratrone, **5**, α–(3-methoxy-2-hydroxyphenyl)acetoveratrone, **6** (Scheme 16.3). In the CDNIP steady-state spectrum, reactive intermediates such as the enol of acetoveratrone (**3a**) and the cyclohexadione precursors (**5a** and **6a**) to α–(3-methoxy-4-hydroxyphenyl)acetoveratrone (**5**) and α–(3-methoxy-2-hydroxy-phenyl)acetoveratrone (**6**) were observed [26,27]. The sign of the polarization of the reactive intermediates indicates that the cleavage of the β–O–4 aryl ether bond proceeds in part through the singlet excited state.

When α-GAV was irradiated in 40% water in dioxane or acetonitrile, the triplet lifetimes and disappearance quantum yield, ϕ_d increased [24,25]. The addition of 0.1 mol/L of cyclohexadiene to ethanol solutions of α-GAV decreased the disappearance quantum yield from 0.34 to 0.10 [25]; whereas, the addition of cyclohexadiene to α-GAV (**1**) in water-free acetonitrile or dioxane had no effect on ϕ_d (Figure 16.4). However, the addition of water to aprotic solvents has a profound effect on decrease in ϕ_d caused by cyclohexadiene. In acetonitrile/40% water and dioxane/40% water ϕ_d was decreased with the addition of 0.1 mol/L of cyclohexadiene, as seen in Figure 16.5. Since 99% of the triplet excited state is quenched by 0.1 mol/L of cyclohexadiene, a significant amount of α-GAV is degraded via the triplet excited state in protic solvents such as ethanol, or in aprotic solvents containing water. In aprotic solvents such as acetonitrile and dioxane cyclohexadiene had no effect on ϕ_d, indicating exclusive singlet state photochemistry [25].

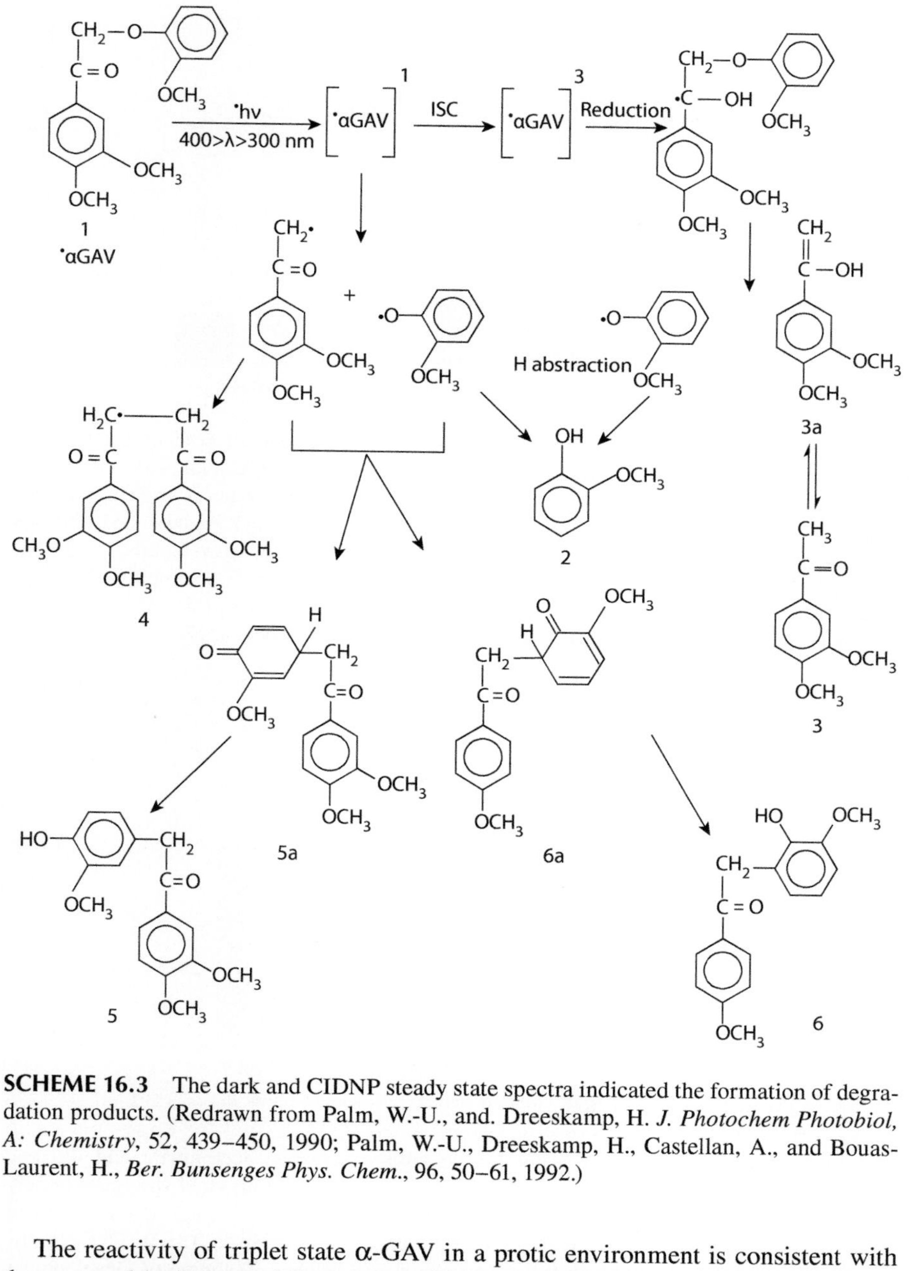

SCHEME 16.3 The dark and CIDNP steady state spectra indicated the formation of degradation products. (Redrawn from Palm, W.-U., and. Dreeskamp, H. *J. Photochem Photobiol, A: Chemistry*, 52, 439–450, 1990; Palm, W.-U., Dreeskamp, H., Castellan, A., and Bouas-Laurent, H., *Ber. Bunsenges Phys. Chem.*, 96, 50–61, 1992.)

The reactivity of triplet state α-GAV in a protic environment is consistent with the proposal by Palm and Dreeskamp [26,27] that triplet photochemistry proceeds through photoreduction to the ketyl radical followed by a facile cleavage of the β–O–4 aryl ether bond (Scheme 16.3).

The decomposition of α-GAV from the triplet excited state was monitored in 1:3 ethanol:benzene by time resolved ESR spectroscopy. A broad emissive resonance

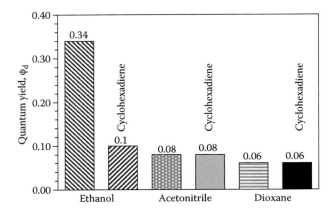

FIGURE 16.4 The addition of triplet quencher cyclohexadiene to ethanol solutions of α-GAV decreased the diappearance quantaum yield by one-third. Cyclohexadiene did not affect the quantum yields of α-GAV in acetonitrile or dioxane. (Redrawn from Schmidt, J.A., Berinstain, A.B., de Rege, F., Heitner, C., and Johnston, L.J. in JF Kennedy, G.O. Phillips and P.A. Williams, Eds. Ligno-Cellulosics, Science, Technology, Development and Use. Chichester: Ellis Horwood, 587, 1992; Schmidt, J.A., Goldszmidt, E. Heitner, C. Scaiano, J.C. Berinstain, A.B., and Johnston, L.J., in C. Heitner and J.C. Scaiano, Eds. Photochemistry of Lignocellulosic Materials, Washington: American Chemical Society, 122, 1993.)

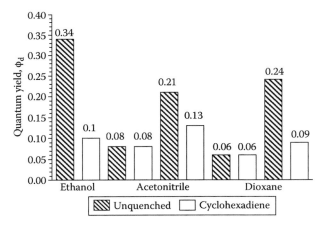

FIGURE 16.5 The addition of water to the aprotic solvents, acetonitrile and dioxane increased the disappearance quantum yield. This increased quantum yield was quenched by cyclohexadiene indicating triplet state production in the presence of water. (Redrawn from Schmidt, J.A., Berinstain, A.B., de Rege, F., Heitner, C., and Johnston, L.J. in JF Kennedy, G.O. Phillips and P.A. Williams, Eds. Ligno-Cellulosics, Science, Technology, Development and Use. Chichester: Ellis Horwood, 587, 1992; Schmidt, J.A., Goldszmidt, E. Heitner, C. Scaiano, J.C. Berinstain, A.B., and Johnston, L.J., in C. Heitner and J.C. Scaiano, Eds. Photochemistry of Lignocellulosic Materials, Washington: American Chemical Society, 122, 1993.)

initially appearing has been attributed to the formation of a ketyl radical of α-GAV [28–30]. Very soon after the detection of the ketyl radical, an emissive resonance attributed to the phenacyl radical was observed. This is in contrast to the results of Palm and Dreeskamp, where it was proposed that an enol was the product of cleavage of the β–O–4 aryl ether bond [27]. In aprotic solvents such as dioxane, benzene and acetonitrile no significant polarized signals are observed, indicating the absence of a triplet excited state. The reactivity of the triplet excited state of α-GAV in protic environments was confirmed by the observation of a polarized emissive resonance when α-GAV was irradiated in these solvents containing ethanol or water [28–30].

The two time-resolved spectroscopic methods CIDNP [26,27] and ESR [28–30] as well as classical triplet quenching experiments all indicate that α-GAV decomposes through both triplet and singlet states in protic solvents and exclusively through the singlet state in aprotic solvents.

β–O–4-arylether Group

Another potential source of ketyl radicals is the abstraction of the hydrogen atom from the benzylic carbon atom of a guaiacylglycerol-β–O–4-arylether group. Almost 40% of the interunit linkages in lignin contain guaiacylglycerol-β–O–4-arylether bonds. Abstraction of the hydrogen atom from the benzylic carbon of the guaiacylglycerol-β-arylether group by triplet excited state molecules or by peroxyl radicals produce ketyl radicals. These radicals are known to undergo cleavage of the β–O–4 aryl ether bond [31–35] to produce an aromatic ketone and a phenol.

The reaction pathways leading to ketyl radicals, and subsequently to aromatic ketones and phenoxyl radicals, were determined in some detail [35]. Abstraction of the hydrogen atom from the benzylic carbon atom of 1-phenyl-2-phenoxyethanol with t-butoxy radicals produced from the decomposition of di-t-butylhyponitrite gave only one product, α-phenoxyacetophenone. None of the expected products of β–O–4-arylether cleavage, acetophenone and phenol, were formed. However, thermal decomposition of di-t-butylperoxide in the presence of 1-phenyl-2-phenoxyethanol produced both acetophenone and phenol, in addition to α-phenoxyacetophenone. These results indicate at least two pathways leading to the β–O–4-arylether cleavage, as shown in Scheme 16.4. The first is reaction of the ketyl radical with oxygen to produce the phenacyl aryl ether, which in turn undergoes photochemical induced cleavage of the aryl ether bond to form the phenoxy radical. This mechanism had been proposed previously by Francis et al. [36]. The second is β–O–4-arylether cleavage of the ketyl radical to produce the phenoxy radical.

The operative mechanism depends on the lifetime of the ketyl radical. Scaiano and coworkers proposed a short living phenoxyketyl free radical with the primary reaction being fragmentation of the β–O–4 arylether bond to enol (tautomerizing to ketone) and the phenoxy radical [34]. The reaction with oxygen would be minor. The latest research on the reactivity of lignin model ketyl radical indicated a long-lived radical quenched by oxygen to produce the corresponding α ketone, which in turn

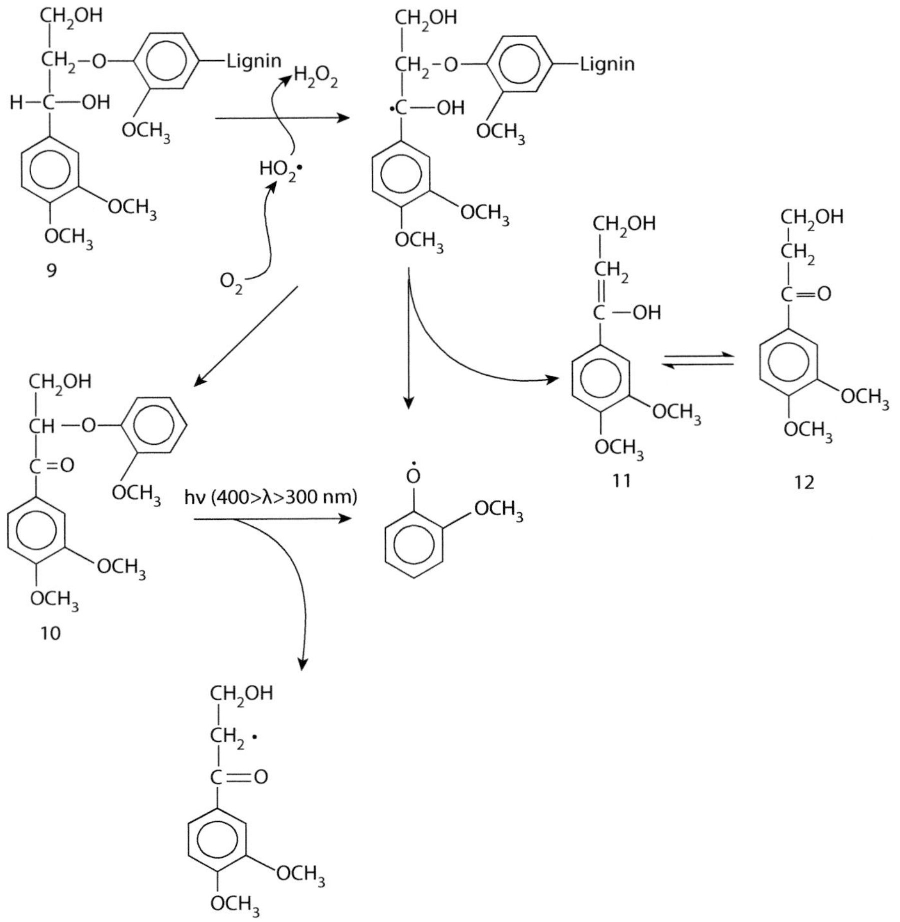

SCHEME 16.4 Reaction pathways leading to β–O–4-arylether cleavage. (Adapted from Scaiano, J.C., Netto-Ferrira, J.C., and Wintgens, V., *J Photochem Photobiol A: Chem*, 59, 265–268, 1991; Huang, Y., Pagé, D., Wayner, D.D.M., and Mulder, P., *Can J Chem*, 73, 2079–2085, 1995; Fabbri, C., Bietti M., and Lanzalunga, O., *J Org Chem*, 70, 2720–2728, 2005.)

absorbs UV light to undergo fragmentation of the β–O–4 aryl ether bond [35,36]. Either of these pathways is consistent with the observation that there was an increase in carbonyl and phenolic hydroxide group after irradiating lignin-containing paper sheets [37].

Coniferyl Alcohol Group

The coniferyl alcohol end group in lignin survives alkaline hydrogen peroxide bleaching [13,38] and is therefore a potential photoactive group in both bleached and unbleached lignin-containing pulps and in wood. The initial research on the photolysis of coniferyl alcohol indicated the formation of the corresponding quinone methide [39–42]. It was thought that quinone methide was produced by UV irradiation of coniferyl alcohol through photoionization and heterolysis of the hydroxyl group to form the phenoxy radical [39,40]. This was reported by Jaeger, Nourmamode, and Castellan who determined the photochemistry of coniferyl alcohol in CH_2Cl_2 or tBuOMe [44]. They found that irradiation for four hours produced a 17 to 18% yield of photoproducts. An analysis of the photoproducts was explained by formation of the phenoxy radical followed by tautomerization to the quionone methide. However, attempts to observe the phenoxy radical by EPR spectroscopy failed [43]. If phenoxy radicals were indeed initially formed, the inability to detect them by EPR may due to their rapid conversion to quinonemethide. In contrast to earlier studies [43], UV irradiation of coniferyl alcohol produced a ligninlike polymer similar to the dehydrogenative polymer, DHP [46].

Since coniferyl alcohol groups are primarily linked to the macromolecule by alkyl-O-4 bonds, the photochemistry of lignin models such as coniferyl alcohol methyl ether is more relevant. In early work on lignin model compounds, Gellerstedt and Pettersson [47] observed that coniferyl alcohol methyl ether, **11** and the analogous structure isoeugonol methyl ether, **12** underwent addition of oxygen to the ethylenic double bond to give the glycol structures **13** and **14** through the addition of $^1O_2$, as seen in Scheme 16.5. Also, $^1O_2$ addition produced the peroxirane intermediate that rearranges with cleavage of the Cα–Cβ bond to produce veratraldehyde, **15** and veratric acid, **16** as well as aldehydes **17** and **18** and acidic fragments **19** and **20**. This was supported by observation made on isolated lignin and lignin in thermomechanical pulp fiber [13]. In addition to the cleavage of the Cα–Cβ bond the alkyl–O–4 coniferyl alcohol moiety also undergoes photooxidation to coniferadehyde [13].

Stilbenes

Although native lignin does not have significant amounts of stilbene groups, the processing of wood fiber into paper causes the formation of stilbenes in lignin-containing papermaking fiber [47,48]. Gellerstedt and Zhang have developed a method of mild acid hydrolysis that releases monomeric and oligermeric end-groups of lignin [38]. Using this method, they demonstrated that the conversion of β-1 structures to stilbene groups occured during wood grinding and refining. However, alkaline hydrogen peroxide bleaching induced the greatest conversion of β-1 structure to stilbenes. Therefore, the stilbene group is a viable leucochronmophore participating-ing in light-induced yellowing of high brightness lignin-containing papers. It was observed that the absorption of light at 350 nm in the acid hydrolysis extracts of spruce wood was attributed to the coniferaldehyde group, whereas the 350 nm light absorption of acid hydrolysis extracts of bleached chemithermomechanical pulp (CTMP) were dominated by the diguaiacyl stilbene group. In solution, the model for the diguaiacyl stilbene groups underwent rapid isomerization to the cis, trans photo stationary state [49]. However, this was not observed when the diguaiacyl stilbene was deposited onto cellulose fibers [50]. In addition, the photochemistry of stilbenes both in solution and deposited onto cellulose is characterized by dimerization to

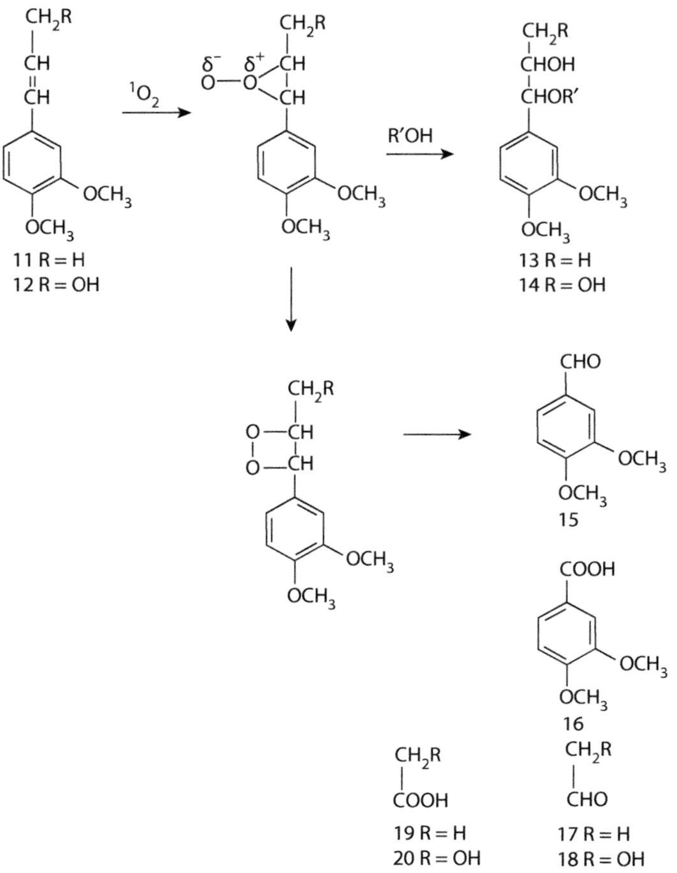

SCHEME 16.5   Photooxidation of coniferyl alcohol. (Adapted from Gellerstedt, G., and Pettersson E.-L., *Acta Chem Scand*, B29, 1005–1010, 1975.)

cyclobutane adducts, and the formation of radical dimer adducts as well as colored quinonoid oxidation products as shown in Scheme 16.6.

Phenyl Coumarone

Phenyl coumarones have been proposed as a product of alkaline hydrogen peroxide bleaching of mechanical and high-yield pulps. Irradiation of this chromophore in solution or on cellulose rapidly produces three colored products [51]. Two are colored dimeric stilbene compounds, one a ketone and the other a conjugated quinone methide, as shown in Scheme 16.7. The third compound isolated was a catechol analog of the starting material, the result of demethylation according to Scheme 16.8. It was proposed that the dimeric products were formed by coupling of the phenoxy radical to produce an unstable cyclobutane dimer, which undergoes rearrangement to the ketone and quinone methide (Scheme 16.7) [52]. Although the dimeric products are produced from starting material adsorbed onto cellulose fibers, one would question

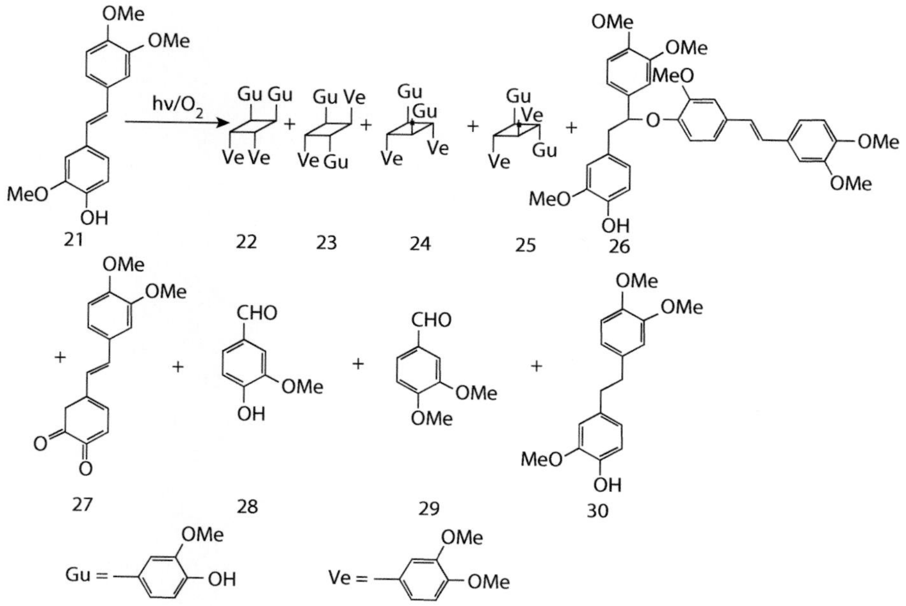

SCHEME 16.6 The photochemistry stilbene phenols in lignin (Redrawn from Ruffin, B., and Castellan, A., *Can J Chem* 78, 73–83, 2000.)

whether the phenyl coumarone moiety was in close enough proximity in the lignin macromolecule to dimerize. The demethylation to catechol is the more likely product in lignin.

Dibenzodioxocin

Brunow et al. have shown that most biphenyl structures in lignin are etherified by the α and β carbons of another phenyl propane unit to give the eight-membered dibenzodioxocin ring, the main branching point in lignins [53,54]. This bonding provides a reactive center for photoinduced radical cleavage.

Nonphenolic dibenzodioxocin, **31** (Scheme 16.9) had an order of magnitude higher disappearance quantum yield than the phenolic analogue, **35** (Scheme 16.10) [56]. This was explained by the efficient cleavage of the α–O–4 in the first step followed by the less efficient β–O–4 bond cleavage. The presence of a phenolic group in the dibenzodioxocin molecule provides a deactivation pathway to quinonemethide (Scheme 16.10) that competes with bond cleavage. This rational is supported by the lower fluorescence quantum yield of the phenolic dibenzodioxocin compared to that of the nonphenolic analogue.

As shown in Schemes 16.9 and 16.10, the products of UV irradiation of the nonphenolic dibenzodioxocin model compound, **31** are different from those of a phenolic dibenzodioxocin model, **35**. The nonphenolic dibenzodioxocin model produced styrene derivative, **32,** which underwent further oxidation to the vanillin, **33** and vanillic acid, **34**. It was proposed that the biphenyl *o*-diketone formed from the biphenyl moiety was converted into uncharacterized colored oligomeric material.

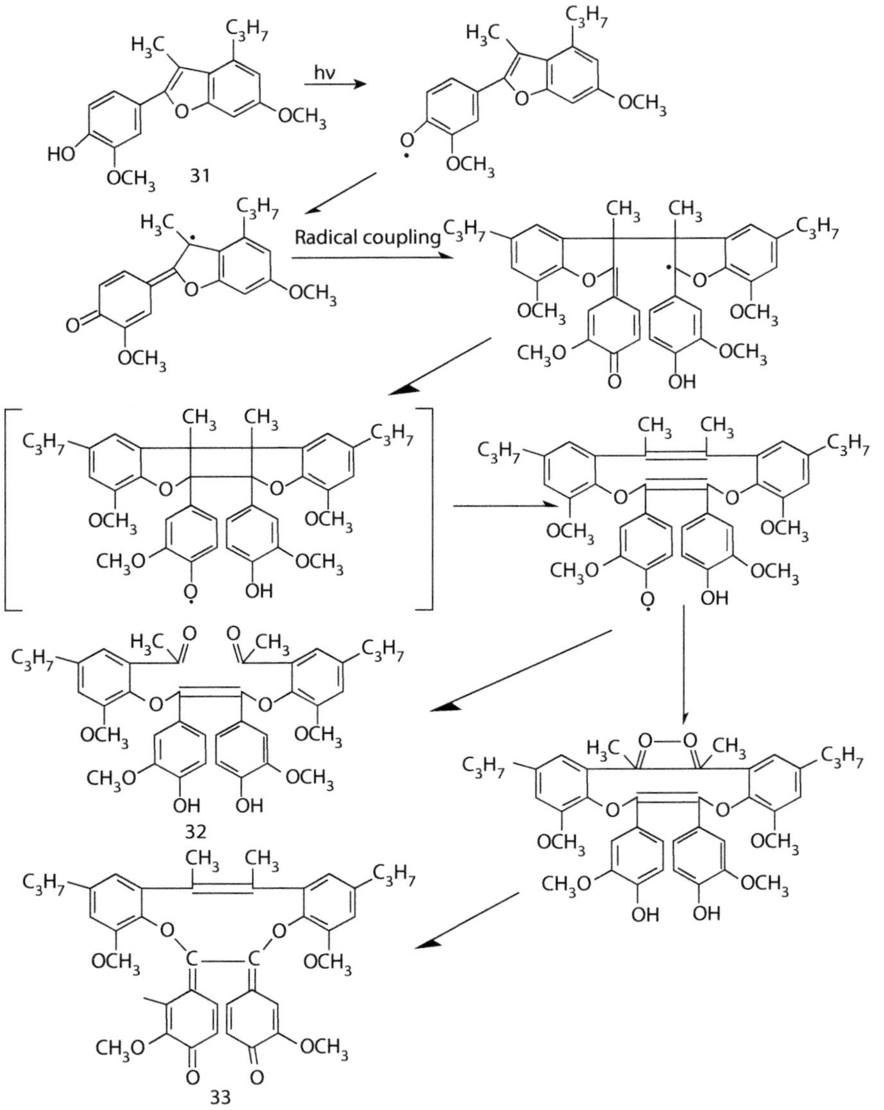

SCHEME 16.7 Dimeric products of the photochemistry of the phenyl coumarone group. (Redrawn from Noutary, C., Fornier de Violet, P., Vercauteren J., and Castellan, A., *Res. Chem. Intermed.*, 21, 223–245, 1995.)

The phenolic dibenzodioxocin model produced an isolatable biphenylbiphenol, **36** as well as compound **37**, the product of the initial α–O–4 cleavage according to Scheme 16.10. The cleavage of the β–O–4 bond is inhibited by the addition of ethanol to the quinonemethide intermediate. The phenolic moiety forms a quinonemethide radical that is thought to form the uncharacterized oligomers.

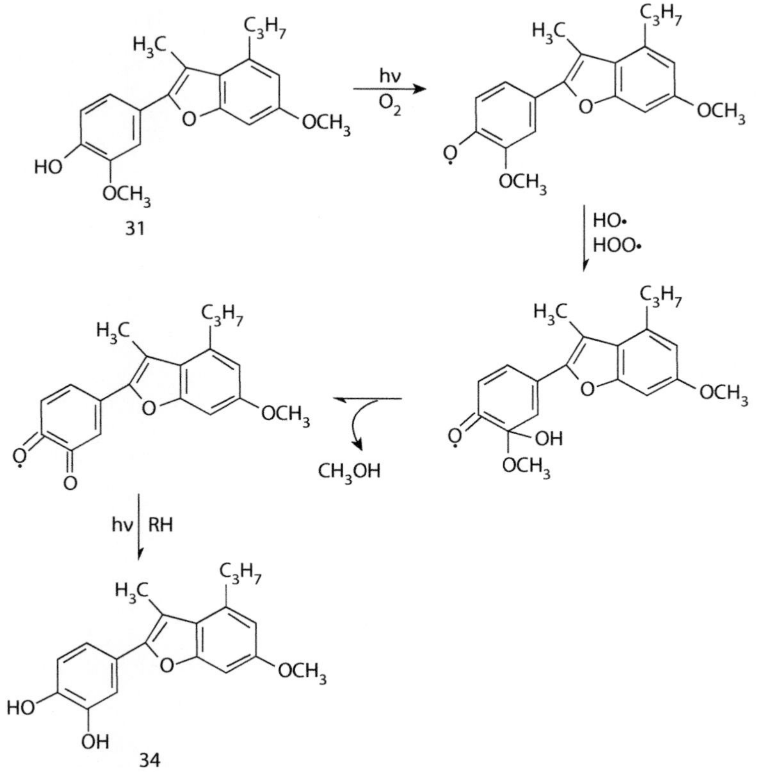

SCHEME 16.8 Light-induced demethylation of phenylcoumarone. (Redrawn from Noutary, C., Fornier de Violet, P., Vercauteren J., and Castellan, A., *Res. Chem. Intermed.*, 21, 247–262, 1995.)

Since dibezodioxocins are considered for the most part nonphenolic, the results of the nonphenolic model would be considered more relevant.

THE ROLE OF SINGLET OXYGEN

DESCRIPTION OF $^1O_2$.

The paramagnetic properties of ground state triplet oxygen are attributed to the two outer electrons with parallel spins. Energy may be transferred from photo-induced excited triplet state molecules to triplet oxygen to produce two high-energy states of singlet oxygen due to the outer electron pair being antiparallel. The lower energy singlet oxygen, the $^1\Delta_g$ state, has a lifetime of one hour in the absence molecular collision and 10^{-6} to 10^{-3} seconds in the liquid state. The higher energy $^1\Sigma_g^-$ state has a much shorter lifetime, seven to 12 seconds in the gas phase and $10^{-11} - 10^{-9}$ seconds in the liquid phase. Therefore, the longer lived $^1\Delta_g$ state oxygen has a long enough lifetime to react with organic and polymeric materials [12].

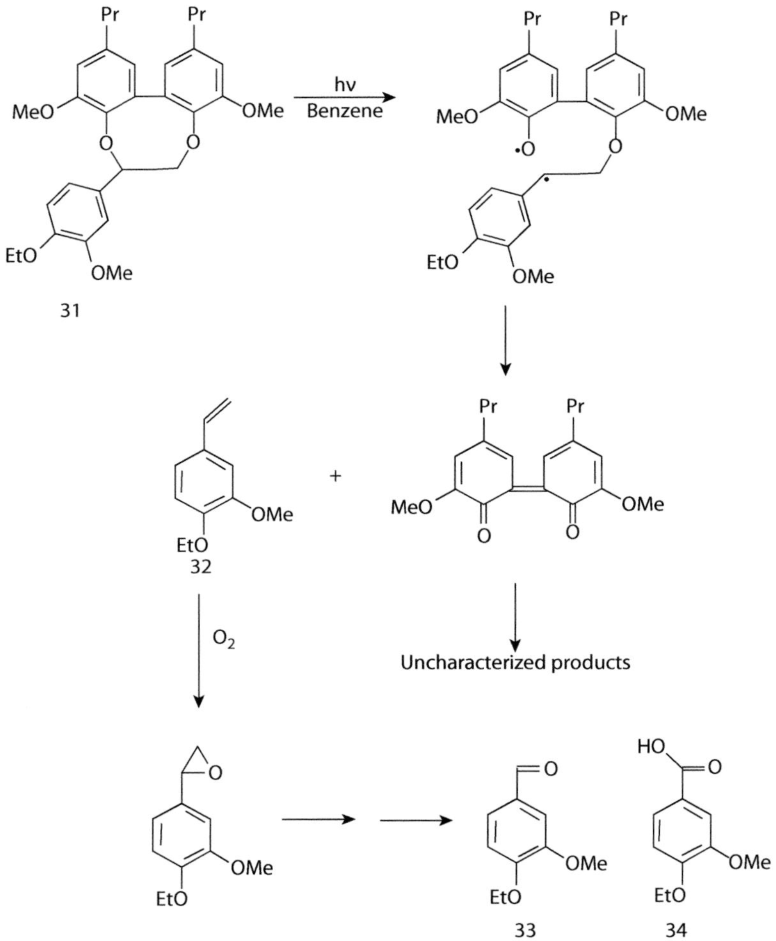

SCHEME 16.9 The photochemistry of nonphenolic dibenzodioxocin. (Redrawn from Gardat, C., Ruggiero, R., Hoareau, W., Nourmamode, A., Grelier S., Siegmund, B., and Castellan, A., *J Photochem Photobiol, A: Chem*, 167, 111–120, 2004.)

REACTIONS AND REACTIVITY

The $^1\Delta_g$ state of oxygen reacts with organic compounds and polymers in five characteristic reaction pathways: hydrogen abstraction and addition (the ene reaction, equation 1), 1,4-cycloaddition (equation 2), 1,2 cyclo-addition (equation 3), oxygenation (equation 4) and hydrogen abstraction from phenolic hydroxyl groups (equation 5) in Figure 16.6 [58].

The lignin macromolecule contains structures that can both sensitize the formation of singlet oxygen [60] and react with singlet oxygen by either the ene, 1,4 or 1,2 cycloaddition or hydrogen abstraction pathways [58].

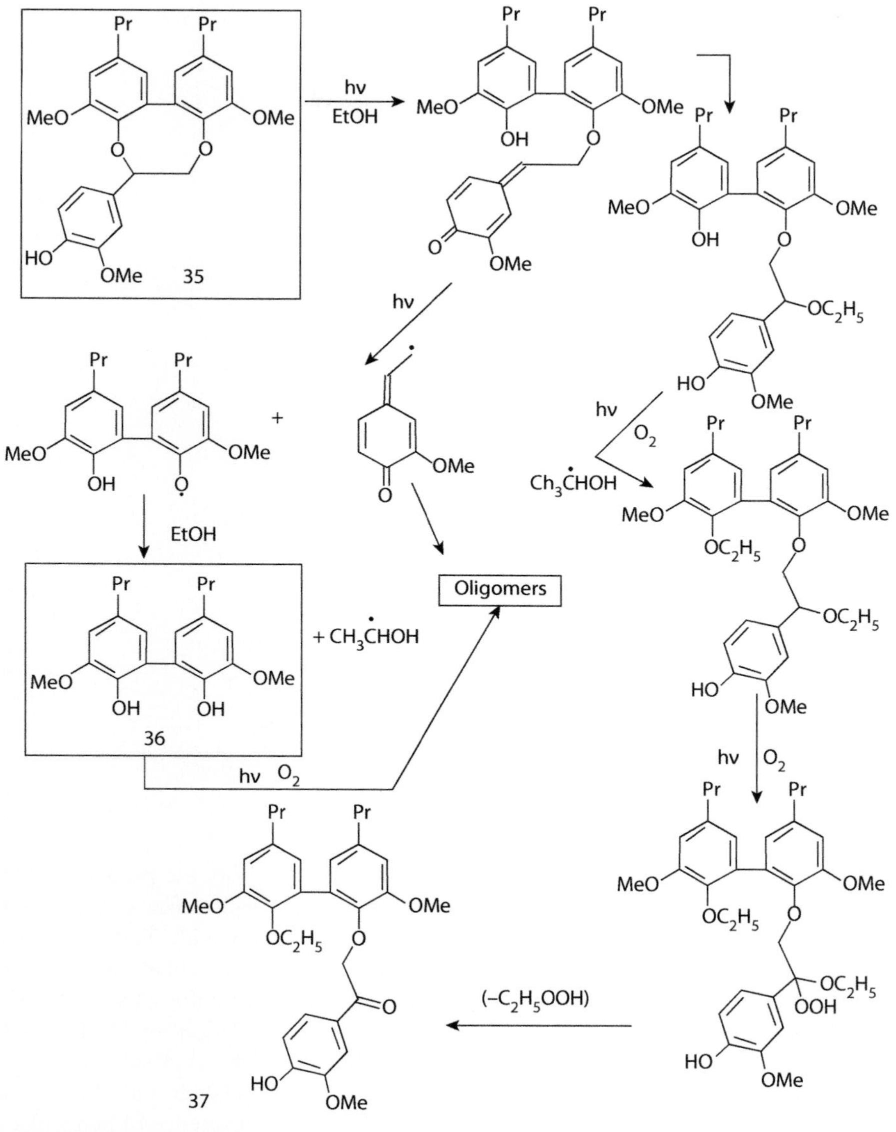

SCHEME 16.10 The photochemistry of phenolic dibenzodioxocin. (Redrawn from Gardat, C., Ruggiero, R.,. Hoareau, W., Damigo, L., Nourmamode, A., Grelier, S., and Castellan, A., *J Photochem Photobiol, A:Chem*, 169, 261–269, 2005.)

Reactions with Lignin Model Compounds

The first example of singlet oxygen reactions with lignin model phenols was reported by Matsuura et al. [59,60]. Singlet oxygen produced by photosensitization oxidizes 4,6-di-*tert*-butyl guaiacol, **38** to the methyl ester of the lactonic acid, **39**, as seen in Scheme 16.11 [59]. These authors proposed that ground state triplet oxygen reacts

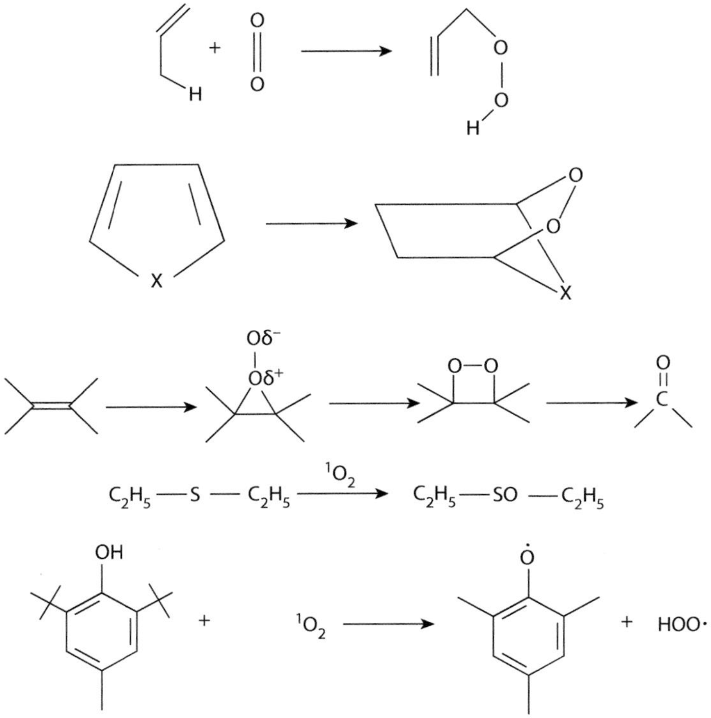

FIGURE 16.6 Equations 1 to 5. Reactions of singlet oxygen with vinylic, dienic and aromatic functional groups.

with the phenoxyl radical. However, it was shown [61] that only the product of the reaction of carbon-based radicals and oxygen, that is, peroxyl radicals, react with phenoxy radicals. Ground state triplet oxygen does not react with phenoxyl radicals. Therefore, Scheme 16.11 includes the reaction of peroxyl radical, the product of hydrogen abstraction by singlet oxygen, in the oxidation of the phenoxyl radical. When there is an alkyl side chain on the phenol, the oxidation by $^1O_2$ does not appear to degrade the aromatic ring, but rather displaces the alkyl side chain, as seen in Schemes 16.12 and 16.13 [62,63]. For example Gellerstedt and Pettersson [63] found that apocynol, **40**, irradiated with UV light in the presence of benzophenone and O_2, produced acetaldehyde and methoxy-p-quinone, **41,** which is unstable and possibly is oxidized to 2-hydroxy-3-methoxy-p-quinone, **42**. Nimz and Turznick obtained similar results with the reaction of singlet oxygen, produced by the reaction of sodium hypochlorite with hydrogen peroxide, with several guaiacyl lignin model compounds shown in Scheme 16.13 [64].

Side chain displacement with the formation of p-quinones and hydroquinones were the primary reactions observed. The product of oxidation by singlet oxygen was a function of the structure of the lignin model compound and the pH, as seen in Scheme 16.13. For example, vanillin, **43** produced methoxy-p-hydroquinone, **50**

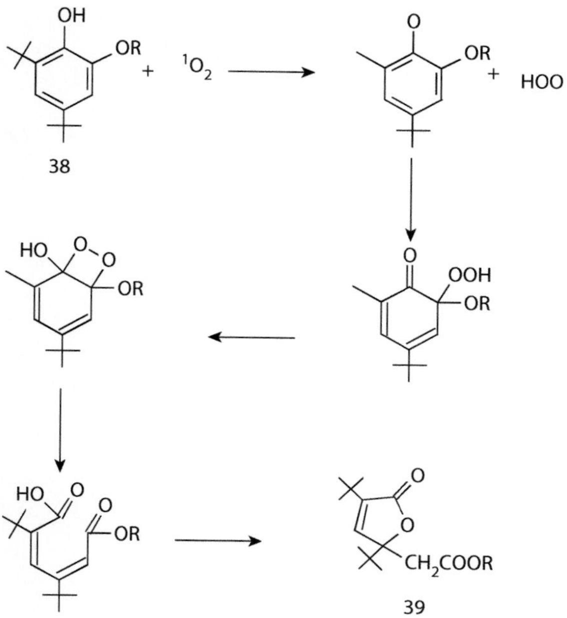

SCHEME 16.11 Singlet oxygen oxidation of aromatic ring. (Adapted from Golnick, K., In: Rånby, B., and Rabek, J.F., Eds. *Singlet Oxygen Reactions with Organic Compounds and Polymers*. New York, John Wiley and Sons, 111–134, 1978.)

at pH 5 and 3-methoxy-5,6-epoxy-*p*-quinone, **51** at pH 9. It was suggested that compound **51** was produced through the oxidation of methoxy-*p*-quinone by residual hydrogen peroxide. Vanillin was degraded by singlet oxygen produced by sensitization with methylene blue [65]. Unfortunately, oxidation products were not determined in this work. Vanillin **43** degradation was determined by the decrease in light absorption at 270 nm. The rate law for the consumption of vanillin, **43** was expressed as first order dependence on singlet oxygen shown in equation 6.

$$\text{Rate of oxidation} = d[\text{Vanillin}]/dt = k_d[\text{Vanillin}][^1O_2] \qquad 6$$

Singlet oxygen oxidation of acetoguaicone, **44** at pH 3 produced only methoxy-*p*-quinone, **52**. Epoxidation of compound **52** by residual hydrogen peroxide does not occur at pH 3. Vanillyl alcohol, **45**, and apocynol, **46** both produced methoxy-*p*-quinone, **52** and 3-methoxy-5,6-epoxy-*p*-quinone, **51**. Apocynol methylether, **47** produced the same quinones (**50** and **52**) as did compounds **45** and **46**, in addition to compound **51**.

Side-chain displacement was not observed for the 4-alkylguaiacols **48** and **49**. The only product was the quinol, **53**. Also, veratraldehyde, in which the phenolic hydroxyl group is methylated, did not react with singlet oxygen. Apparently, both an oxygen atom and a phenolic hydroxyl group are required for side-chain displacement by singlet oxygen.

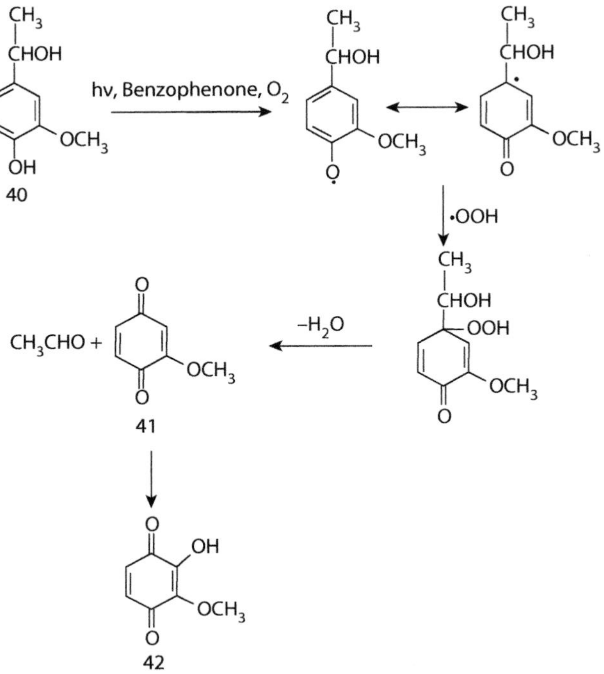

SCHEME 16.12 Singlet oxygen reaction with apocynol. (Redrawn from Gellerstedt, G., and Pettersson, E.- L., *Svensk Papperstn*, 80, 15–21, 1978.)

Scheme 16.14 shows the reactions of two dimeric cyclic ether lignin models with $^1O_2$ [64] Oxidation of the dimeric cyclic ether pinoresinol, **54** with $^1O_2$ from NaOCl/ H_2O_2 at pH 7 underwent cleavage of the phenyl-aliphatic bond to produce methoxy-hydroquinone, **50** and 2,6-dihydroxy-3,7-dioxabicyclo [3.3.0] octane, **55**. When treated under the same conditions as pinoresinol, **54**, dihydrodehydrodiisoeugenol, **56** did not undergo any bond cleavage but rather an oxidation of the phenolic moiety to form the quinol, **57** (Scheme 16.14).

Scheme 16.5 shows that the lignin models for the coniferyl alcohol end group, isoeugenol methyl ether, **11** and the aryl methyl ether of coniferyl alcohol, **12** react with triplet sensitized singlet oxygen to form the aldehydic degradation products **16** or **17** and **18** [47]. The glycol products, **13** and **14** from the addition solvent, water or ethanol, and the peroxirane intermediate were also isolated. Quenching with β-carotene or DABCO indicated participation by singlet oxygen. This reaction pathway of isoeugenol alkyl ether has been confirmed by the production of the same products after irradiating in the presence of singlet oxygen sensitizers methylene blue or rose bengal [64].

An alternate source of singlet oxygen, H_2O_2, and NaOCl, was used to oxidize isoeugenol methyl ether, **11** to products identical to those obtained by direct irradiation [64]. However, Nimz and Turznik [64] reported that isoeugenol methyl ether, **11** treated with singlet oxygen produced by H_2O_2 and NaOCl did not produce aldehydes,

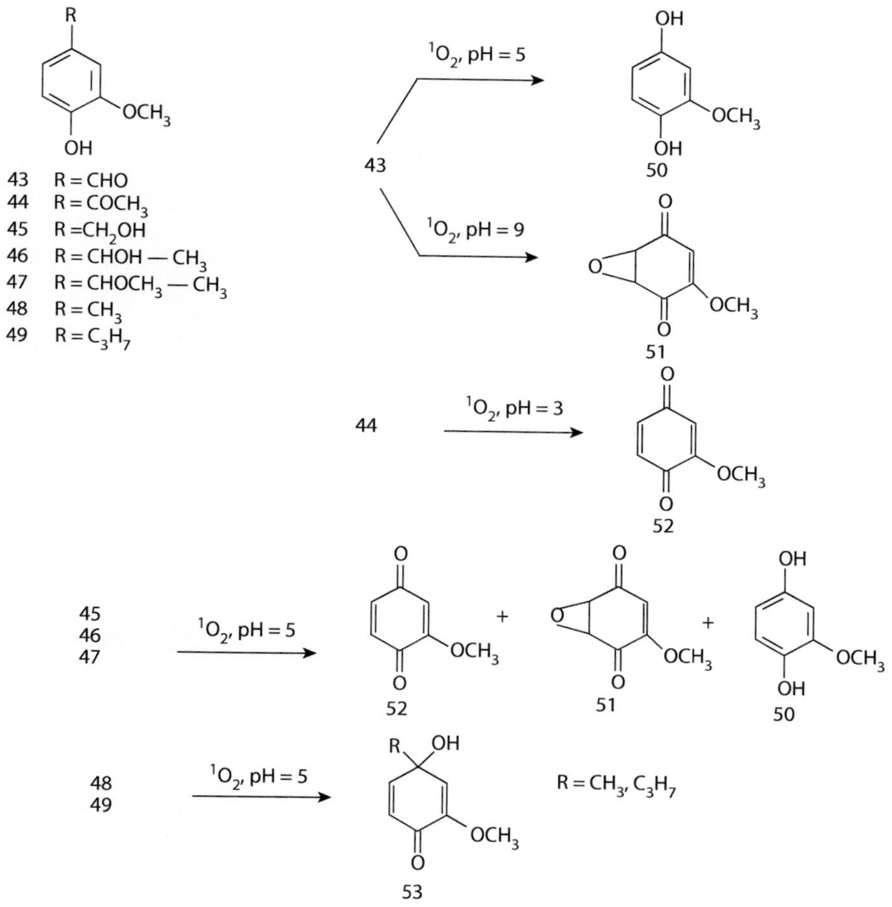

SCHEME 16.13 The reaction of singlet oxygen with guaiacyl based lignin model compounds. (Adapted from Nimz, H.H., and Turznik, G., *Cell. Chem. Technol.*, 14, 727–742, 1980.)

18 and **16**; only small a yield (~1%) the glycol, **13** could be isolated. Quite clearly oxidation of the coniferyl type end-group by irradiation with UV light/O_2 and with $^1O_2$ from H_2O_2 and NaOCl occur by different mechanisms.

As shown in Scheme 16.15, $^1O_2$ produced by the irradiation of oxygen in the presence of methylene blue reacted with dehydro-diisoeugenol, **58** produced a significant yield (22%) of the aldehyde **59** [64].

Lignin models containing the two- or three-carbon side chain react with $^1O_2$ differently than they do with oxygen-based radicals. When 4-ethoxy -3-methoxy-2-(2-methoxyphenoxy)acetophenone, **60a** or 4-ethoxy -3-methoxy-2-(2,3-dimethoxyphenoxy)acetophenone, **60b** was irradiated at wavelengths greater than 400 nm, guaiacol, **2**, or syringol, **62**, 3-methoxy-4-ethoxy-acetophenone, **63** and (3-methoxy-4-ethoxy-2-phenyl)-2-oxoacetaldhyde, **64** were formed [66,67] as shown in Scheme 16.16. It

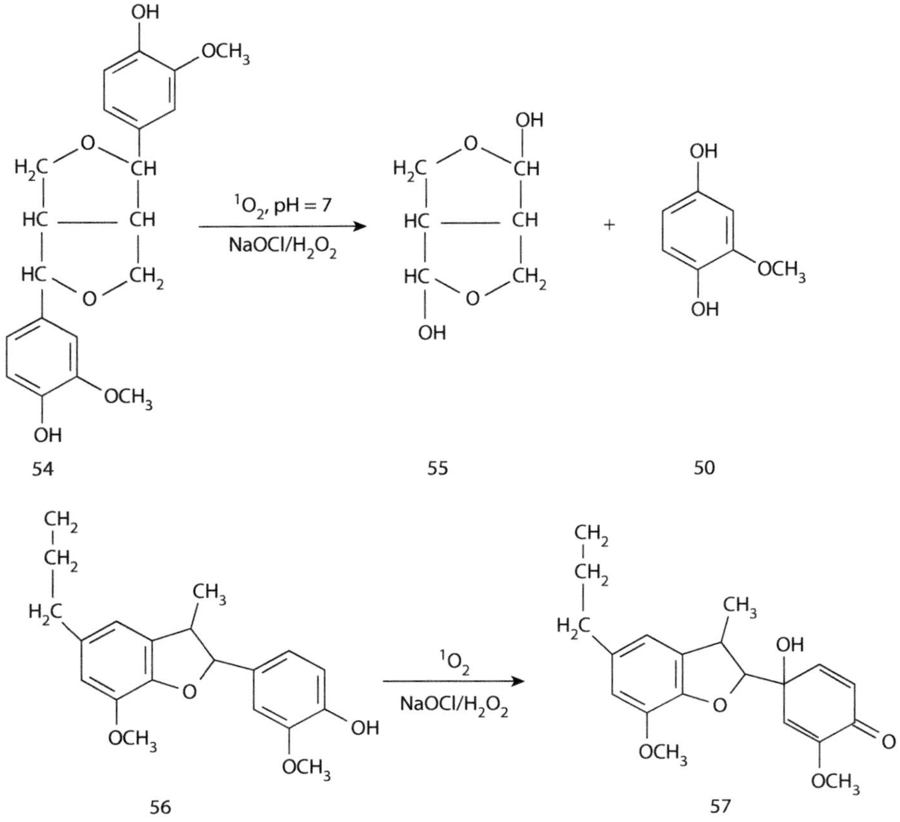

SCHEME 16.14 Singlet oxygen reactions with pinoresinol and dihydrodehydrodi-isoeugenol. (Adapted from Nimz, H.H., and Turznik, G., *Cell. Chem. Technol.*, 14, 727–742, 1980.)

was proposed in Scheme 16.17 that the interaction of singlet oxygen reacted with the phenoxy moiety compounds **60a** and **60b** to form an 1,2-endoperoxide, which in turn decomposes to compounds **2**, **62**, **63**, and **64** [68]. It is interesting to note that the endoperoxide intermediate was used by Matsuura et al. [60,61] to explain the oxidation of di-*t*-butyl phenol to the muconic acid.

Reactions with Isolated Lignin

The fact that DABCO inhibits the photodegradation of MWL by about 50% suggests a significant role of singlet oxygen [8,11]. Irradiation of lignin isolated from steam-exploded straw, beech, and pine with visible light in the presence of rose bengal decreased the degree of polymerization [69]. The UV band at 270 nm, attributed phenolic structures, was substantially decreased. Analysis of the products from steam exploded lignin from beech showed the presence of small molecules such as vanillin and sinapyl alcohol in the presence of highly oxidized phenols in the residue

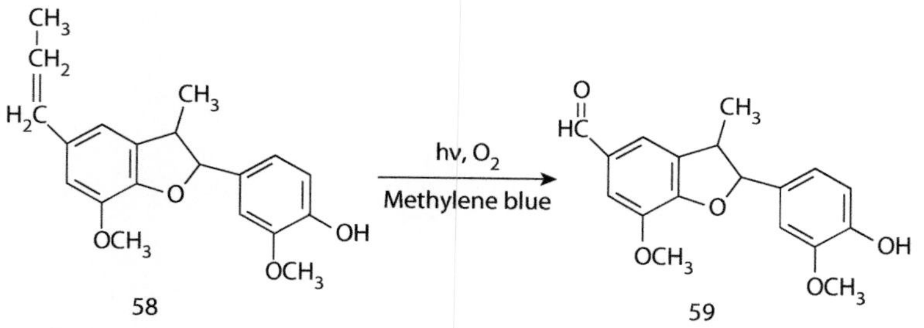

SCHEME 16.15 Methylene blue induced singlet oxygen oxidatively cleaves the propylene chain of dehydro-diisoeugenol to form the aldehyde. (Adapted from Nimz, H.H., and Turznik, G., *Cell. Chem. Technol.*, 14, 727–742, 1980.)

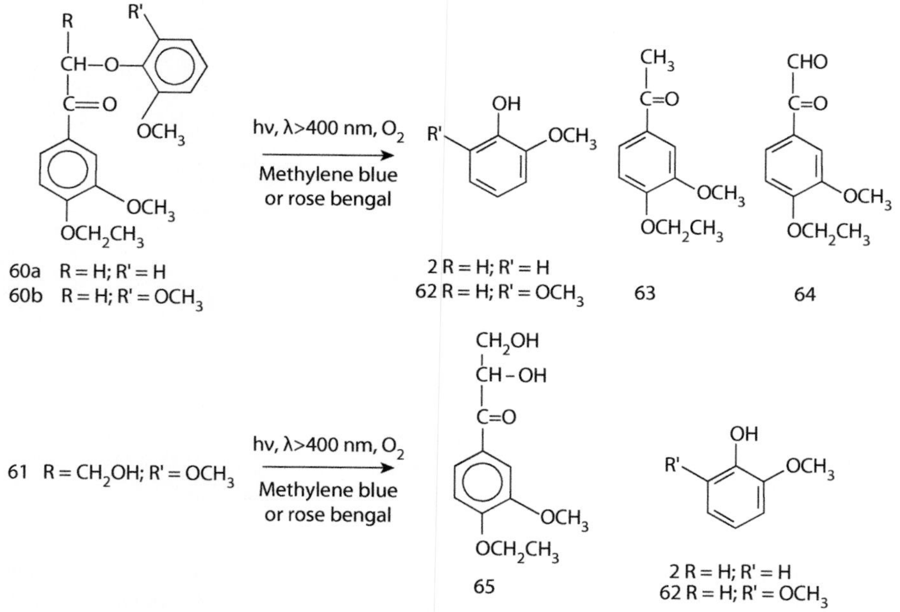

SCHEME 16.16 Reaction of singlet oxygen with nonphenolic methoxyphenoxyguaiacone. (Redrawn from Kutsuki, H., Enoki, A., and Gold, M.H., *Photochem. and Photobiol.*, 32, 1–7, 1983; Crestini, C., and D'Auria, M., *J Photochem Photobiol A: Chem*, 101, 69–73, 1996.)

obtained from silica gel chromatography [70]. In addition, singlet oxygen generated by hydrogen peroxide and sodium hypochlorite similarly degraded Klason lignin isolated from pine and beech woods [71]. One-hour irradiation of steam exploded lignin in the presence of singlet oxygen gave products showing high molecular weights. Two to three hours irradiation produced sinapyl alcohol, 4-hydroxy-3,5-dimethoxy-benzaldehyde, and 2,4-dioctylphenol [72]. It is interesting to note the isolation of a dehydroabietic acid derivative only after the treatment of lignin with singlet oxygen for five hours. After irradiation of the same sample for 12 hours,

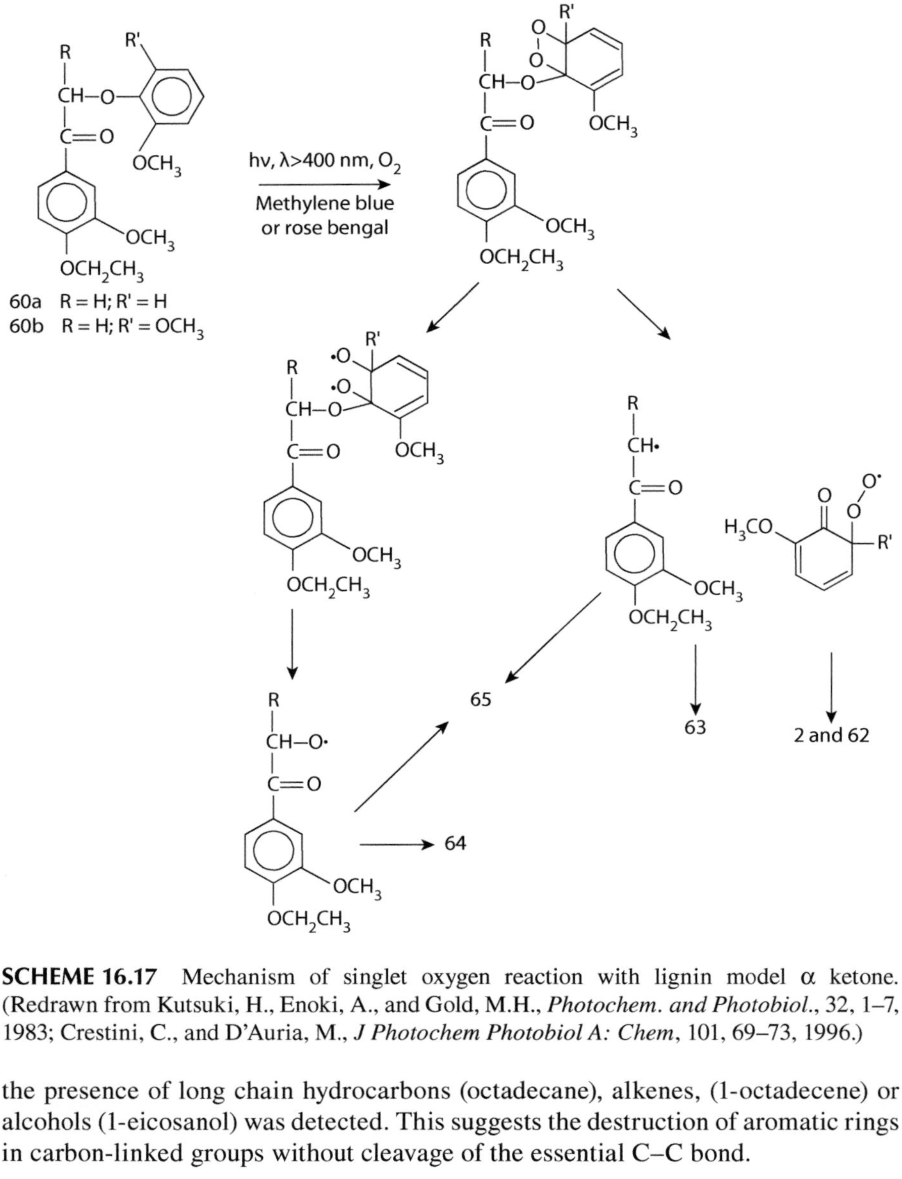

SCHEME 16.17 Mechanism of singlet oxygen reaction with lignin model α ketone. (Redrawn from Kutsuki, H., Enoki, A., and Gold, M.H., *Photochem. and Photobiol.*, 32, 1–7, 1983; Crestini, C., and D'Auria, M., *J Photochem Photobiol A: Chem*, 101, 69–73, 1996.)

the presence of long chain hydrocarbons (octadecane), alkenes, (1-octadecene) or alcohols (1-eicosanol) was detected. This suggests the destruction of aromatic rings in carbon-linked groups without cleavage of the essential C–C bond.

Reactions with Lignin-Containing Wood Fiber

The reactivity of singlet oxygen with lignin model compounds and extracted lignin could be duplicated on lignin-containing fibers, groundwood and chemithermome-chanical pulp under specific conditions. Forsskåhl and coworkers exposed these pulps to microwave generated singlet oxygen [73]. The only significant changes were a small increase in brightness. The pulps were not washed after treatment to determine yields. Subsequently, Forsskåhl and another group [74] observed similar results with singlet oxygen generated from 1,4-dimethylnaphthalene endoperoxide or irradiation of oxygen in the presence of rose bengal. Since only a small amount of bleaching occurred with no evidence of yellowing, it appears that the singlet oxygen reacted only with chromophores on the fiber surface and deactivated to the ground-state triplet before penetration into the fiber wall. The lifetime of singlet oxygen passing through a water layer such as on a wood fiber is in the order of microseconds [12]. The use of methylene blue as a singlet oxygen sensitizer and NaOH to dissolve the lignin decreased the residual lignin content measured by a decrease in kappa number from 13.3 to 8.1 [75,76] compared to 9.3 for UV light alone. rose bengal impregnated into the fiber as singlet oxygen sensitizer was used produce singlet oxygen *In situ* [77], to degrade the lignin steam-exploded pulps beech, straw and cardboard. The delignification using rose bengal, oxygen and irradiation with visible light appears to be more effective than that of methylene blue and UV light [76,77]. Singlet oxygen produced by irradiation of kraft pulp in the presence of oxygen and rose bengal removed about 70% of the residual lignin. Adding alkali increased the delignification to 90%. In contrast to other forms of oxidative delignification, singlet oxygen reduced the levels of condensed lignin (C5) structure [78,79].

Singlet oxygen delignification appears at first sight to be a viable chemical pulp bleaching stage when scaled-up to mill production. However, the usual *caveat* applies regarding the effect of such a stage on cellulose quality. A plot of intrinsic cellulose viscosity against degree of delignification showed that the cellulose degraded after greater than 60% of the residual lignin was removed [78]. This was attributed to the radical by-products of singlet oxygen reactions with the residual lignin.

SUMMARY

The lignin macromolecule undergoes photooxidative degradation through two mechanisms. One involves excitation followed by abstraction of hydrogen, which is followed by addition of ground-state triplet oxygen to carbon-based radicals to form hydroperoxides. The other mechanism is sensitizing the formation of singlet oxygen that undergoes electrophilic addition to the aromatic rings. Lignin contains a multitude of sites that can absorb light and produce excited states that sensitize the formation of free-radicals and singlet oxygen. Both free radicals and singlet oxygen produced inside the fiber wall degrade the macromolecule [9,11]. It was shown that singlet oxygen produced outside the fiber wall does not diffuse into the fiber wall before decomposing and therefore has no significant effect [73,74]. This facile photo-toxidation yields a large number of products that after isolation may yet be useful components and feed-stock in a biorefinery.

REFERENCES

1. RH Turner. Delignification and Bleaching of Lignocellulosic Pulp Via Photo-Oxygenation. U.S. Patent Number 4 292 654, 1981.
2. B Marcoccia, DAI Goring, and DW Reeve. Photo-enhanced Oxygen Delignification of Softwood Kraft Pulp: Effect of Some Process Variables. *J Pulp Paper Sci* 17:J34–J38, 1991.

3. B Marcoccia, DW Reeve, and DAI Goring. Photo-enhanced Oxygen Delignification of Softwood Kraft Pulp. *J. Pulp Paper Sci* 19:J97–J101, 1993.

4. AE de H Machado, R Ruggiero, MGH Terrones, A Nourmamode, S Grelier, and A Castellan. Photodelignification of *Eucalyptis Grandis* Organosolv Pulp. *J Photochem Photobiol A: Chemistry* 94:253–262, 1996.

5. MG Neumann, RAMC de Groote, and AEH Machado. Flash Photolysis of Lignin. Part 1. Dearated Solutions of Dioxane-Lignin. *Polym Photochem* 7:461–469, 1986.

6. H Koch, K Hübner, and K Fischer. The Influence of Light on the Molecular Mass of Lignin. *J Wood Chem Technol* 14:339–349, 1994.

7. K Fischer, M Beyer, and H Koch. Photo-Induced Yellowing of High Yield Pulps. *Holzforschung* 49:203–210, 1995.

8. J-N Destiné, J Wang, C Heitner, and R St. John Manley. The Photodegradation of Milled Wood Lignin. Part I. The Role of Oxygen. *J Pulp Paper Sci* 22:J24–J30, 1996.

9. J Wang, C Heitner, and R St. John Manley. The Photodegradation of Milled Wood Lignin. Part III. The Effect of Time and Media. *J Pulp Paper Sci* 24:337–340, 1998.

10. DS Argyropoulos and Y Sun. Photochemically Induced Solid-State Degradation, Condensation, and Rearrangement Reactions in Lignin Model Compounds and Milled Wood Lignin. *Photochem Phorobiol* 64:510–517, 1996.

11. J Wang, C Heitner, and R St. John Manley. The Photodegradation of Milled Wood Lignin. Part II. The Effect of Inhibitors. *J Pulp Paper Sci* 22:J58–J62, 1996.

12. D Beluš. Quenchers of Singlet Oxygen: A Critical Review. In: B Rånby and JF Rabeck, Eds. *Singlet Oxygen, Reactions with Organic Compounds and Polymers.* New York: John Wiley and Sons, 1978, p. 61.

13. (a) X Pan, D Lachenal, C Lapierre, and B Monties. Structure and Reactivity of Spruce Mechanical Pulp Lignins Part I. Bleaching and Photoyellowing of *in situ* Lignins. *J Wood ChemTechnol* 12:135–147, 1992. (b) X Pan, D Lachenal, C Lapierre, and B Monties. Structure and Reactivity of Spruce Mechanical Pulp Lignins. Part III. Bleaching and Photoyellowing of Isolated Lignin Fractions, *J Wood ChemTechnol* 13: 145–165, 1993.

14. J Gierer and SY Lin. Photodegradation of Lignin: A Contribution to the Mechanism of Chromophore Formation. *Svensk Papperstidn* 75:233–239, 1971.

15. JC Scaiano and JC Netto-Ferreira. Photochemistry of β-phenoxy-acetophenone: An Interesting Case of Intramolecular Triplet Deactivation. *J Photochem* 32:253–259, 1986.

16. C Vanucci, P Fornier de Violet, H Bouas-Laurent, and A Castellan. Photodegradation of Lignin. A Photophysical and Photochemical Study of a Non-Phenolic α-Carbonyl β-O-4 Lignin Model Dimer: 3,4-Dimethoxy-α-(2′-methoxyphenoxy)acetophenone. *J Photochem Photobiol A: Chemistry* 41:251–265, 1988.

17. JC Netto-Ferreira and JC Scaiano. Photochemistry of α-Phenoxy-p-methoxy-acetophenone. *Tetrahedran Letters* 30:443–446, 1980.

18. A Castellan, N Colombo, C Cucuphat, and P Fornier, de Violet, Photo-degradation of Lignin: A Photochemical Study of a Phenolic α-Carbonyl β-O-4 Lignin Model Dimer 4-Hydroxy-3-Methoxy-α-(2′-Methoxyphenoxy)–Acetophenone. *Holzforschung* 43:179–185, 1989.

19. A Castellan, N Colombo, P Fornier, de Violet, A Nourmamode, and H Bouas-Laurent. Photodegradation of Lignin: A Photochemical Study of Bleached CTMP and Lignin Model Compounds. Proceedings International Symposium on Wood and Pulping Chemistry, Raleigh, NC, 1989, pp. 421–430.

20. JC Netto-Ferreira, IGJ Avellar, and JC Scaiano. Effect of Ring Substitution on the Photochemistry of α-(Aryloxy)acetophenones. *J Org Chem* 55:89–92, 1990.

21. N Fukagawa and A. Ishizu. Photoreaction of Phenacyl Aryl Ether Type Lignols. *J Wood Chem Technol* 11:263–289, 1991.

22. DN-S Hon. Thermomechanical Pulp and Light Photoactivity of α-Carbonyl Group in Solid Lignin. *J Wood Chem Technol* 12:179–196, 1992.

23. JA Schmidt, AB Berinstain, F de Rege, C Heitner, LJ Johnston, and JC Scaiano. Photodegradation of Lignin Model α-Guaiacoxyacetoveratrone. Unusual Effects of Solvent, Oxygen and Singlet State Participation. *Can J Chem* 69:104–107, 1991.

24. JA Schmidt, AB Berinstain F de Rege, C Heitner, and LJ Johnston, Effect of Solvent on the Photodegradation of α-Guaiacoxylacetoveratrone, a Phenacyl-β-aryl Ether Lignin Model, in JF Kennedy, GO Phillips, and PA Williams, Eds. *Ligno-Cellulosics, Science, Technology, Development and Use.* Chichester: Ellis Horwood, 1992, p. 587.

25. JA Schmidt, E Goldszmidt, C Heitner, JC Scaiano, AB Berinstain, and LJ Johnston. Photodegradation of α-Guaiacoxyacetoveratrone: Triplet-State Reactivity Induced by Protic Solvents, in C Heitner and JC Scaiano, Eds. *Photochemistry of Lignocellulosic Materials,* Washington, DC: American Chemical Society, 1993, p. 122.

26. WU Palm and H Dreeskamp. Evidence for Singlet State β Cleavage in the Photoreaction of α-(2,6-Dimethoxyphenoxy)-acetophenone Inferred from Time-Resolved CIDNP Spectroscopy. *J Photochem Photobiol, A: Chemistry,* 52:439–450, 1990.

27. W-U Palm, H Dreeskamp, A Castellan, and H Bouas-Laurent. The Photochemistry of β-Phenoxyacetophenones Investigated by Flash-CIDNP spectroscopy. *Ber Bunsenges Phys Chem* 96:50–61, 1992.

28. JKS Wan, MY Tse, and MC Depew. CDIEP Studies of the Photolysis of the Lignin model Compound α-Guaiacoxyacetoveratrone: The Role of Triplet Reactions in Aqueous and Hydroxylic Solvents. *Res Chem Intermed* 17:59–75, 1992.

29. B-J Zhao, MC Depew, NA Weir, and JK-S Wan. Some Mechanistic Aspects of the Light-Induced Yellowing of Lignin: A Further Study of Photochemical Reactions of α-Guaiacoxy-β-hydroxyproprioveratrone, and Some Substituted Methoxybenzenes. *Res Chem Intermed* 19:449–461, 1993.

30. JKS Wan, IA Shkrob, and MC Depew. Primary Photophysical and Photochemical Processes of α-Guaiacoxyacetoveratrone. In C Heitner and JC Scaiano, Eds. *Photochemistry of Lignocellulosic Materials.* Washington, DC: American Chemical Society, 1993, p. 99.

31. Y Saburi, T Yoshimoto, and K Minami. Photochemical Reaction of 2-Phenoxy-1-phenylethanol in the Presence of 1,4-Naphthoquinone. *Nihon Kagaku Zasshi* 91:(4) 371–373, 1970.

32. Y Saburi, T Yoshimoto, and K Minami. Photoreaction of 2-Phenoxy-1-phenylethanol in the Presence of Carbonyl Compound. *Nihon Kagaku. Zasshi* 90:587–590, 1969.

33. Y Saburi, T Yoshimoto, and K Minami. Effect of Substituent on the Photochemical Reaction of 2-Phenoxy-1-Phenylethanol in the Presence of Benzophenone or 1,4-Naphthoquinone. *Nihon Kagaku Zasshi* 91:462–466, 1970.

34. JC Scaiano, JC Netto-Ferrira, and V Wintgens. Fragmentation of Ketyl Radicals Derived From α-Phenoxyacetophenone: An Important Mode of Decay for Lignin-Related Radicals? *J Photochem Photobiol A: Chem* 59:265–268, 1991.

35. Y Huang, D Pagé, DDM Wayner, and P Mulder. Radical-Induced Degradation of a Lignin Model Compound. Decomposition of 1-phenyl-2-phenoxyethanol. *Can J Chem* 73:2079–2085, 1995.

36. C Fabbri, M Bietti, and O Lanzalunga. Generation and Reactivity of Ketyl Radicals with Lignin Related Structures. On the Importance of the Ketyl Pathway in the Photoyellowing of Lignin Containing Pulps and Papers. *J Org Chem* 70:2720–2728, 2005.

37. JA Schmidt and C Heitner. Light-Induced Yellowing of Mechanical and Ultra-High Yield Pulps. Part 2. Radical-Induced Cleavage of Etherified Guaiacylglycerol-β-arylether Groups Is the Main Degradative Pathway. *J Wood Chem Technol* 13(3):309, 1993.

38. G Gellerstedt and L Zhang. Formation and Reactions of Leucochromophoric Structures in High Yield Pulping. *J Wood Chem Technol* 12(4):384–412, 1992.

39. RC Francis, CW Dence, TC Alexander, R Agnemo, S Omori. Photostabilization of Thermomechanical Pulp by Alkylation and Borohydride Reduction. *Tappi J* 74:127–133, 1991.

40. G Leary, Flash Photolysis of Coniferyl Alcohol. *Chem. Comm.* 13:688–689, 1971.

41. G Leary, The Chemistry of Reactive Lignin Intermediates. Part I. Transients in Coniferyl Alcohol Photolysis. *J Chem Soc, Perkin Trans II,* 640, 1972.

42. IJ Miller and GJ Smith, The Photochemical Formation of Quinone Methide from Coniferyl Alcohol. *Aust J Chem* 28(4):825–830, 1975.

43. GJ Smith, IJ Miller, and WH Melhuish. Excitation Spectra of the Triplet States of Some Model Lignin Compounds. *Aust J Chem* 29(9):2073–2076, 1976.

44. C Jaeger, A Nourmamode, and A Castellan. Photodegradation of Lignin: A Photochemical Study of Phenolic Coniferyl Alcohol Lignin Model Molecules. *Holzforschung* 47(5):375–390, 1993.

45. K Radotic, J Zakrzewska, D. Sladic, and M. Jeremic. Study of Photochemical Reactions of Coniferyl Alcohol I. Mechanism and Intermediate Products of UV Radiation-induced Polymerization of Coniferyl Alcohol. *Photochem Photobiol* 65(2):284–291, 1997.

46. K Radotic, S Todorovic, J Zakrzewska, and M. Jeremic. Study of Photochemical Reactions of Coniferyl Alcohol I. Comparative Structural of a Photochemical and Enzymic Polymer of Coniferyl Alcohol. *Photochem Photobiol* 68(5):703–709, 1998.

47. G Gellerstedt and E-L Pettersson. Light-Induced Oxidation of Lignin: The behaviour of Structural Units Containing a Ring-Conjugated Double Bond. *Acta Chem Scand* B29:1005–1010, 1975.

48. G Gellerstedt and L Zhang. Formation of Leucochromophores during High-Yield Pulping and H_2O_2 Bleaching. ACS Symposium Series 531, pp. 129–146, 1993.

49. WJ Leigh, TJ Lewis, V Lin, and JA Postigo. The Photochemistry of 3,3′,4,4′-Tetramethoxy and 4-hydroxy-3,3′,4′-Trimetoxystilbene models for Stilbene Chromophores in Peroxide-Bleached High Yield Wood Pulps. *Can J Chem* 74:263–275, 1996.

50. B Ruffin and A Castellan. Photoyellowing of Peroxide-Bleached Lignin-Rich Pulps: A Photochemical Study on Stilbene-Hydroquinone Chromophores Issued From β-5 Units of Lignin during Refining and (or) Bleaching. *Can J Chem* 78(1):73–83, 2000.

51. C Noutary, P Fornier de Violet, J Vercauteren, and A Castellan. Photochemical Sudies on a Phenoliuc Phenylcoumarone Lignin Model in Relation to the Photodegradation of Lignocellulosic Materials. Part 1. Structure of Photoproducts. *Res Chem Intermed* 21(3–5):223–245, 1995.

52. C Noutary, P Fornier de Violet, J Vercauteren, and A Castellan. Photochemical Sudies on a Phenoliuc Phenylcoumarone Lignin Model in Relation to the Photodegradation of Lignocellulosic Materials. Part 2. Photochemical and Photophysical Studies. *Res Chem Intermed* 21(3–5):247–262, 1995.

53. P Karhunen, P Rummakko, J Sipilä, G Brunow, and I Kilpeläinen. Dibenzodioxocins: A Novel Type of Linkage in Softwood Lignins. *Tetrahedron Lett* 36(1):169–170, 1995.

54. P Karhunen, P Rummakko, J Sipila, and G brunow. The Formation of Dibenzodioxocin Structures by Oxidative Coupling: A Model Reaction for Lignin Biosynthesis. *Tetrahedron Lett* 36(25):4501–4504, 1995.

55. C Gardat, R Ruggiero, W Hoareau, A Nourmamode, S Grelier, B Siegmund, and A Castellan. Photochemical Study of an o-Ethyl Dibenzodioxocin Molecule as a Model for the Photodegradation of Non-phenolic Lignin Units of Lignocellulosics. *J Photochem Photobiol, A: Chem* 167:111–120, 2004.

56. C Gardat, R Ruggiero, W Hoareau, L Damigo, A Nourmamode, S Grelier, and A Castellan. Photochemical Study of 4-(4,9-Dimethoxy-2,11-n-dipropyl-6,7-dihydro-5,8-dioxa-dibenzo[a,c]cycloocten-6-yl)-2-methoxyphenol, a Lignin Model of Phenolic Dibenzodioxocin Unit. *J Photochem Photobiol, A:Chem* 169:261–269, 2005.

57. AE de Hora Machado, R De Paula, R Ruggiero, C Gardat, and A Castellan. Photophysics of Dibenzodioxocins. *J Photochem Photobiol, A: Chem* 180:165–174, 2006.

58. LRC Barclay, JK Grandy, HD MacKinnon, HC Nichol, and MR Vinqvist. Peroxidations Initiated by Lignin Model Compounds: Investigating the Role of Singlet Oxygen in Photo-Yellowing. *Can J Chem* 76:1805–1816, 1998.

59. K Golnick. Mechanism and Kinetics of Chemical Reactions of Singlet Oxygen with Organic Compounds. In: B Rånby and JF Rabek, Eds. *Singlet Oxygen Reactions with Organic Compounds and Polymers.* New York: John Wiley and Sons, 1978, pp. 111–134.

60. T Matsuura, N Yoshimura, A Nishinaga, and I Saito. Photoinduced Reactions. LVI. Participation of Singlet Oxygen in the Hydrogen Abstraction from a Phenol in the Photosensitized Oxygenation. *Tetrahedron* 28:4933–4938, 1972.

61. T Matsuura, N Yoshimura, A Nishinaga, and I Saito. Photoinduced Reactions. LVII. Photosensitized Oxygenation of Catechol and Hydroquinone Derivatives: Non enzymatic Models for the Enzymatic Cleavage of Phenolic Rings. *Tetrahedron* 28:5119–5129, 1972.

62. IA Shkrob, MC Depew, and JKS Wan. Free-Radical Induced Oxidation of Alkoxyphenols: Some Insights into the Processes of Photoyellowing of Papers. *Res Chem Intermed* 17:271–285, 1992.

63. G Gellerstedt and E-L Pettersson. Light-Induced Oxidation of Lignin. Part 2. The Oxidative Degradation of Aromatic Rings. *Svensk Papperstn* 80(1):15–21, 1978.

64. HH Nimz and G Turznik. Reactions of Lignin with Singlet Oxygen. I. Oxidation of Monomeric and Dimeric Model Compounds with Sodium Hypochlorite: Hydrogen Peroxide. *Cell Chem Technol* 14:727–742, 1980.

65. AEH Machado, AJ Gomes, CMF Campos, MGH Terrones, DS Perez, R Ruggiero, and A Castellan. Photoreactivity of Lignin Model Compounds in the Photobleaching of Chemical Pulps. 2. Study of the Degradation of 4-Hydroxy 3-methoxybenzaldehyde and Two Lignin Fragments Induced by Singlet Oxygen. *J Photochem Photobiol* 110:99–106, 1997.

66. H Kutsuki, A Enoki, and MH Gold. Riboflavin-Photosensitized Oxidative Degradation of a Variety of Lignin Model Compounds. *Photochem and Photobiol* 32:1–7, 1983.

67. C Crestini and M D'Auria. Photodegradation of Lignin: The Role of Singlet Oxygen. *J Photochem Photobiol A: Chem* 101:69–73, 1996.

68. C Crestini and M D'Auria. Singlet Oxygen Photodegradation of Lignin Models. *Tetrahedron* 53:7877–7888, 1997.

69. C Bonini, M D'Auria, L D'Alessio, G Mauriello, D Tofani, D Viggiano, and F Zimbardi. Singlet Oxygen Degradation of Lignin. *J Photochem Photobiol A: Chem* 113:119–124, 1998.

70. G Bentivenga, C Bonini, M D'Auria, and A de Bona. Singlet Oxygen Degradation of Lignin: A GC-MS Study on the Residual Products of the Singlet Oxygen Degradation of Steam Explode Lignin from Beech. *J Photochem Photbiol A: Chem* 128:139–143, 1999.

71. G Bentivenga, C Bonini, M D'Auria, A de Bona, and G. Mauriello. Singlet Oxygen Mediated Degradation of Klason Lignin. *Chemosphere* 39(14):2409–2417, 1999.

72. G Bentivenga, C Bonini, M D'Auria, A de Bona, and G Mauriello. Fine Chemicals from Singlet-Oxygen-Mediated Degradation of Lignin: A GC/MS Study at Different Irradiation Times on a Steam-Exploded Lignin. *J Photochem Photobiol A: Chem* 135(2–3):203–206, 2000.

73. I Forsskåhl, C Olkkonen, and H Tylli. Singlet Oxygen-Induced Bleaching of High Yield Pulp Sheets. *J Photochem Photobiol A: Chem* 43:337–344, 1988.

74. H Takagi, I. Foirsskåhl, H. Perakyla, S Omori, and CW Dence. Studies on the Mechanism of the Photoyellowing of bleached Mechanical and Chemimechanical Pulps. *Holzforschung* 44(3):217–222, 1990.

75. A Castellan, S Grelier, AEH Machado, and R Ruggiero. Photodelignification of *Eucalyptus grandis* Organosolv Chemical Pulp. Fourth European Workshop on Lignocellulosics and Pulp: Advances in Characterization and Processing of Wood, Nonwoody, and Secondary Fibers, Stressa, Italy, 1994.

76. D Da Silva Perez, A Castallan, A Nourmamode, S Grelier, R Ruggiero, and A Machado. Photosensitized Delignification of Residual Lignin and Chemical Pulp from *Eucalyptus grandis* Wood. *Holzforschung* 56(6):595–600, 2002.

77. C Bonini, M D'Auria, G Mauriello, D Viggiano, and F Zimbardi. Singlet Oxygen Degradation of Lignin in Pulp. *J Photochem Photobiol A: Chem* 118:107–110, 1998.

78. K-O Hwang. Novel Exploration of the Ability of Singlet Oxygen to Photobleach *P. taeda* Pulps. *Spectrum* 14(2):8–14, 2001.

79. K-O Hwang and LA Lucia. Fundamental Insights into the Oxidation of Lignocellulosics Obtained From Singlet Oxygen Photochemistry. *J Photochem Photobiol A: Chem* 168:205–209, 2004.

17 Pharmacological Properties of Lignans

Takeshi Deyama and Sansei Nishibe

CONTENTS

INTRODUCTION

Lignans and neolignans are natural products derived from the same phenylpropane (C_6-C_3) units that make up the lignin polymer. Lignans have two units linked by a bond between the β-carbons of the side chains. The sesquilignans and dilignans have three and four units, respectively. Neolignans have two units linked by a carbon–carbon bond other than the one between the two β-carbons (although a β-carbon may be involved, e.g., β–5) or by a carbon–oxygen bond. A number of lignans that exhibit a variety of biological activities have been isolated from medicinal plants that are used in traditional and folk medicines. Lignans have been identified in human urine and blood, and they affect human physiology [1–10]. With the resurgent interest in traditional medicines and so-called nutriceuticals, it is appropriate to review the pharmacological properties of lignans in this book.

ANITUMOR ACTIVITY

NATURALLY OCCURRING ANTITUMOR LIGNANS

Lignans with the podophyllotoxin (1) structure have antitumor activity. These lignans were first isolated from plants of the genus *Podophyllum*, such as the American Mayflower, *Podophyllum peltatum*. Subsequently, they have been isolated from other species as well. As bio-assay methods have improved, there has been marked progress in both the isolation of antitumor lignans from natural products and in their synthesis [11–13]. The activities of a number of naturally occurring antitumor lignans against a variety of cancer cell lines are summarized in Table 17.1.

Podophyllotoxin (1), 4′-demethyldeoxypodophyllotoxin (2) and β-apopicropodophyllin (3) strongly inhibited P-388 (murine lymphocytic leukemia); the IC_{50} values (concentration at which 50% of cells were killed) were 0.003, 0.005, and 0.002 μg/ml, respectively. Compounds 1-3 and (–)-yatein (4) showed antimitotic potency [13]. Phyllanthostatin A (5) [14], wikstromol (6) [15–16], (±)-5′-methoxylariciresinol (7), (±)-lariciresinol (8), and (±)-syringaresinol (9) [17] also showed inhibitory activity against P-388. Justiciresinol (10) exhibited cytotoxicity against A-549 (human lung carcinoma), MCF-7 (human breast carcinoma) and HT-29 (human colon adenocarcinoma) cell lines [16]. Justicidin A (11) and diphyllin (12) inhibited 9-KB (human nasopharyngeal carcinoma), while justicidin C-E (13–15) did not [19]. Justicidin B (16) showed inhibition against P-388 and NSCLCN6 (human bronchial epidermoid carcinoma) cells [20]. Picropolygamain (17) inhibited A-549, MCF-7, and HT-29 cells [21]. Deoxypodophyllotoxin (18), β-peltatin methyl ether (19), picro-β-peltatin methyl ether (20) and dehydro-β-peltatin methyl ether (21) showed cytotoxic activities against eleven cell lines: A431 (human epidermoid carcinoma), BC1 (human breast cancer), Col2 (human colon cancer), HT (human fibrocarcinoma), KB (human nasopharyngeal carcinoma), KB-V1 (vinblastine resistant KB), LNCaP (hormone-dependent prostate

TABLE 17.1

Antitumor Activity of Naturally Occurring Lignans

No.	Name	P-388	A431	BC1	Col2	HT	KB	KB-V1	LNCap	Lu1	Mel2	U373	ZR75-1	ASK	BST	Ref.
1	podophyllotoxin	‡	‡	‡	‡	‡	‡	‡	‡	‡		‡	+	‡	‡	[13,17,22, 24,33]
2	4'-demethyldeoxypodophyll otoxine	‡	‡	‡	‡	‡	‡	‡	‡	‡		‡	+	‡	‡	[13]
3	β-apopicropodophyllin	‡		‡	‡	‡	‡	‡	‡	‡		‡	+	‡	‡	[13]
4	(−)-yetin	+		‡	‡	+	‡	+	+	‡		+	+	+	+	[13]
18	deoxypicropodophyllin	‡	‡	‡	‡	‡	+	+	+	‡		‡	+	+		[22]
19	deoxypodophyllotoxin		‡	‡	‡	‡	+	+	‡	‡		‡	‡	‡		[22]
20	β-peltatin methyl ether		‡	‡	‡	+	+	+	‡	‡		+	‡	‡		[22]
21	picroβ-peltatin methyl ether		‡	+	+	+	+		+	+		+		+		[22]
	dehydro-β-peltatin methyl ether		+	+	+	+	+	+	+	+		+		+		[22]
23	liriodendrin	+					6.0									[26]
28	steganangin				+	+	+	+	+	+		+	+	38.1		[30]
29	episteganangin				+	7.4	7.4	+		6.9			13.9	78.1		[30]
30	steganangin		+	+	+	7.4	‡	+	+	+		+	5.0	+		[30]
31	steganoate A													76.9		[30]
32	steganoate B		6.7	17.3	9.9	11.5	9.2	+		5.5	14.9	7.8	4.7	59.3		[30]
33	steganolide A		5.5	8.0	5.1	6.5	4.9	+	9.0	+	9.2	+	8.9			[30]
34	(−)-steganone			+	+	+	+	7.1	+	+		+		68.2		[30]
35	neoisostegane				+	+	6.6									[31]

Note: Results are expressed as ED$_{50}$ values(μg/ml); ‡<0.1, +<4.

P-388, murine lymphocytic leukomia; A431, human epidermoid carcinoma; BC1, human breast cancer; Col2, human colon cancer; HT,human fibrosarcoma; KB, human nasopharyngeal carcinoma; KB-V1, vinblastine resistant KB; LNCap, hormone-dependent prostate cancer; Lu1, human lung cancer; Mel2, human melanoma; U373, human glioblastoma; ZR75-1, human breast cancer; ASK, astrocytoma; BST, brine shrimp (*Artemia salina Leach*) lethality test.

(*Continued*)

TABLE 17.1
Antitumor Activity of Naturally Occurring Lignans (Continued)

No.	Name	P-388	BC1	HT1080	L929	Lu1	L1210	L5178y	Mel2	U373	ZR75-1	HeLa	Col-1	LNCap	BST	Ref.
36	arctiin						4.4									[32]
38	(−)-arctigenin						+	+								[33–35]
39	(−)-trachelogenin							+								[33–35]
40	matairesinol				20											[36]
41	(−)-lirioresinol A				9											[36]
42	(−)-lirioresinol B				8.5											[36]
43	(+)-sesamin											10-100				[33,37]
	(−)-sesamin							42.4							+	[48]
44	anolignan A		4.3	13.7		15.8			7.6	13.5	6.1		16.8	10.8		[38]
45	anolignan B	+	5.1	4.7		11.1			19.0	5.7			18.0	8.7		[38]
46	anolignan C										9.5					[38]
48	hattalin										+					[39]

Note: HT1080, human fibrosarcoma; L1210, mouse leukemia; L5178y, mouse lymphoma.

No.	Name	P-388	A-549	MCF-7	HT-29	KB	NSCLCCN-6	L929	EBVEA	BST	Ref.
5	phyllanthostatin A	+						10			[14]
6	wikstromol	+									[15,16]
7	(±)-5'-methoxylaricirtesinol	+									[17]
8	(±)- lariciresinol	+									[17]
9	(±)-syringaresinol	+									[17]
	(−)-syringaresinol	+									[27]
	(−)-syringaresinol diacetate	+									[27]
10	justiciresinol		31.3	22.3	18.3						[18]
11	justicidin A					+					[19]
12	diphyllin					+					[19]
13	justicidin C					>10.5					[19]
14	justicidin D					9.0					[19]
15	justicidin E					>10.5					[19]
16	justicidin B	+				++					[20]
17	picropolygamain						28			52.2	[21]
25	magnolol		++	++	++				+	+	[29]
26	honokiol								+		[29]
27	monoterpenyl magnolol								+		[29]

Note: A-549 human lung carcinoma; MCF-7, human breast carcinoma; HT-29, human colon adenocarcinoma; NSCLCN6, human bronchial epidermoid carcinoma; KB, nasopharyngeal carcinoma; NSCLCN-6, human bronchial epidermoid carcinoma; L929, murine; EBVEA, Epstein-Barr virus early antigen.

cancer), Lu1 (human lung cancer), Mel2 (human melanoma), U373 (human glioblastoma), ZR75-1 (human breast cancer) and ASK (astrocytoma) [22]. Guaiacylglycerol-β–O–6′-(2-methoxy) cinnamyl alcohol ether (**22**) weakly inhibited HL-60 (human promyelocytic leukomia) cells [23]. Compounds **1** and **18** showed strong inhibitory activity against KB cells [24,25]. Liriodendrin (syringaresinol di-O–β–D- glucopyranoside, **23**) exhibited potent cytotoxic activity against P-388, KB and lung cancer cells [26]. (–)-Syringaresinol and (–)-syringaresinol diacetate showed significant inhibitory activity against P-388 [27]. On the other hand, (–)-syringaresinol O–β–D-glucopyranoside (**24**) was inactive against eleven cancer cell lines: A431, BC1, Col2, HT, KB, KB-V1, LNCaP, Lu1, Mel2, ZR-75-1, and P-388 [28]. Three neolignans isolated from *Magnolia officinalis*, magnolol (**25**), honokiol (**26**) and monoterpenylmagnolol (**27**), inhibited Epstein-Barr virus early antigen (EBV-EA) activation induced by

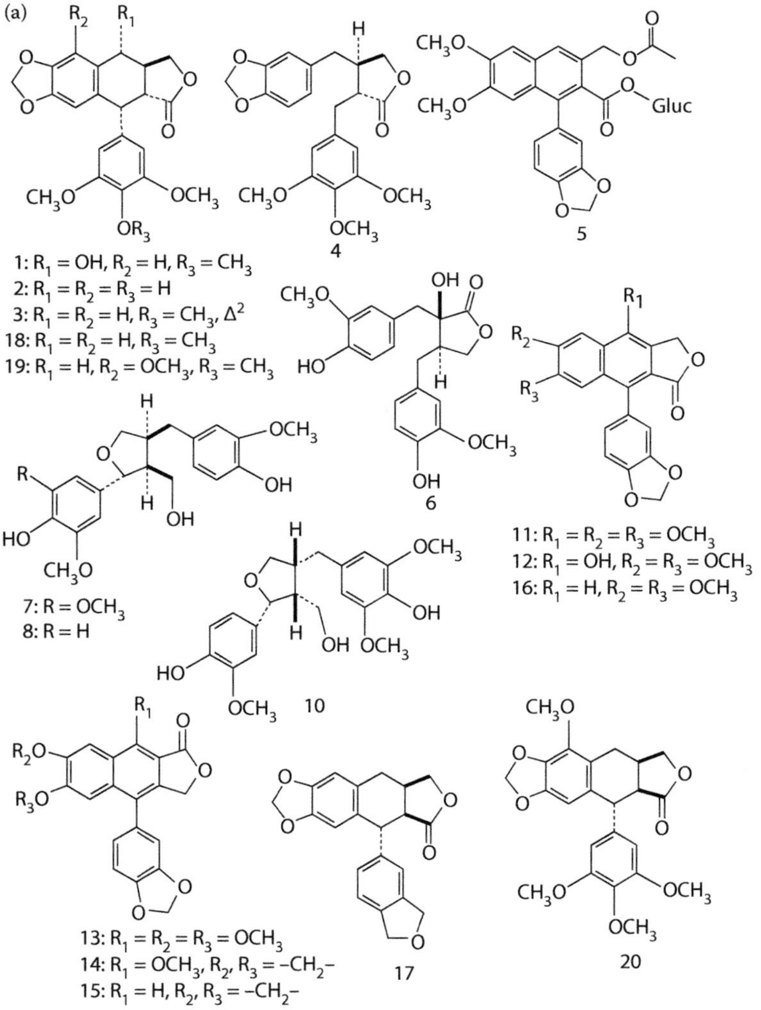

(a)

1: R_1 = OH, R_2 = H, R_3 = CH_3
2: R_1 = R_2 = R_3 = H
3: R_1 = R_2 = H, R_3 = CH_3, Δ^2
18: R_1 = R_2 = H, R_3 = CH_3
19: R_1 = H, R_2 = OCH_3, R_3 = CH_3

4

5

6

7: R = OCH_3
8: R = H

10

11: R_1 = R_2 = R_3 = OCH_3
12: R_1 = OH, R_2 = R_3 = OCH_3
16: R_1 = H, R_2 = R_3 = OCH_3

13: R_1 = R_2 = R_3 = OCH_3
14: R_1 = OCH_3, R_2, R_3 = $-CH_2-$
15: R_1 = H, R_2, R_3 = $-CH_2-$

17

20

(b)

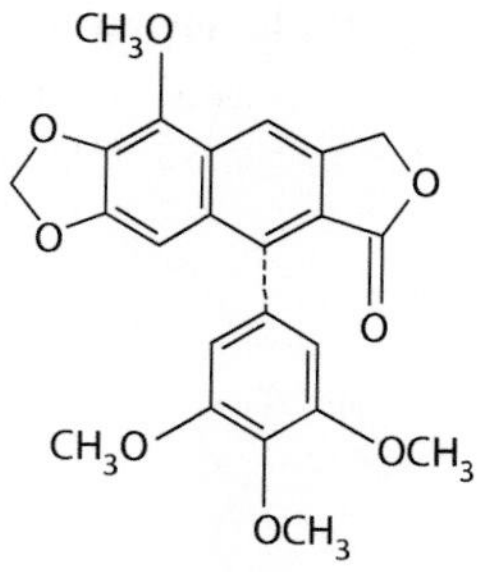

21

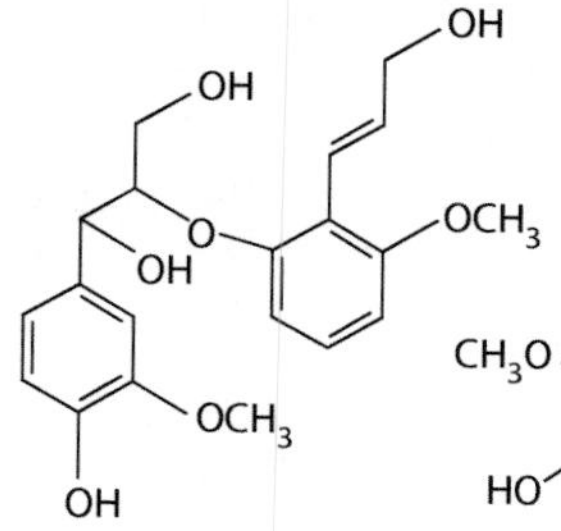

22

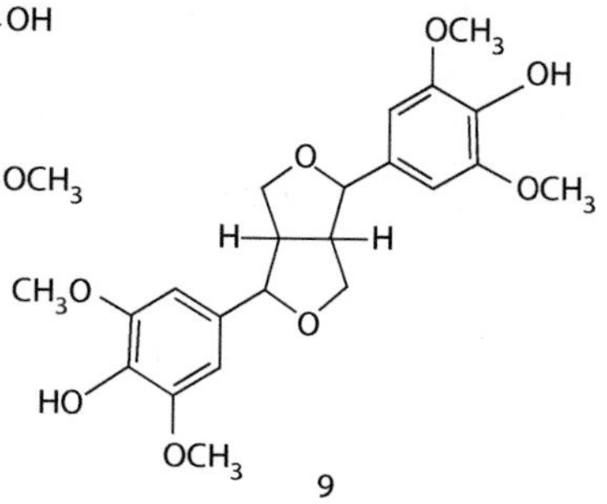

9

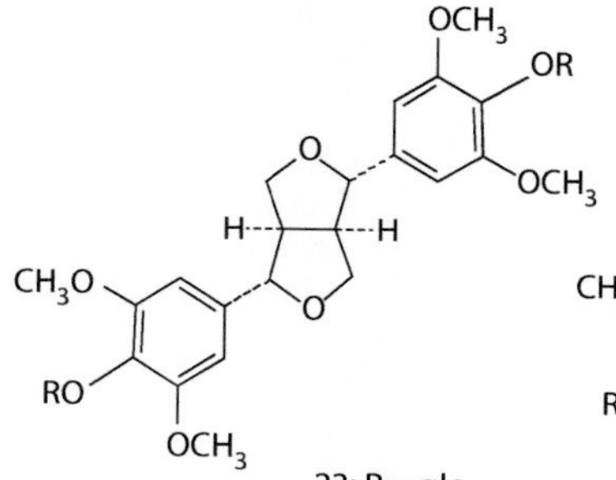

23: R = glc

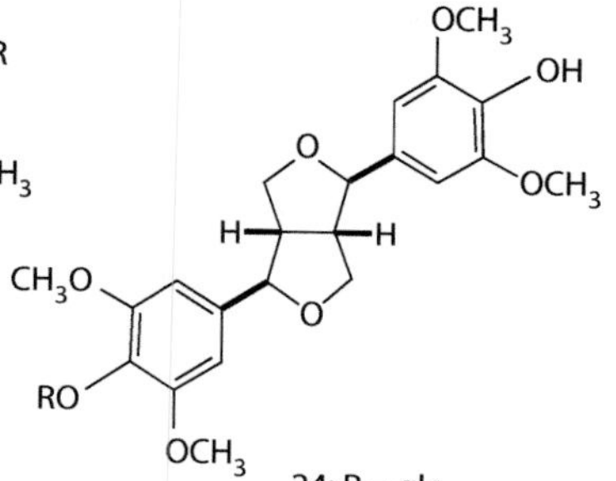

24: R = glc

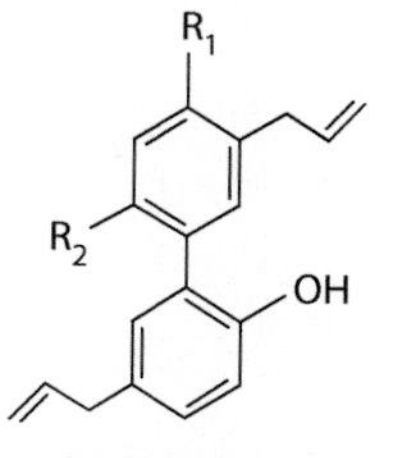

25: R₁ = H, R₂ = OH
26: R₁ = OH, R₂ = H

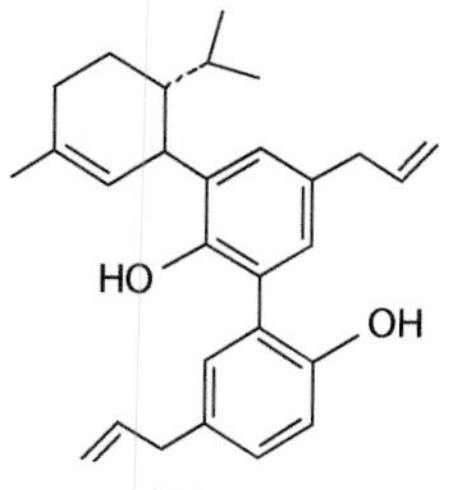

27

12-*O*-tetra decanoylphorbol-13-acetate (TPA), which is a strong tumor promoter in mouse skin. The inhibitory activities of 25 and 26 were more than ten times higher than glycyrrhetinic acid, a strong antitumor agent. Compound 25 delayed the formation of papillomas in mouse skin when it was applied continuously before each TPA treatment [27]. Steganangin (28), episteganangin (29), steganacin (30), steganoate A (31), steganoate B (32), steganolide A (33), and (–)-steganone (34) also inhibited TPA. The magnitude of this activity tended to correlate with the antimiotic activity observed with ASK and in vitro inhibition of microtubule assembly [30]. Neoisostegane (35) showed weak cytotoxic activity against KB cells [31]. Arctiin (36) showed potent inhibitory activity against L1210 (mouse leukemia), IC₅₀ 4.4 μg/ml, while matairesinoside (37) showed no activity [32]. (–)-Arctigenin (38), (–)-trachelogenin (39), and its

2,3-benzylbutyramide derivatives showed remarkable cytotoxic activity against L5718y (mouse lymphoma) cells

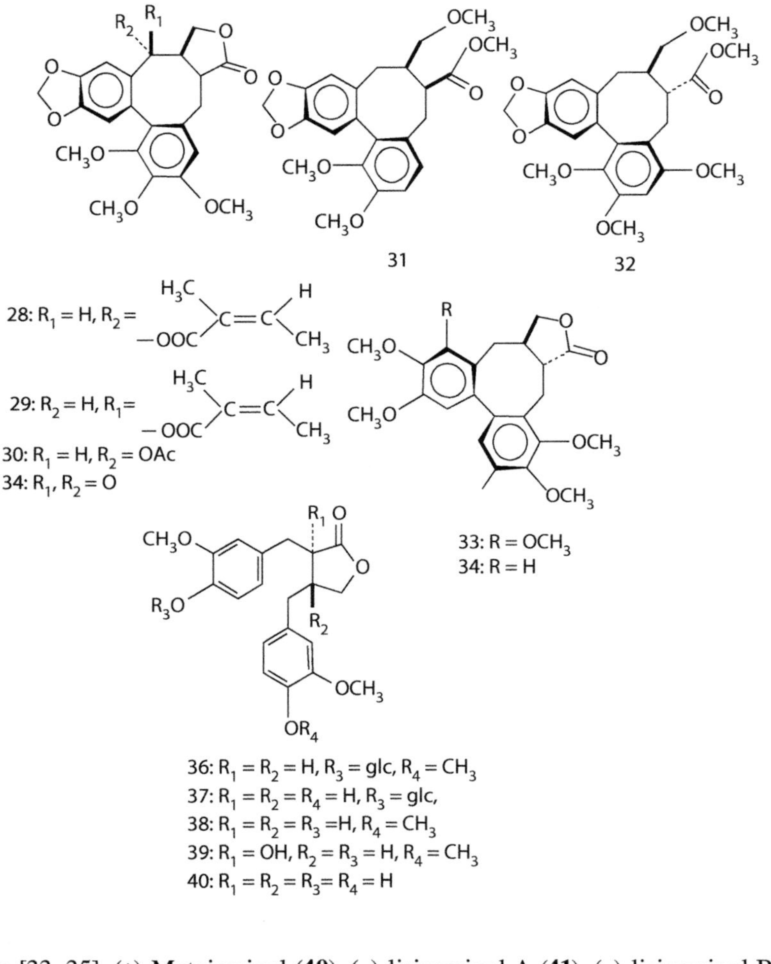
31
32
28: R_1 = H, R_2 = $-OOC(H_3C)C{=}C(H)CH_3$
29: R_2 = H, R_1 = $-OOC(H_3C)C{=}C(H)CH_3$
30: R_1 = H, R_2 = OAc
34: R_1, R_2 = O
33: R = OCH_3
34: R = H
36: $R_1 = R_2$ = H, R_3 = glc, $R_4 = CH_3$
37: $R_1 = R_2 = R_4$ = H, R_3 = glc,
38: $R_1 = R_2 = R_3$ =H, $R_4 = CH_3$
39: R_1 = OH, $R_2 = R_3$ = H, $R_4 = CH_3$
40: $R_1 = R_2 = R_3 = R_4$ = H

in vitro [33–35]. (+)-Matairesinol (**40**), (–)-lirioresinol A (**41**), (–)-lirioresinol B (**42**), and **5** showed potent cytotoxic activity against L929 (murine) cells [36]. (+)-Sesamin (**43**) showed cytotoxic potency against HeLa (human Hela cernix uteri tumor) cell [37]. Anolignan A (**44**) and anolignan B (**45**) showed moderate cytoxicity against BC1, HT1080 (human fibro sarcoma), Lu1, Mel2, Col1 (human colon cancer), LNCaP, and U373 cell lines. Compound **44** showed significant cytoxicity against P-388. Anolignan C (**46**) showed specific, but only moderate cytotoxicity against ZR-75-1. (–)-Secoisolariciresinol (**47**) did not show any cytotoxicity [38]. Hattalin (2,3-dibenzylbutane-1,4-diol) (**48**) showed the strongest inhibitory activity against ZR-75-1 [39]. Gomisin A (**49**) inhibited early-stage hepatocarcinogenesis [41,42], and suppressed tumor promotion [42–44].

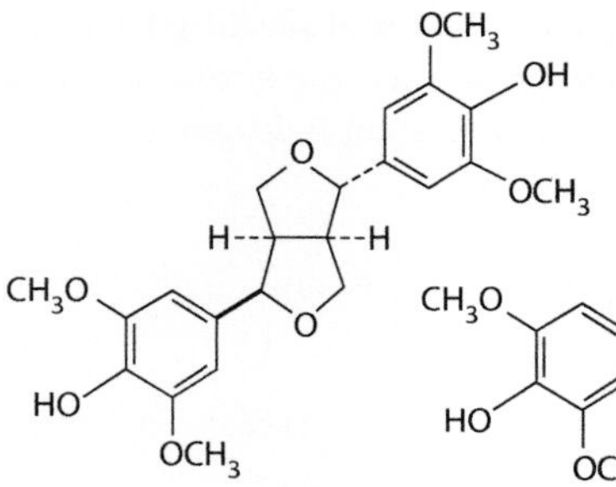

41 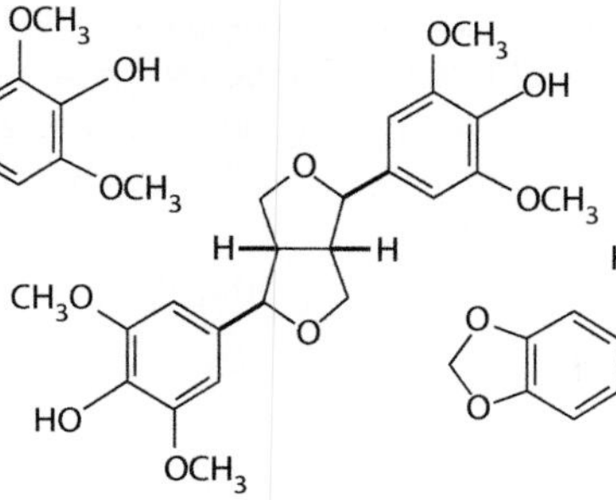   42 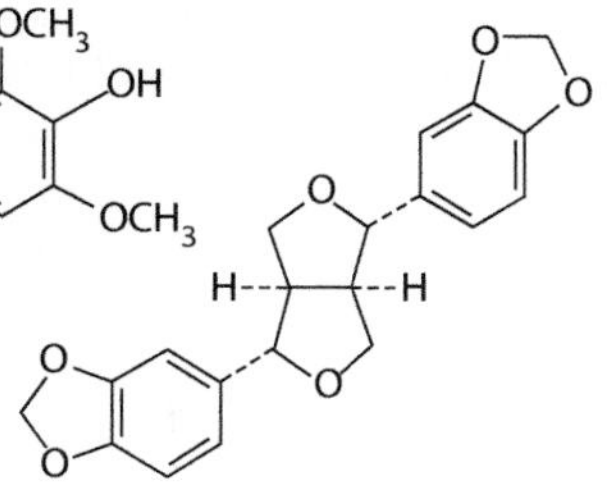   43

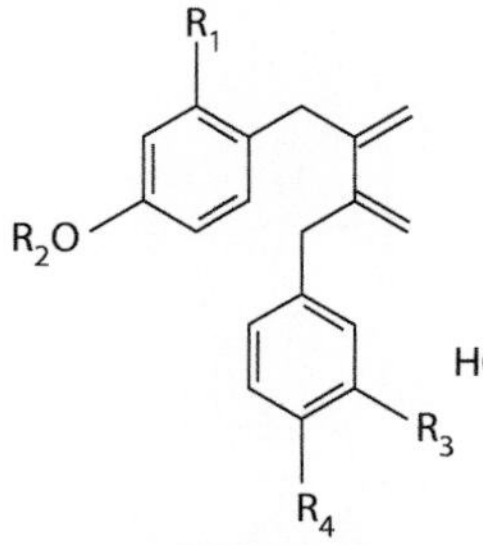

44: R_1 = OH, R_2 = H, R_3, R_4 = -O-CH$_2$-O-
45: R_1 = R_2 = H, R_3 = R_4 = OH

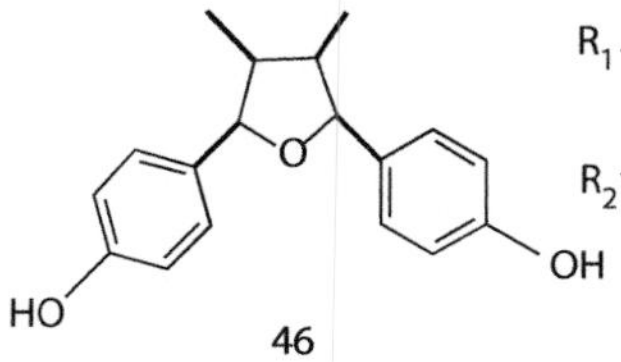

46

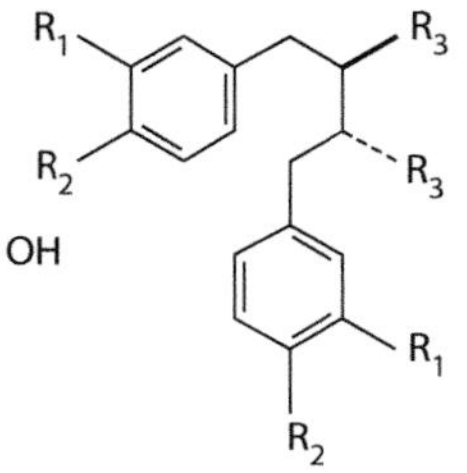

47: R_1 = OCH$_3$, R_2 = OH, R_3 = CH$_2$OH
48: R_1 = R_2 = H, R_3 = CH$_2$OH

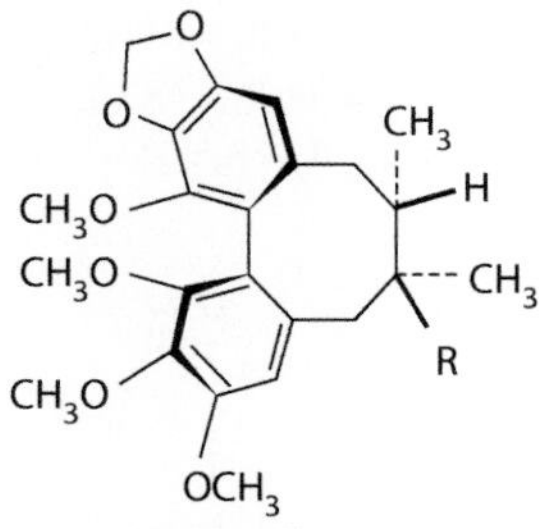

49: R = OH

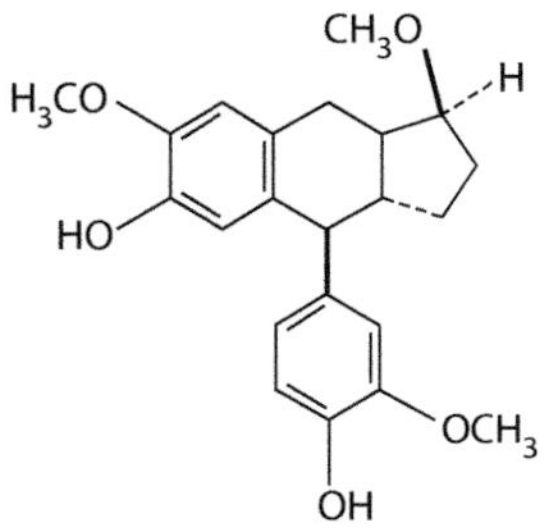

50

The brine shrimp test (BST), the piscidal activity against *Artemia salina*, is related to antitumor activity and is recommended by the National Cancer Institute (NCI) as an in-house test for promising antitumor compounds [45–47]. (–)-Sesamin (enantiomer of **43**) and **16** showed activity against BST [48,49]. Compound **16** had an ichtyotoxic effect on guppies and strongly inhibited swimming activity [50]. (+)-Tsugacetal (**50**) showed weak activity in the BST [51].

A number of antiproliferative drugs, such as anticancer agents, have been reported to promote the terminal differentiation of certain tumor cell lines, and consequently suppressed the tumor phenotype. Thus differentiation inducers seem to be potent agents for the treatment of human cancer. Sixteen lignans isolated from *Arctium lappa* L (Compositae) were differentiation inducers toward mouse meyloid leukemia (Ml) cells: two lignans: **38** and **40**, eight sesquilignans: arctignan A-C (**51–53**), lappaol A-C (**59–61**), lappaol E (**62**) and isolappaol A (**64**), and six dilignans: arctignan D-H (**54–58**) and lappaol F (**63**)]. Compound **38** showed the strongest inducing activity. The activity followed the order lignan > sesquilignan > dilignan [52,53].

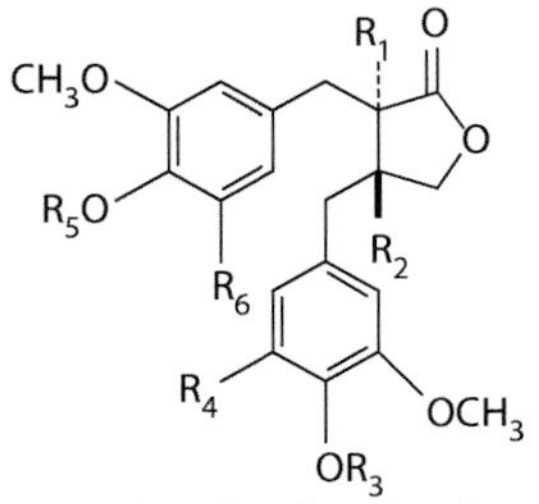

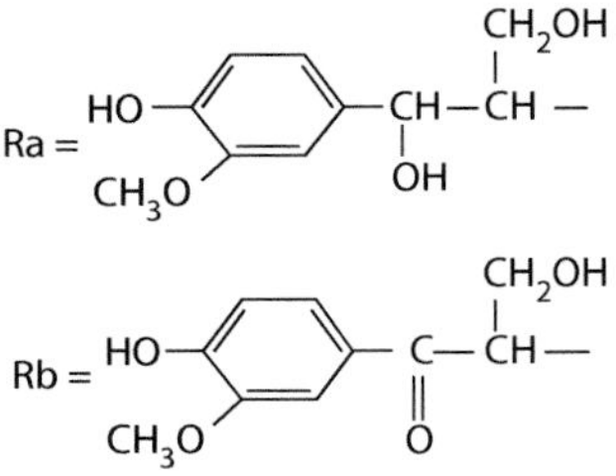

51: $R_1 = R_2 = R_3 = R_4 = R_6 = H, R_5 = Ra$
52: $R_1 = R_2 = R_3 = R_5 = R_6 = H, R_4 = Rb$
53: $R_1 = R_2 = R_3 = R_4 = R_5 = H, R_6 = Rb$
61: $R_1 = R_2 = R_3 = R_5 = R_6 = H, R_4 = Ra$
62: $R_1 = R_2 = R_4 = R_5 = R_6 = H, R_3 = Ra$

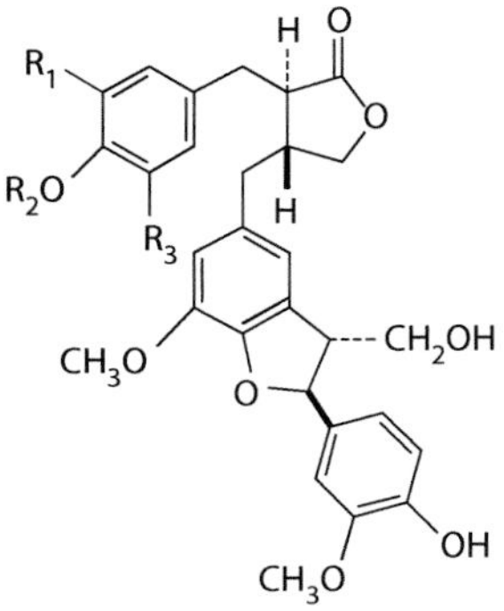

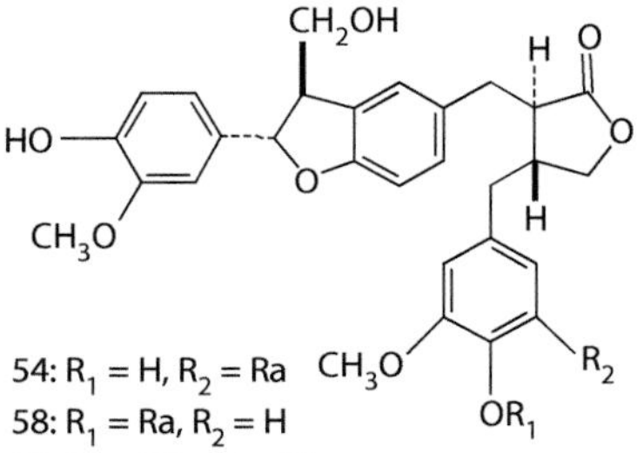

54: $R_1 = H, R_2 = Ra$
58: $R_1 = Ra, R_2 = H$
60: $R_1 = CH_3, R_2 = H$
64: $R_1 = R_2 = H$

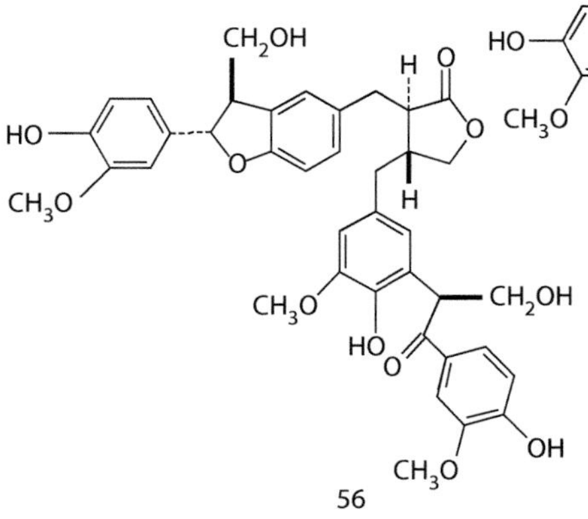

55: $R_1 = Ra, R_2 = H, R_3 = OCH_3$
57: $R_1 = H, R_2 = Ra, R_3 = OCH_3$
59: $R_1 = OCH_3, R_2 = R_3 = H$

56

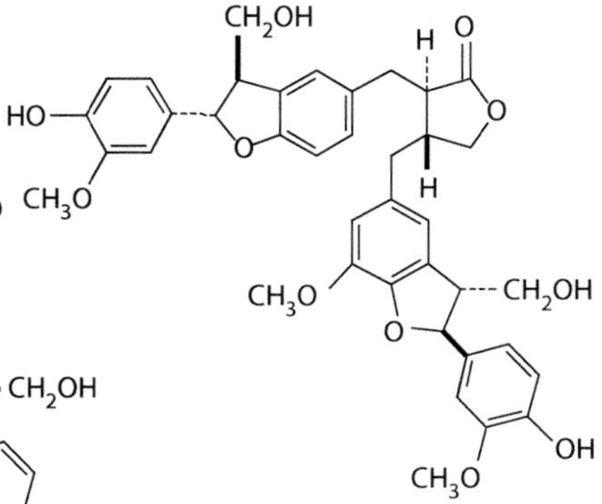

63

Lignans that are present in the serum, urine, bile and seminal fluid of humans, chimpanzees and vervet monkeys are called mammalian lignans. Mammalian lignans inhibit cell membrane Na$^+$/K$^+$-ATPase, the enzyme that maintains the Na$^+$/K$^+$ balance in cells. Hattalin (**48**) inhibited Na$^+$/K$^+$-ATPase of the plasma membrane fraction from both cultured cells and a section of human breast cancer tissue. It also inhibited ATPase from human gastrin cancer tissue (C-ATPase) more than it inhibited ATPase from normal gastric mucosa (N-ATPase). The target ATPase of hattalin was other than sodium and potassium-dependent, ouabain-sensitive ATPase [54].

SYNTHESIZED ANTITUMOR LIGNANS

Podophyllotoxin (**1**) inhibits the assembly of microtubulin and the activity of topo-isomerase II, and exhibits strong antitumor activity. It is considered as the critical structure of antitumor compounds. Since the 1940s, many podophyllotoxin deriva-tives have been synthesized. Among them, etoposide (**VP-16-213, 65**) and teniposide (**VM-26, 66**) exhibit no gastrointestinal toxicity, and have been used with cisplatin (*cis*-diaminodichloroplatinum) in human chemotherapy. Compound **66** is more potent than **65** in L-1210 and HeLa tumor cell lines, and shows strong activity in hematological malignancies.

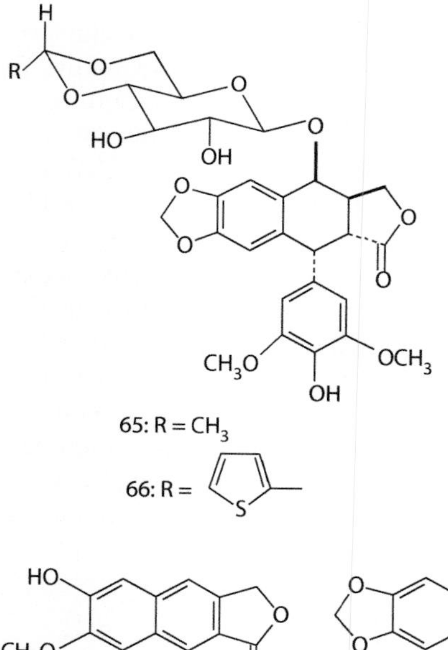

65: R = CH$_3$

66: R =

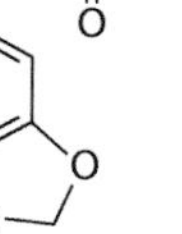

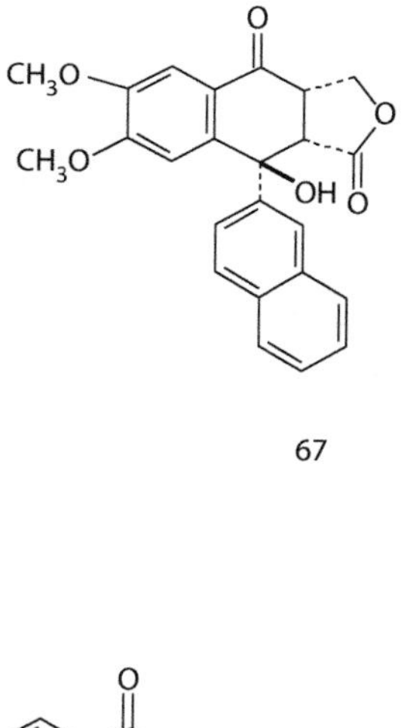

67

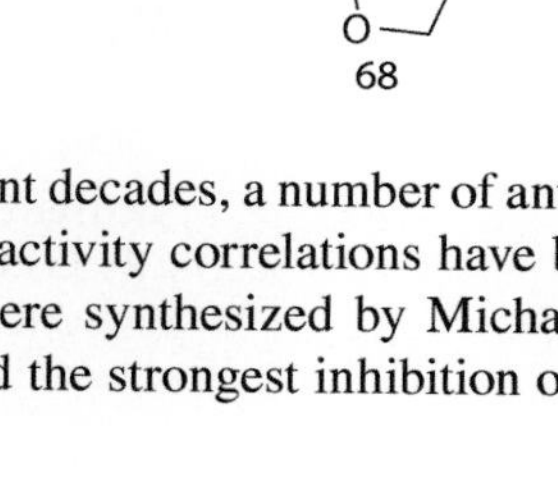

68 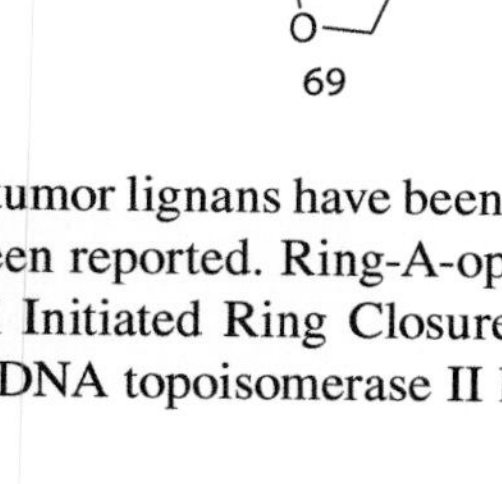   69

In recent decades, a number of antitumor lignans have been synthesized, and their structure-activity correlations have been reported. Ring-A-opened podophyllotoxin lignans were synthesized by Michael Initiated Ring Closure (MIRC). Compound **67** showed the strongest inhibition of DNA topoisomerase II like **65** [55]. Naturally occurring ring-A-opened podophyllin, **16** was synthesized from vertraldehyde [56]. Daurinol (**68**) and retrochinensin (**69**) were synthesized from isovanillin. Methylation of **68** yielded **16** [57]. Other quinone analogs of podophyllotoxin possessing various C–4–β-aniline moieties (R group in Series A, Series B, and Series C below) were synthesized. Some of these were stronger inhibitors of topoisomerase II than **65**. The activity order was series C > series B > series A [58].

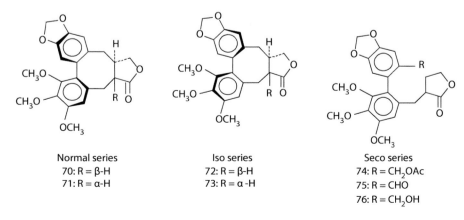

A number of podophyllotoxin congeners with an opened methylenedioxy ring were synthesized. Among them, 3,4,5-trimethoxyphenyl-2-naphthyl podophyllotoxin and 3,5-dimethoxy-4-benzyloxyphenyl podophyllotoxin significantly inhibited DNA topoisomerase II [59]. Many aminoglycosidic variants of 4'-demethyl-1-epipodophyllotoxin showed significant antitumor activity [60,61]. 2',6'-Dibromoyatein and 2'-bromodeoxy-podophyllotoxin showed antitumor activity against P-388 leukomia, A549 and HT-29 cells [62,63].

A series of synthetic analogues of naturally occurring antitumor lignans, the steganacins, were tested for inhibition of microtubule assembly in vitro. Spatial variation exists for both the biaryl junction [normal series: stegane (**70**) and picrostegane (**71**), iso series: isostegane (**72**) and isopicrostegane (**73**)]. Hydroxylation or ketonization at C-5 had little effect on the activity of the stegane skelton. A 5-acetoxyl group (steganacin) considerably increased the activity, while a 5-epiacetoxyl group decreased the activity. Opening the octane ring [secoseries; secosteganacin (**74**), secosteganone(**75**), seco steganol (**76**)] resulted in the loss of activity [64].

C-AMP PHOSPHODIESTERASE INHIBITORY ACTIVITY

Cyclic adenosine monophosphate (c-AMP) is a secondary messenger within cells. The c-AMP phosphodiesterase inhibition test is a useful means for screening biologically active compounds [65]. Correlations of structure with inhibitory activity for lignans and their glucosides are shown in Tables 17.2-1 through 17.2-5. For

TABLE 17.2-1
Inhibitory Activity of c-AMP PDE

No.	R_1	R_2	R_3	R_4	R_5	IC_{50} ($\times 10^{-5}$M)	Ref.
77	H	H	H	H	H	7.5	[67]
78	H	Glc	H	H	H	14.2	[67]
79	H	Glc	H	Glc	H	8.9	[67]
80	H	Glc	H	CH_3	H	21.0	[66]
81	H	CH_3	H	H	H	9.7	[66]
82	H	CH_3	H	CH_3	H	48.0	[66]
83	H	H	OCH_3	H	H	12.1	[67]
84	H	Glc	OCH_3	H	H	29.7	[67]
85	H	Glc	OCH_3	Glc	H	6.3	[67]
86	H	H	OCH_3	H	OCH_3	17.5	[67]
87	H	Glc	OCH_3	H	OCH_3	>50	[67]
23	H	Glc	OCH_3	Glc	OCH_3	12.7	[67]
88	OH	H	H	H	H	21.3	[67]
89	OH	Glc	H	H	H	28.6	[67]
90	OH	H	H	Glc	H	33.2	[67]
91	OH	Glc	H	Glc	H	10.0	[66]
92	OH	H	H	CH_3	H	40.7	[66]
93	OH	CH_3	H	CH_3	H	>50	[66]
94	OAc	H	H	H	H	3.2	[67]
95	OAc	Glc	H	H	H	4.4	[67]
96	OAc	Glc	H	Glc	H	1.1	[67]
97	OAc	Glc	H	H	H	11.5	[66]
98	OAc	Glc	H	CH_3	H	>50	[66]
99	OAc	Glc	H	CH_3	H	>50	[66]

TABLE 17.2-2
Inhibitory Activity of c-AMP PDE

No.	R_1	R_2	R_3	R_4	IC_{50} ($\times 10^{-5}$M)	Ref.
100	H	H	H	H	20.1	[67]
101	H	H	Glc	H	40.1	[67]
102	H	H	H	Glc	35.8	[67]
103	H	H	Glx	Glc	23.7	[67]
104	H	Ac	Glc	H	>50	[108]
105	Ac	H	H	H	>50	[67]
106	Ac	H	Glc	Glc	16.5	[67]
107	Ac	Ac	H	H	>50	[67]

TABLE 17.2-3
Inhibitory Activity of c-AMP PDE

No.	R_1	R_2	R_3	IC_{50} ($\times 10^{-5}$M)	Ref.
40	H	H	H	9.8	[67]
37	H	Glc	H	>50	[67]
108	H	Glc	Glc	11.1	[67]
109	OH	H	H	19.5	[67]
110	OH	Glc	H	>50	[67]
111	OH	Glc	Glc	14.3	[67]

TABLE 17.2-4
Inhibitory Activity of c-AMP PDE

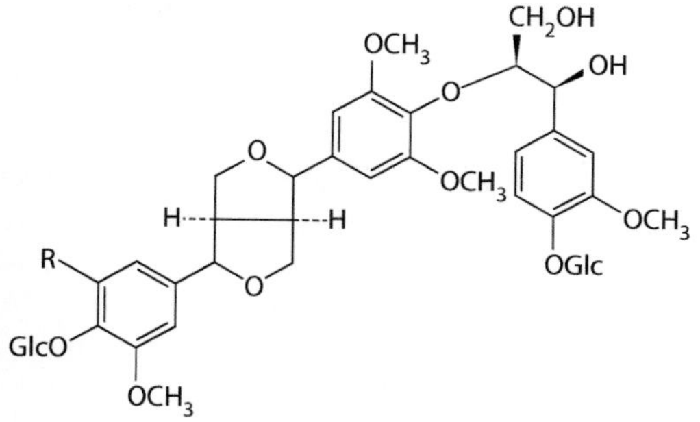

Compound No.	R	IC$_{50}$ ($\times 10^{-5}$M)	Ref.
112	H	>50	[67]
113	OCH3	>50	[67]

Source: Reproduced from Deyama, T., Nishibe, S., Kitagawa, S., Ogihara, Y., Takeda, T., Ohmoto, T., Nikaido, T., and Sankaw, U., *Chem Pharm Bull*, 36, 435–439, 1988. With permission.

TABLE 17.2-5
Inhibitory Activity of c-AMP PDE

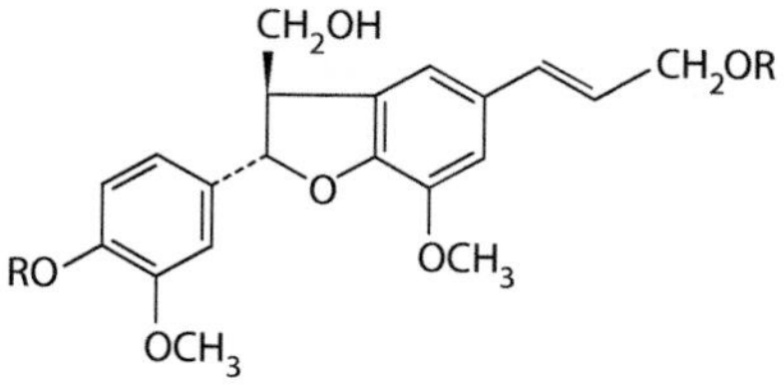

Compound No.	R	IC$_{50}$ ($\times 10^{-5}$M)	Ref.
114	H	>50	[67]
115	Glc	>50	[67]

Source: Reproduced from Deyama, T., Nishibe, S., Kitagawa, S., Ogihara, Y., Takeda, T., Ohmoto, T., Nikaido, T., and Sankaw, U., *Chem Pharm Bull*, 36, 435–439, 1988. With permission.

diaryl-3,7-dioxabicyclo[3,3,0]octane, diarylbutyrolactone and diaryltetrahydrofuran rings the order of inhibitory activity was diglucoside = aglycone > monoglucoside.

(+)-1-Acetoxypinoresinol (**94**) and (+)-1-acetoxypinoresinol 4′,4″-di-*O*–β–D-glucoside (**97**) showed the strongest inhibitory activity, IC$_{50}$ 3.2 × 10^{-5} and 1.1 × 10^{-5}M, respectively [66,67]. The sesquilignans hedyotol C ((guaiacylglycerol-β-medioresinol ether diglucoside (**112**)) and guaiacylglycerol-β-syringaresinol ether diglucoside (**113**) and the neolignans dihydrodehydrodiconiferyl alcohol (**114**) and its 4, γ′-di-*O*–β–D-glucoside (**115**), showed weak inhibitory activity [65]. Pinoresinol lignans were more potent inhibitions than the epipinoresinol type [68].

STRESS REDUCING ACTIVITY

A number of Asian medicinal plants are used as tonics and general restoratives and for enhancing resistance to stress. Early studies on extracts of *Eleutherococcus senticosus* Maxim(Araliaceae) and its component, eleutheroside E[(−)-syringa resinol-di-*O*–β–D-glucoside] (**116**) showed antistress activity [69]. Oral administration of liriodendrin[(+)-syringaresinol-di-*O*–β–D-glucoside] (**23**), a component of *Eucommia ulmoides* Oliv(Eucommiaceae) [70] and of *Acanthopanax senticosus* Harms(*Eleutherococcus senticosus* Maxim) prolonged the exercise time to exhaustion in chronic swimming-stress tests in rats, increased the β-endorphin content in plasma [71], prevented stress-induced decrease in movement and led to accelerated recoveries [72].

116   117

ANTICOMPLEMENTARY ACTIVITY

The complementary system is a major immunity and is activated by a cascade mechanism *via* an antigen–antibody mediated process. The cascade allows for a high amplification rate. The complementary system is normally beneficial for the host, but can also cause adverse effects. Some diseases are related to high complementary activity. Anticomplementary reagents inhibit the activity and prevent the diseases. The anticomplementary effect of nineteen lignans isolated from *Eucommia ulmoides* Oliver [70,73,74–77] are shown in Table 17.3. (+)-Syringaresinol *O*–β–D-glucoside (**87**), (+)-medioresinol *O*–β–D-glucoside(eucommin A) (**84**), and (+)-epipinoresinol (**117**) showed moderate anticomplementary activity. The other lignans exhibited weak activity or none. The lignan glucosides were more active than their corresponding aglycones [78].

TABLE 17.3

Anticomplementary Activity of Lignans from _Eucommia Ulmoides_[a]

No.	Name	Inhibition(%)
	(±)-guaiacylglycerol[b]	19.3 ± 6.6
100	(−)-olivil	4.3 ± 1.9
101	(−)-olivil 4′-O–β–D-glucoside	18.0 ± 5.0
103	(−)-olivil 4′,4″-di-O–β–D-glucoside	18.7 ± 1.5
77	(+)-pinoresinol	3.3 ± 0.9
78	(+)-pinoresinol O–β–D-glucoside	9.0 ± 1.5
79	(+)-pinoresinol di-O–β–D-glucoside	12.0 ± 0.6
86	(+)-syringaresinol	2.7 ± 0.9
87	(+)-syringaresinol O–β–D-glucoside	27.7 ± 4.8
23	(+)-syringaresinol di-O–β–D-glucoside	18.3 ± 2.7
84	(+)-medioresinol 4′-O–β–D-glucoside(eucommin A)	27.7 ± 1.2
85	(+)-medioresinol di-O–β–D-glucoside	13.7 ± 0.7
91	(+)-1-hydroxypinoresinol 4′,4″-di-O–β–D-glucoside	15.0 ± 1.2
117	(+)-epipinorersinol	24.7 ± 6.2
112	guaiacylglycerol-β-medioresinol ether 4″,4‴- di-O–β–D-glucoside (hedytol C 4″,4‴- di-O–β–D-glucoside)	15.0 ± 3.0
	guaiacylglycerol-β-coniferyl aldehyde ether	8.3 ± 1.2
	guaiacylglycerol-β-synapylalcohol ether -4- O–β–D-glucoside (citrusin B)	1.7 ± 1.2
114	dihydrodehydrodiconiferyl alcohol	11.7 ± 1.2

Source: Reproduced from Oshima, Y., Tanaka, S., Hikino, H., Deyama, T., and Kinoshita, G., _J Ethnopharmacol_, 23, 159–164, 1988. With permission.

[a] Data are expressed as mean ±S.E. of three experiments cited from Oshima, Y., Tanaka, S., Hikino, H., Deyama, T., and Kinoshita, G. _J Ethnopharmacol_, 23, 159–164, 1988.

[b] Mixture of erythro- and threo-compounds

CA²⁺ ANTAGONIST ACTIVITY

Ca^{2+} antagonists are drugs that inhibit slow trans-sarcolenmal inward Ca^{2+} current without affecting the Na^+-dependent excitatory process in cardiac and smooth muscle contraction [79]. They are widely used as therapeutic agents for coronary heart diseases and hypertension. Endogeneous digitalis-like factors possess biological properties similar to those of digitalis, such as Na^+, K^+-ATPase inhibition and cross-reaction with antidigitoxin antibodies. Thirty-three compounds from four groups of lignans—dioxabicyclooctanes, butanolides, arylnaphthalenes, tetrahydrofurans and steganes—were tested and 17 compounds exhibited significant inhibition. The potency order was butanolides > dioxabicyclooctanes = steganes.

Trachelogenin (**39**) showed the most potent activity, IC$_{50}$ 1.1×10^{-6} M while arylnaphthalenes showed no significant activity [80]. Several mammalian lignans containing a butyrolactone group—enterolactone (**118**), 3–_O_-methylenterolactone

(119) and prestegane B (120)—inhibited Na$^+$ and K$^+$ pump activity in human blood red cells. The inhibition was much lower than that of oubain, which was used as a reference. Na$^+$ and K$^+$ pump inhibition by lignans did not appear to be competitive with that of oubain, suggesting that they do not act at the digitalis receptor site [81]. Compound 118 inhibited dose-dependent Na$^+$, K$^+$-ATPase activity of human and guinea pig heart [82]. Sixteen mammalian-type lignans were tested for endogeneous digitoxin-like activity, cross-reactivity to antidigitoxin antibody, inhibition of dog kidney Na$^+$, K$^+$-ATPase and oubain displacing activity against [^{3}H] oubain binding to human erythrocytes. Compound 48 showed three types of the activity and may be an endogeneous digitalis-like substance [83]. Dried flower buds of *Magnolia fargesii* Chang (Magnoliaceae) have been used for nasal empyema, allergic rhinitis, sinusitis, and headache. This extract showed a Ca^{2+} antagonist activity [84,85]. Eight Ca^{2+} antagonistic lignans were isolated. Three benzofuran neolignans, fargesone A (121), fargesone B (122), and denudatin B (123) showed high activity. Four dioxabicyclooctadiene lignans, pinoresinol dimethyl ether (82), lirioresinol B dimethyl ether [(+)-yangambin;(+)-syringaresinol dimethyl ether,124], magnolin (125), and fargesin (126) showed lower activity [84]. Compound 23, a lignan component of *Boerhaavia diffusa* L [86] and of *Eucommia ulmoides* Oliv [70], showed a significant Ca^{2+} antagonist effect in frog heart single cells [86].

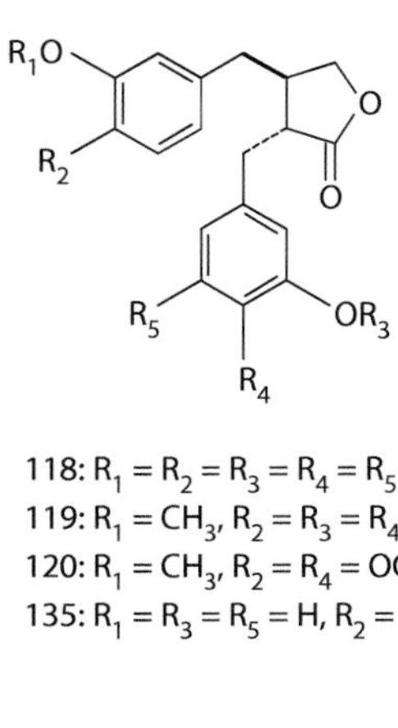

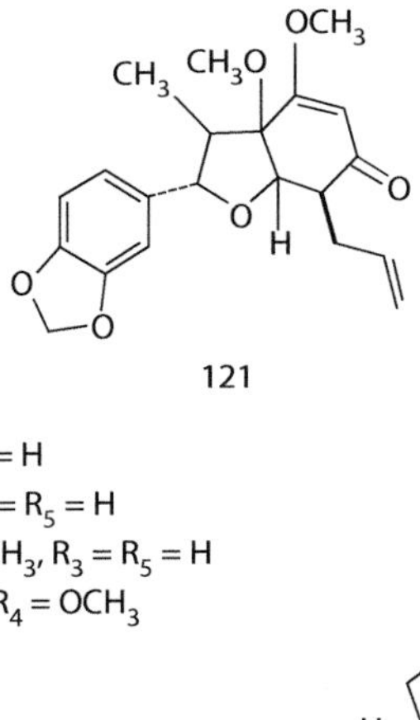

118: $R_1 = R_2 = R_3 = R_4 = R_5 = H$
119: $R_1 = CH_3, R_2 = R_3 = R_4 = R_5 = H$
120: $R_1 = CH_3, R_2 = R_4 = OCH_3, R_3 = R_5 = H$
135: $R_1 = R_3 = R_5 = H, R_2 = R_4 = OCH_3$

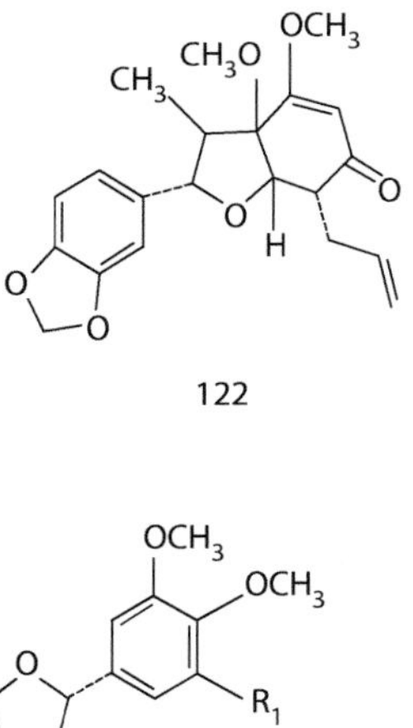

121

122

123

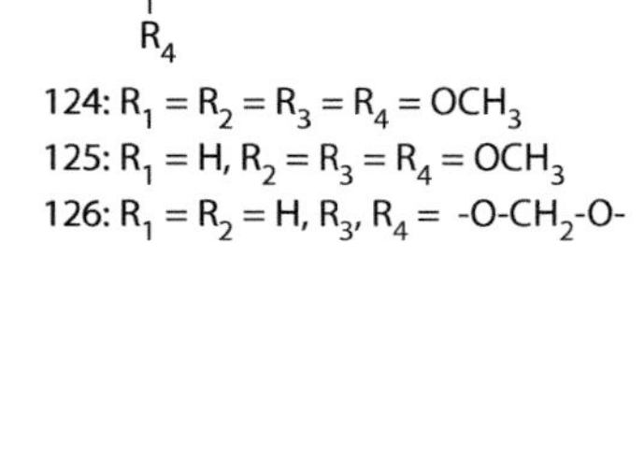

124: $R_1 = R_2 = R_3 = R_4 = OCH_3$
125: $R_1 = H, R_2 = R_3 = R_4 = OCH_3$
126: $R_1 = R_2 = H, R_3, R_4 = -O-CH_2-O-$

ANTIALLERGY EFFECT

The flower buds of *Magnolia salicifolia* (Japanese name Shin-i) have been used for nasal problems in traditional Kampo Medicine. Magnosalicin (**127**), isolated from the extracts, showed potent inhibitory activity on histamin release from rat mast cells. A number of lignans were tested for this inhibitory activity. The dioxabicyclooctane and butanolide lignans were active while tetrahydrofuran lignans were not active. The most active class of lignans showed IC_{50} values less than 20 μM. Compounds **39** and **40** had IC_{50} values of 25 and 18 μM, respectively, and acted as Ca^{2+} antagonists. The inhibitory activity of lignans on histamin release may be closely related to Ca^{2+} antagonist activity [87].

PLATELET ACTIVATING FACTOR ANTAGONIST ACTIVITY

Platelet activating factor (PAF) is a bioactive phospholipid, 1−*O*-hexadecyl or octadecyl -2-acetyl-*syn*-glyceryl-3-phosphorylcholine.

PAF is released from platelets, basophiles, neutrophiles, macropharges, and mast cells when they are stimulated. It causes hypertensive, inflammatory and allergic responses and increases vascular permeability. PAF receptor antagonists are expected to be useful as antiallergic, antiasthma and anti-inflammatory drugs. Reviews of lignans and PAF have been published [4,10,88].

Kadsurenone (128), veraguensin (129), galbelgin (130), galgravin (131), nectandrin A (132) and nectandrin B (133), burseran (134), prestegane A (135) and prestegane B (120) showed inhibitory activity against PAF-induced rabbit platelet aggregation [89]. The fruit of *Forsythia suspense* (Oleaceae) and the seeds of *Arctium lappa* (Compositae) exhibited significant anti-PAF activity. Their active components were dioxabicyclooctane and butanolide lignans. In dioxabicyclooctane lignans, the presence of 3,4-dimethoxyphenyl or 3,4,5-trimethoxyphenyl groups was essential for high PAF antagonistic activity, while hydrophilic groups(hydroxyl and glucosyl) lowered the activity [90]. Two butanolide lignans, arctigenin (38) and trachelogenin (39), inhibited the PAF receptor. On the other hand, their glucosides, arctiin (36) and tracheloside, and the related glucosides matairesinoside (37) and nor-tracheloside (110) showed no inhibitory effect [91]. Compound 124 showed specific inhibitory activity against PAF-induced platelet aggregation in platelet-rich plasma (PRP) [92], and was sensitive to antigen and PAF-induced pleural neutraphil and eosinophil [93]. (+)-Epiyangambin (136) dose-dependently inhibited PAF-induced platelet aggregation with an IC_{50} value of 6.1×10^{-7} M without modifying the amplitude of the maximal response, but had no effect on the platelet aggregation induced by collagen, thrombin or ADP. Thus compound 136 was a potent selective antagonist of PAF [94]. Cinnamophilin (137), its *O,O*-dimethyl ether(1b) and *O,O*-diacetyl derivatives showed significant antiplatelet aggregation activity. The IC_{50} values of 137 toward arachidoic acid (100 μM/*ml*) and collagen-induced (10 μM/*ml*) platelet aggregation were 16 and 100 μM, respectively [95]. Compound 16, neojusticin A (14), taiwanin E (138), and taiwanin E methyl ether showed significant antiplatelet aggregation activity with IC_{50} values of 1.1–8.0 μM. This activity is less than that of indometacin but more than aspirin (IC_{50} values, 20.3 μM) [96]. The synthetic bisphenyltetrahydro furan lignan, *trans*-2,5-bis(3,4,5-trimethoxyphenyl) tetrahydrofuran (L-652,731, 139), was a potent and orally active PAF-specific and competitive receptor antagonist. It inhibited [^{3}H] PAF binding to a receptor site on rabbit platelet membranes, and was more potent than CV-3998, 128 and ginkgolide B as a reported antagonist [97].

136   137   14

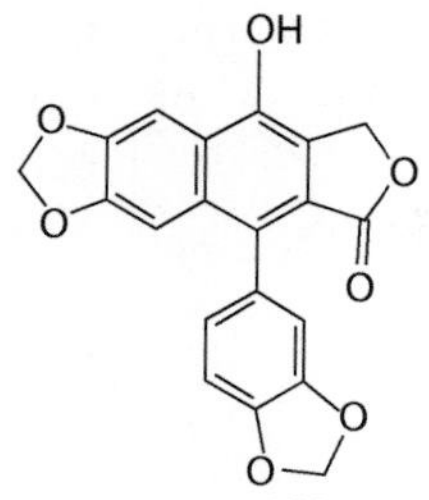

138

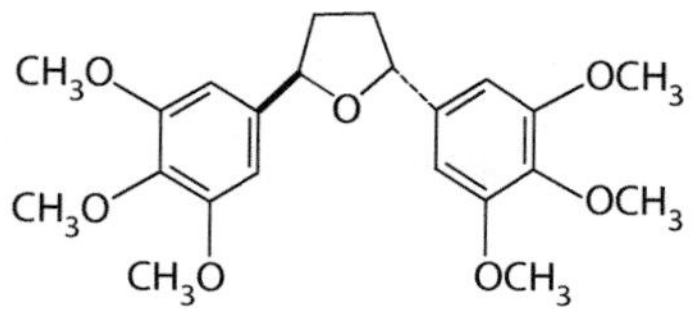

139

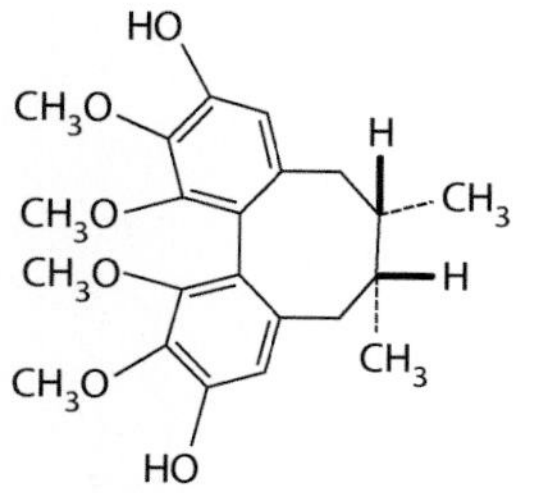

140

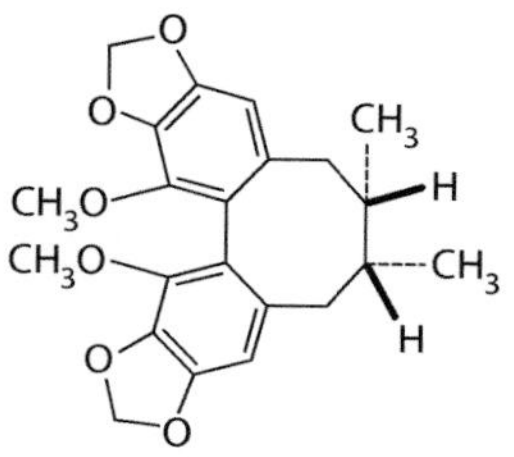

141

ANTI-INFLAMMATORY ACTIVITY

Carragenin-induced paw edema and 12-O-tetradecanoylphorbol 13-acetate (TPA)-induced edema have been used as experimental models of acute inflammation. Gomisin A (**49**), gomisin J (**140**), and wuweizisu C (**141**) inhibited inflammation induced by TPA in mice [44]. Diphyllin acetyl apioside (**142**) and diphyllin apioside (tuberculatin) (**143**) also showed an anti-inflammatory effect with ID_{50} values of 0.27 and 1.23 μM/ ear, respectively, in rabbit [98].

TNF-α INHIBITORY ACTIVITY

Tumor necrosis factor (TNF)-α is a preinflammatory cytokine that is produced by activated macrophages. TNF-α enhances the production of other cytokines by autocrine stimulation and induces the production of prostaglandin E_2 (PG E_2) by synovial fibroblast-type cells and of prostaglandin I_2 (PG I_2) by endothelial cells. Compound **77**, lariciresinol glycoside (**144**), woorenoside IV (**145**) and woorenoside V (**146**) significantly inhibit TNF-α production. The inhibitory activity of **77** is more potent than that of c-AMP PDE inhibitors (theophylline, pentoxifyllin), but less than steroid drugs (prenisolone, dexamethasone, hydrocortisone). These lignans may partly participate in the antiallergic and anti-inflammatory effect of *Coptis japonica* though inhibition of TNF-α production [99,100].

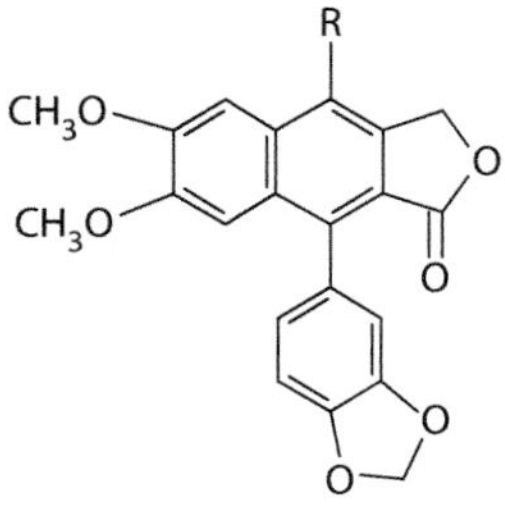

142: R = acetylapiosyl
143: R = apiosyl

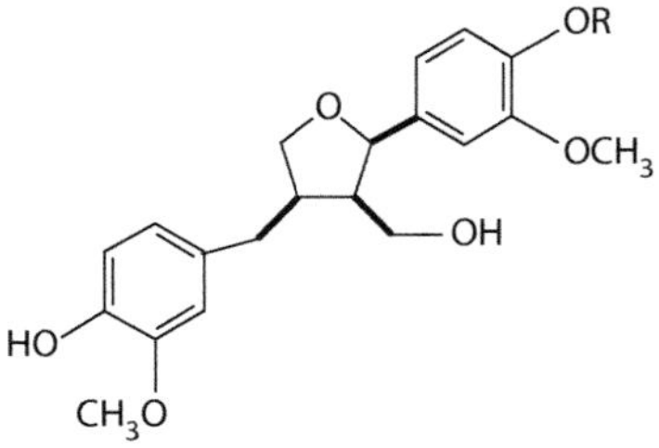

144: R = glc

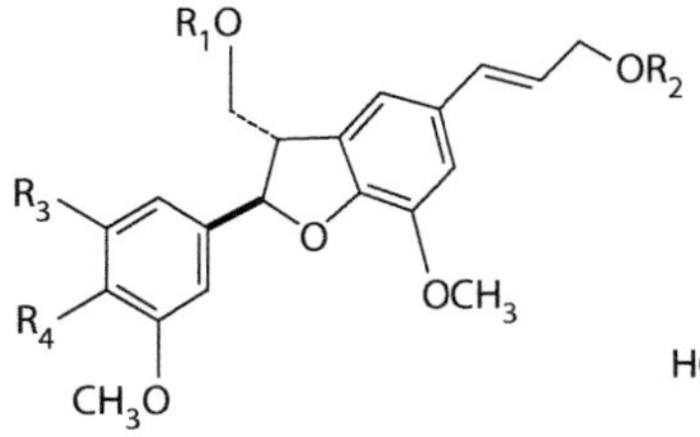

145: R_1 = Ac, R_2 = glc^{6-}A, R_3 = R_4 = OCH$_3$
146: R_1 = H, R_2 = glc^{6-}A, R_3 = H, R_4 = OH

147

$$A = -CO - \overset{\overset{\displaystyle CH_2}{||}}{C} - CH_2 - CH_2OH$$

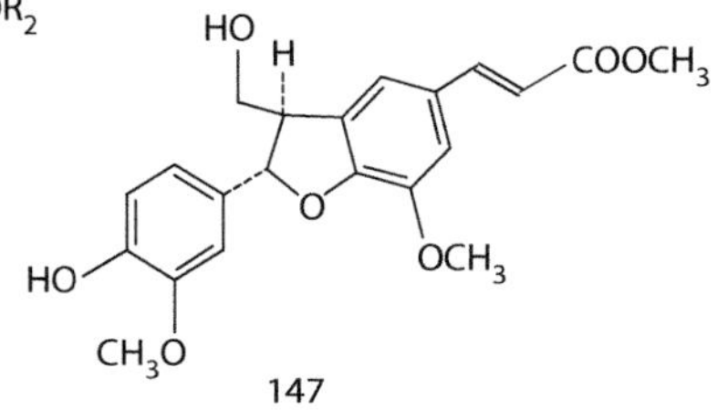

PG I$_2$ INDUCTIVE ACTIVITY

PG I$_2$ is a potent natural inhibitor of platelet aggregation and a powerful vasodilater, and is therefore expected to have potent clinical effects on the cardiovascular system. A neolignan (147), isolated from *Zizyphus jujuba* Mill(Rhammaceae), increased the release of endogeneous PG I$_2$ from rat aorta [101].

RELAXATION EFFECT

Mammalian lignan 48 inhibited the high KCl- and CaCl$_2$-induced contraction in partially depolarized muscle strips in noradrenaline- and angiotensin II-induced contraction, and seemed to induce the relaxation of vascular smooth muscle by inhibition of Ca^{2+} entry. However, compound 48 did not inhibit release of Ca^{2+} from intracellular stores [102]. Compounds 38–40 and 109 relaxed the histamine-induced contraction of tracheal muscle in guinea pig [91]. Todopan Puok (*Fargraceae racemosa* Jack ex Wall) (Loganiceae) is a medicinal plant in Borneo that contains (+)-pinoresinol (77) and (+)-epipinoresinol (117). These compounds relaxed norepinepherine-induced contraction in rat aortic strips without endothelium [103].

ANALGESIC EFFECT

Compound **77,** isolated from *Eucommia ulmoides* Oliv [71] and from *Fagracea racemosa* Jack ex Wall, showed a dose-dependent inhibition of acetic acid-induced writhing in mice [101]. A dihydrofuran lignan (**148**) and an aryl tetrahydronaphthalene lignan (**149**) isolated from *Stauntonia chinensis* (Lardizabalaceae) showed an analgesic effect in mice [104,105]. Dihydrodehydrodiconiferyl alcohol-β–D–(2′–O–hydroxybenzoyl) glucoside (**150**) showed writhing inhibition following oral administration at a dose of 50 mg/kg [106].

HYPOTENSIVE EFFECT

Compounds **39, 78,** and **79** showed a hypotensive effect in spontaneously hypertensive rat (SHR) [80,107,108]. Compound **43** showed an antihypertensive effect in deoxycorticosterone acetate (DOCA)-salt hypertensive rat. In this model, a diet containing **43** markedly suppressed the increase of blood pressure [109]. Compound **43** showed a weak inhibitory activity on angiotensin converting enzyme (ACE) [110]. Graminone B (**151**) inhibited the contractile response of rabbit isolated aorta [111].

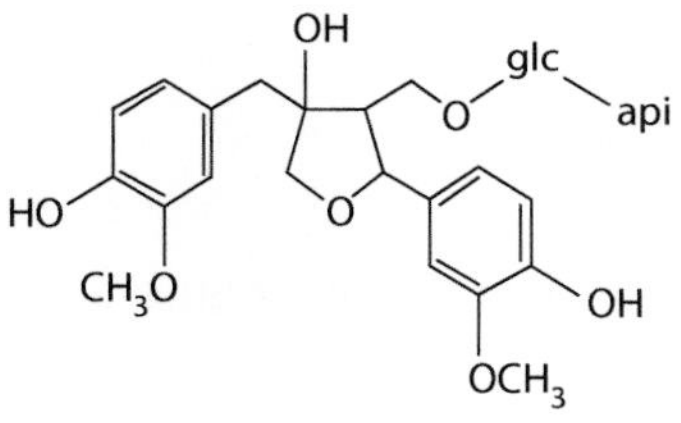

148

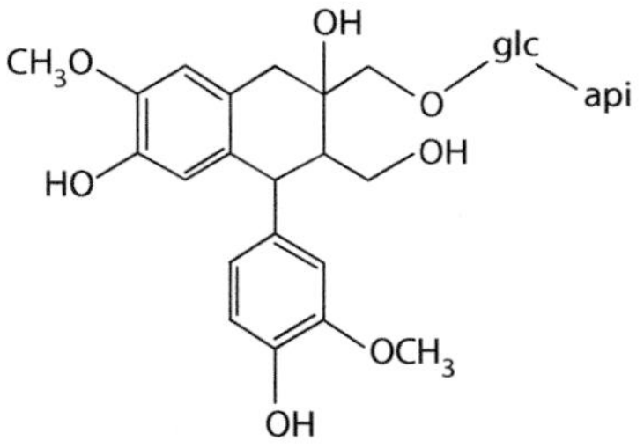

149

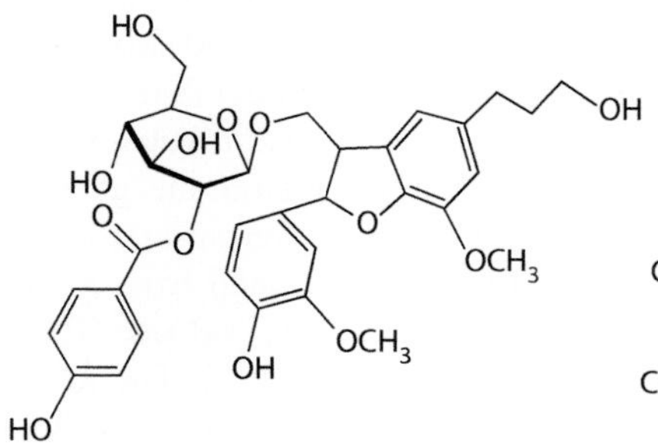

150

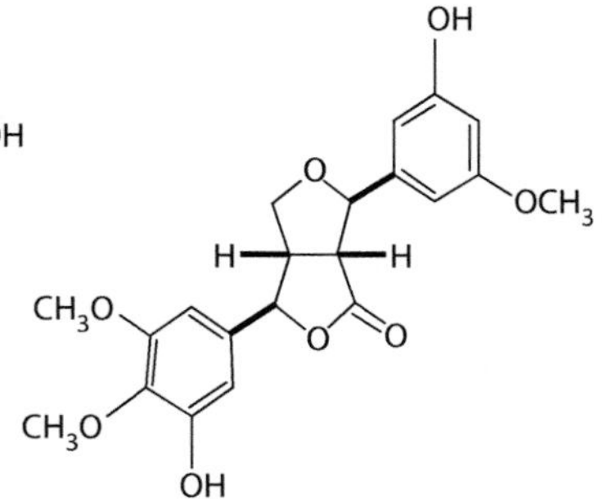

151

5-LIPOXYGENASE INHIBITORY ACTIVITY

Justicidin E (**15**) selectively inhibited 5-lipoxygenase (LO), but did not inhibit the activity of human15-LO nor porcine 12-LO [112]. Naphthalenic lignan lactones were synthesized and tested for inhibition of arachidoic acid oxidation by 5-LO and LTB4 formation in human peripheral blood polymorphonuclear leukocyte and human whole blood. Compounds **152** and **153** showed potent inhibitory activity [113,114].

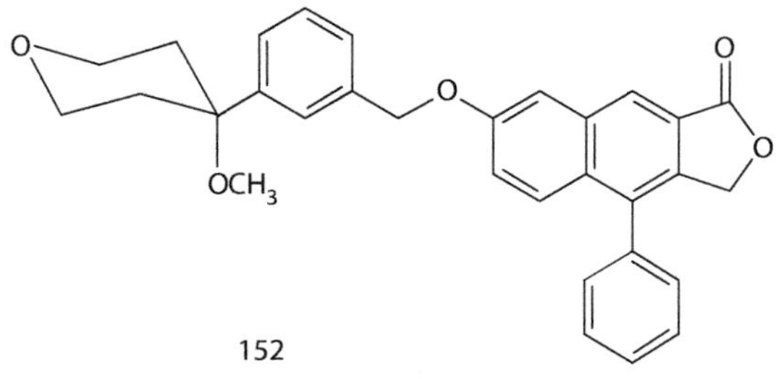

152

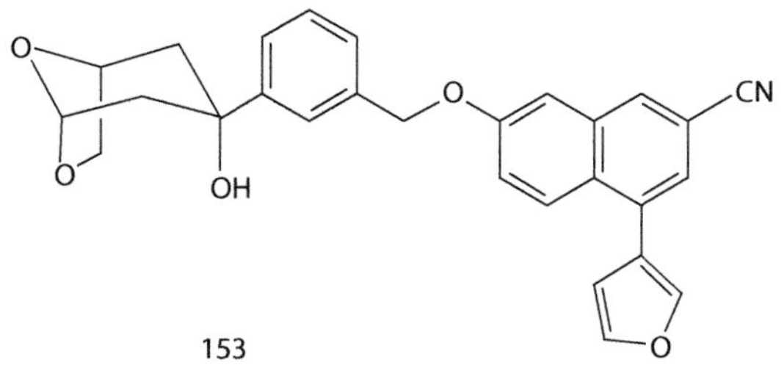

153

ANTIOXIDANT ACTIVITY

A number of phenolic compounds exhibit antioxidant activity. Lignans also show such activity, and foods rich in lignans appear to be effective in preventing diseases linked to active oxygen species [8]. Compound **43**, a main component of sesame lignans, reduced the concentration of lipoperoxide in plasma, liver and tumors as thiobarbituriic acids (TBA) reacting substance [115]. Sesaminol (**154**) was formed in high concentration during purification of unroasted sesame oil and showed a strong antioxidant activity [116]. Compound **154** and sesamol (**155**) exhibited strong anti-oxidant effects in the autoxidation of linoleic acid [117]. Compound **154** significantly inhibited peroxidation of the rabbit erythrocyte membrane ghost system and lipid peroxidation in the liver of CCl_4-administered senescence accelerated mouse (SAM) [118]. Five lignan glucosides (**156–160**) obtained from germinated sesame seeds showed a less potent antioxidant effect (ID_{50} values, 20–30 μM) than α-tocopherol [119], and scavenged hydroxyl radicals [120]. The antioxidant pinoresinol glucosides (**161–163**) were isolated from sesame seeds. Compounds **161** and **162** showed almost the same antioxidant activity as α-tocopherol at a concentration of 100 μM in the erythrocyte membrane ghost system. Compounds **161–163** can produce the antioxidant **77** via hydrolysis by β-glucosidase from intestinal bacteria [121,122]. Similarly, sesaminol glucosides, isolated from sesame seeds, can be hydrolyzed to the strong antioxidant **154** by intestinal β-glucosidase or bacteria [123].

(a)

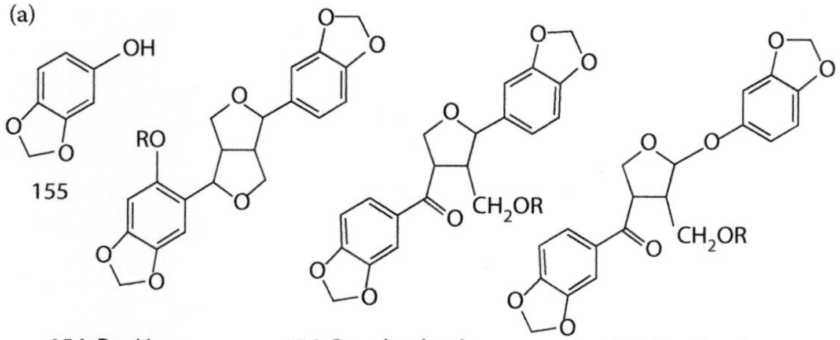

155

154: R = H
157: R = glc-glc-glc
159: R = glc-glc

156: R = glc-glc-glc
158: R = glc-glc

160: R = glc-glc

(b)

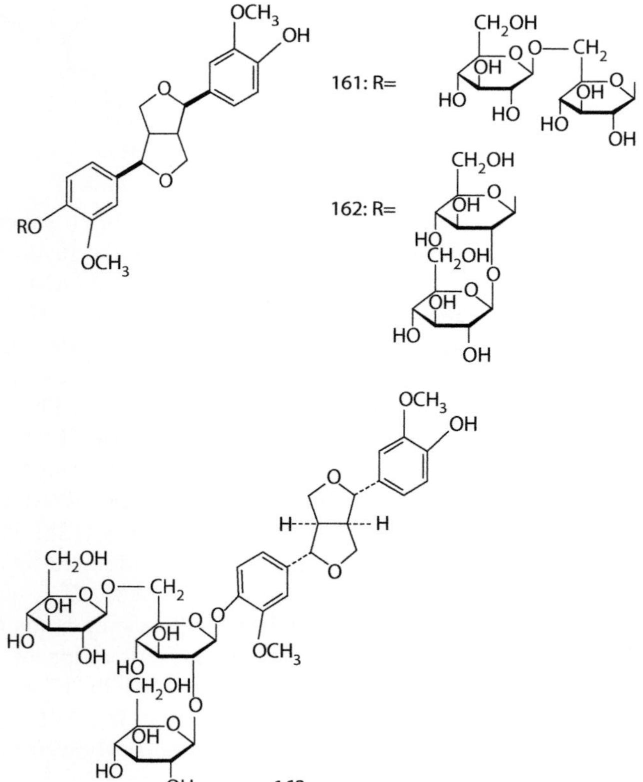

161: R=

162: R=

163

The naturally occurring benzofuran lignan schizotenuin A (**164**) was isolated from *Schizonepeta tenuifolia* Briq. Compound **164** and the synthesized benzofuran lignans **165** and **166** inhibited lipid peroxidation in rat brain homogenate and in rat liver microsomes. The IC_{50} values for **164–166**, 36.3, 3.7, and 4.5 μM, respectively, were more potent than α-tochopherol, IC_{50} value 976 μM. Benzofuran lignans synthesized from methyl ferulate also showed antioxidant activity [124].

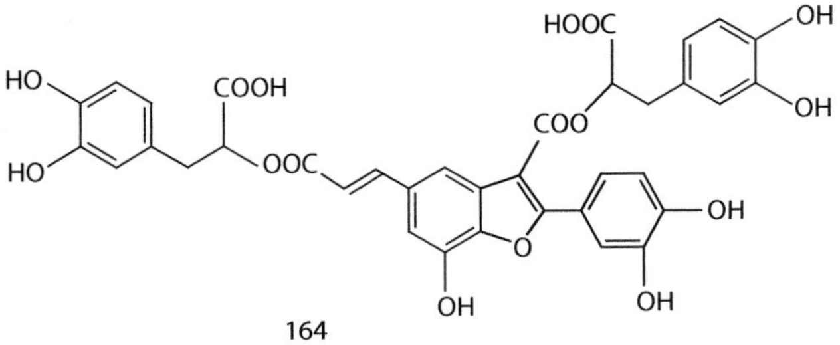

164

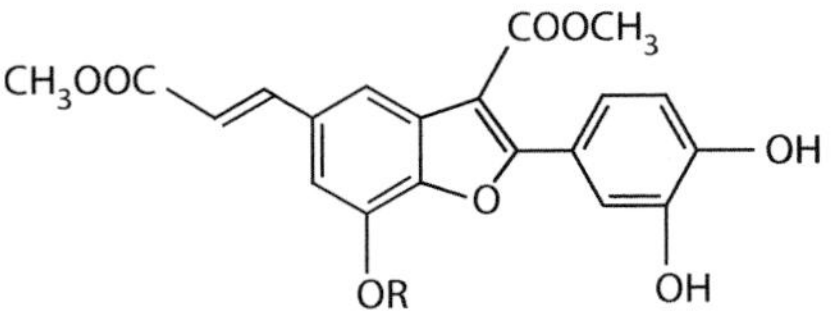

165: R = CH$_3$
166: R = H

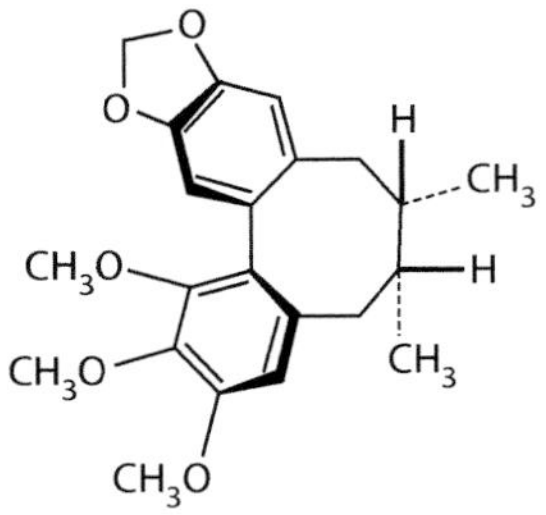

167

(–)-Arctigenin (**38**) and (–)-nortrachelogenin (**109**) inhibited superoxide production [91]. (+)-Lariciresinol (**9**) scavenged 50% diphenylpicrylhydrazyl (DPPH) radical at a concentration of 57.1 µM and showed superoxide dismutase (SOD) like activity with an EC$_{50}$ value of 1.89 µM [125]. Gomisin N (**167**) has antioxidant activity [126]. Compounds **49** and **141** significantly inhibited lipid peroxidation due to ADP/Fe^{3+} or ascorbate/Fe^{2+}-induced CCl$_4$ [127]. Gomisin C (**168**) reduced formation of superoxide by the peptide formyl-Met-Leu-Phe (FMLP) and phorbol myristate acetate (PMA)- and NaF-induced superoxide formation and O$_2$ consumption in rat peripheral nutrophils in vitro. It can be attributed to inhibition of the activity of nicotinamide adenine dinucleotide phosphate (NADPH) oxidase and to reduced release of Ca^{2+} from intracellular Ca^{2+} stores [128]. Heteroclitins A-G (**169–175**), kadsurin (**176**) and interiorin (**177**) also inhibited lipid peroxidation in rat liver homogenate stimulated by Fe^{2+}-ascorbic acid, CCl$_4$-reduced form of NADPH and adenosine 5′-diphosphate ADP-NADPH. Compounds **169** and **172** showed more potent inhibitory activity [129]. Preadministration of **176** caused significant recovery of the SOD activity that is reduced by CCl$_4$-intoxication [130]. The principal lignan of flax, *Linum usitatissimum*, is (–)-secoisolariciresinol diglucoside (**178**). Both of **178** and its (+)-enatiomer showed antioxidative activity [131].

LIVER PROTECTIVE EFFECT

CCl$_4$ or galactosamine elevated the activities of glutamic oxaloacetic transaminase (GOT), glutamic pyruvic transaminase (GPT), lactate dehydrogenase

(LDH), total bile, and total cholesterol in the serum. Pretreatment with **49** markedly reduced the activities. Some lignans from Schizandra fruits, dimethylgomisin J (**179**),(+)-deoxyschizandrin (**180**), schizandrin (**181**), and deoxygomisin A (**182**), showed the same activities with the order of potency **49** >> **179** > **180** > 1 **81** = **182** [132].

(a)

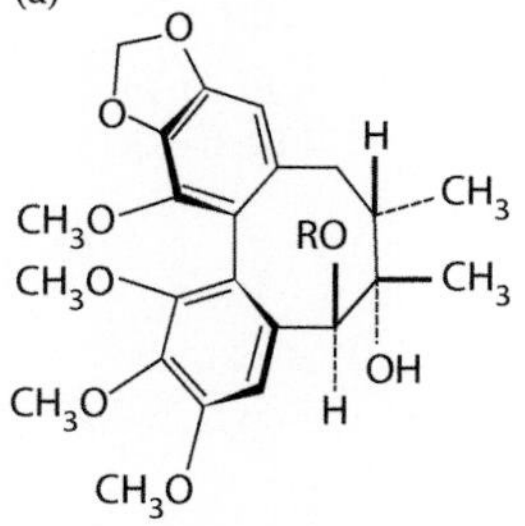

168: R= <chem>benzene-CO—</chem>

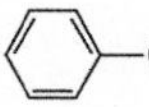

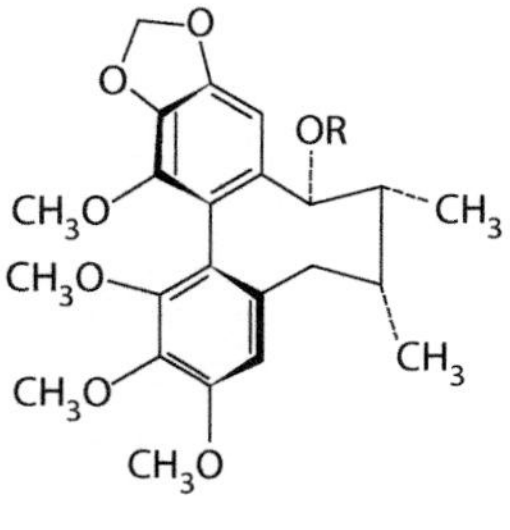

169: R= $-COCH(CH_3)-CH_2-CH_3$

170: R= <chem>-OC\ /CH_3 ; C=C ; H_3C/ \H</chem>

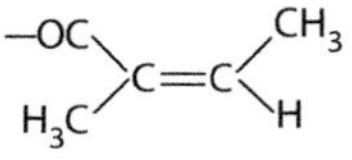

171: R= <chem>-OC\ /H ; C=C ; H_3C/ \CH_3</chem>

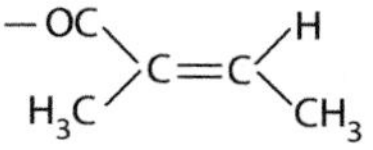

176: R= $-COCH_3$

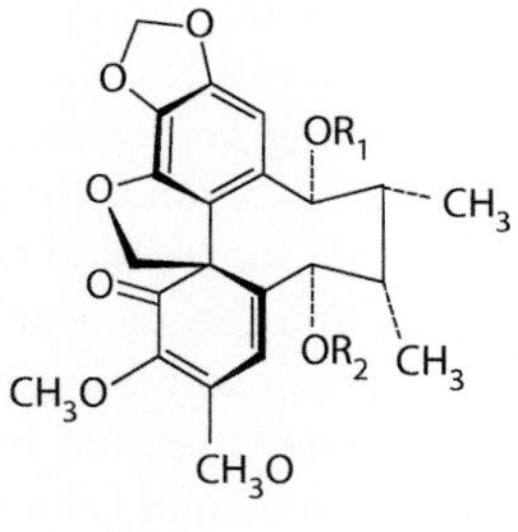

172: R$_1$= <chem>-OC\ /CH_3 ; C=C ; H_3C/ \H</chem> , R$_2$ = H

173: R$_1$= <chem>-OC\ /CH_3 ; C=C ; H_3C/ \H</chem> , R$_2$ = OH

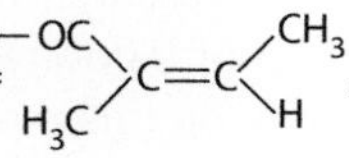

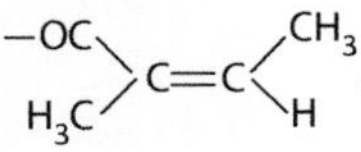

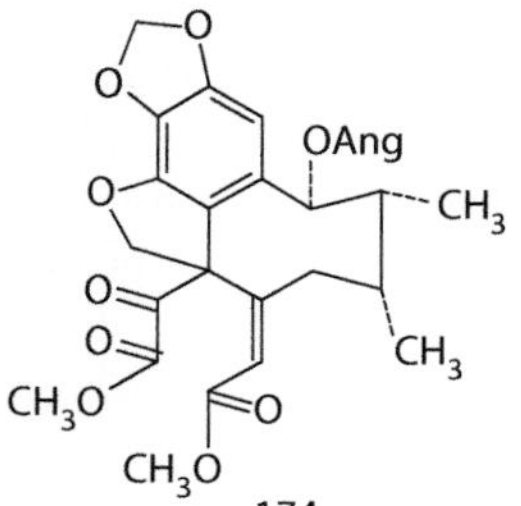

174

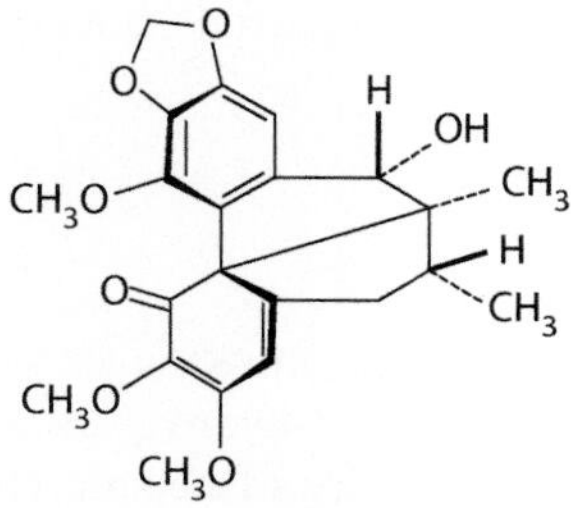

175

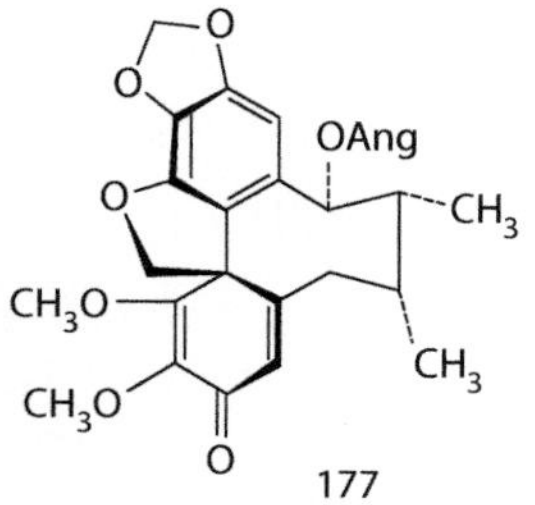

177

(b)

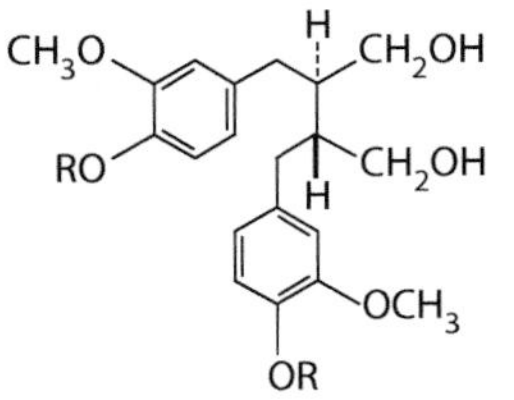

47: R = H
178: R = glc

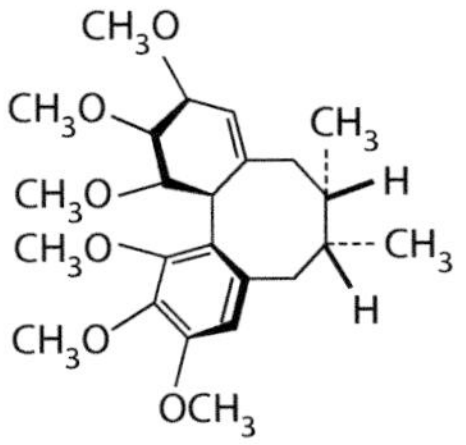

179

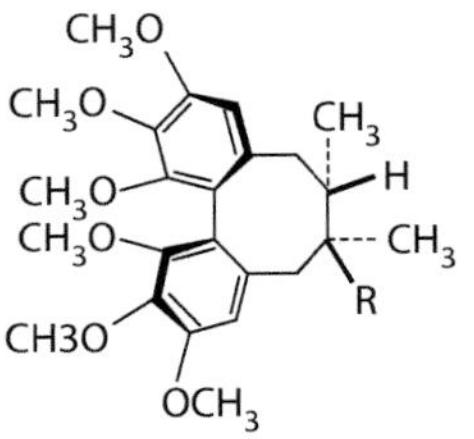

180: R = H
181: R = OH

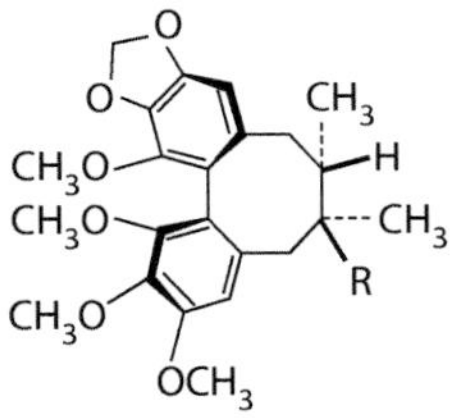

49: R=OH
182: R=H

Compound **49** showed the following liver function-facilitating properties in normal and liver-injured rats [133]: antihepatotoxic, hypolipidemic, and liver protein synthesis-facilitating actions by oral application [134,135], suppression of fibrosis proliferation and acceleration of both liver regeneration and the recovery of liver functions after partial hepatoctomy [136–138]. The acute hepatic failure model— intravenous injection of heat-killed *Propionibacterium acnes* followed by a small amount of lipopolysaccharide seven days later—is similar to human fulminant hepatitis in severity and is considered an effective model in research on the therapy for fulminant hepatitis. Administration of food containing 0.06% **49** remarkably improved the survival rate and serum transaminase levels on this model [139], and on the immunological liver injury model [140,141]. Compound **49** dose-dependently increased the bile flow and enlarged the liver [142].

Sesame oil, which contains a relatively large quantity of **43** and related lignans, has also been used for improvement of health and longevity. Compound **43** showed a significant protective effect against the accumulation of fat droplets and an ability to improve liver functions [143].

HYPOLIPIDEMIC ACTIVITY

A diet containing 0.5% **43** significantly reduced the concentration of serum and liver cholesterol, and the activity of liver microsomal 3-hydroxy 3-methylglutaryl coenzyme A reductase, but did not affect the activity of hepatic cholesterol 7-α-hydroxylase [144]. Hypercholesterolemia patients received 3.6 g of **43**, three times per day. After eight weeks, serum total cholesterol (TC), triglyceride (TG), high-density lipoproteins-cholesterol (HDL-C), apoproteins and lipoproteins were markedly lowered [145]. A series of arylnaphthalene lignans were synthesized and evaluated for hypolipidemic activity. Compound **183** lowered the level of serum cholesterol and elevated HDL-C in rat [146]. The 2-pyridylmethyl derivatives of **183** exhibited more activity than **183** [147].

Obovatol (**184**), **25** and **26** inhibited rat liver cholesterol acyltransferase with IC_{50} values of 42, 86, and 71 μM, respectively [148].

METABOLISM OF LIGNANS AND THEIR PHYTO-ESTROGENIC ACTIVITY

METABOLISM OF LIGNAN

Components of orally administered herbal medicines are often converted to pharmacologically active compounds by intestinal flora. Some lignans in plants and natural medicines are called phytoestrogens, because they are transformed by intestinal microflora to enterodiol (**185**) and enterolactone (**118**), which show estrogen-like biological activity [149–152]. The metabolism by intestinal bacteria of the following lignan compounds and their glycosides to **185** and **118** has been confirmed by GC/MS and HPLC/MS analysis: pinoresinol (**77**), syringaresinol (**9**), lariciresinol (**8**), isolariciresinol (**186**), secoisolariciresinol (**47**), matairesinol (**40**), arctigenin (**4**), phillygenin (**187**) [153–156]. It is interesting that (+)-pinoresinol 4′,4″-di-O–β–D-glucoside (**79**) was converted to (−)-enterolactone (**118**) [156]. During metabolism, an *Enterococcus faecalis* strain, tentatively named PDG-1, transformed (+)-pinoresinol (**77**) to (+)-lariciresinol (**9**) *via* intermediate metabolites. These were isolated, their structures elucidated and the total pathway shown in Figure 17.1 was proposed [157]. (-)-Olivil 4′,4″-di-O–β–D-glucoside (**103**) and (+)-1-hydroxypinoresinol 4′,4″-di-O-β-D-glucoside (**91**) were transformed to (−)-enterolactone (**118**) by human intestinal bacteria via 2-hydroxyenterodiol (**188**). The structure of this newly isolated compound was elucidated by spectral analysis. (+)-Sesamin (**43**) was also converted to **118** in the same way, showing that the methylene dioxy ring was cleaved by intestinal bacteria [158].

Nose et al. have reported the result of incubating arctiin (**36**) and tracheloside (**189**) with rat intestinal flora. Arctiin (**36**) was first converted to arctigenin (**38**) and subsequently metabolized to 2-(3″,4″-dihydroxybenzyl)-3-(3′,4′-dimethoxy-benzyl)-butyrolactone (**190**). Similarly, tracheloside (**189**) was converted to trachelogenin (**39**) and then metabolized to 2-(3″,4″-dihydroxybenzyl)-3-(3′,4′-dimethoxybenzyl)-2-hydroxybutyrolactone (**191**) [159,160]. After oral administration of **36**, **38** appeared as a metabolite in the serum, but **190** did not. It was considered that **190** was rapidly converted to **38** by catechol-O-transferase in the liver. Similarly, after oral administration of **189**, **39** appeared in the serum but **191** was not detected [159,160]. Apart from those, we found that arctiin (**36**) and tracheloside (**189**) were stable to incubation with rat gastric juice (pH 1.2–1.5) and were metabolized to enterolactone (**118**) by the incubation with human and rat intestinal microflora [158]. It seems that the metabolizing activity of the used intestinal bacteria may be different. In our experiments, the intestinal microflora of human or rats often showed different metabolizing activities. Compound **38** was metabolized to a significant extent to **190** in the cell line expressing rat P450 2B1 [161].

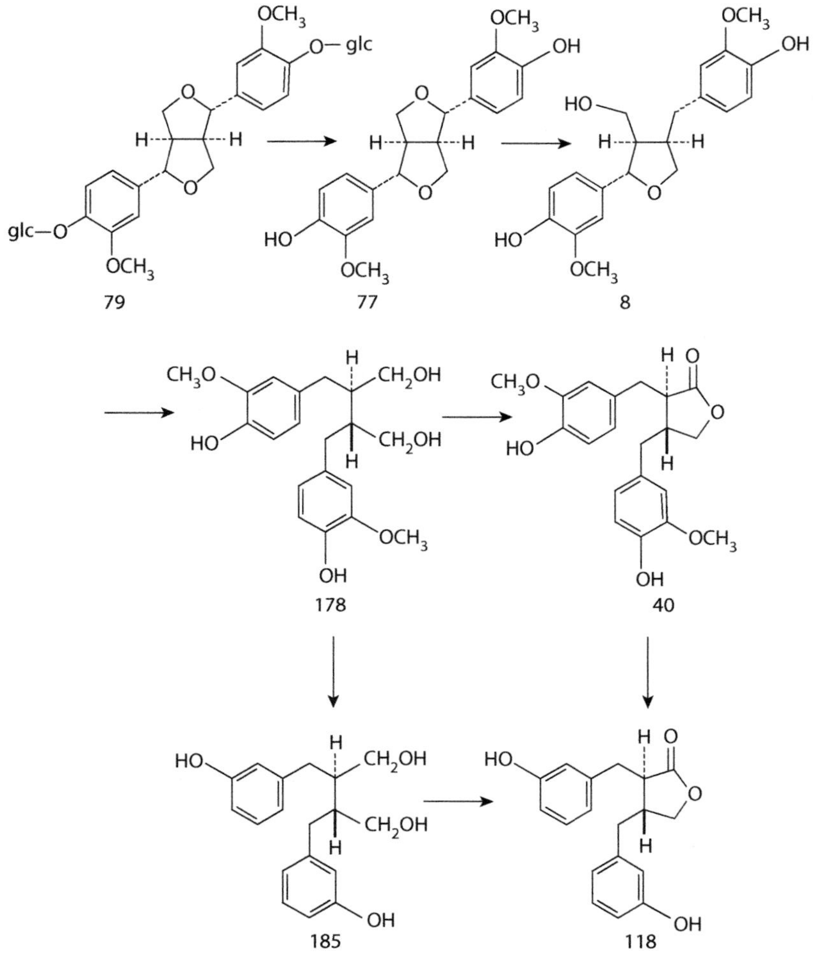

FIGURE 17.1 Proposed metabolic pathway of (+)-pinoresinol diglucoside to enterolactone.

PHYTO-ESTROGENIC ACTIVITY

The important mammalian lignans enterolactone (**118**) and enterodiol (**185**) are present in the serum, urine, bile and seminal fluid, and have a potential protective activity against cancer. Fecal organisms produced **118** and **185** from matairesinol (**40**) and secoisolariciresinol (**47**), which is the major lignan of flax [162]. The effect of diet on urinary excretion of lignans in the adult male chimpanzee has been studied. Compound **118** was the predominate metabolite of the regular diet, while **185** was predominate in a carbohydrate-rich diet [163]. The glucuronized conjugates of **118** and **185** were detected in human urine. The urinary excretion of the dibenzylbutyrolactone increased significantly during the mid-lutal phase of the menstrual cycle and showed a marked increase in early pregnancy in humans as well as in vervet monkeys [164].

Stimulating Activity of Synthesis of Sex Hormone Binding Globulin

It has been reported that vegetarian women excrete more lignans than nonvegetarian women. The lowest mean excretion was found in breast cancer patients. The lignans, formed in the intestinal tract from precursors present in fiber-rich food, stimulate synthesis of sex hormone binding globulin (SHBG) and cause a reduction in hormone bioavailability. Lignans may protect against certain hormone-dependent cancers (breast, endometrium, and prostate).

Compounds **40** and **47** were precursors of the mammalian lignans, **185** and **118**, respectively [161,165]. When healthy young men ate whole wheat and flaxseed bread for six weeks, the content of **185** and **118** in the urine was 7–28 times that of control subjects. Urinary excretion of **118** was 5–10 fold higher than for **174**. No significant change of total testosterone, free testosterone or SHBG in plasma was observed [166]. *Urtica dioica* root, which contains **47**, was successfully applied in the treatment of early stage of benign prostatic hyperplasia. Compound **47** interfered with SHBG [167]. Other lignans of *U. dioica*, **114**, isolariciresinol (**186**), 3,4-divanillyltetrahydrofuran (**192**) and (+)-neoolivil (**193**), showed binding affinity to SHBG. [168].

189: $R_1 = OH, R_2 = R_4 = CH_3, R_3 = glc$
190: $R_1 = R_2 = R_3 = H, R_4 = CH_3$
191: $R_1 = OH, R_2 = R_3 = H, R_4 = CH_3$

185: $R_1 = R_3 = R_5 = OH, R_2 = R_4 = R_6 = H$
188: $R_1 = R_2 = R_3 = R_5 = OH, R_4 = R_6 = H$
194: $R_1 = R_2 = H, R_3 = R_4 = R_5 = R_6 = OH$
198: $R_1 = R_3 = R_4 = R_6 = OH, R_2 = H, R_5 = OCH_3$
199: $R_1 = R_4 = R_6 = OH, R_2 = R_3 = H, R_5 = OCH_3$
200: $R_1 = R_3 = R_4 = R_5 = R_6 = OH, R_2 = H$

Nordihydroguaiaretic acid (**194**) interacts dose-dependently with steroid binding proteins (SBP) by reducing significantly the number of binding sites without changing the association constant (Ka) for estradiol and teststerone. Compound **194** induces conformational changes in steroid binding proteins and causes a reduction or loss of immunorecognition of SBP by anti-SBP antibodies. SBP can carry phytoestrogens into target cells, where they may compete with endogeneous estrogens for receptor sites and interfere with estrogen-mediated processes.

The phytoestrogens could inhibit aromatase activity, and so decrease intracellular estrogen production [169]. Compound **192** showed the highest SHBG binding affinity. In lignans, the 8-8' coupled structure seems to be a prerequisite for an effective binding to SHBG. The following structure-activity relationships between were shown: 1) (±)- Diastereoisomers are more active than meso compounds, 2) The 4-hydroxy-3-methoxy(guaiacyl) substitution pattern in the aromatic part is most effective. The activity increases with the declining porlarity of the aliphatic part of the molecule [170].

AROMATASE (ESTROGEN SYNTHETASE) INHIBITORY ACTIVITY

Aromatase, human estrogen synthetase, catalyzes the conversion of androgens to estrogens in many tissues. The inhibition of human preadipocyte aromatase activity by lignans suggests a mechanism by which consumption of lignan-rich plant foods may contribute to reduction of estrogen-dependent diseases, such as breast cancer. Compounds **118**, **194**, 4,4'-dihydroxyenterolactone (**195**), and 4,4'-dideoxyenterolactone (**196**) inhibited human placental aromatase. Compounds **195** and **196** are theoretical metabolites of **40**, which is most likely a plant precursor of **118** [171]. Compound **118** and its theroretical precursors, 3'-demethoxy-3-*O*-demethyl matairesinol (**197**) and **196** decreased aromatase activity. A smaller decrease was observed with **185** and its theoretical precursors, *O*-demethylsecoisolariciresinol (**198**), demethoxysecoisolariciresinol (**199**) and didemethylsecoisolariciresinol (**200**) [172].

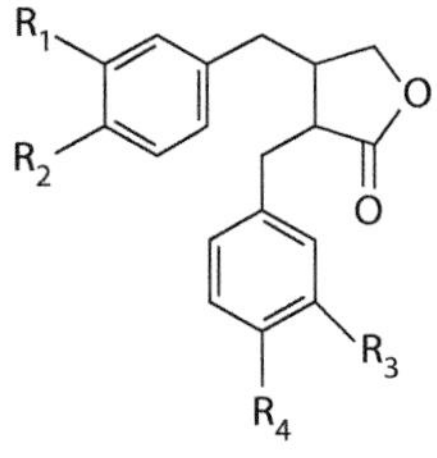

195: $R_1 = R_2 = R_3 = R_4 = OH$
196: $R_1 = R_3 = H, R_2 = R_4 = OH$
197: $R_1 = R_2 = R_4 = OH, R_3 = H$

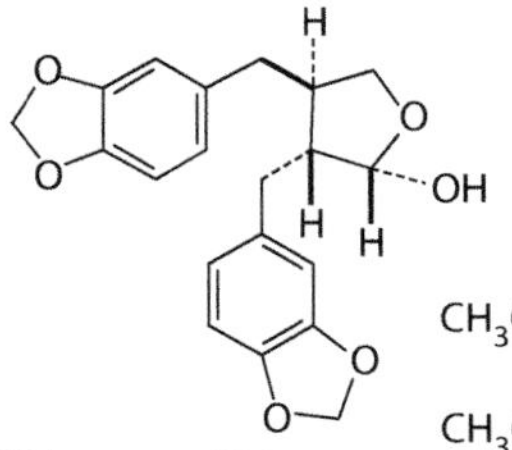

201

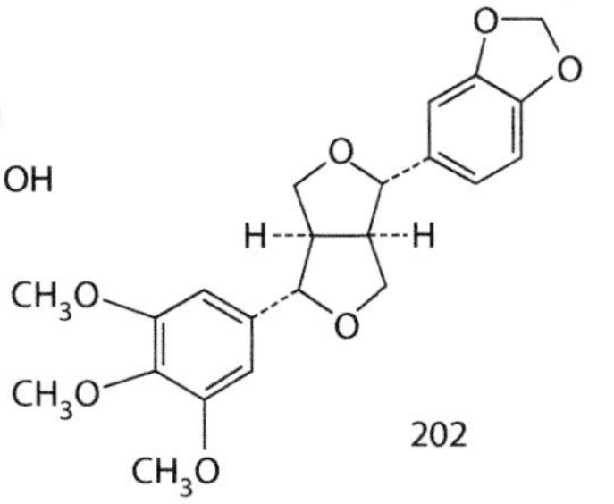

202

ANTI–HUMAN IMMUNODEFICIENCY VIRUS ACTIVITY

(–)-Arctigenin (**38**), (–)-tracheloside (**189**) and etoposide (VP-16-213, **65**) showed a strong cyto-protective effect against human immunodeficiency virus type-1(HIV-1) in vitro, and efficiently inhibited both cellular topoisomerase II activity and the

HIV-1 integrase reaction. The retro-viral effect displayed by these lignans is mainly due to an inhibition of integration of proviral DNA into the cellular DNA genome rather than inhibition of topoisomerase II. No activity was observed from the structurally related lignans arctiin (**36**), cubebin (**201**), podophyllotoxin (**1**), (+)-sesamin (**43**), aschantin (**202**), 2,4-dihydroxybenzyl-2-(4-hydroxy-3-methoxybenzyl)-3-(3,4-methoxybenzyl) butyramide, picropodophyllotoxin (**203**), 4'-demethylpodophyllotoxin (**204**), 4'-demethylpicropodophyllotoxin (**205**), deoxypodophyllotoxin (**18**) and deoxypicropodophyllotoxin (**206**) [33,35]. Anolignan A (**44**) and anolignan B (**45**) were identified as active HIV-1 reverse transcriptase inhibitors. Compound **45** showed very weak activity when tested alone, but showed strong activity when combined with **44**. Compound **44** also showed similar behavior [38]. Interiotherin A (**207**), interiotherin B (**208**), angeloylgomisin R (**209**) and schisantherin D (**210**) were isolated from *Kadsura interior*. Compound **207** and **210** showed potent inhibitory activity against HIV replication in H9 lymphocytes with EC_{50} values of 0.5 and 3.1 µg/ml, respectively. Compounds **208** and **209** showed weak anti-HIV activity [173].

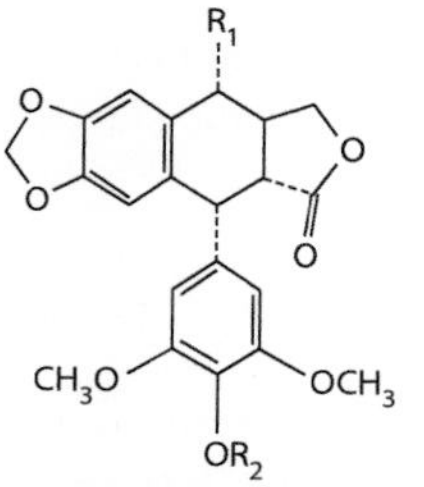

1: R_1 = OH R_2 = CH_3
204: R_1 = OH R_2 = H

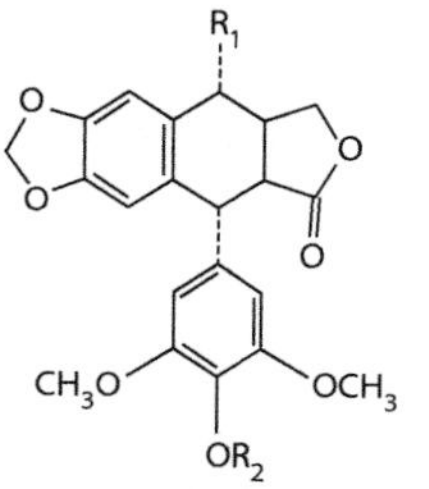

203: R_1 = OH R_2 = CH_3
205: R_1 = OH R_2 = H
206: R_1 = H R_2 = CH_3

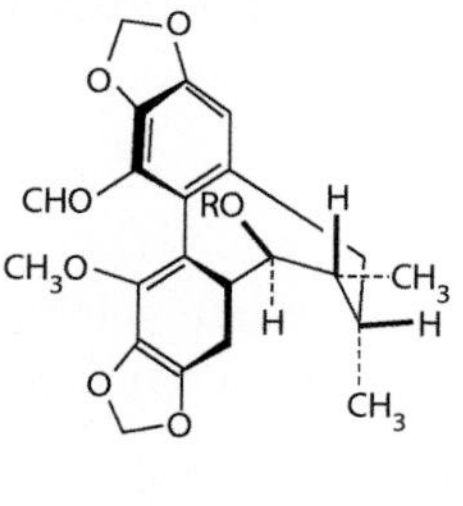

207: R = Ra
209: R = Rb

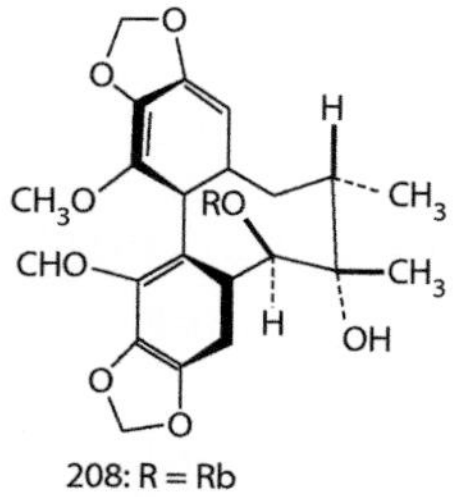

208: R = Rb
210: R = Ra

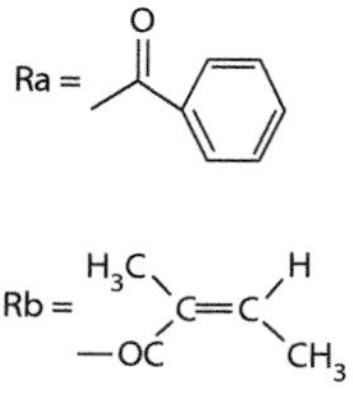

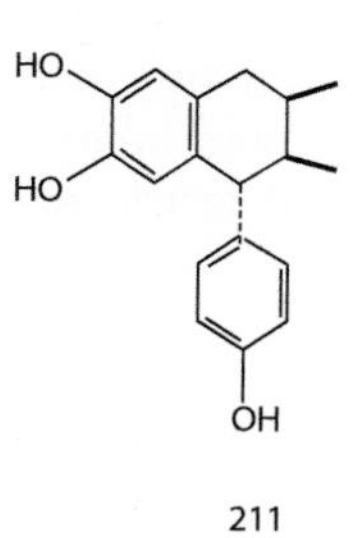

211

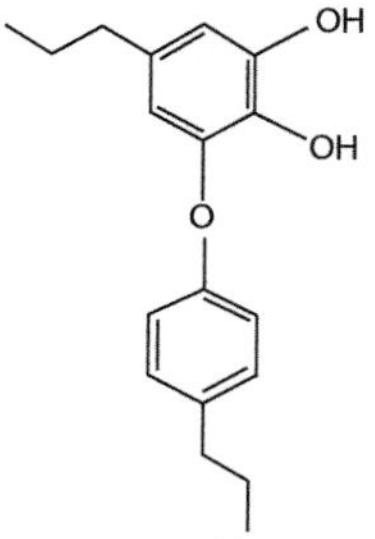

212

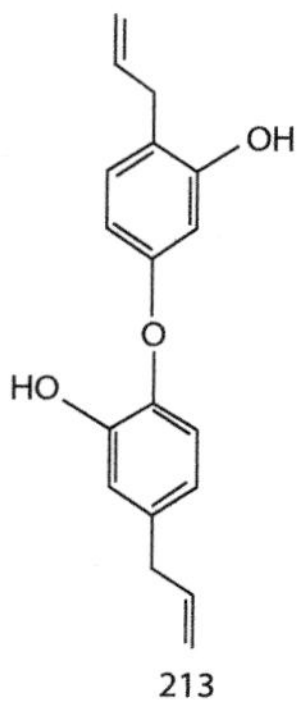

213

OTHER PHARMACOLOGICAL ACTIVITIES

SYNERGISM EFFECT

γ-Tocopherol has strong antioxidant and vitamin E activities. A diet containing sesame seeds showed a marked retardation of aging on the SAM. Sesaminol (**154**) exhibited a remarkable synergism with γ-tocopherol. Sesamin (**43**) also showed potent activity [174].

ANTIEMETIC ACTIVITY

Magnol (**25**) and honokiol (**26**) showed antiemetic activity at a concentration of 20 mg/kg body weight [175].

CONTRACEPTIVE ACTIVITY

Larrea tridentate (DC) Coville(Zygophyllaceae) has been used as contraceptive agent in Mexico, and reported to display uterine relaxation activity in vitro. 3'-Demethoxy-6-*O*-demethylisoguaiacin (**211**) was isolated and showed orally active anti-implantative activity in rats [176].

ANTIBACTERIAL ACTIVITY

Dental caries is an infectious disease caused by plaque forming organisms such as *Streptococcus mutans*. Compounds **25** and **26**, obovatol (**184**) and tetrahydrobovatol (**212**) showed antibacterial activity against *S. mutans* [177,178]. 4,4'-Diallyl-2,3'-dihydroxybiphenyl ether (**213**) and its mono and dimethyl ethers also showed antibacterial activity [179].

ANTIHYPERGLYCEMIC ACTIVITY

Masoprocol (nordihydroguaiaretic acid (**194**)), a lipoxygenase inhibitor, lowered plasma glucose in mice without any increase in insulin concentration [180].

PROTECTIVE EFFECT AGAINST THE STRESS-INDUCED GASTRIC ULCER

The hot water extract of *Acanthpanax senticosus* Harms from Hokkaido, Japan showed a protective effect against stress-induced gastric ulcers in restrained, cold water stressed rats during a two week administration. The activity-guided fractionation of the extract revealed that the n-butanol extract and its main component syringaresinol diglucoside (liriodendrin; **23**) significantly inhibited gastric ulcers [181].

PROTECTIVE EFFECT AGAINST PARKINSON'S DISEASE

Sesamin, a component of sesame seeds and of the stem bark of *Acanthpanax senticosus* Harms showed a protective effect against Parkinson's Disease and its related depressive behavior in rats that were orally administered sesamin (3.30 mg/kg) once a day for two weeks before an intraperitoneal injection of rotenone (2.5 mg/kg) [182].

SUMMARY

Lignans are a large group of natural products. A number of lignans that exhibit various biological activities have been isolated from medicinal plants that are used in traditional and folk medicines. Aryltetralin-type lignans, podophyllotoxin (1) and its congeners, have been isolated from plants of genus *Podophyllum* and exhibit strong antiviral and antitumor activities. The podophyllotoxin derivatives, etoposide (65) and teniposide (66) are successfully synthesized and used in the treatment of malignant conditions. Lignans are a group of polyphenols and exhibit antioxidant activity, and food rich in lignans appear to be effective in preventing diseases linked to active oxygen species. Pinoresinol type lignan glucosides (161–163) and sesaminol glucoside were also identified in sesame seeds and hydrolyzed to pinoresinol (77) and sesaminol (154) by intestinal β-gluciosidase or bacteria. Compound 154 exhibits a strong antioxidant activity.

Pinoresinol (77), arctiigenin (38) and trachelogenin (39) were metabolized to enterolactone (118) and enterodiol (185) by intestinal flora. The important mammalian lignans, 118 and 185 have been reported in the serum, urine, bile and seminal fluid of humans, chimpanzees and vervet monkeys. Compounds 118 and 185 exhibit potential protective activity against cancer. It has been reported that vegetarian women excrete more lignans than nonvegetarian women and that the lowest mean excretion was found in breast cancer patients. The lignans are formed in the intestinal tract from precursors present in fiber-rich food, and stimulate synthesis of sex hormone binding globulin (SHBG). This causes a reduction in hormone bioavailability and may protect against certain hormone -dependent cancers (breast, endometrium, and prostate).

It is clear that lignans have a rich diversity of biological activities, which we have only begun to understand. As further studies reveal the mechanisms of their effects, benefits to human health should follow.

REFERENCES

1. CBS Rao. *Chemistry of Lignans*. Andhra, India: Andhra University Press, 1978.
2. WD MacRae, GHN Towers. Biological activity of lignans. *Phytochem* 23:1207–1200, 1984.
3. S Nishibe. Structural elucidation and biological activities of phenylpropanoids, coumarins and lignans from medicinal plants. In: AU Rahman, ed. *Studies in Natural Products Chemistry, vol. 5. Structure Elucidation* (Part B). Amsterdam: Elsevier, 1989, pp. 505–548.
4. DC Ayres, JD Loike. *Lignans. Clinical, Biological and Clinical Properties*. Cambridge: Cambridge University Press, 1990, pp. 85–132.
5. S Nishibe. Lignans as Bioactive Components in Traditional Medicines. In: A Scalleert ed. *Polyphenolic Phenomena*. Paris: INRA Editions, 1993, pp. 247–255.
6. S Nishibe. Bioactive Lignans and Flavonoids from Traditional Medicines. In: R Brouillord, M Jay, A Scalbert, eds. *Polyphenols 94*, Paris: INRA Editions, 1994, pp. 113–122.

7. J Bruneton. *Pharmacognosy, Phytochemistry, Medicinal Plants.* Paris:Technique & Documentation-Lavoisier, 1995, pp. 241–254.
8. CT Ho, T Osawa, MT Huang, RT Rosan. ACS Symposium Series 547. *Food Phytochemicals for Cancer Prevention II.* Washington, DC: American Chemical Society, 1994, pp. 264–280.
9. SD Sarker. Biological activity of magnolol: A review. *Fitoterapia* 58:3–8, 1997.
10. T Biftu, R Stevenson. Natural 2,5-bisaryltetrahydrofuran lignans: Platelet-activating factor antagonists. *Phytotherapy Research* 1:97–106, 1987.
11. Y Damayanthi, JW Lown. Podophyllotoxins: Current status and recent developments. *Curr Med Chem* 5:205–252, 1998.
12. C Canel, RM Moraes, FE Dayan, D Ferreira. Podophyllotoxin. *Phytochem* 54:115–120, 2000.
13. M Novelo, JG Cruz, L Hernandez, R Pereda-Misanda, H Chai, W Mar, JM Pezzuto. Cytotoxic constituents from *Hyptis verticillata. J Nat Prod* 56:728–1736, 1993.
14. GR Pettit, DE Sahaufelberger. Isolation and structure of the cytostatic lignan glycoside phyllanthostatin A. *J Nat Prod* 51:1104–1112, 1988.
15. SJ Torrance, JJ Hoffmann, JR Cole. Wikstromol, antitumor lignan from Wikstroemia foetida var. oahuensis Gray and Wikstroemia uva-ursi Gray (Thymelaceae). *J Pharmaceu Sci* 68:664–665, 1979.
16. KH Lee, K Tagahara, H Suzuki, RY Wu, M Haruna, IH Hall, HC Huang. Antitumor agents. 49. Tricin, Kaempherol-3-O–β–D-glucopyranoside and (+)-nortrachelogenin, antileukemic principles from Wikstroemia indica. *J Nat Prod* 44:530–535, 1981.
17. CY Duh, CH Phoebe, Jr, JM Pezzuto, AD Kinghorn, NR Fansworth. Plant anticancer agents XLII. Cytotoxic constituents from Wikstromia elliptica. *Nat Prod* 49:706–709, 1986.
18. GV Subbaraju, KKK Kumar, BL Raju, KR Pillai, MC Reddy. Justiciresinol, a new furanoid lignan from Justica glauca. *J Nat Prod* 54:1639–1641, 991.
19. N Fukamiya, KH Lee. Antitumor agents, 81. Justicidin A and diphyllin, two cytotoxic principles from Justica procumbens. *J Nat Prod* 49:348–350, 1986.
20. H Joseph, J Gleye, C Moulis, LT Mensah, C Roussakis, C Gratas. Justicidin B, a cytotoxic principle from Justica retoralis. *J Nat Prod* 51:599–600, 1988.
21. SR Peraza-Sanches, LM Pena-Podriguez. Isolation of picropolygamain from the resin of Bursera simaruba. *J Nat Prod* 55:1788–1771, 1992.
22. DBM Wickramaratne, W Mar, H Chai, JJ Castillo, NR Farmsworth, DD Soejarto, GA Cordell, JM Pezzuto, AD Kinhhorn. Cytotoxic constituents of Bursera permollis. *Planta Medica* 61:80–81, 1995.
23. L Luyengi, N Suh, HHS Fong, JM Pezzuto, AD. Kinghorn. A lignan and four terpenoids from Brucea javanica that induce differentiation with cultured HL-60 promyelocytic leukemia cells. *Phytochem* 43:409–412, 1996.
24. MM Anderson, MJ O'Neill, JD Phillipson, DC Warhurst. In vitro cytotoxicity of a series of quassinoids from Brucera javanica fruits against KB cells. *Planta Medica* 57:62–64, 1991.
25. A Montagnac, JP Provost, M Litaudon, M Pais. Antimitotic and cytotoxic constituents of Myodocarpus gracilis. *Planta Medica* 63:365–366, 1997.
26. LBS Kardono, S Tsauri, K Padmawinata, JM Pezzuto, AD Kinghorn. Cytotoxic constituents of the bark of Plumeria rubra collected in Indonesia. *J Nat Prod* 53:1447–1455, 1990.
27. YC Wu, GY Chang, FN Ko, CM Teng. Bioactive constituents from the stem of Annona montana. *Planta Medica* 61:146–149, 1995.
28. N Kaneda, H Chai, JM Pezzuto, AD Kinghorn, NR Farnsworth, P Tuchinda, J Udchachon, T Sunstisuk, V Reutrakul. Cytotoxic activity of cardenolides from Beaumontia brevituba stems. *Planta Medica* 58:429–431, 1992.

29. T Konoshima, M Kozuka, H Tokuda, H Nishino, A Iwashima, M Haruna, K Ito, M Tanabe. Studies on inhibitors of skin tumor promotion. IX. Neolignans from Magnolia officinalis. *J Nat Prod* 54:816–822, 1991.

30. DBM Wickramaratne, T Pengsuparp, W Mar, HB Chai, TE Chagwedera, CWW Beecher, NR Farnsworth, AD Kinghorn, JM Pezzuto, GA Cordell. Novel antimitotic dibenzocyclo-octadiene lignan constituents of the stem bark of Steganotaenia araliacea. *J Nat Prod* 56:2083–2093, 1993.

31. M Taafrout, F Roussac, JP Robin, RP Hicks, DD Shillady, AT Sneden. Neoisostegane, a new bisbenzoylcyclooctadiene lignan lactone from *Steganotaenia araliacea. J Nat Prod* 47:600–606, 1984.

32. T Suzuki, T Aota, K Endo. The chemical constituents of the leave of Forsythia viridis-sima Lindley and their cytotoxicity. Proceedings of the 118th Annual Meeting of the Japanese Society of Pharmacology, Kyoto, 1998, p. 162.

33. HC Schröder, H Merz, R Steffen, WEG Müller, PS Sarin, S Trumm, J Schulz, E Eich. Differential in vitro anti-HIV activity of natural lignans. *Z Naturforsch* 45c:1215–1221, 1990.

34. S Trumm, E Eich. Cytostatic activities of lignanolides from Ipomoea cairica. *Planta Medica* 55:658–659, 1989.

35. K Pfeifer, H Merz, R Stefen, WEG Müller, S Trumm, J Schulz, E Eich, HC Schröder. In-vitro anti-HIV activity of lignans- differential inhibition of HIV-1 integrase reaction, topoisomerase activity and cellular microtubules. *J Pharm Med* 2:75–97, 1992.

36. RC Lin, AL Skaltsounis, E Seguin, F Tillequin, M Koch. Phenolic constituents of Selaginella doederleinii. *Planta Medica* 60:168–170, 1994.

37. V Darias, L Bravo, CCS Mateo, DAM Herrera. Cytostatic and antibcterial activity of some compounds isolated from several Lamiaceae species from the Canary islands. *Planta Medica* 56:70–72, 1990.

38. AM Rimando, JM Pezzuto, NR Farnsworth, T Suntisuk, V Reutrakul, K Kawanishi. New lignans from Anogeissus acuminata with HIV-1 reverse transcriptase inhibitory activity. *J Nat Prod* 57:896–904, 1994.

39. T Hirano, K Fukuoka, K Oka, T Naito, K Hosaka, H Mitsuhashi, Y Matsumoto. Antiproliferative activity against the human breast carcinoma cell line, ZR-75-1. *Cancer Invest* 8:595–602, 1990.

40. K Miyamoto, K Hiramatsu, Y Ohtani, M Kanitani, M Nomura, M Aburada. Effects of gomisin A on the promotor action and serum bile acid concentration in hepatocarcino-genesis induced by 3′-methyl-4-dimethy aminoazobenzene. *Biol Pharm Bull* 18:1443–1445, 1994.

41. Y Ohtani, M nomura, T Hida, K Miyamoto, M Kanitani, T Aizawa, M Aburada. Inhibition by gomisin A, a lignan compound, of hepatocarcinogenesis by 3′-methyl-4-dimethylaminoazobenzene in rat. *Biol Pharm Bull* 17:808–814,1994.

42. M Nomura, M Nakachiyama, T Hida, Y Ohtaki, K Sudo, T Aizawa, M Aburada, K Miyamoto. Gomisin A, a lignan component of Schizandra fruits, inhibits develope-ment of preneoplastic lesions in rat liver by 3′-methyl-4- dimethylamino-azobenzene. *Cancer Letters* 76:11–18, 1994.

43. K Miyamoto, S Wakusawa, M Nomura, F Sanae, R Sakai, K Sudo, Y Ohtani, S Takeda, Y Fujii. Effect of gomisin A on hepatocarcinogenesis by 3′-methyl-4- dimethylamino-azobenzene in rats. *Japan J Pharmacol* 57:71–77, 1991.

44. K Yasukawa, Y Ikeya, H Mitsuhashi, M Iwasaki, M Aburada, S Nakagawa, M Takeuchi, M Takido. Gomisin A inhibits tumor promotion by 12-O-tetra-decanoylphorbol-13-acetate in two-stage carcinogenesis in mouse skin. *Oncology* 49:68–71, 1992.

45. NR Ferringi, JL McLaughlin, RG Powel, CR Smith. Use of potato disk and brine shrimp bioassay to detect activity and isolate piceatanol as the anti-leukemic principle from the seeds of Euphobia lagascae. *J Nat Prod* 47: 347–352, 1984.

46. JE Anderson, CM Goetz, JL MacLaughlin, M Suffness. A blinde comparison of simple bench-top bioassays and human tumor cell cytotoxicities as anti-tumor prescreenes. *Phytochem Anal* 2:107–111, 1991.

47. PN Soils, CW Wright, MM Anderson, MPG Gupta, JD Phillipson. A microwell cytotoxicity assay using Artemia salina (Brine shrimp). *Planta Medica* 59: 250–252, 1993.

48. C Spatafora, C Tringali. Bioactive metabolites from the bark of Fagara macrophylla. *Phytochem Anal* 8:139–142, 1997.

49. YH Hui, CJ Chang, JL McLaughlin, RG Powell. Justicidin B, a bioactive trace lignan from the seeds of Sesbania drummondii. *J Nat Prod* 46:1175–1178, 1986.

50. TL Bachmann, F Ghia, KBG Tarssell. Lignans and lactones from Phyllanthus anisolobus. *Phytochem* 33:189–191, 1993.

51. K He, G Shi, L Zeng, Q Ye, JL McLaughlin. Konishiol, a new sesquiterpene, and bioactive components from Cunninghamia konishii. *Planta Medica* 63: 58–160, 1997.

52. K Umehara, A Sugawa, M Kuroyanagi, A Ueno, T Taki. Studies on differentiation-inducers from Arctium fructus. *Chem Pharm Bull* 41:1774–1779, 1993.

53. K Umehara, M Nakamura, T Miyase, M Kuroyanagi, A Ueno. Studies on differentiation inducersVI. Lignan derivatives from Arctium fructus (2). *Chem Pharm Bull* 44: 2300–2304, 1996.

54. T Hirano, K Fukuoka, K Oka, Y Matsumoto. Differential sensitivity of human gastric cancer ATPase and normal gastric mucosa ATPase to the synthetic mammalian lignan analogue 2,3-dibenzylbutane-1,4-diol (hattalin). *Cancer Invest* 9:145–150, 1991.

55. A Kamal, M Daneshtalab, K Atchison, RG Micetich. Synthesis of ring-A-opened isopicropodophyllins as potential DNA topoisomerase II inhibitors. *Bioorg Medic Chem Lett* 4:1513–1518, 1994.

56. A Kamal, M Daneshtalab, RG Micetich. A rapid entry into podophyllotoxin congeners: Synthesis of justicidin B. *Tetrahedron Letters* 35:3879–3882, 1994.

57. PT Anastas, R Stevenson. Synthesis of natural lignan arylnaphthalene-lactones, daurinol and retrochinensin. *J Nat Prod* 54:1687-1691, 1991.

58. YL Zhang, YC Shen, ZQ Wang, HX Chen, X Guo, YC Chen, KH Lee. Antitumor agents, 130. Novel 4β-arylamino derivatives of 3′,4′-didemethoxy-3′,4′-dioxo-4-deoxypodophyllotoxin as potent inhibitors of human DNA topoisomerase II. *J Nat Prod* 55:1100–1111, 1992.

59. A Kamal, K Atchison, M Daneshtalab, RG Micetich. Synthesis of podophyllotoxin congeners as potential DNA topoisomeraseII inhibitors. *Anti-Cancer Drug Design* 10:545–554, 1995.

60. H Saito, H Yoshikawa, Y Nishimura, S Kondo, T Takeuchi, H Umezawa. Studies on lignan lactone antitumor agents.I. Synthesis of aminoglycosidic lignan variants related to podophyllotoxin. *Chem Pharm Bull* 34: 3733–3740, 1986.

61. H Saito, H Yoshikawa, Y nishimura, S Kondo, T Takeuchi, H Umezawa. Studies on lignan lactone antitumor agents. II. Synthesis of N-alkylamino-and 2,6-dideoxy-2-aminoglycosidic lignan variants related to podophyllotoxin. *Chem Pharm Bull* 34:3741–3746, 1986.

62. Y Hitotsuyanagi, K Yamagami, A Fujii, Y Naka, Y Ito, T Tahara. Synthesis and biological properties of novel aza-podophyllotoxin analogs possessing pronounced antitumor activity. *Bioorg Medic Chem Lett* 5:1039–1042, 1995.

63. AS Feliciano, M Medarde, RP Lamamie de Clairac, JL Lopez, P Puebla, MDG Gravalos, PR Lazaro, MT Garcia de Quesada. Synthesis and biological activity of bromolignans and cyclolignans. *Arch Pharm* (Weinheim) 326:421–426, 1993.

64. F Zavala, D Guenard, JP Robin, E Brown. Structure-antitubulin activity relationships in steganacin congeners and analogues. Inhibition of tubulin polymerization in vitro by (±)-isodeoxypodophyllotoxin. *J Med Chem* 23:546–549, 1980.

65. EW Sutherland, TW Rall. Fractionation & characterization of a cyclic adenine ribonucleotide formed by tissue particles. *J Bio Chem* 232:1077–1091, 1958.

66. T Nikaido, T Ohmoto, T Kinoshita, U Sankawa, S Nishibe, S Hisada. Inhibition of cyclic AMP phosphodiesterase by lignans. *Chem Pharm Bull* 29:3586–3592, 1981.

67. T Deyama, S Nishibe, S Kitagawa, Y Ogihara, T Takeda, T Ohmoto, T Nikaido, U Sankawa. Inhibition of adenosine 3′,5′-cyclic monophosphate phosphodiesterase by lignan glucosides of Eucommia bark. *Chem Pharm Bull* 36:435–439, 1988.

68. S Nishibe H Tsukamoto, S Hisada, T Nikaido, T Ohmoto, U sankawa. Inhibition of cyclic AMP phosphodiesterase by lignans and coumarins of Olea and Fraxinus barks. *Shoyakugaku Zasshi* 40:89–94, 1986.

69. II Brekhman, IV Dardymov. Pharmacological investigation of glycosides from Ginseng and Eleutherococcus. *Lloydia* 32:46–51, 1969.

70. T Deyama. The constituents of Eucommia ulmoides Oliv. I. Isolation of (+)-medioresinol di-O-β–D-glucopyranoside. *Chem Pharm Bull* 31:2993–2997, 1983.

71. S Nishibe, H Kinosita, H Takeda, G Okano. Phenolic compounds from stem bark of Acanthopanax senticosus and their pharmacological effect in chronic swimming stressed rats. *Chem Pharm Bull* 38:1783–1785, 1990.

72. N Takasugi, T moriguchi, T Fuwa, S Sanada, Y Ida, J Shoji, H Saito. Effect of Eleutherococcus senticosus and its components on rectal temperature, body and grip tones, motor coordination, and exploratory and spontaneous movements in acute stressed mice. *Shoyakugaku Zasshi* 39:232–237, 1985.

73. T Deyama, T Ikawa, S Kitagawa, S Nishibe. The constituents of Eucommia ulmoides Oliv. V. Isolation of dihydroxydehydrodiconiferyl alcohol isomers and phenolic compounds. *Chem Pharm Bull* 35:1785–1789, 1987.

74. T Deyama, T Ikawa, S Nishibe. The constituents of Eucommia ulmoides Oliv II. Isolation and structures of three new lignan glycosides. *Chem Pharm Bull* 33:3651–3657, 1985.

75. T Deyama, T Ikawa, S Kitagawa, S Nishibe. The constituents of Eucommia ulmoides Oliv III. Isolation and structure of a new lignan glycoside. *Chem Pharm Bull* 34:523–527, 1986.

76. T Deyama, T Ikawa, S Kitagawa, S Nishibe. The constituents of Eucommia ulmoides Oliv IV. Isolation of a new sesquilignan glycoside and iridoids. *Chem Pharm Bull* 34:4933–4938, 1986.

77. T Deyama, T Ikawa, S Kitagawa, S Nishibe. The constituents of Eucommia ulmoides Oliv VI. Isolation of a new sesquilignan and neolignan glycosides *Chem Pharm Bull* 35:1803–1807, 1987.

78. Y Oshima, S Tanaka, H Hikino, T Deyama, G Kinoshita. Anticomplementary activity of the constituents of Eucommia ulmoides bark. *J Ethnopharmacol* 23:159–164, 1988.

79. A Fleckenstein. Specific pharmacology of calcium in myocardium pacemakers, and vascular smooth muscle. *Ann Rev Pharmacol Toxicol* 17:149–166, 1977.

80. K Ichikawa, T Kinosita, S Nishibe, U Sankawa. The Ca^{2+} antagonist activity of lignans. *Chem Pharm Bull* 34:3514–3517, 1986.

81. P Braquet, N Senn, JP Robin, A Esanu, T Godfraind, R Garay. Inhibition of the erythrocyte Na^+, K^+-pump by mammalian lignans. *Pharmacol Res Commun* 18:227–239, 1986.

82. M Fagoo, P Braquet, JP Robin, A Esanu, T Godfraind. Biochem. *Biophys Res Commun* 134:1064–1070, 1986.

83. T Hirano, K Oka, T Naitoh, K Hosaka, H Mitsuhashi. Endogeneous digoxin-like activity of mammalian-lignans and their derivatives. *Res Commun Chem Pathol Pharmacol* 64:227–240, 1989.

84. CC Chen, YL Huang, HT Chen, YP Chen, HY Hsu. On the Ca^{2+} antagonistic principle of the flower buds of Magnolia fargesii. *Planta Medica:* 438–440, 1988.

85. YL Huang, CC Chen, YP Chen, HY Hsu, YH Kuo. (−)-Fargesol, a new lignan from the flower buds of Magnolia fargesii. *Planta Medica* 56:237–238, 1990.
86. N Lami, S Kadota, T Kikuchi, Y Momose. Constituents of the roots of Boerhaavia diffuse L. III. Identification of Ca^{2+} channel antagonistic compound from the methanol extract. *Chem Pharm Bull* 39:1551–1555, 1991.
87. T Tsuruga, Y Ebizuka, J Nakajima, YT Chun, H Noguchi, Y Iitaka, U Sankawa. Biologically active constituents of Magnolia salicifolia: Inhibitors of induced histamin release from rat mast cells. *Chem Pharm Bull* 39:3265–3271, 1991.
88. TY Shen. Chemical and biological characterization of lignan analogs as novel PAF receptor antagonists. *Lipids* 26:1154–1156, 1991.
89. P Braquet, JJ Goldfroid. PAF-acether specific binding sites:2. Design of specific antagonists. *Trends in Pharmacol Sci* 397–403, 1986.
90. S Iwakami, Y Ebizuka, U Sankawa. Lignans and sesquiterpenoids as PAF antagonist. *Heterocycle* 30:795–798, 1990.
91. T Fujimoto, M Nose, T Takeda, Y Ogihara, S Nishibe, M Minami. Studies on the Chinese crude drug "Luoshiteng"(II) On the biologically active components in the stem part of Luoshiteng originating from rachelospermum jasminoides.*Shoyakugaku Zasshi* 46:224–249, 1992.
92. HCCF Neto, PT Bozza, HN Crurz, CLM Silva, FA Violante, JMB Filho, G Thomas, MA Martins, EV Tibirica, F Noel, RSB Cordeiro. Yangambin: A new naturally occurring platelet-activating factor receptor antagonist: Binding and in vitro functional studies. *Planta Medica* 61:101–105, 1995.
93. MF Serra, BL Diaz, EO Barreto, APB P antagonist yangambin. *Planta Medica* 63:207–212, 1997.
94. HCCF Neto, MA Martins, PMR Silva, PT Bozza, HN Cruz, MQ Paulo, MAC Kaplan, RSB Cordeiro. Pharmacological profile of epiyangambin: A frofrun lignan with PAF antagonist activity. *J Lipid Mediators* 7:1–9, 1993.
95. TS Wu, YL Leu, YY Chan, SM Yu, CM Teng, JD Su. Lignans and an aromatic acid from Cinnamomum philippinense. *Phytochem* 36:785–788, 1994.
96. CC Chen, WC Hsin, FN Ko, YL Huang, JC Ou, CM Teng. Antiplatelet arylnaphthalide lignans from Justicia procumbens. *J Nat Prod* 59:1149–1150, 1996.
97. SB Hwang, MH Lam, T Biftu,TR Beattie, TY Shen. Trans-2,5-Bis(3,4,5-trimethoxyphenyl) tetrahydrofuran. *J Biol Chem* 260:15639–15645, 1985.
98. JM Prieto, MC Recio, RM Giner, S Manez, A Massmanian, PG Waterman, JL Rios. Topical anti-inflammatory lignans from Haplophyllum hispanicum. *Z Naturforsch* 51C:618–622, 1996.
99. K Yoshikawa, H Kinoshita, Y Kan, S Arihara. Neolignans and phenylpropanoids from the rhizomes of Coptis japonica var dissecta. *Chem Pharm Bull* 43:578–581, 1995.
100. JY Cho, J Park, ES Yoo, K Yoshikawa, KU Baik, J Lee, MH Park. Inhibitory effect of lignans from the rhizomes of Coptis japonica var dissecta on tumor necrosis factor-α production in lipopolysaccharide-stimulated RAW 264.7 cells. *Arch Pharm Res* 21:12–16, 1998.
101. Y Fukuyama, K Mizuta, K Nakagawa, Q Wenjuan, W Xiue. A new neo-lignan, a prostaglandin I2 inducer from the leaves of Zizyphus jujuba. *Planta Medica,* 501–502, 1986.
102. M Abe, M Morikawa, M Inoue, A Nakajima, M Tsuboi, T Naito, K Hosaka, HMitsuhashi. Effect of 2,3-dibenzylbutane-1,4-diol on vascular smooth muscle of rabbit aorta. *Arch Int Pharmacodyn* 301:40–50, 1989.
103. E Okuyama, K Suzumura, M Yamazaki. Pharmacologically active components of Todopon Puok(Fargracea racemosa), a medicinal plant from Borneo. *Chem Pharm Bull* 32:2200–2204, 1995.
104. HB Wan, DC Ju, ST Lian, S Watanabe, M Tamai, S Okuyama, S Kitsukawa, S Omura. JP Patent 01242596, 1989.

105. S Watanabe, M Tamai, S Okuyama, S Kitsukawa, S Omura. JP Patent 01290693, 1989.
106. E Okuyama, S Fujimori, M Yamazaki, T Deyama. Pharmacologically active components of Viticis fructus (Vitex rotundifolia).II. The components having analgestic effect. *Chem Pharm Bull* 46:655–662, 1998.
107. CJ Sih, PR Ravikumar, FC Huang, C Buchner, H Whitelock Jr. Isolation and synthesis of pinoresinol diglucoside, a new antihypertensive principle of Tu-Chang (Eucommia ulmoides Oliver). *J Am Chem Soc* 98:5412–5413, 1976.
108. T Deyama. Studies on the constituents of Eucommia ulmoides Oliver. Doctoral thesis of Nagoya City University, 1987.
109. Y Matsumura, S Kita, S Morimoto, K Akimoto, M Furuya, N Oka, T Tanaka. Antihypertensive effect of sesamine. I. Protection against deoxycorticosterone acetate-salt-induced hypertension and cardiovascular hypertrophy. *Biol Pharm Bull* 18:1016–1019, 1995.
110. S Kita, Y Matsumura, S Morimoto, K Akimoto, M Furuya, N Oka, T Tanaka. Antihypertensive effect of sesamine II. Protection against two-kidney, one-clip renal hypertension and cardiovascular hypertrophy. *Biol Pharm Bull* 18:1283–1285, 1995.
111. K Matsunaga, M Shibuya, Y Ohizumi. Graminone B, a novel lignan with vasodilative activity from *Imperata cylindrica. J Nat Prod* 57:1734–1736, 1994.
112. M Therien, BJ Fitzsimmons, J Scheigetz, D Macdonald, LY Choo, J Guay, JP Falgueyret, D Riendeau. Justicidin E: A new leukotriene biosynthesis inhibitor. *Bioorg Medic Chem Lett* 3:2063–2066, 1993.
113. Y Ducharme, C Brideau, D Dube', CC Chan, JP Falgueyret, JW Gillard, J Guay, JH Hutchinson, CS McFarlane, D Riendeau, J Scheigetz, Y Girad. Naphthalenic lignan lactones as selective, nonredox 5-lipoxygenase inhibitors. Synthesis and biological activity of (methylalkyl)thiazole and methoxytetrahydrofuran hybrids. *J Med Chem* 37: 512–518, 1994.
114. D Delorme, Y Ducharme, C Brideau, CC Chan, N Chauret, S Desmarais, D Dubé, JP Falgueyret, R Fortin, J Guay, P Hamel, TR Jones, C Lépine, C Li, M McAuliffe, CS McFarlane, DA Nicoll-Griffith, D Riendeau, JA Yergey, Y Girad. Dioxabicyclooctanyl naphthalenenitriles as nonredox 5-lipoxygenase inhibitors: Structure-activity relationship study directed toward the improvement of metabolic stability. *J Med Chem* 39:3951–3970, 1996.
115. N Hirose, F Doi, T Ueki, K Akazawa, K Chijiiwa, M Sugano, K Akimoto, S Shimizu, H Yamada. Suppressive effect of sesamin against 7,12-dimethyl-benz[a]-anthracene induced rat mammary carcinogensis. *Anticancer Res* 12:1259–1265, 1992.
116. Y Fukuda, M Nagata, T Osawa, M Namiki. Contribution of lignan analogues to antioxidative activity of refined unroasted sesame seed oil. *J Am Oil Chem Soc* 63:1027–1031, 1986.
117. Y Fukuda, M Namiki. Recent studies on sesame seed and oil. *Nippon Shokuhin Kogyo Gakkaishi* 35:525–562, 1988.
118. T Osawa. Aging, investigation from anti-oxidants. *Genndai Iryo* 25:3327–3332, 1993.
119. K Kuriyama, T Murai. Inhibition effects of lignan glucosides on lipid peroxidation. *Nippon Nogei Kagaku Kaishi* 70:161–167, 1996.
120. K Kuriyama, T Murai. Scavenging of hydroxy radicals by lignan glucosides in germinated sesame seeds. *Nippon Nogei Kagaku Kaishi* 69:703–705, 1995.
121. H Katsuzaki, M Kawasumi, S Kawakishi, T Osawa. Structure of novel antioxidative lignan glucosides isolated from sesame seeds. *Biosci Biotech Biochem* 56:2087–2088, 1992.
122. H Katsuzaki, S Kawakishi, T Osawa. Structure of novel antioxidative lignan triglucoside isolated from sesame seeds. *Heterocycles* 36:933–936, 1993.
123. H Katsuzaki, S Kawakishi, T Osawa. Sesaminol glucosides in sesame seeds. *Phytochem* 35:773–776, 1994.

124. S Maeda, H Masuda, T Tokoroyama. Studies on preparation of bioactive lignans by oxidative coupling reaction. I. Preparatiom and lipid peroxidation inhibitory effect of benzofuran lignans related to schizotenuins. *Chem Pharm Bull* 42:2500–2505, 1994.

125. H Hirano, T Tokuhira, T Yokoi, T Shingu. Isolation of free radical scavenger from Coptidis rhizoma. *Natural Medicines* 51:539–540, 1997.

126. S Toda, M Kimura, M Ohnishi, K Nakashima, Y Ikeya, H Taguchi, H Mitsuhashi. Natural Antioxidant (IV). Antioxidative components isolated from Schizandra fruit. *Shoyakugaku Zasshi* 42:156–159, 1988.

127. Y Kiso, M Tohkin, H Hikino, Y Ikeya, H Taguchi. Mechanism of antihepatotoxic activity of wuweizisu C and gomisin A. *Planta Medica* 331–334, 1985.

128. JP Wang, SL Raung, MF Hsu, CC Chen. Inhibition by gomisin C (a lignan from Schizandra chinensis) of the respiratory burst of rat neutrophils. *Br J Pharmacol* 113:945–953, 1994.

129. XW Yang, H Miyashiro, M Hattori, T Namba, Y Tezuka, T Kikuchi, DF Chen, GJ Xu, T Hori, M Extine, H Mizuno. Isolation of novel lignans, heteroclitins F and G, from the stem of Kadsura heteroclita, and anti-lipid peroxidative actions of heteroclitins A-G and related compounds in the in vitro rat liver homogenate system. *Chem Pharm Bull* 40:1510–1516, 1992.

130. XW Yang, M Hattori, T Namba, DF Chen, GJ Xu. Anti-lipid peroxidative effect of an exract of the stem of Kadsura heteroclita and its major constituent, kadsurin, in mice. *Chem Pharm Bull* 40:406–409, 1992.

131. AD Muir, ND Westcott. Flax seed lignans, stereochemistry, quantitation and biological activity. Proceedings of Phytochemical Society of North America for "Phytochemicals in Human Health Protection, Nutrition and Plant Defense." Pullman, USA, 1998, pp. 29.

132. S Maeda, K Sudo, Y Miyamoto, S Takeda, M Shinbo, M Aburada, Y Ikeya, H Taguchi, M Harada. Pharmacological studies on Schizandra fruits. II. Effect of constituents of Schizandra fruits on drugs hepatic damage in rats. *Yakugaku Zasshi* 102:579–588, 1982.

133. S Maeda, S Takeda, Y Miyamoto, M Aburada, M Harada. Effect of gomisin A on liver function in hepatotoxic chemicals-treated rats. *Japan J Pharmacol* 38:347–353, 1985.

134. S Takeda, S Maemura, K Sudo, Y Kase, I Arai, Y Ohkura, S Funo, Y Fujii, M Aburada, E Hosoya. Effects of gomisin A, a lignan component of Schizandra fruits, on experimental liver injuries and liver microsomal drug-metabolizing enzymes. *Folia Pharmacol Japan* 87:169–187, 1986.

135. S Takeda, I Arai, Y Kase, Y Ohkura, M Hasegawa, Y Sekiguchi, K Sudo, M Aburada, E Hosoya. Pharmacological studies on antihepatotoxic action of (+)-(6S,7S,R-biar)-5,6,7,8-tetrahydro-1,2,3,12-tetramethoxy-6,7-dimethyl-10,11-methylenedioxy-6-dibenzo[a,c] cyclooctenol(TJN-101), a lignan component of Schizandra fruits. Influence of resolvents on the efficacy of TJN-101 in the experimental acute hepatic injuries. *Yakugaku Zasshi* 107:517–524, 1987.

136. S Takeda, Y Kase, I Arai, M Hasegawa, Y Sekiguchi, S Funo, M Aburada, E Hosoya, Y Mizoguchi, S Morisawa. Effects of TJN-101((+)-(6S,7S,R-biar)-5,6,7,8-tetrahydro-1,2,3,12- tetramethoxy-6,7-dimethyl-10,11-methylenedioxy-6-dibenzo[a,c] cyclooctenol) on liver regeneration after partial hepatectomy, and on regional hepatic blood flow and fine structure of the liver in normal rats. *Folia Pharmacol Japan* 88:321–330, 1986.

137. S Takeda, Y Kase, I Arai, Y Ohkura, M Hasegawa, Y Sekiguchi, A Tatsugi, S Funo, M Aburada, E Hosoya. Effect of TJN-101, a lignan compound isolated from Schizandra fruits, on liver fibrosis and on liver regeneration after partial hepatectomy in rats with chronic liver injury induced by CCl4. *Folia Pharmacol Japan* 90:51–65, 1987.

138. S Kubo, Y Ohkura, Y Mizoguchi, I Matsui-Yuasa, S Otani, S Morisawa, H Kinoshita, S Takeda, M Aburada, E Hosoya. Effect of gomisin A(TJN-101) on liver regeneration. *Planta Medica* 58:489–492, 1992.

139. Y Mizoguchi, N Kawada, Y Ichikawa, H Tsutsui. Effect of gomisin A in the prevention of acute hepatic failure induction. *Planta Medica* 57:320–324, 1991.

140. H Nagai, I Yakuo, M Aoki, K Teshima, Y Ono, T Sengoku, T Shimazawa, M Aburada, A Koda. The effect of gomisin A on immunologic liver injury in mice. *Planta Medica* 55:13–17, 1989.

141. Y Ohkura, Y Mizoguchi, Y Sakagami, K Kobayashi, S Yamamoto, S Morisawa, S Takeda, M Aburada. Inhibitory effect of TJN-101((+)-(6S,7S,R-biar)-5,6,7,8-tetrahydro-1,2,3,12- tetramethoxy-6,7-dimethyl-10,11-methylenedioxy-6-dibenzo[a,c] cyclooctenol) on immunologically induced liver injuries. *Japan J Pharmacol* 44:179–185, 1987.

142. S Takeda, I Arai, M Hasegawa, A Tatsugi, M Aburada, E Hosoya. Effect of gomisin A(TJN-101), a lignan compound isolated from Schizandra fruits, on liver function in rats. *Folia Pharmacol Japan* 91:237–244, 1988.

143. K Akimoto, Y Kitagawa, T Akamatsu, N Hirose, M Sugano, S Shimizu, H Yamada. Protective effects of sesamin against liver damage caused by alcohol or carbon tetra-chloride in rodents. *Ann Nutr Metab* 37:218–224, 1993.

144. N Hirose, T Inoue, K Nishihara, M Sugano, K Akimoto, S Shimizu, H Yamada. Inhibition of cholesterol absorption and synthesis in rat by sesamin. *J Lipid Res* 32:629–638, 1991.

145. F Hirata, K Fujita, Y Ishikura, K Hosoda, T Ishikawa, H Nakamura. Hypocholesterolemic effect of sesame lignan in humans. *Atherosclerosis* 122:135–136, 1996.

146. T Iwasaki, K Kondo, T Nishitani, T Kuroda, K Hirakoso, A Ohtani, K Takashima. Arylnaphthalene lignans as novel series of hypolipidemic agents raising high-density lipoprotein level. *Chem Pharm Bull* 43:1701–1705, 1995.

147. T Kuroda, K Kondo, T Iwasaki, A Ohtani, K Takashima. Synthesis and hypolipidemic activity of diesters of arylnaphthalene lignan and their heteroaromatic analogs. *Chem Pharm Bull* 45:678–684, 1997.

148. BM Kwon, MK Kim, SH Lee, JA Kim, IR Lee, YK Kim, SH Bok. Acyl-CoA: Cholesterol acyltransferase inhibitors from Magnolia obovata. *Planta Medica* 63:550–551, 1997.

149. RJ Fletcher. Food sources of phytoestrogens and their precursors in Europe. *Br J Nutr* 89:Supple 1, S39–S43, 2003.

150. J Liggins, R Grimwood, SA Bingham. Extraction and quantification of lignan phytoestrogens in food and human samples. *Anal Biochem* 287:102–109, 2000.

151. T Nurmi, S Heinonen, W Mazur, T Deyama, S Nishibe, H Adlercreutz. Analytical, nutritional and clinical methods. Lignans in selected wines. *Food Chem* 83:303–309, 2003.

152. S Heinonen, T Nurmi, K Liukkonen, K Potanen, K Wähälä, T Deyama, S Nishibe, Adlercreutz. In Vitro Metabolism of plant lignans: New precursors of mammalian lignans enterolactone and enterodiol. *J Agric Food Chem* 49:3178–3186, 2001.

153. NM Saarinen, A Smeds, SI Mäkelä, J Ämmälä, K Hakala, JM Pihlava, EL Ryhänen, R Sjöholm, R Santti. Structural determination of plant lignans for the formation of enterolactone in vivo. *J Chromatogr B* 25:311–319, 2002.

154. L Valentin-Blasini, BC Blount, HS Rogers, LL Needham. HPLC-MS/MS method for the measurement of seven phytoestrogens in human serum and urine. *Journal of Exposure Analysis and Environmental Epidemiology* 10:799–807, 2000.

155. LH Xie, EM Ahn, T Akao, AA Abdel-Hafez, N Nakamura, M Hattori. Transformation of arctiin to estragenic and antiestragenic substances by human intestinal bacteria. *Chem Pharm Bull* 51:378–384, 2003.

156. LH Xie, T Akao, K Hamasaki, T Deyama, M Hattori. Biotransformation of pinoresinol diglucoside to mammalian lignans by human intestinal microflora, and Isolation of *enterococcus faecalis* strain PDG-1 responsible for the transformation of (+)-pinoresinol to (+)-lariciresinol. *Chem Pharm Bull* 51:508–515, 2003.

157. K Hamasaki, K Furukawa, M Hattori, S Nishibe, H Adlercreutz, T Deyama. Metabolism of lignans by intestinal bacteria II. Metabolism of lignans from Eucomiae Cortex and other herbal medicines. Proceeding of the 50th Annual Meeting of the Japanese Society of Pharmacognosy, Tokyo, 2003, p. 69.

158. S Nishibe, T Deyama, K Hamasaki, T Kawamura, T Tanaka, E Sakai, S Heinonen, H Adlercreutz. Use of Forsythia leaves as phytoestrogen source. Abstract books of the 1st internatuional conference on polyphenols and health, Vichy-France, 2003, p. 225.

159. M Nose, T Fujimoto, T Takeda, S Nishibe, Y Ogihara. Structural transformation of lignan compounds in rat gastrointestinal tract. *Planta Medica* 58:520–523, 1992.

160. M Nose, T Fujimoto, S Nishibe, Y Ogihara. Structural transformation of lignan compounds in rat gastrointestinal tract II. Serum concentration of lignans and their metabolites. *Planta Medica* 59:131–134, 1993.

161. R Kasper, G Ganßer, J Doehmer. Biotransformation of the naturally occurring lignan (−)-arctigenin in mammalian cell lines genetically engineered for expression of single cytochrome P450 isoforms. *Planta Medica* 60:441–444, 1994.

162. SP Borriello, KDR Setchell, M Axelson, AM Lawson. Production and metabolism of lignans by human faecal flora. *J Appl Bacteriol* 58:37–43, 1985.

163. PI Musey, H Adlercreutz, KG Gould, DC Collins, T Fotsis, C Bannwart, T Mäkelä, K Wähälä, G Brunow, T Hase. Effect of diet lignans and isoflavonoid phytoestrogens in chimpanzees. *Life Sciences* 57:655–664, 1995.

164. KDR Setchell, AM Lawson, FL Mitchell, H Adlercreutz, DN Kirk, M Axelson. Lignans in man and in animal species. *Nature* 287:740–742, 1980.

165. M Axelson, J Sjövall, BE Gustafsson, KDR Setchell. Origin of lignans in mammals and identification of a precursor from plants. *Nature* 298:659–660, 1982.

166. TD Shultz, WR Bonorden, WR Seaman. Effect of short-term flaxseed comsumption on lignan and sex hormone metabolism in men. *Nutrition Res* 11:1089–1100, 1991.

167. D Ganßer, G Spiteller. Plant constituents interfering with human sex hormone-binding globulin. Evaluation of a test method and its application to Urtica dioica root extracts. *Z Naturforsch C Biosci* 50:98–104, 1995.

168. M Schöttner, D Ganßer, G Spiteller. Lignans from the root of Urtica dioica and their metabolites bind to human sex hormone binding globulin (SHBG). *Planta Medica* 63:529–632, 1997.

169. ME Martin, M Haourigui, C Pelissero, C Benassayag, EA Nunez. Interacxtion between phytoestrogens and human sex steroid binding protein. *Life Sciences* 58:429–436, 1996.

170. M Schöttner, D Ganßer, G Spiteller. Interaction of lignans with human sex hormone binding Globulin (SHBG). *J Biosci* 52:834–843, 1997.

171. H Adlercreutz, C Bannwart, K Wähälä, T Mäkelä, G Brunow, T Hase, PJ Arosemena, JT Kellis Jr, LE Vickery. Inhibition of human aromatase by mammalian lignans and isoflavonoid phytoestrogens. *J Steroid Biochem Molec Biol* 44:147–153, 1993.

172. C Wang, T Mäkelä, T Hase, H Adlercreutz, MS Kurzer. Lignans and flavonoids inhibit aromatase enzyme in human preadipocytes. *J Steroid Biochem Molec Biol* 50:205–212, 1994.

173. DF Chen, SX Zhang, K Chen, BN Zhou, P Wang, LM Cosentino, KH Lee. Two new lignans, interiotherins A and B, as anti-HIV principles from Kadsura interior. *J Nat Prod* 59:1066–1068, 1996.

174. K Yamashita, Y Nohara, K Katayama, M Namiki. Sesame seed lignans and γ-tochopherol act synergistically to produce vitamine E activity in rats. *Nutr* 122:2240–2246, 1992.

175. T Kawai, K Kinoshita, K Koyama, K Takahashi. Anti-emetic principles of Magnolia obovata bark and Zingiber officinale rhizome. *Planta Medica* 60: 17–20, 1994.
176. C Konno, ZZ Lu, HZ Xue, CAJ Erdelmeier, D Meksuriyen, CT Che, GA Cordell, DD Soejarto, DP Waller, HHS Fong. Furanoid lignans from *Larrea tridentata. J Nat Prod* 53:396–406, 1990.
177. T Namba, M Tsunezuka, KH Bae, M Hattori. Studies on dental caries prevention by traditional Chinese medicines(Part I) Screening of crude drugs for antibacterial action against Streptcoccus mutans. *Shoyakugaku Zasshi* 35:295–302, 1981.
178. K Ito, T Iida, K Ichino, M Tsunezuka, M Hattori, T Namba. Obovatol and obovatal, novel biphenyl ether lignans from leaves of Magnolia obovata Thunb. *Chem Pharm Bull* 30:3347–3353, 1982.
179. JK Nitao, MG Nair, DL Thorogood, KS Johnson, JM Scriber. Bioactive neolignans from the leaves of Magnolia virginiana. *Phytochem* 30:2193–2195, 1991.
180. J Luo, T Chuang, J Cheung, J Quan, J Tsai, C Sullivan, RF Hector, MJ Reed, K Meszaros, SR King, TJ Carlson, GM Reaven. Masoprocol (nordihydroguaiaretic acid): A new antihyperglycemic agent isolated from the creosote bush(Larrea tridentata). *Eur J Pharmacol* 346:77–79, 1998.
181. T Fujikawa, A Yamaguchi, I Morita, H Takeda, S Nishibe. Protective effects of *Acanthopanax sencticosus* Harms from Hokkaido and its components on gastric ulcer in restrained cold water stressed rats. *Biol Pharm Bull* 19:1227–1230, 1996.
182. T Fujikawa, N Kanda, A Shimada, M Ogata, I Suzuki, I Hayashi, K Nakashima. Effects of sesamin in *Acanthopanax sencticosus* Harms on behavioral dysfunction in rotenone-induced Parkinsonian rats. *Biol Pharm Bull* 28:169–172, 2005.

Index

A

ABAP, *see* 2,2′-Azobis(amidinopropane)-
  dihydrochloride (ABAP)
Absorption of substituted benzenes related to
  lignin, 51
  effect of substituents, 52–54
  extinction coefficient, 52
ABTS, *see* 2,2′-Azobis-3-ethylbenzthiazoline-6-
  sulfonate (ABTS)
Acanthpanax senticosus, 600, 618–619
ACE, *see* Angiotensin converting enzyme (ACE)
Acetoguaiacone, 332, 410, 472
Acetoguaicone, 573
Acetophenone and phenol chromophores, 84
Acetoxypinoresinol, 599
Acetylated lignins
  ^{13}C-NMR chemical shift assignments for,
    166–169
  2D correlation chemical shifts, 213
  hardwood, 1D ^{13}C NMR spectra, 188
  HSQC spectra of, 194
  softwood, 1D ^{13}C NMR spectra, 187
  volume integrals for, 202
Acetyl bromide treatment, 38
Acidolysis, 19
  phenyl courmaran by, 61
Acid-soluble lignin, 57
Acousto-optic tuneable filter (AOTF), 125
Action spectrum, 75
Activation energy by Arrhenius equation, 315
AHQ ion radicals and dianions, redox reactions
  of, 364
Alcoholysis lignin (AL)
  thermal stability
    DSC heating curve of, 304–305
    by thermogravimetry (TG), 302–303
Aldehydes, 183, 574–575
Alfalfa samples
  control (wild-type) and C3H-deficient
    ^{13}C–^{1}H correlation spectra, 195
    HMBC spectra, 196
  NIR for remote sensing, 128
Alkali darkening, 73–74
Alkaline extraction (DE), 405
Alkaline hydrolysis
  of lignin, 396–397
  stilbene by, 61
Alkaline pulping of wood, *see* Chemical pulping
  process

Alkaline sulfite anthraquinone methanol
  (ASAM) process, 350
Amineboranes, 481, 487
  bleaching efficiency, 485
  PC number, 488–489
  selectivity of, 482
Aminolysis, 59
Angeloylgomisin R, 617
Angiosperm lignins, 30
  thioacidolysis and desulfurization
    relative frequencies of dimers, 37
Angiotensin converting enzyme (ACE), 607
Anolignan
  anolignan A and B, 592, 617
  anolignan C, 592
Anthraquinone (AQ), 331, 350, 362–365
  pulping, 211
Antibacterial activity, dental caries, 618
Antiemetic activity, magnol and honokiol, 618
Anti hyperglycemic activity, masoprocol, 618
Antitumor lignans
  analgesic effect, 607
  antiallergy effect
    dioxabicyclooctane and butanolide
      lignans, 603
    Magnolia salicifolia, flower buds of, 603
  anticomplementary activity
    complementary system, 600
    Eucommia ulmoides, 601
  anti-HIV activity, 616
    retro-viral effect, 617
  anti-inflammatory activity, 605
  antioxidant activity
    benzofuran lignans, 609
    DPPH and SOD, 610
    FMLP and PMA, 610
    NADPH, 610
    TBA and SMA, 608
  aromatase, 616
  Ca^{2}+ antagonists activity
    endogeneous digitalis-like factors, 601
    Magnolia fargesii, dried flower buds of, 602
    trachelogenin, 601–602
  hypolipidemic activity
    hypercholesterolemia patients, 612–613
  hypotensive effect
    DOCA and ACE, 607
  5-lipoxygenase (LO) inhibitory activity
    naphthalenic lignan lactones, 607–608